SINGLE VARIABLE
CALCULUS

SEVENTH EDITION

HYBRID

JAMES STEWART

McMASTER UNIVERSITY

AND

UNIVERSITY OF TORONTO

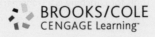

BROOKS/COLE
CENGAGE Learning™

Australia · Brazil · Japan · Korea · Mexico · Singapore · Spain · United Kingdom · United States

BROOKS/COLE
CENGAGE Learning·

Single Variable Calculus, **Seventh Edition, Hybrid**
James Stewart

Executive Editor: Liz Covello

Assistant Editor: Liza Neustaetter

Editorial Assistant: Jennifer Staller

Media Editor : Maureen Ross

Marketing Manager: Jennifer Jones

Marketing Coordinator: Michael Ledesma

Marketing Communications Manager: Mary Anne Payumo

Content Project Manager: Cheryll Linthicum

Art Director: Vernon T. Boes

Print Buyer: Becky Cross

Rights Acquisitions Specialist: Don Schlotman

Production Service: TECH·arts

Text Designer: TECH·arts

Photo Researcher: Terri Wright, www.terriwright.com

Text Researcher: Sarah D'Stair

Copy Editor: Kathi Townes

Cover Designer: Denise Davidson

Cover Illustration: Irene Morris

Compositor: Stephanie Kuhns, TECH·arts

For product information and technology assistance, contact us at
Cengage Learning Customer & Sales Support, 1-800-354-9706.

For permission to use material from this text or product, submit all requests online at **www.cengage.com/permissions**.

Further permissions questions can be e-mailed to **permissionrequest@cengage.com**.

Library of Congress Control Number: 2010936608

ISBN-13: 978-1-111-42675-0
ISBN-10: 1-111-42675-9

Brooks/Cole
20 Davis Drive
Belmont, CA 94002-3098
USA

Cengage Learning is a leading provider of customized learning solutions with office locations around the globe, including Singapore, the United Kingdom, Australia, Mexico, Brazil, and Japan. Locate your local office at **www.cengage.com/global**.

Cengage Learning products are represented in Canada by Nelson Education, Ltd.

To learn more about Brooks/Cole, visit **www.cengage.com/brookscole**.

Purchase any of our products at your local college store or at our preferred online store **www.cengagebrain.com**.

Trademarks

ExamView® and ExamViewPro® are registered trademarks of FSCreations, Inc.
Windows is a registered trademark of the Microsoft Corporation and used herein under license.
Macintosh and Power Macintosh are registered trademarks of Apple Computer, Inc. Used herein under license.
Derive is a registered trademark of Soft Warehouse, Inc.
Maple is a registered trademark of Waterloo Maple, Inc.
Mathematica is a registered trademark of Wolfram Research, Inc.
Tools for Enriching is a trademark used herein under license.

K01T11

Printed in the United States of America
1 2 3 4 5 6 7 15 14 13 12 11

To Bill Ralph and Bruce Thompson

Contents

Preface   *xi*

To the Student   *xxiii*

Diagnostic Tests   *Available in Enhanced WebAssign*

A Preview of Calculus   1

1   Functions and Limits   9

1.1   Four Ways to Represent a Function   10

1.2   Mathematical Models: A Catalog of Essential Functions   19

1.3   New Functions from Old Functions   29

1.4   The Tangent and Velocity Problems   35

1.5   The Limit of a Function   39

1.6   Calculating Limits Using the Limit Laws   48

1.7   The Precise Definition of a Limit   56

1.8   Continuity   64

    Review   73

Principles of Problem Solving   77

2   Derivatives   83

2.1   Derivatives and Rates of Change   84

    Writing Project ▪ Early Methods for Finding Tangents   91

2.2   The Derivative as a Function   91

2.3   Differentiation Formulas   99

    Applied Project ▪ Building a Better Roller Coaster   109

2.4   Derivatives of Trigonometric Functions   110

2.5   The Chain Rule   116

    Applied Project ▪ Where Should a Pilot Start Descent?   122

2.6   Implicit Differentiation   122

    Laboratory Project ▪ Families of Implicit Curves   127

2.7   Rates of Change in the Natural and Social Sciences   128

2.8   Related Rates   137

2.9   Linear Approximations and Differentials   141

      Laboratory Project ▪ Taylor Polynomials   145

      Review   146

Problems Plus   151

3   Applications of Differentiation   155

3.1   Maximum and Minimum Values   156

      Applied Project ▪ The Calculus of Rainbows   162

3.2   The Mean Value Theorem   163

3.3   How Derivatives Affect the Shape of a Graph   168

3.4   Limits at Infinity; Horizontal Asymptotes   175

3.5   Summary of Curve Sketching   186

3.6   Graphing with Calculus *and* Calculators   191

3.7   Optimization Problems   197

      Applied Project ▪ The Shape of a Can   203

3.8   Newton's Method   204

3.9   Antiderivatives   208

      Review   213

Problems Plus   217

4   Integrals   221

4.1   Areas and Distances   222

4.2   The Definite Integral   231

      Discovery Project ▪ Area Functions   242

4.3   The Fundamental Theorem of Calculus   243

4.4   Indefinite Integrals and the Net Change Theorem   251

      Writing Project ▪ Newton, Leibniz, and the Invention of Calculus   257

4.5   The Substitution Rule   258

      Review   263

Problems Plus   268

5   Applications of Integration   271

5.1   Areas Between Curves   272

     Applied Project ▪ The Gini Index   277

5.2   Volumes   278

5.3   Volumes by Cylindrical Shells   286

5.4   Work   290

5.5   Average Value of a Function   294

     Applied Project ▪ Calculus and Baseball   296

     Review   297

Problems Plus   299

6   Inverse Functions: 303

Exponential, Logarithmic, and Inverse Trigonometric Functions

6.1   Inverse Functions   304

Instructors may cover either Sections 6.2–6.4 or Sections 6.2*–6.4*. See the Preface.

6.2 Exponential Functions and Their Derivatives 310	6.2* The Natural Logarithmic Function 332
6.3 Logarithmic Functions 319	6.3* The Natural Exponential Function 339
6.4 Derivatives of Logarithmic Functions 324	6.4* General Logarithmic and Exponential Functions 343

6.5   Exponential Growth and Decay   351

6.6   Inverse Trigonometric Functions   356

     Applied Project ▪ Where to Sit at the Movies   363

6.7   Hyperbolic Functions   364

6.8   Indeterminate Forms and l'Hospital's Rule   369

     Writing Project ▪ The Origins of l'Hospital's Rule   377

     Review   378

Problems Plus   382

7   **Techniques of Integration   385**

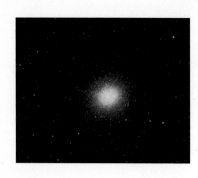

7.1   Integration by Parts   386

7.2   Trigonometric Integrals   390

7.3   Trigonometric Substitution   396

7.4   Integration of Rational Functions by Partial Fractions   402

7.5   Strategy for Integration   410

7.6   Integration Using Tables and Computer Algebra Systems   414
        Discovery Project ▪ Patterns in Integrals   419

7.7   Approximate Integration   420

7.8   Improper Integrals   429
        Review   437

Problems Plus   440

8   **Further Applications of Integration   443**

8.1   Arc Length   444
        Discovery Project ▪ Arc Length Contest   449

8.2   Area of a Surface of Revolution   450
        Discovery Project ▪ Rotating on a Slant   454

8.3   Applications to Physics and Engineering   456
        Discovery Project ▪ Complementary Coffee Cups   464

8.4   Applications to Economics and Biology   465

8.5   Probability   468
        Review   474

Problems Plus   476

9   **Differential Equations   479**

9.1   Modeling with Differential Equations   480

9.2   Direction Fields and Euler's Method   484

9.3   Separable Equations   491
        Applied Project ▪ How Fast Does a Tank Drain?   497
        Applied Project ▪ Which Is Faster, Going Up or Coming Down?   498

9.4   Models for Population Growth   499

9.5   Linear Equations   507

9.6   Predator-Prey Systems   511

Review   516

Problems Plus   520

10   Parametric Equations and Polar Coordinates   523

10.1   Curves Defined by Parametric Equations   524

Laboratory Project ▪ Running Circles around Circles   529

10.2   Calculus with Parametric Curves   530

Laboratory Project ▪ Bézier Curves   536

10.3   Polar Coordinates   536

Laboratory Project ▪ Families of Polar Curves   545

10.4   Areas and Lengths in Polar Coordinates   545

10.5   Conic Sections   549

10.6   Conic Sections in Polar Coordinates   555

Review   560

Problems Plus   564

11   Infinite Sequences and Series   565

11.1   Sequences   566

Laboratory Project ▪ Logistic Sequences   576

11.2   Series   576

11.3   The Integral Test and Estimates of Sums   584

11.4   The Comparison Tests   590

11.5   Alternating Series   594

11.6   Absolute Convergence and the Ratio and Root Tests   598

11.7   Strategy for Testing Series   604

11.8   Power Series   605

11.9   Representations of Functions as Power Series   610

11.10   Taylor and Maclaurin Series   615

Laboratory Project ▪ An Elusive Limit   627

Writing Project ▪ How Newton Discovered the Binomial Series   627

11.11   Applications of Taylor Polynomials   628

Applied Project ▪ Radiation from the Stars   634

Review   635

Problems Plus   639

Appendixes   A1

A   Numbers, Inequalities, and Absolute Values   A2

B   Coordinate Geometry and Lines   A9

C   Graphs of Second-Degree Equations   A14

D   Trigonometry   A21

E   Sigma Notation   A29

F   Proofs of Theorems   A34

G   Graphing Calculators and Computers   A41

H   Complex Numbers   A46

I   Answers to Odd-Numbered Exercises   A55

Index   A63

Preface

> A great discovery solves a great problem but there is a grain of discovery in the solution of any problem. Your problem may be modest; but if it challenges your curiosity and brings into play your inventive faculties, and if you solve it by your own means, you may experience the tension and enjoy the triumph of discovery.
>
> GEORGE POLYA

Many traditional lecture-based Calculus courses are evolving into lecture-lab courses or into courses in which all homework, even tests, is delivered online. Moreover, distance learning is growing rapidly. Thus instructors are increasingly faced with the dual challenge of integrating technology and teaching the concepts and skills of calculus.

In this hybrid text, as in all editions of my hardcover *Calculus* books, my aim is to provide instructors and students with the tools they can use to meet this challenge.

What do I mean by *hybrid*? A hybrid text involves an integration of several products and can be used best in a course with a strong blend of lecture time and online course work.

For this hybrid edition, the diagnostic tests, the end-of-section exercises, and the exercises in the appendixes have been removed from the text and are available exclusively in Enhanced WebAssign, an easy-to-use online homework system. I believe that a slightly smaller book will be more manageable for the student who is spending his or her homework time on a computer.

Although this text has been tailored for instructors and students who use online homework, the focus of the exposition remains unchanged.

 ## Philosophy of the Book

The art of teaching, Mark Van Doren said, is the art of assisting discovery. I have tried to write a book that assists students in discovering calculus—both for its practical power and its surprising beauty. In this edition, as in the first six editions, I aim to convey to the student a sense of the utility of calculus and develop technical competence, but I also strive to give some appreciation for the intrinsic beauty of the subject. Newton undoubtedly experienced a sense of triumph when he made his great discoveries. I want students to share some of that excitement.

The emphasis is on understanding concepts. I think that nearly everybody agrees that this should be the primary goal of calculus instruction. In fact, the impetus for the current calculus reform movement came from the Tulane Conference in 1986, which formulated as their first recommendation:

Focus on conceptual understanding.

I have tried to implement this goal through the *Rule of Three:* "Topics should be presented geometrically, numerically, and algebraically." Visualization, numerical and graphical experimentation, and other approaches have changed how we teach conceptual reasoning in fundamental ways. The Rule of Three has been expanded to become the *Rule of Four* by emphasizing the verbal, or descriptive, point of view as well.

In writing the seventh edition my premise has been that it is possible to achieve conceptual understanding and still retain the best traditions of traditional calculus. The book contains elements of reform, but within the context of a traditional curriculum.

Alternative Versions

I have written several other calculus textbooks that might be preferable for some instructors. Most of them also come in single variable and multivariable versions.

- *Calculus,* Seventh Edition, is similar to the present textbook in content and coverage except that it includes all end-of-section exercises, as well as all end-of-chapter review material.

- *Calculus: Early Transcendentals,* Seventh Edition, is similar to *Calculus,* Seventh Edition, except that the exponential, logarithmic, and inverse trigonometric functions are covered in the first semester.

- *Calculus: Early Transcendentals*, Seventh Edition, Hybrid Version, is similar to *Calculus: Early Transcendentals*, Seventh Edition, in content and coverage except that all end-of-section exercises are available only in Enhanced WebAssign. The printed text includes all end-of-chapter review material.

- *Essential Calculus* is a much briefer book (800 pages), though it contains almost all of the topics in *Calculus,* Seventh Edition. The relative brevity is achieved through briefer exposition of some topics and putting some features on the website.

- *Essential Calculus: Early Transcendentals* resembles *Essential Calculus,* but the exponential, logarithmic, and inverse trigonometric functions are covered in Chapter 3.

- *Calculus: Concepts and Contexts*, Fourth Edition, emphasizes conceptual understanding even more strongly than this book. The coverage of topics is not encyclopedic and the material on transcendental functions and on parametric equations is woven throughout the book instead of being treated in separate chapters.

- *Calculus: Early Vectors* introduces vectors and vector functions in the first semester and integrates them throughout the book. It is suitable for students taking Engineering and Physics courses concurrently with calculus.

- *Brief Applied Calculus* is intended for students in business, the social sciences, and the life sciences.

What's New in the Seventh Edition?

The changes have resulted from talking with my colleagues and students at the University of Toronto and from reading journals, as well as suggestions from users and reviewers. Here are some of the many improvements that I've incorporated into this edition:

- Some material has been rewritten for greater clarity or for better motivation. See, for instance, the introduction to maximum and minimum values on page 156 and the introduction to series on page 576.

- New examples have been added, and the solutions to some of the existing examples have been amplified. A case in point: I added details to the solution of Example 1.6.11

because when I taught Section 1.6 from the sixth edition I realized that students need more guidance when setting up inequalities for the Squeeze Theorem.

- Chapter 1, *Functions and Limits*, consists of most of the material from Chapters 1 and 2 of the sixth edition. The section on Graphing Calculators and Computers is now Appendix G.

- The art program has been revamped: New figures have been incorporated and a substantial percentage of the existing figures have been redrawn.

- The data in examples and exercises have been updated to be more timely.

- Three new projects have been added: *The Gini Index* (page 277) explores how to measure income distribution among inhabitants of a given country and is a nice application of areas between curves. (I thank Klaus Volpert for suggesting this project.) *Families of Implicit Curves* (page 127) investigates the changing shapes of implicitly defined curves as parameters in a family are varied. *Families of Polar Curves* (page 545) exhibits the fascinating shapes of polar curves and how they evolve within a family.

- More than 25% of the exercises are new.

Technology Enhancements

- The media and technology to support the text have been enhanced to give professors greater control over their course, to provide extra help to deal with the varying levels of student preparedness for the calculus course, and to improve support for conceptual understanding. New Enhanced WebAssign features including a customizable Cengage YouBook, *Just in Time* review, *Show Your Work,* Answer Evaluator, Personalized Study Plan, Master Its, solution videos, lecture video clips (with associated questions), and *Visualizing Calculus* (TEC animations with associated questions) have been developed to facilitate improved student learning and flexible classroom teaching.

- *Tools for Enriching Calculus* (TEC) has been completely redesigned and is accessible in Enhanced WebAssign, CourseMate, and PowerLecture. Selected Visuals and Modules are available at www.stewartcalculus.com.

Features

ENHANCED WEBASSIGN Technology is having an impact on the way homework is assigned to students, particularly in large classes. The use of online homework is growing and its appeal depends on ease of use, grading precision, and reliability. With the seventh edition we have been working with the calculus community and WebAssign to develop a more robust online homework system. Exercises are assignable as online homework in free response, multiple choice, and multi-part formats.

    The system also includes Active Examples, in which students are guided in step-by-step tutorials through text examples, with links to the textbook and to video solutions. New enhancements to the system include a customizable eBook, a *Show Your Work* feature, *Just in Time* review of precalculus prerequisites, an improved Assignment Editor, and an Answer Evaluator that accepts more mathematically equivalent answers and allows for homework grading in much the same way that an instructor grades.

CONCEPTUAL EXERCISES The most important way to foster conceptual understanding is through the problems that we assign. To that end I have devised various types of problems. Some exercise sets begin with requests to explain the meanings of the basic concepts of the section. (See, for instance, the first few exercises in Sections 1.5, 1.8, 11.2.) Similarly, all the review sections begin with a Concept Check and a True-False Quiz. Other exercises test conceptual understanding through graphs or tables.

Another type of exercise uses verbal description to test conceptual understanding. I particularly value problems that combine and compare graphical, numerical, and algebraic approaches.

GRADED EXERCISE SETS End-of-section exercises within Enhanced WebAssign have been carefully graded, progressing from basic conceptual exercises and skill-development problems to more challenging problems involving applications and proofs.

REAL-WORLD DATA My assistants and I spent a great deal of time looking in libraries, contacting companies and government agencies, and searching the Internet for interesting real-world data to introduce, motivate, and illustrate the concepts of calculus. As a result, many of the examples and exercises deal with functions defined by such numerical data or graphs. See, for instance, Figure 1 in Section 1.1 (seismograms from the Northridge earthquake) and Figure 4 in Section 4.4 (San Francisco power consumption).

PROJECTS One way of involving students and making them active learners is to have them work (perhaps in groups) on extended projects that give a feeling of substantial accomplishment when completed. I have included four kinds of projects: *Applied Projects* involve applications that are designed to appeal to the imagination of students. The project after Section 9.3 asks whether a ball thrown upward takes longer to reach its maximum height or to fall back to its original height. (The answer might surprise you.) *Laboratory Projects* involve technology; the one following Section 10.2 shows how to use Bézier curves to design shapes that represent letters for a laser printer. *Writing Projects* ask students to compare present-day methods with those of the founders of calculus—Fermat's method for finding tangents, for instance. Suggested references are supplied. *Discovery Projects* anticipate results to be discussed later or encourage discovery through pattern recognition (see the one following Section 7.6). Additional projects can be found in the *Instructor's Guide* (see, for instance, Group Exercise 4.1: Position from Samples).

PROBLEM SOLVING Students usually have difficulties with problems for which there is no single well-defined procedure for obtaining the answer. I think nobody has improved very much on George Polya's four-stage problem-solving strategy and, accordingly, I have included a version of his problem-solving principles following Chapter 1. They are applied, both explicitly and implicitly, throughout the book. After the other chapters I have placed sections called *Problems Plus*, which feature examples of how to tackle challenging calculus problems. In selecting the varied problems for these sections I kept in mind the following advice from David Hilbert: "A mathematical problem should be difficult in order to entice us, yet not inaccessible lest it mock our efforts." When I put these challenging problems on assignments and tests I grade them in a different way. Here I reward a student significantly for ideas toward a solution and for recognizing which problem-solving principles are relevant.

DUAL TREATMENT OF EXPONENTIAL AND LOGARITHMIC FUNCTIONS There are two possible ways of treating the exponential and logarithmic functions and each method has its passionate advocates. Because one often finds advocates of both approaches teaching the same course, I include full treatments of both methods. In Sections 6.2, 6.3, and 6.4 the exponential function is defined first, followed by the logarithmic function as its inverse. (Students have seen these functions introduced this way since high school.) In the alternative approach, presented in Sections 6.2*, 6.3*, and 6.4*, the logarithm is defined as an integral and the exponential function is its inverse. This latter method is, of course, less intuitive but more elegant. You can use whichever treatment you prefer.

If the first approach is taken, then much of Chapter 6 can be covered before Chapters 4 and 5, if desired. This order of presentation allows a faster-paced course to teach the transcendental functions and the definite integral in the first semester of the course.

For instructors who would like to go even further in this direction I have prepared an alternate edition of this book, called *Calculus, Early Transcendentals,* Seventh Edition, in which the exponential and logarithmic functions are introduced in the first chapter. Their limits and derivatives are found in the second and third chapters at the same time as polynomials and the other elementary functions.

TOOLS FOR ENRICHING™ CALCULUS TEC is a companion to the text and is intended to enrich and complement its contents. (It is now accessible in Enhanced WebAssign, CourseMate, and PowerLecture. Selected Visuals and Modules are available at www.stewartcalculus.com.) Developed by Harvey Keynes, Dan Clegg, Hubert Hohn, and myself, TEC uses a discovery and exploratory approach. In sections of the book where technology is particularly appropriate, marginal icons direct students to TEC modules that provide a laboratory environment in which they can explore the topic in different ways and at different levels. **Visuals are animations of figures in text; Modules are more elaborate activities and include exercises.** Instructors can choose to become involved at several different levels, ranging from simply encouraging students to use the Visuals and Modules for independent exploration, to assigning specific exercises from those included with each Module, or to creating additional exercises, labs, and projects that make use of the Visuals and Modules.

HOMEWORK HINTS Homework Hints presented in the form of questions try to imitate an effective teaching assistant by functioning as a silent tutor. They are constructed so as not to reveal any more of the actual solution than is minimally necessary to make further progress. Hints for representative exercises are available to students at stewartcalculus.com and in CourseMate and Enhanced WebAssign.

www.stewartcalculus.com This site includes the following.

- Homework Hints
- Algebra Review
- Lies My Calculator and Computer Told Me
- History of Mathematics, with links to the better historical websites
- Additional Topics (complete with exercise sets): Fourier Series, Formulas for the Remainder Term in Taylor Series, Rotation of Axes
- Archived Problems (Drill exercises that appeared in previous editions, together with their solutions)
- Challenge Problems (some from the Problems Plus sections from prior editions)
- Links, for particular topics, to outside web resources
- Selected Tools for Enriching Calculus (TEC) Modules and Visuals

Content

Diagnostic Tests Diagnostic tests in Basic Algebra, Analytic Geometry, Functions, and Trigonometry are available online in Enhanced WebAssign.

A Preview of Calculus This is an overview of the subject and includes a list of questions to motivate the study of calculus.

1 Functions and Limits   From the beginning, multiple representations of functions are stressed: verbal, numerical, visual, and algebraic. A discussion of mathematical models leads to a review of the standard functions from these four points of view. The material on limits is motivated by a prior discussion of the tangent and velocity problems. Limits are treated from descriptive, graphical, numerical, and algebraic points of view. Section 1.7, on the precise epsilon-delta definition of a limit, is an optional section.

2 Derivatives   The material on derivatives is covered in two sections in order to give students more time to get used to the idea of a derivative as a function. The examples and exercises explore the meanings of derivatives in various contexts. Higher derivatives are introduced in Section 2.2.

3 Applications of Differentiation   The basic facts concerning extreme values and shapes of curves are deduced from the Mean Value Theorem. Graphing with technology emphasizes the interaction between calculus and calculators and the analysis of families of curves. Some substantial optimization problems are provided, including an explanation of why you need to raise your head 42° to see the top of a rainbow.

4 Integrals   The area problem and the distance problem serve to motivate the definite integral, with sigma notation introduced as needed. (Full coverage of sigma notation is provided in Appendix E.) Emphasis is placed on explaining the meanings of integrals in various contexts and on estimating their values from graphs and tables.

5 Applications of Integration   Here I present the applications of integration—area, volume, work, average value—that can reasonably be done without specialized techniques of integration. General methods are emphasized. The goal is for students to be able to divide a quantity into small pieces, estimate with Riemann sums, and recognize the limit as an integral.

6 Inverse Functions: Exponential, Logarithmic, and Inverse Trigonometric Functions   As discussed more fully on page xiv, only one of the two treatments of these functions need be covered. Exponential growth and decay are covered in this chapter.

7 Techniques of Integration   All the standard methods are covered but, of course, the real challenge is to be able to recognize which technique is best used in a given situation. Accordingly, in Section 7.5, I present a strategy for integration. The use of computer algebra systems is discussed in Section 7.6.

8 Further Applications of Integration   Here are the applications of integration—arc length and surface area—for which it is useful to have available all the techniques of integration, as well as applications to biology, economics, and physics (hydrostatic force and centers of mass). I have also included a section on probability. There are more applications here than can realistically be covered in a given course. Instructors should select applications suitable for their students and for which they themselves have enthusiasm.

9 Differential Equations   Modeling is the theme that unifies this introductory treatment of differential equations. Direction fields and Euler's method are studied before separable and linear equations are solved explicitly, so that qualitative, numerical, and analytic approaches are given equal consideration. These methods are applied to the exponential, logistic, and other models for population growth. The first four or five sections of this chapter serve as a good introduction to first-order differential equations. An optional final section uses predator-prey models to illustrate systems of differential equations.

10 Parametric Equations and Polar Coordinates

This chapter introduces parametric and polar curves and applies the methods of calculus to them. Parametric curves are well suited to laboratory projects; the three presented here involve families of curves and Bézier curves. A brief treatment of conic sections in polar coordinates prepares the way for Kepler's Laws in Chapter 13.

11 Infinite Sequences and Series

The convergence tests have intuitive justifications (see page 584) as well as formal proofs. Numerical estimates of sums of series are based on which test was used to prove convergence. The emphasis is on Taylor series and polynomials and their applications to physics. Error estimates include those from graphing devices.

 ## Ancillaries

Single Variable Calculus, Seventh Edition, Hybrid Version, is supported by a complete set of ancillaries developed under my direction. Each piece has been designed to enhance student understanding and to facilitate creative instruction. With this edition, new media and technologies have been developed that help students to visualize calculus and instructors to customize content to better align with the way they teach their course. The tables on pages xxi–xxii describe each of these ancillaries.

 ## Acknowledgments

The preparation of this and previous editions has involved much time spent reading the reasoned (but sometimes contradictory) advice from a large number of astute reviewers. I greatly appreciate the time they spent to understand my motivation for the approach taken. I have learned something from each of them.

SEVENTH EDITION REVIEWERS

Amy Austin, *Texas A&M University*
Anthony J. Bevelacqua, *University of North Dakota*
Zhen-Qing Chen, *University of Washington—Seattle*
Jenna Carpenter, *Louisiana Tech University*
Le Baron O. Ferguson, *University of California—Riverside*
Shari Harris, *John Wood Community College*
Amer Iqbal, *University of Washington—Seattle*
Akhtar Khan, *Rochester Institute of Technology*
Marianne Korten, *Kansas State University*
Joyce Longman, *Villanova University*

Richard Millspaugh, *University of North Dakota*
Lon H. Mitchell, *Virginia Commonwealth University*
Ho Kuen Ng, *San Jose State University*
Norma Ortiz-Robinson, *Virginia Commonwealth University*
Qin Sheng, *Baylor University*
Magdalena Toda, *Texas Tech University*
Ruth Trygstad, *Salt Lake Community College*
Klaus Volpert, *Villanova University*
Peiyong Wang, *Wayne State University*

FOCUS GROUP PARTICIPANTS

Edward Bonan-Hamada, *Mesa State College*
Mark Brenneman, *Mesa Community College*
Barry Brunson, *Western Kentucky University*
Daria Filippova, *Bowling Green State University*
Lee R. Gibson, *University of Louisville*
Eric Hsu, *San Francisco State University*
Jonathan Lee, *University of Oklahoma*
Mary Margarita Legner, *Riverside City College*

Joseph McCollum, *Siena College*
Hien Nguyen, *University of Wisconsin–Whitewater*
John O'Neill, *Siena College*
David Price, *Tarrant County College, South East Campus*
Joe Rody, *Arizona State University*
Raymond Sadeghi, *University of Texas at San Antonio*
Paul Seeburger, *Monroe Community College*
David Torain, *Hampton University*

TECHNOLOGY REVIEWERS

Maria Andersen, *Muskegon Community College*
Eric Aurand, *Eastfield College*
Joy Becker, *University of Wisconsin–Stout*
Przemyslaw Bogacki, *Old Dominion University*
Amy Elizabeth Bowman, *University of Alabama in Huntsville*
Monica Brown, *University of Missouri–St. Louis*
Roxanne Byrne, *University of Colorado at Denver and Health Sciences Center*
Teri Christiansen, *University of Missouri–Columbia*
Bobby Dale Daniel, *Lamar University*
Jennifer Daniel, *Lamar University*
Andras Domokos, *California State University, Sacramento*
Timothy Flaherty, *Carnegie Mellon University*
Lee Gibson, *University of Louisville*
Jane Golden, *Hillsborough Community College*
Semion Gutman, *University of Oklahoma*
Diane Hoffoss, *University of San Diego*
Lorraine Hughes, *Mississippi State University*
Jay Jahangiri, *Kent State University*
John Jernigan, *Community College of Philadelphia*

Brian Karasek, *South Mountain Community College*
Jason Kozinski, *University of Florida*
Carole Krueger, *The University of Texas at Arlington*
Ken Kubota, *University of Kentucky*
John Mitchell, *Clark College*
Donald Paul, *Tulsa Community College*
Chad Pierson, *University of Minnesota, Duluth*
Lanita Presson, *University of Alabama in Huntsville*
Karin Reinhold, *State University of New York at Albany*
Thomas Riedel, *University of Louisville*
Christopher Schroeder, *Morehead State University*
Angela Sharp, *University of Minnesota, Duluth*
Patricia Shaw, *Mississippi State University*
Carl Spitznagel, *John Carroll University*
Mohammad Tabanjeh, *Virginia State University*
Capt. Koichi Takagi, *United States Naval Academy*
Lorna TenEyck, *Chemeketa Community College*
Roger Werbylo, *Pima Community College*
David Williams, *Clayton State University*
Zhuan Ye, *Northern Illinois University*

PREVIOUS EDITION REVIEWERS

B. D. Aggarwala, *University of Calgary*
John Alberghini, *Manchester Community College*
Michael Albert, *Carnegie-Mellon University*
Daniel Anderson, *University of Iowa*
Donna J. Bailey, *Northeast Missouri State University*
Wayne Barber, *Chemeketa Community College*
Marilyn Belkin, *Villanova University*
Neil Berger, *University of Illinois, Chicago*
David Berman, *University of New Orleans*
Richard Biggs, *University of Western Ontario*
Robert Blumenthal, *Oglethorpe University*
Martina Bode, *Northwestern University*
Barbara Bohannon, *Hofstra University*
Philip L. Bowers, *Florida State University*
Amy Elizabeth Bowman, *University of Alabama in Huntsville*
Jay Bourland, *Colorado State University*
Stephen W. Brady, *Wichita State University*
Michael Breen, *Tennessee Technological University*
Robert N. Bryan, *University of Western Ontario*
David Buchthal, *University of Akron*
Jorge Cassio, *Miami-Dade Community College*
Jack Ceder, *University of California, Santa Barbara*
Scott Chapman, *Trinity University*
James Choike, *Oklahoma State University*
Barbara Cortzen, *DePaul University*
Carl Cowen, *Purdue University*
Philip S. Crooke, *Vanderbilt University*
Charles N. Curtis, *Missouri Southern State College*
Daniel Cyphert, *Armstrong State College*
Robert Dahlin
M. Hilary Davies, *University of Alaska Anchorage*
Gregory J. Davis, *University of Wisconsin–Green Bay*
Elias Deeba, *University of Houston–Downtown*

Daniel DiMaria, *Suffolk Community College*
Seymour Ditor, *University of Western Ontario*
Greg Dresden, *Washington and Lee University*
Daniel Drucker, *Wayne State University*
Kenn Dunn, *Dalhousie University*
Dennis Dunninger, *Michigan State University*
Bruce Edwards, *University of Florida*
David Ellis, *San Francisco State University*
John Ellison, *Grove City College*
Martin Erickson, *Truman State University*
Garret Etgen, *University of Houston*
Theodore G. Faticoni, *Fordham University*
Laurene V. Fausett, *Georgia Southern University*
Norman Feldman, *Sonoma State University*
Newman Fisher, *San Francisco State University*
José D. Flores, *The University of South Dakota*
William Francis, *Michigan Technological University*
James T. Franklin, *Valencia Community College, East*
Stanley Friedlander, *Bronx Community College*
Patrick Gallagher, *Columbia University–New York*
Paul Garrett, *University of Minnesota–Minneapolis*
Frederick Gass, *Miami University of Ohio*
Bruce Gilligan, *University of Regina*
Matthias K. Gobbert, *University of Maryland, Baltimore County*
Gerald Goff, *Oklahoma State University*
Stuart Goldenberg, *California Polytechnic State University*
John A. Graham, *Buckingham Browne & Nichols School*
Richard Grassl, *University of New Mexico*
Michael Gregory, *University of North Dakota*
Charles Groetsch, *University of Cincinnati*
Paul Triantafilos Hadavas, *Armstrong Atlantic State University*
Salim M. Haïdar, *Grand Valley State University*

D. W. Hall, *Michigan State University*

Robert L. Hall, *University of Wisconsin–Milwaukee*

Howard B. Hamilton, *California State University, Sacramento*

Darel Hardy, *Colorado State University*

Gary W. Harrison, *College of Charleston*

Melvin Hausner, *New York University/Courant Institute*

Curtis Herink, *Mercer University*

Russell Herman, *University of North Carolina at Wilmington*

Allen Hesse, *Rochester Community College*

Randall R. Holmes, *Auburn University*

James F. Hurley, *University of Connecticut*

Matthew A. Isom, *Arizona State University*

Gerald Janusz, *University of Illinois at Urbana-Champaign*

John H. Jenkins, *Embry-Riddle Aeronautical University, Prescott Campus*

Clement Jeske, *University of Wisconsin, Platteville*

Carl Jockusch, *University of Illinois at Urbana-Champaign*

Jan E. H. Johansson, *University of Vermont*

Jerry Johnson, *Oklahoma State University*

Zsuzsanna M. Kadas, *St. Michael's College*

Nets Katz, *Indiana University Bloomington*

Matt Kaufman

Matthias Kawski, *Arizona State University*

Frederick W. Keene, *Pasadena City College*

Robert L. Kelley, *University of Miami*

Virgil Kowalik, *Texas A&I University*

Kevin Kreider, *University of Akron*

Leonard Krop, *DePaul University*

Mark Krusemeyer, *Carleton College*

John C. Lawlor, *University of Vermont*

Christopher C. Leary, *State University of New York at Geneseo*

David Leeming, *University of Victoria*

Sam Lesseig, *Northeast Missouri State University*

Phil Locke, *University of Maine*

Joan McCarter, *Arizona State University*

Phil McCartney, *Northern Kentucky University*

James McKinney, *California State Polytechnic University, Pomona*

Igor Malyshev, *San Jose State University*

Larry Mansfield, *Queens College*

Mary Martin, *Colgate University*

Nathaniel F. G. Martin, *University of Virginia*

Gerald Y. Matsumoto, *American River College*

Tom Metzger, *University of Pittsburgh*

Michael Montaño, *Riverside Community College*

Teri Jo Murphy, *University of Oklahoma*

Martin Nakashima, *California State Polytechnic University, Pomona*

Richard Nowakowski, *Dalhousie University*

Hussain S. Nur, *California State University, Fresno*

Wayne N. Palmer, *Utica College*

Vincent Panico, *University of the Pacific*

F. J. Papp, *University of Michigan–Dearborn*

Mike Penna, *Indiana University–Purdue University Indianapolis*

Mark Pinsky, *Northwestern University*

Lothar Redlin, *The Pennsylvania State University*

Joel W. Robbin, *University of Wisconsin–Madison*

Lila Roberts, *Georgia College and State University*

E. Arthur Robinson, Jr., *The George Washington University*

Richard Rockwell, *Pacific Union College*

Rob Root, *Lafayette College*

Richard Ruedemann, *Arizona State University*

David Ryeburn, *Simon Fraser University*

Richard St. Andre, *Central Michigan University*

Ricardo Salinas, *San Antonio College*

Robert Schmidt, *South Dakota State University*

Eric Schreiner, *Western Michigan University*

Mihr J. Shah, *Kent State University–Trumbull*

Theodore Shifrin, *University of Georgia*

Wayne Skrapek, *University of Saskatchewan*

Larry Small, *Los Angeles Pierce College*

Teresa Morgan Smith, *Blinn College*

William Smith, *University of North Carolina*

Donald W. Solomon, *University of Wisconsin–Milwaukee*

Edward Spitznagel, *Washington University*

Joseph Stampfli, *Indiana University*

Kristin Stoley, *Blinn College*

M. B. Tavakoli, *Chaffey College*

Paul Xavier Uhlig, *St. Mary's University, San Antonio*

Stan Ver Nooy, *University of Oregon*

Andrei Verona, *California State University–Los Angeles*

Russell C. Walker, *Carnegie Mellon University*

William L. Walton, *McCallie School*

Jack Weiner, *University of Guelph*

Alan Weinstein, *University of California, Berkeley*

Theodore W. Wilcox, *Rochester Institute of Technology*

Steven Willard, *University of Alberta*

Robert Wilson, *University of Wisconsin–Madison*

Jerome Wolbert, *University of Michigan–Ann Arbor*

Dennis H. Wortman, *University of Massachusetts, Boston*

Mary Wright, *Southern Illinois University–Carbondale*

Paul M. Wright, *Austin Community College*

Xian Wu, *University of South Carolina*

In addition, I would like to thank Jordan Bell, George Bergman, Leon Gerber, Mary Pugh, and Simon Smith for their suggestions; Al Shenk and Dennis Zill for permission to use exercises from their calculus texts; COMAP for permission to use project material; George Bergman, David Bleecker, Dan Clegg, Victor Kaftal, Anthony Lam, Jamie Lawson, Ira Rosenholtz, Paul Sally, Lowell Smylie, and Larry Wallen for ideas for exercises; Dan Drucker for the roller derby project; Thomas Banchoff, Tom Farmer, Fred Gass, John Ramsay, Larry Riddle, Philip Straffin, and Klaus Volpert for ideas for projects; Dan Anderson, Dan Clegg, Jeff Cole, Dan Drucker, and Barbara Frank for solving the new exercises and suggesting ways to improve them; Marv Riedesel and Mary Johnson for accuracy in proofreading; and Jeff Cole and Dan Clegg for their careful preparation and proofreading of the answer manuscript.

In addition, I thank those who have contributed to past editions: Ed Barbeau, Fred Brauer, Andy Bulman-Fleming, Bob Burton, David Cusick, Tom DiCiccio, Garret Etgen, Chris Fisher, Stuart Goldenberg, Arnold Good, Gene Hecht, Harvey Keynes, E.L. Koh, Zdislav Kovarik, Kevin Kreider, Emile LeBlanc, David Leep, Gerald Leibowitz, Larry Peterson, Lothar Redlin, Carl Riehm, John Ringland, Peter Rosenthal, Doug Shaw, Dan Silver, Norton Starr, Saleem Watson, Alan Weinstein, and Gail Wolkowicz.

I also thank Kathi Townes and Stephanie Kuhns, TECHarts for their production services and the following Brooks/Cole staff: Cheryll Linthicum, content project manager; Liza Neustaetter, assistant editor; Maureen Ross, media editor; Sam Subity, managing media editor; Jennifer Jones, marketing manager; and Vernon Boes, art director. They have all done an outstanding job.

I have been very fortunate to have worked with some of the best mathematics editors in the business over the past three decades: Ron Munro, Harry Campbell, Craig Barth, Jeremy Hayhurst, Gary Ostedt, Bob Pirtle, Richard Stratton, and now Liz Covello. All of them have contributed greatly to the success of this book.

JAMES STEWART

■ Ancillaries for Instructors

PowerLecture
ISBN 0-8400-5414-9

This comprehensive DVD contains all art from the text in both jpeg and PowerPoint formats, key equations and tables from the text, complete pre-built PowerPoint lectures, an electronic version of the Instructor's Guide, Solution Builder, ExamView testing software, Tools for Enriching Calculus, video instruction, and JoinIn on TurningPoint clicker content.

Instructor's Guide
by Douglas Shaw
ISBN 0-8400-5407-6

Each section of the text is discussed from several viewpoints. The Instructor's Guide contains suggested time to allot, points to stress, text discussion topics, core materials for lecture, workshop/discussion suggestions, group work exercises in a form suitable for handout, and suggested homework assignments. An electronic version of the Instructor's Guide is available on the PowerLecture DVD.

Complete Solutions Manual

Single Variable
By Daniel Anderson, Jeffery A. Cole, and Daniel Drucker
ISBN 0-8400-5302-9

Multivariable
By Dan Clegg and Barbara Frank
ISBN 0-8400-4947-1

Includes worked-out solutions to all exercises in the text.

Solution Builder
www.cengage.com/solutionbuilder

This online instructor database offers complete worked out solutions to all exercises in the text. Solution Builder allows you to create customized, secure solutions printouts (in PDF format) matched exactly to the problems you assign in class.

Printed Test Bank
By William Steven Harmon
ISBN 0-8400-5408-4

Contains text-specific multiple-choice and free response test items.

ExamView Testing

Create, deliver, and customize tests in print and online formats with ExamView, an easy-to-use assessment and tutorial software. ExamView contains hundreds of multiple-choice and free response test items. ExamView testing is available on the PowerLecture DVD.

■ Ancillaries for Instructors and Students

Stewart Website
www.stewartcalculus.com

Contents: *Homework Hints* ■ *Algebra Review* ■ *Additional Topics* ■ *Drill exercises* ■ *Challenge Problems* ■ *Web Links* ■ *History of Mathematics* ■ *Tools for Enriching Calculus (TEC)*

TEC Tools for Enriching™ Calculus
By James Stewart, Harvey Keynes, Dan Clegg, and developer Hu Hohn

Tools for Enriching Calculus (TEC) functions as both a powerful tool for instructors, as well as a tutorial environment in which students can explore and review selected topics. The Flash simulation modules in TEC include instructions, written and audio explanations of the concepts, and exercises. TEC is accessible in CourseMate, WebAssign, and PowerLecture. Selected Visuals and Modules are available at www.stewartcalculus.com.

WebAssign Enhanced WebAssign
www.webassign.net

WebAssign's homework delivery system lets instructors deliver, collect, grade, and record assignments via the web. Enhanced WebAssign for Stewart's Calculus *now includes opportunities for students to review prerequisite skills and content both at the start of the course and at the beginning of each section. In addition, for selected problems, students can get extra help in the form of "enhanced feedback" (rejoinders) and video solutions.* **Other key features include:** *thousands of problems from Stewart's* Calculus, *a customizable Cengage YouBook, Personal Study Plans, Show Your Work, Just in Time Review, Answer Evaluator, Visualizing Calculus animations and modules, quizzes, lecture videos (with associated questions), and more!*

Cengage Customizable YouBook

YouBook is a Flash-based eBook that is interactive and customizable! Containing all the content from Stewart's Calculus, *YouBook features a text edit tool that allows instructors to modify the textbook narrative as needed. With YouBook, instructors can quickly re-order entire sections and chapters or hide any content they don't teach to create an eBook that perfectly matches their syllabus. Instructors can further customize the text by adding instructor-created or YouTube video links. Additional media assets include: animated figures, video clips, highlighting, notes, and more! YouBook is available in Enhanced WebAssign.*

■ Electronic items   ■ Printed items

(Table continues on page xxii.)

CourseMate CourseMate

www.cengagebrain.com

CourseMate is a perfect self-study tool for students, and requires no set up from instructors. CourseMate brings course concepts to life with interactive learning, study, and exam preparation tools that support the printed textbook. CourseMate for Stewart's Calculus *includes: an interactive eBook, Tools for Enriching Calculus, videos, quizzes, flashcards, and more! For instructors, CourseMate includes Engagement Tracker, a first-of-its-kind tool that monitors student engagement.*

Maple CD-ROM

Maple provides an advanced, high performance mathematical computation engine with fully integrated numerics & symbolics, all accessible from a WYSIWYG technical document environment.

CengageBrain.com

To access additional course materials and companion resources, please visit www.cengagebrain.com. At the CengageBrain.com home page, search for the ISBN of your title (from the back cover of your book) using the search box at the top of the page. This will take you to the product page where free companion resources can be found.

Ancillaries for Students

Student Solutions Manual

Single Variable
By Daniel Anderson, Jeffery A. Cole, and Daniel Drucker
ISBN 0-8400-4949-8

Multivariable
By Dan Clegg and Barbara Frank
ISBN 0-8400-4945-5

Provides completely worked-out solutions to all odd-numbered exercises in the text, giving students a chance to check their answers and ensure they took the correct steps to arrive at an answer.

Study Guide

Single Variable
By Richard St. Andre
ISBN 0-8400-5409-2

Multivariable
By Richard St. Andre
ISBN 0-8400-5410-6

For each section of the text, the Study Guide provides students with a brief introduction, a short list of concepts to master, as well as summary and focus questions with explained answers. The Study Guide also contains "Technology Plus" questions, and multiple-choice "On Your Own" exam-style questions.

CalcLabs with Maple

Single Variable By Philip B. Yasskin and Robert Lopez
ISBN 0-8400-5811-X

Multivariable By Philip B. Yasskin and Robert Lopez
ISBN 0-8400-5812-8

CalcLabs with Mathematica

Single Variable By Selwyn Hollis
ISBN 0-8400-5814-4

Multivariable By Selwyn Hollis
ISBN 0-8400-5813-6

Each of these comprehensive lab manuals will help students learn to use the technology tools available to them. CalcLabs contain clearly explained exercises and a variety of labs and projects to accompany the text.

A Companion to Calculus

By Dennis Ebersole, Doris Schattschneider, Alicia Sevilla, and Kay Somers
ISBN 0-495-01124-X

Written to improve algebra and problem-solving skills of students taking a Calculus course, every chapter in this companion is keyed to a calculus topic, providing conceptual background and specific algebra techniques needed to understand and solve calculus problems related to that topic. It is designed for calculus courses that integrate the review of precalculus concepts or for individual use.

Linear Algebra for Calculus

by Konrad J. Heuvers, William P. Francis, John H. Kuisti, Deborah F. Lockhart, Daniel S. Moak, and Gene M. Ortner
ISBN 0-534-25248-6

This comprehensive book, designed to supplement the calculus course, provides an introduction to and review of the basic ideas of linear algebra.

■ Electronic items   ■ Printed items

To the Student

Reading a calculus textbook is different from reading a newspaper or a novel, or even a physics book. Don't be discouraged if you have to read a passage more than once in order to understand it. You should have pencil and paper and calculator at hand to sketch a diagram or make a calculation.

Some students start by trying their homework problems and read the text only if they get stuck on an exercise. I suggest that a far better plan is to read and understand a section of the text before attempting the exercises. In particular, you should look at the definitions to see the exact meanings of the terms. And before you read each example, I suggest that you cover up the solution and try solving the problem yourself. You'll get a lot more from looking at the solution if you do so.

Part of the aim of this course is to train you to think logically. Learn to write the solutions of the exercises in a connected, step-by-step fashion with explanatory sentences—not just a string of disconnected equations or formulas.

The answers to the odd-numbered review exercises appear at the back of the book, in Appendix I. Some exercises ask for a verbal explanation or interpretation or description. In such cases there is no single correct way of expressing the answer, so don't worry that you haven't found the definitive answer. In addition, there are often several different forms in which to express a numerical or algebraic answer, so if your answer differs from mine, don't immediately assume you're wrong. For example, if the answer given in the back of the book is $\sqrt{2} - 1$ and you obtain $1/(1 + \sqrt{2})$, then you're right and rationalizing the denominator will show that the answers are equivalent.

The icon ⌂ indicates an exercise that definitely requires the use of either a graphing calculator or a computer with graphing software. (Appendix G discusses the use of these graphing devices and some of the pitfalls that you may encounter.) But that doesn't mean that graphing devices can't be used to check your work on the other exercises as well. The symbol CAS is

reserved for problems in which the full resources of a computer algebra system (like Derive, Maple, Mathematica, or the TI-89/92) are required.

You will also encounter the symbol ⊘, which warns you against committing an error. I have placed this symbol in the margin in situations where I have observed that a large proportion of my students tend to make the same mistake.

Tools for Enriching Calculus, which is a companion to this text, is referred to by means of the symbol TEC and can be accessed in Enhanced WebAssign and CourseMate (selected Visuals and Modules are available at www.stewartcalculus.com). It directs you to modules in which you can explore aspects of calculus for which the computer is particularly useful.

Homework Hints for representative exercises can be found on stewartcalculus.com as well as Enhanced WebAssign and CourseMate. The homework hints ask you questions that allow you to make progress toward a solution without actually giving you the answer. You need to pursue each hint in an active manner with pencil and paper to work out the details. If a particular hint doesn't enable you to solve the problem, you can click to reveal the next hint.

I recommend that you keep this book for reference purposes after you finish the course. Because you will likely forget some of the specific details of calculus, the book will serve as a useful reminder when you need to use calculus in subsequent courses. And, because this book contains more material than can be covered in any one course, it can also serve as a valuable resource for a working scientist or engineer.

Calculus is an exciting subject, justly considered to be one of the greatest achievements of the human intellect. I hope you will discover that it is not only useful but also intrinsically beautiful.

JAMES STEWART

A Preview of Calculus

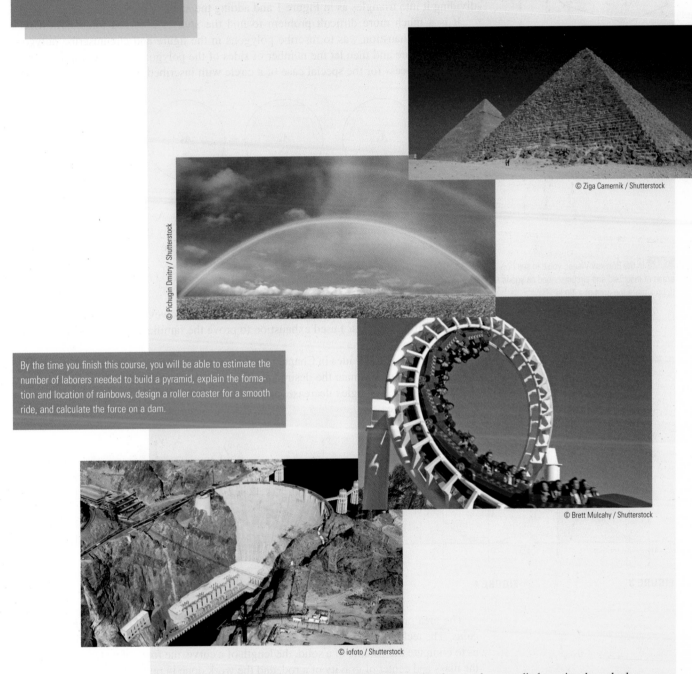

© Ziga Camernik / Shutterstock

© Pichugin Dmitry / Shutterstock

By the time you finish this course, you will be able to estimate the number of laborers needed to build a pyramid, explain the formation and location of rainbows, design a roller coaster for a smooth ride, and calculate the force on a dam.

© Brett Mulcahy / Shutterstock

© iofoto / Shutterstock

Calculus is fundamentally different from the mathematics that you have studied previously: calculus is less static and more dynamic. It is concerned with change and motion; it deals with quantities that approach other quantities. For that reason it may be useful to have an overview of the subject before beginning its intensive study. Here we give a glimpse of some of the main ideas of calculus by showing how the concept of a limit arises when we attempt to solve a variety of problems.

$$A = A_1 + A_2 + A_3 + A_4 + A_5$$

FIGURE 1

The Area Problem

The origins of calculus go back at least 2500 years to the ancient Greeks, who found areas using the "method of exhaustion." They knew how to find the area A of any polygon by dividing it into triangles as in Figure 1 and adding the areas of these triangles.

It is a much more difficult problem to find the area of a curved figure. The Greek method of exhaustion was to inscribe polygons in the figure and circumscribe polygons about the figure and then let the number of sides of the polygons increase. Figure 2 illustrates this process for the special case of a circle with inscribed regular polygons.

   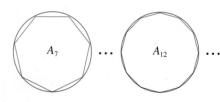

FIGURE 2

Let A_n be the area of the inscribed polygon with n sides. As n increases, it appears that A_n becomes closer and closer to the area of the circle. We say that the area of the circle is the *limit* of the areas of the inscribed polygons, and we write

$$A = \lim_{n \to \infty} A_n$$

The Greeks themselves did not use limits explicitly. However, by indirect reasoning, Eudoxus (fifth century BC) used exhaustion to prove the familiar formula for the area of a circle: $A = \pi r^2$.

We will use a similar idea in Chapter 4 to find areas of regions of the type shown in Figure 3. We will approximate the desired area A by areas of rectangles (as in Figure 4), let the width of the rectangles decrease, and then calculate A as the limit of these sums of areas of rectangles.

FIGURE 3   **FIGURE 4**

The area problem is the central problem in the branch of calculus called *integral calculus*. The techniques that we will develop in Chapter 4 for finding areas will also enable us to compute the volume of a solid, the length of a curve, the force of water against a dam, the mass and center of gravity of a rod, and the work done in pumping water out of a tank.

The Tangent Problem

Consider the problem of trying to find an equation of the tangent line t to a curve with equation $y = f(x)$ at a given point P. (We will give a precise definition of a tangent line in

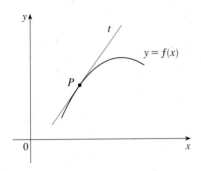

FIGURE 5
The tangent line at P

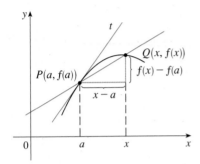

FIGURE 6
The secant line PQ

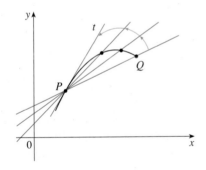

FIGURE 7
Secant lines approaching the
tangent line

Chapter 1. For now you can think of it as a line that touches the curve at P as in Figure 5.) Since we know that the point P lies on the tangent line, we can find the equation of t if we know its slope m. The problem is that we need two points to compute the slope and we know only one point, P, on t. To get around the problem we first find an approximation to m by taking a nearby point Q on the curve and computing the slope m_{PQ} of the secant line PQ. From Figure 6 we see that

$$\boxed{1} \qquad m_{PQ} = \frac{f(x) - f(a)}{x - a}$$

Now imagine that Q moves along the curve toward P as in Figure 7. You can see that the secant line rotates and approaches the tangent line as its limiting position. This means that the slope m_{PQ} of the secant line becomes closer and closer to the slope m of the tangent line. We write

$$m = \lim_{Q \to P} m_{PQ}$$

and we say that m is the limit of m_{PQ} as Q approaches P along the curve. Since x approaches a as Q approaches P, we could also use Equation 1 to write

$$\boxed{2} \qquad m = \lim_{x \to a} \frac{f(x) - f(a)}{x - a}$$

Specific examples of this procedure will be given in Chapter 1.

The tangent problem has given rise to the branch of calculus called *differential calculus*, which was not invented until more than 2000 years after integral calculus. The main ideas behind differential calculus are due to the French mathematician Pierre Fermat (1601–1665) and were developed by the English mathematicians John Wallis (1616–1703), Isaac Barrow (1630–1677), and Isaac Newton (1642–1727) and the German mathematician Gottfried Leibniz (1646–1716).

The two branches of calculus and their chief problems, the area problem and the tangent problem, appear to be very different, but it turns out that there is a very close connection between them. The tangent problem and the area problem are inverse problems in a sense that will be described in Chapter 4.

Velocity

When we look at the speedometer of a car and read that the car is traveling at 48 mi/h, what does that information indicate to us? We know that if the velocity remains constant, then after an hour we will have traveled 48 mi. But if the velocity of the car varies, what does it mean to say that the velocity at a given instant is 48 mi/h?

In order to analyze this question, let's examine the motion of a car that travels along a straight road and assume that we can measure the distance traveled by the car (in feet) at 1-second intervals as in the following chart:

t = Time elapsed (s)	0	1	2	3	4	5
d = Distance (ft)	0	2	9	24	42	71

As a first step toward finding the velocity after 2 seconds have elapsed, we find the average velocity during the time interval $2 \leqslant t \leqslant 4$:

$$\text{average velocity} = \frac{\text{change in position}}{\text{time elapsed}}$$

$$= \frac{42 - 9}{4 - 2}$$

$$= 16.5 \text{ ft/s}$$

Similarly, the average velocity in the time interval $2 \leqslant t \leqslant 3$ is

$$\text{average velocity} = \frac{24 - 9}{3 - 2} = 15 \text{ ft/s}$$

We have the feeling that the velocity at the instant $t = 2$ can't be much different from the average velocity during a short time interval starting at $t = 2$. So let's imagine that the distance traveled has been measured at 0.1-second time intervals as in the following chart:

t	2.0	2.1	2.2	2.3	2.4	2.5
d	9.00	10.02	11.16	12.45	13.96	15.80

Then we can compute, for instance, the average velocity over the time interval $[2, 2.5]$:

$$\text{average velocity} = \frac{15.80 - 9.00}{2.5 - 2} = 13.6 \text{ ft/s}$$

The results of such calculations are shown in the following chart:

Time interval	[2, 3]	[2, 2.5]	[2, 2.4]	[2, 2.3]	[2, 2.2]	[2, 2.1]
Average velocity (ft/s)	15.0	13.6	12.4	11.5	10.8	10.2

The average velocities over successively smaller intervals appear to be getting closer to a number near 10, and so we expect that the velocity at exactly $t = 2$ is about 10 ft/s. In Chapter 1 we will define the instantaneous velocity of a moving object as the limiting value of the average velocities over smaller and smaller time intervals.

In Figure 8 we show a graphical representation of the motion of the car by plotting the distance traveled as a function of time. If we write $d = f(t)$, then $f(t)$ is the number of feet traveled after t seconds. The average velocity in the time interval $[2, t]$ is

$$\text{average velocity} = \frac{\text{change in position}}{\text{time elapsed}} = \frac{f(t) - f(2)}{t - 2}$$

which is the same as the slope of the secant line PQ in Figure 8. The velocity v when $t = 2$ is the limiting value of this average velocity as t approaches 2; that is,

$$v = \lim_{t \to 2} \frac{f(t) - f(2)}{t - 2}$$

and we recognize from Equation 2 that this is the same as the slope of the tangent line to the curve at P.

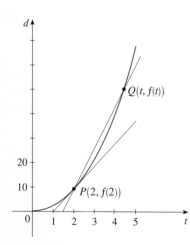

FIGURE 8

Thus, when we solve the tangent problem in differential calculus, we are also solving problems concerning velocities. The same techniques also enable us to solve problems involving rates of change in all of the natural and social sciences.

The Limit of a Sequence

In the fifth century BC the Greek philosopher Zeno of Elea posed four problems, now known as *Zeno's paradoxes,* that were intended to challenge some of the ideas concerning space and time that were held in his day. Zeno's second paradox concerns a race between the Greek hero Achilles and a tortoise that has been given a head start. Zeno argued, as follows, that Achilles could never pass the tortoise: Suppose that Achilles starts at position a_1 and the tortoise starts at position t_1. (See Figure 9.) When Achilles reaches the point $a_2 = t_1$, the tortoise is farther ahead at position t_2. When Achilles reaches $a_3 = t_2$, the tortoise is at t_3. This process continues indefinitely and so it appears that the tortoise will always be ahead! But this defies common sense.

FIGURE 9

One way of explaining this paradox is with the idea of a *sequence*. The successive positions of Achilles $(a_1, a_2, a_3, \ldots)$ or the successive positions of the tortoise $(t_1, t_2, t_3, \ldots)$ form what is known as a sequence.

In general, a sequence $\{a_n\}$ is a set of numbers written in a definite order. For instance, the sequence

$$\left\{1, \tfrac{1}{2}, \tfrac{1}{3}, \tfrac{1}{4}, \tfrac{1}{5}, \ldots\right\}$$

can be described by giving the following formula for the nth term:

$$a_n = \frac{1}{n}$$

(a)

(b)

FIGURE 10

We can visualize this sequence by plotting its terms on a number line as in Figure 10(a) or by drawing its graph as in Figure 10(b). Observe from either picture that the terms of the sequence $a_n = 1/n$ are becoming closer and closer to 0 as n increases. In fact, we can find terms as small as we please by making n large enough. We say that the limit of the sequence is 0, and we indicate this by writing

$$\lim_{n \to \infty} \frac{1}{n} = 0$$

In general, the notation

$$\lim_{n \to \infty} a_n = L$$

is used if the terms a_n approach the number L as n becomes large. This means that the numbers a_n can be made as close as we like to the number L by taking n sufficiently large.

The concept of the limit of a sequence occurs whenever we use the decimal representation of a real number. For instance, if

$$a_1 = 3.1$$

$$a_2 = 3.14$$

$$a_3 = 3.141$$

$$a_4 = 3.1415$$

$$a_5 = 3.14159$$

$$a_6 = 3.141592$$

$$a_7 = 3.1415926$$

$$\vdots$$

then

$$\lim_{n \to \infty} a_n = \pi$$

The terms in this sequence are rational approximations to π.

Let's return to Zeno's paradox. The successive positions of Achilles and the tortoise form sequences $\{a_n\}$ and $\{t_n\}$, where $a_n < t_n$ for all n. It can be shown that both sequences have the same limit:

$$\lim_{n \to \infty} a_n = p = \lim_{n \to \infty} t_n$$

It is precisely at this point p that Achilles overtakes the tortoise.

The Sum of a Series

Another of Zeno's paradoxes, as passed on to us by Aristotle, is the following: "A man standing in a room cannot walk to the wall. In order to do so, he would first have to go half the distance, then half the remaining distance, and then again half of what still remains. This process can always be continued and can never be ended." (See Figure 11.)

FIGURE 11

Of course, we know that the man can actually reach the wall, so this suggests that perhaps the total distance can be expressed as the sum of infinitely many smaller distances as follows:

[3]
$$1 = \frac{1}{2} + \frac{1}{4} + \frac{1}{8} + \frac{1}{16} + \cdots + \frac{1}{2^n} + \cdots$$

Zeno was arguing that it doesn't make sense to add infinitely many numbers together. But there are other situations in which we implicitly use infinite sums. For instance, in decimal notation, the symbol $0.\overline{3} = 0.3333\ldots$ means

$$\frac{3}{10} + \frac{3}{100} + \frac{3}{1000} + \frac{3}{10,000} + \cdots$$

and so, in some sense, it must be true that

$$\frac{3}{10} + \frac{3}{100} + \frac{3}{1000} + \frac{3}{10,000} + \cdots = \frac{1}{3}$$

More generally, if d_n denotes the nth digit in the decimal representation of a number, then

$$0.d_1 d_2 d_3 d_4 \ldots = \frac{d_1}{10} + \frac{d_2}{10^2} + \frac{d_3}{10^3} + \cdots + \frac{d_n}{10^n} + \cdots$$

Therefore some infinite sums, or infinite series as they are called, have a meaning. But we must define carefully what the sum of an infinite series is.

Returning to the series in Equation 3, we denote by s_n the sum of the first n terms of the series. Thus

$$s_1 = \tfrac{1}{2} = 0.5$$

$$s_2 = \tfrac{1}{2} + \tfrac{1}{4} = 0.75$$

$$s_3 = \tfrac{1}{2} + \tfrac{1}{4} + \tfrac{1}{8} = 0.875$$

$$s_4 = \tfrac{1}{2} + \tfrac{1}{4} + \tfrac{1}{8} + \tfrac{1}{16} = 0.9375$$

$$s_5 = \tfrac{1}{2} + \tfrac{1}{4} + \tfrac{1}{8} + \tfrac{1}{16} + \tfrac{1}{32} = 0.96875$$

$$s_6 = \tfrac{1}{2} + \tfrac{1}{4} + \tfrac{1}{8} + \tfrac{1}{16} + \tfrac{1}{32} + \tfrac{1}{64} = 0.984375$$

$$s_7 = \tfrac{1}{2} + \tfrac{1}{4} + \tfrac{1}{8} + \tfrac{1}{16} + \tfrac{1}{32} + \tfrac{1}{64} + \tfrac{1}{128} = 0.9921875$$

$$\vdots$$

$$s_{10} = \tfrac{1}{2} + \tfrac{1}{4} + \cdots + \tfrac{1}{1024} \approx 0.99902344$$

$$\vdots$$

$$s_{16} = \frac{1}{2} + \frac{1}{4} + \cdots + \frac{1}{2^{16}} \approx 0.99998474$$

Observe that as we add more and more terms, the partial sums become closer and closer to 1. In fact, it can be shown that by taking n large enough (that is, by adding sufficiently many terms of the series), we can make the partial sum s_n as close as we please to the number 1. It therefore seems reasonable to say that the sum of the infinite series is 1 and to write

$$\frac{1}{2} + \frac{1}{4} + \frac{1}{8} + \cdots + \frac{1}{2^n} + \cdots = 1$$

In other words, the reason the sum of the series is 1 is that

$$\lim_{n \to \infty} s_n = 1$$

In Chapter 11 we will discuss these ideas further. We will then use Newton's idea of combining infinite series with differential and integral calculus.

Summary

We have seen that the concept of a limit arises in trying to find the area of a region, the slope of a tangent to a curve, the velocity of a car, or the sum of an infinite series. In each case the common theme is the calculation of a quantity as the limit of other, easily calculated quantities. It is this basic idea of a limit that sets calculus apart from other areas of mathematics. In fact, we could define calculus as the part of mathematics that deals with limits.

After Sir Isaac Newton invented his version of calculus, he used it to explain the motion of the planets around the sun. Today calculus is used in calculating the orbits of satellites and spacecraft, in predicting population sizes, in estimating how fast oil prices rise or fall, in forecasting weather, in measuring the cardiac output of the heart, in calculating life insurance premiums, and in a great variety of other areas. We will explore some of these uses of calculus in this book.

In order to convey a sense of the power of the subject, we end this preview with a list of some of the questions that you will be able to answer using calculus:

1. How can we explain the fact, illustrated in Figure 12, that the angle of elevation from an observer up to the highest point in a rainbow is 42°? (See page 162.)

2. How can we explain the shapes of cans on supermarket shelves? (See page 203.)

3. Where is the best place to sit in a movie theater? (See page 363.)

4. How can we design a roller coaster for a smooth ride? (See page 109.)

5. How far away from an airport should a pilot start descent? (See page 122.)

6. How can we fit curves together to design shapes to represent letters on a laser printer? (See page 536.)

7. Where should an infielder position himself to catch a baseball thrown by an outfielder and relay it to home plate? (See page 521.)

8. Does a ball thrown upward take longer to reach its maximum height or to fall back to its original height? (See page 498.)

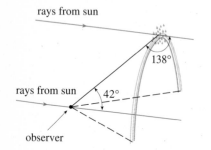

rays from sun

138°

rays from sun   42°

observer

FIGURE 12

Functions and Limits

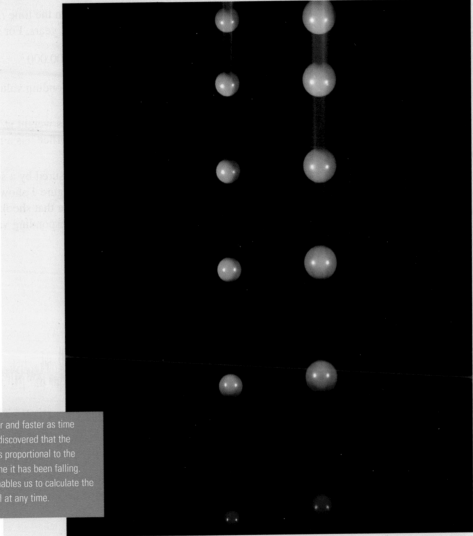

© 1986 Peticolas / Megna, Fundamental Photographs, NYC

A ball falls faster and faster as time passes. Galileo discovered that the distance fallen is proportional to the square of the time it has been falling. Calculus then enables us to calculate the speed of the ball at any time.

The fundamental objects that we deal with in calculus are functions. We stress that a function can be represented in different ways: by an equation, in a table, by a graph, or in words. We look at the main types of functions that occur in calculus and describe the process of using these functions as mathematical models of real-world phenomena.

In *A Preview of Calculus* (page 1) we saw how the idea of a limit underlies the various branches of calculus. It is therefore appropriate to begin our study of calculus by investigating limits of functions and their properties.

1.1 Four Ways to Represent a Function

Functions arise whenever one quantity depends on another. Consider the following four situations.

A. The area A of a circle depends on the radius r of the circle. The rule that connects r and A is given by the equation $A = \pi r^2$. With each positive number r there is associated one value of A, and we say that A is a *function* of r.

B. The human population of the world P depends on the time t. The table gives estimates of the world population $P(t)$ at time t, for certain years. For instance,

$$P(1950) \approx 2{,}560{,}000{,}000$$

But for each value of the time t there is a corresponding value of P, and we say that P is a function of t.

C. The cost C of mailing an envelope depends on its weight w. Although there is no simple formula that connects w and C, the post office has a rule for determining C when w is known.

D. The vertical acceleration a of the ground as measured by a seismograph during an earthquake is a function of the elapsed time t. Figure 1 shows a graph generated by seismic activity during the Northridge earthquake that shook Los Angeles in 1994. For a given value of t, the graph provides a corresponding value of a.

Year	Population (millions)
1900	1650
1910	1750
1920	1860
1930	2070
1940	2300
1950	2560
1960	3040
1970	3710
1980	4450
1990	5280
2000	6080
2010	6870

Calif. Dept. of Mines and Geology

FIGURE 1
Vertical ground acceleration during the Northridge earthquake

Each of these examples describes a rule whereby, given a number (r, t, w, or t), another number (A, P, C, or a) is assigned. In each case we say that the second number is a function of the first number.

> A **function** f is a rule that assigns to each element x in a set D exactly one element, called $f(x)$, in a set E.

We usually consider functions for which the sets D and E are sets of real numbers. The set D is called the **domain** of the function. The number $f(x)$ is the **value of f at x** and is read "f of x." The **range** of f is the set of all possible values of $f(x)$ as x varies throughout the domain. A symbol that represents an arbitrary number in the *domain* of a function f is called an **independent variable**. A symbol that represents a number in the *range* of f is called a **dependent variable**. In Example A, for instance, r is the independent variable and A is the dependent variable.

FIGURE 2
Machine diagram for a function f

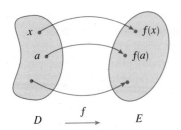

FIGURE 3
Arrow diagram for f

It's helpful to think of a function as a **machine** (see Figure 2). If x is in the domain of the function f, then when x enters the machine, it's accepted as an input and the machine produces an output $f(x)$ according to the rule of the function. Thus we can think of the domain as the set of all possible inputs and the range as the set of all possible outputs.

The preprogrammed functions in a calculator are good examples of a function as a machine. For example, the square root key on your calculator computes such a function. You press the key labeled $\sqrt{}$ (or $\sqrt{x}$) and enter the input x. If $x < 0$, then x is not in the domain of this function; that is, x is not an acceptable input, and the calculator will indicate an error. If $x \geqslant 0$, then an *approximation* to $\sqrt{x}$ will appear in the display. Thus the $\sqrt{x}$ key on your calculator is not quite the same as the exact mathematical function f defined by $f(x) = \sqrt{x}$.

Another way to picture a function is by an **arrow diagram** as in Figure 3. Each arrow connects an element of D to an element of E. The arrow indicates that $f(x)$ is associated with x, $f(a)$ is associated with a, and so on.

The most common method for visualizing a function is its graph. If f is a function with domain D, then its **graph** is the set of ordered pairs

$$\{(x, f(x)) \mid x \in D\}$$

(Notice that these are input-output pairs.) In other words, the graph of f consists of all points (x, y) in the coordinate plane such that $y = f(x)$ and x is in the domain of f.

The graph of a function f gives us a useful picture of the behavior or "life history" of a function. Since the y-coordinate of any point (x, y) on the graph is $y = f(x)$, we can read the value of $f(x)$ from the graph as being the height of the graph above the point x (see Figure 4). The graph of f also allows us to picture the domain of f on the x-axis and its range on the y-axis as in Figure 5.

FIGURE 4

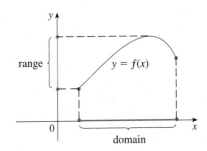

FIGURE 5

EXAMPLE 1 The graph of a function f is shown in Figure 6.
(a) Find the values of $f(1)$ and $f(5)$.
(b) What are the domain and range of f?

SOLUTION
(a) We see from Figure 6 that the point $(1, 3)$ lies on the graph of f, so the value of f at 1 is $f(1) = 3$. (In other words, the point on the graph that lies above $x = 1$ is 3 units above the x-axis.)

When $x = 5$, the graph lies about 0.7 unit below the x-axis, so we estimate that $f(5) \approx -0.7$.

(b) We see that $f(x)$ is defined when $0 \leqslant x \leqslant 7$, so the domain of f is the closed interval $[0, 7]$. Notice that f takes on all values from -2 to 4, so the range of f is

$$\{y \mid -2 \leqslant y \leqslant 4\} = [-2, 4]$$

FIGURE 6

The notation for intervals is given in Appendix A.

FIGURE 7

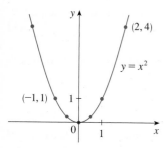

FIGURE 8

The expression

$$\frac{f(a + h) - f(a)}{h}$$

in Example 3 is called a **difference quotient** and occurs frequently in calculus. As we will see in Chapter 2, it represents the average rate of change of $f(x)$ between $x = a$ and $x = a + h$.

EXAMPLE 2   Sketch the graph and find the domain and range of each function.

(a) $f(x) = 2x - 1$ (b) $g(x) = x^2$

SOLUTION

(a) The equation of the graph is $y = 2x - 1$, and we recognize this as being the equation of a line with slope 2 and y-intercept -1. (Recall the slope-intercept form of the equation of a line: $y = mx + b$. See Appendix B.) This enables us to sketch a portion of the graph of f in Figure 7. The expression $2x - 1$ is defined for all real numbers, so the domain of f is the set of all real numbers, which we denote by $\mathbb{R}$. The graph shows that the range is also $\mathbb{R}$.

(b) Since $g(2) = 2^2 = 4$ and $g(-1) = (-1)^2 = 1$, we could plot the points $(2, 4)$ and $(-1, 1)$, together with a few other points on the graph, and join them to produce the graph (Figure 8). The equation of the graph is $y = x^2$, which represents a parabola (see Appendix C). The domain of g is $\mathbb{R}$. The range of g consists of all values of $g(x)$, that is, all numbers of the form x^2. But $x^2 \geqslant 0$ for all numbers x and any positive number y is a square. So the range of g is $\{y \mid y \geqslant 0\} = [0, \infty)$. This can also be seen from Figure 8.

EXAMPLE 3   If $f(x) = 2x^2 - 5x + 1$ and $h \neq 0$, evaluate $\dfrac{f(a + h) - f(a)}{h}$.

SOLUTION   We first evaluate $f(a + h)$ by replacing x by $a + h$ in the expression for $f(x)$:

$$f(a + h) = 2(a + h)^2 - 5(a + h) + 1$$

$$= 2(a^2 + 2ah + h^2) - 5(a + h) + 1$$

$$= 2a^2 + 4ah + 2h^2 - 5a - 5h + 1$$

Then we substitute into the given expression and simplify:

$$\frac{f(a + h) - f(a)}{h} = \frac{(2a^2 + 4ah + 2h^2 - 5a - 5h + 1) - (2a^2 - 5a + 1)}{h}$$

$$= \frac{2a^2 + 4ah + 2h^2 - 5a - 5h + 1 - 2a^2 + 5a - 1}{h}$$

$$= \frac{4ah + 2h^2 - 5h}{h} = 4a + 2h - 5$$

Representations of Functions

There are four possible ways to represent a function:

- verbally   (by a description in words)
- numerically   (by a table of values)
- visually   (by a graph)
- algebraically   (by an explicit formula)

If a single function can be represented in all four ways, it's often useful to go from one representation to another to gain additional insight into the function. (In Example 2, for instance, we started with algebraic formulas and then obtained the graphs.) But certain functions are described more naturally by one method than by another. With this in mind, let's reexamine the four situations that we considered at the beginning of this section.

A. The most useful representation of the area of a circle as a function of its radius is probably the algebraic formula $A(r) = \pi r^2$, though it is possible to compile a table of values or to sketch a graph (half a parabola). Because a circle has to have a positive radius, the domain is $\{r \mid r > 0\} = (0, \infty)$, and the range is also $(0, \infty)$.

B. We are given a description of the function in words: $P(t)$ is the human population of the world at time t. Let's measure t so that $t = 0$ corresponds to the year 1900. The table of values of world population provides a convenient representation of this function. If we plot these values, we get the graph (called a *scatter plot*) in Figure 9. It too is a useful representation; the graph allows us to absorb all the data at once. What about a formula? Of course, it's impossible to devise an explicit formula that gives the exact human population $P(t)$ at any time t. But it is possible to find an expression for a function that *approximates* $P(t)$. In fact, using methods explained in Section 1.2, we obtain the approximation

$$P(t) \approx f(t) = (1.43653 \times 10^9) \cdot (1.01395)^t$$

Figure 10 shows that it is a reasonably good "fit." The function f is called a *mathematical model* for population growth. In other words, it is a function with an explicit formula that approximates the behavior of our given function. We will see, however, that the ideas of calculus can be applied to a table of values; an explicit formula is not necessary.

t	Population (millions)
0	1650
10	1750
20	1860
30	2070
40	2300
50	2560
60	3040
70	3710
80	4450
90	5280
100	6080
110	6870

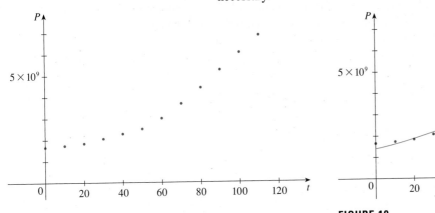

FIGURE 9

FIGURE 10

A function defined by a table of values is called a *tabular* function.

The function P is typical of the functions that arise whenever we attempt to apply calculus to the real world. We start with a verbal description of a function. Then we may be able to construct a table of values of the function, perhaps from instrument readings in a scientific experiment. Even though we don't have complete knowledge of the values of the function, we will see throughout the book that it is still possible to perform the operations of calculus on such a function.

C. Again the function is described in words: Let $C(w)$ be the cost of mailing a large envelope with weight w. The rule that the US Postal Service used as of 2010 is as follows: The cost is 88 cents for up to 1 oz, plus 17 cents for each additional ounce (or less) up to 13 oz. The table of values shown in the margin is the most convenient representation for this function, though it is possible to sketch a graph (see Example 10).

D. The graph shown in Figure 1 is the most natural representation of the vertical acceleration function $a(t)$. It's true that a table of values could be compiled, and it is even possible to devise an approximate formula. But everything a geologist needs to know—amplitudes and patterns—can be seen easily from the graph. (The same is true for the patterns seen in electrocardiograms of heart patients and polygraphs for lie-detection.)

w (ounces)	$C(w)$ (dollars)
$0 < w \leq 1$	0.88
$1 < w \leq 2$	1.05
$2 < w \leq 3$	1.22
$3 < w \leq 4$	1.39
$4 < w \leq 5$	1.56
$\vdots$	$\vdots$

FIGURE 11

In the next example we sketch the graph of a function that is defined verbally.

EXAMPLE 4 When you turn on a hot-water faucet, the temperature T of the water depends on how long the water has been running. Draw a rough graph of T as a function of the time t that has elapsed since the faucet was turned on.

SOLUTION The initial temperature of the running water is close to room temperature because the water has been sitting in the pipes. When the water from the hot-water tank starts flowing from the faucet, T increases quickly. In the next phase, T is constant at the temperature of the heated water in the tank. When the tank is drained, T decreases to the temperature of the water supply. This enables us to make the rough sketch of T as a function of t in Figure 11. ▬

In the following example we start with a verbal description of a function in a physical situation and obtain an explicit algebraic formula. The ability to do this is a useful skill in solving calculus problems that ask for the maximum or minimum values of quantities.

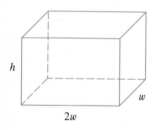

FIGURE 12

V EXAMPLE 5 A rectangular storage container with an open top has a volume of 10 m³. The length of its base is twice its width. Material for the base costs \$10 per square meter; material for the sides costs \$6 per square meter. Express the cost of materials as a function of the width of the base.

SOLUTION We draw a diagram as in Figure 12 and introduce notation by letting w and $2w$ be the width and length of the base, respectively, and h be the height.

The area of the base is $(2w)w = 2w^2$, so the cost, in dollars, of the material for the base is $10(2w^2)$. Two of the sides have area wh and the other two have area $2wh$, so the cost of the material for the sides is $6[2(wh) + 2(2wh)]$. The total cost is therefore

$$C = 10(2w^2) + 6[2(wh) + 2(2wh)] = 20w^2 + 36wh$$

To express C as a function of w alone, we need to eliminate h and we do so by using the fact that the volume is 10 m³. Thus

$$w(2w)h = 10$$

which gives

$$h = \frac{10}{2w^2} = \frac{5}{w^2}$$

Substituting this into the expression for C, we have

$$C = 20w^2 + 36w\left(\frac{5}{w^2}\right) = 20w^2 + \frac{180}{w}$$

Therefore the equation

$$C(w) = 20w^2 + \frac{180}{w} \qquad w > 0$$

expresses C as a function of w. ▬

PS In setting up applied functions as in Example 5, it may be useful to review the principles of problem solving as discussed on page 77, particularly *Step 1: Understand the Problem*.

EXAMPLE 6 Find the domain of each function.

(a) $f(x) = \sqrt{x + 2}$
(b) $g(x) = \dfrac{1}{x^2 - x}$

Domain Convention

If a function is given by a formula and the domain is not stated explicitly, the convention is that the domain is the set of all numbers for which the formula makes sense and defines a real number.

SOLUTION

(a) Because the square root of a negative number is not defined (as a real number), the domain of f consists of all values of x such that $x + 2 \geq 0$. This is equivalent to $x \geq -2$, so the domain is the interval $[-2, \infty)$.

(b) Since

$$g(x) = \frac{1}{x^2 - x} = \frac{1}{x(x - 1)}$$

and division by 0 is not allowed, we see that $g(x)$ is not defined when $x = 0$ or $x = 1$. Thus the domain of g is

$$\{x \mid x \neq 0, x \neq 1\}$$

which could also be written in interval notation as

$$(-\infty, 0) \cup (0, 1) \cup (1, \infty)$$

The graph of a function is a curve in the xy-plane. But the question arises: Which curves in the xy-plane are graphs of functions? This is answered by the following test.

> **The Vertical Line Test** A curve in the xy-plane is the graph of a function of x if and only if no vertical line intersects the curve more than once.

The reason for the truth of the Vertical Line Test can be seen in Figure 13. If each vertical line $x = a$ intersects a curve only once, at (a, b), then exactly one functional value is defined by $f(a) = b$. But if a line $x = a$ intersects the curve twice, at (a, b) and (a, c), then the curve can't represent a function because a function can't assign two different values to a.

FIGURE 13

For example, the parabola $x = y^2 - 2$ shown in Figure 14(a) is not the graph of a function of x because, as you can see, there are vertical lines that intersect the parabola twice. The parabola, however, does contain the graphs of *two* functions of x. Notice that the equation $x = y^2 - 2$ implies $y^2 = x + 2$, so $y = \pm\sqrt{x + 2}$. Thus the upper and lower halves of the parabola are the graphs of the functions $f(x) = \sqrt{x + 2}$ [from Example 6(a)] and $g(x) = -\sqrt{x + 2}$. [See Figures 14(b) and (c).] We observe that if we reverse the roles of x and y, then the equation $x = h(y) = y^2 - 2$ *does* define x as a function of y (with y as the independent variable and x as the dependent variable) and the parabola now appears as the graph of the function h.

FIGURE 14   (a) $x = y^2 - 2$ (b) $y = \sqrt{x + 2}$ (c) $y = -\sqrt{x + 2}$

Piecewise Defined Functions

The functions in the following four examples are defined by different formulas in different parts of their domains. Such functions are called **piecewise defined functions**.

V **EXAMPLE 7** A function f is defined by

$$f(x) = \begin{cases} 1 - x & \text{if } x \leq -1 \\ x^2 & \text{if } x > -1 \end{cases}$$

Evaluate $f(-2)$, $f(-1)$, and $f(0)$ and sketch the graph.

SOLUTION Remember that a function is a rule. For this particular function the rule is the following: First look at the value of the input x. If it happens that $x \leq -1$, then the value of $f(x)$ is $1 - x$. On the other hand, if $x > -1$, then the value of $f(x)$ is x^2.

$$\text{Since } -2 \leq -1, \text{ we have } f(-2) = 1 - (-2) = 3.$$

$$\text{Since } -1 \leq -1, \text{ we have } f(-1) = 1 - (-1) = 2.$$

$$\text{Since } 0 > -1, \text{ we have } f(0) = 0^2 = 0.$$

How do we draw the graph of f? We observe that if $x \leq -1$, then $f(x) = 1 - x$, so the part of the graph of f that lies to the left of the vertical line $x = -1$ must coincide with the line $y = 1 - x$, which has slope -1 and y-intercept 1. If $x > -1$, then $f(x) = x^2$, so the part of the graph of f that lies to the right of the line $x = -1$ must coincide with the graph of $y = x^2$, which is a parabola. This enables us to sketch the graph in Figure 15. The solid dot indicates that the point $(-1, 2)$ is included on the graph; the open dot indicates that the point $(-1, 1)$ is excluded from the graph.

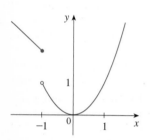

FIGURE 15

The next example of a piecewise defined function is the absolute value function. Recall that the **absolute value** of a number a, denoted by $|a|$, is the distance from a to 0 on the real number line. Distances are always positive or 0, so we have

For a more extensive review of absolute values, see Appendix A.

$$|a| \geq 0 \qquad \text{for every number } a$$

For example,

$$|3| = 3 \qquad |-3| = 3 \qquad |0| = 0 \qquad |\sqrt{2} - 1| = \sqrt{2} - 1 \qquad |3 - \pi| = \pi - 3$$

In general, we have

$$\begin{aligned} |a| &= a \qquad \text{if } a \geq 0 \\ |a| &= -a \qquad \text{if } a < 0 \end{aligned}$$

(Remember that if a is negative, then $-a$ is positive.)

EXAMPLE 8 Sketch the graph of the absolute value function $f(x) = |x|$.

SOLUTION From the preceding discussion we know that

$$|x| = \begin{cases} x & \text{if } x \geq 0 \\ -x & \text{if } x < 0 \end{cases}$$

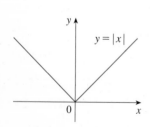

FIGURE 16

Using the same method as in Example 7, we see that the graph of f coincides with the line $y = x$ to the right of the y-axis and coincides with the line $y = -x$ to the left of the y-axis (see Figure 16).

EXAMPLE 9 Find a formula for the function f graphed in Figure 17.

FIGURE 17

SOLUTION The line through $(0, 0)$ and $(1, 1)$ has slope $m = 1$ and y-intercept $b = 0$, so its equation is $y = x$. Thus, for the part of the graph of f that joins $(0, 0)$ to $(1, 1)$, we have

$$f(x) = x \qquad \text{if } 0 \leqslant x \leqslant 1$$

Point-slope form of the equation of a line:

$$y - y_1 = m(x - x_1)$$

See Appendix B.

The line through $(1, 1)$ and $(2, 0)$ has slope $m = -1$, so its point-slope form is

$$y - 0 = (-1)(x - 2) \qquad \text{or} \qquad y = 2 - x$$

So we have $\qquad f(x) = 2 - x \qquad \text{if } 1 < x \leqslant 2$

We also see that the graph of f coincides with the x-axis for $x > 2$. Putting this information together, we have the following three-piece formula for f:

$$f(x) = \begin{cases} x & \text{if } 0 \leqslant x \leqslant 1 \\ 2 - x & \text{if } 1 < x \leqslant 2 \\ 0 & \text{if } x > 2 \end{cases}$$

EXAMPLE 10 In Example C at the beginning of this section we considered the cost $C(w)$ of mailing a large envelope with weight w. In effect, this is a piecewise defined function because, from the table of values on page 13, we have

$$C(w) = \begin{cases} 0.88 & \text{if } 0 < w \leqslant 1 \\ 1.05 & \text{if } 1 < w \leqslant 2 \\ 1.22 & \text{if } 2 < w \leqslant 3 \\ 1.39 & \text{if } 3 < w \leqslant 4 \\ \vdots \end{cases}$$

FIGURE 18

The graph is shown in Figure 18. You can see why functions similar to this one are called **step functions**—they jump from one value to the next. Such functions will be studied in Chapter 2.

Symmetry

If a function f satisfies $f(-x) = f(x)$ for every number x in its domain, then f is called an **even function**. For instance, the function $f(x) = x^2$ is even because

$$f(-x) = (-x)^2 = x^2 = f(x)$$

The geometric significance of an even function is that its graph is symmetric with respect to the y-axis (see Figure 19). This means that if we have plotted the graph of f for $x \geqslant 0$, we obtain the entire graph simply by reflecting this portion about the y-axis.

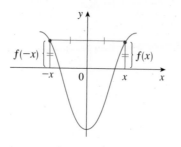

FIGURE 19 An even function   **FIGURE 20** An odd function

If f satisfies $f(-x) = -f(x)$ for every number x in its domain, then f is called an **odd function**. For example, the function $f(x) = x^3$ is odd because

$$f(-x) = (-x)^3 = -x^3 = -f(x)$$

The graph of an odd function is symmetric about the origin (see Figure 20). If we already have the graph of f for $x \geqslant 0$, we can obtain the entire graph by rotating this portion through $180°$ about the origin.

V EXAMPLE 11 Determine whether each of the following functions is even, odd, or neither even nor odd.

(a) $f(x) = x^5 + x$ (b) $g(x) = 1 - x^4$ (c) $h(x) = 2x - x^2$

SOLUTION

(a)
$$f(-x) = (-x)^5 + (-x) = (-1)^5 x^5 + (-x)$$
$$= -x^5 - x = -(x^5 + x)$$
$$= -f(x)$$

Therefore f is an odd function.

(b)
$$g(-x) = 1 - (-x)^4 = 1 - x^4 = g(x)$$

So g is even.

(c)
$$h(-x) = 2(-x) - (-x)^2 = -2x - x^2$$

Since $h(-x) \neq h(x)$ and $h(-x) \neq -h(x)$, we conclude that h is neither even nor odd.

The graphs of the functions in Example 11 are shown in Figure 21. Notice that the graph of h is symmetric neither about the y-axis nor about the origin.

FIGURE 21   (a)   (b)   (c)

FIGURE 22

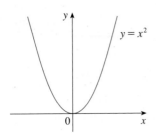

FIGURE 23

Increasing and Decreasing Functions

The graph shown in Figure 22 rises from A to B, falls from B to C, and rises again from C to D. The function f is said to be increasing on the interval $[a, b]$, decreasing on $[b, c]$, and increasing again on $[c, d]$. Notice that if x_1 and x_2 are any two numbers between a and b with $x_1 < x_2$, then $f(x_1) < f(x_2)$. We use this as the defining property of an increasing function.

A function f is called **increasing** on an interval I if

$$f(x_1) < f(x_2) \qquad \text{whenever } x_1 < x_2 \text{ in } I$$

It is called **decreasing** on I if

$$f(x_1) > f(x_2) \qquad \text{whenever } x_1 < x_2 \text{ in } I$$

In the definition of an increasing function it is important to realize that the inequality $f(x_1) < f(x_2)$ must be satisfied for *every* pair of numbers x_1 and x_2 in I with $x_1 < x_2$.

You can see from Figure 23 that the function $f(x) = x^2$ is decreasing on the interval $(-\infty, 0]$ and increasing on the interval $[0, \infty)$.

1.1 Exercises

Access end-of-section exercises and tools for visualizing calculus online at **www.webassign.net**

1.2 Mathematical Models: A Catalog of Essential Functions

A **mathematical model** is a mathematical description (often by means of a function or an equation) of a real-world phenomenon such as the size of a population, the demand for a product, the speed of a falling object, the concentration of a product in a chemical reaction, the life expectancy of a person at birth, or the cost of emission reductions. The purpose of the model is to understand the phenomenon and perhaps to make predictions about future behavior.

Figure 1 (on page 20) illustrates the process of mathematical modeling. Given a real-world problem, our first task is to formulate a mathematical model by identifying and naming the independent and dependent variables and making assumptions that simplify the phenomenon enough to make it mathematically tractable. We use our knowledge of the physical situation and our mathematical skills to obtain equations that relate the variables. In situations where there is no physical law to guide us, we may need to collect data (either from a library or the Internet or by conducting our own experiments) and examine the data in the form of a table in order to discern patterns. From this numerical representation of a function we may wish to obtain a graphical representation by plotting the data. The graph might even suggest a suitable algebraic formula in some cases.

FIGURE 1   The modeling process

The second stage is to apply the mathematics that we know (such as the calculus that will be developed throughout this book) to the mathematical model that we have formulated in order to derive mathematical conclusions. Then, in the third stage, we take those mathematical conclusions and interpret them as information about the original real-world phenomenon by way of offering explanations or making predictions. The final step is to test our predictions by checking against new real data. If the predictions don't compare well with reality, we need to refine our model or to formulate a new model and start the cycle again.

A mathematical model is never a completely accurate representation of a physical situation—it is an *idealization*. A good model simplifies reality enough to permit mathematical calculations but is accurate enough to provide valuable conclusions. It is important to realize the limitations of the model. In the end, Mother Nature has the final say.

There are many different types of functions that can be used to model relationships observed in the real world. In what follows, we discuss the behavior and graphs of these functions and give examples of situations appropriately modeled by such functions.

Linear Models

The coordinate geometry of lines is reviewed in Appendix B.

When we say that y is a **linear function** of x, we mean that the graph of the function is a line, so we can use the slope-intercept form of the equation of a line to write a formula for the function as

$$y = f(x) = mx + b$$

where m is the slope of the line and b is the y-intercept.

A characteristic feature of linear functions is that they grow at a constant rate. For instance, Figure 2 shows a graph of the linear function $f(x) = 3x - 2$ and a table of sample values. Notice that whenever x increases by 0.1, the value of $f(x)$ increases by 0.3. So $f(x)$ increases three times as fast as x. Thus the slope of the graph $y = 3x - 2$, namely 3, can be interpreted as the rate of change of y with respect to x.

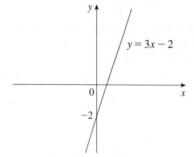

x	$f(x) = 3x - 2$
1.0	1.0
1.1	1.3
1.2	1.6
1.3	1.9
1.4	2.2
1.5	2.5

FIGURE 2

V   EXAMPLE 1

(a) As dry air moves upward, it expands and cools. If the ground temperature is 20°C and the temperature at a height of 1 km is 10°C, express the temperature T (in °C) as a function of the height h (in kilometers), assuming that a linear model is appropriate.

(b) Draw the graph of the function in part (a). What does the slope represent?

(c) What is the temperature at a height of 2.5 km?

SOLUTION

(a) Because we are assuming that T is a linear function of h, we can write

$$T = mh + b$$

We are given that $T = 20$ when $h = 0$, so

$$20 = m \cdot 0 + b = b$$

In other words, the y-intercept is $b = 20$.

We are also given that $T = 10$ when $h = 1$, so

$$10 = m \cdot 1 + 20$$

The slope of the line is therefore $m = 10 - 20 = -10$ and the required linear function is

$$T = -10h + 20$$

(b) The graph is sketched in Figure 3. The slope is $m = -10°\text{C/km}$, and this represents the rate of change of temperature with respect to height.

(c) At a height of $h = 2.5$ km, the temperature is

$$T = -10(2.5) + 20 = -5°\text{C}$$

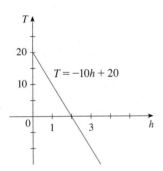

FIGURE 3

If there is no physical law or principle to help us formulate a model, we construct an **empirical model**, which is based entirely on collected data. We seek a curve that "fits" the data in the sense that it captures the basic trend of the data points.

V **EXAMPLE 2** Table 1 lists the average carbon dioxide level in the atmosphere, measured in parts per million at Mauna Loa Observatory from 1980 to 2008. Use the data in Table 1 to find a model for the carbon dioxide level.

SOLUTION We use the data in Table 1 to make the scatter plot in Figure 4, where t represents time (in years) and C represents the CO_2 level (in parts per million, ppm).

TABLE 1

Year	CO_2 level (in ppm)	Year	CO_2 level (in ppm)
1980	338.7	1996	362.4
1982	341.2	1998	366.5
1984	344.4	2000	369.4
1986	347.2	2002	373.2
1988	351.5	2004	377.5
1990	354.2	2006	381.9
1992	356.3	2008	385.6
1994	358.6		

FIGURE 4 Scatter plot for the average CO_2 level

Notice that the data points appear to lie close to a straight line, so it's natural to choose a linear model in this case. But there are many possible lines that approximate these data points, so which one should we use? One possibility is the line that passes through the first and last data points. The slope of this line is

$$\frac{385.6 - 338.7}{2008 - 1980} = \frac{46.9}{28} = 1.675$$

and its equation is

$$C - 338.7 = 1.675(t - 1980)$$

or

1 $$C = 1.675t - 2977.8$$

Equation 1 gives one possible linear model for the carbon dioxide level; it is graphed in Figure 5.

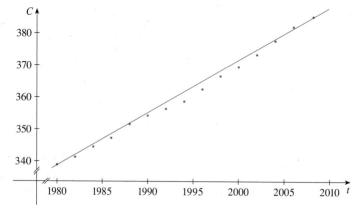

FIGURE 5
Linear model through
first and last data points

A computer or graphing calculator finds the regression line by the method of **least squares**, which is to minimize the sum of the squares of the vertical distances between the data points and the line. The details are explained in Section 14.7.

Notice that our model gives values higher than most of the actual CO_2 levels. A better linear model is obtained by a procedure from statistics called *linear regression*. If we use a graphing calculator, we enter the data from Table 1 into the data editor and choose the linear regression command. (With Maple we use the fit[leastsquare] command in the stats package; with Mathematica we use the Fit command.) The machine gives the slope and *y*-intercept of the regression line as

$$m = 1.65429 \qquad b = -2938.07$$

So our least squares model for the CO_2 level is

2 $$C = 1.65429t - 2938.07$$

In Figure 6 we graph the regression line as well as the data points. Comparing with Figure 5, we see that it gives a better fit than our previous linear model.

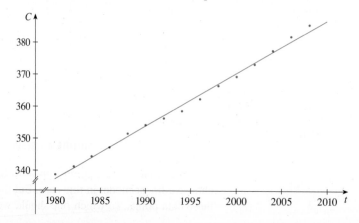

FIGURE 6
The regression line

V **EXAMPLE 3** Use the linear model given by Equation 2 to estimate the average CO_2 level for 1987 and to predict the level for the year 2015. According to this model, when will the CO_2 level exceed 420 parts per million?

SOLUTION Using Equation 2 with $t = 1987$, we estimate that the average CO_2 level in 1987 was

$$C(1987) = (1.65429)(1987) - 2938.07 \approx 349.00$$

This is an example of *interpolation* because we have estimated a value *between* observed values. (In fact, the Mauna Loa Observatory reported that the average CO_2 level in 1987 was 348.93 ppm, so our estimate is quite accurate.)

With $t = 2015$, we get

$$C(2015) = (1.65429)(2015) - 2938.07 \approx 395.32$$

So we predict that the average CO_2 level in the year 2015 will be 395.3 ppm. This is an example of *extrapolation* because we have predicted a value *outside* the region of observations. Consequently, we are far less certain about the accuracy of our prediction.

Using Equation 2, we see that the CO_2 level exceeds 420 ppm when

$$1.65429t - 2938.07 > 420$$

Solving this inequality, we get

$$t > \frac{3358.07}{1.65429} \approx 2029.92$$

We therefore predict that the CO_2 level will exceed 420 ppm by the year 2030. This prediction is risky because it involves a time quite remote from our observations. In fact, we see from Figure 6 that the trend has been for CO_2 levels to increase rather more rapidly in recent years, so the level might exceed 420 ppm well before 2030. ▬

Polynomials

A function P is called a **polynomial** if

$$P(x) = a_n x^n + a_{n-1} x^{n-1} + \cdots + a_2 x^2 + a_1 x + a_0$$

where n is a nonnegative integer and the numbers $a_0, a_1, a_2, \ldots, a_n$ are constants called the **coefficients** of the polynomial. The domain of any polynomial is $\mathbb{R} = (-\infty, \infty)$. If the leading coefficient $a_n \neq 0$, then the **degree** of the polynomial is n. For example, the function

$$P(x) = 2x^6 - x^4 + \tfrac{2}{5}x^3 + \sqrt{2}$$

is a polynomial of degree 6.

A polynomial of degree 1 is of the form $P(x) = mx + b$ and so it is a linear function. A polynomial of degree 2 is of the form $P(x) = ax^2 + bx + c$ and is called a **quadratic function**. Its graph is always a parabola obtained by shifting the parabola $y = ax^2$, as we will see in the next section. The parabola opens upward if $a > 0$ and downward if $a < 0$. (See Figure 7.)

FIGURE 7
The graphs of quadratic functions are parabolas.

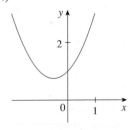

(a) $y = x^2 + x + 1$

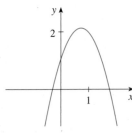

(b) $y = -2x^2 + 3x + 1$

A polynomial of degree 3 is of the form

$$P(x) = ax^3 + bx^2 + cx + d \qquad a \neq 0$$

and is called a **cubic function**. Figure 8 shows the graph of a cubic function in part (a) and graphs of polynomials of degrees 4 and 5 in parts (b) and (c). We will see later why the graphs have these shapes.

  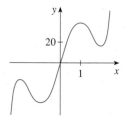

FIGURE 8

(a) $y = x^3 - x + 1$ (b) $y = x^4 - 3x^2 + x$ (c) $y = 3x^5 - 25x^3 + 60x$

Polynomials are commonly used to model various quantities that occur in the natural and social sciences. For instance, in Section 2.7 we will explain why economists often use a polynomial $P(x)$ to represent the cost of producing x units of a commodity. In the following example we use a quadratic function to model the fall of a ball.

TABLE 2

Time (seconds)	Height (meters)
0	450
1	445
2	431
3	408
4	375
5	332
6	279
7	216
8	143
9	61

EXAMPLE 4 A ball is dropped from the upper observation deck of the CN Tower, 450 m above the ground, and its height h above the ground is recorded at 1-second intervals in Table 2. Find a model to fit the data and use the model to predict the time at which the ball hits the ground.

SOLUTION We draw a scatter plot of the data in Figure 9 and observe that a linear model is inappropriate. But it looks as if the data points might lie on a parabola, so we try a quadratic model instead. Using a graphing calculator or computer algebra system (which uses the least squares method), we obtain the following quadratic model:

$$\boxed{3} \qquad h = 449.36 + 0.96t - 4.90t^2$$

 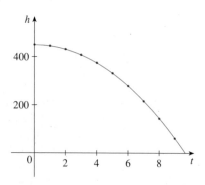

FIGURE 9

Scatter plot for a falling ball

FIGURE 10

Quadratic model for a falling ball

In Figure 10 we plot the graph of Equation 3 together with the data points and see that the quadratic model gives a very good fit.

The ball hits the ground when $h = 0$, so we solve the quadratic equation

$$-4.90t^2 + 0.96t + 449.36 = 0$$

The quadratic formula gives

$$t = \frac{-0.96 \pm \sqrt{(0.96)^2 - 4(-4.90)(449.36)}}{2(-4.90)}$$

The positive root is $t \approx 9.67$, so we predict that the ball will hit the ground after about 9.7 seconds.

Power Functions

A function of the form $f(x) = x^a$, where a is a constant, is called a **power function**. We consider several cases.

(i) $a = n$, where n is a positive integer

The graphs of $f(x) = x^n$ for $n = 1, 2, 3, 4,$ and 5 are shown in Figure 11. (These are polynomials with only one term.) We already know the shape of the graphs of $y = x$ (a line through the origin with slope 1) and $y = x^2$ [a parabola, see Example 2(b) in Section 1.1].

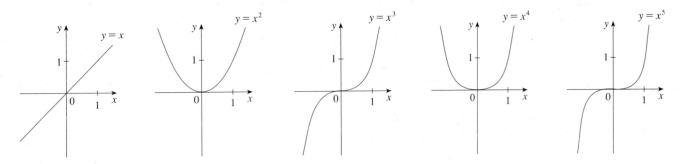

FIGURE 11 Graphs of $f(x) = x^n$ for $n = 1, 2, 3, 4, 5$

The general shape of the graph of $f(x) = x^n$ depends on whether n is even or odd. If n is even, then $f(x) = x^n$ is an even function and its graph is similar to the parabola $y = x^2$. If n is odd, then $f(x) = x^n$ is an odd function and its graph is similar to that of $y = x^3$. Notice from Figure 12, however, that as n increases, the graph of $y = x^n$ becomes flatter near 0 and steeper when $|x| \geq 1$. (If x is small, then x^2 is smaller, x^3 is even smaller, x^4 is smaller still, and so on.)

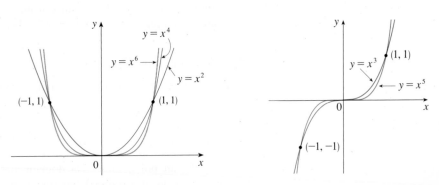

FIGURE 12

Families of power functions

(ii) $a = 1/n$, where n is a positive integer

The function $f(x) = x^{1/n} = \sqrt[n]{x}$ is a **root function**. For $n = 2$ it is the square root function $f(x) = \sqrt{x}$, whose domain is $[0, \infty)$ and whose graph is the upper half of the parabola $x = y^2$. [See Figure 13(a).] For other even values of n, the graph of $y = \sqrt[n]{x}$ is similar to that of $y = \sqrt{x}$. For $n = 3$ we have the cube root function $f(x) = \sqrt[3]{x}$ whose domain is $\mathbb{R}$ (recall that every real number has a cube root) and whose graph is shown in Figure 13(b). The graph of $y = \sqrt[n]{x}$ for n odd ($n > 3$) is similar to that of $y = \sqrt[3]{x}$.

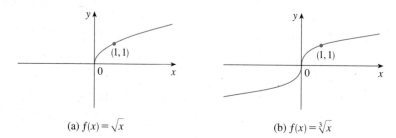

FIGURE 13

Graphs of root functions

(a) $f(x) = \sqrt{x}$

(b) $f(x) = \sqrt[3]{x}$

(iii) $a = -1$

The graph of the **reciprocal function** $f(x) = x^{-1} = 1/x$ is shown in Figure 14. Its graph has the equation $y = 1/x$, or $xy = 1$, and is a hyperbola with the coordinate axes as its asymptotes. This function arises in physics and chemistry in connection with Boyle's Law, which says that, when the temperature is constant, the volume V of a gas is inversely proportional to the pressure P:

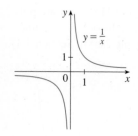

FIGURE 14
The reciprocal function

$$V = \frac{C}{P}$$

where C is a constant. Thus the graph of V as a function of P (see Figure 15) has the same general shape as the right half of Figure 14.

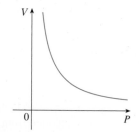

FIGURE 15
Volume as a function of pressure
at constant temperature

Power functions are also used to model species-area relationships (Exercises 26–27), illumination as a function of a distance from a light source (Exercise 25), and the period of revolution of a planet as a function of its distance from the sun (Exercise 28).

▮ Rational Functions

A **rational function** f is a ratio of two polynomials:

$$f(x) = \frac{P(x)}{Q(x)}$$

where P and Q are polynomials. The domain consists of all values of x such that $Q(x) \neq 0$. A simple example of a rational function is the function $f(x) = 1/x$, whose domain is

FIGURE 16
$$f(x) = \frac{2x^4 - x^2 + 1}{x^2 - 4}$$

$\{x \mid x \neq 0\}$; this is the reciprocal function graphed in Figure 14. The function

$$f(x) = \frac{2x^4 - x^2 + 1}{x^2 - 4}$$

is a rational function with domain $\{x \mid x \neq \pm 2\}$. Its graph is shown in Figure 16.

Algebraic Functions

A function f is called an **algebraic function** if it can be constructed using algebraic operations (such as addition, subtraction, multiplication, division, and taking roots) starting with polynomials. Any rational function is automatically an algebraic function. Here are two more examples:

$$f(x) = \sqrt{x^2 + 1} \qquad g(x) = \frac{x^4 - 16x^2}{x + \sqrt{x}} + (x - 2)\sqrt[3]{x + 1}$$

When we sketch algebraic functions in Chapter 3, we will see that their graphs can assume a variety of shapes. Figure 17 illustrates some of the possibilities.

FIGURE 17
(a) $f(x) = x\sqrt{x + 3}$ (b) $g(x) = \sqrt[4]{x^2 - 25}$ (c) $h(x) = x^{2/3}(x - 2)^2$

An example of an algebraic function occurs in the theory of relativity. The mass of a particle with velocity v is

$$m = f(v) = \frac{m_0}{\sqrt{1 - v^2/c^2}}$$

where m_0 is the rest mass of the particle and $c = 3.0 \times 10^5$ km/s is the speed of light in a vacuum.

Trigonometric Functions

The Reference Pages are located at the front and back of the book.

Trigonometry and the trigonometric functions are reviewed on Reference Page 2 and also in Appendix D. In calculus the convention is that radian measure is always used (except when otherwise indicated). For example, when we use the function $f(x) = \sin x$, it is understood that $\sin x$ means the sine of the angle whose radian measure is x. Thus the graphs of the sine and cosine functions are as shown in Figure 18.

(a) $f(x) = \sin x$ (b) $g(x) = \cos x$

FIGURE 18

Notice that for both the sine and cosine functions the domain is $(-\infty, \infty)$ and the range is the closed interval $[-1, 1]$. Thus, for all values of x, we have

$$-1 \le \sin x \le 1 \qquad -1 \le \cos x \le 1$$

or, in terms of absolute values,

$$|\sin x| \le 1 \qquad |\cos x| \le 1$$

Also, the zeros of the sine function occur at the integer multiples of π; that is,

$$\sin x = 0 \qquad \text{when} \qquad x = n\pi \quad n \text{ an integer}$$

An important property of the sine and cosine functions is that they are periodic functions and have period 2π. This means that, for all values of x,

$$\sin(x + 2\pi) = \sin x \qquad \cos(x + 2\pi) = \cos x$$

The periodic nature of these functions makes them suitable for modeling repetitive phenomena such as tides, vibrating springs, and sound waves. For instance, in Example 4 in Section 1.3 we will see that a reasonable model for the number of hours of daylight in Philadelphia t days after January 1 is given by the function

$$L(t) = 12 + 2.8 \sin\left[\frac{2\pi}{365}(t - 80)\right]$$

The tangent function is related to the sine and cosine functions by the equation

$$\tan x = \frac{\sin x}{\cos x}$$

and its graph is shown in Figure 19. It is undefined whenever $\cos x = 0$, that is, when $x = \pm\pi/2, \pm3\pi/2, \ldots$. Its range is $(-\infty, \infty)$. Notice that the tangent function has period π:

$$\tan(x + \pi) = \tan x \qquad \text{for all } x$$

The remaining three trigonometric functions (cosecant, secant, and cotangent) are the reciprocals of the sine, cosine, and tangent functions. Their graphs are shown in Appendix D.

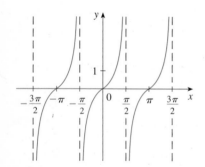

FIGURE 19

$y = \tan x$

▆ Exponential Functions

The **exponential functions** are the functions of the form $f(x) = a^x$, where the base a is a positive constant. The graphs of $y = 2^x$ and $y = (0.5)^x$ are shown in Figure 20. In both cases the domain is $(-\infty, \infty)$ and the range is $(0, \infty)$.

Exponential functions will be studied in detail in Chapter 6, and we will see that they are useful for modeling many natural phenomena, such as population growth (if $a > 1$) and radioactive decay (if $a < 1$).

(a) $y = 2^x$ (b) $y = (0.5)^x$

FIGURE 20

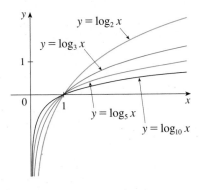

FIGURE 21

Logarithmic Functions

The **logarithmic functions** $f(x) = \log_a x$, where the base a is a positive constant, are the inverse functions of the exponential functions. They will be studied in Chapter 6. Figure 21 shows the graphs of four logarithmic functions with various bases. In each case the domain is $(0, \infty)$, the range is $(-\infty, \infty)$, and the function increases slowly when $x > 1$.

EXAMPLE 5 Classify the following functions as one of the types of functions that we have discussed.

(a) $f(x) = 5^x$

(b) $g(x) = x^5$

(c) $h(x) = \dfrac{1+x}{1-\sqrt{x}}$

(d) $u(t) = 1 - t + 5t^4$

SOLUTION

(a) $f(x) = 5^x$ is an exponential function. (The x is the exponent.)

(b) $g(x) = x^5$ is a power function. (The x is the base.) We could also consider it to be a polynomial of degree 5.

(c) $h(x) = \dfrac{1+x}{1-\sqrt{x}}$ is an algebraic function.

(d) $u(t) = 1 - t + 5t^4$ is a polynomial of degree 4.

1.2 Exercises

Access end-of-section exercises and tools for visualizing calculus online at **www.webassign.net**   **WebAssign**

1.3 New Functions from Old Functions

In this section we start with the basic functions we discussed in Section 1.2 and obtain new functions by shifting, stretching, and reflecting their graphs. We also show how to combine pairs of functions by the standard arithmetic operations and by composition.

Transformations of Functions

By applying certain transformations to the graph of a given function we can obtain the graphs of certain related functions. This will give us the ability to sketch the graphs of many functions quickly by hand. It will also enable us to write equations for given graphs. Let's first consider **translations**. If c is a positive number, then the graph of $y = f(x) + c$ is just the graph of $y = f(x)$ shifted upward a distance of c units (because each y-coordinate is increased by the same number c). Likewise, if $g(x) = f(x - c)$, where $c > 0$, then the value of g at x is the same as the value of f at $x - c$ (c units to the left of x). There-fore the graph of $y = f(x - c)$ is just the graph of $y = f(x)$ shifted c units to the right (see Figure 1 on page 30).

> **Vertical and Horizontal Shifts**   Suppose $c > 0$. To obtain the graph of
>
> $y = f(x) + c$, shift the graph of $y = f(x)$ a distance c units upward
>
> $y = f(x) - c$, shift the graph of $y = f(x)$ a distance c units downward
>
> $y = f(x - c)$, shift the graph of $y = f(x)$ a distance c units to the right
>
> $y = f(x + c)$, shift the graph of $y = f(x)$ a distance c units to the left

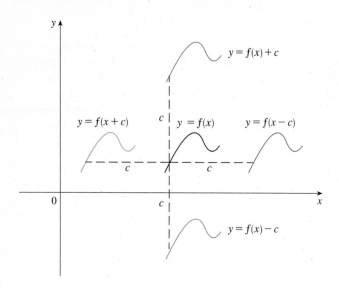

FIGURE 1

Translating the graph of f

FIGURE 2

Stretching and reflecting the graph of f

Now let's consider the **stretching** and **reflecting** transformations. If $c > 1$, then the graph of $y = cf(x)$ is the graph of $y = f(x)$ stretched by a factor of c in the vertical direction (because each y-coordinate is multiplied by the same number c). The graph of $y = -f(x)$ is the graph of $y = f(x)$ reflected about the x-axis because the point (x, y) is replaced by the point $(x, -y)$. (See Figure 2 and the following chart, where the results of other stretching, shrinking, and reflecting transformations are also given.)

Vertical and Horizontal Stretching and Reflecting Suppose $c > 1$. To obtain the graph of

$y = cf(x)$, stretch the graph of $y = f(x)$ vertically by a factor of c

$y = (1/c)f(x)$, shrink the graph of $y = f(x)$ vertically by a factor of c

$y = f(cx)$, shrink the graph of $y = f(x)$ horizontally by a factor of c

$y = f(x/c)$, stretch the graph of $y = f(x)$ horizontally by a factor of c

$y = -f(x)$, reflect the graph of $y = f(x)$ about the x-axis

$y = f(-x)$, reflect the graph of $y = f(x)$ about the y-axis

---

Figure 3 illustrates these stretching transformations when applied to the cosine function with $c = 2$. For instance, in order to get the graph of $y = 2 \cos x$ we multiply the y-coordinate of each point on the graph of $y = \cos x$ by 2. This means that the graph of $y = \cos x$ gets stretched vertically by a factor of 2.

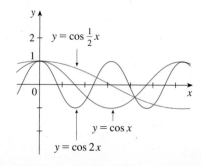

FIGURE 3

V **EXAMPLE 1** Given the graph of $y = \sqrt{x}$, use transformations to graph $y = \sqrt{x} - 2$, $y = \sqrt{x - 2}$, $y = -\sqrt{x}$, $y = 2\sqrt{x}$, and $y = \sqrt{-x}$.

SOLUTION The graph of the square root function $y = \sqrt{x}$, obtained from Figure 13(a) in Section 1.2, is shown in Figure 4(a). In the other parts of the figure we sketch $y = \sqrt{x} - 2$ by shifting 2 units downward, $y = \sqrt{x - 2}$ by shifting 2 units to the right, $y = -\sqrt{x}$ by reflecting about the x-axis, $y = 2\sqrt{x}$ by stretching vertically by a factor of 2, and $y = \sqrt{-x}$ by reflecting about the y-axis.

(a) $y = \sqrt{x}$

(b) $y = \sqrt{x} - 2$

(c) $y = \sqrt{x - 2}$

(d) $y = -\sqrt{x}$

(e) $y = 2\sqrt{x}$

(f) $y = \sqrt{-x}$

FIGURE 4

EXAMPLE 2 Sketch the graph of the function $f(x) = x^2 + 6x + 10$.

SOLUTION Completing the square, we write the equation of the graph as

$$y = x^2 + 6x + 10 = (x + 3)^2 + 1$$

This means we obtain the desired graph by starting with the parabola $y = x^2$ and shifting 3 units to the left and then 1 unit upward (see Figure 5).

(a) $y = x^2$

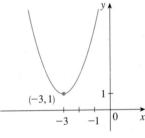
(b) $y = (x + 3)^2 + 1$

FIGURE 5

EXAMPLE 3 Sketch the graphs of the following functions.
(a) $y = \sin 2x$ 　　　　　　　　(b) $y = 1 - \sin x$

SOLUTION
(a) We obtain the graph of $y = \sin 2x$ from that of $y = \sin x$ by compressing horizontally by a factor of 2. (See Figures 6 and 7.) Thus, whereas the period of $y = \sin x$ is 2π, the period of $y = \sin 2x$ is $2\pi/2 = \pi$.

FIGURE 6

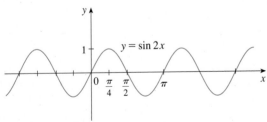

FIGURE 7

(b) To obtain the graph of $y = 1 - \sin x$, we again start with $y = \sin x$. We reflect about the x-axis to get the graph of $y = -\sin x$ and then we shift 1 unit upward to get $y = 1 - \sin x$. (See Figure 8.)

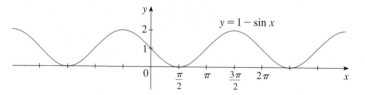

FIGURE 8

EXAMPLE 4 Figure 9 shows graphs of the number of hours of daylight as functions of the time of the year at several latitudes. Given that Philadelphia is located at approximately 40°N latitude, find a function that models the length of daylight at Philadelphia.

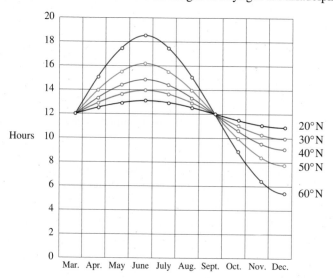

FIGURE 9

Graph of the length of daylight
from March 21 through December 21
at various latitudes

Lucia C. Harrison, *Daylight, Twilight, Darkness and Time*
(New York, 1935) page 40.

SOLUTION Notice that each curve resembles a shifted and stretched sine function. By looking at the blue curve we see that, at the latitude of Philadelphia, daylight lasts about 14.8 hours on June 21 and 9.2 hours on December 21, so the amplitude of the curve (the factor by which we have to stretch the sine curve vertically) is $\frac{1}{2}(14.8 - 9.2) = 2.8$.

By what factor do we need to stretch the sine curve horizontally if we measure the time t in days? Because there are about 365 days in a year, the period of our model should be 365. But the period of $y = \sin t$ is 2π, so the horizontal stretching factor is $c = 2\pi/365$.

We also notice that the curve begins its cycle on March 21, the 80th day of the year, so we have to shift the curve 80 units to the right. In addition, we shift it 12 units upward. Therefore we model the length of daylight in Philadelphia on the tth day of the year by the function

$$L(t) = 12 + 2.8 \sin\left[\frac{2\pi}{365}(t - 80)\right]$$

Another transformation of some interest is taking the *absolute value* of a function. If $y = |f(x)|$, then according to the definition of absolute value, $y = f(x)$ when $f(x) \geq 0$ and $y = -f(x)$ when $f(x) < 0$. This tells us how to get the graph of $y = |f(x)|$ from the graph of $y = f(x)$: The part of the graph that lies above the x-axis remains the same; the part that lies below the x-axis is reflected about the x-axis.

(a) $y = x^2 - 1$

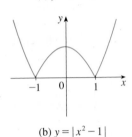

(b) $y = |x^2 - 1|$

FIGURE 10

V EXAMPLE 5 Sketch the graph of the function $y = |x^2 - 1|$.

SOLUTION We first graph the parabola $y = x^2 - 1$ in Figure 10(a) by shifting the parabola $y = x^2$ downward 1 unit. We see that the graph lies below the x-axis when $-1 < x < 1$, so we reflect that part of the graph about the x-axis to obtain the graph of $y = |x^2 - 1|$ in Figure 10(b).

Combinations of Functions

Two functions f and g can be combined to form new functions $f + g$, $f - g$, fg, and f/g in a manner similar to the way we add, subtract, multiply, and divide real numbers. The sum and difference functions are defined by

$$(f + g)(x) = f(x) + g(x) \qquad (f - g)(x) = f(x) - g(x)$$

If the domain of f is A and the domain of g is B, then the domain of $f + g$ is the intersection $A \cap B$ because both $f(x)$ and $g(x)$ have to be defined. For example, the domain of $f(x) = \sqrt{x}$ is $A = [0, \infty)$ and the domain of $g(x) = \sqrt{2 - x}$ is $B = (-\infty, 2]$, so the domain of $(f + g)(x) = \sqrt{x} + \sqrt{2 - x}$ is $A \cap B = [0, 2]$.

Similarly, the product and quotient functions are defined by

$$(fg)(x) = f(x)g(x) \qquad \left(\frac{f}{g}\right)(x) = \frac{f(x)}{g(x)}$$

The domain of fg is $A \cap B$, but we can't divide by 0 and so the domain of f/g is $\{x \in A \cap B \mid g(x) \neq 0\}$. For instance, if $f(x) = x^2$ and $g(x) = x - 1$, then the domain of the rational function $(f/g)(x) = x^2/(x - 1)$ is $\{x \mid x \neq 1\}$, or $(-\infty, 1) \cup (1, \infty)$.

There is another way of combining two functions to obtain a new function. For example, suppose that $y = f(u) = \sqrt{u}$ and $u = g(x) = x^2 + 1$. Since y is a function of u and u is, in turn, a function of x, it follows that y is ultimately a function of x. We compute this by substitution:

$$y = f(u) = f(g(x)) = f(x^2 + 1) = \sqrt{x^2 + 1}$$

The procedure is called *composition* because the new function is *composed* of the two given functions f and g.

In general, given any two functions f and g, we start with a number x in the domain of g and find its image $g(x)$. If this number $g(x)$ is in the domain of f, then we can calculate the value of $f(g(x))$. Notice that the output of one function is used as the input to the next function. The result is a new function $h(x) = f(g(x))$ obtained by substituting g into f. It is called the *composition* (or *composite*) of f and g and is denoted by $f \circ g$ ("f circle g").

Definition Given two functions f and g, the **composite function** $f \circ g$ (also called the **composition** of f and g) is defined by

$$(f \circ g)(x) = f(g(x))$$

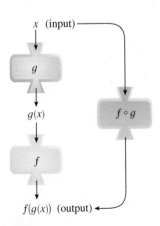

FIGURE 11
The $f \circ g$ machine is composed of the g machine (first) and then the f machine.

The domain of $f \circ g$ is the set of all x in the domain of g such that $g(x)$ is in the domain of f. In other words, $(f \circ g)(x)$ is defined whenever both $g(x)$ and $f(g(x))$ are defined. Figure 11 shows how to picture $f \circ g$ in terms of machines.

EXAMPLE 6 If $f(x) = x^2$ and $g(x) = x - 3$, find the composite functions $f \circ g$ and $g \circ f$.

SOLUTION We have

$$(f \circ g)(x) = f(g(x)) = f(x - 3) = (x - 3)^2$$

$$(g \circ f)(x) = g(f(x)) = g(x^2) = x^2 - 3$$

⊘   **NOTE** You can see from Example 6 that, in general, $f \circ g \neq g \circ f$. Remember, the notation $f \circ g$ means that the function g is applied first and then f is applied second. In Example 6, $f \circ g$ is the function that *first* subtracts 3 and *then* squares; $g \circ f$ is the function that *first* squares and *then* subtracts 3.

V EXAMPLE 7 If $f(x) = \sqrt{x}$ and $g(x) = \sqrt{2 - x}$, find each function and its domain.
(a) $f \circ g$ (b) $g \circ f$ (c) $f \circ f$ (d) $g \circ g$

SOLUTION

(a) $$(f \circ g)(x) = f(g(x)) = f(\sqrt{2 - x}) = \sqrt{\sqrt{2 - x}} = \sqrt[4]{2 - x}$$

The domain of $f \circ g$ is $\{x \mid 2 - x \geq 0\} = \{x \mid x \leq 2\} = (-\infty, 2]$.

(b) $$(g \circ f)(x) = g(f(x)) = g(\sqrt{x}) = \sqrt{2 - \sqrt{x}}$$

If $0 \leq a \leq b$, then $a^2 \leq b^2$.

For $\sqrt{x}$ to be defined we must have $x \geq 0$. For $\sqrt{2 - \sqrt{x}}$ to be defined we must have $2 - \sqrt{x} \geq 0$, that is, $\sqrt{x} \leq 2$, or $x \leq 4$. Thus we have $0 \leq x \leq 4$, so the domain of $g \circ f$ is the closed interval $[0, 4]$.

(c) $$(f \circ f)(x) = f(f(x)) = f(\sqrt{x}) = \sqrt{\sqrt{x}} = \sqrt[4]{x}$$

The domain of $f \circ f$ is $[0, \infty)$.

(d) $$(g \circ g)(x) = g(g(x)) = g(\sqrt{2 - x}) = \sqrt{2 - \sqrt{2 - x}}$$

This expression is defined when both $2 - x \geq 0$ and $2 - \sqrt{2 - x} \geq 0$. The first inequality means $x \leq 2$, and the second is equivalent to $\sqrt{2 - x} \leq 2$, or $2 - x \leq 4$, or $x \geq -2$. Thus $-2 \leq x \leq 2$, so the domain of $g \circ g$ is the closed interval $[-2, 2]$.

It is possible to take the composition of three or more functions. For instance, the composite function $f \circ g \circ h$ is found by first applying h, then g, and then f as follows:

$$(f \circ g \circ h)(x) = f(g(h(x)))$$

EXAMPLE 8 Find $f \circ g \circ h$ if $f(x) = x/(x + 1)$, $g(x) = x^{10}$, and $h(x) = x + 3$.

SOLUTION $$(f \circ g \circ h)(x) = f(g(h(x))) = f(g(x + 3))$$

$$= f((x + 3)^{10}) = \frac{(x + 3)^{10}}{(x + 3)^{10} + 1}$$

So far we have used composition to build complicated functions from simpler ones. But in calculus it is often useful to be able to *decompose* a complicated function into simpler ones, as in the following example.

EXAMPLE 9 Given $F(x) = \cos^2(x + 9)$, find functions f, g, and h such that $F = f \circ g \circ h$.

SOLUTION Since $F(x) = [\cos(x + 9)]^2$, the formula for F says: First add 9, then take the cosine of the result, and finally square. So we let

$$h(x) = x + 9 \qquad g(x) = \cos x \qquad f(x) = x^2$$

Then $\qquad (f \circ g \circ h)(x) = f(g(h(x))) = f(g(x + 9)) = f(\cos(x + 9))$

$$= [\cos(x + 9)]^2 = F(x)$$

1.3 Exercises

Access end-of-section exercises and tools for visualizing calculus online at **www.webassign.net**

1.4 The Tangent and Velocity Problems

In this section we see how limits arise when we attempt to find the tangent to a curve or the velocity of an object.

The Tangent Problem

The word *tangent* is derived from the Latin word *tangens*, which means "touching." Thus a tangent to a curve is a line that touches the curve. In other words, a tangent line should have the same direction as the curve at the point of contact. How can this idea be made precise?

For a circle we could simply follow Euclid and say that a tangent is a line that intersects the circle once and only once, as in Figure 1(a). For more complicated curves this definition is inadequate. Figure l(b) shows two lines l and t passing through a point P on a curve C. The line l intersects C only once, but it certainly does not look like what we think of as a tangent. The line t, on the other hand, looks like a tangent but it intersects C twice.

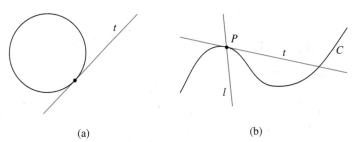

FIGURE 1   (a)   (b)

To be specific, let's look at the problem of trying to find a tangent line t to the parabola $y = x^2$ in the following example.

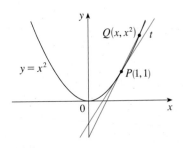

FIGURE 2

V EXAMPLE 1 Find an equation of the tangent line to the parabola $y = x^2$ at the point $P(1, 1)$.

SOLUTION We will be able to find an equation of the tangent line t as soon as we know its slope m. The difficulty is that we know only one point, P, on t, whereas we need two points to compute the slope. But observe that we can compute an approximation to m by choosing a nearby point $Q(x, x^2)$ on the parabola (as in Figure 2) and computing the slope m_{PQ} of the secant line PQ. [A **secant line**, from the Latin word *secans*, meaning cutting, is a line that cuts (intersects) a curve more than once.]

We choose $x \neq 1$ so that $Q \neq P$. Then

$$m_{PQ} = \frac{x^2 - 1}{x - 1}$$

For instance, for the point $Q(1.5, 2.25)$ we have

$$m_{PQ} = \frac{2.25 - 1}{1.5 - 1} = \frac{1.25}{0.5} = 2.5$$

x	m_{PQ}
2	3
1.5	2.5
1.1	2.1
1.01	2.01
1.001	2.001

The tables in the margin show the values of m_{PQ} for several values of x close to 1. The closer Q is to P, the closer x is to 1 and, it appears from the tables, the closer m_{PQ} is to 2. This suggests that the slope of the tangent line t should be $m = 2$.

We say that the slope of the tangent line is the *limit* of the slopes of the secant lines, and we express this symbolically by writing

$$\lim_{Q \to P} m_{PQ} = m \qquad \text{and} \qquad \lim_{x \to 1} \frac{x^2 - 1}{x - 1} = 2$$

x	m_{PQ}
0	1
0.5	1.5
0.9	1.9
0.99	1.99
0.999	1.999

Assuming that the slope of the tangent line is indeed 2, we use the point-slope form of the equation of a line (see Appendix B) to write the equation of the tangent line through $(1, 1)$ as

$$y - 1 = 2(x - 1) \qquad \text{or} \qquad y = 2x - 1$$

Figure 3 illustrates the limiting process that occurs in this example. As Q approaches P along the parabola, the corresponding secant lines rotate about P and approach the tangent line t.

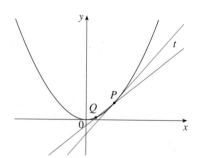

Q approaches P from the right

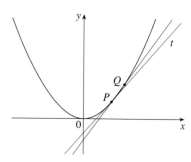

Q approaches P from the left

FIGURE 3

TEC In Visual 1.4 you can see how the process in Figure 3 works for additional functions.

t	Q
0.00	100.00
0.02	81.87
0.04	67.03
0.06	54.88
0.08	44.93
0.10	36.76

Many functions that occur in science are not described by explicit equations; they are defined by experimental data. The next example shows how to estimate the slope of the tangent line to the graph of such a function.

▼ EXAMPLE 2 The flash unit on a camera operates by storing charge on a capacitor and releasing it suddenly when the flash is set off. The data in the table describe the charge Q remaining on the capacitor (measured in microcoulombs) at time t (measured in seconds after the flash goes off). Use the data to draw the graph of this function and estimate the slope of the tangent line at the point where $t = 0.04$. [*Note:* The slope of the tangent line represents the electric current flowing from the capacitor to the flash bulb (measured in microamperes).]

SOLUTION In Figure 4 we plot the given data and use them to sketch a curve that approximates the graph of the function.

FIGURE 4

Given the points $P(0.04, 67.03)$ and $R(0.00, 100.00)$ on the graph, we find that the slope of the secant line PR is

$$m_{PR} = \frac{100.00 - 67.03}{0.00 - 0.04} = -824.25$$

R	m_{PR}
(0.00, 100.00)	−824.25
(0.02, 81.87)	−742.00
(0.06, 54.88)	−607.50
(0.08, 44.93)	−552.50
(0.10, 36.76)	−504.50

The table at the left shows the results of similar calculations for the slopes of other secant lines. From this table we would expect the slope of the tangent line at $t = 0.04$ to lie somewhere between -742 and -607.5. In fact, the average of the slopes of the two closest secant lines is

$$\tfrac{1}{2}(-742 - 607.5) = -674.75$$

So, by this method, we estimate the slope of the tangent line to be -675.

Another method is to draw an approximation to the tangent line at P and measure the sides of the triangle ABC, as in Figure 4. This gives an estimate of the slope of the tangent line as

$$-\frac{|AB|}{|BC|} \approx -\frac{80.4 - 53.6}{0.06 - 0.02} = -670$$

The physical meaning of the answer in Example 2 is that the electric current flowing from the capacitor to the flash bulb after 0.04 second is about −670 microamperes.

The Velocity Problem

If you watch the speedometer of a car as you travel in city traffic, you see that the needle doesn't stay still for very long; that is, the velocity of the car is not constant. We assume from watching the speedometer that the car has a definite velocity at each moment, but how is the "instantaneous" velocity defined? Let's investigate the example of a falling ball.

The CN Tower in Toronto was the tallest free-standing building in the world for 32 years.

�V EXAMPLE 3 Suppose that a ball is dropped from the upper observation deck of the CN Tower in Toronto, 450 m above the ground. Find the velocity of the ball after 5 seconds.

SOLUTION Through experiments carried out four centuries ago, Galileo discovered that the distance fallen by any freely falling body is proportional to the square of the time it has been falling. (This model for free fall neglects air resistance.) If the distance fallen after t seconds is denoted by $s(t)$ and measured in meters, then Galileo's law is expressed by the equation

$$s(t) = 4.9t^2$$

The difficulty in finding the velocity after 5 s is that we are dealing with a single instant of time ($t = 5$), so no time interval is involved. However, we can approximate the desired quantity by computing the average velocity over the brief time interval of a tenth of a second from $t = 5$ to $t = 5.1$:

$$\text{average velocity} = \frac{\text{change in position}}{\text{time elapsed}}$$

$$= \frac{s(5.1) - s(5)}{0.1}$$

$$= \frac{4.9(5.1)^2 - 4.9(5)^2}{0.1} = 49.49 \text{ m/s}$$

The following table shows the results of similar calculations of the average velocity over successively smaller time periods.

Time interval	Average velocity (m/s)
$5 \leqslant t \leqslant 6$	53.9
$5 \leqslant t \leqslant 5.1$	49.49
$5 \leqslant t \leqslant 5.05$	49.245
$5 \leqslant t \leqslant 5.01$	49.049
$5 \leqslant t \leqslant 5.001$	49.0049

It appears that as we shorten the time period, the average velocity is becoming closer to 49 m/s. The **instantaneous velocity** when $t = 5$ is defined to be the limiting value of these average velocities over shorter and shorter time periods that start at $t = 5$. Thus the (instantaneous) velocity after 5 s is

$$v = 49 \text{ m/s} \qquad \blacksquare$$

You may have the feeling that the calculations used in solving this problem are very similar to those used earlier in this section to find tangents. In fact, there is a close connection between the tangent problem and the problem of finding velocities. If we draw the graph of the distance function of the ball (as in Figure 5) and we consider the points $P(a, 4.9a^2)$ and $Q(a + h, 4.9(a + h)^2)$ on the graph, then the slope of the secant line PQ is

$$m_{PQ} = \frac{4.9(a + h)^2 - 4.9a^2}{(a + h) - a}$$

which is the same as the average velocity over the time interval $[a, a + h]$. Therefore the velocity at time $t = a$ (the limit of these average velocities as h approaches 0) must be equal to the slope of the tangent line at P (the limit of the slopes of the secant lines).

FIGURE 5

Examples 1 and 3 show that in order to solve tangent and velocity problems we must be able to find limits. After studying methods for computing limits in the next four sections, we will return to the problems of finding tangents and velocities in Chapter 2.

1.4   Exercises

Access end-of-section exercises and tools for visualizing calculus online at **www.webassign.net**

1.5   The Limit of a Function

Having seen in the preceding section how limits arise when we want to find the tangent to a curve or the velocity of an object, we now turn our attention to limits in general and numerical and graphical methods for computing them.

Let's investigate the behavior of the function f defined by $f(x) = x^2 - x + 2$ for values of x near 2. The following table gives values of $f(x)$ for values of x close to 2 but not equal to 2.

x	$f(x)$	x	$f(x)$
1.0	2.000000	3.0	8.000000
1.5	2.750000	2.5	5.750000
1.8	3.440000	2.2	4.640000
1.9	3.710000	2.1	4.310000
1.95	3.852500	2.05	4.152500
1.99	3.970100	2.01	4.030100
1.995	3.985025	2.005	4.015025
1.999	3.997001	2.001	4.003001

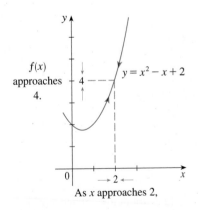

As x approaches 2,

FIGURE 1

From the table and the graph of f (a parabola) shown in Figure 1 we see that when x is close to 2 (on either side of 2), $f(x)$ is close to 4. In fact, it appears that we can make the values of $f(x)$ as close as we like to 4 by taking x sufficiently close to 2. We express this by saying "the limit of the function $f(x) = x^2 - x + 2$ as x approaches 2 is equal to 4." The

notation for this is

$$\lim_{x \to 2} (x^2 - x + 2) = 4$$

In general, we use the following notation.

---

1 **Definition** Suppose $f(x)$ is defined when x is near the number a. (This means that f is defined on some open interval that contains a, except possibly at a itself.) Then we write

$$\lim_{x \to a} f(x) = L$$

and say   "the limit of $f(x)$, as x approaches a, equals L"

if we can make the values of $f(x)$ arbitrarily close to L (as close to L as we like) by taking x to be sufficiently close to a (on either side of a) but not equal to a.

Roughly speaking, this says that the values of $f(x)$ approach L as x approaches a. In other words, the values of $f(x)$ tend to get closer and closer to the number L as x gets closer and closer to the number a (from either side of a) but $x \ne a$. (A more precise definition will be given in Section 1.7.)

An alternative notation for

$$\lim_{x \to a} f(x) = L$$

is $f(x) \to L$ as $x \to a$

which is usually read "$f(x)$ approaches L as x approaches a."

Notice the phrase "but $x \ne a$" in the definition of limit. This means that in finding the limit of $f(x)$ as x approaches a, we never consider $x = a$. In fact, $f(x)$ need not even be defined when $x = a$. The only thing that matters is how f is defined *near a*.

Figure 2 shows the graphs of three functions. Note that in part (c), $f(a)$ is not defined and in part (b), $f(a) \ne L$. But in each case, regardless of what happens at a, it is true that $\lim_{x \to a} f(x) = L$.

(a)

(b)

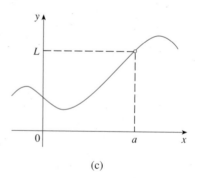

(c)

FIGURE 2 $\lim_{x \to a} f(x) = L$ in all three cases

EXAMPLE 1 Guess the value of $\displaystyle\lim_{x \to 1} \frac{x - 1}{x^2 - 1}$.

SOLUTION Notice that the function $f(x) = (x - 1)/(x^2 - 1)$ is not defined when $x = 1$, but that doesn't matter because the definition of $\lim_{x \to a} f(x)$ says that we consider values of x that are close to a but not equal to a.

$x < 1$	$f(x)$
0.5	0.666667
0.9	0.526316
0.99	0.502513
0.999	0.500250
0.9999	0.500025

$x > 1$	$f(x)$
1.5	0.400000
1.1	0.476190
1.01	0.497512
1.001	0.499750
1.0001	0.499975

The tables at the left give values of $f(x)$ (correct to six decimal places) for values of x that approach 1 (but are not equal to 1). On the basis of the values in the tables, we make the guess that

$$\lim_{x \to 1} \frac{x - 1}{x^2 - 1} = 0.5$$

Example 1 is illustrated by the graph of f in Figure 3. Now let's change f slightly by giving it the value 2 when $x = 1$ and calling the resulting function g:

$$g(x) = \begin{cases} \dfrac{x - 1}{x^2 - 1} & \text{if } x \neq 1 \\ 2 & \text{if } x = 1 \end{cases}$$

This new function g still has the same limit as x approaches 1. (See Figure 4.)

FIGURE 3

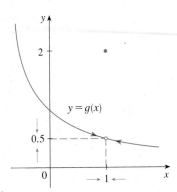

FIGURE 4

EXAMPLE 2   Estimate the value of $\displaystyle\lim_{t \to 0} \frac{\sqrt{t^2 + 9} - 3}{t^2}$.

SOLUTION   The table lists values of the function for several values of t near 0.

t	$\dfrac{\sqrt{t^2 + 9} - 3}{t^2}$
± 1.0	0.16228
± 0.5	0.16553
± 0.1	0.16662
± 0.05	0.16666
± 0.01	0.16667

As t approaches 0, the values of the function seem to approach $0.1666666\ldots$ and so we guess that

$$\lim_{t \to 0} \frac{\sqrt{t^2 + 9} - 3}{t^2} = \frac{1}{6}$$

t	$\dfrac{\sqrt{t^2 + 9} - 3}{t^2}$
± 0.0005	0.16800
± 0.0001	0.20000
± 0.00005	0.00000
± 0.00001	0.00000

In Example 2 what would have happened if we had taken even smaller values of t? The table in the margin shows the results from one calculator; you can see that something strange seems to be happening.

If you try these calculations on your own calculator you might get different values, but eventually you will get the value 0 if you make t sufficiently small. Does this mean that the answer is really 0 instead of $\frac{1}{6}$? No, the value of the limit is $\frac{1}{6}$, as we will show in the next section. The problem is that the calculator gave false values because $\sqrt{t^2+9}$ is very close to 3 when t is small. (In fact, when t is sufficiently small, a calculator's value for $\sqrt{t^2+9}$ is 3.000... to as many digits as the calculator is capable of carrying.)

Something similar happens when we try to graph the function

$$f(t) = \frac{\sqrt{t^2+9}-3}{t^2}$$

of Example 2 on a graphing calculator or computer. Parts (a) and (b) of Figure 5 show quite accurate graphs of f, and when we use the trace mode (if available) we can estimate easily that the limit is about $\frac{1}{6}$. But if we zoom in too much, as in parts (c) and (d), then we get inaccurate graphs, again because of problems with subtraction.

www.stewartcalculus.com

For a further explanation of why calculators sometimes give false values, click on *Lies My Calculator and Computer Told Me*. In particular, see the section called *The Perils of Subtraction*.

(a) $[-5, 5]$ by $[-0.1, 0.3]$

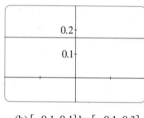

(b) $[-0.1, 0.1]$ by $[-0.1, 0.3]$

(c) $[-10^{-6}, 10^{-6}]$ by $[-0.1, 0.3]$

(d) $[-10^{-7}, 10^{-7}]$ by $[-0.1, 0.3]$

FIGURE 5

V EXAMPLE 3 Guess the value of $\displaystyle\lim_{x\to 0}\frac{\sin x}{x}$.

SOLUTION The function $f(x) = (\sin x)/x$ is not defined when $x = 0$. Using a calculator (and remembering that, if $x \in \mathbb{R}$, $\sin x$ means the sine of the angle whose *radian* measure is x), we construct a table of values correct to eight decimal places. From the table at the left and the graph in Figure 6 we guess that

$$\lim_{x\to 0}\frac{\sin x}{x} = 1$$

This guess is in fact correct, as will be proved in Chapter 2 using a geometric argument.

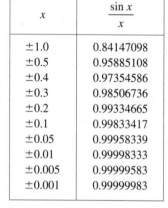

x	$\dfrac{\sin x}{x}$
± 1.0	0.84147098
± 0.5	0.95885108
± 0.4	0.97354586
± 0.3	0.98506736
± 0.2	0.99334665
± 0.1	0.99833417
± 0.05	0.99958339
± 0.01	0.99998333
± 0.005	0.99999583
± 0.001	0.99999983

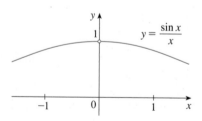

FIGURE 6

V EXAMPLE 4 Investigate $\displaystyle\lim_{x\to 0}\sin\frac{\pi}{x}$.

SOLUTION Again the function $f(x) = \sin(\pi/x)$ is undefined at 0. Evaluating the function for some small values of x, we get

$$f(1) = \sin \pi = 0 \qquad\qquad f\left(\tfrac{1}{2}\right) = \sin 2\pi = 0$$

$$f\left(\tfrac{1}{3}\right) = \sin 3\pi = 0 \qquad\qquad f\left(\tfrac{1}{4}\right) = \sin 4\pi = 0$$

$$f(0.1) = \sin 10\pi = 0 \qquad\qquad f(0.01) = \sin 100\pi = 0$$

Similarly, $f(0.001) = f(0.0001) = 0$. On the basis of this information we might be tempted to guess that

$$\lim_{x \to 0} \sin \frac{\pi}{x} = 0$$

⊘ but this time our guess is wrong. Note that although $f(1/n) = \sin n\pi = 0$ for any integer n, it is also true that $f(x) = 1$ for infinitely many values of x that approach 0. You can see this from the graph of f shown in Figure 7.

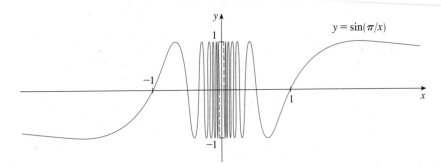

FIGURE 7

The dashed lines near the y-axis indicate that the values of $\sin(\pi/x)$ oscillate between 1 and -1 infinitely often as x approaches 0. (See Exercise 43.)

Since the values of $f(x)$ do not approach a fixed number as x approaches 0,

$$\lim_{x \to 0} \sin \frac{\pi}{x} \quad \text{does not exist} \qquad\qquad \blacksquare$$

x	$x^3 + \dfrac{\cos 5x}{10{,}000}$
1	1.000028
0.5	0.124920
0.1	0.001088
0.05	0.000222
0.01	0.000101

x	$x^3 + \dfrac{\cos 5x}{10{,}000}$
0.005	0.00010009
0.001	0.00010000

EXAMPLE 5 Find $\displaystyle\lim_{x \to 0}\left(x^3 + \frac{\cos 5x}{10{,}000}\right)$.

SOLUTION As before, we construct a table of values. From the first table in the margin it appears that

$$\lim_{x \to 0}\left(x^3 + \frac{\cos 5x}{10{,}000}\right) = 0$$

But if we persevere with smaller values of x, the second table suggests that

$$\lim_{x \to 0}\left(x^3 + \frac{\cos 5x}{10{,}000}\right) = 0.000100 = \frac{1}{10{,}000}$$

Later we will see that $\lim_{x \to 0} \cos 5x = 1$; then it follows that the limit is 0.0001. ▬

⊘ Examples 4 and 5 illustrate some of the pitfalls in guessing the value of a limit. It is easy to guess the wrong value if we use inappropriate values of x, but it is difficult to know when to stop calculating values. And, as the discussion after Example 2 shows, sometimes calculators and computers give the wrong values. In the next section, however, we will develop foolproof methods for calculating limits.

V **EXAMPLE 6** The Heaviside function H is defined by

$$H(t) = \begin{cases} 0 & \text{if } t < 0 \\ 1 & \text{if } t \geqslant 0 \end{cases}$$

[This function is named after the electrical engineer Oliver Heaviside (1850–1925) and can be used to describe an electric current that is switched on at time $t = 0$.] Its graph is shown in Figure 8.

As t approaches 0 from the left, $H(t)$ approaches 0. As t approaches 0 from the right, $H(t)$ approaches 1. There is no single number that $H(t)$ approaches as t approaches 0. Therefore $\lim_{t \to 0} H(t)$ does not exist.

FIGURE 8
The Heaviside function

One-Sided Limits

We noticed in Example 6 that $H(t)$ approaches 0 as t approaches 0 from the left and $H(t)$ approaches 1 as t approaches 0 from the right. We indicate this situation symbolically by writing

$$\lim_{t \to 0^-} H(t) = 0 \qquad \text{and} \qquad \lim_{t \to 0^+} H(t) = 1$$

The symbol "$t \to 0^-$" indicates that we consider only values of t that are less than 0. Likewise, "$t \to 0^+$" indicates that we consider only values of t that are greater than 0.

---

2 **Definition** We write

$$\lim_{x \to a^-} f(x) = L$$

and say the **left-hand limit of $f(x)$ as x approaches a** [or the **limit of $f(x)$ as x approaches a from the left**] is equal to L if we can make the values of $f(x)$ arbitrarily close to L by taking x to be sufficiently close to a and x less than a.

Notice that Definition 2 differs from Definition 1 only in that we require x to be less than a. Similarly, if we require that x be greater than a, we get "the **right-hand limit of $f(x)$ as x approaches a** is equal to L" and we write

$$\lim_{x \to a^+} f(x) = L$$

Thus the symbol "$x \to a^+$" means that we consider only $x > a$. These definitions are illustrated in Figure 9.

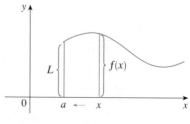

FIGURE 9

(a) $\displaystyle\lim_{x \to a^-} f(x) = L$

(b) $\displaystyle\lim_{x \to a^+} f(x) = L$

By comparing Definition 1 with the definitions of one-sided limits, we see that the following is true.

$$\boxed{3} \quad \lim_{x \to a} f(x) = L \quad \text{if and only if} \quad \lim_{x \to a^-} f(x) = L \quad \text{and} \quad \lim_{x \to a^+} f(x) = L$$

FIGURE 10

V   EXAMPLE 7   The graph of a function g is shown in Figure 10. Use it to state the values (if they exist) of the following:

(a) $\lim\limits_{x \to 2^-} g(x)$ (b) $\lim\limits_{x \to 2^+} g(x)$ (c) $\lim\limits_{x \to 2} g(x)$

(d) $\lim\limits_{x \to 5^-} g(x)$ (e) $\lim\limits_{x \to 5^+} g(x)$ (f) $\lim\limits_{x \to 5} g(x)$

SOLUTION   From the graph we see that the values of $g(x)$ approach 3 as x approaches 2 from the left, but they approach 1 as x approaches 2 from the right. Therefore

$$\text{(a)} \ \lim_{x \to 2^-} g(x) = 3 \quad \text{and} \quad \text{(b)} \ \lim_{x \to 2^+} g(x) = 1$$

(c) Since the left and right limits are different, we conclude from $\boxed{3}$ that $\lim_{x \to 2} g(x)$ does not exist.

The graph also shows that

$$\text{(d)} \ \lim_{x \to 5^-} g(x) = 2 \quad \text{and} \quad \text{(e)} \ \lim_{x \to 5^+} g(x) = 2$$

(f) This time the left and right limits are the same and so, by $\boxed{3}$, we have

$$\lim_{x \to 5} g(x) = 2$$

Despite this fact, notice that $g(5) \neq 2$.   ∎

Infinite Limits

EXAMPLE 8   Find $\lim\limits_{x \to 0} \dfrac{1}{x^2}$ if it exists.

x	$\dfrac{1}{x^2}$
± 1	1
± 0.5	4
± 0.2	25
± 0.1	100
± 0.05	400
± 0.01	10,000
± 0.001	1,000,000

SOLUTION   As x becomes close to 0, x^2 also becomes close to 0, and $1/x^2$ becomes very large. (See the table in the margin.) In fact, it appears from the graph of the function $f(x) = 1/x^2$ shown in Figure 11 that the values of $f(x)$ can be made arbitrarily large by taking x close enough to 0. Thus the values of $f(x)$ do not approach a number, so $\lim_{x \to 0} (1/x^2)$ does not exist.   ∎

To indicate the kind of behavior exhibited in Example 8, we use the notation

$$\lim_{x \to 0} \frac{1}{x^2} = \infty$$

⊘ This does not mean that we are regarding ∞ as a number. Nor does it mean that the limit exists. It simply expresses the particular way in which the limit does not exist: $1/x^2$ can be made as large as we like by taking x close enough to 0.

In general, we write symbolically

$$\lim_{x \to a} f(x) = \infty$$

FIGURE 11

to indicate that the values of $f(x)$ tend to become larger and larger (or "increase without bound") as x becomes closer and closer to a.

> **4 Definition** Let f be a function defined on both sides of a, except possibly at a itself. Then
>
> $$\lim_{x \to a} f(x) = \infty$$
>
> means that the values of $f(x)$ can be made arbitrarily large (as large as we please) by taking x sufficiently close to a, but not equal to a.

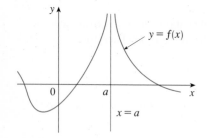

FIGURE 12
$\lim\limits_{x \to a} f(x) = \infty$

When we say a number is "large negative," we mean that it is negative but its magnitude (absolute value) is large.

Another notation for $\lim_{x \to a} f(x) = \infty$ is

$$f(x) \to \infty \qquad \text{as} \qquad x \to a$$

Again, the symbol ∞ is not a number, but the expression $\lim_{x \to a} f(x) = \infty$ is often read as

"the limit of $f(x)$, as x approaches a, is infinity"

or   "$f(x)$ becomes infinite as x approaches a"

or   "$f(x)$ increases without bound as x approaches a"

This definition is illustrated graphically in Figure 12.

A similar sort of limit, for functions that become large negative as x gets close to a, is defined in Definition 5 and is illustrated in Figure 13.

> **5 Definition** Let f be defined on both sides of a, except possibly at a itself. Then
>
> $$\lim_{x \to a} f(x) = -\infty$$
>
> means that the values of $f(x)$ can be made arbitrarily large negative by taking x sufficiently close to a, but not equal to a.

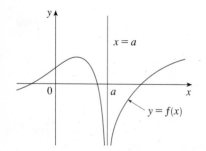

FIGURE 13
$\lim\limits_{x \to a} f(x) = -\infty$

The symbol $\lim_{x \to a} f(x) = -\infty$ can be read as "the limit of $f(x)$, as x approaches a, is negative infinity" or "$f(x)$ decreases without bound as x approaches a." As an example we have

$$\lim_{x \to 0} \left(-\frac{1}{x^2} \right) = -\infty$$

Similar definitions can be given for the one-sided infinite limits

$$\lim_{x \to a^-} f(x) = \infty \qquad\qquad \lim_{x \to a^+} f(x) = \infty$$

$$\lim_{x \to a^-} f(x) = -\infty \qquad\qquad \lim_{x \to a^+} f(x) = -\infty$$

remembering that "$x \to a^-$" means that we consider only values of x that are less than a, and similarly "$x \to a^+$" means that we consider only $x > a$. Illustrations of these four cases are given in Figure 14.

(a) $\lim\limits_{x \to a^-} f(x) = \infty$

(b) $\lim\limits_{x \to a^+} f(x) = \infty$

(c) $\lim\limits_{x \to a^-} f(x) = -\infty$

(d) $\lim\limits_{x \to a^+} f(x) = -\infty$

FIGURE 14

> **6 Definition** The line $x = a$ is called a **vertical asymptote** of the curve $y = f(x)$ if at least one of the following statements is true:
>
> $$\lim_{x \to a} f(x) = \infty \qquad \lim_{x \to a^-} f(x) = \infty \qquad \lim_{x \to a^+} f(x) = \infty$$
>
> $$\lim_{x \to a} f(x) = -\infty \qquad \lim_{x \to a^-} f(x) = -\infty \qquad \lim_{x \to a^+} f(x) = -\infty$$

For instance, the y-axis is a vertical asymptote of the curve $y = 1/x^2$ because $\lim_{x \to 0} (1/x^2) = \infty$. In Figure 14 the line $x = a$ is a vertical asymptote in each of the four cases shown. In general, knowledge of vertical asymptotes is very useful in sketching graphs.

EXAMPLE 9 Find $\lim\limits_{x \to 3^+} \dfrac{2x}{x-3}$ and $\lim\limits_{x \to 3^-} \dfrac{2x}{x-3}$.

SOLUTION If x is close to 3 but larger than 3, then the denominator $x - 3$ is a small positive number and $2x$ is close to 6. So the quotient $2x/(x - 3)$ is a large *positive* number. Thus, intuitively, we see that

$$\lim_{x \to 3^+} \frac{2x}{x-3} = \infty$$

Likewise, if x is close to 3 but smaller than 3, then $x - 3$ is a small negative number but $2x$ is still a positive number (close to 6). So $2x/(x - 3)$ is a numerically large *negative* number. Thus

$$\lim_{x \to 3^-} \frac{2x}{x-3} = -\infty$$

The graph of the curve $y = 2x/(x - 3)$ is given in Figure 15. The line $x = 3$ is a vertical asymptote.

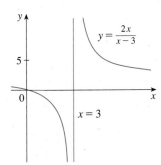

$y = \dfrac{2x}{x-3}$

$x = 3$

FIGURE 15

EXAMPLE 10 Find the vertical asymptotes of $f(x) = \tan x$.

SOLUTION Because

$$\tan x = \frac{\sin x}{\cos x}$$

there are potential vertical asymptotes where $\cos x = 0$. In fact, since $\cos x \to 0^+$ as $x \to (\pi/2)^-$ and $\cos x \to 0^-$ as $x \to (\pi/2)^+$, whereas $\sin x$ is positive when x is near

$\pi/2$, we have

$$\lim_{x \to (\pi/2)^-} \tan x = \infty \qquad \text{and} \qquad \lim_{x \to (\pi/2)^+} \tan x = -\infty$$

This shows that the line $x = \pi/2$ is a vertical asymptote. Similar reasoning shows that the lines $x = (2n + 1)\pi/2$, where n is an integer, are all vertical asymptotes of $f(x) = \tan x$. The graph in Figure 16 confirms this.

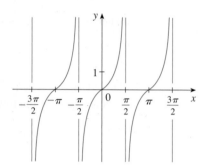

FIGURE 16

$y = \tan x$

1.5   Exercises

Access end-of-section exercises and tools for visualizing calculus online at **www.webassign.net**

1.6   Calculating Limits Using the Limit Laws

In Section 1.5 we used calculators and graphs to guess the values of limits, but we saw that such methods don't always lead to the correct answer. In this section we use the following properties of limits, called the *Limit Laws,* to calculate limits.

Limit Laws   Suppose that c is a constant and the limits

$$\lim_{x \to a} f(x) \qquad \text{and} \qquad \lim_{x \to a} g(x)$$

exist. Then

1. $\displaystyle \lim_{x \to a} [f(x) + g(x)] = \lim_{x \to a} f(x) + \lim_{x \to a} g(x)$

2. $\displaystyle \lim_{x \to a} [f(x) - g(x)] = \lim_{x \to a} f(x) - \lim_{x \to a} g(x)$

3. $\displaystyle \lim_{x \to a} [cf(x)] = c \lim_{x \to a} f(x)$

4. $\displaystyle \lim_{x \to a} [f(x) g(x)] = \lim_{x \to a} f(x) \cdot \lim_{x \to a} g(x)$

5. $\displaystyle \lim_{x \to a} \frac{f(x)}{g(x)} = \frac{\displaystyle \lim_{x \to a} f(x)}{\displaystyle \lim_{x \to a} g(x)} \quad \text{if } \lim_{x \to a} g(x) \neq 0$

These five laws can be stated verbally as follows:

Sum Law

1. The limit of a sum is the sum of the limits.

Difference Law

2. The limit of a difference is the difference of the limits.

Constant Multiple Law

3. The limit of a constant times a function is the constant times the limit of the function.

Product Law

4. The limit of a product is the product of the limits.

Quotient Law

5. The limit of a quotient is the quotient of the limits (provided that the limit of the denominator is not 0).

It is easy to believe that these properties are true. For instance, if $f(x)$ is close to L and $g(x)$ is close to M, it is reasonable to conclude that $f(x) + g(x)$ is close to $L + M$. This gives us an intuitive basis for believing that Law 1 is true. In Section 1.7 we give a precise definition of a limit and use it to prove this law. The proofs of the remaining laws are given in Appendix F.

EXAMPLE 1 Use the Limit Laws and the graphs of f and g in Figure 1 to evaluate the following limits, if they exist.

(a) $\displaystyle\lim_{x \to -2} [f(x) + 5g(x)]$ (b) $\displaystyle\lim_{x \to 1} [f(x)g(x)]$ (c) $\displaystyle\lim_{x \to 2} \frac{f(x)}{g(x)}$

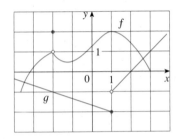

FIGURE 1

SOLUTION

(a) From the graphs of f and g we see that

$$\lim_{x \to -2} f(x) = 1 \quad \text{and} \quad \lim_{x \to -2} g(x) = -1$$

Therefore we have

$$\lim_{x \to -2} [f(x) + 5g(x)] = \lim_{x \to -2} f(x) + \lim_{x \to -2} [5g(x)] \quad \text{(by Law 1)}$$

$$= \lim_{x \to -2} f(x) + 5 \lim_{x \to -2} g(x) \quad \text{(by Law 3)}$$

$$= 1 + 5(-1) = -4$$

(b) We see that $\lim_{x \to 1} f(x) = 2$. But $\lim_{x \to 1} g(x)$ does not exist because the left and right limits are different:

$$\lim_{x \to 1^-} g(x) = -2 \qquad \lim_{x \to 1^+} g(x) = -1$$

So we can't use Law 4 for the desired limit. But we *can* use Law 4 for the one-sided limits:

$$\lim_{x \to 1^-} [f(x)g(x)] = 2 \cdot (-2) = -4 \qquad \lim_{x \to 1^+} [f(x)g(x)] = 2 \cdot (-1) = -2$$

The left and right limits aren't equal, so $\lim_{x \to 1} [f(x)g(x)]$ does not exist.

(c) The graphs show that

$$\lim_{x \to 2} f(x) \approx 1.4 \quad \text{and} \quad \lim_{x \to 2} g(x) = 0$$

Because the limit of the denominator is 0, we can't use Law 5. The given limit does not exist because the denominator approaches 0 while the numerator approaches a nonzero number.

If we use the Product Law repeatedly with $g(x) = f(x)$, we obtain the following law.

Power Law

6. $\lim\limits_{x \to a} [f(x)]^n = \left[\lim\limits_{x \to a} f(x) \right]^n$ where n is a positive integer

In applying these six limit laws, we need to use two special limits:

7. $\lim\limits_{x \to a} c = c$ **8.** $\lim\limits_{x \to a} x = a$

These limits are obvious from an intuitive point of view (state them in words or draw graphs of $y = c$ and $y = x$), but proofs based on the precise definition are requested in the exercises for Section 1.7.

If we now put $f(x) = x$ in Law 6 and use Law 8, we get another useful special limit.

9. $\lim\limits_{x \to a} x^n = a^n$ where n is a positive integer

A similar limit holds for roots as follows. (For square roots the proof is outlined in Exercise 37 in Section 1.7.)

10. $\lim\limits_{x \to a} \sqrt[n]{x} = \sqrt[n]{a}$ where n is a positive integer

(If n is even, we assume that $a > 0$.)

More generally, we have the following law, which is proved in Section 1.8 as a consequence of Law 10.

Root Law

11. $\lim\limits_{x \to a} \sqrt[n]{f(x)} = \sqrt[n]{\lim\limits_{x \to a} f(x)}$ where n is a positive integer

$\left[\text{If } n \text{ is even, we assume that } \lim\limits_{x \to a} f(x) > 0. \right]$

EXAMPLE 2 Evaluate the following limits and justify each step.

(a) $\lim\limits_{x \to 5} (2x^2 - 3x + 4)$ (b) $\lim\limits_{x \to -2} \dfrac{x^3 + 2x^2 - 1}{5 - 3x}$

SOLUTION

(a) $\lim\limits_{x \to 5} (2x^2 - 3x + 4) = \lim\limits_{x \to 5} (2x^2) - \lim\limits_{x \to 5} (3x) + \lim\limits_{x \to 5} 4$ (by Laws 2 and 1)

$= 2 \lim\limits_{x \to 5} x^2 - 3 \lim\limits_{x \to 5} x + \lim\limits_{x \to 5} 4$ (by 3)

$= 2(5^2) - 3(5) + 4$ (by 9, 8, and 7)

$= 39$

Newton and Limits

Isaac Newton was born on Christmas Day in 1642, the year of Galileo's death. When he entered Cambridge University in 1661 Newton didn't know much mathematics, but he learned quickly by reading Euclid and Descartes and by attending the lectures of Isaac Barrow. Cambridge was closed because of the plague in 1665 and 1666, and Newton returned home to reflect on what he had learned. Those two years were amazingly productive for at that time he made four of his major discoveries: (1) his representation of functions as sums of infinite series, including the binomial theorem; (2) his work on differential and integral calculus; (3) his laws of motion and law of universal gravitation; and (4) his prism experiments on the nature of light and color. Because of a fear of controversy and criticism, he was reluctant to publish his discoveries and it wasn't until 1687, at the urging of the astronomer Halley, that Newton published *Principia Mathematica*. In this work, the greatest scientific treatise ever written, Newton set forth his version of calculus and used it to investigate mechanics, fluid dynamics, and wave motion, and to explain the motion of planets and comets.

The beginnings of calculus are found in the calculations of areas and volumes by ancient Greek scholars such as Eudoxus and Archimedes. Although aspects of the idea of a limit are implicit in their "method of exhaustion," Eudoxus and Archimedes never explicitly formulated the concept of a limit. Likewise, mathematicians such as Cavalieri, Fermat, and Barrow, the immediate precursors of Newton in the development of calculus, did not actually use limits. It was Isaac Newton who was the first to talk explicitly about limits. He explained that the main idea behind limits is that quantities "approach nearer than by any given difference." Newton stated that the limit was the basic concept in calculus, but it was left to later mathematicians like Cauchy to clarify his ideas about limits.

(b) We start by using Law 5, but its use is fully justified only at the final stage when we see that the limits of the numerator and denominator exist and the limit of the denominator is not 0.

$$\lim_{x\to-2}\frac{x^3+2x^2-1}{5-3x}=\frac{\lim_{x\to-2}(x^3+2x^2-1)}{\lim_{x\to-2}(5-3x)} \quad \text{(by Law 5)}$$

$$=\frac{\lim_{x\to-2}x^3+2\lim_{x\to-2}x^2-\lim_{x\to-2}1}{\lim_{x\to-2}5-3\lim_{x\to-2}x} \quad \text{(by 1, 2, and 3)}$$

$$=\frac{(-2)^3+2(-2)^2-1}{5-3(-2)} \quad \text{(by 9, 8, and 7)}$$

$$=-\frac{1}{11}$$

NOTE If we let $f(x)=2x^2-3x+4$, then $f(5)=39$. In other words, we would have gotten the correct answer in Example 2(a) by substituting 5 for x. Similarly, direct substitution provides the correct answer in part (b). The functions in Example 2 are a polynomial and a rational function, respectively, and similar use of the Limit Laws proves that direct substitution always works for such functions (see Exercises 55 and 56). We state this fact as follows.

Direct Substitution Property If f is a polynomial or a rational function and a is in the domain of f, then

$$\lim_{x\to a}f(x)=f(a)$$

Functions with the Direct Substitution Property are called *continuous at a* and will be studied in Section 1.8. However, not all limits can be evaluated by direct substitution, as the following examples show.

EXAMPLE 3 Find $\lim_{x\to1}\dfrac{x^2-1}{x-1}$.

SOLUTION Let $f(x)=(x^2-1)/(x-1)$. We can't find the limit by substituting $x=1$ because $f(1)$ isn't defined. Nor can we apply the Quotient Law, because the limit of the denominator is 0. Instead, we need to do some preliminary algebra. We factor the numerator as a difference of squares:

$$\frac{x^2-1}{x-1}=\frac{(x-1)(x+1)}{x-1}$$

The numerator and denominator have a common factor of $x-1$. When we take the limit as x approaches 1, we have $x\neq1$ and so $x-1\neq0$. Therefore we can cancel the common factor and compute the limit as follows:

$$\lim_{x\to1}\frac{x^2-1}{x-1}=\lim_{x\to1}\frac{(x-1)(x+1)}{x-1}$$

$$=\lim_{x\to1}(x+1)$$

$$=1+1=2$$

The limit in this example arose in Section 1.4 when we were trying to find the tangent to the parabola $y = x^2$ at the point $(1, 1)$.   ▬

   NOTE   In Example 3 we were able to compute the limit by replacing the given function $f(x) = (x^2 - 1)/(x - 1)$ by a simpler function, $g(x) = x + 1$, with the same limit. This is valid because $f(x) = g(x)$ except when $x = 1$, and in computing a limit as x approaches 1 we don't consider what happens when x is actually *equal* to 1. In general, we have the following useful fact.

If $f(x) = g(x)$ when $x \neq a$, then $\lim\limits_{x \to a} f(x) = \lim\limits_{x \to a} g(x)$, provided the limits exist.

EXAMPLE 4   Find $\lim\limits_{x \to 1} g(x)$ where

$$g(x) = \begin{cases} x + 1 & \text{if } x \neq 1 \\ \pi & \text{if } x = 1 \end{cases}$$

SOLUTION   Here g is defined at $x = 1$ and $g(1) = \pi$, but the value of a limit as x approaches 1 does not depend on the value of the function at 1. Since $g(x) = x + 1$ for $x \neq 1$, we have

$$\lim\limits_{x \to 1} g(x) = \lim\limits_{x \to 1} (x + 1) = 2$$ ▬

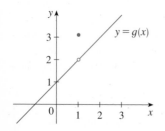

FIGURE 2
The graphs of the functions f (from Example 3) and g (from Example 4)

   Note that the values of the functions in Examples 3 and 4 are identical except when $x = 1$ (see Figure 2) and so they have the same limit as x approaches 1.

V **EXAMPLE 5**   Evaluate $\lim\limits_{h \to 0} \dfrac{(3 + h)^2 - 9}{h}$.

SOLUTION   If we define

$$F(h) = \frac{(3 + h)^2 - 9}{h}$$

then, as in Example 3, we can't compute $\lim_{h \to 0} F(h)$ by letting $h = 0$ since $F(0)$ is undefined. But if we simplify $F(h)$ algebraically, we find that

$$F(h) = \frac{(9 + 6h + h^2) - 9}{h} = \frac{6h + h^2}{h} = 6 + h$$

(Recall that we consider only $h \neq 0$ when letting h approach 0.) Thus

$$\lim\limits_{h \to 0} \frac{(3 + h)^2 - 9}{h} = \lim\limits_{h \to 0} (6 + h) = 6$$ ▬

EXAMPLE 6   Find $\lim\limits_{t \to 0} \dfrac{\sqrt{t^2 + 9} - 3}{t^2}$.

SOLUTION We can't apply the Quotient Law immediately, since the limit of the denominator is 0. Here the preliminary algebra consists of rationalizing the numerator:

$$\lim_{t \to 0} \frac{\sqrt{t^2 + 9} - 3}{t^2} = \lim_{t \to 0} \frac{\sqrt{t^2 + 9} - 3}{t^2} \cdot \frac{\sqrt{t^2 + 9} + 3}{\sqrt{t^2 + 9} + 3}$$

$$= \lim_{t \to 0} \frac{(t^2 + 9) - 9}{t^2 \left(\sqrt{t^2 + 9} + 3\right)}$$

$$= \lim_{t \to 0} \frac{t^2}{t^2 \left(\sqrt{t^2 + 9} + 3\right)}$$

$$= \lim_{t \to 0} \frac{1}{\sqrt{t^2 + 9} + 3}$$

$$= \frac{1}{\sqrt{\lim_{t \to 0} (t^2 + 9)} + 3}$$

$$= \frac{1}{3 + 3} = \frac{1}{6}$$

This calculation confirms the guess that we made in Example 2 in Section 1.5. ▬

Some limits are best calculated by first finding the left- and right-hand limits. The following theorem is a reminder of what we discovered in Section 1.5. It says that a two-sided limit exists if and only if both of the one-sided limits exist and are equal.

---

1 Theorem $\lim_{x \to a} f(x) = L$ if and only if $\lim_{x \to a^-} f(x) = L = \lim_{x \to a^+} f(x)$

When computing one-sided limits, we use the fact that the Limit Laws also hold for one-sided limits.

EXAMPLE 7 Show that $\lim_{x \to 0} |x| = 0$.

SOLUTION Recall that

$$|x| = \begin{cases} x & \text{if } x \geqslant 0 \\ -x & \text{if } x < 0 \end{cases}$$

Since $|x| = x$ for $x > 0$, we have

$$\lim_{x \to 0^+} |x| = \lim_{x \to 0^+} x = 0$$

For $x < 0$ we have $|x| = -x$ and so

$$\lim_{x \to 0^-} |x| = \lim_{x \to 0^-} (-x) = 0$$

Therefore, by Theorem 1,

$$\lim_{x \to 0} |x| = 0$$

The result of Example 7 looks plausible from Figure 3.

FIGURE 3

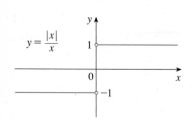

$y = \dfrac{|x|}{x}$

FIGURE 4

V **EXAMPLE 8** Prove that $\lim\limits_{x \to 0} \dfrac{|x|}{x}$ does not exist.

SOLUTION
$$\lim_{x \to 0^+} \frac{|x|}{x} = \lim_{x \to 0^+} \frac{x}{x} = \lim_{x \to 0^+} 1 = 1$$

$$\lim_{x \to 0^-} \frac{|x|}{x} = \lim_{x \to 0^-} \frac{-x}{x} = \lim_{x \to 0^-} (-1) = -1$$

Since the right- and left-hand limits are different, it follows from Theorem 1 that $\lim_{x \to 0} |x|/x$ does not exist. The graph of the function $f(x) = |x|/x$ is shown in Figure 4 and supports the one-sided limits that we found.

EXAMPLE 9 If
$$f(x) = \begin{cases} \sqrt{x - 4} & \text{if } x > 4 \\ 8 - 2x & \text{if } x < 4 \end{cases}$$

determine whether $\lim_{x \to 4} f(x)$ exists.

It is shown in Example 3 in Section 1.7 that $\lim_{x \to 0^+} \sqrt{x} = 0$.

SOLUTION Since $f(x) = \sqrt{x - 4}$ for $x > 4$, we have
$$\lim_{x \to 4^+} f(x) = \lim_{x \to 4^+} \sqrt{x - 4} = \sqrt{4 - 4} = 0$$

Since $f(x) = 8 - 2x$ for $x < 4$, we have
$$\lim_{x \to 4^-} f(x) = \lim_{x \to 4^-} (8 - 2x) = 8 - 2 \cdot 4 = 0$$

The right- and left-hand limits are equal. Thus the limit exists and
$$\lim_{x \to 4} f(x) = 0$$

FIGURE 5

The graph of f is shown in Figure 5.

Other notations for $[\![x]\!]$ are $[x]$ and $\lfloor x \rfloor$. The greatest integer function is sometimes called the *floor function*.

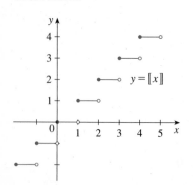

$y = [\![x]\!]$

FIGURE 6
Greatest integer function

EXAMPLE 10 The **greatest integer function** is defined by $[\![x]\!] = $ the largest integer that is less than or equal to x. (For instance, $[\![4]\!] = 4$, $[\![4.8]\!] = 4$, $[\![\pi]\!] = 3$, $[\![\sqrt{2}\,]\!] = 1$, $[\![-\frac{1}{2}]\!] = -1$.) Show that $\lim_{x \to 3} [\![x]\!]$ does not exist.

SOLUTION The graph of the greatest integer function is shown in Figure 6. Since $[\![x]\!] = 3$ for $3 \le x < 4$, we have
$$\lim_{x \to 3^+} [\![x]\!] = \lim_{x \to 3^+} 3 = 3$$

Since $[\![x]\!] = 2$ for $2 \le x < 3$, we have
$$\lim_{x \to 3^-} [\![x]\!] = \lim_{x \to 3^-} 2 = 2$$

Because these one-sided limits are not equal, $\lim_{x \to 3} [\![x]\!]$ does not exist by Theorem 1.

The next two theorems give two additional properties of limits. Their proofs can be found in Appendix F.

---

2 **Theorem** If $f(x) \leqslant g(x)$ when x is near a (except possibly at a) and the limits of f and g both exist as x approaches a, then

$$\lim_{x \to a} f(x) \leqslant \lim_{x \to a} g(x)$$

---

3 **The Squeeze Theorem** If $f(x) \leqslant g(x) \leqslant h(x)$ when x is near a (except possibly at a) and

$$\lim_{x \to a} f(x) = \lim_{x \to a} h(x) = L$$

then

$$\lim_{x \to a} g(x) = L$$

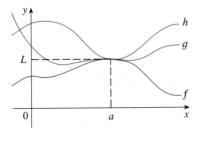

FIGURE 7

The Squeeze Theorem, which is sometimes called the Sandwich Theorem or the Pinching Theorem, is illustrated by Figure 7. It says that if $g(x)$ is squeezed between $f(x)$ and $h(x)$ near a, and if f and h have the same limit L at a, then g is forced to have the same limit L at a.

V **EXAMPLE 11** Show that $\displaystyle\lim_{x \to 0} x^2 \sin \frac{1}{x} = 0$.

SOLUTION First note that we **cannot** use

$\varnothing$

$$\lim_{x \to 0} x^2 \sin \frac{1}{x} = \lim_{x \to 0} x^2 \cdot \lim_{x \to 0} \sin \frac{1}{x}$$

because $\lim_{x \to 0} \sin(1/x)$ does not exist (see Example 4 in Section 1.5).

Instead we apply the Squeeze Theorem, and so we need to find a function f smaller than $g(x) = x^2 \sin(1/x)$ and a function h bigger than g such that both $f(x)$ and $h(x)$ approach 0. To do this we use our knowledge of the sine function. Because the sine of any number lies between -1 and 1, we can write

4
$$-1 \leqslant \sin \frac{1}{x} \leqslant 1$$

Any inequality remains true when multiplied by a positive number. We know that $x^2 \geqslant 0$ for all x and so, multiplying each side of the inequalities in **4** by x^2, we get

$$-x^2 \leqslant x^2 \sin \frac{1}{x} \leqslant x^2$$

as illustrated by Figure 8. We know that

$$\lim_{x \to 0} x^2 = 0 \qquad \text{and} \qquad \lim_{x \to 0} (-x^2) = 0$$

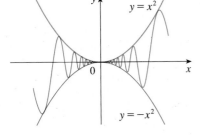

FIGURE 8
$y = x^2 \sin(1/x)$

Taking $f(x) = -x^2$, $g(x) = x^2 \sin(1/x)$, and $h(x) = x^2$ in the Squeeze Theorem, we obtain

$$\lim_{x \to 0} x^2 \sin \frac{1}{x} = 0$$

1.6   Exercises

Access end-of-section exercises and tools for visualizing calculus online at **www.webassign.net**

1.7   The Precise Definition of a Limit

The intuitive definition of a limit given in Section 1.5 is inadequate for some purposes because such phrases as "x is close to 2" and "$f(x)$ gets closer and closer to L" are vague. In order to be able to prove conclusively that

$$\lim_{x \to 0} \left(x^3 + \frac{\cos 5x}{10{,}000} \right) = 0.0001 \qquad \text{or} \qquad \lim_{x \to 0} \frac{\sin x}{x} = 1$$

we must make the definition of a limit precise.

To motivate the precise definition of a limit, let's consider the function

$$f(x) = \begin{cases} 2x - 1 & \text{if } x \ne 3 \\ 6 & \text{if } x = 3 \end{cases}$$

Intuitively, it is clear that when x is close to 3 but $x \ne 3$, then $f(x)$ is close to 5, and so $\lim_{x \to 3} f(x) = 5$.

To obtain more detailed information about how $f(x)$ varies when x is close to 3, we ask the following question:

How close to 3 does x have to be so that $f(x)$ differs from 5 by less than 0.1?

It is traditional to use the Greek letter δ (delta) in this situation.

The distance from x to 3 is $|x - 3|$ and the distance from $f(x)$ to 5 is $|f(x) - 5|$, so our problem is to find a number δ such that

$$|f(x) - 5| < 0.1 \qquad \text{if} \qquad |x - 3| < \delta \quad \text{but } x \ne 3$$

If $|x - 3| > 0$, then $x \ne 3$, so an equivalent formulation of our problem is to find a number δ such that

$$|f(x) - 5| < 0.1 \qquad \text{if} \qquad 0 < |x - 3| < \delta$$

Notice that if $0 < |x - 3| < (0.1)/2 = 0.05$, then

$$|f(x) - 5| = |(2x - 1) - 5| = |2x - 6| = 2|x - 3| < 2(0.05) = 0.1$$

that is, $$|f(x) - 5| < 0.1 \qquad \text{if} \qquad 0 < |x - 3| < 0.05$$

Thus an answer to the problem is given by $\delta = 0.05$; that is, if x is within a distance of 0.05 from 3, then $f(x)$ will be within a distance of 0.1 from 5.

If we change the number 0.1 in our problem to the smaller number 0.01, then by using the same method we find that $f(x)$ will differ from 5 by less than 0.01 provided that x differs from 3 by less than $(0.01)/2 = 0.005$:

$$|f(x) - 5| < 0.01 \qquad \text{if} \qquad 0 < |x - 3| < 0.005$$

Similarly,

$$|f(x) - 5| < 0.001 \qquad \text{if} \qquad 0 < |x - 3| < 0.0005$$

The numbers $0.1, 0.01$, and 0.001 that we have considered are *error tolerances* that we might allow. For 5 to be the precise limit of $f(x)$ as x approaches 3, we must not only be able to bring the difference between $f(x)$ and 5 below each of these three numbers; we must be able to bring it below *any* positive number. And, by the same reasoning, we can! If we write ε (the Greek letter epsilon) for an arbitrary positive number, then we find as before that

$$\boxed{1} \qquad\qquad |f(x) - 5| < \varepsilon \qquad \text{if} \qquad 0 < |x - 3| < \delta = \frac{\varepsilon}{2}$$

This is a precise way of saying that $f(x)$ is close to 5 when x is close to 3 because $\boxed{1}$ says that we can make the values of $f(x)$ within an arbitrary distance ε from 5 by taking the values of x within a distance $\varepsilon/2$ from 3 (but $x \neq 3$).

Note that $\boxed{1}$ can be rewritten as follows:

$$\text{if} \quad 3 - \delta < x < 3 + \delta \quad (x \neq 3) \qquad \text{then} \qquad 5 - \varepsilon < f(x) < 5 + \varepsilon$$

and this is illustrated in Figure 1. By taking the values of $x\ (\neq 3)$ to lie in the interval $(3 - \delta, 3 + \delta)$ we can make the values of $f(x)$ lie in the interval $(5 - \varepsilon, 5 + \varepsilon)$.

Using $\boxed{1}$ as a model, we give a precise definition of a limit.

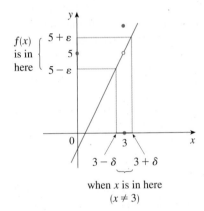

$f(x)$ is in here

$5 + \varepsilon$
5
$5 - \varepsilon$

0

3

$3 - \delta \qquad 3 + \delta$

when x is in here
$(x \neq 3)$

FIGURE 1

$\boxed{2}$ **Definition**   Let f be a function defined on some open interval that contains the number a, except possibly at a itself. Then we say that the **limit of $f(x)$ as x approaches a is L**, and we write

$$\lim_{x \to a} f(x) = L$$

if for every number $\varepsilon > 0$ there is a number $\delta > 0$ such that

$$\text{if} \quad 0 < |x - a| < \delta \qquad \text{then} \qquad |f(x) - L| < \varepsilon$$

Since $|x - a|$ is the distance from x to a and $|f(x) - L|$ is the distance from $f(x)$ to L, and since ε can be arbitrarily small, the definition of a limit can be expressed in words as follows:

   $\lim_{x \to a} f(x) = L$ means that the distance between $f(x)$ and L can be made arbitrarily small
   by taking the distance from x to a sufficiently small (but not 0).

Alternatively,

   $\lim_{x \to a} f(x) = L$ means that the values of $f(x)$ can be made as close as we please to L
   by taking x close enough to a (but not equal to a).

We can also reformulate Definition 2 in terms of intervals by observing that the inequality $|x - a| < \delta$ is equivalent to $-\delta < x - a < \delta$, which in turn can be written as $a - \delta < x < a + \delta$. Also $0 < |x - a|$ is true if and only if $x - a \neq 0$, that is, $x \neq a$. Similarly, the inequality $|f(x) - L| < \varepsilon$ is equivalent to the pair of inequalities

$L - \varepsilon < f(x) < L + \varepsilon$. Therefore, in terms of intervals, Definition 2 can be stated as follows:

$\lim_{x \to a} f(x) = L$ means that for every $\varepsilon > 0$ (no matter how small ε is) we can find $\delta > 0$ such that if x lies in the open interval $(a - \delta, a + \delta)$ and $x \neq a$, then $f(x)$ lies in the open interval $(L - \varepsilon, L + \varepsilon)$.

We interpret this statement geometrically by representing a function by an arrow diagram as in Figure 2, where f maps a subset of $\mathbb{R}$ onto another subset of $\mathbb{R}$.

FIGURE 2

The definition of limit says that if any small interval $(L - \varepsilon, L + \varepsilon)$ is given around L, then we can find an interval $(a - \delta, a + \delta)$ around a such that f maps all the points in $(a - \delta, a + \delta)$ (except possibly a) into the interval $(L - \varepsilon, L + \varepsilon)$. (See Figure 3.)

FIGURE 3

Another geometric interpretation of limits can be given in terms of the graph of a function. If $\varepsilon > 0$ is given, then we draw the horizontal lines $y = L + \varepsilon$ and $y = L - \varepsilon$ and the graph of f. (See Figure 4.) If $\lim_{x \to a} f(x) = L$, then we can find a number $\delta > 0$ such that if we restrict x to lie in the interval $(a - \delta, a + \delta)$ and take $x \neq a$, then the curve $y = f(x)$ lies between the lines $y = L - \varepsilon$ and $y = L + \varepsilon$. (See Figure 5.) You can see that if such a δ has been found, then any smaller δ will also work.

It is important to realize that the process illustrated in Figures 4 and 5 must work for *every* positive number ε, no matter how small it is chosen. Figure 6 shows that if a smaller ε is chosen, then a smaller δ may be required.

FIGURE 4

FIGURE 5

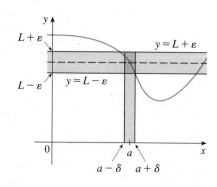

FIGURE 6

EXAMPLE 1 Use a graph to find a number δ such that

$$\text{if} \qquad |x - 1| < \delta \qquad \text{then} \qquad |(x^3 - 5x + 6) - 2| < 0.2$$

In other words, find a number δ that corresponds to $\varepsilon = 0.2$ in the definition of a limit for the function $f(x) = x^3 - 5x + 6$ with $a = 1$ and $L = 2$.

SOLUTION A graph of f is shown in Figure 7; we are interested in the region near the point $(1, 2)$. Notice that we can rewrite the inequality

$$|(x^3 - 5x + 6) - 2| < 0.2$$

as

$$1.8 < x^3 - 5x + 6 < 2.2$$

FIGURE 7

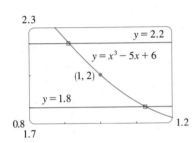

FIGURE 8

So we need to determine the values of x for which the curve $y = x^3 - 5x + 6$ lies between the horizontal lines $y = 1.8$ and $y = 2.2$. Therefore we graph the curves $y = x^3 - 5x + 6$, $y = 1.8$, and $y = 2.2$ near the point $(1, 2)$ in Figure 8. Then we use the cursor to estimate that the x-coordinate of the point of intersection of the line $y = 2.2$ and the curve $y = x^3 - 5x + 6$ is about 0.911. Similarly, $y = x^3 - 5x + 6$ intersects the line $y = 1.8$ when $x \approx 1.124$. So, rounding to be safe, we can say that

$$\text{if} \qquad 0.92 < x < 1.12 \qquad \text{then} \qquad 1.8 < x^3 - 5x + 6 < 2.2$$

This interval $(0.92, 1.12)$ is not symmetric about $x = 1$. The distance from $x = 1$ to the left endpoint is $1 - 0.92 = 0.08$ and the distance to the right endpoint is 0.12. We can choose δ to be the smaller of these numbers, that is, $\delta = 0.08$. Then we can rewrite our inequalities in terms of distances as follows:

$$\text{if} \qquad |x - 1| < 0.08 \qquad \text{then} \qquad |(x^3 - 5x + 6) - 2| < 0.2$$

This just says that by keeping x within 0.08 of 1, we are able to keep $f(x)$ within 0.2 of 2.

Although we chose $\delta = 0.08$, any smaller positive value of δ would also have worked.

The graphical procedure in Example 1 gives an illustration of the definition for $\varepsilon = 0.2$, but it does not *prove* that the limit is equal to 2. A proof has to provide a δ for *every* ε.

In proving limit statements it may be helpful to think of the definition of limit as a challenge. First it challenges you with a number ε. Then you must be able to produce a suitable δ. You have to be able to do this for *every* $\varepsilon > 0$, not just a particular ε.

Imagine a contest between two people, A and B, and imagine yourself to be B. Person A stipulates that the fixed number L should be approximated by the values of $f(x)$ to within a degree of accuracy ε (say, 0.01). Person B then responds by finding a number δ such that if $0 < |x - a| < \delta$, then $|f(x) - L| < \varepsilon$. Then A may become more exacting and challenge B with a smaller value of ε (say, 0.0001). Again B has to respond by finding a corresponding δ. Usually the smaller the value of ε, the smaller the corresponding value of δ must be. If B always wins, no matter how small A makes ε, then $\lim_{x \to a} f(x) = L$.

TEC In Module 1.7/3.4 you can explore the precise definition of a limit both graphically and numerically.

V EXAMPLE 2 Prove that $\lim_{x \to 3} (4x - 5) = 7$.

SOLUTION

1. *Preliminary analysis of the problem* (*guessing a value for* δ). Let ε be a given positive number. We want to find a number δ such that

$$\text{if} \qquad 0 < |x - 3| < \delta \qquad \text{then} \qquad |(4x - 5) - 7| < \varepsilon$$

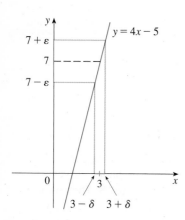

FIGURE 9

Cauchy and Limits

After the invention of calculus in the 17th century, there followed a period of free development of the subject in the 18th century. Mathematicians like the Bernoulli brothers and Euler were eager to exploit the power of calculus and boldly explored the consequences of this new and wonderful mathematical theory without worrying too much about whether their proofs were completely correct.

The 19th century, by contrast, was the Age of Rigor in mathematics. There was a movement to go back to the foundations of the subject—to provide careful definitions and rigorous proofs. At the forefront of this movement was the French mathematician Augustin-Louis Cauchy (1789–1857), who started out as a military engineer before becoming a mathematics professor in Paris. Cauchy took Newton's idea of a limit, which was kept alive in the 18th century by the French mathematician Jean d'Alembert, and made it more precise. His definition of a limit reads as follows: "When the successive values attributed to a variable approach indefinitely a fixed value so as to end by differing from it by as little as one wishes, this last is called the *limit* of all the others." But when Cauchy used this definition in examples and proofs, he often employed delta-epsilon inequalities similar to the ones in this section. A typical Cauchy proof starts with: "Designate by δ and ε two very small numbers; . . ." He used ε because of the correspondence between epsilon and the French word *erreur* and δ because delta corresponds to *différence*. Later, the German mathematician Karl Weierstrass (1815–1897) stated the definition of a limit exactly as in our Definition 2.

But $|(4x - 5) - 7| = |4x - 12| = |4(x - 3)| = 4|x - 3|$. Therefore we want δ such that

$$\text{if} \qquad 0 < |x - 3| < \delta \qquad \text{then} \qquad 4|x - 3| < \varepsilon$$

that is,

$$\text{if} \qquad 0 < |x - 3| < \delta \qquad \text{then} \qquad |x - 3| < \frac{\varepsilon}{4}$$

This suggests that we should choose $\delta = \varepsilon/4$.

2. *Proof* (*showing that this δ works*).   Given $\varepsilon > 0$, choose $\delta = \varepsilon/4$. If $0 < |x - 3| < \delta$, then

$$|(4x - 5) - 7| = |4x - 12| = 4|x - 3| < 4\delta = 4\left(\frac{\varepsilon}{4}\right) = \varepsilon$$

Thus

$$\text{if} \qquad 0 < |x - 3| < \delta \qquad \text{then} \qquad |(4x - 5) - 7| < \varepsilon$$

Therefore, by the definition of a limit,

$$\lim_{x \to 3} (4x - 5) = 7$$

This example is illustrated by Figure 9.

Note that in the solution of Example 2 there were two stages—guessing and proving. We made a preliminary analysis that enabled us to guess a value for δ. But then in the second stage we had to go back and prove in a careful, logical fashion that we had made a correct guess. This procedure is typical of much of mathematics. Sometimes it is necessary to first make an intelligent guess about the answer to a problem and then prove that the guess is correct.

The intuitive definitions of one-sided limits that were given in Section 1.5 can be precisely reformulated as follows.

---

3 **Definition of Left-Hand Limit**

$$\lim_{x \to a^-} f(x) = L$$

if for every number $\varepsilon > 0$ there is a number $\delta > 0$ such that

$$\text{if} \qquad a - \delta < x < a \qquad \text{then} \qquad |f(x) - L| < \varepsilon$$

---

4 **Definition of Right-Hand Limit**

$$\lim_{x \to a^+} f(x) = L$$

if for every number $\varepsilon > 0$ there is a number $\delta > 0$ such that

$$\text{if} \qquad a < x < a + \delta \qquad \text{then} \qquad |f(x) - L| < \varepsilon$$

Notice that Definition 3 is the same as Definition 2 except that x is restricted to lie in the *left* half $(a - \delta, a)$ of the interval $(a - \delta, a + \delta)$. In Definition 4, x is restricted to lie in the *right* half $(a, a + \delta)$ of the interval $(a - \delta, a + \delta)$.

V **EXAMPLE 3** Use Definition 4 to prove that $\lim\limits_{x \to 0^+} \sqrt{x} = 0$.

SOLUTION

1. *Guessing a value for* δ. Let ε be a given positive number. Here $a = 0$ and $L = 0$, so we want to find a number δ such that

$$\text{if} \quad 0 < x < \delta \quad \text{then} \quad |\sqrt{x} - 0| < \varepsilon$$

that is,
$$\text{if} \quad 0 < x < \delta \quad \text{then} \quad \sqrt{x} < \varepsilon$$

or, squaring both sides of the inequality $\sqrt{x} < \varepsilon$, we get

$$\text{if} \quad 0 < x < \delta \quad \text{then} \quad x < \varepsilon^2$$

This suggests that we should choose $\delta = \varepsilon^2$.

2. *Showing that this* δ *works.* Given $\varepsilon > 0$, let $\delta = \varepsilon^2$. If $0 < x < \delta$, then

$$\sqrt{x} < \sqrt{\delta} = \sqrt{\varepsilon^2} = \varepsilon$$

so
$$|\sqrt{x} - 0| < \varepsilon$$

According to Definition 4, this shows that $\lim_{x \to 0^+} \sqrt{x} = 0$.

EXAMPLE 4 Prove that $\lim\limits_{x \to 3} x^2 = 9$.

SOLUTION

1. *Guessing a value for* δ. Let $\varepsilon > 0$ be given. We have to find a number $\delta > 0$ such that

$$\text{if} \quad 0 < |x - 3| < \delta \quad \text{then} \quad |x^2 - 9| < \varepsilon$$

To connect $|x^2 - 9|$ with $|x - 3|$ we write $|x^2 - 9| = |(x + 3)(x - 3)|$. Then we want

$$\text{if} \quad 0 < |x - 3| < \delta \quad \text{then} \quad |x + 3||x - 3| < \varepsilon$$

Notice that if we can find a positive constant C such that $|x + 3| < C$, then

$$|x + 3||x - 3| < C|x - 3|$$

and we can make $C|x - 3| < \varepsilon$ by taking $|x - 3| < \varepsilon/C = \delta$.

We can find such a number C if we restrict x to lie in some interval centered at 3. In fact, since we are interested only in values of x that are close to 3, it is reasonable to assume that x is within a distance 1 from 3, that is, $|x - 3| < 1$. Then $2 < x < 4$, so $5 < x + 3 < 7$. Thus we have $|x + 3| < 7$, and so $C = 7$ is a suitable choice for the constant.

But now there are two restrictions on $|x - 3|$, namely

$$|x - 3| < 1 \quad \text{and} \quad |x - 3| < \frac{\varepsilon}{C} = \frac{\varepsilon}{7}$$

To make sure that both of these inequalities are satisfied, we take δ to be the smaller of the two numbers 1 and $\varepsilon/7$. The notation for this is $\delta = \min\{1, \varepsilon/7\}$.

2. *Showing that this δ works.* Given $\varepsilon > 0$, let $\delta = \min\{1, \varepsilon/7\}$. If $0 < |x - 3| < \delta$, then $|x - 3| < 1 \Rightarrow 2 < x < 4 \Rightarrow |x + 3| < 7$ (as in part 1). We also have $|x - 3| < \varepsilon/7$, so

$$|x^2 - 9| = |x + 3||x - 3| < 7 \cdot \frac{\varepsilon}{7} = \varepsilon$$

This shows that $\lim_{x \to 3} x^2 = 9$.

As Example 4 shows, it is not always easy to prove that limit statements are true using the ε, δ definition. In fact, if we had been given a more complicated function such as $f(x) = (6x^2 - 8x + 9)/(2x^2 - 1)$, a proof would require a great deal of ingenuity. Fortunately this is unnecessary because the Limit Laws stated in Section 1.6 can be proved using Definition 2, and then the limits of complicated functions can be found rigorously from the Limit Laws without resorting to the definition directly.

For instance, we prove the Sum Law: If $\lim_{x \to a} f(x) = L$ and $\lim_{x \to a} g(x) = M$ both exist, then

$$\lim_{x \to a} [f(x) + g(x)] = L + M$$

The remaining laws are proved in the exercises and in Appendix F.

PROOF OF THE SUM LAW Let $\varepsilon > 0$ be given. We must find $\delta > 0$ such that

$$\text{if} \quad 0 < |x - a| < \delta \quad \text{then} \quad |f(x) + g(x) - (L + M)| < \varepsilon$$

Using the Triangle Inequality we can write

Triangle Inequality:

$$|a + b| \leqslant |a| + |b|$$

(See Appendix A.)

$$\boxed{5} \qquad |f(x) + g(x) - (L + M)| = |(f(x) - L) + (g(x) - M)|$$
$$\leqslant |f(x) - L| + |g(x) - M|$$

We make $|f(x) + g(x) - (L + M)|$ less than ε by making each of the terms $|f(x) - L|$ and $|g(x) - M|$ less than $\varepsilon/2$.

Since $\varepsilon/2 > 0$ and $\lim_{x \to a} f(x) = L$, there exists a number $\delta_1 > 0$ such that

$$\text{if} \quad 0 < |x - a| < \delta_1 \quad \text{then} \quad |f(x) - L| < \frac{\varepsilon}{2}$$

Similarly, since $\lim_{x \to a} g(x) = M$, there exists a number $\delta_2 > 0$ such that

$$\text{if} \quad 0 < |x - a| < \delta_2 \quad \text{then} \quad |g(x) - M| < \frac{\varepsilon}{2}$$

Let $\delta = \min\{\delta_1, \delta_2\}$, the smaller of the numbers δ_1 and δ_2. Notice that

$$\text{if} \quad 0 < |x - a| < \delta \quad \text{then} \quad 0 < |x - a| < \delta_1 \quad \text{and} \quad 0 < |x - a| < \delta_2$$

and so

$$|f(x) - L| < \frac{\varepsilon}{2} \quad \text{and} \quad |g(x) - M| < \frac{\varepsilon}{2}$$

Therefore, by $\boxed{5}$,

$$|f(x) + g(x) - (L + M)| \leq |f(x) - L| + |g(x) - M|$$

$$< \frac{\varepsilon}{2} + \frac{\varepsilon}{2} = \varepsilon$$

To summarize,

$$\text{if} \quad 0 < |x - a| < \delta \quad \text{then} \quad |f(x) + g(x) - (L + M)| < \varepsilon$$

Thus, by the definition of a limit,

$$\lim_{x \to a} [f(x) + g(x)] = L + M$$

Infinite Limits

Infinite limits can also be defined in a precise way. The following is a precise version of Definition 4 in Section 1.5.

$\boxed{6}$ **Definition** Let f be a function defined on some open interval that contains the number a, except possibly at a itself. Then

$$\lim_{x \to a} f(x) = \infty$$

means that for every positive number M there is a positive number δ such that

$$\text{if} \quad 0 < |x - a| < \delta \quad \text{then} \quad f(x) > M$$

This says that the values of $f(x)$ can be made arbitrarily large (larger than any given number M) by taking x close enough to a (within a distance δ, where δ depends on M, but with $x \neq a$). A geometric illustration is shown in Figure 10.

Given any horizontal line $y = M$, we can find a number $\delta > 0$ such that if we restrict x to lie in the interval $(a - \delta, a + \delta)$ but $x \neq a$, then the curve $y = f(x)$ lies above the line $y = M$. You can see that if a larger M is chosen, then a smaller δ may be required.

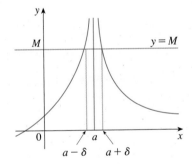

FIGURE 10

V EXAMPLE 5 Use Definition 6 to prove that $\lim_{x \to 0} \dfrac{1}{x^2} = \infty$.

SOLUTION Let M be a given positive number. We want to find a number δ such that

$$\text{if} \quad 0 < |x| < \delta \quad \text{then} \quad 1/x^2 > M$$

But

$$\frac{1}{x^2} > M \quad \Longleftrightarrow \quad x^2 < \frac{1}{M} \quad \Longleftrightarrow \quad |x| < \frac{1}{\sqrt{M}}$$

So if we choose $\delta = 1/\sqrt{M}$ and $0 < |x| < \delta = 1/\sqrt{M}$, then $1/x^2 > M$. This shows that $1/x^2 \to \infty$ as $x \to 0$.

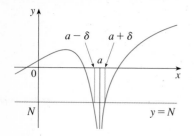

FIGURE 11

Similarly, the following is a precise version of Definition 5 in Section 1.5. It is illustrated by Figure 11.

---

7 **Definition** Let f be a function defined on some open interval that contains the number a, except possibly at a itself. Then

$$\lim_{x \to a} f(x) = -\infty$$

means that for every negative number N there is a positive number δ such that

$$\text{if} \quad 0 < |x - a| < \delta \quad \text{then} \quad f(x) < N$$

1.7 Exercises

Access end-of-section exercises and tools for visualizing calculus online at **www.webassign.net**

1.8 Continuity

We noticed in Section 1.6 that the limit of a function as x approaches a can often be found simply by calculating the value of the function at a. Functions with this property are called *continuous at a*. We will see that the mathematical definition of continuity corresponds closely with the meaning of the word *continuity* in everyday language. (A continuous process is one that takes place gradually, without interruption or abrupt change.)

As illustrated in Figure 1, if f is continuous, then the points $(x, f(x))$ on the graph of f approach the point $(a, f(a))$ on the graph. So there is no gap in the curve.

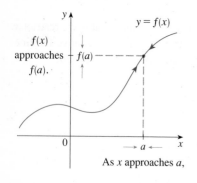

As x approaches a,

FIGURE 1

---

1 **Definition** A function f is **continuous at a number a** if

$$\lim_{x \to a} f(x) = f(a)$$

Notice that Definition 1 implicitly requires three things if f is continuous at a:

1. $f(a)$ is defined (that is, a is in the domain of f)

2. $\lim_{x \to a} f(x)$ exists

3. $\lim_{x \to a} f(x) = f(a)$

The definition says that f is continuous at a if $f(x)$ approaches $f(a)$ as x approaches a. Thus a continuous function f has the property that a small change in x produces only a small change in $f(x)$. In fact, the change in $f(x)$ can be kept as small as we please by keeping the change in x sufficiently small.

If f is defined near a (in other words, f is defined on an open interval containing a, except perhaps at a), we say that f is **discontinuous at a** (or f has a **discontinuity** at a) if f is not continuous at a.

Physical phenomena are usually continuous. For instance, the displacement or velocity of a vehicle varies continuously with time, as does a person's height. But discontinuities do occur in such situations as electric currents. [See Example 6 in Section 1.5, where the Heaviside function is discontinuous at 0 because $\lim_{t \to 0} H(t)$ does not exist.]

Geometrically, you can think of a function that is continuous at every number in an interval as a function whose graph has no break in it. The graph can be drawn without removing your pen from the paper.

EXAMPLE 1 Figure 2 shows the graph of a function f. At which numbers is f discontinuous? Why?

SOLUTION It looks as if there is a discontinuity when $a = 1$ because the graph has a break there. The official reason that f is discontinuous at 1 is that $f(1)$ is not defined.

The graph also has a break when $a = 3$, but the reason for the discontinuity is different. Here, $f(3)$ is defined, but $\lim_{x \to 3} f(x)$ does not exist (because the left and right limits are different). So f is discontinuous at 3.

What about $a = 5$? Here, $f(5)$ is defined and $\lim_{x \to 5} f(x)$ exists (because the left and right limits are the same). But

$$\lim_{x \to 5} f(x) \neq f(5)$$

So f is discontinuous at 5.

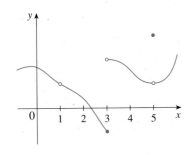

FIGURE 2

Now let's see how to detect discontinuities when a function is defined by a formula.

V **EXAMPLE 2** Where are each of the following functions discontinuous?

(a) $f(x) = \dfrac{x^2 - x - 2}{x - 2}$

(b) $f(x) = \begin{cases} \dfrac{1}{x^2} & \text{if } x \neq 0 \\ 1 & \text{if } x = 0 \end{cases}$

(c) $f(x) = \begin{cases} \dfrac{x^2 - x - 2}{x - 2} & \text{if } x \neq 2 \\ 1 & \text{if } x = 2 \end{cases}$

(d) $f(x) = [\![x]\!]$

SOLUTION

(a) Notice that $f(2)$ is not defined, so f is discontinuous at 2. Later we'll see why f is continuous at all other numbers.

(b) Here $f(0) = 1$ is defined but

$$\lim_{x \to 0} f(x) = \lim_{x \to 0} \frac{1}{x^2}$$

does not exist. (See Example 8 in Section 1.5.) So f is discontinuous at 0.

(c) Here $f(2) = 1$ is defined and

$$\lim_{x \to 2} f(x) = \lim_{x \to 2} \frac{x^2 - x - 2}{x - 2} = \lim_{x \to 2} \frac{(x - 2)(x + 1)}{x - 2} = \lim_{x \to 2} (x + 1) = 3$$

exists. But

$$\lim_{x \to 2} f(x) \neq f(2)$$

so f is not continuous at 2.

(d) The greatest integer function $f(x) = [\![x]\!]$ has discontinuities at all of the integers because $\lim_{x \to n} [\![x]\!]$ does not exist if n is an integer. (See Example 10 and Exercise 51 in Section 1.6.)

Figure 3 shows the graphs of the functions in Example 2. In each case the graph can't be drawn without lifting the pen from the paper because a hole or break or jump occurs in the graph. The kind of discontinuity illustrated in parts (a) and (c) is called **removable** because we could remove the discontinuity by redefining f at just the single number 2. [The function $g(x) = x + 1$ is continuous.] The discontinuity in part (b) is called an **infinite discontinuity**. The discontinuities in part (d) are called **jump discontinuities** because the function "jumps" from one value to another.

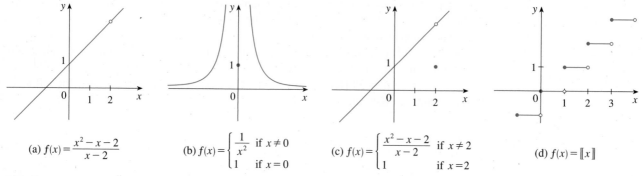

(a) $f(x) = \dfrac{x^2 - x - 2}{x - 2}$ (b) $f(x) = \begin{cases} \dfrac{1}{x^2} & \text{if } x \neq 0 \\ 1 & \text{if } x = 0 \end{cases}$ (c) $f(x) = \begin{cases} \dfrac{x^2 - x - 2}{x - 2} & \text{if } x \neq 2 \\ 1 & \text{if } x = 2 \end{cases}$ (d) $f(x) = [\![x]\!]$

FIGURE 3
Graphs of the functions in Example 2

---

2 **Definition** A function f is **continuous from the right at a number a** if

$$\lim_{x \to a^+} f(x) = f(a)$$

and f is **continuous from the left at a** if

$$\lim_{x \to a^-} f(x) = f(a)$$

EXAMPLE 3 At each integer n, the function $f(x) = [\![x]\!]$ [see Figure 3(d)] is continuous from the right but discontinuous from the left because

$$\lim_{x \to n^+} f(x) = \lim_{x \to n^+} [\![x]\!] = n = f(n)$$

but

$$\lim_{x \to n^-} f(x) = \lim_{x \to n^-} [\![x]\!] = n - 1 \neq f(n)$$ ∎

---

3 **Definition** A function f is **continuous on an interval** if it is continuous at every number in the interval. (If f is defined only on one side of an endpoint of the interval, we understand *continuous* at the endpoint to mean *continuous from the right* or *continuous from the left*.)

EXAMPLE 4 Show that the function $f(x) = 1 - \sqrt{1 - x^2}$ is continuous on the interval $[-1, 1]$.

SOLUTION If $-1 < a < 1$, then using the Limit Laws, we have

$$\lim_{x \to a} f(x) = \lim_{x \to a} \left(1 - \sqrt{1 - x^2}\right)$$

$$= 1 - \lim_{x \to a} \sqrt{1 - x^2} \qquad \text{(by Laws 2 and 7)}$$

$$= 1 - \sqrt{\lim_{x \to a} (1 - x^2)} \qquad \text{(by 11)}$$

$$= 1 - \sqrt{1 - a^2} \qquad \text{(by 2, 7, and 9)}$$

$$= f(a)$$

Thus, by Definition 1, f is continuous at a if $-1 < a < 1$. Similar calculations show that

$$\lim_{x \to -1^+} f(x) = 1 = f(-1) \qquad \text{and} \qquad \lim_{x \to 1^-} f(x) = 1 = f(1)$$

so f is continuous from the right at -1 and continuous from the left at 1. Therefore, according to Definition 3, f is continuous on $[-1, 1]$.

The graph of f is sketched in Figure 4. It is the lower half of the circle

$$x^2 + (y - 1)^2 = 1 \qquad \blacksquare$$

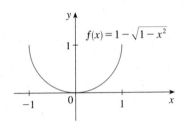

$f(x) = 1 - \sqrt{1 - x^2}$

FIGURE 4

Instead of always using Definitions 1, 2, and 3 to verify the continuity of a function as we did in Example 4, it is often convenient to use the next theorem, which shows how to build up complicated continuous functions from simple ones.

---

4 Theorem If f and g are continuous at a and c is a constant, then the following functions are also continuous at a:

1. $f + g$ **2.** $f - g$ **3.** cf

4. fg **5.** $\dfrac{f}{g}$ if $g(a) \neq 0$

PROOF Each of the five parts of this theorem follows from the corresponding Limit Law in Section 1.6. For instance, we give the proof of part 1. Since f and g are continuous at a, we have

$$\lim_{x \to a} f(x) = f(a) \qquad \text{and} \qquad \lim_{x \to a} g(x) = g(a)$$

Therefore

$$\lim_{x \to a} (f + g)(x) = \lim_{x \to a} [f(x) + g(x)]$$

$$= \lim_{x \to a} f(x) + \lim_{x \to a} g(x) \qquad \text{(by Law 1)}$$

$$= f(a) + g(a)$$

$$= (f + g)(a)$$

This shows that $f + g$ is continuous at a. $\blacksquare$

It follows from Theorem 4 and Definition 3 that if f and g are continuous on an interval, then so are the functions $f + g$, $f - g$, cf, fg, and (if g is never 0) f/g. The following theorem was stated in Section 1.6 as the Direct Substitution Property.

> **5** **Theorem**
> (a) Any polynomial is continuous everywhere; that is, it is continuous on $\mathbb{R} = (-\infty, \infty)$.
> (b) Any rational function is continuous wherever it is defined; that is, it is continuous on its domain.

PROOF

(a) A polynomial is a function of the form

$$P(x) = c_n x^n + c_{n-1} x^{n-1} + \cdots + c_1 x + c_0$$

where $c_0, c_1, \ldots, c_n$ are constants. We know that

$$\lim_{x \to a} c_0 = c_0 \qquad \text{(by Law 7)}$$

and

$$\lim_{x \to a} x^m = a^m \qquad m = 1, 2, \ldots, n \qquad \text{(by 9)}$$

This equation is precisely the statement that the function $f(x) = x^m$ is a continuous function. Thus, by part 3 of Theorem 4, the function $g(x) = cx^m$ is continuous. Since P is a sum of functions of this form and a constant function, it follows from part 1 of Theorem 4 that P is continuous.

(b) A rational function is a function of the form

$$f(x) = \frac{P(x)}{Q(x)}$$

where P and Q are polynomials. The domain of f is $D = \{x \in \mathbb{R} \mid Q(x) \neq 0\}$. We know from part (a) that P and Q are continuous everywhere. Thus, by part 5 of Theorem 4, f is continuous at every number in D. ∎

As an illustration of Theorem 5, observe that the volume of a sphere varies continuously with its radius because the formula $V(r) = \frac{4}{3}\pi r^3$ shows that V is a polynomial function of r. Likewise, if a ball is thrown vertically into the air with a velocity of 50 ft/s, then the height of the ball in feet t seconds later is given by the formula $h = 50t - 16t^2$. Again this is a polynomial function, so the height is a continuous function of the elapsed time.

Knowledge of which functions are continuous enables us to evaluate some limits very quickly, as the following example shows. Compare it with Example 2(b) in Section 1.6.

EXAMPLE 5 Find $\lim_{x \to -2} \dfrac{x^3 + 2x^2 - 1}{5 - 3x}$.

SOLUTION The function

$$f(x) = \frac{x^3 + 2x^2 - 1}{5 - 3x}$$

is rational, so by Theorem 5 it is continuous on its domain, which is $\{x \mid x \neq \frac{5}{3}\}$. Therefore

$$\lim_{x \to -2} \frac{x^3 + 2x^2 - 1}{5 - 3x} = \lim_{x \to -2} f(x) = f(-2)$$

$$= \frac{(-2)^3 + 2(-2)^2 - 1}{5 - 3(-2)} = -\frac{1}{11}$$

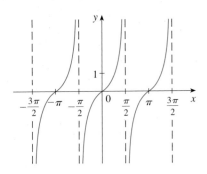

FIGURE 5

It turns out that most of the familiar functions are continuous at every number in their domains. For instance, Limit Law 10 (page 50) is exactly the statement that root functions are continuous.

From the appearance of the graphs of the sine and cosine functions (Figure 18 in Section 1.2), we would certainly guess that they are continuous. We know from the definitions of $\sin \theta$ and $\cos \theta$ that the coordinates of the point P in Figure 5 are $(\cos \theta, \sin \theta)$. As $\theta \to 0$, we see that P approaches the point $(1, 0)$ and so $\cos \theta \to 1$ and $\sin \theta \to 0$. Thus

Another way to establish the limits in $\boxed{6}$ is to use the Squeeze Theorem with the inequality $\sin \theta < \theta$ (for $\theta > 0$), which is proved in Section 2.4.

$$\boxed{6} \qquad \lim_{\theta \to 0} \cos \theta = 1 \qquad \lim_{\theta \to 0} \sin \theta = 0$$

Since $\cos 0 = 1$ and $\sin 0 = 0$, the equations in $\boxed{6}$ assert that the cosine and sine functions are continuous at 0. The addition formulas for cosine and sine can then be used to deduce that these functions are continuous everywhere (see Exercises 60 and 61).

It follows from part 5 of Theorem 4 that

$$\tan x = \frac{\sin x}{\cos x}$$

is continuous except where $\cos x = 0$. This happens when x is an odd integer multiple of $\pi/2$, so $y = \tan x$ has infinite discontinuities when $x = \pm\pi/2, \pm3\pi/2, \pm5\pi/2$, and so on (see Figure 6).

FIGURE 6 $y = \tan x$

$\boxed{7}$ **Theorem** The following types of functions are continuous at every number in their domains:

polynomials	rational functions
root functions	trigonometric functions

EXAMPLE 6 On what intervals is each function continuous?

(a) $f(x) = x^{100} - 2x^{37} + 75$

(b) $g(x) = \dfrac{x^2 + 2x + 17}{x^2 - 1}$

(c) $h(x) = \sqrt{x} + \dfrac{x + 1}{x - 1} - \dfrac{x + 1}{x^2 + 1}$

SOLUTION

(a) f is a polynomial, so it is continuous on $(-\infty, \infty)$ by Theorem 5(a).

(b) g is a rational function, so by Theorem 5(b), it is continuous on its domain, which is $D = \{x \mid x^2 - 1 \neq 0\} = \{x \mid x \neq \pm 1\}$. Thus g is continuous on the intervals $(-\infty, -1)$, $(-1, 1)$, and $(1, \infty)$.

(c) We can write $h(x) = F(x) + G(x) - H(x)$, where

$$F(x) = \sqrt{x} \qquad G(x) = \frac{x + 1}{x - 1} \qquad H(x) = \frac{x + 1}{x^2 + 1}$$

F is continuous on $[0, \infty)$ by Theorem 7. G is a rational function, so it is continuous everywhere except when $x - 1 = 0$, that is, $x = 1$. H is also a rational function, but its denominator is never 0, so H is continuous everywhere. Thus, by parts 1 and 2 of Theorem 4, h is continuous on the intervals $[0, 1)$ and $(1, \infty)$.

EXAMPLE 7 Evaluate $\displaystyle\lim_{x \to \pi} \frac{\sin x}{2 + \cos x}$.

SOLUTION Theorem 7 tells us that $y = \sin x$ is continuous. The function in the denominator, $y = 2 + \cos x$, is the sum of two continuous functions and is therefore continuous. Notice that this function is never 0 because $\cos x \geq -1$ for all x and so $2 + \cos x > 0$ everywhere. Thus the ratio

$$f(x) = \frac{\sin x}{2 + \cos x}$$

is continuous everywhere. Hence, by the definition of a continuous function,

$$\lim_{x \to \pi} \frac{\sin x}{2 + \cos x} = \lim_{x \to \pi} f(x) = f(\pi) = \frac{\sin \pi}{2 + \cos \pi} = \frac{0}{2 - 1} = 0$$ ▬

Another way of combining continuous functions f and g to get a new continuous function is to form the composite function $f \circ g$. This fact is a consequence of the following theorem.

---

8 **Theorem** If f is continuous at b and $\displaystyle\lim_{x \to a} g(x) = b$, then $\displaystyle\lim_{x \to a} f(g(x)) = f(b)$. In other words,

$$\lim_{x \to a} f(g(x)) = f\left(\lim_{x \to a} g(x)\right)$$

This theorem says that a limit symbol can be moved through a function symbol if the function is continuous and the limit exists. In other words, the order of these two symbols can be reversed.

Intuitively, Theorem 8 is reasonable because if x is close to a, then $g(x)$ is close to b, and since f is continuous at b, if $g(x)$ is close to b, then $f(g(x))$ is close to $f(b)$. A proof of Theorem 8 is given in Appendix F.

Let's now apply Theorem 8 in the special case where $f(x) = \sqrt[n]{x}$, with n being a positive integer. Then

$$f(g(x)) = \sqrt[n]{g(x)}$$

and

$$f\left(\lim_{x \to a} g(x)\right) = \sqrt[n]{\lim_{x \to a} g(x)}$$

If we put these expressions into Theorem 8, we get

$$\lim_{x \to a} \sqrt[n]{g(x)} = \sqrt[n]{\lim_{x \to a} g(x)}$$

and so Limit Law 11 has now been proved. (We assume that the roots exist.)

---

9 **Theorem** If g is continuous at a and f is continuous at $g(a)$, then the composite function $f \circ g$ given by $(f \circ g)(x) = f(g(x))$ is continuous at a.

This theorem is often expressed informally by saying "a continuous function of a continuous function is a continuous function."

PROOF Since g is continuous at a, we have

$$\lim_{x \to a} g(x) = g(a)$$

Since f is continuous at $b = g(a)$, we can apply Theorem 8 to obtain

$$\lim_{x \to a} f(g(x)) = f(g(a))$$

which is precisely the statement that the function $h(x) = f(g(x))$ is continuous at a; that is, $f \circ g$ is continuous at a. ■

V **EXAMPLE 8** Where are the following functions continuous?

(a) $h(x) = \sin(x^2)$
(b) $F(x) = \dfrac{1}{\sqrt{x^2 + 7} - 4}$

SOLUTION

(a) We have $h(x) = f(g(x))$, where

$$g(x) = x^2 \qquad \text{and} \qquad f(x) = \sin x$$

Now g is continuous on $\mathbb{R}$ since it is a polynomial, and f is also continuous everywhere. Thus $h = f \circ g$ is continuous on $\mathbb{R}$ by Theorem 9.

(b) Notice that F can be broken up as the composition of four continuous functions:

$$F = f \circ g \circ h \circ k \qquad \text{or} \qquad F(x) = f(g(h(k(x))))$$

where $\quad f(x) = \dfrac{1}{x} \qquad g(x) = x - 4 \qquad h(x) = \sqrt{x} \qquad k(x) = x^2 + 7$

We know that each of these functions is continuous on its domain (by Theorems 5 and 7), so by Theorem 9, F is continuous on its domain, which is

$$\{x \in \mathbb{R} \mid \sqrt{x^2 + 7} \neq 4\} = \{x \mid x \neq \pm 3\} = (-\infty, -3) \cup (-3, 3) \cup (3, \infty) \quad ■$$

An important property of continuous functions is expressed by the following theorem, whose proof is found in more advanced books on calculus.

---

10 **The Intermediate Value Theorem** Suppose that f is continuous on the closed interval $[a, b]$ and let N be any number between $f(a)$ and $f(b)$, where $f(a) \neq f(b)$. Then there exists a number c in (a, b) such that $f(c) = N$.

The Intermediate Value Theorem states that a continuous function takes on every intermediate value between the function values $f(a)$ and $f(b)$. It is illustrated by Figure 7. Note that the value N can be taken on once [as in part (a)] or more than once [as in part (b)].

(a)

(b)

FIGURE 7

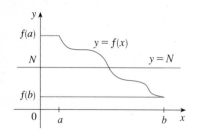

FIGURE 8

If we think of a continuous function as a function whose graph has no hole or break, then it is easy to believe that the Intermediate Value Theorem is true. In geometric terms it says that if any horizontal line $y = N$ is given between $y = f(a)$ and $y = f(b)$ as in Figure 8, then the graph of f can't jump over the line. It must intersect $y = N$ somewhere.

It is important that the function f in Theorem 10 be continuous. The Intermediate Value Theorem is not true in general for discontinuous functions (see Exercise 48).

One use of the Intermediate Value Theorem is in locating roots of equations as in the following example.

V   EXAMPLE 9 Show that there is a root of the equation

$$4x^3 - 6x^2 + 3x - 2 = 0$$

between 1 and 2.

SOLUTION Let $f(x) = 4x^3 - 6x^2 + 3x - 2$. We are looking for a solution of the given equation, that is, a number c between 1 and 2 such that $f(c) = 0$. Therefore we take $a = 1$, $b = 2$, and $N = 0$ in Theorem 10. We have

$$f(1) = 4 - 6 + 3 - 2 = -1 < 0$$

and

$$f(2) = 32 - 24 + 6 - 2 = 12 > 0$$

Thus $f(1) < 0 < f(2)$; that is, $N = 0$ is a number between $f(1)$ and $f(2)$. Now f is continuous since it is a polynomial, so the Intermediate Value Theorem says there is a number c between 1 and 2 such that $f(c) = 0$. In other words, the equation $4x^3 - 6x^2 + 3x - 2 = 0$ has at least one root c in the interval $(1, 2)$.

In fact, we can locate a root more precisely by using the Intermediate Value Theorem again. Since

$$f(1.2) = -0.128 < 0 \qquad \text{and} \qquad f(1.3) = 0.548 > 0$$

a root must lie between 1.2 and 1.3. A calculator gives, by trial and error,

$$f(1.22) = -0.007008 < 0 \qquad \text{and} \qquad f(1.23) = 0.056068 > 0$$

so a root lies in the interval $(1.22, 1.23)$. ▬

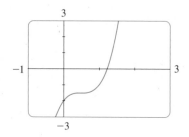

FIGURE 9

We can use a graphing calculator or computer to illustrate the use of the Intermediate Value Theorem in Example 9. Figure 9 shows the graph of f in the viewing rectangle $[-1, 3]$ by $[-3, 3]$ and you can see that the graph crosses the x-axis between 1 and 2. Figure 10 shows the result of zooming in to the viewing rectangle $[1.2, 1.3]$ by $[-0.2, 0.2]$.

In fact, the Intermediate Value Theorem plays a role in the very way these graphing devices work. A computer calculates a finite number of points on the graph and turns on the pixels that contain these calculated points. It assumes that the function is continuous and takes on all the intermediate values between two consecutive points. The computer therefore connects the pixels by turning on the intermediate pixels.

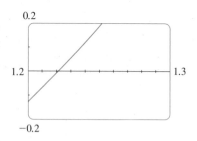

FIGURE 10

 1.8   **Exercises**

Access end-of-section exercises and tools for visualizing calculus online at **www.webassign.net**   ^{ENHANCED} **WebAssign**

1 Review

Concept Check

1. (a) What is a function? What are its domain and range?
   (b) What is the graph of a function?
   (c) How can you tell whether a given curve is the graph of a function?

2. Discuss four ways of representing a function. Illustrate your discussion with examples.

3. (a) What is an even function? How can you tell if a function is even by looking at its graph? Give three examples of an even function.
   (b) What is an odd function? How can you tell if a function is odd by looking at its graph? Give three examples of an odd function.

4. What is an increasing function?

5. What is a mathematical model?

6. Give an example of each type of function.
   (a) Linear function   (b) Power function
   (c) Exponential function   (d) Quadratic function
   (e) Polynomial of degree 5   (f) Rational function

7. Sketch by hand, on the same axes, the graphs of the following functions.
   (a) $f(x) = x$ (b) $g(x) = x^2$
   (c) $h(x) = x^3$ (d) $j(x) = x^4$

8. Draw, by hand, a rough sketch of the graph of each function.
   (a) $y = \sin x$ (b) $y = \tan x$
   (c) $y = 2^x$ (d) $y = 1/x$
   (e) $y = |x|$ (f) $y = \sqrt{x}$

9. Suppose that f has domain A and g has domain B.
   (a) What is the domain of $f + g$?
   (b) What is the domain of fg?
   (c) What is the domain of f/g?

10. How is the composite function $f \circ g$ defined? What is its domain?

11. Suppose the graph of f is given. Write an equation for each of the graphs that are obtained from the graph of f as follows.
    (a) Shift 2 units upward.
    (b) Shift 2 units downward.
    (c) Shift 2 units to the right.
    (d) Shift 2 units to the left.
    (e) Reflect about the x-axis.
    (f) Reflect about the y-axis.
    (g) Stretch vertically by a factor of 2.
    (h) Shrink vertically by a factor of 2.
    (i) Stretch horizontally by a factor of 2.
    (j) Shrink horizontally by a factor of 2.

12. Explain what each of the following means and illustrate with a sketch.
    (a) $\lim_{x \to a^-} f(x) = L$ (b) $\lim_{x \to a^+} f(x) = L$
    (c) $\lim_{x \to a} f(x) = L$ (d) $\lim_{x \to a} f(x) = \infty$
    (e) $\lim_{x \to a} f(x) = -\infty$

13. Describe several ways in which a limit can fail to exist. Illustrate with sketches.

14. What does it mean to say that the line $x = a$ is a vertical asymptote of the curve $y = f(x)$? Draw curves to illustrate the various possibilities.

15. State the following Limit Laws.
    (a) Sum Law   (b) Difference Law
    (c) Constant Multiple Law (d) Product Law
    (e) Quotient Law   (f) Power Law
    (g) Root Law

16. What does the Squeeze Theorem say?

17. (a) What does it mean for f to be continuous at a?
    (b) What does it mean for f to be continuous on the interval $(-\infty, \infty)$? What can you say about the graph of such a function?

18. What does the Intermediate Value Theorem say?

True-False Quiz

Determine whether the statement is true or false. If it is true, explain why. If it is false, explain why or give an example that disproves the statement.

1. If f is a function, then $f(s + t) = f(s) + f(t)$.

2. If $f(s) = f(t)$, then $s = t$.

3. If f is a function, then $f(3x) = 3f(x)$.

4. If $x_1 < x_2$ and f is a decreasing function, then $f(x_1) > f(x_2)$.

5. A vertical line intersects the graph of a function at most once.

6. If x is any real number, then $\sqrt{x^2} = x$.

7. $\lim_{x \to 4} \left(\dfrac{2x}{x - 4} - \dfrac{8}{x - 4} \right) = \lim_{x \to 4} \dfrac{2x}{x - 4} - \lim_{x \to 4} \dfrac{8}{x - 4}$

8. $\lim_{x \to 1} \dfrac{x^2 + 6x - 7}{x^2 + 5x - 6} = \dfrac{\lim_{x \to 1} (x^2 + 6x - 7)}{\lim_{x \to 1} (x^2 + 5x - 6)}$

9. $\lim\limits_{x\to 1} \dfrac{x-3}{x^2+2x-4} = \dfrac{\lim\limits_{x\to 1}(x-3)}{\lim\limits_{x\to 1}(x^2+2x-4)}$

10. If $\lim_{x\to 5} f(x) = 2$ and $\lim_{x\to 5} g(x) = 0$, then $\lim_{x\to 5}\left[f(x)/g(x)\right]$ does not exist.

11. If $\lim_{x\to 5} f(x) = 0$ and $\lim_{x\to 5} g(x) = 0$, then $\lim_{x\to 5}\left[f(x)/g(x)\right]$ does not exist.

12. If neither $\lim_{x\to a} f(x)$ nor $\lim_{x\to a} g(x)$ exists, then $\lim_{x\to a}\left[f(x)+g(x)\right]$ does not exist.

13. If $\lim_{x\to a} f(x)$ exists but $\lim_{x\to a} g(x)$ does not exist, then $\lim_{x\to a}\left[f(x)+g(x)\right]$ does not exist.

14. If $\lim_{x\to 6}\left[f(x)\,g(x)\right]$ exists, then the limit must be $f(6)\,g(6)$.

15. If p is a polynomial, then $\lim_{x\to b} p(x) = p(b)$.

16. If $\lim_{x\to 0} f(x) = \infty$ and $\lim_{x\to 0} g(x) = \infty$, then $\lim_{x\to 0}\left[f(x)-g(x)\right] = 0$.

17. If the line $x = 1$ is a vertical asymptote of $y = f(x)$, then f is not defined at 1.

18. If $f(1) > 0$ and $f(3) < 0$, then there exists a number c between 1 and 3 such that $f(c) = 0$.

19. If f is continuous at 5 and $f(5) = 2$ and $f(4) = 3$, then $\lim_{x\to 2} f(4x^2 - 11) = 2$.

20. If f is continuous on $[-1, 1]$ and $f(-1) = 4$ and $f(1) = 3$, then there exists a number r such that $|r| < 1$ and $f(r) = \pi$.

21. Let f be a function such that $\lim_{x\to 0} f(x) = 6$. Then there exists a number δ such that if $0 < |x| < \delta$, then $|f(x) - 6| < 1$.

22. If $f(x) > 1$ for all x and $\lim_{x\to 0} f(x)$ exists, then $\lim_{x\to 0} f(x) > 1$.

23. The equation $x^{10} - 10x^2 + 5 = 0$ has a root in the interval $(0, 2)$.

24. If f is continuous at a, so is $|f|$.

25. If $|f|$ is continuous at a, so is f.

Exercises

1. Let f be the function whose graph is given.
(a) Estimate the value of $f(2)$.
(b) Estimate the values of x such that $f(x) = 3$.
(c) State the domain of f.
(d) State the range of f.
(e) On what interval is f increasing?
(f) Is f even, odd, or neither even nor odd? Explain.

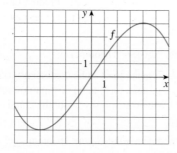

2. Determine whether each curve is the graph of a function of x. If it is, state the domain and range of the function.
(a)   (b)

3. If $f(x) = x^2 - 2x + 3$, evaluate the difference quotient
$$\frac{f(a+h) - f(a)}{h}$$

4. Sketch a rough graph of the yield of a crop as a function of the amount of fertilizer used.

5–8 Find the domain and range of the function. Write your answer in interval notation.

5. $f(x) = 2/(3x - 1)$ **6.** $g(x) = \sqrt{16 - x^4}$

7. $y = 1 + \sin x$ **8.** $F(t) = 3 + \cos 2t$

9. Suppose that the graph of f is given. Describe how the graphs of the following functions can be obtained from the graph of f.
(a) $y = f(x) + 8$ (b) $y = f(x + 8)$
(c) $y = 1 + 2f(x)$ (d) $y = f(x - 2) - 2$
(e) $y = -f(x)$ (f) $y = 3 - f(x)$

10. The graph of f is given. Draw the graphs of the following functions.
(a) $y = f(x - 8)$ (b) $y = -f(x)$
(c) $y = 2 - f(x)$ (d) $y = \frac{1}{2}f(x) - 1$

11–16 Use transformations to sketch the graph of the function.

11. $y = -\sin 2x$

12. $y = (x - 2)^2$

13. $y = 1 + \frac{1}{2}x^3$

14. $y = 2 - \sqrt{x}$

15. $f(x) = \dfrac{1}{x + 2}$

16. $f(x) = \begin{cases} 1 + x & \text{if } x < 0 \\ 1 + x^2 & \text{if } x \geq 0 \end{cases}$

---

17. Determine whether f is even, odd, or neither even nor odd.
(a) $f(x) = 2x^5 - 3x^2 + 2$
(b) $f(x) = x^3 - x^7$
(c) $f(x) = \cos(x^2)$
(d) $f(x) = 1 + \sin x$

18. Find an expression for the function whose graph consists of the line segment from the point $(-2, 2)$ to the point $(-1, 0)$ together with the top half of the circle with center the origin and radius 1.

19. If $f(x) = \sqrt{x}$ and $g(x) = \sin x$, find the functions (a) $f \circ g$, (b) $g \circ f$, (c) $f \circ f$, (d) $g \circ g$, and their domains.

20. Express the function $F(x) = 1/\sqrt{x + \sqrt{x}}$ as a composition of three functions.

21. Life expectancy improved dramatically in the 20th century. The table gives the life expectancy at birth (in years) of males born in the United States. Use a scatter plot to choose an appropriate type of model. Use your model to predict the life span of a male born in the year 2010.

Birth year	Life expectancy	Birth year	Life expectancy
1900	48.3	1960	66.6
1910	51.1	1970	67.1
1920	55.2	1980	70.0
1930	57.4	1990	71.8
1940	62.5	2000	73.0
1950	65.6		

22. A small-appliance manufacturer finds that it costs \$9000 to produce 1000 toaster ovens a week and \$12,000 to produce 1500 toaster ovens a week.
(a) Express the cost as a function of the number of toaster ovens produced, assuming that it is linear. Then sketch the graph.
(b) What is the slope of the graph and what does it represent?
(c) What is the y-intercept of the graph and what does it represent?

23. The graph of f is given.
(a) Find each limit, or explain why it does not exist.
(i) $\lim\limits_{x \to 2^+} f(x)$
(ii) $\lim\limits_{x \to -3^+} f(x)$

(iii) $\lim\limits_{x \to -3} f(x)$
(iv) $\lim\limits_{x \to 4} f(x)$
(v) $\lim\limits_{x \to 0} f(x)$
(vi) $\lim\limits_{x \to 2^-} f(x)$
(b) State the equations of the vertical asymptotes.
(c) At what numbers is f discontinuous? Explain.

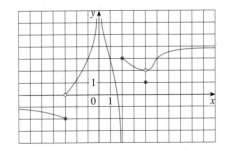

24. Sketch the graph of an example of a function f that satisfies all of the following conditions:
$$\lim_{x \to 0^+} f(x) = -2, \quad \lim_{x \to 0^-} f(x) = 1, \quad f(0) = -1,$$
$$\lim_{x \to 2^-} f(x) = \infty, \quad \lim_{x \to 2^+} f(x) = -\infty$$

25–38 Find the limit.

25. $\lim\limits_{x \to 0} \cos(x + \sin x)$

26. $\lim\limits_{x \to 3} \dfrac{x^2 - 9}{x^2 + 2x - 3}$

27. $\lim\limits_{x \to -3} \dfrac{x^2 - 9}{x^2 + 2x - 3}$

28. $\lim\limits_{x \to 1^+} \dfrac{x^2 - 9}{x^2 + 2x - 3}$

29. $\lim\limits_{h \to 0} \dfrac{(h - 1)^3 + 1}{h}$

30. $\lim\limits_{t \to 2} \dfrac{t^2 - 4}{t^3 - 8}$

31. $\lim\limits_{r \to 9} \dfrac{\sqrt{r}}{(r - 9)^4}$

32. $\lim\limits_{v \to 4^+} \dfrac{4 - v}{|4 - v|}$

33. $\lim\limits_{u \to 1} \dfrac{u^4 - 1}{u^3 + 5u^2 - 6u}$

34. $\lim\limits_{x \to 3} \dfrac{\sqrt{x + 6} - x}{x^3 - 3x^2}$

35. $\lim\limits_{s \to 16} \dfrac{4 - \sqrt{s}}{s - 16}$

36. $\lim\limits_{v \to 2} \dfrac{v^2 + 2v - 8}{v^4 - 16}$

37. $\lim\limits_{x \to 0} \dfrac{1 - \sqrt{1 - x^2}}{x}$

38. $\lim\limits_{x \to 1} \left(\dfrac{1}{x - 1} + \dfrac{1}{x^2 - 3x + 2} \right)$

---

39. If $2x - 1 \leq f(x) \leq x^2$ for $0 < x < 3$, find $\lim\limits_{x \to 1} f(x)$.

40. Prove that $\lim\limits_{x \to 0} x^2 \cos(1/x^2) = 0$.

41–44 Prove the statement using the precise definition of a limit.

41. $\lim\limits_{x \to 2} (14 - 5x) = 4$

42. $\lim\limits_{x \to 0} \sqrt[3]{x} = 0$

43. $\lim\limits_{x \to 2} (x^2 - 3x) = -2$

44. $\lim\limits_{x \to 4^+} \dfrac{2}{\sqrt{x - 4}} = \infty$

---

45. Let

$$f(x) = \begin{cases} \sqrt{-x} & \text{if } x < 0 \\ 3 - x & \text{if } 0 \le x < 3 \\ (x - 3)^2 & \text{if } x > 3 \end{cases}$$

(a) Evaluate each limit, if it exists.

   (i) $\lim\limits_{x \to 0^+} f(x)$ (ii) $\lim\limits_{x \to 0^-} f(x)$ (iii) $\lim\limits_{x \to 0} f(x)$

   (iv) $\lim\limits_{x \to 3^-} f(x)$ (v) $\lim\limits_{x \to 3^+} f(x)$ (vi) $\lim\limits_{x \to 3} f(x)$

(b) Where is f discontinuous?

(c) Sketch the graph of f.

46. Let

$$g(x) = \begin{cases} 2x - x^2 & \text{if } 0 \le x \le 2 \\ 2 - x & \text{if } 2 < x \le 3 \\ x - 4 & \text{if } 3 < x < 4 \\ \pi & \text{if } x \ge 4 \end{cases}$$

(a) For each of the numbers 2, 3, and 4, discover whether g is continuous from the left, continuous from the right, or continuous at the number.

(b) Sketch the graph of g.

47–48 Show that the function is continuous on its domain. State the domain.

47. $h(x) = \sqrt[4]{x} + x^3 \cos x$

48. $g(x) = \dfrac{\sqrt{x^2 - 9}}{x^2 - 2}$

49–50 Use the Intermediate Value Theorem to show that there is a root of the equation in the given interval.

49. $x^5 - x^3 + 3x - 5 = 0,$ $(1, 2)$

50. $2 \sin x = 3 - 2x,$ $(0, 1)$

Principles of Problem Solving

There are no hard and fast rules that will ensure success in solving problems. However, it is possible to outline some general steps in the problem-solving process and to give some principles that may be useful in the solution of certain problems. These steps and principles are just common sense made explicit. They have been adapted from George Polya's book *How To Solve It*.

1 UNDERSTAND THE PROBLEM

The first step is to read the problem and make sure that you understand it clearly. Ask yourself the following questions:

> *What is the unknown?*
> *What are the given quantities?*
> *What are the given conditions?*

For many problems it is useful to

> *draw a diagram*

and identify the given and required quantities on the diagram.
   Usually it is necessary to

> *introduce suitable notation*

In choosing symbols for the unknown quantities we often use letters such as a, b, c, m, n, x, and y, but in some cases it helps to use initials as suggestive symbols; for instance, V for volume or t for time.

2 THINK OF A PLAN

Find a connection between the given information and the unknown that will enable you to calculate the unknown. It often helps to ask yourself explicitly: "How can I relate the given to the unknown?" If you don't see a connection immediately, the following ideas may be helpful in devising a plan.

Try to Recognize Something Familiar   Relate the given situation to previous knowledge. Look at the unknown and try to recall a more familiar problem that has a similar unknown.

Try to Recognize Patterns   Some problems are solved by recognizing that some kind of pattern is occurring. The pattern could be geometric, or numerical, or algebraic. If you can see regularity or repetition in a problem, you might be able to guess what the continuing pattern is and then prove it.

Use Analogy   Try to think of an analogous problem, that is, a similar problem, a related problem, but one that is easier than the original problem. If you can solve the similar, simpler problem, then it might give you the clues you need to solve the original, more difficult problem. For instance, if a problem involves very large numbers, you could first try a similar problem with smaller numbers. Or if the problem involves three-dimensional geometry, you could look for a similar problem in two-dimensional geometry. Or if the problem you start with is a general one, you could first try a special case.

Introduce Something Extra   It may sometimes be necessary to introduce something new, an auxiliary aid, to help make the connection between the given and the unknown. For instance, in a problem where a diagram is useful the auxiliary aid could be a new line drawn in a diagram. In a more algebraic problem it could be a new unknown that is related to the original unknown.

Take Cases We may sometimes have to split a problem into several cases and give a different argument for each of the cases. For instance, we often have to use this strategy in dealing with absolute value.

Work Backward Sometimes it is useful to imagine that your problem is solved and work backward, step by step, until you arrive at the given data. Then you may be able to reverse your steps and thereby construct a solution to the original problem. This procedure is commonly used in solving equations. For instance, in solving the equation $3x - 5 = 7$, we suppose that x is a number that satisfies $3x - 5 = 7$ and work backward. We add 5 to each side of the equation and then divide each side by 3 to get $x = 4$. Since each of these steps can be reversed, we have solved the problem.

Establish Subgoals In a complex problem it is often useful to set subgoals (in which the desired situation is only partially fulfilled). If we can first reach these subgoals, then we may be able to build on them to reach our final goal.

Indirect Reasoning Sometimes it is appropriate to attack a problem indirectly. In using proof by contradiction to prove that P implies Q, we assume that P is true and Q is false and try to see why this can't happen. Somehow we have to use this information and arrive at a contradiction to what we absolutely know is true.

Mathematical Induction In proving statements that involve a positive integer n, it is frequently helpful to use the following principle.

Principle of Mathematical Induction Let S_n be a statement about the positive integer n. Suppose that

1. S_1 is true.

2. S_{k+1} is true whenever S_k is true.

Then S_n is true for all positive integers n.

This is reasonable because, since S_1 is true, it follows from condition 2 (with $k = 1$) that S_2 is true. Then, using condition 2 with $k = 2$, we see that S_3 is true. Again using condition 2, this time with $k = 3$, we have that S_4 is true. This procedure can be followed indefinitely.

3 CARRY OUT THE PLAN

In Step 2 a plan was devised. In carrying out that plan we have to check each stage of the plan and write the details that prove that each stage is correct.

4 LOOK BACK

Having completed our solution, it is wise to look back over it, partly to see if we have made errors in the solution and partly to see if we can think of an easier way to solve the problem. Another reason for looking back is that it will familiarize us with the method of solution and this may be useful for solving a future problem. Descartes said, "Every problem that I solved became a rule which served afterwards to solve other problems."

These principles of problem solving are illustrated in the following examples. Before you look at the solutions, try to solve these problems yourself, referring to these Principles of Problem Solving if you get stuck. You may find it useful to refer to this section from time to time as you solve the exercises in the remaining chapters of this book.

As the first example illustrates, it is often necessary to use the problem-solving principle of *taking cases* when dealing with absolute values.

EXAMPLE 1 Solve the inequality $|x - 3| + |x + 2| < 11$.

SOLUTION Recall the definition of absolute value:

$$|x| = \begin{cases} x & \text{if } x \geqslant 0 \\ -x & \text{if } x < 0 \end{cases}$$

It follows that

$$|x - 3| = \begin{cases} x - 3 & \text{if } x - 3 \geqslant 0 \\ -(x - 3) & \text{if } x - 3 < 0 \end{cases}$$

$$= \begin{cases} x - 3 & \text{if } x \geqslant 3 \\ -x + 3 & \text{if } x < 3 \end{cases}$$

Similarly

$$|x + 2| = \begin{cases} x + 2 & \text{if } x + 2 \geqslant 0 \\ -(x + 2) & \text{if } x + 2 < 0 \end{cases}$$

$$= \begin{cases} x + 2 & \text{if } x \geqslant -2 \\ -x - 2 & \text{if } x < -2 \end{cases}$$

PS Take cases

These expressions show that we must consider three cases:

$$x < -2 \qquad -2 \leqslant x < 3 \qquad x \geqslant 3$$

CASE I If $x < -2$, we have

$$|x - 3| + |x + 2| < 11$$
$$-x + 3 - x - 2 < 11$$
$$-2x < 10$$
$$x > -5$$

CASE II If $-2 \leqslant x < 3$, the given inequality becomes

$$-x + 3 + x + 2 < 11$$
$$5 < 11 \qquad \text{(always true)}$$

CASE III If $x \geqslant 3$, the inequality becomes

$$x - 3 + x + 2 < 11$$
$$2x < 12$$
$$x < 6$$

Combining cases I, II, and III, we see that the inequality is satisfied when $-5 < x < 6$. So the solution is the interval $(-5, 6)$. ▬

79

In the following example we first guess the answer by looking at special cases and recognizing a pattern. Then we prove our conjecture by mathematical induction.

In using the Principle of Mathematical Induction, we follow three steps:

Step 1 Prove that S_n is true when $n = 1$.

Step 2 Assume that S_n is true when $n = k$ and deduce that S_n is true when $n = k + 1$.

Step 3 Conclude that S_n is true for all n by the Principle of Mathematical Induction.

EXAMPLE 2 If $f_0(x) = x/(x + 1)$ and $f_{n+1} = f_0 \circ f_n$ for $n = 0, 1, 2, \ldots$, find a formula for $f_n(x)$.

PS Analogy: Try a similar, simpler problem

SOLUTION We start by finding formulas for $f_n(x)$ for the special cases $n = 1, 2,$ and 3.

$$f_1(x) = (f_0 \circ f_0)(x) = f_0(f_0(x)) = f_0\left(\frac{x}{x + 1}\right)$$

$$= \frac{\dfrac{x}{x + 1}}{\dfrac{x}{x + 1} + 1} = \frac{\dfrac{x}{x + 1}}{\dfrac{2x + 1}{x + 1}} = \frac{x}{2x + 1}$$

$$f_2(x) = (f_0 \circ f_1)(x) = f_0(f_1(x)) = f_0\left(\frac{x}{2x + 1}\right)$$

$$= \frac{\dfrac{x}{2x + 1}}{\dfrac{x}{2x + 1} + 1} = \frac{\dfrac{x}{2x + 1}}{\dfrac{3x + 1}{2x + 1}} = \frac{x}{3x + 1}$$

$$f_3(x) = (f_0 \circ f_2)(x) = f_0(f_2(x)) = f_0\left(\frac{x}{3x + 1}\right)$$

PS Look for a pattern

$$= \frac{\dfrac{x}{3x + 1}}{\dfrac{x}{3x + 1} + 1} = \frac{\dfrac{x}{3x + 1}}{\dfrac{4x + 1}{3x + 1}} = \frac{x}{4x + 1}$$

We notice a pattern: The coefficient of x in the denominator of $f_n(x)$ is $n + 1$ in the three cases we have computed. So we make the guess that, in general,

$$\boxed{4} \qquad f_n(x) = \frac{x}{(n + 1)x + 1}$$

To prove this, we use the Principle of Mathematical Induction. We have already verified that $\boxed{4}$ is true for $n = 1$. Assume that it is true for $n = k$, that is,

$$f_k(x) = \frac{x}{(k + 1)x + 1}$$

Then $\quad f_{k+1}(x) = (f_0 \circ f_k)(x) = f_0(f_k(x)) = f_0\left(\dfrac{x}{(k+1)x+1}\right)$

$$= \frac{\dfrac{x}{(k+1)x+1}}{\dfrac{x}{(k+1)x+1}+1} = \frac{\dfrac{x}{(k+1)x+1}}{\dfrac{(k+2)x+1}{(k+1)x+1}} = \frac{x}{(k+2)x+1}$$

This expression shows that $\boxed{4}$ is true for $n = k + 1$. Therefore, by mathematical induction, it is true for all positive integers n. ▬

In the following example we show how the problem solving strategy of *introducing something extra* is sometimes useful when we evaluate limits. The idea is to change the variable—to introduce a new variable that is related to the original variable—in such a way as to make the problem simpler. Later, in Section 4.5, we will make more extensive use of this general idea.

EXAMPLE 3 Evaluate $\lim\limits_{x \to 0} \dfrac{\sqrt[3]{1+cx}-1}{x}$, where c is a constant.

SOLUTION As it stands, this limit looks challenging. In Section 1.6 we evaluated several limits in which both numerator and denominator approached 0. There our strategy was to perform some sort of algebraic manipulation that led to a simplifying cancellation, but here it's not clear what kind of algebra is necessary.

So we introduce a new variable t by the equation

$$t = \sqrt[3]{1+cx}$$

We also need to express x in terms of t, so we solve this equation:

$$t^3 = 1 + cx \qquad x = \frac{t^3 - 1}{c} \quad \text{(if } c \neq 0\text{)}$$

Notice that $x \to 0$ is equivalent to $t \to 1$. This allows us to convert the given limit into one involving the variable t:

$$\lim_{x \to 0} \frac{\sqrt[3]{1+cx}-1}{x} = \lim_{t \to 1} \frac{t-1}{(t^3-1)/c} = \lim_{t \to 1} \frac{c(t-1)}{t^3-1}$$

The change of variable allowed us to replace a relatively complicated limit by a simpler one of a type that we have seen before. Factoring the denominator as a difference of cubes, we get

$$\lim_{t \to 1} \frac{c(t-1)}{t^3-1} = \lim_{t \to 1} \frac{c(t-1)}{(t-1)(t^2+t+1)}$$

$$= \lim_{t \to 1} \frac{c}{t^2+t+1} = \frac{c}{3}$$

In making the change of variable we had to rule out the case $c = 0$. But if $c = 0$, the function is 0 for all nonzero x and so its limit is 0. Therefore, in all cases, the limit is $c/3$. ▬

The following problems are meant to test and challenge your problem-solving skills. Some of them require a considerable amount of time to think through, so don't be discouraged if you can't solve them right away. If you get stuck, you might find it helpful to refer to the discussion of the principles of problem solving.

1. Draw the graph of the equation $x + |x| = y + |y|$.

2. Sketch the region in the plane consisting of all points (x, y) such that $|x - y| + |x| - |y| \leq 2$.

3. If $f_0(x) = x^2$ and $f_{n+1}(x) = f_0(f_n(x))$ for $n = 0, 1, 2, \ldots$, find a formula for $f_n(x)$.

4. (a) If $f_0(x) = \dfrac{1}{2 - x}$ and $f_{n+1} = f_0 \circ f_n$ for $n = 0, 1, 2, \ldots$, find an expression for $f_n(x)$ and use mathematical induction to prove it.

    (b) Graph f_0, f_1, f_2, f_3 on the same screen and describe the effects of repeated composition.

5. Evaluate $\displaystyle\lim_{x \to 1} \frac{\sqrt[3]{x} - 1}{\sqrt{x} - 1}$.

6. Find numbers a and b such that $\displaystyle\lim_{x \to 0} \frac{\sqrt{ax + b} - 2}{x} = 1$.

7. Evaluate $\displaystyle\lim_{x \to 0} \frac{|2x - 1| - |2x + 1|}{x}$.

8. The figure shows a point P on the parabola $y = x^2$ and the point Q where the perpendicular bisector of OP intersects the y-axis. As P approaches the origin along the parabola, what happens to Q? Does it have a limiting position? If so, find it.

9. Evaluate the following limits, if they exist, where $\llbracket x \rrbracket$ denotes the greatest integer function.

   (a) $\displaystyle\lim_{x \to 0} \frac{\llbracket x \rrbracket}{x}$ (b) $\displaystyle\lim_{x \to 0} x \llbracket 1/x \rrbracket$

10. Sketch the region in the plane defined by each of the following equations.

    (a) $\llbracket x \rrbracket^2 + \llbracket y \rrbracket^2 = 1$ (b) $\llbracket x \rrbracket^2 - \llbracket y \rrbracket^2 = 3$ (c) $\llbracket x + y \rrbracket^2 = 1$ (d) $\llbracket x \rrbracket + \llbracket y \rrbracket = 1$

11. Find all values of a such that f is continuous on $\mathbb{R}$:

$$f(x) = \begin{cases} x + 1 & \text{if } x \leq a \\ x^2 & \text{if } x > a \end{cases}$$

12. A **fixed point** of a function f is a number c in its domain such that $f(c) = c$. (The function doesn't move c; it stays fixed.)
    (a) Sketch the graph of a continuous function with domain $[0, 1]$ whose range also lies in $[0, 1]$. Locate a fixed point of f.
    (b) Try to draw the graph of a continuous function with domain $[0, 1]$ and range in $[0, 1]$ that does *not* have a fixed point. What is the obstacle?
    (c) Use the Intermediate Value Theorem to prove that any continuous function with domain $[0, 1]$ and range in $[0, 1]$ must have a fixed point.

13. If $\lim_{x \to a} [f(x) + g(x)] = 2$ and $\lim_{x \to a} [f(x) - g(x)] = 1$, find $\lim_{x \to a} [f(x) g(x)]$.

14. (a) The figure shows an isosceles triangle ABC with $\angle B = \angle C$. The bisector of angle B intersects the side AC at the point P. Suppose that the base BC remains fixed but the altitude $|AM|$ of the triangle approaches 0, so A approaches the midpoint M of BC. What happens to P during this process? Does it have a limiting position? If so, find it.
    (b) Try to sketch the path traced out by P during this process. Then find an equation of this curve and use this equation to sketch the curve.

15. (a) If we start from $0°$ latitude and proceed in a westerly direction, we can let $T(x)$ denote the temperature at the point x at any given time. Assuming that T is a continuous function of x, show that at any fixed time there are at least two diametrically opposite points on the equator that have exactly the same temperature.
    (b) Does the result in part (a) hold for points lying on any circle on the earth's surface?
    (c) Does the result in part (a) hold for barometric pressure and for altitude above sea level?

FIGURE FOR PROBLEM 8

FIGURE FOR PROBLEM 14

 Graphing calculator or computer required

2 Derivatives

For a roller coaster ride to be smooth, the straight stretches of the track need to be connected to the curved segments so that there are no abrupt changes in direction. In the project on page 109 you will see how to design the first ascent and drop of a new coaster for a smooth ride.

© Brett Mulcahy / Shutterstock

In this chapter we begin our study of differential calculus, which is concerned with how one quantity changes in relation to another quantity. The central concept of differential calculus is the *derivative*, which is an outgrowth of the velocities and slopes of tangents that we considered in Chapter 1. After learning how to calculate derivatives, we use them to solve problems involving rates of change and the approximation of functions.

2.1 Derivatives and Rates of Change

The problem of finding the tangent line to a curve and the problem of finding the velocity of an object both involve finding the same type of limit, as we saw in Section 1.4. This special type of limit is called a *derivative* and we will see that it can be interpreted as a rate of change in any of the sciences or engineering.

Tangents

If a curve C has equation $y = f(x)$ and we want to find the tangent line to C at the point $P(a, f(a))$, then we consider a nearby point $Q(x, f(x))$, where $x \neq a$, and compute the slope of the secant line PQ:

$$m_{PQ} = \frac{f(x) - f(a)}{x - a}$$

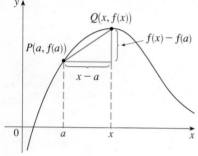

Then we let Q approach P along the curve C by letting x approach a. If m_{PQ} approaches a number m, then we define the *tangent t* to be the line through P with slope m. (This amounts to saying that the tangent line is the limiting position of the secant line PQ as Q approaches P. See Figure 1.)

---

1 Definition The **tangent line** to the curve $y = f(x)$ at the point $P(a, f(a))$ is the line through P with slope

$$m = \lim_{x \to a} \frac{f(x) - f(a)}{x - a}$$

provided that this limit exists.

FIGURE 1

In our first example we confirm the guess we made in Example 1 in Section 1.4.

V EXAMPLE 1 Find an equation of the tangent line to the parabola $y = x^2$ at the point $P(1, 1)$.

SOLUTION Here we have $a = 1$ and $f(x) = x^2$, so the slope is

$$m = \lim_{x \to 1} \frac{f(x) - f(1)}{x - 1} = \lim_{x \to 1} \frac{x^2 - 1}{x - 1}$$

$$= \lim_{x \to 1} \frac{(x - 1)(x + 1)}{x - 1}$$

$$= \lim_{x \to 1} (x + 1) = 1 + 1 = 2$$

Point-slope form for a line through the point (x_1, y_1) with slope m:

$$y - y_1 = m(x - x_1)$$

Using the point-slope form of the equation of a line, we find that an equation of the tangent line at $(1, 1)$ is

$$y - 1 = 2(x - 1) \qquad \text{or} \qquad y = 2x - 1$$

We sometimes refer to the slope of the tangent line to a curve at a point as the **slope of the curve** at the point. The idea is that if we zoom in far enough toward the point, the curve looks almost like a straight line. Figure 2 illustrates this procedure for the curve $y = x^2$ in

TEC Visual 2.1 shows an animation of Figure 2.

Example 1. The more we zoom in, the more the parabola looks like a line. In other words, the curve becomes almost indistinguishable from its tangent line.

FIGURE 2 Zooming in toward the point $(1, 1)$ on the parabola $y = x^2$

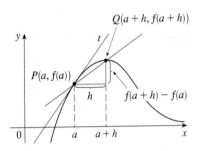

FIGURE 3

There is another expression for the slope of a tangent line that is sometimes easier to use. If $h = x - a$, then $x = a + h$ and so the slope of the secant line PQ is

$$m_{PQ} = \frac{f(a + h) - f(a)}{h}$$

(See Figure 3 where the case $h > 0$ is illustrated and Q is to the right of P. If it happened that $h < 0$, however, Q would be to the left of P.)

Notice that as x approaches a, h approaches 0 (because $h = x - a$) and so the expression for the slope of the tangent line in Definition 1 becomes

2
$$m = \lim_{h \to 0} \frac{f(a + h) - f(a)}{h}$$

EXAMPLE 2 Find an equation of the tangent line to the hyperbola $y = 3/x$ at the point $(3, 1)$.

SOLUTION Let $f(x) = 3/x$. Then the slope of the tangent at $(3, 1)$ is

$$m = \lim_{h \to 0} \frac{f(3 + h) - f(3)}{h} = \lim_{h \to 0} \frac{\frac{3}{3 + h} - 1}{h} = \lim_{h \to 0} \frac{\frac{3 - (3 + h)}{3 + h}}{h}$$

$$= \lim_{h \to 0} \frac{-h}{h(3 + h)} = \lim_{h \to 0} -\frac{1}{3 + h} = -\frac{1}{3}$$

Therefore an equation of the tangent at the point $(3, 1)$ is

$$y - 1 = -\tfrac{1}{3}(x - 3)$$

which simplifies to

$$x + 3y - 6 = 0$$

The hyperbola and its tangent are shown in Figure 4.

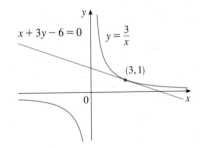

FIGURE 4

Velocities

In Section 1.4 we investigated the motion of a ball dropped from the CN Tower and defined its velocity to be the limiting value of average velocities over shorter and shorter time periods.

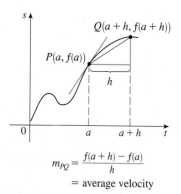

FIGURE 5

$$m_{PQ} = \frac{f(a+h) - f(a)}{h}$$

$$= \text{average velocity}$$

FIGURE 6

Recall from Section 1.4: The distance
(in meters) fallen after t seconds is $4.9t^2$.

In general, suppose an object moves along a straight line according to an equation of motion $s = f(t)$, where s is the displacement (directed distance) of the object from the origin at time t. The function f that describes the motion is called the **position function** of the object. In the time interval from $t = a$ to $t = a + h$ the change in position is $f(a + h) - f(a)$. (See Figure 5.) The average velocity over this time interval is

$$\text{average velocity} = \frac{\text{displacement}}{\text{time}} = \frac{f(a+h) - f(a)}{h}$$

which is the same as the slope of the secant line PQ in Figure 6.

Now suppose we compute the average velocities over shorter and shorter time intervals $[a, a + h]$. In other words, we let h approach 0. As in the example of the falling ball, we define the **velocity** (or **instantaneous velocity**) $v(a)$ at time $t = a$ to be the limit of these average velocities:

$$\boxed{3} \qquad v(a) = \lim_{h \to 0} \frac{f(a+h) - f(a)}{h}$$

This means that the velocity at time $t = a$ is equal to the slope of the tangent line at P (compare Equations 2 and 3).

Now that we know how to compute limits, let's reconsider the problem of the falling ball.

▶ EXAMPLE 3 Suppose that a ball is dropped from the upper observation deck of the CN Tower, 450 m above the ground.
(a) What is the velocity of the ball after 5 seconds?
(b) How fast is the ball traveling when it hits the ground?

SOLUTION We will need to find the velocity both when $t = 5$ and when the ball hits the ground, so it's efficient to start by finding the velocity at a general time $t = a$. Using the equation of motion $s = f(t) = 4.9t^2$, we have

$$v(a) = \lim_{h \to 0} \frac{f(a+h) - f(a)}{h} = \lim_{h \to 0} \frac{4.9(a+h)^2 - 4.9a^2}{h}$$

$$= \lim_{h \to 0} \frac{4.9(a^2 + 2ah + h^2 - a^2)}{h} = \lim_{h \to 0} \frac{4.9(2ah + h^2)}{h}$$

$$= \lim_{h \to 0} 4.9(2a + h) = 9.8a$$

(a) The velocity after 5 s is $v(5) = (9.8)(5) = 49$ m/s.

(b) Since the observation deck is 450 m above the ground, the ball will hit the ground at the time t_1 when $s(t_1) = 450$, that is,

$$4.9t_1^2 = 450$$

This gives

$$t_1^2 = \frac{450}{4.9} \qquad \text{and} \qquad t_1 = \sqrt{\frac{450}{4.9}} \approx 9.6 \text{ s}$$

The velocity of the ball as it hits the ground is therefore

$$v(t_1) = 9.8t_1 = 9.8\sqrt{\frac{450}{4.9}} \approx 94 \text{ m/s}$$

Derivatives

We have seen that the same type of limit arises in finding the slope of a tangent line (Equation 2) or the velocity of an object (Equation 3). In fact, limits of the form

$$\lim_{h \to 0} \frac{f(a + h) - f(a)}{h}$$

arise whenever we calculate a rate of change in any of the sciences or engineering, such as a rate of reaction in chemistry or a marginal cost in economics. Since this type of limit occurs so widely, it is given a special name and notation.

---

4 Definition The **derivative of a function f at a number a**, denoted by $f'(a)$, is

$$f'(a) = \lim_{h \to 0} \frac{f(a + h) - f(a)}{h}$$

if this limit exists.

$f'(a)$ is read "f prime of a."

If we write $x = a + h$, then we have $h = x - a$ and h approaches 0 if and only if x approaches a. Therefore an equivalent way of stating the definition of the derivative, as we saw in finding tangent lines, is

5

$$f'(a) = \lim_{x \to a} \frac{f(x) - f(a)}{x - a}$$

V **EXAMPLE 4** Find the derivative of the function $f(x) = x^2 - 8x + 9$ at the number a.

SOLUTION From Definition 4 we have

$$f'(a) = \lim_{h \to 0} \frac{f(a + h) - f(a)}{h}$$

$$= \lim_{h \to 0} \frac{[(a + h)^2 - 8(a + h) + 9] - [a^2 - 8a + 9]}{h}$$

$$= \lim_{h \to 0} \frac{a^2 + 2ah + h^2 - 8a - 8h + 9 - a^2 + 8a - 9}{h}$$

$$= \lim_{h \to 0} \frac{2ah + h^2 - 8h}{h} = \lim_{h \to 0} (2a + h - 8)$$

$$= 2a - 8$$

We defined the tangent line to the curve $y = f(x)$ at the point $P(a, f(a))$ to be the line that passes through P and has slope m given by Equation 1 or 2. Since, by Definition 4, this is the same as the derivative $f'(a)$, we can now say the following.

The tangent line to $y = f(x)$ at $(a, f(a))$ is the line through $(a, f(a))$ whose slope is equal to $f'(a)$, the derivative of f at a.

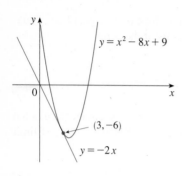

FIGURE 7

If we use the point-slope form of the equation of a line, we can write an equation of the tangent line to the curve $y = f(x)$ at the point $(a, f(a))$:

$$y - f(a) = f'(a)(x - a)$$

V **EXAMPLE 5** Find an equation of the tangent line to the parabola $y = x^2 - 8x + 9$ at the point $(3, -6)$.

SOLUTION From Example 4 we know that the derivative of $f(x) = x^2 - 8x + 9$ at the number a is $f'(a) = 2a - 8$. Therefore the slope of the tangent line at $(3, -6)$ is $f'(3) = 2(3) - 8 = -2$. Thus an equation of the tangent line, shown in Figure 7, is

$$y - (-6) = (-2)(x - 3) \qquad \text{or} \qquad y = -2x$$

Rates of Change

Suppose y is a quantity that depends on another quantity x. Thus y is a function of x and we write $y = f(x)$. If x changes from x_1 to x_2, then the change in x (also called the **increment** of x) is

$$\Delta x = x_2 - x_1$$

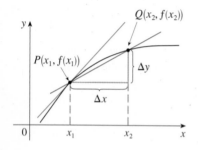

average rate of change $= m_{PQ}$

instantaneous rate of change =
    slope of tangent at P

FIGURE 8

and the corresponding change in y is

$$\Delta y = f(x_2) - f(x_1)$$

The difference quotient

$$\frac{\Delta y}{\Delta x} = \frac{f(x_2) - f(x_1)}{x_2 - x_1}$$

is called the **average rate of change of y with respect to x** over the interval $[x_1, x_2]$ and can be interpreted as the slope of the secant line PQ in Figure 8.

By analogy with velocity, we consider the average rate of change over smaller and smaller intervals by letting x_2 approach x_1 and therefore letting Δx approach 0. The limit of these average rates of change is called the (**instantaneous**) **rate of change of y with respect to x** at $x = x_1$, which is interpreted as the slope of the tangent to the curve $y = f(x)$ at $P(x_1, f(x_1))$:

$$\boxed{6} \qquad \text{instantaneous rate of change} = \lim_{\Delta x \to 0} \frac{\Delta y}{\Delta x} = \lim_{x_2 \to x_1} \frac{f(x_2) - f(x_1)}{x_2 - x_1}$$

We recognize this limit as being the derivative $f'(x_1)$.

We know that one interpretation of the derivative $f'(a)$ is as the slope of the tangent line to the curve $y = f(x)$ when $x = a$. We now have a second interpretation:

The derivative $f'(a)$ is the instantaneous rate of change of $y = f(x)$ with respect to x when $x = a$.

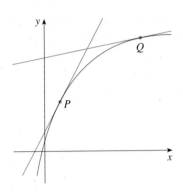

FIGURE 9
The y-values are changing rapidly at P and slowly at Q.

The connection with the first interpretation is that if we sketch the curve $y = f(x)$, then the instantaneous rate of change is the slope of the tangent to this curve at the point where $x = a$. This means that when the derivative is large (and therefore the curve is steep, as at the point P in Figure 9), the y-values change rapidly. When the derivative is small, the curve is relatively flat (as at point Q) and the y-values change slowly.

In particular, if $s = f(t)$ is the position function of a particle that moves along a straight line, then $f'(a)$ is the rate of change of the displacement s with respect to the time t. In other words, $f'(a)$ *is the velocity of the particle at time* $t = a$. The **speed** of the particle is the absolute value of the velocity, that is, $|f'(a)|$.

In the next example we discuss the meaning of the derivative of a function that is defined verbally.

V EXAMPLE 6 A manufacturer produces bolts of a fabric with a fixed width. The cost of producing x yards of this fabric is $C = f(x)$ dollars.
(a) What is the meaning of the derivative $f'(x)$? What are its units?
(b) In practical terms, what does it mean to say that $f'(1000) = 9$?
(c) Which do you think is greater, $f'(50)$ or $f'(500)$? What about $f'(5000)$?

SOLUTION
(a) The derivative $f'(x)$ is the instantaneous rate of change of C with respect to x; that is, $f'(x)$ means the rate of change of the production cost with respect to the number of yards produced. (Economists call this rate of change the *marginal cost*. This idea is discussed in more detail in Sections 2.7 and 3.7.)

Because

$$f'(x) = \lim_{\Delta x \to 0} \frac{\Delta C}{\Delta x}$$

the units for $f'(x)$ are the same as the units for the difference quotient $\Delta C/\Delta x$. Since ΔC is measured in dollars and Δx in yards, it follows that the units for $f'(x)$ are dollars per yard.

(b) The statement that $f'(1000) = 9$ means that, after 1000 yards of fabric have been manufactured, the rate at which the production cost is increasing is \$9/yard. (When $x = 1000$, C is increasing 9 times as fast as x.)

Since $\Delta x = 1$ is small compared with $x = 1000$, we could use the approximation

Here we are assuming that the cost function is well behaved; in other words, $C(x)$ doesn't oscillate rapidly near $x = 1000$.

$$f'(1000) \approx \frac{\Delta C}{\Delta x} = \frac{\Delta C}{1} = \Delta C$$

and say that the cost of manufacturing the 1000th yard (or the 1001st) is about \$9.

(c) The rate at which the production cost is increasing (per yard) is probably lower when $x = 500$ than when $x = 50$ (the cost of making the 500th yard is less than the cost of the 50th yard) because of economies of scale. (The manufacturer makes more efficient use of the fixed costs of production.) So

$$f'(50) > f'(500)$$

But, as production expands, the resulting large-scale operation might become inefficient and there might be overtime costs. Thus it is possible that the rate of increase of costs will eventually start to rise. So it may happen that

$$f'(5000) > f'(500)$$

In the following example we estimate the rate of change of the national debt with respect to time. Here the function is defined not by a formula but by a table of values.

t	$D(t)$
1980	930.2
1985	1945.9
1990	3233.3
1995	4974.0
2000	5674.2
2005	7932.7

V EXAMPLE 7 Let $D(t)$ be the US national debt at time t. The table in the margin gives approximate values of this function by providing end of year estimates, in billions of dollars, from 1980 to 2005. Interpret and estimate the value of $D'(1990)$.

SOLUTION The derivative $D'(1990)$ means the rate of change of D with respect to t when $t = 1990$, that is, the rate of increase of the national debt in 1990.

According to Equation 5,

$$D'(1990) = \lim_{t \to 1990} \frac{D(t) - D(1990)}{t - 1990}$$

t	$\dfrac{D(t) - D(1990)}{t - 1990}$
1980	230.31
1985	257.48
1995	348.14
2000	244.09
2005	313.29

So we compute and tabulate values of the difference quotient (the average rates of change) as shown in the table at the left. From this table we see that $D'(1990)$ lies somewhere between 257.48 and 348.14 billion dollars per year. [Here we are making the reasonable assumption that the debt didn't fluctuate wildly between 1980 and 2000.] We estimate that the rate of increase of the national debt of the United States in 1990 was the average of these two numbers, namely

$$D'(1990) \approx 303 \text{ billion dollars per year}$$

A Note on Units

The units for the average rate of change $\Delta D/\Delta t$ are the units for ΔD divided by the units for Δt, namely, billions of dollars per year. The instantaneous rate of change is the limit of the average rates of change, so it is measured in the same units: billions of dollars per year.

Another method would be to plot the debt function and estimate the slope of the tangent line when $t = 1990$. ▬

In Examples 3, 6, and 7 we saw three specific examples of rates of change: the velocity of an object is the rate of change of displacement with respect to time; marginal cost is the rate of change of production cost with respect to the number of items produced; the rate of change of the debt with respect to time is of interest in economics. Here is a small sample of other rates of change: In physics, the rate of change of work with respect to time is called *power*. Chemists who study a chemical reaction are interested in the rate of change in the concentration of a reactant with respect to time (called the *rate of reaction*). A biologist is interested in the rate of change of the population of a colony of bacteria with respect to time. In fact, the computation of rates of change is important in all of the natural sciences, in engineering, and even in the social sciences. Further examples will be given in Section 2.7.

All these rates of change are derivatives and can therefore be interpreted as slopes of tangents. This gives added significance to the solution of the tangent problem. Whenever we solve a problem involving tangent lines, we are not just solving a problem in geometry. We are also implicitly solving a great variety of problems involving rates of change in science and engineering.

2.1   Exercises

Access end-of-section exercises and tools for visualizing calculus online at **www.webassign.net**   ᴱᴺᴴᴬᴺᴄᴇᴰ **WebAssign**

WRITING PROJECT   **EARLY METHODS FOR FINDING TANGENTS**

The first person to formulate explicitly the ideas of limits and derivatives was Sir Isaac Newton in the 1660s. But Newton acknowledged that "If I have seen further than other men, it is because I have stood on the shoulders of giants." Two of those giants were Pierre Fermat (1601–1665) and Newton's mentor at Cambridge, Isaac Barrow (1630–1677). Newton was familiar with the methods that these men used to find tangent lines, and their methods played a role in Newton's eventual formulation of calculus.

The following references contain explanations of these methods. Read one or more of the references and write a report comparing the methods of either Fermat or Barrow to modern methods. In particular, use the method of Section 2.1 to find an equation of the tangent line to the curve $y = x^3 + 2x$ at the point $(1, 3)$ and show how either Fermat or Barrow would have solved the same problem. Although you used derivatives and they did not, point out similarities between the methods.

1. Carl Boyer and Uta Merzbach, *A History of Mathematics* (New York: Wiley, 1989), pp. 389, 432.

2. C. H. Edwards, *The Historical Development of the Calculus* (New York: Springer-Verlag, 1979), pp. 124, 132.

3. Howard Eves, *An Introduction to the History of Mathematics*, 6th ed. (New York: Saunders, 1990), pp. 391, 395.

4. Morris Kline, *Mathematical Thought from Ancient to Modern Times* (New York: Oxford University Press, 1972), pp. 344, 346.

2.2   The Derivative as a Function

In the preceding section we considered the derivative of a function f at a fixed number a:

$$\boxed{1} \qquad f'(a) = \lim_{h \to 0} \frac{f(a + h) - f(a)}{h}$$

Here we change our point of view and let the number a vary. If we replace a in Equation 1 by a variable x, we obtain

$$\boxed{2} \qquad \boxed{f'(x) = \lim_{h \to 0} \frac{f(x + h) - f(x)}{h}}$$

Given any number x for which this limit exists, we assign to x the number $f'(x)$. So we can regard f' as a new function, called the **derivative of f** and defined by Equation 2. We know that the value of f' at x, $f'(x)$, can be interpreted geometrically as the slope of the tangent line to the graph of f at the point $(x, f(x))$.

The function f' is called the derivative of f because it has been "derived" from f by the limiting operation in Equation 2. The domain of f' is the set $\{x \mid f'(x) \text{ exists}\}$ and may be smaller than the domain of f.

FIGURE 1

V **EXAMPLE 1** The graph of a function f is given in Figure 1. Use it to sketch the graph of the derivative f'.

SOLUTION We can estimate the value of the derivative at any value of x by drawing the tangent at the point $(x, f(x))$ and estimating its slope. For instance, for $x = 5$ we draw the tangent at P in Figure 2(a) and estimate its slope to be about $\frac{3}{2}$, so $f'(5) \approx 1.5$. This allows us to plot the point $P'(5, 1.5)$ on the graph of f' directly beneath P. Repeating this procedure at several points, we get the graph shown in Figure 2(b). Notice that the tangents at A, B, and C are horizontal, so the derivative is 0 there and the graph of f' crosses the x-axis at the points A', B', and C', directly beneath A, B, and C. Between A and B the tangents have positive slope, so $f'(x)$ is positive there. But between B and C the tangents have negative slope, so $f'(x)$ is negative there.

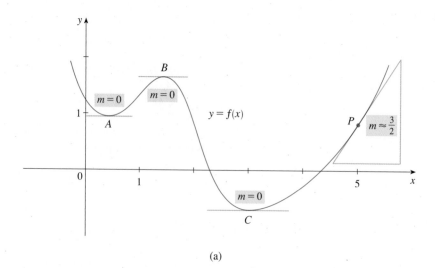

(a)

TEC Visual 2.2 shows an animation of Figure 2 for several functions.

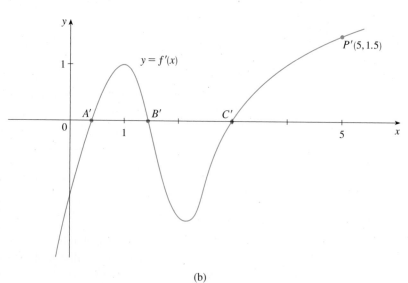

FIGURE 2

(b)

V **EXAMPLE 2**
(a) If $f(x) = x^3 - x$, find a formula for $f'(x)$.
(b) Illustrate by comparing the graphs of f and f'.

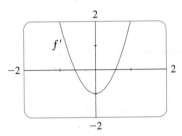

FIGURE 3

SOLUTION

(a) When using Equation 2 to compute a derivative, we must remember that the variable is h and that x is temporarily regarded as a constant during the calculation of the limit.

$$f'(x) = \lim_{h \to 0} \frac{f(x+h) - f(x)}{h} = \lim_{h \to 0} \frac{[(x+h)^3 - (x+h)] - [x^3 - x]}{h}$$

$$= \lim_{h \to 0} \frac{x^3 + 3x^2h + 3xh^2 + h^3 - x - h - x^3 + x}{h}$$

$$= \lim_{h \to 0} \frac{3x^2h + 3xh^2 + h^3 - h}{h}$$

$$= \lim_{h \to 0} (3x^2 + 3xh + h^2 - 1) = 3x^2 - 1$$

(b) We use a graphing device to graph f and f' in Figure 3. Notice that $f'(x) = 0$ when f has horizontal tangents and $f'(x)$ is positive when the tangents have positive slope. So these graphs serve as a check on our work in part (a). ▬

EXAMPLE 3 If $f(x) = \sqrt{x}$, find the derivative of f. State the domain of f'.

SOLUTION

$$f'(x) = \lim_{h \to 0} \frac{f(x+h) - f(x)}{h} = \lim_{h \to 0} \frac{\sqrt{x+h} - \sqrt{x}}{h}$$

$$= \lim_{h \to 0} \left(\frac{\sqrt{x+h} - \sqrt{x}}{h} \cdot \frac{\sqrt{x+h} + \sqrt{x}}{\sqrt{x+h} + \sqrt{x}} \right)$$

$$= \lim_{h \to 0} \frac{(x+h) - x}{h(\sqrt{x+h} + \sqrt{x})} = \lim_{h \to 0} \frac{1}{\sqrt{x+h} + \sqrt{x}}$$

$$= \frac{1}{\sqrt{x} + \sqrt{x}} = \frac{1}{2\sqrt{x}}$$

Here we rationalize the numerator.

We see that $f'(x)$ exists if $x > 0$, so the domain of f' is $(0, \infty)$. This is smaller than the domain of f, which is $[0, \infty)$. ▬

Let's check to see that the result of Example 3 is reasonable by looking at the graphs of f and f' in Figure 4. When x is close to 0, $\sqrt{x}$ is also close to 0, so $f'(x) = 1/(2\sqrt{x})$ is very large and this corresponds to the steep tangent lines near $(0, 0)$ in Figure 4(a) and the large values of $f'(x)$ just to the right of 0 in Figure 4(b). When x is large, $f'(x)$ is very small and this corresponds to the flatter tangent lines at the far right of the graph of f and the horizontal asymptote of the graph of f'.

FIGURE 4

(a) $f(x) = \sqrt{x}$

(b) $f'(x) = \dfrac{1}{2\sqrt{x}}$

EXAMPLE 4 Find f' if $f(x) = \dfrac{1-x}{2+x}$.

SOLUTION

$$\dfrac{\dfrac{a}{b} - \dfrac{c}{d}}{e} = \dfrac{ad - bc}{bd} \cdot \dfrac{1}{e}$$

$$f'(x) = \lim_{h \to 0} \frac{f(x+h) - f(x)}{h} = \lim_{h \to 0} \frac{\dfrac{1-(x+h)}{2+(x+h)} - \dfrac{1-x}{2+x}}{h}$$

$$= \lim_{h \to 0} \frac{(1-x-h)(2+x) - (1-x)(2+x+h)}{h(2+x+h)(2+x)}$$

$$= \lim_{h \to 0} \frac{(2-x-2h-x^2-xh) - (2-x+h-x^2-xh)}{h(2+x+h)(2+x)}$$

$$= \lim_{h \to 0} \frac{-3h}{h(2+x+h)(2+x)}$$

$$= \lim_{h \to 0} \frac{-3}{(2+x+h)(2+x)} = -\frac{3}{(2+x)^2}$$

Other Notations

If we use the traditional notation $y = f(x)$ to indicate that the independent variable is x and the dependent variable is y, then some common alternative notations for the derivative are as follows:

$$f'(x) = y' = \frac{dy}{dx} = \frac{df}{dx} = \frac{d}{dx} f(x) = Df(x) = D_x f(x)$$

The symbols D and d/dx are called **differentiation operators** because they indicate the operation of **differentiation**, which is the process of calculating a derivative.

The symbol dy/dx, which was introduced by Leibniz, should not be regarded as a ratio (for the time being); it is simply a synonym for $f'(x)$. Nonetheless, it is a very useful and suggestive notation, especially when used in conjunction with increment notation. Referring to Equation 2.1.6, we can rewrite the definition of derivative in Leibniz notation in the form

$$\frac{dy}{dx} = \lim_{\Delta x \to 0} \frac{\Delta y}{\Delta x}$$

If we want to indicate the value of a derivative dy/dx in Leibniz notation at a specific number a, we use the notation

$$\frac{dy}{dx}\bigg|_{x=a} \qquad \text{or} \qquad \frac{dy}{dx}\bigg]_{x=a}$$

which is a synonym for $f'(a)$.

---

3 Definition A function f is **differentiable at a** if $f'(a)$ exists. It is **differentiable on an open interval** (a, b) [or (a, ∞) or $(-\infty, a)$ or $(-\infty, \infty)$] if it is differentiable at every number in the interval.

Leibniz

Gottfried Wilhelm Leibniz was born in Leipzig in 1646 and studied law, theology, philosophy, and mathematics at the university there, graduating with a bachelor's degree at age 17. After earning his doctorate in law at age 20, Leibniz entered the diplomatic service and spent most of his life traveling to the capitals of Europe on political missions. In particular, he worked to avert a French military threat against Germany and attempted to reconcile the Catholic and Protestant churches.

His serious study of mathematics did not begin until 1672 while he was on a diplomatic mission in Paris. There he built a calculating machine and met scientists, like Huygens, who directed his attention to the latest developments in mathematics and science. Leibniz sought to develop a symbolic logic and system of notation that would simplify logical reasoning. In particular, the version of calculus that he published in 1684 established the notation and the rules for finding derivatives that we use today.

Unfortunately, a dreadful priority dispute arose in the 1690s between the followers of Newton and those of Leibniz as to who had invented calculus first. Leibniz was even accused of plagiarism by members of the Royal Society in England. The truth is that each man invented calculus independently. Newton arrived at his version of calculus first but, because of his fear of controversy, did not publish it immediately. So Leibniz's 1684 account of calculus was the first to be published.

V **EXAMPLE 5** Where is the function $f(x) = |x|$ differentiable?

SOLUTION If $x > 0$, then $|x| = x$ and we can choose h small enough that $x + h > 0$ and hence $|x + h| = x + h$. Therefore, for $x > 0$, we have

$$f'(x) = \lim_{h \to 0} \frac{|x + h| - |x|}{h} = \lim_{h \to 0} \frac{(x + h) - x}{h}$$

$$= \lim_{h \to 0} \frac{h}{h} = \lim_{h \to 0} 1 = 1$$

and so f is differentiable for any $x > 0$.

Similarly, for $x < 0$ we have $|x| = -x$ and h can be chosen small enough that $x + h < 0$ and so $|x + h| = -(x + h)$. Therefore, for $x < 0$,

$$f'(x) = \lim_{h \to 0} \frac{|x + h| - |x|}{h} = \lim_{h \to 0} \frac{-(x + h) - (-x)}{h}$$

$$= \lim_{h \to 0} \frac{-h}{h} = \lim_{h \to 0} (-1) = -1$$

and so f is differentiable for any $x < 0$.

For $x = 0$ we have to investigate

$$f'(0) = \lim_{h \to 0} \frac{f(0 + h) - f(0)}{h}$$

$$= \lim_{h \to 0} \frac{|0 + h| - |0|}{h} \qquad \text{(if it exists)}$$

Let's compute the left and right limits separately:

$$\lim_{h \to 0^+} \frac{|0 + h| - |0|}{h} = \lim_{h \to 0^+} \frac{|h|}{h} = \lim_{h \to 0^+} \frac{h}{h} = \lim_{h \to 0^+} 1 = 1$$

and $\qquad \lim_{h \to 0^-} \frac{|0 + h| - |0|}{h} = \lim_{h \to 0^-} \frac{|h|}{h} = \lim_{h \to 0^-} \frac{-h}{h} = \lim_{h \to 0^-} (-1) = -1$

Since these limits are different, $f'(0)$ does not exist. Thus f is differentiable at all x except 0.

A formula for f' is given by

$$f'(x) = \begin{cases} 1 & \text{if } x > 0 \\ -1 & \text{if } x < 0 \end{cases}$$

and its graph is shown in Figure 5(b). The fact that $f'(0)$ does not exist is reflected geometrically in the fact that the curve $y = |x|$ does not have a tangent line at $(0, 0)$. [See Figure 5(a).]

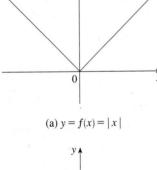

(a) $y = f(x) = |x|$

(b) $y = f'(x)$

FIGURE 5

Both continuity and differentiability are desirable properties for a function to have. The following theorem shows how these properties are related.

> **4** **Theorem** If f is differentiable at a, then f is continuous at a.

PROOF To prove that f is continuous at a, we have to show that $\lim_{x \to a} f(x) = f(a)$. We do this by showing that the difference $f(x) - f(a)$ approaches 0.

The given information is that f is differentiable at a, that is,

$$f'(a) = \lim_{x \to a} \frac{f(x) - f(a)}{x - a}$$

PS An important aspect of problem solving is trying to find a connection between the given and the unknown. See Step 2 (Think of a Plan) in *Principles of Problem Solving* on page 77.

exists (see Equation 2.1.5). To connect the given and the unknown, we divide and multiply $f(x) - f(a)$ by $x - a$ (which we can do when $x \neq a$):

$$f(x) - f(a) = \frac{f(x) - f(a)}{x - a} (x - a)$$

Thus, using the Product Law and (2.1.5), we can write

$$\lim_{x \to a} [f(x) - f(a)] = \lim_{x \to a} \frac{f(x) - f(a)}{x - a} (x - a)$$

$$= \lim_{x \to a} \frac{f(x) - f(a)}{x - a} \cdot \lim_{x \to a} (x - a)$$

$$= f'(a) \cdot 0 = 0$$

To use what we have just proved, we start with $f(x)$ and add and subtract $f(a)$:

$$\lim_{x \to a} f(x) = \lim_{x \to a} [f(a) + (f(x) - f(a))]$$

$$= \lim_{x \to a} f(a) + \lim_{x \to a} [f(x) - f(a)]$$

$$= f(a) + 0 = f(a)$$

Therefore f is continuous at a.   ▬

NOTE The converse of Theorem 4 is false; that is, there are functions that are continuous but not differentiable. For instance, the function $f(x) = |x|$ is continuous at 0 because

$$\lim_{x \to 0} f(x) = \lim_{x \to 0} |x| = 0 = f(0)$$

(See Example 7 in Section 1.6.) But in Example 5 we showed that f is not differentiable at 0.

▬ How Can a Function Fail to Be Differentiable?

We saw that the function $y = |x|$ in Example 5 is not differentiable at 0 and Figure 5(a) shows that its graph changes direction abruptly when $x = 0$. In general, if the graph of a function f has a "corner" or "kink" in it, then the graph of f has no tangent at this point and f is not differentiable there. [In trying to compute $f'(a)$, we find that the left and right limits are different.]

FIGURE 6

Theorem 4 gives another way for a function not to have a derivative. It says that if f is not continuous at a, then f is not differentiable at a. So at any discontinuity (for instance, a jump discontinuity) f fails to be differentiable.

A third possibility is that the curve has a **vertical tangent line** when $x = a$; that is, f is continuous at a and

$$\lim_{x \to a} \left| f'(x) \right| = \infty$$

This means that the tangent lines become steeper and steeper as $x \to a$. Figure 6 shows one way that this can happen; Figure 7(c) shows another. Figure 7 illustrates the three possibilities that we have discussed.

FIGURE 7
Three ways for f not to be differentiable at a

(a) A corner

(b) A discontinuity

(c) A vertical tangent

A graphing calculator or computer provides another way of looking at differentiability. If f is differentiable at a, then when we zoom in toward the point $(a, f(a))$ the graph straightens out and appears more and more like a line. (See Figure 8. We saw a specific example of this in Figure 2 in Section 2.1.) But no matter how much we zoom in toward a point like the ones in Figures 6 and 7(a), we can't eliminate the sharp point or corner (see Figure 9).

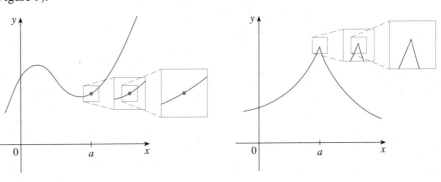

FIGURE 8
f is differentiable at a.

FIGURE 9
f is not differentiable at a.

■ Higher Derivatives

If f is a differentiable function, then its derivative f' is also a function, so f' may have a derivative of its own, denoted by $(f')' = f''$. This new function f'' is called the **second derivative** of f because it is the derivative of the derivative of f. Using Leibniz notation, we write the second derivative of $y = f(x)$ as

$$\frac{d}{dx}\left(\frac{dy}{dx}\right) = \frac{d^2 y}{dx^2}$$

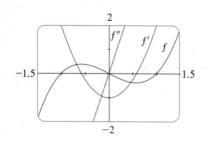

FIGURE 10

TEC In Module 2.2 you can see how changing the coefficients of a polynomial f affects the appearance of the graphs of f, f', and f''.

EXAMPLE 6 If $f(x) = x^3 - x$, find and interpret $f''(x)$.

SOLUTION In Example 2 we found that the first derivative is $f'(x) = 3x^2 - 1$. So the second derivative is

$$f''(x) = (f')'(x) = \lim_{h \to 0} \frac{f'(x + h) - f'(x)}{h} = \lim_{h \to 0} \frac{[3(x + h)^2 - 1] - [3x^2 - 1]}{h}$$

$$= \lim_{h \to 0} \frac{3x^2 + 6xh + 3h^2 - 1 - 3x^2 + 1}{h} = \lim_{h \to 0} (6x + 3h) = 6x$$

The graphs of f, f', and f'' are shown in Figure 10.

We can interpret $f''(x)$ as the slope of the curve $y = f'(x)$ at the point $(x, f'(x))$. In other words, it is the rate of change of the slope of the original curve $y = f(x)$.

Notice from Figure 10 that $f''(x)$ is negative when $y = f'(x)$ has negative slope and positive when $y = f'(x)$ has positive slope. So the graphs serve as a check on our calculations.

In general, we can interpret a second derivative as a rate of change of a rate of change. The most familiar example of this is *acceleration,* which we define as follows.

If $s = s(t)$ is the position function of an object that moves in a straight line, we know that its first derivative represents the velocity $v(t)$ of the object as a function of time:

$$v(t) = s'(t) = \frac{ds}{dt}$$

The instantaneous rate of change of velocity with respect to time is called the **acceleration** $a(t)$ of the object. Thus the acceleration function is the derivative of the velocity function and is therefore the second derivative of the position function:

$$a(t) = v'(t) = s''(t)$$

or, in Leibniz notation,

$$a = \frac{dv}{dt} = \frac{d^2s}{dt^2}$$

The **third derivative** f''' is the derivative of the second derivative: $f''' = (f'')'$. So $f'''(x)$ can be interpreted as the slope of the curve $y = f''(x)$ or as the rate of change of $f''(x)$. If $y = f(x)$, then alternative notations for the third derivative are

$$y''' = f'''(x) = \frac{d}{dx}\left(\frac{d^2y}{dx^2}\right) = \frac{d^3y}{dx^3}$$

The process can be continued. The fourth derivative f'''' is usually denoted by $f^{(4)}$. In general, the nth derivative of f is denoted by $f^{(n)}$ and is obtained from f by differentiating n times. If $y = f(x)$, we write

$$y^{(n)} = f^{(n)}(x) = \frac{d^n y}{dx^n}$$

EXAMPLE 7 If $f(x) = x^3 - x$, find $f'''(x)$ and $f^{(4)}(x)$.

SOLUTION In Example 6 we found that $f''(x) = 6x$. The graph of the second derivative has equation $y = 6x$ and so it is a straight line with slope 6. Since the derivative $f'''(x)$ is the slope of $f''(x)$, we have

$$f'''(x) = 6$$

for all values of x. So f''' is a constant function and its graph is a horizontal line. Therefore, for all values of x,

$$f^{(4)}(x) = 0$$

We can also interpret the third derivative physically in the case where the function is the position function $s = s(t)$ of an object that moves along a straight line. Because $s''' = (s'')' = a'$, the third derivative of the position function is the derivative of the acceleration function and is called the **jerk**:

$$j = \frac{da}{dt} = \frac{d^3s}{dt^3}$$

Thus the jerk j is the rate of change of acceleration. It is aptly named because a large jerk means a sudden change in acceleration, which causes an abrupt movement in a vehicle.

We have seen that one application of second and third derivatives occurs in analyzing the motion of objects using acceleration and jerk. We will investigate another application of second derivatives in Section 3.3, where we show how knowledge of f'' gives us information about the shape of the graph of f. In Chapter 11 we will see how second and higher derivatives enable us to represent functions as sums of infinite series.

2.2   Exercises

Access end-of-section exercises and tools for visualizing calculus online at **www.webassign.net**

2.3   Differentiation Formulas

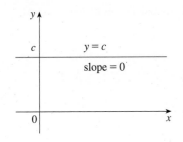

FIGURE 1
The graph of $f(x) = c$ is the line $y = c$, so $f'(x) = 0$.

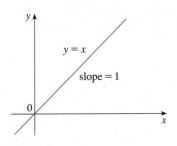

FIGURE 2
The graph of $f(x) = x$ is the line $y = x$, so $f'(x) = 1$.

If it were always necessary to compute derivatives directly from the definition, as we did in the preceding section, such computations would be tedious and the evaluation of some limits would require ingenuity. Fortunately, several rules have been developed for finding derivatives without having to use the definition directly. These formulas greatly simplify the task of differentiation.

Let's start with the simplest of all functions, the constant function $f(x) = c$. The graph of this function is the horizontal line $y = c$, which has slope 0, so we must have $f'(x) = 0$. (See Figure 1.) A formal proof, from the definition of a derivative, is also easy:

$$f'(x) = \lim_{h \to 0} \frac{f(x + h) - f(x)}{h} = \lim_{h \to 0} \frac{c - c}{h} = \lim_{h \to 0} 0 = 0$$

In Leibniz notation, we write this rule as follows.

Derivative of a Constant Function
$$\frac{d}{dx}(c) = 0$$

Power Functions

We next look at the functions $f(x) = x^n$, where n is a positive integer. If $n = 1$, the graph of $f(x) = x$ is the line $y = x$, which has slope 1. (See Figure 2.) So

1 $$\frac{d}{dx}(x) = 1$$

(You can also verify Equation 1 from the definition of a derivative.) We have already investigated the cases $n = 2$ and $n = 3$. In fact, in Section 2.2 (Exercises 17 and 18) we found that

$$\boxed{2} \qquad \frac{d}{dx}(x^2) = 2x \qquad \frac{d}{dx}(x^3) = 3x^2$$

For $n = 4$ we find the derivative of $f(x) = x^4$ as follows:

$$f'(x) = \lim_{h \to 0} \frac{f(x + h) - f(x)}{h} = \lim_{h \to 0} \frac{(x + h)^4 - x^4}{h}$$

$$= \lim_{h \to 0} \frac{x^4 + 4x^3h + 6x^2h^2 + 4xh^3 + h^4 - x^4}{h}$$

$$= \lim_{h \to 0} \frac{4x^3h + 6x^2h^2 + 4xh^3 + h^4}{h}$$

$$= \lim_{h \to 0} (4x^3 + 6x^2h + 4xh^2 + h^3) = 4x^3$$

Thus

$$\boxed{3} \qquad \frac{d}{dx}(x^4) = 4x^3$$

Comparing the equations in $\boxed{1}$, $\boxed{2}$, and $\boxed{3}$, we see a pattern emerging. It seems to be a reasonable guess that, when n is a positive integer, $(d/dx)(x^n) = nx^{n-1}$. This turns out to be true. We prove it in two ways; the second proof uses the Binomial Theorem.

The Power Rule   If n is a positive integer, then

$$\frac{d}{dx}(x^n) = nx^{n-1}$$

FIRST PROOF   The formula

$$x^n - a^n = (x - a)(x^{n-1} + x^{n-2}a + \cdots + xa^{n-2} + a^{n-1})$$

can be verified simply by multiplying out the right-hand side (or by summing the second factor as a geometric series). If $f(x) = x^n$, we can use Equation 2.1.5 for $f'(a)$ and the equation above to write

$$f'(a) = \lim_{x \to a} \frac{f(x) - f(a)}{x - a} = \lim_{x \to a} \frac{x^n - a^n}{x - a}$$

$$= \lim_{x \to a} (x^{n-1} + x^{n-2}a + \cdots + xa^{n-2} + a^{n-1})$$

$$= a^{n-1} + a^{n-2}a + \cdots + aa^{n-2} + a^{n-1}$$

$$= na^{n-1}$$

SECOND PROOF

$$f'(x) = \lim_{h \to 0} \frac{f(x+h) - f(x)}{h} = \lim_{h \to 0} \frac{(x+h)^n - x^n}{h}$$

The Binomial Theorem is given on Reference Page 1.

In finding the derivative of x^4 we had to expand $(x+h)^4$. Here we need to expand $(x+h)^n$ and we use the Binomial Theorem to do so:

$$f'(x) = \lim_{h \to 0} \frac{\left[x^n + nx^{n-1}h + \dfrac{n(n-1)}{2}x^{n-2}h^2 + \cdots + nxh^{n-1} + h^n \right] - x^n}{h}$$

$$= \lim_{h \to 0} \frac{nx^{n-1}h + \dfrac{n(n-1)}{2}x^{n-2}h^2 + \cdots + nxh^{n-1} + h^n}{h}$$

$$= \lim_{h \to 0} \left[nx^{n-1} + \frac{n(n-1)}{2}x^{n-2}h + \cdots + nxh^{n-2} + h^{n-1} \right]$$

$$= nx^{n-1}$$

because every term except the first has h as a factor and therefore approaches 0. ▬

We illustrate the Power Rule using various notations in Example 1.

EXAMPLE 1

(a) If $f(x) = x^6$, then $f'(x) = 6x^5$.

(b) If $y = x^{1000}$, then $y' = 1000x^{999}$.

(c) If $y = t^4$, then $\dfrac{dy}{dt} = 4t^3$.

(d) $\dfrac{d}{dr}(r^3) = 3r^2$ ▬

▬ New Derivatives from Old

When new functions are formed from old functions by addition, subtraction, or multiplication by a constant, their derivatives can be calculated in terms of derivatives of the old functions. In particular, the following formula says that *the derivative of a constant times a function is the constant times the derivative of the function.*

GEOMETRIC INTERPRETATION OF THE CONSTANT MULTIPLE RULE

Multiplying by $c = 2$ stretches the graph vertically by a factor of 2. All the rises have been doubled but the runs stay the same. So the slopes are doubled too.

The Constant Multiple Rule If c is a constant and f is a differentiable function, then

$$\frac{d}{dx}[cf(x)] = c\frac{d}{dx}f(x)$$

PROOF Let $g(x) = cf(x)$. Then

$$g'(x) = \lim_{h \to 0} \frac{g(x+h) - g(x)}{h} = \lim_{h \to 0} \frac{cf(x+h) - cf(x)}{h}$$

$$= \lim_{h \to 0} c\left[\frac{f(x+h) - f(x)}{h} \right]$$

$$= c \lim_{h \to 0} \frac{f(x+h) - f(x)}{h} \qquad \text{(by Law 3 of limits)}$$

$$= cf'(x)$$
▬

EXAMPLE 2

(a) $\dfrac{d}{dx}(3x^4) = 3\dfrac{d}{dx}(x^4) = 3(4x^3) = 12x^3$

(b) $\dfrac{d}{dx}(-x) = \dfrac{d}{dx}[(-1)x] = (-1)\dfrac{d}{dx}(x) = -1(1) = -1$

The next rule tells us that *the derivative of a sum of functions is the sum of the derivatives.*

Using prime notation, we can write the Sum Rule as

$$(f + g)' = f' + g'$$

The Sum Rule If f and g are both differentiable, then

$$\frac{d}{dx}[f(x) + g(x)] = \frac{d}{dx}f(x) + \frac{d}{dx}g(x)$$

PROOF   Let $F(x) = f(x) + g(x)$. Then

$$F'(x) = \lim_{h \to 0} \frac{F(x + h) - F(x)}{h}$$

$$= \lim_{h \to 0} \frac{[f(x + h) + g(x + h)] - [f(x) + g(x)]}{h}$$

$$= \lim_{h \to 0} \left[\frac{f(x + h) - f(x)}{h} + \frac{g(x + h) - g(x)}{h} \right]$$

$$= \lim_{h \to 0} \frac{f(x + h) - f(x)}{h} + \lim_{h \to 0} \frac{g(x + h) - g(x)}{h} \qquad \text{(by Law 1)}$$

$$= f'(x) + g'(x)$$

The Sum Rule can be extended to the sum of any number of functions. For instance, using this theorem twice, we get

$$(f + g + h)' = [(f + g) + h]' = (f + g)' + h' = f' + g' + h'$$

By writing $f - g$ as $f + (-1)g$ and applying the Sum Rule and the Constant Multiple Rule, we get the following formula.

The Difference Rule If f and g are both differentiable, then

$$\frac{d}{dx}[f(x) - g(x)] = \frac{d}{dx}f(x) - \frac{d}{dx}g(x)$$

The Constant Multiple Rule, the Sum Rule, and the Difference Rule can be combined with the Power Rule to differentiate any polynomial, as the following examples demonstrate.

EXAMPLE 3

$$\frac{d}{dx}(x^8 + 12x^5 - 4x^4 + 10x^3 - 6x + 5)$$

$$= \frac{d}{dx}(x^8) + 12\frac{d}{dx}(x^5) - 4\frac{d}{dx}(x^4) + 10\frac{d}{dx}(x^3) - 6\frac{d}{dx}(x) + \frac{d}{dx}(5)$$

$$= 8x^7 + 12(5x^4) - 4(4x^3) + 10(3x^2) - 6(1) + 0$$

$$= 8x^7 + 60x^4 - 16x^3 + 30x^2 - 6$$

V EXAMPLE 4   Find the points on the curve $y = x^4 - 6x^2 + 4$ where the tangent line is horizontal.

SOLUTION   Horizontal tangents occur where the derivative is zero. We have

$$\frac{dy}{dx} = \frac{d}{dx}(x^4) - 6\frac{d}{dx}(x^2) + \frac{d}{dx}(4)$$

$$= 4x^3 - 12x + 0 = 4x(x^2 - 3)$$

Thus $dy/dx = 0$ if $x = 0$ or $x^2 - 3 = 0$, that is, $x = \pm\sqrt{3}$. So the given curve has horizontal tangents when $x = 0$, $\sqrt{3}$, and $-\sqrt{3}$. The corresponding points are $(0, 4)$, $(\sqrt{3}, -5)$, and $(-\sqrt{3}, -5)$. (See Figure 3.)

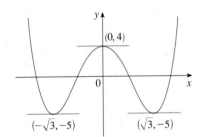

FIGURE 3
The curve $y = x^4 - 6x^2 + 4$ and
its horizontal tangents

EXAMPLE 5   The equation of motion of a particle is $s = 2t^3 - 5t^2 + 3t + 4$, where s is measured in centimeters and t in seconds. Find the acceleration as a function of time. What is the acceleration after 2 seconds?

SOLUTION   The velocity and acceleration are

$$v(t) = \frac{ds}{dt} = 6t^2 - 10t + 3$$

$$a(t) = \frac{dv}{dt} = 12t - 10$$

The acceleration after 2 s is $a(2) = 14$ cm/s².

Next we need a formula for the derivative of a product of two functions. By analogy with the Sum and Difference Rules, one might be tempted to guess, as Leibniz did three centuries ago, that the derivative of a product is the product of the derivatives. We can see, however, that this guess is wrong by looking at a particular example. Let $f(x) = x$ and $g(x) = x^2$. Then the Power Rule gives $f'(x) = 1$ and $g'(x) = 2x$. But $(fg)(x) = x^3$, so $(fg)'(x) = 3x^2$. Thus $(fg)' \neq f'g'$. The correct formula was discovered by Leibniz (soon after his false start) and is called the Product Rule.

We can write the Product Rule in prime
notation as

$$(fg)' = fg' + gf'$$

The Product Rule   If f and g are both differentiable, then

$$\frac{d}{dx}[f(x)g(x)] = f(x)\frac{d}{dx}[g(x)] + g(x)\frac{d}{dx}[f(x)]$$

PROOF Let $F(x) = f(x)g(x)$. Then

$$F'(x) = \lim_{h \to 0} \frac{F(x + h) - F(x)}{h}$$

$$= \lim_{h \to 0} \frac{f(x + h)g(x + h) - f(x)g(x)}{h}$$

In order to evaluate this limit, we would like to separate the functions f and g as in the proof of the Sum Rule. We can achieve this separation by subtracting and adding the term $f(x + h)g(x)$ in the numerator:

$$F'(x) = \lim_{h \to 0} \frac{f(x + h)g(x + h) - f(x + h)g(x) + f(x + h)g(x) - f(x)g(x)}{h}$$

$$= \lim_{h \to 0} \left[f(x + h) \frac{g(x + h) - g(x)}{h} + g(x) \frac{f(x + h) - f(x)}{h} \right]$$

$$= \lim_{h \to 0} f(x + h) \cdot \lim_{h \to 0} \frac{g(x + h) - g(x)}{h} + \lim_{h \to 0} g(x) \cdot \lim_{h \to 0} \frac{f(x + h) - f(x)}{h}$$

$$= f(x)g'(x) + g(x)f'(x)$$

Note that $\lim_{h \to 0} g(x) = g(x)$ because $g(x)$ is a constant with respect to the variable h. Also, since f is differentiable at x, it is continuous at x by Theorem 2.2.4, and so $\lim_{h \to 0} f(x + h) = f(x)$. (See Exercise 59 in Section 1.8.)

In words, the Product Rule says that *the derivative of a product of two functions is the first function times the derivative of the second function plus the second function times the derivative of the first function.*

EXAMPLE 6 Find $F'(x)$ if $F(x) = (6x^3)(7x^4)$.

SOLUTION By the Product Rule, we have

$$F'(x) = (6x^3) \frac{d}{dx}(7x^4) + (7x^4) \frac{d}{dx}(6x^3)$$

$$= (6x^3)(28x^3) + (7x^4)(18x^2)$$

$$= 168x^6 + 126x^6 = 294x^6$$

Notice that we could verify the answer to Example 6 directly by first multiplying the factors:

$$F(x) = (6x^3)(7x^4) = 42x^7 \quad \Rightarrow \quad F'(x) = 42(7x^6) = 294x^6$$

But later we will meet functions, such as $y = x^2 \sin x$, for which the Product Rule is the only possible method.

V **EXAMPLE 7** If $h(x) = xg(x)$ and it is known that $g(3) = 5$ and $g'(3) = 2$, find $h'(3)$.

SOLUTION Applying the Product Rule, we get

$$h'(x) = \frac{d}{dx}[xg(x)] = x\frac{d}{dx}[g(x)] + g(x)\frac{d}{dx}[x]$$

$$= xg'(x) + g(x)$$

Therefore $\qquad h'(3) = 3g'(3) + g(3) = 3 \cdot 2 + 5 = 11$ ▬

In prime notation we can write the Quotient Rule as

$$\left(\frac{f}{g}\right)' = \frac{gf' - fg'}{g^2}$$

The Quotient Rule If f and g are differentiable, then

$$\frac{d}{dx}\left[\frac{f(x)}{g(x)}\right] = \frac{g(x)\dfrac{d}{dx}[f(x)] - f(x)\dfrac{d}{dx}[g(x)]}{[g(x)]^2}$$

PROOF Let $F(x) = f(x)/g(x)$. Then

$$F'(x) = \lim_{h \to 0} \frac{F(x+h) - F(x)}{h} = \lim_{h \to 0} \frac{\dfrac{f(x+h)}{g(x+h)} - \dfrac{f(x)}{g(x)}}{h}$$

$$= \lim_{h \to 0} \frac{f(x+h)g(x) - f(x)g(x+h)}{hg(x+h)g(x)}$$

We can separate f and g in this expression by subtracting and adding the term $f(x)g(x)$ in the numerator:

$$F'(x) = \lim_{h \to 0} \frac{f(x+h)g(x) - f(x)g(x) + f(x)g(x) - f(x)g(x+h)}{hg(x+h)g(x)}$$

$$= \lim_{h \to 0} \frac{g(x)\dfrac{f(x+h) - f(x)}{h} - f(x)\dfrac{g(x+h) - g(x)}{h}}{g(x+h)g(x)}$$

$$= \frac{\displaystyle\lim_{h \to 0} g(x) \cdot \lim_{h \to 0} \frac{f(x+h) - f(x)}{h} - \lim_{h \to 0} f(x) \cdot \lim_{h \to 0} \frac{g(x+h) - g(x)}{h}}{\displaystyle\lim_{h \to 0} g(x+h) \cdot \lim_{h \to 0} g(x)}$$

$$= \frac{g(x)f'(x) - f(x)g'(x)}{[g(x)]^2}$$

Again g is continuous by Theorem 2.2.4, so $\lim_{h \to 0} g(x+h) = g(x)$. ▬

In words, the Quotient Rule says that the *derivative of a quotient is the denominator times the derivative of the numerator minus the numerator times the derivative of the denominator, all divided by the square of the denominator.*

The theorems of this section show that any polynomial is differentiable on $\mathbb{R}$ and any rational function is differentiable on its domain. Furthermore, the Quotient Rule and the other differentiation formulas enable us to compute the derivative of any rational function, as the next example illustrates.

We can use a graphing device to check that the answer to Example 8 is plausible. Figure 4 shows the graphs of the function of Example 8 and its derivative. Notice that when y grows rapidly (near -2), y' is large. And when y grows slowly, y' is near 0.

FIGURE 4

V **EXAMPLE 8** Let $y = \dfrac{x^2 + x - 2}{x^3 + 6}$. Then

$$y' = \frac{(x^3 + 6)\dfrac{d}{dx}(x^2 + x - 2) - (x^2 + x - 2)\dfrac{d}{dx}(x^3 + 6)}{(x^3 + 6)^2}$$

$$= \frac{(x^3 + 6)(2x + 1) - (x^2 + x - 2)(3x^2)}{(x^3 + 6)^2}$$

$$= \frac{(2x^4 + x^3 + 12x + 6) - (3x^4 + 3x^3 - 6x^2)}{(x^3 + 6)^2}$$

$$= \frac{-x^4 - 2x^3 + 6x^2 + 12x + 6}{(x^3 + 6)^2}$$

NOTE Don't use the Quotient Rule *every* time you see a quotient. Sometimes it's easier to rewrite a quotient first to put it in a form that is simpler for the purpose of differentiation. For instance, although it is possible to differentiate the function

$$F(x) = \frac{3x^2 + 2\sqrt{x}}{x}$$

using the Quotient Rule, it is much easier to perform the division first and write the function as

$$F(x) = 3x + 2x^{-1/2}$$

before differentiating.

General Power Functions

The Quotient Rule can be used to extend the Power Rule to the case where the exponent is a negative integer.

If n is a positive integer, then

$$\frac{d}{dx}(x^{-n}) = -nx^{-n-1}$$

PROOF
$$\frac{d}{dx}(x^{-n}) = \frac{d}{dx}\left(\frac{1}{x^n}\right)$$

$$= \frac{x^n \dfrac{d}{dx}(1) - 1 \cdot \dfrac{d}{dx}(x^n)}{(x^n)^2} = \frac{x^n \cdot 0 - 1 \cdot nx^{n-1}}{x^{2n}}$$

$$= \frac{-nx^{n-1}}{x^{2n}} = -nx^{n-1-2n} = -nx^{-n-1}$$

EXAMPLE 9

(a) If $y = \dfrac{1}{x}$, then $\dfrac{dy}{dx} = \dfrac{d}{dx}(x^{-1}) = -x^{-2} = -\dfrac{1}{x^2}$

(b) $\dfrac{d}{dt}\left(\dfrac{6}{t^3}\right) = 6\dfrac{d}{dt}(t^{-3}) = 6(-3)t^{-4} = -\dfrac{18}{t^4}$

So far we know that the Power Rule holds if the exponent n is a positive or negative integer. If $n = 0$, then $x^0 = 1$, which we know has a derivative of 0. Thus the Power Rule holds for any integer n. What if the exponent is a fraction? In Example 3 in Section 2.2 we found that

$$\frac{d}{dx}\sqrt{x} = \frac{1}{2\sqrt{x}}$$

which can be written as

$$\frac{d}{dx}(x^{1/2}) = \tfrac{1}{2}x^{-1/2}$$

This shows that the Power Rule is true even when $n = \tfrac{1}{2}$. In fact, it also holds for *any real number* n, as we will prove in Chapter 6. (A proof for rational values of n is indicated in Exercise 48 in Section 2.6.) In the meantime we state the general version and use it in the examples and exercises.

The Power Rule (General Version) If n is any real number, then

$$\frac{d}{dx}(x^n) = nx^{n-1}$$

EXAMPLE 10

(a) If $f(x) = x^\pi$, then $f'(x) = \pi x^{\pi - 1}$.

(b) Let

$$y = \frac{1}{\sqrt[3]{x^2}}$$

Then

$$\frac{dy}{dx} = \frac{d}{dx}(x^{-2/3}) = -\tfrac{2}{3}x^{-(2/3)-1}$$

$$= -\tfrac{2}{3}x^{-5/3}$$

In Example 11, a and b are constants. It is customary in mathematics to use letters near the beginning of the alphabet to represent constants and letters near the end of the alphabet to represent variables.

EXAMPLE 11 Differentiate the function $f(t) = \sqrt{t}\,(a + bt)$.

SOLUTION 1 Using the Product Rule, we have

$$f'(t) = \sqrt{t}\,\frac{d}{dt}(a + bt) + (a + bt)\frac{d}{dt}\left(\sqrt{t}\right)$$

$$= \sqrt{t} \cdot b + (a + bt) \cdot \tfrac{1}{2}t^{-1/2}$$

$$= b\sqrt{t} + \frac{a + bt}{2\sqrt{t}} = \frac{a + 3bt}{2\sqrt{t}}$$

SOLUTION 2 If we first use the laws of exponents to rewrite $f(t)$, then we can proceed directly without using the Product Rule.

$$f(t) = a\sqrt{t} + bt\sqrt{t} = at^{1/2} + bt^{3/2}$$

$$f'(t) = \tfrac{1}{2}at^{-1/2} + \tfrac{3}{2}bt^{1/2}$$

which is equivalent to the answer given in Solution 1.

The differentiation rules enable us to find tangent lines without having to resort to the definition of a derivative. They also enable us to find *normal lines*. The **normal line** to a curve C at point P is the line through P that is perpendicular to the tangent line at P. (In the study of optics, one needs to consider the angle between a light ray and the normal line to a lens.)

EXAMPLE 12 Find equations of the tangent line and normal line to the curve $y = \sqrt{x}/(1 + x^2)$ at the point $\left(1, \tfrac{1}{2}\right)$.

SOLUTION According to the Quotient Rule, we have

$$\frac{dy}{dx} = \frac{(1 + x^2)\dfrac{d}{dx}\left(\sqrt{x}\right) - \sqrt{x}\,\dfrac{d}{dx}(1 + x^2)}{(1 + x^2)^2}$$

$$= \frac{(1 + x^2)\dfrac{1}{2\sqrt{x}} - \sqrt{x}\,(2x)}{(1 + x^2)^2}$$

$$= \frac{(1 + x^2) - 4x^2}{2\sqrt{x}\,(1 + x^2)^2} = \frac{1 - 3x^2}{2\sqrt{x}\,(1 + x^2)^2}$$

So the slope of the tangent line at $\left(1, \tfrac{1}{2}\right)$ is

$$\left.\frac{dy}{dx}\right|_{x=1} = \frac{1 - 3 \cdot 1^2}{2\sqrt{1}\,(1 + 1^2)^2} = -\frac{1}{4}$$

We use the point-slope form to write an equation of the tangent line at $\left(1, \tfrac{1}{2}\right)$:

$$y - \tfrac{1}{2} = -\tfrac{1}{4}(x - 1) \qquad \text{or} \qquad y = -\tfrac{1}{4}x + \tfrac{3}{4}$$

The slope of the normal line at $\left(1, \tfrac{1}{2}\right)$ is the negative reciprocal of $-\tfrac{1}{4}$, namely 4, so an equation is

$$y - \tfrac{1}{2} = 4(x - 1) \qquad \text{or} \qquad y = 4x - \tfrac{7}{2}$$

The curve and its tangent and normal lines are graphed in Figure 5.

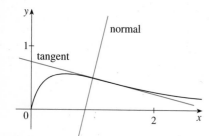

FIGURE 5

EXAMPLE 13 At what points on the hyperbola $xy = 12$ is the tangent line parallel to the line $3x + y = 0$?

SOLUTION Since $xy = 12$ can be written as $y = 12/x$, we have

$$\frac{dy}{dx} = 12\,\frac{d}{dx}\,(x^{-1}) = 12(-x^{-2}) = -\frac{12}{x^2}$$

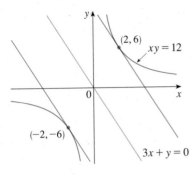

FIGURE 6

Let the x-coordinate of one of the points in question be a. Then the slope of the tangent line at that point is $-12/a^2$. This tangent line will be parallel to the line $3x + y = 0$, or $y = -3x$, if it has the same slope, that is, -3. Equating slopes, we get

$$-\frac{12}{a^2} = -3 \qquad \text{or} \qquad a^2 = 4 \qquad \text{or} \qquad a = \pm 2$$

Therefore the required points are $(2, 6)$ and $(-2, -6)$. The hyperbola and the tangents are shown in Figure 6.

We summarize the differentiation formulas we have learned so far as follows.

Table of Differentiation Formulas

$$\frac{d}{dx}(c) = 0 \qquad\qquad \frac{d}{dx}(x^n) = nx^{n-1}$$

$$(cf)' = cf' \qquad\qquad (f + g)' = f' + g' \qquad\qquad (f - g)' = f' - g'$$

$$(fg)' = fg' + gf' \qquad\qquad \left(\frac{f}{g}\right)' = \frac{gf' - fg'}{g^2}$$

2.3 Exercises

Access end-of-section exercises and tools for visualizing calculus online at **www.webassign.net**   ENHANCED **WebAssign**

APPLIED PROJECT   BUILDING A BETTER ROLLER COASTER

Suppose you are asked to design the first ascent and drop for a new roller coaster. By studying photographs of your favorite coasters, you decide to make the slope of the ascent 0.8 and the slope of the drop -1.6. You decide to connect these two straight stretches $y = L_1(x)$ and $y = L_2(x)$ with part of a parabola $y = f(x) = ax^2 + bx + c$, where x and $f(x)$ are measured in feet. For the track to be smooth there can't be abrupt changes in direction, so you want the linear segments L_1 and L_2 to be tangent to the parabola at the transition points P and Q. (See the figure.) To simplify the equations, you decide to place the origin at P.

1. (a) Suppose the horizontal distance between P and Q is 100 ft. Write equations in a, b, and c that will ensure that the track is smooth at the transition points.
   (b) Solve the equations in part (a) for a, b, and c to find a formula for $f(x)$.
   (c) Plot L_1, f, and L_2 to verify graphically that the transitions are smooth.
   (d) Find the difference in elevation between P and Q.

2. The solution in Problem 1 might *look* smooth, but it might not *feel* smooth because the piecewise defined function [consisting of $L_1(x)$ for $x < 0$, $f(x)$ for $0 \leq x \leq 100$, and $L_2(x)$ for $x > 100$] doesn't have a continuous second derivative. So you decide to improve

---

⊞ Graphing calculator or computer required

CAS Computer algebra system required

the design by using a quadratic function $q(x) = ax^2 + bx + c$ only on the interval $10 \leqslant x \leqslant 90$ and connecting it to the linear functions by means of two cubic functions:

$$g(x) = kx^3 + lx^2 + mx + n \quad 0 \leqslant x < 10$$

$$h(x) = px^3 + qx^2 + rx + s \quad 90 < x \leqslant 100$$

(a) Write a system of equations in 11 unknowns that ensure that the functions and their first two derivatives agree at the transition points.

CAS (b) Solve the equations in part (a) with a computer algebra system to find formulas for $q(x)$, $g(x)$, and $h(x)$.

(c) Plot L_1, g, q, h, and L_2, and compare with the plot in Problem 1(c).

2.4   Derivatives of Trigonometric Functions

A review of trigonometric functions is given in Appendix D.

Before starting this section, you might need to review the trigonometric functions. In particular, it is important to remember that when we talk about the function f defined for all real numbers x by

$$f(x) = \sin x$$

it is understood that $\sin x$ means the sine of the angle whose *radian* measure is x. A similar convention holds for the other trigonometric functions cos, tan, csc, sec, and cot. Recall from Section 1.8 that all of the trigonometric functions are continuous at every number in their domains.

If we sketch the graph of the function $f(x) = \sin x$ and use the interpretation of $f'(x)$ as the slope of the tangent to the sine curve in order to sketch the graph of f' (see Exercise 16 in Section 2.2), then it looks as if the graph of f' may be the same as the cosine curve (see Figure 1).

TEC Visual 2.4 shows an animation of Figure 1.

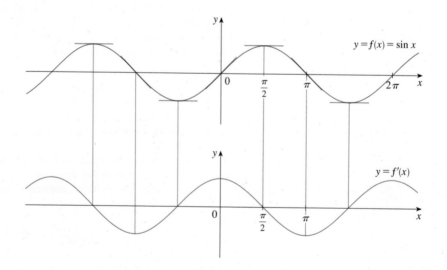

FIGURE 1

Let's try to confirm our guess that if $f(x) = \sin x$, then $f'(x) = \cos x$. From the definition of a derivative, we have

$$f'(x) = \lim_{h \to 0} \frac{f(x + h) - f(x)}{h} = \lim_{h \to 0} \frac{\sin(x + h) - \sin x}{h}$$

We have used the addition formula for sine. See Appendix D.

$$= \lim_{h \to 0} \frac{\sin x \cos h + \cos x \sin h - \sin x}{h}$$

$$= \lim_{h \to 0} \left[\frac{\sin x \cos h - \sin x}{h} + \frac{\cos x \sin h}{h} \right]$$

$$= \lim_{h \to 0} \left[\sin x \left(\frac{\cos h - 1}{h} \right) + \cos x \left(\frac{\sin h}{h} \right) \right]$$

$$\boxed{1} \qquad = \lim_{h \to 0} \sin x \cdot \lim_{h \to 0} \frac{\cos h - 1}{h} + \lim_{h \to 0} \cos x \cdot \lim_{h \to 0} \frac{\sin h}{h}$$

Two of these four limits are easy to evaluate. Since we regard x as a constant when computing a limit as $h \to 0$, we have

$$\lim_{h \to 0} \sin x = \sin x \qquad \text{and} \qquad \lim_{h \to 0} \cos x = \cos x$$

The limit of $(\sin h)/h$ is not so obvious. In Example 3 in Section 1.5 we made the guess, on the basis of numerical and graphical evidence, that

$$\boxed{2} \qquad \boxed{\lim_{\theta \to 0} \frac{\sin \theta}{\theta} = 1}$$

We now use a geometric argument to prove Equation 2. Assume first that θ lies between 0 and $\pi/2$. Figure 2(a) shows a sector of a circle with center O, central angle θ, and radius 1. BC is drawn perpendicular to OA. By the definition of radian measure, we have arc $AB = \theta$. Also $|BC| = |OB| \sin \theta = \sin \theta$. From the diagram we see that

$$|BC| < |AB| < \text{arc } AB$$

Therefore $\qquad\qquad \sin \theta < \theta \qquad \text{so} \qquad \frac{\sin \theta}{\theta} < 1$

Let the tangent lines at A and B intersect at E. You can see from Figure 2(b) that the circumference of a circle is smaller than the length of a circumscribed polygon, and so arc $AB < |AE| + |EB|$. Thus

$$\theta = \text{arc } AB < |AE| + |EB|$$

$$< |AE| + |ED|$$

$$= |AD| = |OA| \tan \theta$$

$$= \tan \theta$$

(a)

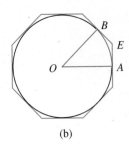

(b)

FIGURE 2

(In Appendix F the inequality $\theta \leqslant \tan \theta$ is proved directly from the definition of the length of an arc without resorting to geometric intuition as we did here.) Therefore we have

$$\theta < \frac{\sin \theta}{\cos \theta}$$

so

$$\cos \theta < \frac{\sin \theta}{\theta} < 1$$

We know that $\lim_{\theta \to 0} 1 = 1$ and $\lim_{\theta \to 0} \cos \theta = 1$, so by the Squeeze Theorem, we have

$$\lim_{\theta \to 0^+} \frac{\sin \theta}{\theta} = 1$$

But the function $(\sin \theta)/\theta$ is an even function, so its right and left limits must be equal. Hence, we have

$$\lim_{\theta \to 0} \frac{\sin \theta}{\theta} = 1$$

so we have proved Equation 2.

We can deduce the value of the remaining limit in $\boxed{1}$ as follows:

We multiply numerator and denominator by $\cos \theta + 1$ in order to put the function in a form in which we can use the limits we know.

$$\lim_{\theta \to 0} \frac{\cos \theta - 1}{\theta} = \lim_{\theta \to 0} \left(\frac{\cos \theta - 1}{\theta} \cdot \frac{\cos \theta + 1}{\cos \theta + 1} \right) = \lim_{\theta \to 0} \frac{\cos^2 \theta - 1}{\theta \,(\cos \theta + 1)}$$

$$= \lim_{\theta \to 0} \frac{-\sin^2 \theta}{\theta \,(\cos \theta + 1)} = -\lim_{\theta \to 0} \left(\frac{\sin \theta}{\theta} \cdot \frac{\sin \theta}{\cos \theta + 1} \right)$$

$$= -\lim_{\theta \to 0} \frac{\sin \theta}{\theta} \cdot \lim_{\theta \to 0} \frac{\sin \theta}{\cos \theta + 1}$$

$$= -1 \cdot \left(\frac{0}{1 + 1} \right) = 0 \qquad \text{(by Equation 2)}$$

$\boxed{3}$

$$\lim_{\theta \to 0} \frac{\cos \theta - 1}{\theta} = 0$$

If we now put the limits $\boxed{2}$ and $\boxed{3}$ in $\boxed{1}$, we get

$$f'(x) = \lim_{h \to 0} \sin x \cdot \lim_{h \to 0} \frac{\cos h - 1}{h} + \lim_{h \to 0} \cos x \cdot \lim_{h \to 0} \frac{\sin h}{h}$$

$$= (\sin x) \cdot 0 + (\cos x) \cdot 1 = \cos x$$

So we have proved the formula for the derivative of the sine function:

$\boxed{4}$

$$\frac{d}{dx} (\sin x) = \cos x$$

V **EXAMPLE 1** Differentiate $y = x^2 \sin x$.

SOLUTION Using the Product Rule and Formula 4, we have

$$\frac{dy}{dx} = x^2 \frac{d}{dx}(\sin x) + \sin x \frac{d}{dx}(x^2) = x^2 \cos x + 2x \sin x$$

Figure 3 shows the graphs of the function of Example 1 and its derivative. Notice that $y' = 0$ whenever y has a horizontal tangent.

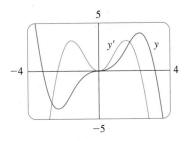

FIGURE 3

Using the same methods as in the proof of Formula 4, one can prove (see Exercise 20) that

$$\boxed{5} \qquad \frac{d}{dx}(\cos x) = -\sin x$$

The tangent function can also be differentiated by using the definition of a derivative, but it is easier to use the Quotient Rule together with Formulas 4 and 5:

$$\frac{d}{dx}(\tan x) = \frac{d}{dx}\left(\frac{\sin x}{\cos x}\right)$$

$$= \frac{\cos x \frac{d}{dx}(\sin x) - \sin x \frac{d}{dx}(\cos x)}{\cos^2 x}$$

$$= \frac{\cos x \cdot \cos x - \sin x(-\sin x)}{\cos^2 x}$$

$$= \frac{\cos^2 x + \sin^2 x}{\cos^2 x}$$

$$= \frac{1}{\cos^2 x} = \sec^2 x$$

$$\boxed{6} \qquad \frac{d}{dx}(\tan x) = \sec^2 x$$

The derivatives of the remaining trigonometric functions, csc, sec , and cot , can also be found easily using the Quotient Rule (see Exercises 17–19). We collect all the differentiation formulas for trigonometric functions in the following table. Remember that they are valid only when x is measured in radians.

When you memorize this table, it is helpful to notice that the minus signs go with the derivatives of the "cofunctions," that is, cosine, cosecant, and cotangent.

Derivatives of Trigonometric Functions

$$\frac{d}{dx}(\sin x) = \cos x \qquad\qquad \frac{d}{dx}(\csc x) = -\csc x \cot x$$

$$\frac{d}{dx}(\cos x) = -\sin x \qquad\qquad \frac{d}{dx}(\sec x) = \sec x \tan x$$

$$\frac{d}{dx}(\tan x) = \sec^2 x \qquad\qquad \frac{d}{dx}(\cot x) = -\csc^2 x$$

EXAMPLE 2 Differentiate $f(x) = \dfrac{\sec x}{1 + \tan x}$. For what values of x does the graph of f have a horizontal tangent?

SOLUTION The Quotient Rule gives

$$f'(x) = \frac{(1 + \tan x)\dfrac{d}{dx}(\sec x) - \sec x \dfrac{d}{dx}(1 + \tan x)}{(1 + \tan x)^2}$$

$$= \frac{(1 + \tan x)\sec x \tan x - \sec x \cdot \sec^2 x}{(1 + \tan x)^2}$$

$$= \frac{\sec x (\tan x + \tan^2 x - \sec^2 x)}{(1 + \tan x)^2}$$

$$= \frac{\sec x (\tan x - 1)}{(1 + \tan x)^2}$$

In simplifying the answer we have used the identity $\tan^2 x + 1 = \sec^2 x$.

Since $\sec x$ is never 0, we see that $f'(x) = 0$ when $\tan x = 1$, and this occurs when $x = n\pi + \pi/4$, where n is an integer (see Figure 4).

FIGURE 4

The horizontal tangents in Example 2

Trigonometric functions are often used in modeling real-world phenomena. In particular, vibrations, waves, elastic motions, and other quantities that vary in a periodic manner can be described using trigonometric functions. In the following example we discuss an instance of simple harmonic motion.

V **EXAMPLE 3** An object at the end of a vertical spring is stretched 4 cm beyond its rest position and released at time $t = 0$. (See Figure 5 and note that the downward direction is positive.) Its position at time t is

$$s = f(t) = 4 \cos t$$

Find the velocity and acceleration at time t and use them to analyze the motion of the object.

SOLUTION The velocity and acceleration are

$$v = \frac{ds}{dt} = \frac{d}{dt}(4 \cos t) = 4\frac{d}{dt}(\cos t) = -4 \sin t$$

$$a = \frac{dv}{dt} = \frac{d}{dt}(-4 \sin t) = -4\frac{d}{dt}(\sin t) = -4 \cos t$$

FIGURE 5

The object oscillates from the lowest point ($s = 4$ cm) to the highest point ($s = -4$ cm). The period of the oscillation is 2π, the period of $\cos t$.

The speed is $|v| = 4|\sin t|$, which is greatest when $|\sin t| = 1$, that is, when $\cos t = 0$. So the object moves fastest as it passes through its equilibrium position ($s = 0$). Its speed is 0 when $\sin t = 0$, that is, at the high and low points.

The acceleration $a = -4 \cos t = 0$ when $s = 0$. It has greatest magnitude at the high and low points. See the graphs in Figure 6.

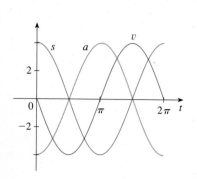

FIGURE 6

EXAMPLE 4 Find the 27th derivative of $\cos x$.

PS Look for a pattern.

SOLUTION The first few derivatives of $f(x) = \cos x$ are as follows:

$$f'(x) = -\sin x$$
$$f''(x) = -\cos x$$
$$f'''(x) = \sin x$$
$$f^{(4)}(x) = \cos x$$
$$f^{(5)}(x) = -\sin x$$

We see that the successive derivatives occur in a cycle of length 4 and, in particular, $f^{(n)}(x) = \cos x$ whenever n is a multiple of 4. Therefore

$$f^{(24)}(x) = \cos x$$

and, differentiating three more times, we have

$$f^{(27)}(x) = \sin x$$

Our main use for the limit in Equation 2 has been to prove the differentiation formula for the sine function. But this limit is also useful in finding certain other trigonometric limits, as the following two examples show.

EXAMPLE 5 Find $\displaystyle\lim_{x \to 0} \frac{\sin 7x}{4x}$.

SOLUTION In order to apply Equation 2, we first rewrite the function by multiplying and dividing by 7:

Note that $\sin 7x \neq 7 \sin x$.

$$\frac{\sin 7x}{4x} = \frac{7}{4}\left(\frac{\sin 7x}{7x}\right)$$

If we let $\theta = 7x$, then $\theta \to 0$ as $x \to 0$, so by Equation 2 we have

$$\lim_{x \to 0} \frac{\sin 7x}{4x} = \frac{7}{4}\lim_{x \to 0}\left(\frac{\sin 7x}{7x}\right)$$

$$= \frac{7}{4}\lim_{\theta \to 0}\frac{\sin \theta}{\theta} = \frac{7}{4} \cdot 1 = \frac{7}{4}$$

V **EXAMPLE 6** Calculate $\displaystyle\lim_{x \to 0} x \cot x$.

SOLUTION Here we divide numerator and denominator by x:

$$\lim_{x \to 0} x \cot x = \lim_{x \to 0} \frac{x \cos x}{\sin x}$$

$$= \lim_{x \to 0} \frac{\cos x}{\dfrac{\sin x}{x}} = \frac{\displaystyle\lim_{x \to 0} \cos x}{\displaystyle\lim_{x \to 0} \dfrac{\sin x}{x}}$$

$$= \frac{\cos 0}{1} \qquad \text{(by the continuity of cosine and Equation 2)}$$

$$= 1$$

2.4 **Exercises**

Access end-of-section exercises and tools for visualizing calculus online at **www.webassign.net** ENHANCED WebAssign

2.5 The Chain Rule

Suppose you are asked to differentiate the function

$$F(x) = \sqrt{x^2 + 1}$$

The differentiation formulas you learned in the previous sections of this chapter do not enable you to calculate $F'(x)$.

See Section 1.3 for a review of composite functions.

Observe that F is a composite function. In fact, if we let $y = f(u) = \sqrt{u}$ and let $u = g(x) = x^2 + 1$, then we can write $y = F(x) = f(g(x))$, that is, $F = f \circ g$. We know how to differentiate both f and g, so it would be useful to have a rule that tells us how to find the derivative of $F = f \circ g$ in terms of the derivatives of f and g.

It turns out that the derivative of the composite function $f \circ g$ is the product of the derivatives of f and g. This fact is one of the most important of the differentiation rules and is called the *Chain Rule*. It seems plausible if we interpret derivatives as rates of change. Regard du/dx as the rate of change of u with respect to x, dy/du as the rate of change of y with respect to u, and dy/dx as the rate of change of y with respect to x. If u changes twice as fast as x and y changes three times as fast as u, then it seems reasonable that y changes six times as fast as x, and so we expect that

$$\frac{dy}{dx} = \frac{dy}{du}\frac{du}{dx}$$

The Chain Rule If g is differentiable at x and f is differentiable at $g(x)$, then the composite function $F = f \circ g$ defined by $F(x) = f(g(x))$ is differentiable at x and F' is given by the product

$$F'(x) = f'(g(x)) \cdot g'(x)$$

In Leibniz notation, if $y = f(u)$ and $u = g(x)$ are both differentiable functions, then

$$\frac{dy}{dx} = \frac{dy}{du}\frac{du}{dx}$$

COMMENTS ON THE PROOF OF THE CHAIN RULE   Let Δu be the change in u corresponding to a change of Δx in x, that is,

$$\Delta u = g(x + \Delta x) - g(x)$$

Then the corresponding change in y is

$$\Delta y = f(u + \Delta u) - f(u)$$

James Gregory

The first person to formulate the Chain Rule was the Scottish mathematician James Gregory (1638–1675), who also designed the first practical reflecting telescope. Gregory discovered the basic ideas of calculus at about the same time as Newton. He became the first Professor of Mathematics at the University of St. Andrews and later held the same position at the University of Edinburgh. But one year after accepting that position he died at the age of 36.

It is tempting to write

$$\frac{dy}{dx} = \lim_{\Delta x \to 0} \frac{\Delta y}{\Delta x}$$

$$\boxed{1} \qquad = \lim_{\Delta x \to 0} \frac{\Delta y}{\Delta u} \cdot \frac{\Delta u}{\Delta x}$$

$$= \lim_{\Delta x \to 0} \frac{\Delta y}{\Delta u} \cdot \lim_{\Delta x \to 0} \frac{\Delta u}{\Delta x}$$

$$= \lim_{\Delta u \to 0} \frac{\Delta y}{\Delta u} \cdot \lim_{\Delta x \to 0} \frac{\Delta u}{\Delta x}$$

$$= \frac{dy}{du} \frac{du}{dx}$$

The only flaw in this reasoning is that in $\boxed{1}$ it might happen that $\Delta u = 0$ (even when $\Delta x \neq 0$) and, of course, we can't divide by 0. Nonetheless, this reasoning does at least *suggest* that the Chain Rule is true. A full proof of the Chain Rule is given at the end of this section.

The Chain Rule can be written either in the prime notation

$$\boxed{2} \qquad (f \circ g)'(x) = f'(g(x)) \cdot g'(x)$$

or, if $y = f(u)$ and $u = g(x)$, in Leibniz notation:

$$\boxed{3} \qquad \frac{dy}{dx} = \frac{dy}{du} \frac{du}{dx}$$

Equation 3 is easy to remember because if dy/du and du/dx were quotients, then we could cancel du. Remember, however, that du has not been defined and du/dx should not be thought of as an actual quotient.

EXAMPLE 1 Find $F'(x)$ if $F(x) = \sqrt{x^2 + 1}$.

SOLUTION 1 (using Equation 2): At the beginning of this section we expressed F as $F(x) = (f \circ g)(x) = f(g(x))$ where $f(u) = \sqrt{u}$ and $g(x) = x^2 + 1$. Since

$$f'(u) = \tfrac{1}{2} u^{-1/2} = \frac{1}{2\sqrt{u}} \qquad \text{and} \qquad g'(x) = 2x$$

we have
$$F'(x) = f'(g(x)) \cdot g'(x)$$
$$= \frac{1}{2\sqrt{x^2 + 1}} \cdot 2x = \frac{x}{\sqrt{x^2 + 1}}$$

SOLUTION 2 (using Equation 3): If we let $u = x^2 + 1$ and $y = \sqrt{u}$, then

$$F'(x) = \frac{dy}{du} \frac{du}{dx} = \frac{1}{2\sqrt{u}}(2x) = \frac{1}{2\sqrt{x^2 + 1}}(2x) = \frac{x}{\sqrt{x^2 + 1}}$$

When using Formula 3 we should bear in mind that dy/dx refers to the derivative of y when y is considered as a function of x (called the *derivative of y with respect to x*), whereas dy/du refers to the derivative of y when considered as a function of u (the derivative of y with respect to u). For instance, in Example 1, y can be considered as a function of x $\left(y = \sqrt{x^2 + 1}\right)$ and also as a function of u $\left(y = \sqrt{u}\right)$. Note that

$$\frac{dy}{dx} = F'(x) = \frac{x}{\sqrt{x^2 + 1}} \qquad \text{whereas} \qquad \frac{dy}{du} = f'(u) = \frac{1}{2\sqrt{u}}$$

NOTE In using the Chain Rule we work from the outside to the inside. Formula 2 says that *we differentiate the outer function f [at the inner function $g(x)$] and then we multiply by the derivative of the inner function.*

$$\frac{d}{dx} \underbrace{f}_{\substack{\text{outer} \\ \text{function}}} \underbrace{(g(x))}_{\substack{\text{evaluated} \\ \text{at inner} \\ \text{function}}} = \underbrace{f'}_{\substack{\text{derivative} \\ \text{of outer} \\ \text{function}}} \underbrace{(g(x))}_{\substack{\text{evaluated} \\ \text{at inner} \\ \text{function}}} \cdot \underbrace{g'(x)}_{\substack{\text{derivative} \\ \text{of inner} \\ \text{function}}}$$

V EXAMPLE 2 Differentiate (a) $y = \sin(x^2)$ and (b) $y = \sin^2 x$.

SOLUTION

(a) If $y = \sin(x^2)$, then the outer function is the sine function and the inner function is the squaring function, so the Chain Rule gives

$$\frac{dy}{dx} = \frac{d}{dx} \underbrace{\sin}_{\substack{\text{outer} \\ \text{function}}} \underbrace{(x^2)}_{\substack{\text{evaluated} \\ \text{at inner} \\ \text{function}}} = \underbrace{\cos}_{\substack{\text{derivative} \\ \text{of outer} \\ \text{function}}} \underbrace{(x^2)}_{\substack{\text{evaluated} \\ \text{at inner} \\ \text{function}}} \cdot \underbrace{2x}_{\substack{\text{derivative} \\ \text{of inner} \\ \text{function}}}$$

$$= 2x \cos(x^2)$$

(b) Note that $\sin^2 x = (\sin x)^2$. Here the outer function is the squaring function and the inner function is the sine function. So

$$\frac{dy}{dx} = \frac{d}{dx} \underbrace{(\sin x)^2}_{\substack{\text{inner} \\ \text{function}}} = \underbrace{2}_{\substack{\text{derivative} \\ \text{of outer} \\ \text{function}}} \cdot \underbrace{(\sin x)}_{\substack{\text{evaluated} \\ \text{at inner} \\ \text{function}}} \cdot \underbrace{\cos x}_{\substack{\text{derivative} \\ \text{of inner} \\ \text{function}}}$$

The answer can be left as $2 \sin x \cos x$ or written as $\sin 2x$ (by a trigonometric identity known as the double-angle formula).

In Example 2(a) we combined the Chain Rule with the rule for differentiating the sine function. In general, if $y = \sin u$, where u is a differentiable function of x, then, by the Chain Rule,

$$\frac{dy}{dx} = \frac{dy}{du} \frac{du}{dx} = \cos u \frac{du}{dx}$$

Thus $$\frac{d}{dx} (\sin u) = \cos u \frac{du}{dx}$$

In a similar fashion, all of the formulas for differentiating trigonometric functions can be combined with the Chain Rule.

See Reference Page 2 or Appendix D.

Let's make explicit the special case of the Chain Rule where the outer function f is a power function. If $y = [g(x)]^n$, then we can write $y = f(u) = u^n$ where $u = g(x)$. By using the Chain Rule and then the Power Rule, we get

$$\frac{dy}{dx} = \frac{dy}{du}\frac{du}{dx} = nu^{n-1}\frac{du}{dx} = n[g(x)]^{n-1}g'(x)$$

---

4 **The Power Rule Combined with the Chain Rule** If n is any real number and $u = g(x)$ is differentiable, then

$$\frac{d}{dx}(u^n) = nu^{n-1}\frac{du}{dx}$$

Alternatively, $\qquad \dfrac{d}{dx}[g(x)]^n = n[g(x)]^{n-1} \cdot g'(x)$

Notice that the derivative in Example 1 could be calculated by taking $n = \frac{1}{2}$ in Rule 4.

EXAMPLE 3 Differentiate $y = (x^3 - 1)^{100}$.

SOLUTION Taking $u = g(x) = x^3 - 1$ and $n = 100$ in $\boxed{4}$, we have

$$\frac{dy}{dx} = \frac{d}{dx}(x^3 - 1)^{100} = 100(x^3 - 1)^{99}\frac{d}{dx}(x^3 - 1)$$

$$= 100(x^3 - 1)^{99} \cdot 3x^2 = 300x^2(x^3 - 1)^{99}$$

V EXAMPLE 4 Find $f'(x)$ if $f(x) = \dfrac{1}{\sqrt[3]{x^2 + x + 1}}$.

SOLUTION First rewrite f: $\qquad f(x) = (x^2 + x + 1)^{-1/3}$

Thus $\qquad\qquad f'(x) = -\tfrac{1}{3}(x^2 + x + 1)^{-4/3}\dfrac{d}{dx}(x^2 + x + 1)$

$$= -\tfrac{1}{3}(x^2 + x + 1)^{-4/3}(2x + 1)$$

EXAMPLE 5 Find the derivative of the function

$$g(t) = \left(\frac{t - 2}{2t + 1}\right)^9$$

SOLUTION Combining the Power Rule, Chain Rule, and Quotient Rule, we get

$$g'(t) = 9\left(\frac{t - 2}{2t + 1}\right)^8 \frac{d}{dt}\left(\frac{t - 2}{2t + 1}\right)$$

$$= 9\left(\frac{t - 2}{2t + 1}\right)^8 \frac{(2t + 1) \cdot 1 - 2(t - 2)}{(2t + 1)^2} = \frac{45(t - 2)^8}{(2t + 1)^{10}}$$

EXAMPLE 6 Differentiate $y = (2x + 1)^5(x^3 - x + 1)^4$.

SOLUTION In this example we must use the Product Rule before using the Chain Rule:

$$\frac{dy}{dx} = (2x + 1)^5 \frac{d}{dx}(x^3 - x + 1)^4 + (x^3 - x + 1)^4 \frac{d}{dx}(2x + 1)^5$$

$$= (2x + 1)^5 \cdot 4(x^3 - x + 1)^3 \frac{d}{dx}(x^3 - x + 1)$$

$$+ (x^3 - x + 1)^4 \cdot 5(2x + 1)^4 \frac{d}{dx}(2x + 1)$$

$$= 4(2x + 1)^5(x^3 - x + 1)^3(3x^2 - 1) + 5(x^3 - x + 1)^4(2x + 1)^4 \cdot 2$$

Noticing that each term has the common factor $2(2x + 1)^4(x^3 - x + 1)^3$, we could factor it out and write the answer as

$$\frac{dy}{dx} = 2(2x + 1)^4(x^3 - x + 1)^3(17x^3 + 6x^2 - 9x + 3) \quad \blacksquare$$

The graphs of the functions y and y' in Example 6 are shown in Figure 1. Notice that y' is large when y increases rapidly and $y' = 0$ when y has a horizontal tangent. So our answer appears to be reasonable.

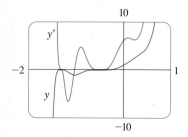

FIGURE 1

The reason for the name "Chain Rule" becomes clear when we make a longer chain by adding another link. Suppose that $y = f(u)$, $u = g(x)$, and $x = h(t)$, where f, g, and h are differentiable functions. Then, to compute the derivative of y with respect to t, we use the Chain Rule twice:

$$\frac{dy}{dt} = \frac{dy}{dx}\frac{dx}{dt} = \frac{dy}{du}\frac{du}{dx}\frac{dx}{dt}$$

V   **EXAMPLE 7** If $f(x) = \sin(\cos(\tan x))$, then

$$f'(x) = \cos(\cos(\tan x)) \frac{d}{dx}\cos(\tan x)$$

$$= \cos(\cos(\tan x))[-\sin(\tan x)] \frac{d}{dx}(\tan x)$$

$$= -\cos(\cos(\tan x))\sin(\tan x)\sec^2 x$$

Notice that we used the Chain Rule twice. $\blacksquare$

EXAMPLE 8 Differentiate $y = \sqrt{\sec x^3}$.

SOLUTION Here the outer function is the square root function, the middle function is the secant function, and the inner function is the cubing function. So we have

$$\frac{dy}{dx} = \frac{1}{2\sqrt{\sec x^3}}\frac{d}{dx}(\sec x^3)$$

$$= \frac{1}{2\sqrt{\sec x^3}}\sec x^3 \tan x^3 \frac{d}{dx}(x^3)$$

$$= \frac{3x^2 \sec x^3 \tan x^3}{2\sqrt{\sec x^3}} \quad \blacksquare$$

How to Prove the Chain Rule

Recall that if $y = f(x)$ and x changes from a to $a + \Delta x$, we define the increment of y as

$$\Delta y = f(a + \Delta x) - f(a)$$

According to the definition of a derivative, we have

$$\lim_{\Delta x \to 0} \frac{\Delta y}{\Delta x} = f'(a)$$

So if we denote by ε the difference between the difference quotient and the derivative, we obtain

$$\lim_{\Delta x \to 0} \varepsilon = \lim_{\Delta x \to 0} \left(\frac{\Delta y}{\Delta x} - f'(a) \right) = f'(a) - f'(a) = 0$$

But $\qquad \varepsilon = \dfrac{\Delta y}{\Delta x} - f'(a) \qquad \Rightarrow \qquad \Delta y = f'(a)\,\Delta x + \varepsilon\,\Delta x$

If we define ε to be 0 when $\Delta x = 0$, then ε becomes a continuous function of Δx. Thus, for a differentiable function f, we can write

$$\boxed{5} \qquad \Delta y = f'(a)\,\Delta x + \varepsilon\,\Delta x \qquad \text{where} \quad \varepsilon \to 0 \text{ as } \Delta x \to 0$$

and ε is a continuous function of Δx. This property of differentiable functions is what enables us to prove the Chain Rule.

PROOF OF THE CHAIN RULE Suppose $u = g(x)$ is differentiable at a and $y = f(u)$ is differentiable at $b = g(a)$. If Δx is an increment in x and Δu and Δy are the corresponding increments in u and y, then we can use Equation 5 to write

$$\boxed{6} \qquad \Delta u = g'(a)\,\Delta x + \varepsilon_1\,\Delta x = [g'(a) + \varepsilon_1]\,\Delta x$$

where $\varepsilon_1 \to 0$ as $\Delta x \to 0$. Similarly

$$\boxed{7} \qquad \Delta y = f'(b)\,\Delta u + \varepsilon_2\,\Delta u = [f'(b) + \varepsilon_2]\,\Delta u$$

where $\varepsilon_2 \to 0$ as $\Delta u \to 0$. If we now substitute the expression for Δu from Equation 6 into Equation 7, we get

$$\Delta y = [f'(b) + \varepsilon_2][g'(a) + \varepsilon_1]\,\Delta x$$

so $\qquad \dfrac{\Delta y}{\Delta x} = [f'(b) + \varepsilon_2][g'(a) + \varepsilon_1]$

As $\Delta x \to 0$, Equation 6 shows that $\Delta u \to 0$. So both $\varepsilon_1 \to 0$ and $\varepsilon_2 \to 0$ as $\Delta x \to 0$. Therefore

$$\frac{dy}{dx} = \lim_{\Delta x \to 0} \frac{\Delta y}{\Delta x} = \lim_{\Delta x \to 0} [f'(b) + \varepsilon_2][g'(a) + \varepsilon_1]$$

$$= f'(b)\,g'(a) = f'(g(a))\,g'(a)$$

This proves the Chain Rule.

2.5   Exercises

Access end-of-section exercises and tools for visualizing calculus online at **www.webassign.net**

APPLIED PROJECT   WHERE SHOULD A PILOT START DESCENT?

$y = P(x)$

An approach path for an aircraft landing is shown in the figure and satisfies the following conditions:

(i) The cruising altitude is h when descent starts at a horizontal distance ℓ from touchdown at the origin.

(ii) The pilot must maintain a constant horizontal speed v throughout descent.

(iii) The absolute value of the vertical acceleration should not exceed a constant k (which is much less than the acceleration due to gravity).

1. Find a cubic polynomial $P(x) = ax^3 + bx^2 + cx + d$ that satisfies condition (i) by imposing suitable conditions on $P(x)$ and $P'(x)$ at the start of descent and at touchdown.

2. Use conditions (ii) and (iii) to show that

$$\frac{6hv^2}{\ell^2} \leq k$$

3. Suppose that an airline decides not to allow vertical acceleration of a plane to exceed $k = 860$ mi/h². If the cruising altitude of a plane is 35,000 ft and the speed is 300 mi/h, how far away from the airport should the pilot start descent?

4. Graph the approach path if the conditions stated in Problem 3 are satisfied.

Graphing calculator or computer required

2.6   Implicit Differentiation

The functions that we have met so far can be described by expressing one variable explicitly in terms of another variable—for example,

$$y = \sqrt{x^3 + 1} \qquad \text{or} \qquad y = x \sin x$$

or, in general, $y = f(x)$. Some functions, however, are defined implicitly by a relation between x and y such as

$$\boxed{1} \qquad\qquad\qquad x^2 + y^2 = 25$$

or

$$\boxed{2} \qquad\qquad\qquad x^3 + y^3 = 6xy$$

In some cases it is possible to solve such an equation for y as an explicit function (or several functions) of x. For instance, if we solve Equation 1 for y, we get $y = \pm\sqrt{25 - x^2}$, so two of the functions determined by the implicit Equation 1 are $f(x) = \sqrt{25 - x^2}$ and $g(x) = -\sqrt{25 - x^2}$. The graphs of f and g are the upper and lower semicircles of the circle $x^2 + y^2 = 25$. (See Figure 1.)

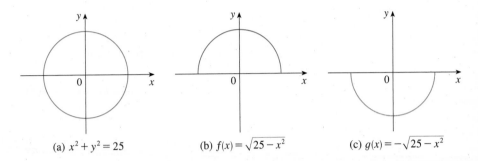

FIGURE 1   (a) $x^2 + y^2 = 25$ (b) $f(x) = \sqrt{25 - x^2}$ (c) $g(x) = -\sqrt{25 - x^2}$

It's not easy to solve Equation 2 for y explicitly as a function of x by hand. (A computer algebra system has no trouble, but the expressions it obtains are very complicated.) Nonetheless, ☐2 is the equation of a curve called the **folium of Descartes** shown in Figure 2 and it implicitly defines y as several functions of x. The graphs of three such functions are shown in Figure 3. When we say that f is a function defined implicitly by Equation 2, we mean that the equation

$$x^3 + [f(x)]^3 = 6xf(x)$$

is true for all values of x in the domain of f.

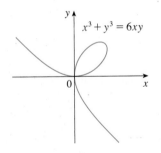

FIGURE 2   The folium of Descartes

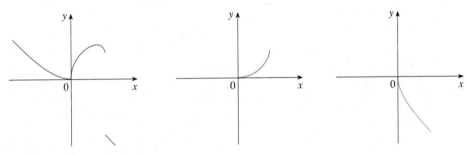

FIGURE 3   Graphs of three functions defined by the folium of Descartes

Fortunately, we don't need to solve an equation for y in terms of x in order to find the derivative of y. Instead we can use the method of **implicit differentiation**. This consists of differentiating both sides of the equation with respect to x and then solving the resulting equation for y'. In the examples and exercises of this section it is always assumed that the given equation determines y implicitly as a differentiable function of x so that the method of implicit differentiation can be applied.

V   EXAMPLE 1

(a) If $x^2 + y^2 = 25$, find $\dfrac{dy}{dx}$.

(b) Find an equation of the tangent to the circle $x^2 + y^2 = 25$ at the point $(3, 4)$.

SOLUTION 1

(a) Differentiate both sides of the equation $x^2 + y^2 = 25$:

$$\frac{d}{dx}(x^2 + y^2) = \frac{d}{dx}(25)$$

$$\frac{d}{dx}(x^2) + \frac{d}{dx}(y^2) = 0$$

Remembering that y is a function of x and using the Chain Rule, we have

$$\frac{d}{dx}(y^2) = \frac{d}{dy}(y^2)\frac{dy}{dx} = 2y\frac{dy}{dx}$$

Thus
$$2x + 2y\frac{dy}{dx} = 0$$

Now we solve this equation for dy/dx:

$$\frac{dy}{dx} = -\frac{x}{y}$$

(b) At the point $(3, 4)$ we have $x = 3$ and $y = 4$, so

$$\frac{dy}{dx} = -\frac{3}{4}$$

An equation of the tangent to the circle at $(3, 4)$ is therefore

$$y - 4 = -\tfrac{3}{4}(x - 3) \qquad \text{or} \qquad 3x + 4y = 25$$

SOLUTION 2

(b) Solving the equation $x^2 + y^2 = 25$, we get $y = \pm\sqrt{25 - x^2}$. The point $(3, 4)$ lies on the upper semicircle $y = \sqrt{25 - x^2}$ and so we consider the function $f(x) = \sqrt{25 - x^2}$. Differentiating f using the Chain Rule, we have

$$f'(x) = \tfrac{1}{2}(25 - x^2)^{-1/2}\frac{d}{dx}(25 - x^2)$$

$$= \tfrac{1}{2}(25 - x^2)^{-1/2}(-2x) = -\frac{x}{\sqrt{25 - x^2}}$$

So
$$f'(3) = -\frac{3}{\sqrt{25 - 3^2}} = -\frac{3}{4}$$

Example 1 illustrates that even when it is possible to solve an equation explicitly for y in terms of x, it may be easier to use implicit differentiation.

and, as in Solution 1, an equation of the tangent is $3x + 4y = 25$. ▬

NOTE 1 The expression $dy/dx = -x/y$ in Solution 1 gives the derivative in terms of both x and y. It is correct no matter which function y is determined by the given equation. For instance, for $y = f(x) = \sqrt{25 - x^2}$ we have

$$\frac{dy}{dx} = -\frac{x}{y} = -\frac{x}{\sqrt{25 - x^2}}$$

whereas for $y = g(x) = -\sqrt{25 - x^2}$ we have

$$\frac{dy}{dx} = -\frac{x}{y} = -\frac{x}{-\sqrt{25 - x^2}} = \frac{x}{\sqrt{25 - x^2}}$$

V EXAMPLE 2

(a) Find y' if $x^3 + y^3 = 6xy$.

(b) Find the tangent to the folium of Descartes $x^3 + y^3 = 6xy$ at the point $(3, 3)$.

(c) At what point in the first quadrant is the tangent line horizontal?

SOLUTION

(a) Differentiating both sides of $x^3 + y^3 = 6xy$ with respect to x, regarding y as a function of x, and using the Chain Rule on the term y^3 and the Product Rule on the term $6xy$, we get

$$3x^2 + 3y^2y' = 6xy' + 6y$$

or

$$x^2 + y^2y' = 2xy' + 2y$$

We now solve for y':

$$y^2y' - 2xy' = 2y - x^2$$

$$(y^2 - 2x)y' = 2y - x^2$$

$$y' = \frac{2y - x^2}{y^2 - 2x}$$

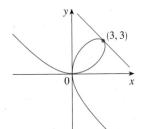

(3, 3)

FIGURE 4

(b) When $x = y = 3$,

$$y' = \frac{2 \cdot 3 - 3^2}{3^2 - 2 \cdot 3} = -1$$

and a glance at Figure 4 confirms that this is a reasonable value for the slope at $(3, 3)$. So an equation of the tangent to the folium at $(3, 3)$ is

$$y - 3 = -1(x - 3) \qquad \text{or} \qquad x + y = 6$$

(c) The tangent line is horizontal if $y' = 0$. Using the expression for y' from part (a), we see that $y' = 0$ when $2y - x^2 = 0$ (provided that $y^2 - 2x \neq 0$). Substituting $y = \frac{1}{2}x^2$ in the equation of the curve, we get

$$x^3 + \left(\tfrac{1}{2}x^2\right)^3 = 6x\left(\tfrac{1}{2}x^2\right)$$

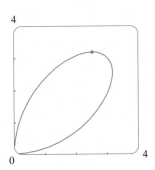

FIGURE 5

which simplifies to $x^6 = 16x^3$. Since $x \neq 0$ in the first quadrant, we have $x^3 = 16$. If $x = 16^{1/3} = 2^{4/3}$, then $y = \frac{1}{2}(2^{8/3}) = 2^{5/3}$. Thus the tangent is horizontal at $(2^{4/3}, 2^{5/3})$, which is approximately $(2.5198, 3.1748)$. Looking at Figure 5, we see that our answer is reasonable. ▬

NOTE 2 There is a formula for the three roots of a cubic equation that is like the quadratic formula but much more complicated. If we use this formula (or a computer algebra system) to solve the equation $x^3 + y^3 = 6xy$ for y in terms of x, we get three functions determined by the equation:

$$y = f(x) = \sqrt[3]{-\tfrac{1}{2}x^3 + \sqrt{\tfrac{1}{4}x^6 - 8x^3}} + \sqrt[3]{-\tfrac{1}{2}x^3 - \sqrt{\tfrac{1}{4}x^6 - 8x^3}}$$

and

$$y = \tfrac{1}{2}\left[-f(x) \pm \sqrt{-3}\left(\sqrt[3]{-\tfrac{1}{2}x^3 + \sqrt{\tfrac{1}{4}x^6 - 8x^3}} - \sqrt[3]{-\tfrac{1}{2}x^3 - \sqrt{\tfrac{1}{4}x^6 - 8x^3}}\right)\right]$$

(These are the three functions whose graphs are shown in Figure 3.) You can see that the method of implicit differentiation saves an enormous amount of work in cases such as this. Moreover, implicit differentiation works just as easily for equations such as

$$y^5 + 3x^2y^2 + 5x^4 = 12$$

for which it is *impossible* to find a similar expression for y in terms of x.

EXAMPLE 3 Find y' if $\sin(x + y) = y^2 \cos x$.

SOLUTION Differentiating implicitly with respect to x and remembering that y is a function of x, we get

$$\cos(x + y) \cdot (1 + y') = y^2(-\sin x) + (\cos x)(2yy')$$

(Note that we have used the Chain Rule on the left side and the Product Rule and Chain Rule on the right side.) If we collect the terms that involve y', we get

$$\cos(x + y) + y^2 \sin x = (2y \cos x)y' - \cos(x + y) \cdot y'$$

So
$$y' = \frac{y^2 \sin x + \cos(x + y)}{2y \cos x - \cos(x + y)}$$

Figure 6, drawn with the implicit-plotting command of a computer algebra system, shows part of the curve $\sin(x + y) = y^2 \cos x$. As a check on our calculation, notice that $y' = -1$ when $x = y = 0$ and it appears from the graph that the slope is approximately -1 at the origin.

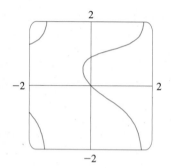

FIGURE 6

Figures 7, 8, and 9 show three more curves produced by a computer algebra system with an implicit-plotting command. In Exercises 41–42 you will have an opportunity to create and examine unusual curves of this nature.

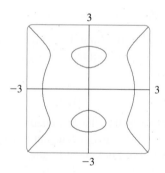

FIGURE 7
$(y^2 - 1)(y^2 - 4) = x^2(x^2 - 4)$

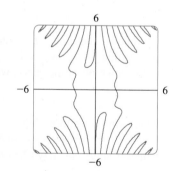

FIGURE 8
$(y^2 - 1)\sin(xy) = x^2 - 4$

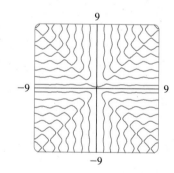

FIGURE 9
$y \sin 3x = x \cos 3y$

The following example shows how to find the second derivative of a function that is defined implicitly.

EXAMPLE 4 Find y'' if $x^4 + y^4 = 16$.

SOLUTION Differentiating the equation implicitly with respect to x, we get

$$4x^3 + 4y^3y' = 0$$

Solving for y' gives

$$\boxed{3} \qquad\qquad y' = -\frac{x^3}{y^3}$$

To find y'' we differentiate this expression for y' using the Quotient Rule and remembering that y is a function of x:

$$y'' = \frac{d}{dx}\left(-\frac{x^3}{y^3}\right) = -\frac{y^3\,(d/dx)(x^3) - x^3\,(d/dx)(y^3)}{(y^3)^2}$$

$$= -\frac{y^3 \cdot 3x^2 - x^3(3y^2 y')}{y^6}$$

If we now substitute Equation 3 into this expression, we get

$$y'' = -\frac{3x^2 y^3 - 3x^3 y^2\left(-\dfrac{x^3}{y^3}\right)}{y^6}$$

$$= -\frac{3(x^2 y^4 + x^6)}{y^7} = -\frac{3x^2(y^4 + x^4)}{y^7}$$

Figure 10 shows the graph of the curve $x^4 + y^4 = 16$ of Example 4. Notice that it's a stretched and flattened version of the circle $x^2 + y^2 = 4$. For this reason it's sometimes called a *fat circle*. It starts out very steep on the left but quickly becomes very flat. This can be seen from the expression

$$y' = -\frac{x^3}{y^3} = -\left(\frac{x}{y}\right)^3$$

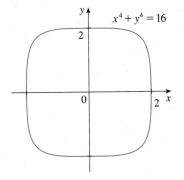

FIGURE 10

But the values of x and y must satisfy the original equation $x^4 + y^4 = 16$. So the answer simplifies to

$$y'' = -\frac{3x^2(16)}{y^7} = -48\,\frac{x^2}{y^7}$$

2.6 Exercises

Access end-of-section exercises and tools for visualizing calculus online at **www.webassign.net**

LABORATORY PROJECT [CAS] **FAMILIES OF IMPLICIT CURVES**

In this project you will explore the changing shapes of implicitly defined curves as you vary the constants in a family, and determine which features are common to all members of the family.

1. Consider the family of curves

$$y^2 - 2x^2(x + 8) = c[(y + 1)^2(y + 9) - x^2]$$

(a) By graphing the curves with $c = 0$ and $c = 2$, determine how many points of intersection there are. (You might have to zoom in to find all of them.)

(b) Now add the curves with $c = 5$ and $c = 10$ to your graphs in part (a). What do you notice? What about other values of c?

[CAS] Computer algebra system required

2. (a) Graph several members of the family of curves

$$x^2 + y^2 + cx^2y^2 = 1$$

Describe how the graph changes as you change the value of c.

(b) What happens to the curve when $c = -1$? Describe what appears on the screen. Can you prove it algebraically?

(c) Find y' by implicit differentiation. For the case $c = -1$, is your expression for y' consistent with what you discovered in part (b)?

Rates of Change in the Natural and Social Sciences

We know that if $y = f(x)$, then the derivative dy/dx can be interpreted as the rate of change of y with respect to x. In this section we examine some of the applications of this idea to physics, chemistry, biology, economics, and other sciences.

Let's recall from Section 2.1 the basic idea behind rates of change. If x changes from x_1 to x_2, then the change in x is

$$\Delta x = x_2 - x_1$$

and the corresponding change in y is

$$\Delta y = f(x_2) - f(x_1)$$

The difference quotient

$$\frac{\Delta y}{\Delta x} = \frac{f(x_2) - f(x_1)}{x_2 - x_1}$$

is the **average rate of change of y with respect to x** over the interval $[x_1, x_2]$ and can be interpreted as the slope of the secant line PQ in Figure 1. Its limit as $\Delta x \to 0$ is the derivative $f'(x_1)$, which can therefore be interpreted as the **instantaneous rate of change of y with respect to x** or the slope of the tangent line at $P(x_1, f(x_1))$. Using Leibniz notation, we write the process in the form

$$\frac{dy}{dx} = \lim_{\Delta x \to 0} \frac{\Delta y}{\Delta x}$$

Whenever the function $y = f(x)$ has a specific interpretation in one of the sciences, its derivative will have a specific interpretation as a rate of change. (As we discussed in Section 2.1, the units for dy/dx are the units for y divided by the units for x.) We now look at some of these interpretations in the natural and social sciences.

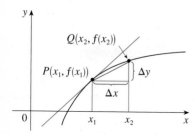

m_{PQ} = average rate of change
$m = f'(x_1)$ = instantaneous rate of change

FIGURE 1

▮ Physics

If $s = f(t)$ is the position function of a particle that is moving in a straight line, then $\Delta s/\Delta t$ represents the average velocity over a time period Δt, and $v = ds/dt$ represents the instantaneous **velocity** (the rate of change of displacement with respect to time). The instantaneous rate of change of velocity with respect to time is **acceleration**: $a(t) = v'(t) = s''(t)$. This was discussed in Sections 2.1 and 2.2, but now that we know the differentiation formulas, we are able to solve problems involving the motion of objects more easily.

V EXAMPLE 1 The position of a particle is given by the equation

$$s = f(t) = t^3 - 6t^2 + 9t$$

where t is measured in seconds and s in meters.
(a) Find the velocity at time t.
(b) What is the velocity after 2 s? After 4 s?
(c) When is the particle at rest?
(d) When is the particle moving forward (that is, in the positive direction)?
(e) Draw a diagram to represent the motion of the particle.
(f) Find the total distance traveled by the particle during the first five seconds.
(g) Find the acceleration at time t and after 4 s.
(h) Graph the position, velocity, and acceleration functions for $0 \leqslant t \leqslant 5$.
(i) When is the particle speeding up? When is it slowing down?

SOLUTION
(a) The velocity function is the derivative of the position function.

$$s = f(t) = t^3 - 6t^2 + 9t$$

$$v(t) = \frac{ds}{dt} = 3t^2 - 12t + 9$$

(b) The velocity after 2 s means the instantaneous velocity when $t = 2$, that is,

$$v(2) = \frac{ds}{dt}\bigg|_{t=2} = 3(2)^2 - 12(2) + 9 = -3 \text{ m/s}$$

The velocity after 4 s is

$$v(4) = 3(4)^2 - 12(4) + 9 = 9 \text{ m/s}$$

(c) The particle is at rest when $v(t) = 0$, that is,

$$3t^2 - 12t + 9 = 3(t^2 - 4t + 3) = 3(t - 1)(t - 3) = 0$$

and this is true when $t = 1$ or $t = 3$. Thus the particle is at rest after 1 s and after 3 s.
(d) The particle moves in the positive direction when $v(t) > 0$, that is,

$$3t^2 - 12t + 9 = 3(t - 1)(t - 3) > 0$$

This inequality is true when both factors are positive ($t > 3$) or when both factors are negative ($t < 1$). Thus the particle moves in the positive direction in the time intervals $t < 1$ and $t > 3$. It moves backward (in the negative direction) when $1 < t < 3$.

(e) Using the information from part (d) we make a schematic sketch in Figure 2 of the motion of the particle back and forth along a line (the s-axis).

(f) Because of what we learned in parts (d) and (e), we need to calculate the distances traveled during the time intervals [0, 1], [1, 3], and [3, 5] separately.
  The distance traveled in the first second is

$$|f(1) - f(0)| = |4 - 0| = 4 \text{ m}$$

From $t = 1$ to $t = 3$ the distance traveled is

$$|f(3) - f(1)| = |0 - 4| = 4 \text{ m}$$

$t = 3$
$s = 0$

$t = 0$ $t = 1$
$s = 0$ $s = 4$

FIGURE 2

From $t = 3$ to $t = 5$ the distance traveled is

$$|f(5) - f(3)| = |20 - 0| = 20 \text{ m}$$

The total distance is $4 + 4 + 20 = 28$ m.

(g) The acceleration is the derivative of the velocity function:

$$a(t) = \frac{d^2s}{dt^2} = \frac{dv}{dt} = 6t - 12$$

$$a(4) = 6(4) - 12 = 12 \text{ m/s}^2$$

(h) Figure 3 shows the graphs of s, v, and a.

(i) The particle speeds up when the velocity is positive and increasing (v and a are both positive) and also when the velocity is negative and decreasing (v and a are both negative). In other words, the particle speeds up when the velocity and acceleration have the same sign. (The particle is pushed in the same direction it is moving.) From Figure 3 we see that this happens when $1 < t < 2$ and when $t > 3$. The particle slows down when v and a have opposite signs, that is, when $0 \le t < 1$ and when $2 < t < 3$. Figure 4 summarizes the motion of the particle.

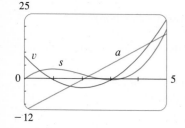

FIGURE 3

TEC In Module 2.7 you can see an animation of Figure 4 with an expression for s that you can choose yourself.

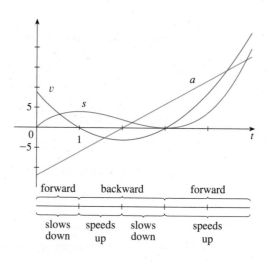

FIGURE 4

EXAMPLE 2 If a rod or piece of wire is homogeneous, then its linear density is uniform and is defined as the mass per unit length ($\rho = m/l$) and measured in kilograms per meter. Suppose, however, that the rod is not homogeneous but that its mass measured from its left end to a point x is $m = f(x)$, as shown in Figure 5.

FIGURE 5   This part of the rod has mass $f(x)$.

The mass of the part of the rod that lies between $x = x_1$ and $x = x_2$ is given by $\Delta m = f(x_2) - f(x_1)$, so the average density of that part of the rod is

$$\text{average density} = \frac{\Delta m}{\Delta x} = \frac{f(x_2) - f(x_1)}{x_2 - x_1}$$

If we now let $\Delta x \to 0$ (that is, $x_2 \to x_1$), we are computing the average density over smaller and smaller intervals. The **linear density** ρ at x_1 is the limit of these average densities as $\Delta x \to 0$; that is, the linear density is the rate of change of mass with respect to length. Symbolically,

$$\rho = \lim_{\Delta x \to 0} \frac{\Delta m}{\Delta x} = \frac{dm}{dx}$$

Thus the linear density of the rod is the derivative of mass with respect to length.

For instance, if $m = f(x) = \sqrt{x}$, where x is measured in meters and m in kilograms, then the average density of the part of the rod given by $1 \le x \le 1.2$ is

$$\frac{\Delta m}{\Delta x} = \frac{f(1.2) - f(1)}{1.2 - 1} = \frac{\sqrt{1.2} - 1}{0.2} \approx 0.48 \text{ kg/m}$$

while the density right at $x = 1$ is

$$\rho = \frac{dm}{dx}\bigg|_{x=1} = \frac{1}{2\sqrt{x}}\bigg|_{x=1} = 0.50 \text{ kg/m}$$

V EXAMPLE 3 A current exists whenever electric charges move. Figure 6 shows part of a wire and electrons moving through a plane surface, shaded red. If ΔQ is the net charge that passes through this surface during a time period Δt, then the average current during this time interval is defined as

FIGURE 6

$$\text{average current} = \frac{\Delta Q}{\Delta t} = \frac{Q_2 - Q_1}{t_2 - t_1}$$

If we take the limit of this average current over smaller and smaller time intervals, we get what is called the **current** I at a given time t_1:

$$I = \lim_{\Delta t \to 0} \frac{\Delta Q}{\Delta t} = \frac{dQ}{dt}$$

Thus the current is the rate at which charge flows through a surface. It is measured in units of charge per unit time (often coulombs per second, called amperes).

Velocity, density, and current are not the only rates of change that are important in physics. Others include power (the rate at which work is done), the rate of heat flow, temperature gradient (the rate of change of temperature with respect to position), and the rate of decay of a radioactive substance in nuclear physics.

Chemistry

EXAMPLE 4 A chemical reaction results in the formation of one or more substances (called *products*) from one or more starting materials (called *reactants*). For instance, the "equation"

$$2H_2 + O_2 \to 2H_2O$$

indicates that two molecules of hydrogen and one molecule of oxygen form two molecules of water. Let's consider the reaction

$$A + B \to C$$

where A and B are the reactants and C is the product. The **concentration** of a reactant A is the number of moles (1 mole $= 6.022 \times 10^{23}$ molecules) per liter and is denoted by [A]. The concentration varies during a reaction, so [A], [B], and [C] are all functions of time (t). The average rate of reaction of the product C over a time interval $t_1 \leqslant t \leqslant t_2$ is

$$\frac{\Delta[C]}{\Delta t} = \frac{[C](t_2) - [C](t_1)}{t_2 - t_1}$$

But chemists are more interested in the **instantaneous rate of reaction**, which is obtained by taking the limit of the average rate of reaction as the time interval Δt approaches 0:

$$\text{rate of reaction} = \lim_{\Delta t \to 0} \frac{\Delta[C]}{\Delta t} = \frac{d[C]}{dt}$$

Since the concentration of the product increases as the reaction proceeds, the derivative $d[C]/dt$ will be positive, and so the rate of reaction of C is positive. The concentrations of the reactants, however, decrease during the reaction, so, to make the rates of reaction of A and B positive numbers, we put minus signs in front of the derivatives $d[A]/dt$ and $d[B]/dt$. Since [A] and [B] each decrease at the same rate that [C] increases, we have

$$\text{rate of reaction} = \frac{d[C]}{dt} = -\frac{d[A]}{dt} = -\frac{d[B]}{dt}$$

More generally, it turns out that for a reaction of the form

$$a\text{A} + b\text{B} \to c\text{C} + d\text{D}$$

we have

$$-\frac{1}{a}\frac{d[A]}{dt} = -\frac{1}{b}\frac{d[B]}{dt} = \frac{1}{c}\frac{d[C]}{dt} = \frac{1}{d}\frac{d[D]}{dt}$$

The rate of reaction can be determined from data and graphical methods. In some cases there are explicit formulas for the concentrations as functions of time, which enable us to compute the rate of reaction (see Exercise 24).

EXAMPLE 5 One of the quantities of interest in thermodynamics is compressibility. If a given substance is kept at a constant temperature, then its volume V depends on its pressure P. We can consider the rate of change of volume with respect to pressure—namely, the derivative dV/dP. As P increases, V decreases, so $dV/dP < 0$. The **compressibility** is defined by introducing a minus sign and dividing this derivative by the volume V:

$$\text{isothermal compressibility} = \beta = -\frac{1}{V}\frac{dV}{dP}$$

Thus β measures how fast, per unit volume, the volume of a substance decreases as the pressure on it increases at constant temperature.

For instance, the volume V (in cubic meters) of a sample of air at 25°C was found to be related to the pressure P (in kilopascals) by the equation

$$V = \frac{5.3}{P}$$

The rate of change of V with respect to P when $P = 50$ kPa is

$$\left. \frac{dV}{dP} \right|_{P=50} = \left. -\frac{5.3}{P^2} \right|_{P=50}$$

$$= -\frac{5.3}{2500} = -0.00212 \ \text{m}^3/\text{kPa}$$

The compressibility at that pressure is

$$\beta = -\frac{1}{V} \left. \frac{dV}{dP} \right|_{P=50} = \frac{0.00212}{\dfrac{5.3}{50}} = 0.02 \ (\text{m}^3/\text{kPa})/\text{m}^3$$

Biology

EXAMPLE 6 Let $n = f(t)$ be the number of individuals in an animal or plant population at time t. The change in the population size between the times $t = t_1$ and $t = t_2$ is $\Delta n = f(t_2) - f(t_1)$, and so the average rate of growth during the time period $t_1 \le t \le t_2$ is

$$\text{average rate of growth} = \frac{\Delta n}{\Delta t} = \frac{f(t_2) - f(t_1)}{t_2 - t_1}$$

The **instantaneous rate of growth** is obtained from this average rate of growth by letting the time period Δt approach 0:

$$\text{growth rate} = \lim_{\Delta t \to 0} \frac{\Delta n}{\Delta t} = \frac{dn}{dt}$$

Strictly speaking, this is not quite accurate because the actual graph of a population function $n = f(t)$ would be a step function that is discontinuous whenever a birth or death occurs and therefore not differentiable. However, for a large animal or plant population, we can replace the graph by a smooth approximating curve as in Figure 7.

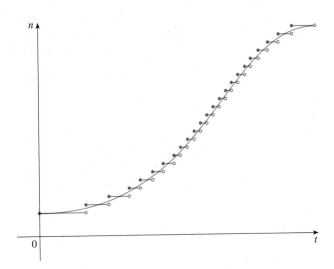

FIGURE 7

A smooth curve approximating
a growth function

E. coli bacteria are about 2 micrometers (μm) long and 0.75 μm wide. The image was produced with a scanning electron microscope.

To be more specific, consider a population of bacteria in a homogeneous nutrient medium. Suppose that by sampling the population at certain intervals it is determined that the population doubles every hour. If the initial population is n_0 and the time t is measured in hours, then

$$f(1) = 2f(0) = 2n_0$$
$$f(2) = 2f(1) = 2^2n_0$$
$$f(3) = 2f(2) = 2^3n_0$$

and, in general,

$$f(t) = 2^t n_0$$

The population function is $n = n_0 2^t$.

This is an example of an exponential function. In Chapter 6 we will discuss exponential functions in general; at that time we will be able to compute their derivatives and thereby determine the rate of growth of the bacteria population.

EXAMPLE 7 When we consider the flow of blood through a blood vessel, such as a vein or artery, we can model the shape of the blood vessel by a cylindrical tube with radius R and length l as illustrated in Figure 8.

FIGURE 8
Blood flow in an artery

Because of friction at the walls of the tube, the velocity v of the blood is greatest along the central axis of the tube and decreases as the distance r from the axis increases until v becomes 0 at the wall. The relationship between v and r is given by the **law of laminar flow** discovered by the French physician Jean-Louis-Marie Poiseuille in 1840. This law states that

$$\boxed{1} \qquad v = \frac{P}{4\eta l}(R^2 - r^2)$$

For more detailed information, see W. Nichols and M. O'Rourke (eds.), *McDonald's Blood Flow in Arteries: Theoretical, Experimental, and Clinical Principles*, 5th ed. (New York, 2005).

where η is the viscosity of the blood and P is the pressure difference between the ends of the tube. If P and l are constant, then v is a function of r with domain $[0, R]$.

The average rate of change of the velocity as we move from $r = r_1$ outward to $r = r_2$ is given by

$$\frac{\Delta v}{\Delta r} = \frac{v(r_2) - v(r_1)}{r_2 - r_1}$$

and if we let $\Delta r \to 0$, we obtain the **velocity gradient**, that is, the instantaneous rate of change of velocity with respect to r:

$$\text{velocity gradient} = \lim_{\Delta r \to 0} \frac{\Delta v}{\Delta r} = \frac{dv}{dr}$$

Using Equation 1, we obtain

$$\frac{dv}{dr} = \frac{P}{4\eta l}(0 - 2r) = -\frac{Pr}{2\eta l}$$

For one of the smaller human arteries we can take $\eta = 0.027$, $R = 0.008$ cm, $l = 2$ cm, and $P = 4000$ dynes/cm^2, which gives

$$v = \frac{4000}{4(0.027)2}(0.000064 - r^2)$$

$$\approx 1.85 \times 10^4 (6.4 \times 10^{-5} - r^2)$$

At $r = 0.002$ cm the blood is flowing at a speed of

$$v(0.002) \approx 1.85 \times 10^4 (64 \times 10^{-6} - 4 \times 10^{-6})$$

$$= 1.11 \text{ cm/s}$$

and the velocity gradient at that point is

$$\left.\frac{dv}{dr}\right|_{r=0.002} = -\frac{4000(0.002)}{2(0.027)2} \approx -74 \text{ (cm/s)/cm}$$

To get a feeling for what this statement means, let's change our units from centimeters to micrometers (1 cm = 10,000 μm). Then the radius of the artery is 80 μm. The velocity at the central axis is 11,850 μm/s, which decreases to 11,110 μm/s at a distance of $r = 20$ μm. The fact that $dv/dr = -74$ (μm/s)/μm means that, when $r = 20$ μm, the velocity is decreasing at a rate of about 74 μm/s for each micrometer that we proceed away from the center.

Economics

V EXAMPLE 8 Suppose $C(x)$ is the total cost that a company incurs in producing x units of a certain commodity. The function C is called a **cost function**. If the number of items produced is increased from x_1 to x_2, then the additional cost is $\Delta C = C(x_2) - C(x_1)$, and the average rate of change of the cost is

$$\frac{\Delta C}{\Delta x} = \frac{C(x_2) - C(x_1)}{x_2 - x_1} = \frac{C(x_1 + \Delta x) - C(x_1)}{\Delta x}$$

The limit of this quantity as $\Delta x \to 0$, that is, the instantaneous rate of change of cost with respect to the number of items produced, is called the **marginal cost** by economists:

$$\text{marginal cost} = \lim_{\Delta x \to 0} \frac{\Delta C}{\Delta x} = \frac{dC}{dx}$$

[Since x often takes on only integer values, it may not make literal sense to let Δx approach 0, but we can always replace $C(x)$ by a smooth approximating function as in Example 6.]

Taking $\Delta x = 1$ and n large (so that Δx is small compared to n), we have

$$C'(n) \approx C(n + 1) - C(n)$$

Thus the marginal cost of producing n units is approximately equal to the cost of producing one more unit [the $(n + 1)$st unit].

It is often appropriate to represent a total cost function by a polynomial

$$C(x) = a + bx + cx^2 + dx^3$$

where a represents the overhead cost (rent, heat, maintenance) and the other terms represent the cost of raw materials, labor, and so on. (The cost of raw materials may be proportional to x, but labor costs might depend partly on higher powers of x because of overtime costs and inefficiencies involved in large-scale operations.)

For instance, suppose a company has estimated that the cost (in dollars) of producing x items is

$$C(x) = 10,000 + 5x + 0.01x^2$$

Then the marginal cost function is

$$C'(x) = 5 + 0.02x$$

The marginal cost at the production level of 500 items is

$$C'(500) = 5 + 0.02(500) = \$15/\text{item}$$

This gives the rate at which costs are increasing with respect to the production level when $x = 500$ and predicts the cost of the 501st item.

The actual cost of producing the 501st item is

$$C(501) - C(500) = [10,000 + 5(501) + 0.01(501)^2]$$
$$- [10,000 + 5(500) + 0.01(500)^2]$$
$$= \$15.01$$

Notice that $C'(500) \approx C(501) - C(500)$. ▬

Economists also study marginal demand, marginal revenue, and marginal profit, which are the derivatives of the demand, revenue, and profit functions. These will be considered in Chapter 3 after we have developed techniques for finding the maximum and minimum values of functions.

▬ Other Sciences

Rates of change occur in all the sciences. A geologist is interested in knowing the rate at which an intruded body of molten rock cools by conduction of heat into surrounding rocks. An engineer wants to know the rate at which water flows into or out of a reservoir. An urban geographer is interested in the rate of change of the population density in a city as the distance from the city center increases. A meteorologist is concerned with the rate of change of atmospheric pressure with respect to height (see Exercise 17 in Section 6.5).

In psychology, those interested in learning theory study the so-called learning curve, which graphs the performance $P(t)$ of someone learning a skill as a function of the training time t. Of particular interest is the rate at which performance improves as time passes, that is, dP/dt.

In sociology, differential calculus is used in analyzing the spread of rumors (or innovations or fads or fashions). If $p(t)$ denotes the proportion of a population that knows a rumor by time t, then the derivative dp/dt represents the rate of spread of the rumor (see Exercise 63 in Section 6.2).

▬ A Single Idea, Many Interpretations

Velocity, density, current, power, and temperature gradient in physics; rate of reaction and compressibility in chemistry; rate of growth and blood velocity gradient in biology; marginal cost and marginal profit in economics; rate of heat flow in geology; rate of improvement of performance in psychology; rate of spread of a rumor in sociology—these are all special cases of a single mathematical concept, the derivative.

This is an illustration of the fact that part of the power of mathematics lies in its abstractness. A single abstract mathematical concept (such as the derivative) can have different interpretations in each of the sciences. When we develop the properties of the mathematical concept once and for all, we can then turn around and apply these results to all of the sciences. This is much more efficient than developing properties of special concepts in each separate science. The French mathematician Joseph Fourier (1768–1830) put it succinctly: "Mathematics compares the most diverse phenomena and discovers the secret analogies that unite them."

2.7   Exercises

Access end-of-section exercises and tools for visualizing calculus online at **www.webassign.net**

2.8   Related Rates

If we are pumping air into a balloon, both the volume and the radius of the balloon are increasing and their rates of increase are related to each other. But it is much easier to measure directly the rate of increase of the volume than the rate of increase of the radius.

In a related rates problem the idea is to compute the rate of change of one quantity in terms of the rate of change of another quantity (which may be more easily measured). The procedure is to find an equation that relates the two quantities and then use the Chain Rule to differentiate both sides with respect to time.

V   EXAMPLE 1   Air is being pumped into a spherical balloon so that its volume increases at a rate of 100 cm³/s. How fast is the radius of the balloon increasing when the diameter is 50 cm?

SOLUTION   We start by identifying two things:

the *given information:*

the rate of increase of the volume of air is 100 cm³/s

and the *unknown:*

the rate of increase of the radius when the diameter is 50 cm

In order to express these quantities mathematically, we introduce some suggestive *notation:*

Let V be the volume of the balloon and let r be its radius.

The key thing to remember is that rates of change are derivatives. In this problem, the volume and the radius are both functions of the time t. The rate of increase of the volume with respect to time is the derivative dV/dt, and the rate of increase of the radius is dr/dt. We can therefore restate the given and the unknown as follows:

$$\text{Given:} \quad \frac{dV}{dt} = 100 \text{ cm}^3/\text{s}$$

$$\text{Unknown:} \quad \frac{dr}{dt} \quad \text{when } r = 25 \text{ cm}$$

In order to connect dV/dt and dr/dt, we first relate V and r by the formula for the volume of a sphere:

$$V = \tfrac{4}{3}\pi r^3$$

PS According to the Principles of Problem Solving discussed on page 77, the first step is to understand the problem. This includes reading the problem carefully, identifying the given and the unknown, and introducing suitable notation.

PS The second stage of problem solving is to think of a plan for connecting the given and the unknown.

In order to use the given information, we differentiate each side of this equation with respect to t. To differentiate the right side, we need to use the Chain Rule:

$$\frac{dV}{dt} = \frac{dV}{dr}\frac{dr}{dt} = 4\pi r^2 \frac{dr}{dt}$$

Now we solve for the unknown quantity:

$$\frac{dr}{dt} = \frac{1}{4\pi r^2}\frac{dV}{dt}$$

If we put $r = 25$ and $dV/dt = 100$ in this equation, we obtain

$$\frac{dr}{dt} = \frac{1}{4\pi(25)^2}100 = \frac{1}{25\pi}$$

The radius of the balloon is increasing at the rate of $1/(25\pi) \approx 0.0127$ cm/s.

EXAMPLE 2 A ladder 10 ft long rests against a vertical wall. If the bottom of the ladder slides away from the wall at a rate of 1 ft/s, how fast is the top of the ladder sliding down the wall when the bottom of the ladder is 6 ft from the wall?

SOLUTION We first draw a diagram and label it as in Figure 1. Let x feet be the distance from the bottom of the ladder to the wall and y feet the distance from the top of the ladder to the ground. Note that x and y are both functions of t (time, measured in seconds).

We are given that $dx/dt = 1$ ft/s and we are asked to find dy/dt when $x = 6$ ft (see Figure 2). In this problem, the relationship between x and y is given by the Pythagorean Theorem:

$$x^2 + y^2 = 100$$

Differentiating each side with respect to t using the Chain Rule, we have

$$2x\frac{dx}{dt} + 2y\frac{dy}{dt} = 0$$

and solving this equation for the desired rate, we obtain

$$\frac{dy}{dt} = -\frac{x}{y}\frac{dx}{dt}$$

When $x = 6$, the Pythagorean Theorem gives $y = 8$ and so, substituting these values and $dx/dt = 1$, we have

$$\frac{dy}{dt} = -\frac{6}{8}(1) = -\frac{3}{4}\text{ ft/s}$$

The fact that dy/dt is negative means that the distance from the top of the ladder to the ground is *decreasing* at a rate of $\frac{3}{4}$ ft/s. In other words, the top of the ladder is sliding down the wall at a rate of $\frac{3}{4}$ ft/s.

EXAMPLE 3 A water tank has the shape of an inverted circular cone with base radius 2 m and height 4 m. If water is being pumped into the tank at a rate of 2 m³/min, find the rate at which the water level is rising when the water is 3 m deep.

SOLUTION We first sketch the cone and label it as in Figure 3. Let V, r, and h be the volume of the water, the radius of the surface, and the height of the water at time t, where t is measured in minutes.

Notice that, although dV/dt is constant, dr/dt is *not* constant.

FIGURE 1

FIGURE 2

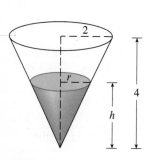

FIGURE 3

We are given that $dV/dt = 2$ m³/min and we are asked to find dh/dt when h is 3 m. The quantities V and h are related by the equation

$$V = \tfrac{1}{3}\pi r^2 h$$

but it is very useful to express V as a function of h alone. In order to eliminate r, we use the similar triangles in Figure 3 to write

$$\frac{r}{h} = \frac{2}{4} \qquad r = \frac{h}{2}$$

and the expression for V becomes

$$V = \frac{1}{3}\pi\left(\frac{h}{2}\right)^2 h = \frac{\pi}{12}h^3$$

Now we can differentiate each side with respect to t:

$$\frac{dV}{dt} = \frac{\pi}{4}h^2\frac{dh}{dt}$$

so

$$\frac{dh}{dt} = \frac{4}{\pi h^2}\frac{dV}{dt}$$

Substituting $h = 3$ m and $dV/dt = 2$ m³/min, we have

$$\frac{dh}{dt} = \frac{4}{\pi(3)^2}\cdot 2 = \frac{8}{9\pi}$$

The water level is rising at a rate of $8/(9\pi) \approx 0.28$ m/min.

PS Look back: What have we learned from Examples 1–3 that will help us solve future problems?

Problem Solving Strategy It is useful to recall some of the problem-solving principles from page 77 and adapt them to related rates in light of our experience in Examples 1–3:

1. Read the problem carefully.
2. Draw a diagram if possible.
3. Introduce notation. Assign symbols to all quantities that are functions of time.
4. Express the given information and the required rate in terms of derivatives.
5. Write an equation that relates the various quantities of the problem. If necessary, use the geometry of the situation to eliminate one of the variables by substitution (as in Example 3).
6. Use the Chain Rule to differentiate both sides of the equation with respect to t.
7. Substitute the given information into the resulting equation and solve for the unknown rate.

⊘ **WARNING** A common error is to substitute the given numerical information (for quantities that vary with time) too early. This should be done only *after* the differentiation. (Step 7 follows Step 6.) For instance, in Example 3 we dealt with general values of h until we finally substituted $h = 3$ at the last stage. (If we had put $h = 3$ earlier, we would have gotten $dV/dt = 0$, which is clearly wrong.)

The following examples are further illustrations of the strategy.

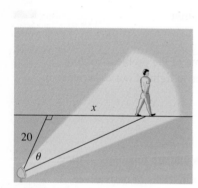

FIGURE 4

V EXAMPLE 4 Car A is traveling west at 50 mi/h and car B is traveling north at 60 mi/h. Both are headed for the intersection of the two roads. At what rate are the cars approaching each other when car A is 0.3 mi and car B is 0.4 mi from the intersection?

SOLUTION We draw Figure 4, where C is the intersection of the roads. At a given time t, let x be the distance from car A to C, let y be the distance from car B to C, and let z be the distance between the cars, where x, y, and z are measured in miles.

We are given that $dx/dt = -50$ mi/h and $dy/dt = -60$ mi/h. (The derivatives are negative because x and y are decreasing.) We are asked to find dz/dt. The equation that relates x, y, and z is given by the Pythagorean Theorem:

$$z^2 = x^2 + y^2$$

Differentiating each side with respect to t, we have

$$2z\frac{dz}{dt} = 2x\frac{dx}{dt} + 2y\frac{dy}{dt}$$

$$\frac{dz}{dt} = \frac{1}{z}\left(x\frac{dx}{dt} + y\frac{dy}{dt}\right)$$

When $x = 0.3$ mi and $y = 0.4$ mi, the Pythagorean Theorem gives $z = 0.5$ mi, so

$$\frac{dz}{dt} = \frac{1}{0.5}[0.3(-50) + 0.4(-60)]$$

$$= -78 \text{ mi/h}$$

The cars are approaching each other at a rate of 78 mi/h.   ▬

V EXAMPLE 5 A man walks along a straight path at a speed of 4 ft/s. A searchlight is located on the ground 20 ft from the path and is kept focused on the man. At what rate is the searchlight rotating when the man is 15 ft from the point on the path closest to the searchlight?

SOLUTION We draw Figure 5 and let x be the distance from the man to the point on the path closest to the searchlight. We let θ be the angle between the beam of the searchlight and the perpendicular to the path.

We are given that $dx/dt = 4$ ft/s and are asked to find $d\theta/dt$ when $x = 15$. The equation that relates x and θ can be written from Figure 5:

$$\frac{x}{20} = \tan\theta \qquad x = 20\tan\theta$$

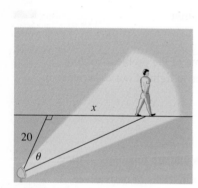

FIGURE 5

Differentiating each side with respect to t, we get

$$\frac{dx}{dt} = 20\sec^2\theta\,\frac{d\theta}{dt}$$

so

$$\frac{d\theta}{dt} = \frac{1}{20}\cos^2\theta\,\frac{dx}{dt}$$

$$= \frac{1}{20}\cos^2\theta\,(4) = \frac{1}{5}\cos^2\theta$$

When $x = 15$, the length of the beam is 25, so $\cos \theta = \frac{4}{5}$ and

$$\frac{d\theta}{dt} = \frac{1}{5}\left(\frac{4}{5}\right)^2 = \frac{16}{125} = 0.128$$

The searchlight is rotating at a rate of 0.128 rad/s.

2.8 Exercises

Access end-of-section exercises and tools for visualizing calculus online at **www.webassign.net**

2.9 Linear Approximations and Differentials

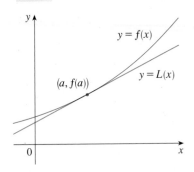

FIGURE 1

We have seen that a curve lies very close to its tangent line near the point of tangency. In fact, by zooming in toward a point on the graph of a differentiable function, we noticed that the graph looks more and more like its tangent line. (See Figure 2 in Section 2.1.) This observation is the basis for a method of finding approximate values of functions.

The idea is that it might be easy to calculate a value $f(a)$ of a function, but difficult (or even impossible) to compute nearby values of f. So we settle for the easily computed values of the linear function L whose graph is the tangent line of f at $(a, f(a))$. (See Figure 1.)

In other words, we use the tangent line at $(a, f(a))$ as an approximation to the curve $y = f(x)$ when x is near a. An equation of this tangent line is

$$y = f(a) + f'(a)(x - a)$$

and the approximation

$$\boxed{1} \qquad f(x) \approx f(a) + f'(a)(x - a)$$

is called the **linear approximation** or **tangent line approximation** of f at a. The linear function whose graph is this tangent line, that is,

$$\boxed{2} \qquad L(x) = f(a) + f'(a)(x - a)$$

is called the **linearization** of f at a.

V **EXAMPLE 1** Find the linearization of the function $f(x) = \sqrt{x + 3}$ at $a = 1$ and use it to approximate the numbers $\sqrt{3.98}$ and $\sqrt{4.05}$. Are these approximations overestimates or underestimates?

SOLUTION The derivative of $f(x) = (x + 3)^{1/2}$ is

$$f'(x) = \tfrac{1}{2}(x + 3)^{-1/2} = \frac{1}{2\sqrt{x + 3}}$$

and so we have $f(1) = 2$ and $f'(1) = \frac{1}{4}$. Putting these values into Equation 2, we see that the linearization is

$$L(x) = f(1) + f'(1)(x - 1) = 2 + \tfrac{1}{4}(x - 1) = \frac{7}{4} + \frac{x}{4}$$

The corresponding linear approximation $\boxed{1}$ is

$$\sqrt{x + 3} \approx \frac{7}{4} + \frac{x}{4} \qquad \text{(when } x \text{ is near 1)}$$

FIGURE 2

In particular, we have

$$\sqrt{3.98} \approx \tfrac{7}{4} + \tfrac{0.98}{4} = 1.995 \qquad \text{and} \qquad \sqrt{4.05} \approx \tfrac{7}{4} + \tfrac{1.05}{4} = 2.0125$$

The linear approximation is illustrated in Figure 2. We see that, indeed, the tangent line approximation is a good approximation to the given function when x is near 1. We also see that our approximations are overestimates because the tangent line lies above the curve.

Of course, a calculator could give us approximations for $\sqrt{3.98}$ and $\sqrt{4.05}$, but the linear approximation gives an approximation *over an entire interval*.   ▬

In the following table we compare the estimates from the linear approximation in Example 1 with the true values. Notice from this table, and also from Figure 2, that the tangent line approximation gives good estimates when x is close to 1 but the accuracy of the approximation deteriorates when x is farther away from 1.

	x	From $L(x)$	Actual value
$\sqrt{3.9}$	0.9	1.975	1.97484176 ...
$\sqrt{3.98}$	0.98	1.995	1.99499373 ...
$\sqrt{4}$	1	2	2.00000000 ...
$\sqrt{4.05}$	1.05	2.0125	2.01246117 ...
$\sqrt{4.1}$	1.1	2.025	2.02484567 ...
$\sqrt{5}$	2	2.25	2.23606797 ...
$\sqrt{6}$	3	2.5	2.44948974 ...

How good is the approximation that we obtained in Example 1? The next example shows that by using a graphing calculator or computer we can determine an interval throughout which a linear approximation provides a specified accuracy.

EXAMPLE 2   For what values of x is the linear approximation

$$\sqrt{x + 3} \approx \frac{7}{4} + \frac{x}{4}$$

accurate to within 0.5? What about accuracy to within 0.1?

SOLUTION   Accuracy to within 0.5 means that the functions should differ by less than 0.5:

$$\left| \sqrt{x + 3} - \left(\frac{7}{4} + \frac{x}{4} \right) \right| < 0.5$$

Equivalently, we could write

$$\sqrt{x + 3} - 0.5 < \frac{7}{4} + \frac{x}{4} < \sqrt{x + 3} + 0.5$$

This says that the linear approximation should lie between the curves obtained by shifting the curve $y = \sqrt{x + 3}$ upward and downward by an amount 0.5. Figure 3 shows the tangent line $y = (7 + x)/4$ intersecting the upper curve $y = \sqrt{x + 3} + 0.5$ at P

FIGURE 3

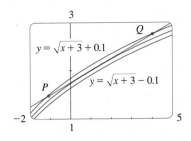

FIGURE 4

and Q. Zooming in and using the cursor, we estimate that the x-coordinate of P is about -2.66 and the x-coordinate of Q is about 8.66. Thus we see from the graph that the approximation

$$\sqrt{x+3} \approx \frac{7}{4} + \frac{x}{4}$$

is accurate to within 0.5 when $-2.6 < x < 8.6$. (We have rounded to be safe.)

Similarly, from Figure 4 we see that the approximation is accurate to within 0.1 when $-1.1 < x < 3.9$. ∎

Applications to Physics

Linear approximations are often used in physics. In analyzing the consequences of an equation, a physicist sometimes needs to simplify a function by replacing it with its linear approximation. For instance, in deriving a formula for the period of a pendulum, physics textbooks obtain the expression $a_T = -g \sin \theta$ for tangential acceleration and then replace $\sin \theta$ by θ with the remark that $\sin \theta$ is very close to θ if θ is not too large. [See, for example, *Physics: Calculus*, 2d ed., by Eugene Hecht (Pacific Grove, CA, 2000), p. 431.] You can verify that the linearization of the function $f(x) = \sin x$ at $a = 0$ is $L(x) = x$ and so the linear approximation at 0 is

$$\sin x \approx x$$

(see Exercise 40). So, in effect, the derivation of the formula for the period of a pendulum uses the tangent line approximation for the sine function.

Another example occurs in the theory of optics, where light rays that arrive at shallow angles relative to the optical axis are called *paraxial rays*. In paraxial (or Gaussian) optics, both $\sin \theta$ and $\cos \theta$ are replaced by their linearizations. In other words, the linear approximations

$$\sin \theta \approx \theta \qquad \text{and} \qquad \cos \theta \approx 1$$

are used because θ is close to 0. The results of calculations made with these approximations became the basic theoretical tool used to design lenses. [See *Optics*, 4th ed., by Eugene Hecht (San Francisco, 2002), p. 154.]

In Section 11.11 we will present several other applications of the idea of linear approximations to physics and engineering.

Differentials

The ideas behind linear approximations are sometimes formulated in the terminology and notation of *differentials*. If $y = f(x)$, where f is a differentiable function, then the **differential** dx is an independent variable; that is, dx can be given the value of any real number. The **differential** dy is then defined in terms of dx by the equation

$$\boxed{3} \qquad\qquad dy = f'(x)\, dx$$

So dy is a dependent variable; it depends on the values of x and dx. If dx is given a specific value and x is taken to be some specific number in the domain of f, then the numerical value of dy is determined.

If $dx \neq 0$, we can divide both sides of Equation 3 by dx to obtain

$$\frac{dy}{dx} = f'(x)$$

We have seen similar equations before, but now the left side can genuinely be interpreted as a ratio of differentials.

FIGURE 5

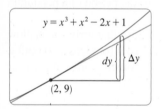

Figure 6 shows the function in Example 3 and a comparison of dy and Δy when $a = 2$. The viewing rectangle is $[1.8, 2.5]$ by $[6, 18]$.

FIGURE 6

The geometric meaning of differentials is shown in Figure 5. Let $P(x, f(x))$ and $Q(x + \Delta x, f(x + \Delta x))$ be points on the graph of f and let $dx = \Delta x$. The corresponding change in y is

$$\Delta y = f(x + \Delta x) - f(x)$$

The slope of the tangent line PR is the derivative $f'(x)$. Thus the directed distance from S to R is $f'(x)\, dx = dy$. Therefore dy represents the amount that the tangent line rises or falls (the change in the linearization), whereas Δy represents the amount that the curve $y = f(x)$ rises or falls when x changes by an amount dx.

EXAMPLE 3   Compare the values of Δy and dy if $y = f(x) = x^3 + x^2 - 2x + 1$ and x changes (a) from 2 to 2.05 and (b) from 2 to 2.01.

SOLUTION
(a) We have

$$f(2) = 2^3 + 2^2 - 2(2) + 1 = 9$$

$$f(2.05) = (2.05)^3 + (2.05)^2 - 2(2.05) + 1 = 9.717625$$

$$\Delta y = f(2.05) - f(2) = 0.717625$$

In general, $dy = f'(x)\, dx = (3x^2 + 2x - 2)\, dx$

When $x = 2$ and $dx = \Delta x = 0.05$, this becomes

$$dy = [3(2)^2 + 2(2) - 2]0.05 = 0.7$$

(b) $f(2.01) = (2.01)^3 + (2.01)^2 - 2(2.01) + 1 = 9.140701$

$$\Delta y = f(2.01) - f(2) = 0.140701$$

When $dx = \Delta x = 0.01$,

$$dy = [3(2)^2 + 2(2) - 2]0.01 = 0.14$$

Notice that the approximation $\Delta y \approx dy$ becomes better as Δx becomes smaller in Example 3. Notice also that dy was easier to compute than Δy. For more complicated functions it may be impossible to compute Δy exactly. In such cases the approximation by differentials is especially useful.

In the notation of differentials, the linear approximation $\boxed{1}$ can be written as

$$f(a + dx) \approx f(a) + dy$$

For instance, for the function $f(x) = \sqrt{x + 3}$ in Example 1, we have

$$dy = f'(x)\, dx = \frac{dx}{2\sqrt{x + 3}}$$

If $a = 1$ and $dx = \Delta x = 0.05$, then

$$dy = \frac{0.05}{2\sqrt{1 + 3}} = 0.0125$$

and $\sqrt{4.05} = f(1.05) \approx f(1) + dy = 2.0125$

just as we found in Example 1.

Our final example illustrates the use of differentials in estimating the errors that occur because of approximate measurements.

V EXAMPLE 4 The radius of a sphere was measured and found to be 21 cm with a possible error in measurement of at most 0.05 cm. What is the maximum error in using this value of the radius to compute the volume of the sphere?

SOLUTION If the radius of the sphere is r, then its volume is $V = \frac{4}{3}\pi r^3$. If the error in the measured value of r is denoted by $dr = \Delta r$, then the corresponding error in the calculated value of V is ΔV, which can be approximated by the differential

$$dV = 4\pi r^2\, dr$$

When $r = 21$ and $dr = 0.05$, this becomes

$$dV = 4\pi(21)^2 0.05 \approx 277$$

The maximum error in the calculated volume is about 277 cm^3.

NOTE Although the possible error in Example 4 may appear to be rather large, a better picture of the error is given by the **relative error**, which is computed by dividing the error by the total volume:

$$\frac{\Delta V}{V} \approx \frac{dV}{V} = \frac{4\pi r^2\, dr}{\frac{4}{3}\pi r^3} = 3\,\frac{dr}{r}$$

Thus the relative error in the volume is about three times the relative error in the radius. In Example 4 the relative error in the radius is approximately $dr/r = 0.05/21 \approx 0.0024$ and it produces a relative error of about 0.007 in the volume. The errors could also be expressed as **percentage errors** of 0.24% in the radius and 0.7% in the volume.

2.9 Exercises

Access end-of-section exercises and tools for visualizing calculus online at **www.webassign.net**

LABORATORY PROJECT   TAYLOR POLYNOMIALS

The tangent line approximation $L(x)$ is the best first-degree (linear) approximation to $f(x)$ near $x = a$ because $f(x)$ and $L(x)$ have the same rate of change (derivative) at a. For a better approximation than a linear one, let's try a second-degree (quadratic) approximation $P(x)$. In other words, we approximate a curve by a parabola instead of by a straight line. To make sure that the approximation is a good one, we stipulate the following:

(i) $P(a) = f(a)$ (P and f should have the same value at a.)

(ii) $P'(a) = f'(a)$ (P and f should have the same rate of change at a.)

(iii) $P''(a) = f''(a)$ (The slopes of P and f should change at the same rate at a.)

1. Find the quadratic approximation $P(x) = A + Bx + Cx^2$ to the function $f(x) = \cos x$ that satisfies conditions (i), (ii), and (iii) with $a = 0$. Graph P, f, and the linear approximation $L(x) = 1$ on a common screen. Comment on how well the functions P and L approximate f.

Graphing calculator or computer with graphing software required

2. Determine the values of x for which the quadratic approximation $f(x) \approx P(x)$ in Problem 1 is accurate to within 0.1. [*Hint:* Graph $y = P(x)$, $y = \cos x - 0.1$, and $y = \cos x + 0.1$ on a common screen.]

3. To approximate a function f by a quadratic function P near a number a, it is best to write P in the form

$$P(x) = A + B(x - a) + C(x - a)^2$$

Show that the quadratic function that satisfies conditions (i), (ii), and (iii) is

$$P(x) = f(a) + f'(a)(x - a) + \tfrac{1}{2}f''(a)(x - a)^2$$

4. Find the quadratic approximation to $f(x) = \sqrt{x + 3}$ near $a = 1$. Graph f, the quadratic approximation, and the linear approximation from Example 2 in Section 2.9 on a common screen. What do you conclude?

5. Instead of being satisfied with a linear or quadratic approximation to $f(x)$ near $x = a$, let's try to find better approximations with higher-degree polynomials. We look for an nth-degree polynomial

$$T_n(x) = c_0 + c_1(x - a) + c_2(x - a)^2 + c_3(x - a)^3 + \cdots + c_n(x - a)^n$$

such that T_n and its first n derivatives have the same values at $x = a$ as f and its first n derivatives. By differentiating repeatedly and setting $x = a$, show that these conditions are satisfied if $c_0 = f(a)$, $c_1 = f'(a)$, $c_2 = \tfrac{1}{2}f''(a)$, and in general

$$c_k = \frac{f^{(k)}(a)}{k!}$$

where $k! = 1 \cdot 2 \cdot 3 \cdot 4 \cdot \cdots \cdot k$. The resulting polynomial

$$T_n(x) = f(a) + f'(a)(x - a) + \frac{f''(a)}{2!}(x - a)^2 + \cdots + \frac{f^{(n)}(a)}{n!}(x - a)^n$$

is called the **nth-degree Taylor polynomial of f centered at a**.

6. Find the 8th-degree Taylor polynomial centered at $a = 0$ for the function $f(x) = \cos x$. Graph f together with the Taylor polynomials T_2, T_4, T_6, T_8 in the viewing rectangle $[-5, 5]$ by $[-1.4, 1.4]$ and comment on how well they approximate f.

2   Review

Concept Check

1. Write an expression for the slope of the tangent line to the curve $y = f(x)$ at the point $(a, f(a))$.

2. Suppose an object moves along a straight line with position $f(t)$ at time t. Write an expression for the instantaneous velocity of the object at time $t = a$. How can you interpret this velocity in terms of the graph of f?

3. If $y = f(x)$ and x changes from x_1 to x_2, write expressions for the following.
(a) The average rate of change of y with respect to x over the interval $[x_1, x_2]$.
(b) The instantaneous rate of change of y with respect to x at $x = x_1$.

4. Define the derivative $f'(a)$. Discuss two ways of interpreting this number.

5. (a) What does it mean for f to be differentiable at a?
(b) What is the relation between the differentiability and continuity of a function?
(c) Sketch the graph of a function that is continuous but not differentiable at $a = 2$.

6. Describe several ways in which a function can fail to be differentiable. Illustrate with sketches.

7. What are the second and third derivatives of a function f? If f is the position function of an object, how can you interpret f'' and f'''?

8. State each differentiation rule both in symbols and in words.
   (a) The Power Rule   (b) The Constant Multiple Rule
   (c) The Sum Rule   (d) The Difference Rule
   (e) The Product Rule   (f) The Quotient Rule
   (g) The Chain Rule

9. State the derivative of each function.
   (a) $y = x^n$ (b) $y = \sin x$ (c) $y = \cos x$
   (d) $y = \tan x$ (e) $y = \csc x$ (f) $y = \sec x$
   (g) $y = \cot x$

10. Explain how implicit differentiation works.

11. Give several examples of how the derivative can be interpreted as a rate of change in physics, chemistry, biology, economics, or other sciences.

12. (a) Write an expression for the linearization of f at a.
   (b) If $y = f(x)$, write an expression for the differential dy.
   (c) If $dx = \Delta x$, draw a picture showing the geometric meanings of Δy and dy.

True-False Quiz

Determine whether the statement is true or false. If it is true, explain why. If it is false, explain why or give an example that disproves the statement.

1. If f is continuous at a, then f is differentiable at a.

2. If f and g are differentiable, then
$$\frac{d}{dx}[f(x) + g(x)] = f'(x) + g'(x)$$

3. If f and g are differentiable, then
$$\frac{d}{dx}[f(x)g(x)] = f'(x)g'(x)$$

4. If f and g are differentiable, then
$$\frac{d}{dx}[f(g(x))] = f'(g(x))g'(x)$$

5. If f is differentiable, then $\dfrac{d}{dx}\sqrt{f(x)} = \dfrac{f'(x)}{2\sqrt{f(x)}}$.

6. If f is differentiable, then $\dfrac{d}{dx}f(\sqrt{x}) = \dfrac{f'(x)}{2\sqrt{x}}$.

7. $\dfrac{d}{dx}|x^2 + x| = |2x + 1|$

8. If $f'(r)$ exists, then $\lim_{x \to r} f(x) = f(r)$.

9. If $g(x) = x^5$, then $\lim_{x \to 2} \dfrac{g(x) - g(2)}{x - 2} = 80$.

10. $\dfrac{d^2y}{dx^2} = \left(\dfrac{dy}{dx}\right)^2$

11. An equation of the tangent line to the parabola $y = x^2$ at $(-2, 4)$ is $y - 4 = 2x(x + 2)$.

12. $\dfrac{d}{dx}(\tan^2 x) = \dfrac{d}{dx}(\sec^2 x)$

Exercises

1. The displacement (in meters) of an object moving in a straight line is given by $s = 1 + 2t + \frac{1}{4}t^2$, where t is measured in seconds.
   (a) Find the average velocity over each time period.
      (i) $[1, 3]$ (ii) $[1, 2]$
      (iii) $[1, 1.5]$ (iv) $[1, 1.1]$
   (b) Find the instantaneous velocity when $t = 1$.

2. The graph of f is shown. State, with reasons, the numbers at which f is not differentiable.

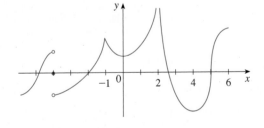

3–4 Trace or copy the graph of the function. Then sketch a graph of its derivative directly beneath.

3.

4.

---

5. The figure shows the graphs of f, f', and f''. Identify each curve, and explain your choices.

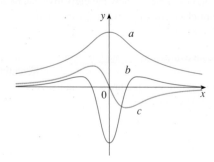

6. Find a function f and a number a such that

$$\lim_{h \to 0} \frac{(2 + h)^6 - 64}{h} = f'(a)$$

7. The total cost of repaying a student loan at an interest rate of $r\%$ per year is $C = f(r)$.
(a) What is the meaning of the derivative $f'(r)$? What are its units?
(b) What does the statement $f'(10) = 1200$ mean?
(c) Is $f'(r)$ always positive or does it change sign?

8. The *total fertility rate* at time t, denoted by $F(t)$, is an estimate of the average number of children born to each woman (assuming that current birth rates remain constant). The graph of the total fertility rate in the United States shows the fluctuations from 1940 to 1990.
(a) Estimate the values of $F'(1950)$, $F'(1965)$, and $F'(1987)$.
(b) What are the meanings of these derivatives?
(c) Can you suggest reasons for the values of these derivatives?

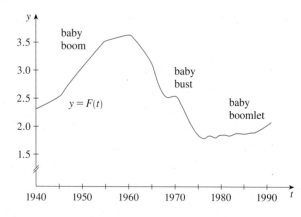

9. Let $C(t)$ be the total value of US currency (coins and banknotes) in circulation at time t. The table gives values of this function from 1980 to 2000, as of September 30, in billions of dollars. Interpret and estimate the value of $C'(1990)$.

t	1980	1985	1990	1995	2000
$C(t)$	129.9	187.3	271.9	409.3	568.6

10–11 Find $f'(x)$ from first principles, that is, directly from the definition of a derivative.

10. $f(x) = \dfrac{4 - x}{3 + x}$

11. $f(x) = x^3 + 5x + 4$

12. (a) If $f(x) = \sqrt{3 - 5x}$, use the definition of a derivative to find $f'(x)$.
(b) Find the domains of f and f'.
(c) Graph f and f' on a common screen. Compare the graphs to see whether your answer to part (a) is reasonable.

13–40 Calculate y'.

13. $y = (x^2 + x^3)^4$

14. $y = \dfrac{1}{\sqrt{x}} - \dfrac{1}{\sqrt[5]{x^3}}$

15. $y = \dfrac{x^2 - x + 2}{\sqrt{x}}$

16. $y = \dfrac{\tan x}{1 + \cos x}$

17. $y = x^2 \sin \pi x$

18. $y = \left(x + \dfrac{1}{x^2}\right)^{\sqrt{7}}$

19. $y = \dfrac{t^4 - 1}{t^4 + 1}$

20. $y = \sin(\cos x)$

21. $y = \tan \sqrt{1 - x}$

22. $y = \dfrac{1}{\sin(x - \sin x)}$

23. $xy^4 + x^2 y = x + 3y$

24. $y = \sec(1 + x^2)$

25. $y = \dfrac{\sec 2\theta}{1 + \tan 2\theta}$

26. $x^2 \cos y + \sin 2y = xy$

27. $y = (1 - x^{-1})^{-1}$

28. $y = 1/\sqrt[3]{x + \sqrt{x}}$

29. $\sin(xy) = x^2 - y$

30. $y = \sqrt{\sin \sqrt{x}}$

31. $y = \cot(3x^2 + 5)$

32. $y = \dfrac{(x + \lambda)^4}{x^4 + \lambda^4}$

33. $y = \sqrt{x} \cos \sqrt{x}$

34. $y = \dfrac{\sin mx}{x}$

35. $y = \tan^2(\sin \theta)$

36. $x \tan y = y - 1$

37. $y = \sqrt[5]{x \tan x}$

38. $y = \dfrac{(x - 1)(x - 4)}{(x - 2)(x - 3)}$

39. $y = \sin(\tan \sqrt{1 + x^3})$

40. $y = \sin^2(\cos \sqrt{\sin \pi x})$

41. If $f(t) = \sqrt{4t + 1}$, find $f''(2)$.

42. If $g(\theta) = \theta \sin \theta$, find $g''(\pi/6)$.

43. Find y'' if $x^6 + y^6 = 1$.

44. Find $f^{(n)}(x)$ if $f(x) = 1/(2 - x)$.

45–46 Find the limit.

45. $\displaystyle\lim_{x\to 0}\frac{\sec x}{1-\sin x}$

46. $\displaystyle\lim_{t\to 0}\frac{t^3}{\tan^3 2t}$

47–48 Find an equation of the tangent to the curve at the given point.

47. $y = 4\sin^2 x$, $(\pi/6, 1)$

48. $y = \dfrac{x^2-1}{x^2+1}$, $(0, -1)$

49–50 Find equations of the tangent line and normal line to the curve at the given point.

49. $y = \sqrt{1+4\sin x}$, $(0, 1)$

50. $x^2 + 4xy + y^2 = 13$, $(2, 1)$

51. (a) If $f(x) = x\sqrt{5-x}$, find $f'(x)$.
   (b) Find equations of the tangent lines to the curve
      $y = x\sqrt{5-x}$ at the points $(1, 2)$ and $(4, 4)$.
    (c) Illustrate part (b) by graphing the curve and tangent lines on the same screen.
    (d) Check to see that your answer to part (a) is reasonable by comparing the graphs of f and f'.

52. (a) If $f(x) = 4x - \tan x$, $-\pi/2 < x < \pi/2$, find f' and f''.
    (b) Check to see that your answers to part (a) are reasonable by comparing the graphs of f, f', and f''.

53. At what points on the curve $y = \sin x + \cos x$, $0 \le x \le 2\pi$, is the tangent line horizontal?

54. Find the points on the ellipse $x^2 + 2y^2 = 1$ where the tangent line has slope 1.

55. Find a parabola $y = ax^2 + bx + c$ that passes through the point $(1, 4)$ and whose tangent lines at $x = -1$ and $x = 5$ have slopes 6 and -2, respectively.

56. How many tangent lines to the curve $y = x/(x+1)$ pass through the point $(1, 2)$? At which points do these tangent lines touch the curve?

57. If $f(x) = (x-a)(x-b)(x-c)$, show that

$$\frac{f'(x)}{f(x)} = \frac{1}{x-a} + \frac{1}{x-b} + \frac{1}{x-c}$$

58. (a) By differentiating the double-angle formula

$$\cos 2x = \cos^2 x - \sin^2 x$$

obtain the double-angle formula for the sine function.
   (b) By differentiating the addition formula

$$\sin(x+a) = \sin x \cos a + \cos x \sin a$$

obtain the addition formula for the cosine function.

59. Suppose that $h(x) = f(x)g(x)$ and $F(x) = f(g(x))$, where $f(2) = 3$, $g(2) = 5$, $g'(2) = 4$, $f'(2) = -2$, and $f'(5) = 11$. Find (a) $h'(2)$ and (b) $F'(2)$.

60. If f and g are the functions whose graphs are shown, let $P(x) = f(x)g(x)$, $Q(x) = f(x)/g(x)$, and $C(x) = f(g(x))$. Find (a) $P'(2)$, (b) $Q'(2)$, and (c) $C'(2)$.

61–68 Find f' in terms of g'.

61. $f(x) = x^2 g(x)$

62. $f(x) = g(x^2)$

63. $f(x) = [g(x)]^2$

64. $f(x) = x^a g(x^b)$

65. $f(x) = g(g(x))$

66. $f(x) = \sin(g(x))$

67. $f(x) = g(\sin x)$

68. $f(x) = g(\tan \sqrt{x})$

69–71 Find h' in terms of f' and g'.

69. $h(x) = \dfrac{f(x)g(x)}{f(x)+g(x)}$

70. $h(x) = \sqrt{\dfrac{f(x)}{g(x)}}$

71. $h(x) = f(g(\sin 4x))$

72. A particle moves along a horizontal line so that its coordinate at time t is $x = \sqrt{b^2 + c^2 t^2}$, $t \ge 0$, where b and c are positive constants.
   (a) Find the velocity and acceleration functions.
   (b) Show that the particle always moves in the positive direction.

73. A particle moves on a vertical line so that its coordinate at time t is $y = t^3 - 12t + 3$, $t \ge 0$.
   (a) Find the velocity and acceleration functions.
   (b) When is the particle moving upward and when is it moving downward?
   (c) Find the distance that the particle travels in the time interval $0 \le t \le 3$.
    (d) Graph the position, velocity, and acceleration functions for $0 \le t \le 3$.
   (e) When is the particle speeding up? When is it slowing down?

74. The volume of a right circular cone is $V = \frac{1}{3}\pi r^2 h$, where r is the radius of the base and h is the height.
   (a) Find the rate of change of the volume with respect to the height if the radius is constant.

(b) Find the rate of change of the volume with respect to the radius if the height is constant.

75. The mass of part of a wire is $x\left(1 + \sqrt{x}\right)$ kilograms, where x is measured in meters from one end of the wire. Find the linear density of the wire when $x = 4$ m.

76. The cost, in dollars, of producing x units of a certain commodity is

$$C(x) = 920 + 2x - 0.02x^2 + 0.00007x^3$$

(a) Find the marginal cost function.
(b) Find $C'(100)$ and explain its meaning.
(c) Compare $C'(100)$ with the cost of producing the 101st item.

77. The volume of a cube is increasing at a rate of 10 cm³/min. How fast is the surface area increasing when the length of an edge is 30 cm?

78. A paper cup has the shape of a cone with height 10 cm and radius 3 cm (at the top). If water is poured into the cup at a rate of 2 cm³/s, how fast is the water level rising when the water is 5 cm deep?

79. A balloon is rising at a constant speed of 5 ft/s. A boy is cycling along a straight road at a speed of 15 ft/s. When he passes under the balloon, it is 45 ft above him. How fast is the distance between the boy and the balloon increasing 3 s later?

80. A waterskier skis over the ramp shown in the figure at a speed of 30 ft/s. How fast is she rising as she leaves the ramp?

4 ft

15 ft

81. The angle of elevation of the sun is decreasing at a rate of 0.25 rad/h. How fast is the shadow cast by a 400-ft-tall building increasing when the angle of elevation of the sun is $\pi/6$?

82. (a) Find the linear approximation to $f(x) = \sqrt{25 - x^2}$ near 3.
(b) Illustrate part (a) by graphing f and the linear approximation.
(c) For what values of x is the linear approximation accurate to within 0.1?

83. (a) Find the linearization of $f(x) = \sqrt[3]{1 + 3x}$ at $a = 0$. State the corresponding linear approximation and use it to give an approximate value for $\sqrt[3]{1.03}$.
(b) Determine the values of x for which the linear approximation given in part (a) is accurate to within 0.1.

84. Evaluate dy if $y = x^3 - 2x^2 + 1$, $x = 2$, and $dx = 0.2$.

85. A window has the shape of a square surmounted by a semicircle. The base of the window is measured as having width 60 cm with a possible error in measurement of 0.1 cm. Use differentials to estimate the maximum error possible in computing the area of the window.

86–88 Express the limit as a derivative and evaluate.

86. $\lim\limits_{x \to 1} \dfrac{x^{17} - 1}{x - 1}$

87. $\lim\limits_{h \to 0} \dfrac{\sqrt[4]{16 + h} - 2}{h}$

88. $\lim\limits_{\theta \to \pi/3} \dfrac{\cos \theta - 0.5}{\theta - \pi/3}$

---

89. Evaluate $\lim\limits_{x \to 0} \dfrac{\sqrt{1 + \tan x} - \sqrt{1 + \sin x}}{x^3}$.

90. Suppose f is a differentiable function such that $f(g(x)) = x$ and $f'(x) = 1 + [f(x)]^2$. Show that $g'(x) = 1/(1 + x^2)$.

91. Find $f'(x)$ if it is known that

$$\frac{d}{dx}[f(2x)] = x^2$$

92. Show that the length of the portion of any tangent line to the astroid $x^{2/3} + y^{2/3} = a^{2/3}$ cut off by the coordinate axes is constant.

Problems Plus

Before you look at the example, cover up the solution and try it yourself first.

EXAMPLE 1 How many lines are tangent to both of the parabolas $y = -1 - x^2$ and $y = 1 + x^2$? Find the coordinates of the points at which these tangents touch the parabolas.

SOLUTION To gain insight into this problem, it is essential to draw a diagram. So we sketch the parabolas $y = 1 + x^2$ (which is the standard parabola $y = x^2$ shifted 1 unit upward) and $y = -1 - x^2$ (which is obtained by reflecting the first parabola about the x-axis). If we try to draw a line tangent to both parabolas, we soon discover that there are only two possibilities, as illustrated in Figure 1.

Let P be a point at which one of these tangents touches the upper parabola and let a be its x-coordinate. (The choice of notation for the unknown is important. Of course we could have used b or c or x_0 or x_1 instead of a. However, it's not advisable to use x in place of a because that x could be confused with the variable x in the equation of the parabola.) Then, since P lies on the parabola $y = 1 + x^2$, its y-coordinate must be $1 + a^2$. Because of the symmetry shown in Figure 1, the coordinates of the point Q where the tangent touches the lower parabola must be $(-a, -(1 + a^2))$.

To use the given information that the line is a tangent, we equate the slope of the line PQ to the slope of the tangent line at P. We have

$$m_{PQ} = \frac{1 + a^2 - (-1 - a^2)}{a - (-a)} = \frac{1 + a^2}{a}$$

If $f(x) = 1 + x^2$, then the slope of the tangent line at P is $f'(a) = 2a$. Thus the condition that we need to use is that

$$\frac{1 + a^2}{a} = 2a$$

Solving this equation, we get $1 + a^2 = 2a^2$, so $a^2 = 1$ and $a = \pm 1$. Therefore the points are $(1, 2)$ and $(-1, -2)$. By symmetry, the two remaining points are $(-1, 2)$ and $(1, -2)$.

FIGURE 1

Problems

1. Find points P and Q on the parabola $y = 1 - x^2$ so that the triangle ABC formed by the x-axis and the tangent lines at P and Q is an equilateral triangle (see the figure).

2. Find the point where the curves $y = x^3 - 3x + 4$ and $y = 3(x^2 - x)$ are tangent to each other, that is, have a common tangent line. Illustrate by sketching both curves and the common tangent.

3. Show that the tangent lines to the parabola $y = ax^2 + bx + c$ at any two points with x-coordinates p and q must intersect at a point whose x-coordinate is halfway between p and q.

4. Show that

$$\frac{d}{dx}\left(\frac{\sin^2 x}{1 + \cot x} + \frac{\cos^2 x}{1 + \tan x}\right) = -\cos 2x$$

5. If $f(x) = \lim_{t \to x} \frac{\sec t - \sec x}{t - x}$, find the value of $f'(\pi/4)$.

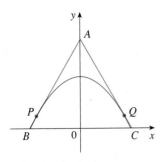

FIGURE FOR PROBLEM 1

Graphing calculator or computer required

CAS Computer algebra system required

6. Find the values of the constants a and b such that

$$\lim_{x \to 0} \frac{\sqrt[3]{ax+b}-2}{x} = \frac{5}{12}$$

7. Prove that $\dfrac{d^n}{dx^n}(\sin^4 x + \cos^4 x) = 4^{n-1}\cos(4x + n\pi/2)$.

8. Find the nth derivative of the function $f(x) = x^n/(1-x)$.

9. The figure shows a circle with radius 1 inscribed in the parabola $y = x^2$. Find the center of the circle.

10. If f is differentiable at a, where $a > 0$, evaluate the following limit in terms of $f'(a)$:

$$\lim_{x \to a} \frac{f(x) - f(a)}{\sqrt{x} - \sqrt{a}}$$

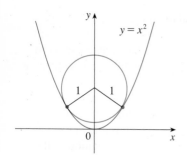

$y = x^2$

FIGURE FOR PROBLEM 9

11. The figure shows a rotating wheel with radius 40 cm and a connecting rod AP with length 1.2 m. The pin P slides back and forth along the x-axis as the wheel rotates counterclockwise at a rate of 360 revolutions per minute.
(a) Find the angular velocity of the connecting rod, $d\alpha/dt$, in radians per second, when $\theta = \pi/3$.
(b) Express the distance $x = |OP|$ in terms of θ.
(c) Find an expression for the velocity of the pin P in terms of θ.

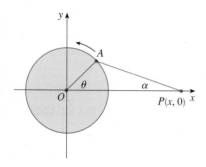

FIGURE FOR PROBLEM 11

12. Tangent lines T_1 and T_2 are drawn at two points P_1 and P_2 on the parabola $y = x^2$ and they intersect at a point P. Another tangent line T is drawn at a point between P_1 and P_2; it intersects T_1 at Q_1 and T_2 at Q_2. Show that

$$\frac{|PQ_1|}{|PP_1|} + \frac{|PQ_2|}{|PP_2|} = 1$$

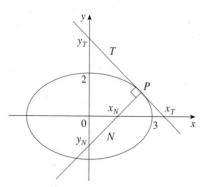

FIGURE FOR PROBLEM 13

13. Let T and N be the tangent and normal lines to the ellipse $x^2/9 + y^2/4 = 1$ at any point P on the ellipse in the first quadrant. Let x_T and y_T be the x- and y-intercepts of T and x_N and y_N be the intercepts of N. As P moves along the ellipse in the first quadrant (but not on the axes), what values can x_T, y_T, x_N, and y_N take on? First try to guess the answers just by looking at the figure. Then use calculus to solve the problem and see how good your intuition is.

14. Evaluate $\displaystyle\lim_{x \to 0} \frac{\sin(3+x)^2 - \sin 9}{x}$.

15. (a) Use the identity for $\tan(x - y)$ (see Equation 14b in Appendix D) to show that if two lines L_1 and L_2 intersect at an angle α, then

$$\tan \alpha = \frac{m_2 - m_1}{1 + m_1 m_2}$$

where m_1 and m_2 are the slopes of L_1 and L_2, respectively.
(b) The **angle between the curves** C_1 and C_2 at a point of intersection P is defined to be the angle between the tangent lines to C_1 and C_2 at P (if these tangent lines exist). Use part (a) to find, correct to the nearest degree, the angle between each pair of curves at each point of intersection.
(i) $y = x^2$ and $y = (x-2)^2$
(ii) $x^2 - y^2 = 3$ and $x^2 - 4x + y^2 + 3 = 0$

FIGURE FOR PROBLEM 16

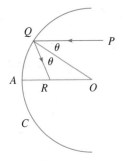

FIGURE FOR PROBLEM 17

16. Let $P(x_1, y_1)$ be a point on the parabola $y^2 = 4px$ with focus $F(p, 0)$. Let α be the angle between the parabola and the line segment FP, and let β be the angle between the horizontal line $y = y_1$ and the parabola as in the figure. Prove that $\alpha = \beta$. (Thus, by a principle of geometrical optics, light from a source placed at F will be reflected along a line parallel to the x-axis. This explains why *paraboloids*, the surfaces obtained by rotating parabolas about their axes, are used as the shape of some automobile headlights and mirrors for telescopes.)

17. Suppose that we replace the parabolic mirror of Problem 16 by a spherical mirror. Although the mirror has no focus, we can show the existence of an *approximate* focus. In the figure, C is a semicircle with center O. A ray of light coming in toward the mirror parallel to the axis along the line PQ will be reflected to the point R on the axis so that $\angle PQO = \angle OQR$ (the angle of incidence is equal to the angle of reflection). What happens to the point R as P is taken closer and closer to the axis?

18. If f and g are differentiable functions with $f(0) = g(0) = 0$ and $g'(0) \neq 0$, show that

$$\lim_{x \to 0} \frac{f(x)}{g(x)} = \frac{f'(0)}{g'(0)}$$

19. Evaluate $\displaystyle\lim_{x \to 0} \frac{\sin(a + 2x) - 2\sin(a + x) + \sin a}{x^2}$.

20. Given an ellipse $x^2/a^2 + y^2/b^2 = 1$, where $a \neq b$, find the equation of the set of all points from which there are two tangents to the curve whose slopes are (a) reciprocals and (b) negative reciprocals.

21. Find the two points on the curve $y = x^4 - 2x^2 - x$ that have a common tangent line.

22. Suppose that three points on the parabola $y = x^2$ have the property that their normal lines intersect at a common point. Show that the sum of their x-coordinates is 0.

23. A *lattice point* in the plane is a point with integer coordinates. Suppose that circles with radius r are drawn using all lattice points as centers. Find the smallest value of r such that any line with slope $\frac{2}{5}$ intersects some of these circles.

24. A cone of radius r centimeters and height h centimeters is lowered point first at a rate of 1 cm/s into a tall cylinder of radius R centimeters that is partially filled with water. How fast is the water level rising at the instant the cone is completely submerged?

25. A container in the shape of an inverted cone has height 16 cm and radius 5 cm at the top. It is partially filled with a liquid that oozes through the sides at a rate proportional to the area of the container that is in contact with the liquid. (The surface area of a cone is $\pi r l$, where r is the radius and l is the slant height.) If we pour the liquid into the container at a rate of 2 cm³/min, then the height of the liquid decreases at a rate of 0.3 cm/min when the height is 10 cm. If our goal is to keep the liquid at a constant height of 10 cm, at what rate should we pour the liquid into the container?

CAS 26. (a) The cubic function $f(x) = x(x - 2)(x - 6)$ has three distinct zeros: 0, 2, and 6. Graph f and its tangent lines at the *average* of each pair of zeros. What do you notice?
    (b) Suppose the cubic function $f(x) = (x - a)(x - b)(x - c)$ has three distinct zeros: a, b, and c. Prove, with the help of a computer algebra system, that a tangent line drawn at the average of the zeros a and b intersects the graph of f at the third zero.

3 Applications of Differentiation

The calculus that you learn in this chapter will enable you to explain the location of rainbows in the sky and why the colors in the secondary rainbow appear in the opposite order to those in the primary rainbow. (See the project on pages 162–163.)

© Pichugin Dmitry / Shutterstock

We have already investigated some of the applications of derivatives, but now that we know the differentiation rules we are in a better position to pursue the applications of differentiation in greater depth. Here we learn how derivatives affect the shape of a graph of a function and, in particular, how they help us locate maximum and minimum values of functions. Many practical problems require us to minimize a cost or maximize an area or somehow find the best possible outcome of a situation. In particular, we will be able to investigate the optimal shape of a can and to explain the location of rainbows in the sky.

3.1 Maximum and Minimum Values

Some of the most important applications of differential calculus are *optimization problems*, in which we are required to find the optimal (best) way of doing something. Here are examples of such problems that we will solve in this chapter:

- What is the shape of a can that minimizes manufacturing costs?

- What is the maximum acceleration of a space shuttle? (This is an important question to the astronauts who have to withstand the effects of acceleration.)

- What is the radius of a contracted windpipe that expels air most rapidly during a cough?

- At what angle should blood vessels branch so as to minimize the energy expended by the heart in pumping blood?

These problems can be reduced to finding the maximum or minimum values of a function. Let's first explain exactly what we mean by maximum and minimum values.

We see that the highest point on the graph of the function f shown in Figure 1 is the point $(3, 5)$. In other words, the largest value of f is $f(3) = 5$. Likewise, the smallest value is $f(6) = 2$. We say that $f(3) = 5$ is the *absolute maximum* of f and $f(6) = 2$ is the *absolute minimum*. In general, we use the following definition.

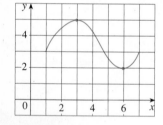

FIGURE 1

---

1   **Definition**   Let c be a number in the domain D of a function f. Then $f(c)$ is the

- **absolute maximum** value of f on D if $f(c) \geqslant f(x)$ for all x in D.
- **absolute minimum** value of f on D if $f(c) \leqslant f(x)$ for all x in D.

An absolute maximum or minimum is sometimes called a **global** maximum or minimum. The maximum and minimum values of f are called **extreme values** of f.

Figure 2 shows the graph of a function f with absolute maximum at d and absolute minimum at a. Note that $(d, f(d))$ is the highest point on the graph and $(a, f(a))$ is the lowest point. In Figure 2, if we consider only values of x near b [for instance, if we restrict our attention to the interval (a, c)], then $f(b)$ is the largest of those values of $f(x)$ and is called a *local maximum value* of f. Likewise, $f(c)$ is called a *local minimum value* of f because $f(c) \leqslant f(x)$ for x near c [in the interval (b, d), for instance]. The function f also has a local minimum at e. In general, we have the following definition.

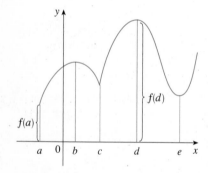

FIGURE 2

Abs min $f(a)$, abs max $f(d)$,
loc min $f(c)$, $f(e)$, loc max $f(b)$, $f(d)$

---

2   **Definition**   The number $f(c)$ is a

- **local maximum** value of f if $f(c) \geqslant f(x)$ when x is near c.
- **local minimum** value of f if $f(c) \leqslant f(x)$ when x is near c.

In Definition 2 (and elsewhere), if we say that something is true **near** c, we mean that it is true on some open interval containing c. For instance, in Figure 3 we see that $f(4) = 5$ is a local minimum because it's the smallest value of f on the interval I. It's not the absolute minimum because $f(x)$ takes smaller values when x is near 12 (in the interval K, for instance). In fact $f(12) = 3$ is both a local minimum and the absolute minimum. Similarly, $f(8) = 7$ is a local maximum, but not the absolute maximum because f takes larger values near 1.

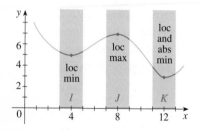

FIGURE 3

EXAMPLE 1 The function $f(x) = \cos x$ takes on its (local and absolute) maximum value of 1 infinitely many times, since $\cos 2n\pi = 1$ for any integer n and $-1 \le \cos x \le 1$ for all x. Likewise, $\cos(2n + 1)\pi = -1$ is its minimum value, where n is any integer. ▪

EXAMPLE 2 If $f(x) = x^2$, then $f(x) \ge f(0)$ because $x^2 \ge 0$ for all x. Therefore $f(0) = 0$ is the absolute (and local) minimum value of f. This corresponds to the fact that the origin is the lowest point on the parabola $y = x^2$. (See Figure 4.) However, there is no highest point on the parabola and so this function has no maximum value. ▪

FIGURE 4

Minimum value 0, no maximum

EXAMPLE 3 From the graph of the function $f(x) = x^3$, shown in Figure 5, we see that this function has neither an absolute maximum value nor an absolute minimum value. In fact, it has no local extreme values either.

FIGURE 5

No minimum, no maximum

V EXAMPLE 4 The graph of the function

$$f(x) = 3x^4 - 16x^3 + 18x^2 \qquad -1 \le x \le 4$$

is shown in Figure 6. You can see that $f(1) = 5$ is a local maximum, whereas the absolute maximum is $f(-1) = 37$. (This absolute maximum is not a local maximum because it occurs at an endpoint.) Also, $f(0) = 0$ is a local minimum and $f(3) = -27$ is both a local and an absolute minimum. Note that f has neither a local nor an absolute maximum at $x = 4$. ▪

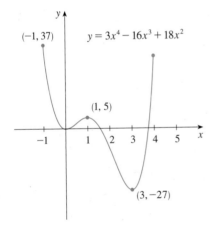

FIGURE 6

We have seen that some functions have extreme values, whereas others do not. The following theorem gives conditions under which a function is guaranteed to possess extreme values.

> **3   The Extreme Value Theorem**   If f is continuous on a closed interval $[a, b]$, then f attains an absolute maximum value $f(c)$ and an absolute minimum value $f(d)$ at some numbers c and d in $[a, b]$.

The Extreme Value Theorem is illustrated in Figure 7. Note that an extreme value can be taken on more than once. Although the Extreme Value Theorem is intuitively very plausible, it is difficult to prove and so we omit the proof.

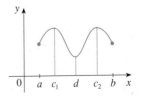

FIGURE 7

Figures 8 and 9 show that a function need not possess extreme values if either hypothesis (continuity or closed interval) is omitted from the Extreme Value Theorem.

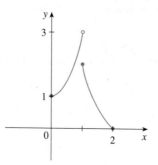

FIGURE 8

This function has minimum value $f(2) = 0$, but no maximum value.

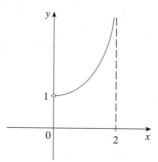

FIGURE 9

This continuous function g has no maximum or minimum.

The function f whose graph is shown in Figure 8 is defined on the closed interval [0, 2] but has no maximum value. (Notice that the range of f is [0, 3). The function takes on values arbitrarily close to 3, but never actually attains the value 3.) This does not contradict the Extreme Value Theorem because f is not continuous. [Nonetheless, a discontinuous function *could* have maximum and minimum values. See Exercise 13(b).]

The function g shown in Figure 9 is continuous on the open interval (0, 2) but has neither a maximum nor a minimum value. [The range of g is (1, ∞). The function takes on arbitrarily large values.] This does not contradict the Extreme Value Theorem because the interval (0, 2) is not closed.

The Extreme Value Theorem says that a continuous function on a closed interval has a maximum value and a minimum value, but it does not tell us how to find these extreme values. We start by looking for local extreme values.

Figure 10 shows the graph of a function f with a local maximum at c and a local minimum at d. It appears that at the maximum and minimum points the tangent lines are horizontal and therefore each has slope 0. We know that the derivative is the slope of the tangent line, so it appears that $f'(c) = 0$ and $f'(d) = 0$. The following theorem says that this is always true for differentiable functions.

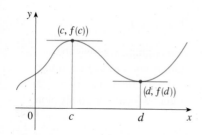

FIGURE 10

4   Fermat's Theorem   If f has a local maximum or minimum at c, and if $f'(c)$ exists, then $f'(c) = 0$.

PROOF   Suppose, for the sake of definiteness, that f has a local maximum at c. Then, according to Definition 2, $f(c) \geqslant f(x)$ if x is sufficiently close to c. This implies that if h is sufficiently close to 0, with h being positive or negative, then

$$f(c) \geqslant f(c + h)$$

and therefore

5
$$f(c + h) - f(c) \leqslant 0$$

We can divide both sides of an inequality by a positive number. Thus, if $h > 0$ and h is sufficiently small, we have

$$\frac{f(c + h) - f(c)}{h} \leqslant 0$$

Taking the right-hand limit of both sides of this inequality (using Theorem 1.6.2), we get

$$\lim_{h \to 0^+} \frac{f(c + h) - f(c)}{h} \leq \lim_{h \to 0^+} 0 = 0$$

But since $f'(c)$ exists, we have

$$f'(c) = \lim_{h \to 0} \frac{f(c + h) - f(c)}{h} = \lim_{h \to 0^+} \frac{f(c + h) - f(c)}{h}$$

and so we have shown that $f'(c) \leq 0$.

If $h < 0$, then the direction of the inequality $\boxed{5}$ is reversed when we divide by h:

$$\frac{f(c + h) - f(c)}{h} \geq 0 \qquad h < 0$$

So, taking the left-hand limit, we have

$$f'(c) = \lim_{h \to 0} \frac{f(c + h) - f(c)}{h} = \lim_{h \to 0^-} \frac{f(c + h) - f(c)}{h} \geq 0$$

We have shown that $f'(c) \geq 0$ and also that $f'(c) \leq 0$. Since both of these inequalities must be true, the only possibility is that $f'(c) = 0$.

We have proved Fermat's Theorem for the case of a local maximum. The case of a local minimum can be proved in a similar manner, or we could use Exercise 70 to deduce it from the case we have just proved (see Exercise 71). ◼

The following examples caution us against reading too much into Fermat's Theorem: We can't expect to locate extreme values simply by setting $f'(x) = 0$ and solving for x.

EXAMPLE 5 If $f(x) = x^3$, then $f'(x) = 3x^2$, so $f'(0) = 0$. But f has no maximum or minimum at 0, as you can see from its graph in Figure 11. (Or observe that $x^3 > 0$ for $x > 0$ but $x^3 < 0$ for $x < 0$.) The fact that $f'(0) = 0$ simply means that the curve $y = x^3$ has a horizontal tangent at $(0, 0)$. Instead of having a maximum or minimum at $(0, 0)$, the curve crosses its horizontal tangent there. ◼

EXAMPLE 6 The function $f(x) = |x|$ has its (local and absolute) minimum value at 0, but that value can't be found by setting $f'(x) = 0$ because, as was shown in Example 5 in Section 2.2, $f'(0)$ does not exist. (See Figure 12.) ◼

⊘ **WARNING** Examples 5 and 6 show that we must be careful when using Fermat's Theorem. Example 5 demonstrates that even when $f'(c) = 0$ there need not be a maximum or minimum at c. (In other words, the converse of Fermat's Theorem is false in general.) Furthermore, there may be an extreme value even when $f'(c)$ does not exist (as in Example 6).

Fermat's Theorem does suggest that we should at least *start* looking for extreme values of f at the numbers c where $f'(c) = 0$ or where $f'(c)$ does not exist. Such numbers are given a special name.

FIGURE 11
If $f(x) = x^3$, then $f'(0) = 0$ but f has no maximum or minimum.

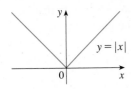

FIGURE 12
If $f(x) = |x|$, then $f(0) = 0$ is a minimum value, but $f'(0)$ does not exist.

$\boxed{6}$ **Definition** A **critical number** of a function f is a number c in the domain of f such that either $f'(c) = 0$ or $f'(c)$ does not exist.

Figure 13 shows a graph of the function f in Example 7. It supports our answer because there is a horizontal tangent when $x = 1.5$ and a vertical tangent when $x = 0$.

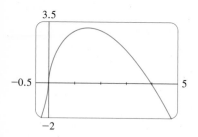

FIGURE 13

V   EXAMPLE 7 Find the critical numbers of $f(x) = x^{3/5}(4 - x)$.

SOLUTION The Product Rule gives

$$f'(x) = x^{3/5}(-1) + (4 - x)(\tfrac{3}{5}x^{-2/5}) = -x^{3/5} + \frac{3(4 - x)}{5x^{2/5}}$$

$$= \frac{-5x + 3(4 - x)}{5x^{2/5}} = \frac{12 - 8x}{5x^{2/5}}$$

[The same result could be obtained by first writing $f(x) = 4x^{3/5} - x^{8/5}$.] Therefore $f'(x) = 0$ if $12 - 8x = 0$, that is, $x = \tfrac{3}{2}$, and $f'(x)$ does not exist when $x = 0$. Thus the critical numbers are $\tfrac{3}{2}$ and 0.   ▬

In terms of critical numbers, Fermat's Theorem can be rephrased as follows (compare Definition 6 with Theorem 4):

---

7 If f has a local maximum or minimum at c, then c is a critical number of f.

To find an absolute maximum or minimum of a continuous function on a closed interval, we note that either it is local [in which case it occurs at a critical number by **7**] or it occurs at an endpoint of the interval. Thus the following three-step procedure always works.

The Closed Interval Method To find the *absolute* maximum and minimum values of a continuous function f on a closed interval $[a, b]$:

1. Find the values of f at the critical numbers of f in (a, b).

2. Find the values of f at the endpoints of the interval.

3. The largest of the values from Steps 1 and 2 is the absolute maximum value; the smallest of these values is the absolute minimum value.

V   EXAMPLE 8 Find the absolute maximum and minimum values of the function

$$f(x) = x^3 - 3x^2 + 1 \qquad -\tfrac{1}{2} \le x \le 4$$

SOLUTION Since f is continuous on $[-\tfrac{1}{2}, 4]$, we can use the Closed Interval Method:

$$f(x) = x^3 - 3x^2 + 1$$

$$f'(x) = 3x^2 - 6x = 3x(x - 2)$$

Since $f'(x)$ exists for all x, the only critical numbers of f occur when $f'(x) = 0$, that is, $x = 0$ or $x = 2$. Notice that each of these critical numbers lies in the interval $\left(-\tfrac{1}{2}, 4\right)$. The values of f at these critical numbers are

$$f(0) = 1 \qquad f(2) = -3$$

The values of f at the endpoints of the interval are

$$f\left(-\tfrac{1}{2}\right) = \tfrac{1}{8} \qquad f(4) = 17$$

Comparing these four numbers, we see that the absolute maximum value is $f(4) = 17$ and the absolute minimum value is $f(2) = -3$.

$$y = x^3 - 3x^2 + 1$$

(4, 17)

(2, −3)

FIGURE 14

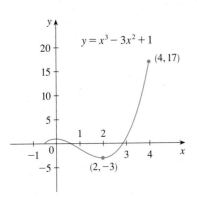

FIGURE 15

Note that in this example the absolute maximum occurs at an endpoint, whereas the absolute minimum occurs at a critical number. The graph of f is sketched in Figure 14.

If you have a graphing calculator or a computer with graphing software, it is possible to estimate maximum and minimum values very easily. But, as the next example shows, calculus is needed to find the *exact* values.

EXAMPLE 9
(a) Use a graphing device to estimate the absolute minimum and maximum values of the function $f(x) = x - 2 \sin x$, $0 \leq x \leq 2\pi$.
(b) Use calculus to find the exact minimum and maximum values.

SOLUTION
(a) Figure 15 shows a graph of f in the viewing rectangle $[0, 2\pi]$ by $[-1, 8]$. By moving the cursor close to the maximum point, we see that the y-coordinates don't change very much in the vicinity of the maximum. The absolute maximum value is about 6.97 and it occurs when $x \approx 5.2$. Similarly, by moving the cursor close to the minimum point, we see that the absolute minimum value is about -0.68 and it occurs when $x \approx 1.0$. It is possible to get more accurate estimates by zooming in toward the maximum and minimum points, but instead let's use calculus.

(b) The function $f(x) = x - 2 \sin x$ is continuous on $[0, 2\pi]$. Since $f'(x) = 1 - 2 \cos x$, we have $f'(x) = 0$ when $\cos x = \frac{1}{2}$ and this occurs when $x = \pi/3$ or $5\pi/3$. The values of f at these critical numbers are

$$f(\pi/3) = \frac{\pi}{3} - 2 \sin \frac{\pi}{3} = \frac{\pi}{3} - \sqrt{3} \approx -0.684853$$

and
$$f(5\pi/3) = \frac{5\pi}{3} - 2 \sin \frac{5\pi}{3} = \frac{5\pi}{3} + \sqrt{3} \approx 6.968039$$

The values of f at the endpoints are

$$f(0) = 0 \quad \text{and} \quad f(2\pi) = 2\pi \approx 6.28$$

Comparing these four numbers and using the Closed Interval Method, we see that the absolute minimum value is $f(\pi/3) = \pi/3 - \sqrt{3}$ and the absolute maximum value is $f(5\pi/3) = 5\pi/3 + \sqrt{3}$. The values from part (a) serve as a check on our work.

EXAMPLE 10 The Hubble Space Telescope was deployed on April 24, 1990, by the space shuttle *Discovery*. A model for the velocity of the shuttle during this mission, from liftoff at $t = 0$ until the solid rocket boosters were jettisoned at $t = 126$ s, is given by

$$v(t) = 0.001302t^3 - 0.09029t^2 + 23.61t - 3.083$$

(in feet per second). Using this model, estimate the absolute maximum and minimum values of the *acceleration* of the shuttle between liftoff and the jettisoning of the boosters.

SOLUTION We are asked for the extreme values not of the given velocity function, but rather of the acceleration function. So we first need to differentiate to find the acceleration:

$$a(t) = v'(t) = \frac{d}{dt}(0.001302t^3 - 0.09029t^2 + 23.61t - 3.083)$$

$$= 0.003906t^2 - 0.18058t + 23.61$$

We now apply the Closed Interval Method to the continuous function a on the interval $0 \leq t \leq 126$. Its derivative is

$$a'(t) = 0.007812t - 0.18058$$

The only critical number occurs when $a'(t) = 0$:

$$t_1 = \frac{0.18058}{0.007812} \approx 23.12$$

Evaluating $a(t)$ at the critical number and at the endpoints, we have

$$a(0) = 23.61 \qquad a(t_1) \approx 21.52 \qquad a(126) \approx 62.87$$

So the maximum acceleration is about 62.87 ft/s² and the minimum acceleration is about 21.52 ft/s².

3.1 Exercises

Access end-of-section exercises and tools for visualizing calculus online at **www.webassign.net**   ENHANCED **WebAssign**

APPLIED PROJECT   **THE CALCULUS OF RAINBOWS**

Formation of the primary rainbow

Rainbows are created when raindrops scatter sunlight. They have fascinated mankind since ancient times and have inspired attempts at scientific explanation since the time of Aristotle. In this project we use the ideas of Descartes and Newton to explain the shape, location, and colors of rainbows.

1. The figure shows a ray of sunlight entering a spherical raindrop at A. Some of the light is reflected, but the line AB shows the path of the part that enters the drop. Notice that the light is refracted toward the normal line AO and in fact Snell's Law says that $\sin \alpha = k \sin \beta$, where α is the angle of incidence, β is the angle of refraction, and $k \approx \frac{4}{3}$ is the index of refraction for water. At B some of the light passes through the drop and is refracted into the air, but the line BC shows the part that is reflected. (The angle of incidence equals the angle of reflection.) When the ray reaches C, part of it is reflected, but for the time being we are more interested in the part that leaves the raindrop at C. (Notice that it is refracted away from the normal line.) The *angle of deviation* $D(\alpha)$ is the amount of clockwise rotation that the ray has undergone during this three-stage process. Thus

$$D(\alpha) = (\alpha - \beta) + (\pi - 2\beta) + (\alpha - \beta) = \pi + 2\alpha - 4\beta$$

Show that the minimum value of the deviation is $D(\alpha) \approx 138°$ and occurs when $\alpha \approx 59.4°$.

The significance of the minimum deviation is that when $\alpha \approx 59.4°$ we have $D'(\alpha) \approx 0$, so $\Delta D/\Delta \alpha \approx 0$. This means that many rays with $\alpha \approx 59.4°$ become deviated by approximately the same amount. It is the *concentration* of rays coming from near the direction of minimum deviation that creates the brightness of the primary rainbow. The figure at the left shows that the angle of elevation from the observer up to the highest point on the rainbow is $180° - 138° = 42°$. (This angle is called the *rainbow angle*.)

2. Problem 1 explains the location of the primary rainbow, but how do we explain the colors? Sunlight comprises a range of wavelengths, from the red range through orange, yellow, green, blue, indigo, and violet. As Newton discovered in his prism experiments of 1666, the

index of refraction is different for each color. (The effect is called *dispersion.*) For red light the refractive index is $k \approx 1.3318$ whereas for violet light it is $k \approx 1.3435$. By repeating the calculation of Problem 1 for these values of k, show that the rainbow angle is about 42.3° for the red bow and 40.6° for the violet bow. So the rainbow really consists of seven individual bows corresponding to the seven colors.

3. Perhaps you have seen a fainter secondary rainbow above the primary bow. That results from the part of a ray that enters a raindrop and is refracted at A, reflected twice (at B and C), and refracted as it leaves the drop at D (see the figure at the left). This time the deviation angle $D(\alpha)$ is the total amount of counterclockwise rotation that the ray undergoes in this four-stage process. Show that

$$D(\alpha) = 2\alpha - 6\beta + 2\pi$$

and $D(\alpha)$ has a minimum value when

$$\cos \alpha = \sqrt{\frac{k^2 - 1}{8}}$$

Taking $k = \frac{4}{3}$, show that the minimum deviation is about 129° and so the rainbow angle for the secondary rainbow is about 51°, as shown in the figure at the left.

4. Show that the colors in the secondary rainbow appear in the opposite order from those in the primary rainbow.

Formation of the secondary rainbow

3.2   The Mean Value Theorem

We will see that many of the results of this chapter depend on one central fact, which is called the Mean Value Theorem. But to arrive at the Mean Value Theorem we first need the following result.

Rolle

Rolle's Theorem was first published in 1691 by the French mathematician Michel Rolle (1652–1719) in a book entitled *Méthode pour resoudre les Egalitez*. He was a vocal critic of the methods of his day and attacked calculus as being a "collection of ingenious fallacies." Later, however, he became convinced of the essential correctness of the methods of calculus.

Rolle's Theorem   Let f be a function that satisfies the following three hypotheses:

1. f is continuous on the closed interval $[a, b]$.
2. f is differentiable on the open interval (a, b).
3. $f(a) = f(b)$

Then there is a number c in (a, b) such that $f'(c) = 0$.

Before giving the proof let's take a look at the graphs of some typical functions that satisfy the three hypotheses. Figure 1 shows the graphs of four such functions. In each case it appears that there is at least one point $(c, f(c))$ on the graph where the tangent is horizontal and therefore $f'(c) = 0$. Thus Rolle's Theorem is plausible.

(a)

(b)

(c)

(d)

FIGURE 1

PS Take cases

PROOF   There are three cases:

CASE I $f(x) = k$, **a constant**
Then $f'(x) = 0$, so the number c can be taken to be *any* number in (a, b).

CASE II $f(x) > f(a)$ **for some x in (a, b)** [as in Figure 1(b) or (c)]
By the Extreme Value Theorem (which we can apply by hypothesis 1), f has a maximum value somewhere in $[a, b]$. Since $f(a) = f(b)$, it must attain this maximum value at a number c in the open interval (a, b). Then f has a *local* maximum at c and, by hypothesis 2, f is differentiable at c. Therefore $f'(c) = 0$ by Fermat's Theorem.

CASE III $f(x) < f(a)$ **for some x in (a, b)** [as in Figure 1(c) or (d)]
By the Extreme Value Theorem, f has a minimum value in $[a, b]$ and, since $f(a) = f(b)$, it attains this minimum value at a number c in (a, b). Again $f'(c) = 0$ by Fermat's Theorem.

EXAMPLE 1   Let's apply Rolle's Theorem to the position function $s = f(t)$ of a moving object. If the object is in the same place at two different instants $t = a$ and $t = b$, then $f(a) = f(b)$. Rolle's Theorem says that there is some instant of time $t = c$ between a and b when $f'(c) = 0$; that is, the velocity is 0. (In particular, you can see that this is true when a ball is thrown directly upward.)

Figure 2 shows a graph of the function $f(x) = x^3 + x - 1$ discussed in Example 2. Rolle's Theorem shows that, no matter how much we enlarge the viewing rectangle, we can never find a second x-intercept.

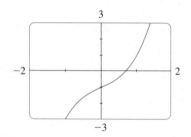

FIGURE 2

EXAMPLE 2   Prove that the equation $x^3 + x - 1 = 0$ has exactly one real root.

SOLUTION   First we use the Intermediate Value Theorem (1.8.10) to show that a root exists. Let $f(x) = x^3 + x - 1$. Then $f(0) = -1 < 0$ and $f(1) = 1 > 0$. Since f is a polynomial, it is continuous, so the Intermediate Value Theorem states that there is a number c between 0 and 1 such that $f(c) = 0$. Thus the given equation has a root.

To show that the equation has no other real root, we use Rolle's Theorem and argue by contradiction. Suppose that it had two roots a and b. Then $f(a) = 0 = f(b)$ and, since f is a polynomial, it is differentiable on (a, b) and continuous on $[a, b]$. Thus, by Rolle's Theorem, there is a number c between a and b such that $f'(c) = 0$. But

$$f'(x) = 3x^2 + 1 \geqslant 1 \qquad \text{for all } x$$

(since $x^2 \geqslant 0$) so $f'(x)$ can never be 0. This gives a contradiction. Therefore the equation can't have two real roots.

Our main use of Rolle's Theorem is in proving the following important theorem, which was first stated by another French mathematician, Joseph-Louis Lagrange.

The Mean Value Theorem is an example of what is called an existence theorem. Like the Intermediate Value Theorem, the Extreme Value Theorem, and Rolle's Theorem, it guarantees that there *exists* a number with a certain property, but it doesn't tell us how to find the number.

The Mean Value Theorem Let f be a function that satisfies the following hypotheses:

1. f is continuous on the closed interval $[a, b]$.

2. f is differentiable on the open interval (a, b).

Then there is a number c in (a, b) such that

$$\boxed{1} \qquad f'(c) = \frac{f(b) - f(a)}{b - a}$$

or, equivalently,

$$\boxed{2} \qquad f(b) - f(a) = f'(c)(b - a)$$

Before proving this theorem, we can see that it is reasonable by interpreting it geometrically. Figures 3 and 4 show the points $A(a, f(a))$ and $B(b, f(b))$ on the graphs of two differentiable functions. The slope of the secant line AB is

$$\boxed{3} \qquad m_{AB} = \frac{f(b) - f(a)}{b - a}$$

which is the same expression as on the right side of Equation 1. Since $f'(c)$ is the slope of the tangent line at the point $(c, f(c))$, the Mean Value Theorem, in the form given by Equation 1, says that there is at least one point $P(c, f(c))$ on the graph where the slope of the tangent line is the same as the slope of the secant line AB. In other words, there is a point P where the tangent line is parallel to the secant line AB. (Imagine a line parallel to AB, starting far away and moving parallel to itself until it touches the graph for the first time.)

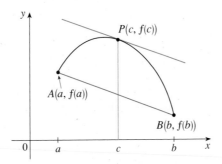

FIGURE 3   **FIGURE 4**

PROOF   We apply Rolle's Theorem to a new function h defined as the difference between f and the function whose graph is the secant line AB. Using Equation 3, we see that the equation of the line AB can be written as

$$y - f(a) = \frac{f(b) - f(a)}{b - a}(x - a)$$

or as

$$y = f(a) + \frac{f(b) - f(a)}{b - a}(x - a)$$

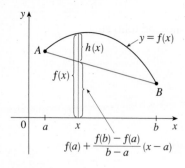

FIGURE 5

So, as shown in Figure 5,

$$\boxed{4} \qquad h(x) = f(x) - f(a) - \frac{f(b) - f(a)}{b - a}(x - a)$$

First we must verify that h satisfies the three hypotheses of Rolle's Theorem.

1. The function h is continuous on $[a, b]$ because it is the sum of f and a first-degree polynomial, both of which are continuous.

2. The function h is differentiable on (a, b) because both f and the first-degree polynomial are differentiable. In fact, we can compute h' directly from Equation 4:

$$h'(x) = f'(x) - \frac{f(b) - f(a)}{b - a}$$

(Note that $f(a)$ and $[f(b) - f(a)]/(b - a)$ are constants.)

3.
$$h(a) = f(a) - f(a) - \frac{f(b) - f(a)}{b - a}(a - a) = 0$$

$$h(b) = f(b) - f(a) - \frac{f(b) - f(a)}{b - a}(b - a)$$

$$= f(b) - f(a) - [f(b) - f(a)] = 0$$

Therefore $h(a) = h(b)$.

Since h satisfies the hypotheses of Rolle's Theorem, that theorem says there is a number c in (a, b) such that $h'(c) = 0$. Therefore

$$0 = h'(c) = f'(c) - \frac{f(b) - f(a)}{b - a}$$

and so

$$f'(c) = \frac{f(b) - f(a)}{b - a} \qquad \blacksquare$$

Lagrange and the Mean Value Theorem

The Mean Value Theorem was first formulated by Joseph-Louis Lagrange (1736–1813), born in Italy of a French father and an Italian mother. He was a child prodigy and became a professor in Turin at the tender age of 19. Lagrange made great contributions to number theory, theory of functions, theory of equations, and analytical and celestial mechanics. In particular, he applied calculus to the analysis of the stability of the solar system. At the invitation of Frederick the Great, he succeeded Euler at the Berlin Academy and, when Frederick died, Lagrange accepted King Louis XVI's invitation to Paris, where he was given apartments in the Louvre and became a professor at the Ecole Polytechnique. Despite all the trappings of luxury and fame, he was a kind and quiet man, living only for science.

EXAMPLE 3 To illustrate the Mean Value Theorem with a specific function, let's consider $f(x) = x^3 - x$, $a = 0$, $b = 2$. Since f is a polynomial, it is continuous and differentiable for all x, so it is certainly continuous on $[0, 2]$ and differentiable on $(0, 2)$. Therefore, by the Mean Value Theorem, there is a number c in $(0, 2)$ such that

$$f(2) - f(0) = f'(c)(2 - 0)$$

Now $f(2) = 6$, $f(0) = 0$, and $f'(x) = 3x^2 - 1$, so this equation becomes

$$6 = (3c^2 - 1)2 = 6c^2 - 2$$

which gives $c^2 = \frac{4}{3}$, that is, $c = \pm 2/\sqrt{3}$. But c must lie in $(0, 2)$, so $c = 2/\sqrt{3}$. Figure 6 illustrates this calculation: The tangent line at this value of c is parallel to the secant line OB.

FIGURE 6

EXAMPLE 4 If an object moves in a straight line with position function $s = f(t)$, then the average velocity between $t = a$ and $t = b$ is

$$\frac{f(b) - f(a)}{b - a}$$

and the velocity at $t = c$ is $f'(c)$. Thus the Mean Value Theorem (in the form of Equation 1) tells us that at some time $t = c$ between a and b the instantaneous velocity $f'(c)$ is equal to that average velocity. For instance, if a car traveled 180 km in 2 hours, then the speedometer must have read 90 km/h at least once.

In general, the Mean Value Theorem can be interpreted as saying that there is a number at which the instantaneous rate of change is equal to the average rate of change over an interval.

The main significance of the Mean Value Theorem is that it enables us to obtain information about a function from information about its derivative. The next example provides an instance of this principle.

V   EXAMPLE 5   Suppose that $f(0) = -3$ and $f'(x) \leq 5$ for all values of x. How large can $f(2)$ possibly be?

SOLUTION   We are given that f is differentiable (and therefore continuous) everywhere. In particular, we can apply the Mean Value Theorem on the interval $[0, 2]$. There exists a number c such that

$$f(2) - f(0) = f'(c)(2 - 0)$$

so

$$f(2) = f(0) + 2f'(c) = -3 + 2f'(c)$$

We are given that $f'(x) \leq 5$ for all x, so in particular we know that $f'(c) \leq 5$. Multiplying both sides of this inequality by 2, we have $2f'(c) \leq 10$, so

$$f(2) = -3 + 2f'(c) \leq -3 + 10 = 7$$

The largest possible value for $f(2)$ is 7.

The Mean Value Theorem can be used to establish some of the basic facts of differential calculus. One of these basic facts is the following theorem. Others will be found in the following sections.

---

5   Theorem   If $f'(x) = 0$ for all x in an interval (a, b), then f is constant on (a, b).

PROOF   Let x_1 and x_2 be any two numbers in (a, b) with $x_1 < x_2$. Since f is differentiable on (a, b), it must be differentiable on (x_1, x_2) and continuous on $[x_1, x_2]$. By applying the Mean Value Theorem to f on the interval $[x_1, x_2]$, we get a number c such that $x_1 < c < x_2$ and

$$\boxed{6} \qquad f(x_2) - f(x_1) = f'(c)(x_2 - x_1)$$

Since $f'(x) = 0$ for all x, we have $f'(c) = 0$, and so Equation 6 becomes

$$f(x_2) - f(x_1) = 0 \qquad \text{or} \qquad f(x_2) = f(x_1)$$

Therefore f has the same value at *any* two numbers x_1 and x_2 in (a, b). This means that f is constant on (a, b).

---

7   Corollary   If $f'(x) = g'(x)$ for all x in an interval (a, b), then $f - g$ is constant on (a, b); that is, $f(x) = g(x) + c$ where c is a constant.

PROOF Let $F(x) = f(x) - g(x)$. Then

$$F'(x) = f'(x) - g'(x) = 0$$

for all x in (a, b). Thus, by Theorem 5, F is constant; that is, $f - g$ is constant.

NOTE Care must be taken in applying Theorem 5. Let

$$f(x) = \frac{x}{|x|} = \begin{cases} 1 & \text{if } x > 0 \\ -1 & \text{if } x < 0 \end{cases}$$

The domain of f is $D = \{x \mid x \neq 0\}$ and $f'(x) = 0$ for all x in D. But f is obviously not a constant function. This does not contradict Theorem 5 because D is not an interval. Notice that f is constant on the interval $(0, \infty)$ and also on the interval $(-\infty, 0)$.

3.3 **How Derivatives Affect the Shape of a Graph**

Many of the applications of calculus depend on our ability to deduce facts about a function f from information concerning its derivatives. Because $f'(x)$ represents the slope of the curve $y = f(x)$ at the point $(x, f(x))$, it tells us the direction in which the curve proceeds at each point. So it is reasonable to expect that information about $f'(x)$ will provide us with information about $f(x)$.

What Does f' Say About f?

To see how the derivative of f can tell us where a function is increasing or decreasing, look at Figure 1. (Increasing functions and decreasing functions were defined in Section 1.1.) Between A and B and between C and D, the tangent lines have positive slope and so $f'(x) > 0$. Between B and C, the tangent lines have negative slope and so $f'(x) < 0$. Thus it appears that f increases when $f'(x)$ is positive and decreases when $f'(x)$ is negative. To prove that this is always the case, we use the Mean Value Theorem.

FIGURE 1

Let's abbreviate the name of this test to the I/D Test.

Increasing/Decreasing Test

(a) If $f'(x) > 0$ on an interval, then f is increasing on that interval.

(b) If $f'(x) < 0$ on an interval, then f is decreasing on that interval.

PROOF

(a) Let x_1 and x_2 be any two numbers in the interval with $x_1 < x_2$. According to the definition of an increasing function (page 19), we have to show that $f(x_1) < f(x_2)$.

Because we are given that $f'(x) > 0$, we know that f is differentiable on $[x_1, x_2]$. So, by the Mean Value Theorem, there is a number c between x_1 and x_2 such that

$$\boxed{1} \qquad f(x_2) - f(x_1) = f'(c)(x_2 - x_1)$$

Now $f'(c) > 0$ by assumption and $x_2 - x_1 > 0$ because $x_1 < x_2$. Thus the right side of Equation 1 is positive, and so

$$f(x_2) - f(x_1) > 0 \quad \text{or} \quad f(x_1) < f(x_2)$$

This shows that f is increasing.

Part (b) is proved similarly.

V EXAMPLE 1 Find where the function $f(x) = 3x^4 - 4x^3 - 12x^2 + 5$ is increasing and where it is decreasing.

SOLUTION $\qquad f'(x) = 12x^3 - 12x^2 - 24x = 12x(x - 2)(x + 1)$

To use the I/D Test we have to know where $f'(x) > 0$ and where $f'(x) < 0$. This depends on the signs of the three factors of $f'(x)$, namely, $12x$, $x - 2$, and $x + 1$. We divide the real line into intervals whose endpoints are the critical numbers $-1, 0$, and 2 and arrange our work in a chart. A plus sign indicates that the given expression is positive, and a minus sign indicates that it is negative. The last column of the chart gives the conclusion based on the I/D Test. For instance, $f'(x) < 0$ for $0 < x < 2$, so f is decreasing on $(0, 2)$. (It would also be true to say that f is decreasing on the closed interval $[0, 2]$.)

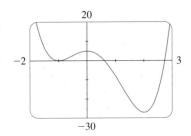

Interval	$12x$	$x - 2$	$x + 1$	$f'(x)$	f
$x < -1$	$-$	$-$	$-$	$-$	decreasing on $(-\infty, -1)$
$-1 < x < 0$	$-$	$-$	$+$	$+$	increasing on $(-1, 0)$
$0 < x < 2$	$+$	$-$	$+$	$-$	decreasing on $(0, 2)$
$x > 2$	$+$	$+$	$+$	$+$	increasing on $(2, \infty)$

The graph of f shown in Figure 2 confirms the information in the chart.

FIGURE 2

Recall from Section 3.1 that if f has a local maximum or minimum at c, then c must be a critical number of f (by Fermat's Theorem), but not every critical number gives rise to a maximum or a minimum. We therefore need a test that will tell us whether or not f has a local maximum or minimum at a critical number.

You can see from Figure 2 that $f(0) = 5$ is a local maximum value of f because f increases on $(-1, 0)$ and decreases on $(0, 2)$. Or, in terms of derivatives, $f'(x) > 0$ for $-1 < x < 0$ and $f'(x) < 0$ for $0 < x < 2$. In other words, the sign of $f'(x)$ changes from positive to negative at 0. This observation is the basis of the following test.

The First Derivative Test Suppose that c is a critical number of a continuous function f.

(a) If f' changes from positive to negative at c, then f has a local maximum at c.

(b) If f' changes from negative to positive at c, then f has a local minimum at c.

(c) If f' does not change sign at c (for example, if f' is positive on both sides of c or negative on both sides), then f has no local maximum or minimum at c.

The First Derivative Test is a consequence of the I/D Test. In part (a), for instance, since the sign of $f'(x)$ changes from positive to negative at c, f is increasing to the left of c and decreasing to the right of c. It follows that f has a local maximum at c.

It is easy to remember the First Derivative Test by visualizing diagrams such as those in Figure 3.

(a) Local maximum   (b) Local minimum   (c) No maximum or minimum   (d) No maximum or minimum

FIGURE 3

V   **EXAMPLE 2**   Find the local minimum and maximum values of the function f in Example 1.

SOLUTION   From the chart in the solution to Example 1 we see that $f'(x)$ changes from negative to positive at -1, so $f(-1) = 0$ is a local minimum value by the First Derivative Test. Similarly, f' changes from negative to positive at 2, so $f(2) = -27$ is also a local minimum value. As previously noted, $f(0) = 5$ is a local maximum value because $f'(x)$ changes from positive to negative at 0. ∎

EXAMPLE 3   Find the local maximum and minimum values of the function

$$g(x) = x + 2 \sin x \qquad 0 \le x \le 2\pi$$

SOLUTION   To find the critical numbers of g, we differentiate:

$$g'(x) = 1 + 2 \cos x$$

So $g'(x) = 0$ when $\cos x = -\frac{1}{2}$. The solutions of this equation are $2\pi/3$ and $4\pi/3$. Because g is differentiable everywhere, the only critical numbers are $2\pi/3$ and $4\pi/3$ and so we analyze g in the following table.

The + signs in the table come from the fact that $g'(x) > 0$ when $\cos x > -\frac{1}{2}$. From the graph of $y = \cos x$, this is true in the indicated intervals.

Interval	$g'(x) = 1 + 2 \cos x$	g
$0 < x < 2\pi/3$	+	increasing on $(0, 2\pi/3)$
$2\pi/3 < x < 4\pi/3$	−	decreasing on $(2\pi/3, 4\pi/3)$
$4\pi/3 < x < 2\pi$	+	increasing on $(4\pi/3, 2\pi)$

Because $g'(x)$ changes from positive to negative at $2\pi/3$, the First Derivative Test tells us that there is a local maximum at $2\pi/3$ and the local maximum value is

$$g(2\pi/3) = \frac{2\pi}{3} + 2 \sin \frac{2\pi}{3} = \frac{2\pi}{3} + 2\left(\frac{\sqrt{3}}{2}\right) = \frac{2\pi}{3} + \sqrt{3} \approx 3.83$$

Likewise, $g'(x)$ changes from negative to positive at $4\pi/3$ and so

$$g(4\pi/3) = \frac{4\pi}{3} + 2 \sin \frac{4\pi}{3} = \frac{4\pi}{3} + 2\left(-\frac{\sqrt{3}}{2}\right) = \frac{4\pi}{3} - \sqrt{3} \approx 2.46$$

FIGURE 4
$g(x) = x + 2 \sin x$

is a local minimum value. The graph of g in Figure 4 supports our conclusion. ∎

■ What Does f'' Say About f?

Figure 5 shows the graphs of two increasing functions on (a, b). Both graphs join point A to point B but they look different because they bend in different directions. How can we distinguish between these two types of behavior? In Figure 6 tangents to these curves have been drawn at several points. In (a) the curve lies above the tangents and f is called *concave upward* on (a, b). In (b) the curve lies below the tangents and g is called *concave downward* on (a, b).

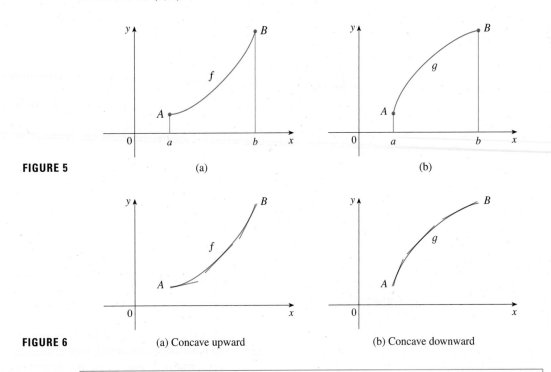

FIGURE 5

(a) (b)

FIGURE 6

(a) Concave upward (b) Concave downward

> **Definition** If the graph of f lies above all of its tangents on an interval I, then it is called **concave upward** on I. If the graph of f lies below all of its tangents on I, it is called **concave downward** on I.

Figure 7 shows the graph of a function that is concave upward (abbreviated CU) on the intervals (b, c), (d, e), and (e, p) and concave downward (CD) on the intervals (a, b), (c, d), and (p, q).

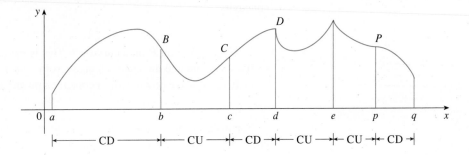

FIGURE 7

Let's see how the second derivative helps determine the intervals of concavity. Looking at Figure 6(a), you can see that, going from left to right, the slope of the tangent increases. This means that the derivative f' is an increasing function and therefore its derivative

f'' is positive. Likewise, in Figure 6(b) the slope of the tangent decreases from left to right, so f' decreases and therefore f'' is negative. This reasoning can be reversed and suggests that the following theorem is true. A proof is given in Appendix F with the help of the Mean Value Theorem.

Concavity Test

(a) If $f''(x) > 0$ for all x in I, then the graph of f is concave upward on I.

(b) If $f''(x) < 0$ for all x in I, then the graph of f is concave downward on I.

EXAMPLE 4   Figure 8 shows a population graph for Cyprian honeybees raised in an apiary. How does the rate of population increase change over time? When is this rate highest? Over what intervals is P concave upward or concave downward?

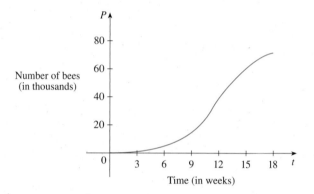

FIGURE 8

SOLUTION   By looking at the slope of the curve as t increases, we see that the rate of increase of the population is initially very small, then gets larger until it reaches a maximum at about $t = 12$ weeks, and decreases as the population begins to level off. As the population approaches its maximum value of about 75,000 (called the *carrying capacity*), the rate of increase, $P'(t)$, approaches 0. The curve appears to be concave upward on $(0, 12)$ and concave downward on $(12, 18)$.   ▬

In Example 4, the population curve changed from concave upward to concave downward at approximately the point $(12, 38,000)$. This point is called an *inflection point* of the curve. The significance of this point is that the rate of population increase has its maximum value there. In general, an inflection point is a point where a curve changes its direction of concavity.

Definition   A point P on a curve $y = f(x)$ is called an **inflection point** if f is continuous there and the curve changes from concave upward to concave downward or from concave downward to concave upward at P.

For instance, in Figure 7, B, C, D, and P are the points of inflection. Notice that if a curve has a tangent at a point of inflection, then the curve crosses its tangent there.

In view of the Concavity Test, there is a point of inflection at any point where the second derivative changes sign.

EXAMPLE 5 Sketch a possible graph of a function f that satisfies the following conditions:

(i) $f(0) = 0$, $f(2) = 3$, $f(4) = 6$, $f'(0) = f'(4) = 0$

(ii) $f'(x) > 0$ for $0 < x < 4$, $f'(x) < 0$ for $x < 0$ and for $x > 4$

(iii) $f''(x) > 0$ for $x < 2$, $f''(x) < 0$ for $x > 2$

SOLUTION   Condition (i) tells us that the graph has horizontal tangents at the points $(0, 0)$ and $(4, 6)$. Condition (ii) says that f is increasing on the interval $(0, 4)$ and decreasing on the intervals $(-\infty, 0)$ and $(4, \infty)$. It follows from the I/D Test that $f(0) = 0$ is a local minimum and $f(4) = 6$ is a local maximum.

Condition (iii) says that the graph is concave upward on the interval $(-\infty, 2)$ and concave downward on $(2, \infty)$. Because the curve changes from concave upward to concave downward when $x = 2$, the point $(2, 3)$ is an inflection point.

We use this information to sketch the graph of f in Figure 9. Notice that we made the curve bend upward when $x < 2$ and bend downward when $x > 2$.

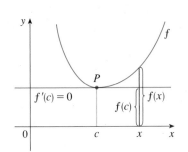

FIGURE 9

Another application of the second derivative is the following test for maximum and minimum values. It is a consequence of the Concavity Test.

The Second Derivative Test   Suppose f'' is continuous near c.

(a) If $f'(c) = 0$ and $f''(c) > 0$, then f has a local minimum at c.

(b) If $f'(c) = 0$ and $f''(c) < 0$, then f has a local maximum at c.

For instance, part (a) is true because $f''(x) > 0$ near c and so f is concave upward near c. This means that the graph of f lies *above* its horizontal tangent at c and so f has a local minimum at c. (See Figure 10.)

FIGURE 10
$f''(c) > 0$, f is concave upward

V **EXAMPLE 6** Discuss the curve $y = x^4 - 4x^3$ with respect to concavity, points of inflection, and local maxima and minima. Use this information to sketch the curve.

SOLUTION   If $f(x) = x^4 - 4x^3$, then

$$f'(x) = 4x^3 - 12x^2 = 4x^2(x - 3)$$
$$f''(x) = 12x^2 - 24x = 12x(x - 2)$$

To find the critical numbers we set $f'(x) = 0$ and obtain $x = 0$ and $x = 3$. To use the Second Derivative Test we evaluate f'' at these critical numbers:

$$f''(0) = 0 \qquad f''(3) = 36 > 0$$

Since $f'(3) = 0$ and $f''(3) > 0$, $f(3) = -27$ is a local minimum. Since $f''(0) = 0$, the Second Derivative Test gives no information about the critical number 0. But since $f'(x) < 0$ for $x < 0$ and also for $0 < x < 3$, the First Derivative Test tells us that f does not have a local maximum or minimum at 0. [In fact, the expression for $f'(x)$ shows that f decreases to the left of 3 and increases to the right of 3.]

Since $f''(x) = 0$ when $x = 0$ or 2, we divide the real line into intervals with these numbers as endpoints and complete the following chart.

Interval	$f''(x) = 12x(x-2)$	Concavity
$(-\infty, 0)$	$+$	upward
$(0, 2)$	$-$	downward
$(2, \infty)$	$+$	upward

The point $(0, 0)$ is an inflection point since the curve changes from concave upward to concave downward there. Also $(2, -16)$ is an inflection point since the curve changes from concave downward to concave upward there.

Using the local minimum, the intervals of concavity, and the inflection points, we sketch the curve in Figure 11.

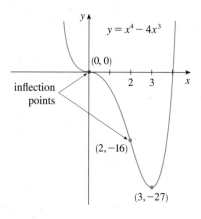

FIGURE 11

NOTE The Second Derivative Test is inconclusive when $f''(c) = 0$. In other words, at such a point there might be a maximum, there might be a minimum, or there might be neither (as in Example 6). This test also fails when $f''(c)$ does not exist. In such cases the First Derivative Test must be used. In fact, even when both tests apply, the First Derivative Test is often the easier one to use.

EXAMPLE 7 Sketch the graph of the function $f(x) = x^{2/3}(6 - x)^{1/3}$.

SOLUTION Calculation of the first two derivatives gives

$$f'(x) = \frac{4 - x}{x^{1/3}(6 - x)^{2/3}} \qquad f''(x) = \frac{-8}{x^{4/3}(6 - x)^{5/3}}$$

Use the differentiation rules to check these calculations.

Since $f'(x) = 0$ when $x = 4$ and $f'(x)$ does not exist when $x = 0$ or $x = 6$, the critical numbers are 0, 4, and 6.

Interval	$4 - x$	$x^{1/3}$	$(6 - x)^{2/3}$	$f'(x)$	f
$x < 0$	$+$	$-$	$+$	$-$	decreasing on $(-\infty, 0)$
$0 < x < 4$	$+$	$+$	$+$	$+$	increasing on $(0, 4)$
$4 < x < 6$	$-$	$+$	$+$	$-$	decreasing on $(4, 6)$
$x > 6$	$-$	$+$	$+$	$-$	decreasing on $(6, \infty)$

TEC In Module 3.3 you can practice using information about f', f'', and asymptotes to determine the shape of the graph of f.

To find the local extreme values we use the First Derivative Test. Since f' changes from negative to positive at 0, $f(0) = 0$ is a local minimum. Since f' changes from positive to negative at 4, $f(4) = 2^{5/3}$ is a local maximum. The sign of f' does not change at 6, so there is no minimum or maximum there. (The Second Derivative Test could be used at 4 but not at 0 or 6 since f'' does not exist at either of these numbers.)

Looking at the expression for $f''(x)$ and noting that $x^{4/3} \geq 0$ for all x, we have $f''(x) < 0$ for $x < 0$ and for $0 < x < 6$ and $f''(x) > 0$ for $x > 6$. So f is concave downward on $(-\infty, 0)$ and $(0, 6)$ and concave upward on $(6, \infty)$, and the only inflection point is $(6, 0)$. The graph is sketched in Figure 12. Note that the curve has vertical tangents at $(0, 0)$ and $(6, 0)$ because $|f'(x)| \to \infty$ as $x \to 0$ and as $x \to 6$.

Try reproducing the graph in Figure 12 with a graphing calculator or computer. Some machines produce the complete graph, some produce only the portion to the right of the y-axis, and some produce only the portion between $x = 0$ and $x = 6$. For an explanation and cure, see Example 7 in Appendix G. An equivalent expression that gives the correct graph is

$$y = (x^2)^{1/3} \cdot \frac{6 - x}{|6 - x|} |6 - x|^{1/3}$$

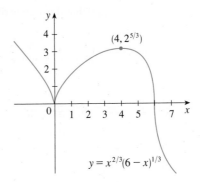

FIGURE 12

Exercises

Access end-of-section exercises and tools for visualizing calculus online at **www.webassign.net**

3.4 **Limits at Infinity; Horizontal Asymptotes**

In Sections 1.5 and 1.7 we investigated infinite limits and vertical asymptotes. There we let x approach a number and the result was that the values of y became arbitrarily large (positive or negative). In this section we let x become arbitrarily large (positive or negative) and see what happens to y. We will find it very useful to consider this so-called *end behavior* when sketching graphs.

Let's begin by investigating the behavior of the function f defined by

$$f(x) = \frac{x^2 - 1}{x^2 + 1}$$

as x becomes large. The table at the left gives values of this function correct to six decimal places, and the graph of f has been drawn by a computer in Figure 1.

x	$f(x)$
0	-1
± 1	0
± 2	0.600000
± 3	0.800000
± 4	0.882353
± 5	0.923077
± 10	0.980198
± 50	0.999200
± 100	0.999800
± 1000	0.999998

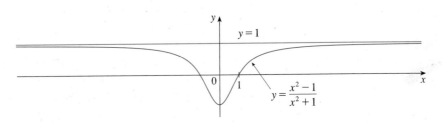

FIGURE 1

As x grows larger and larger you can see that the values of $f(x)$ get closer and closer to 1. In fact, it seems that we can make the values of $f(x)$ as close as we like to 1 by taking x sufficiently large. This situation is expressed symbolically by writing

$$\lim_{x \to \infty} \frac{x^2 - 1}{x^2 + 1} = 1$$

In general, we use the notation

$$\lim_{x \to \infty} f(x) = L$$

to indicate that the values of $f(x)$ approach L as x becomes larger and larger.

---

1   **Definition**   Let f be a function defined on some interval (a, ∞). Then

$$\lim_{x \to \infty} f(x) = L$$

means that the values of $f(x)$ can be made arbitrarily close to L by taking x sufficiently large.

Another notation for $\lim_{x \to \infty} f(x) = L$ is

$$f(x) \to L \quad \text{as} \quad x \to \infty$$

The symbol ∞ does not represent a number. Nonetheless, the expression $\lim_{x \to \infty} f(x) = L$ is often read as

"the limit of $f(x)$, as x approaches infinity, is L"

or   "the limit of $f(x)$, as x becomes infinite, is L"

or   "the limit of $f(x)$, as x increases without bound, is L"

The meaning of such phrases is given by Definition 1. A more precise definition, similar to the ε, δ definition of Section 1.7, is given at the end of this section.

Geometric illustrations of Definition 1 are shown in Figure 2. Notice that there are many ways for the graph of f to approach the line $y = L$ (which is called a *horizontal asymptote*) as we look to the far right of each graph.

FIGURE 2
Examples illustrating $\lim_{x \to \infty} f(x) = L$

Referring back to Figure 1, we see that for numerically large negative values of x, the values of $f(x)$ are close to 1. By letting x decrease through negative values without bound, we can make $f(x)$ as close to 1 as we like. This is expressed by writing

$$\lim_{x \to -\infty} \frac{x^2 - 1}{x^2 + 1} = 1$$

The general definition is as follows.

---

2 **Definition** Let f be a function defined on some interval $(-\infty, a)$. Then

$$\lim_{x \to -\infty} f(x) = L$$

means that the values of $f(x)$ can be made arbitrarily close to L by taking x sufficiently large negative.

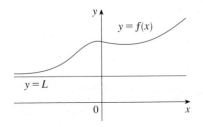

Again, the symbol $-\infty$ does not represent a number, but the expression $\lim_{x \to -\infty} f(x) = L$ is often read as

"the limit of $f(x)$, as x approaches negative infinity, is L"

Definition 2 is illustrated in Figure 3. Notice that the graph approaches the line $y = L$ as we look to the far left of each graph.

---

3 **Definition** The line $y = L$ is called a **horizontal asymptote** of the curve $y = f(x)$ if either

$$\lim_{x \to \infty} f(x) = L \qquad \text{or} \qquad \lim_{x \to -\infty} f(x) = L$$

FIGURE 3
Examples illustrating $\lim_{x \to -\infty} f(x) = L$

For instance, the curve illustrated in Figure 1 has the line $y = 1$ as a horizontal asymptote because

$$\lim_{x \to \infty} \frac{x^2 - 1}{x^2 + 1} = 1$$

The curve $y = f(x)$ sketched in Figure 4 has both $y = -1$ and $y = 2$ as horizontal asymptotes because

$$\lim_{x \to \infty} f(x) = -1 \qquad \text{and} \qquad \lim_{x \to -\infty} f(x) = 2$$

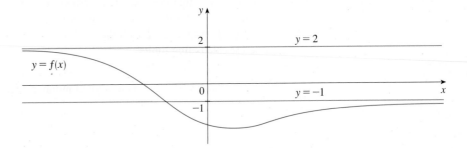

FIGURE 4

EXAMPLE 1 Find the infinite limits, limits at infinity, and asymptotes for the function f whose graph is shown in Figure 5 on page 178.

SOLUTION We see that the values of $f(x)$ become large as $x \to -1$ from both sides, so

$$\lim_{x \to -1} f(x) = \infty$$

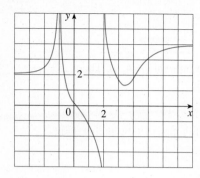

FIGURE 5

Notice that $f(x)$ becomes large negative as x approaches 2 from the left, but large positive as x approaches 2 from the right. So

$$\lim_{x \to 2^-} f(x) = -\infty \quad \text{and} \quad \lim_{x \to 2^+} f(x) = \infty$$

Thus both of the lines $x = -1$ and $x = 2$ are vertical asymptotes.

As x becomes large, it appears that $f(x)$ approaches 4. But as x decreases through negative values, $f(x)$ approaches 2. So

$$\lim_{x \to \infty} f(x) = 4 \quad \text{and} \quad \lim_{x \to -\infty} f(x) = 2$$

This means that both $y = 4$ and $y = 2$ are horizontal asymptotes. ▬

EXAMPLE 2 Find $\lim\limits_{x \to \infty} \dfrac{1}{x}$ and $\lim\limits_{x \to -\infty} \dfrac{1}{x}$.

SOLUTION Observe that when x is large, $1/x$ is small. For instance,

$$\frac{1}{100} = 0.01 \qquad \frac{1}{10,000} = 0.0001 \qquad \frac{1}{1,000,000} = 0.000001$$

In fact, by taking x large enough, we can make $1/x$ as close to 0 as we please. Therefore, according to Definition 1, we have

$$\lim_{x \to \infty} \frac{1}{x} = 0$$

Similar reasoning shows that when x is large negative, $1/x$ is small negative, so we also have

$$\lim_{x \to -\infty} \frac{1}{x} = 0$$

It follows that the line $y = 0$ (the x-axis) is a horizontal asymptote of the curve $y = 1/x$. (This is an equilateral hyperbola; see Figure 6.) ▬

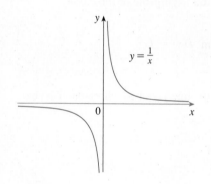

FIGURE 6

$\lim\limits_{x \to \infty} \dfrac{1}{x} = 0$, $\lim\limits_{x \to -\infty} \dfrac{1}{x} = 0$

Most of the Limit Laws that were given in Section 1.6 also hold for limits at infinity. It can be proved that the *Limit Laws listed in Section 1.6 (with the exception of Laws 9 and 10) are also valid if "$x \to a$" is replaced by "$x \to \infty$" or "$x \to -\infty$."* In particular, if we combine Laws 6 and 11 with the results of Example 2, we obtain the following important rule for calculating limits.

---

4   Theorem If $r > 0$ is a rational number, then

$$\lim_{x \to \infty} \frac{1}{x^r} = 0$$

If $r > 0$ is a rational number such that x^r is defined for all x, then

$$\lim_{x \to -\infty} \frac{1}{x^r} = 0$$

V EXAMPLE 3 Evaluate

$$\lim_{x \to \infty} \frac{3x^2 - x - 2}{5x^2 + 4x + 1}$$

and indicate which properties of limits are used at each stage.

SOLUTION As x becomes large, both numerator and denominator become large, so it isn't obvious what happens to their ratio. We need to do some preliminary algebra.

To evaluate the limit at infinity of any rational function, we first divide both the numerator and denominator by the highest power of x that occurs in the denominator. (We may assume that $x \neq 0$, since we are interested only in large values of x.) In this case the highest power of x in the denominator is x^2, so we have

$$\lim_{x \to \infty} \frac{3x^2 - x - 2}{5x^2 + 4x + 1} = \lim_{x \to \infty} \frac{\dfrac{3x^2 - x - 2}{x^2}}{\dfrac{5x^2 + 4x + 1}{x^2}} = \lim_{x \to \infty} \frac{3 - \dfrac{1}{x} - \dfrac{2}{x^2}}{5 + \dfrac{4}{x} + \dfrac{1}{x^2}}$$

$$= \frac{\lim_{x \to \infty} \left(3 - \dfrac{1}{x} - \dfrac{2}{x^2} \right)}{\lim_{x \to \infty} \left(5 + \dfrac{4}{x} + \dfrac{1}{x^2} \right)} \qquad \text{(by Limit Law 5)}$$

$$= \frac{\lim_{x \to \infty} 3 - \lim_{x \to \infty} \dfrac{1}{x} - 2 \lim_{x \to \infty} \dfrac{1}{x^2}}{\lim_{x \to \infty} 5 + 4 \lim_{x \to \infty} \dfrac{1}{x} + \lim_{x \to \infty} \dfrac{1}{x^2}} \qquad \text{(by 1, 2, and 3)}$$

$$= \frac{3 - 0 - 0}{5 + 0 + 0} \qquad \text{(by 7 and Theorem 4)}$$

$$= \frac{3}{5}$$

A similar calculation shows that the limit as $x \to -\infty$ is also $\frac{3}{5}$. Figure 7 illustrates the results of these calculations by showing how the graph of the given rational function approaches the horizontal asymptote $y = \frac{3}{5}$.

FIGURE 7

$$y = \frac{3x^2 - x - 2}{5x^2 + 4x + 1}$$

EXAMPLE 4 Find the horizontal and vertical asymptotes of the graph of the function

$$f(x) = \frac{\sqrt{2x^2 + 1}}{3x - 5}$$

SOLUTION Dividing both numerator and denominator by x and using the properties of limits, we have

$$\lim_{x \to \infty} \frac{\sqrt{2x^2 + 1}}{3x - 5} = \lim_{x \to \infty} \frac{\sqrt{2 + \dfrac{1}{x^2}}}{3 - \dfrac{5}{x}} \qquad \text{(since } \sqrt{x^2} = x \text{ for } x > 0\text{)}$$

$$= \frac{\displaystyle\lim_{x \to \infty} \sqrt{2 + \dfrac{1}{x^2}}}{\displaystyle\lim_{x \to \infty} \left(3 - \dfrac{5}{x}\right)} = \frac{\sqrt{\displaystyle\lim_{x \to \infty} 2 + \lim_{x \to \infty} \dfrac{1}{x^2}}}{\displaystyle\lim_{x \to \infty} 3 - 5 \lim_{x \to \infty} \dfrac{1}{x}}$$

$$= \frac{\sqrt{2 + 0}}{3 - 5 \cdot 0} = \frac{\sqrt{2}}{3}$$

Therefore the line $y = \sqrt{2}/3$ is a horizontal asymptote of the graph of f.

In computing the limit as $x \to -\infty$, we must remember that for $x < 0$, we have $\sqrt{x^2} = |x| = -x$. So when we divide the numerator by x, for $x < 0$ we get

$$\frac{1}{x} \sqrt{2x^2 + 1} = -\frac{1}{\sqrt{x^2}} \sqrt{2x^2 + 1} = -\sqrt{2 + \frac{1}{x^2}}$$

Therefore

$$\lim_{x \to -\infty} \frac{\sqrt{2x^2 + 1}}{3x - 5} = \lim_{x \to -\infty} \frac{-\sqrt{2 + \dfrac{1}{x^2}}}{3 - \dfrac{5}{x}}$$

$$= \frac{-\sqrt{2 + \displaystyle\lim_{x \to -\infty} \dfrac{1}{x^2}}}{3 - 5 \displaystyle\lim_{x \to -\infty} \dfrac{1}{x}} = -\frac{\sqrt{2}}{3}$$

Thus the line $y = -\sqrt{2}/3$ is also a horizontal asymptote.

A vertical asymptote is likely to occur when the denominator, $3x - 5$, is 0, that is, when $x = \frac{5}{3}$. If x is close to $\frac{5}{3}$ and $x > \frac{5}{3}$, then the denominator is close to 0 and $3x - 5$ is positive. The numerator $\sqrt{2x^2 + 1}$ is always positive, so $f(x)$ is positive. Therefore

$$\lim_{x \to (5/3)^+} \frac{\sqrt{2x^2 + 1}}{3x - 5} = \infty$$

If x is close to $\frac{5}{3}$ but $x < \frac{5}{3}$, then $3x - 5 < 0$ and so $f(x)$ is large negative. Thus

$$\lim_{x \to (5/3)^-} \frac{\sqrt{2x^2 + 1}}{3x - 5} = -\infty$$

The vertical asymptote is $x = \frac{5}{3}$. All three asymptotes are shown in Figure 8.

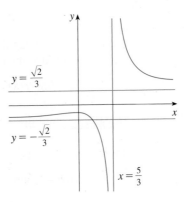

FIGURE 8

$$y = \frac{\sqrt{2x^2 + 1}}{3x - 5}$$

EXAMPLE 5 Compute $\displaystyle\lim_{x \to \infty} \left(\sqrt{x^2 + 1} - x \right)$.

SOLUTION Because both $\sqrt{x^2 + 1}$ and x are large when x is large, it's difficult to see what happens to their difference, so we use algebra to rewrite the function. We first multiply numerator and denominator by the conjugate radical:

We can think of the given function as having a denominator of 1.

$$\lim_{x \to \infty} \left(\sqrt{x^2 + 1} - x \right) = \lim_{x \to \infty} \left(\sqrt{x^2 + 1} - x \right) \frac{\sqrt{x^2 + 1} + x}{\sqrt{x^2 + 1} + x}$$

$$= \lim_{x \to \infty} \frac{(x^2 + 1) - x^2}{\sqrt{x^2 + 1} + x} = \lim_{x \to \infty} \frac{1}{\sqrt{x^2 + 1} + x}$$

Notice that the denominator of this last expression $\left(\sqrt{x^2 + 1} + x \right)$ becomes large as $x \to \infty$ (it's bigger than x). So

$$\lim_{x \to \infty} \left(\sqrt{x^2 + 1} - x \right) = \lim_{x \to \infty} \frac{1}{\sqrt{x^2 + 1} + x} = 0$$

Figure 9 illustrates this result.

$$y = \sqrt{x^2 + 1} - x$$

FIGURE 9

EXAMPLE 6 Evaluate $\displaystyle\lim_{x \to \infty} \sin \frac{1}{x}$.

SOLUTION If we let $t = 1/x$, then $t \to 0^+$ as $x \to \infty$. Therefore

PS The problem-solving strategy for Example 6 is *introducing something extra* (see page 77). Here, the something extra, the auxiliary aid, is the new variable t.

$$\lim_{x \to \infty} \sin \frac{1}{x} = \lim_{t \to 0^+} \sin t = 0$$

(See Exercise 71.)

EXAMPLE 7 Evaluate $\displaystyle\lim_{x \to \infty} \sin x$.

SOLUTION As x increases, the values of $\sin x$ oscillate between 1 and -1 infinitely often and so they don't approach any definite number. Thus $\lim_{x \to \infty} \sin x$ does not exist.

Infinite Limits at Infinity

The notation

$$\lim_{x \to \infty} f(x) = \infty$$

is used to indicate that the values of $f(x)$ become large as x becomes large. Similar meanings are attached to the following symbols:

$$\lim_{x \to -\infty} f(x) = \infty \qquad \lim_{x \to \infty} f(x) = -\infty \qquad \lim_{x \to -\infty} f(x) = -\infty$$

EXAMPLE 8 Find $\lim_{x \to \infty} x^3$ and $\lim_{x \to -\infty} x^3$.

SOLUTION When x becomes large, x^3 also becomes large. For instance,

$$10^3 = 1000 \qquad 100^3 = 1,000,000 \qquad 1000^3 = 1,000,000,000$$

In fact, we can make x^3 as big as we like by taking x large enough. Therefore we can write

$$\lim_{x \to \infty} x^3 = \infty$$

Similarly, when x is large negative, so is x^3. Thus

$$\lim_{x \to -\infty} x^3 = -\infty$$

These limit statements can also be seen from the graph of $y = x^3$ in Figure 10.

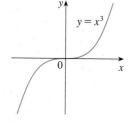

FIGURE 10
$\lim\limits_{x \to \infty} x^3 = \infty$, $\lim\limits_{x \to -\infty} x^3 = -\infty$

EXAMPLE 9 Find $\lim_{x \to \infty} (x^2 - x)$.

SOLUTION It would be **wrong** to write

$$\lim_{x \to \infty} (x^2 - x) = \lim_{x \to \infty} x^2 - \lim_{x \to \infty} x = \infty - \infty$$

The Limit Laws can't be applied to infinite limits because ∞ is not a number ($\infty - \infty$ can't be defined). However, we *can* write

$$\lim_{x \to \infty} (x^2 - x) = \lim_{x \to \infty} x(x - 1) = \infty$$

because both x and $x - 1$ become arbitrarily large and so their product does too.

EXAMPLE 10 Find $\lim_{x \to \infty} \dfrac{x^2 + x}{3 - x}$.

SOLUTION As in Example 3, we divide the numerator and denominator by the highest power of x in the denominator, which is just x:

$$\lim_{x \to \infty} \frac{x^2 + x}{3 - x} = \lim_{x \to \infty} \frac{x + 1}{\dfrac{3}{x} - 1} = -\infty$$

because $x + 1 \to \infty$ and $3/x - 1 \to -1$ as $x \to \infty$.

The next example shows that by using infinite limits at infinity, together with intercepts, we can get a rough idea of the graph of a polynomial without computing derivatives.

V **EXAMPLE 11** Sketch the graph of $y = (x - 2)^4(x + 1)^3(x - 1)$ by finding its intercepts and its limits as $x \to \infty$ and as $x \to -\infty$.

SOLUTION The y-intercept is $f(0) = (-2)^4(1)^3(-1) = -16$ and the x-intercepts are found by setting $y = 0$: $x = 2, -1, 1$. Notice that since $(x - 2)^4$ is positive, the function doesn't change sign at 2; thus the graph doesn't cross the x-axis at 2. The graph crosses the axis at -1 and 1.

When x is large positive, all three factors are large, so

$$\lim_{x \to \infty} (x - 2)^4(x + 1)^3(x - 1) = \infty$$

When x is large negative, the first factor is large positive and the second and third factors are both large negative, so

$$\lim_{x \to -\infty} (x - 2)^4(x + 1)^3(x - 1) = \infty$$

Combining this information, we give a rough sketch of the graph in Figure 11.

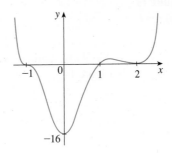

FIGURE 11
$y = (x - 2)^4(x + 1)^3(x - 1)$

Precise Definitions

Definition 1 can be stated precisely as follows.

---

5 **Definition** Let f be a function defined on some interval (a, ∞). Then

$$\lim_{x \to \infty} f(x) = L$$

means that for every $\varepsilon > 0$ there is a corresponding number N such that

$$\text{if} \quad x > N \quad \text{then} \quad |f(x) - L| < \varepsilon$$

In words, this says that the values of $f(x)$ can be made arbitrarily close to L (within a distance ε, where ε is any positive number) by taking x sufficiently large (larger than N, where N depends on ε). Graphically it says that by choosing x large enough (larger than some number N) we can make the graph of f lie between the given horizontal lines

$y = L - \varepsilon$ and $y = L + \varepsilon$ as in Figure 12. This must be true no matter how small we choose ε. Figure 13 shows that if a smaller value of ε is chosen, then a larger value of N may be required.

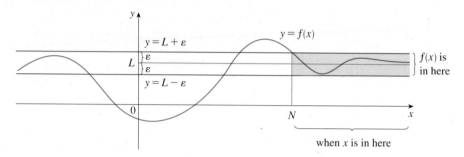

FIGURE 12
$\lim\limits_{x \to \infty} f(x) = L$

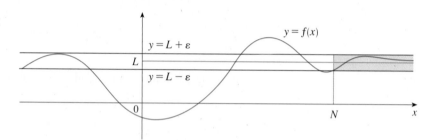

FIGURE 13
$\lim\limits_{x \to \infty} f(x) = L$

Similarly, a precise version of Definition 2 is given by Definition 6, which is illustrated in Figure 14.

---

6   **Definition**   Let f be a function defined on some interval $(-\infty, a)$. Then

$$\lim_{x \to -\infty} f(x) = L$$

means that for every $\varepsilon > 0$ there is a corresponding number N such that

$$\text{if} \quad x < N \qquad \text{then} \qquad |f(x) - L| < \varepsilon$$

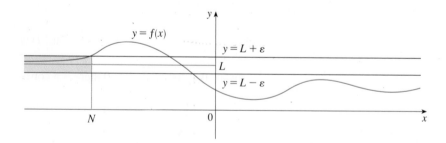

FIGURE 14
$\lim\limits_{x \to -\infty} f(x) = L$

In Example 3 we calculated that

$$\lim_{x \to \infty} \frac{3x^2 - x - 2}{5x^2 + 4x + 1} = \frac{3}{5}$$

In the next example we use a graphing device to relate this statement to Definition 5 with $L = \frac{3}{5}$ and $\varepsilon = 0.1$.

TEC In Module 1.7/3.4 you can explore the precise definition of a limit both graphically and numerically.

EXAMPLE 12 Use a graph to find a number N such that

$$\text{if} \quad x > N \quad \text{then} \quad \left| \frac{3x^2 - x - 2}{5x^2 + 4x + 1} - 0.6 \right| < 0.1$$

SOLUTION We rewrite the given inequality as

$$0.5 < \frac{3x^2 - x - 2}{5x^2 + 4x + 1} < 0.7$$

We need to determine the values of x for which the given curve lies between the horizontal lines $y = 0.5$ and $y = 0.7$. So we graph the curve and these lines in Figure 15. Then we use the cursor to estimate that the curve crosses the line $y = 0.5$ when $x \approx 6.7$. To the right of this number it seems that the curve stays between the lines $y = 0.5$ and $y = 0.7$. Rounding to be safe, we can say that

$$\text{if} \quad x > 7 \quad \text{then} \quad \left| \frac{3x^2 - x - 2}{5x^2 + 4x + 1} - 0.6 \right| < 0.1$$

In other words, for $\varepsilon = 0.1$ we can choose $N = 7$ (or any larger number) in Definition 5.

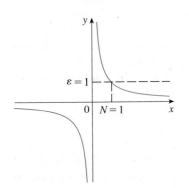

FIGURE 15

EXAMPLE 13 Use Definition 5 to prove that $\lim\limits_{x \to \infty} \dfrac{1}{x} = 0$.

SOLUTION Given $\varepsilon > 0$, we want to find N such that

$$\text{if} \quad x > N \quad \text{then} \quad \left| \frac{1}{x} - 0 \right| < \varepsilon$$

In computing the limit we may assume that $x > 0$. Then $1/x < \varepsilon \iff x > 1/\varepsilon$. Let's choose $N = 1/\varepsilon$. So

$$\text{if} \quad x > N = \frac{1}{\varepsilon} \quad \text{then} \quad \left| \frac{1}{x} - 0 \right| = \frac{1}{x} < \varepsilon$$

Therefore, by Definition 5,

$$\lim_{x \to \infty} \frac{1}{x} = 0$$

Figure 16 illustrates the proof by showing some values of ε and the corresponding values of N.

FIGURE 16

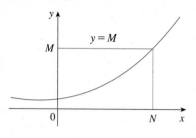

FIGURE 17
$\lim\limits_{x \to \infty} f(x) = \infty$

Finally we note that an infinite limit at infinity can be defined as follows. The geometric illustration is given in Figure 17.

> **7** **Definition** Let f be a function defined on some interval (a, ∞). Then
>
> $$\lim_{x \to \infty} f(x) = \infty$$
>
> means that for every positive number M there is a corresponding positive number N such that
>
> $$\text{if} \quad x > N \quad \text{then} \quad f(x) > M$$

Similar definitions apply when the symbol ∞ is replaced by $-\infty$. (See Exercise 72.)

3.4 Exercises

Access end-of-section exercises and tools for visualizing calculus online at **www.webassign.net**

3.5 Summary of Curve Sketching

So far we have been concerned with some particular aspects of curve sketching: domain, range, symmetry, limits, continuity, and vertical asymptotes in Chapter 1; derivatives and tangents in Chapter 2; and extreme values, intervals of increase and decrease, concavity, points of inflection, and horizontal asymptotes in this chapter. It is now time to put all of this information together to sketch graphs that reveal the important features of functions.

You might ask: Why don't we just use a graphing calculator or computer to graph a curve? Why do we need to use calculus?

It's true that modern technology is capable of producing very accurate graphs. But even the best graphing devices have to be used intelligently. As discussed in Appendix G, it is extremely important to choose an appropriate viewing rectangle to avoid getting a misleading graph. (See especially Examples 1, 3, 4, and 5 in that appendix.) The use of calculus enables us to discover the most interesting aspects of graphs and in many cases to calculate maximum and minimum points and inflection points *exactly* instead of approximately.

For instance, Figure 1 shows the graph of $f(x) = 8x^3 - 21x^2 + 18x + 2$. At first glance it seems reasonable: It has the same shape as cubic curves like $y = x^3$, and it appears to have no maximum or minimum point. But if you compute the derivative, you will see that there is a maximum when $x = 0.75$ and a minimum when $x = 1$. Indeed, if we zoom in to this portion of the graph, we see that behavior exhibited in Figure 2. Without calculus, we could easily have overlooked it.

In the next section we will graph functions by using the interaction between calculus and graphing devices. In this section we draw graphs by first considering the following information. We don't assume that you have a graphing device, but if you do have one you should use it as a check on your work.

FIGURE 1

Guidelines for Sketching a Curve

The following checklist is intended as a guide to sketching a curve $y = f(x)$ by hand. Not every item is relevant to every function. (For instance, a given curve might not have an asymptote or possess symmetry.) But the guidelines provide all the information you need to make a sketch that displays the most important aspects of the function.

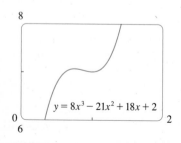

FIGURE 2

A. **Domain** It's often useful to start by determining the domain D of f, that is, the set of values of x for which $f(x)$ is defined.

B. **Intercepts** The y-intercept is $f(0)$ and this tells us where the curve intersects the y-axis. To find the x-intercepts, we set $y = 0$ and solve for x. (You can omit this step if the equation is difficult to solve.)

C. **Symmetry**

   (i) If $f(-x) = f(x)$ for all x in D, that is, the equation of the curve is unchanged when x is replaced by $-x$, then f is an **even function** and the curve is symmetric about the y-axis. This means that our work is cut in half. If we know what the curve looks like for $x \geqslant 0$, then we need only reflect about the y-axis to obtain the complete curve [see Figure 3(a)]. Here are some examples: $y = x^2$, $y = x^4$, $y = |x|$, and $y = \cos x$.

   (ii) If $f(-x) = -f(x)$ for all x in D, then f is an **odd function** and the curve is symmetric about the origin. Again we can obtain the complete curve if we know what it looks like for $x \geqslant 0$. [Rotate 180° about the origin; see Figure 3(b).] Some simple examples of odd functions are $y = x$, $y = x^3$, $y = x^5$, and $y = \sin x$.

   (iii) If $f(x + p) = f(x)$ for all x in D, where p is a positive constant, then f is called a **periodic function** and the smallest such number p is called the **period.** For instance, $y = \sin x$ has period 2π and $y = \tan x$ has period π. If we know what the graph looks like in an interval of length p, then we can use translation to sketch the entire graph (see Figure 4).

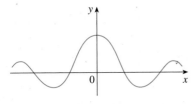

(a) Even function: reflectional symmetry

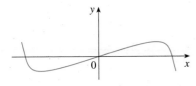

(b) Odd function: rotational symmetry

FIGURE 3

FIGURE 4

Periodic function: translational symmetry

D. **Asymptotes**

   (i) *Horizontal Asymptotes.* Recall from Section 3.4 that if either $\lim_{x \to \infty} f(x) = L$ or $\lim_{x \to -\infty} f(x) = L$, then the line $y = L$ is a horizontal asymptote of the curve $y = f(x)$. If it turns out that $\lim_{x \to \infty} f(x) = \infty$ (or $-\infty$), then we do not have an asymptote to the right, but that is still useful information for sketching the curve.

   (ii) *Vertical Asymptotes.* Recall from Section 1.5 that the line $x = a$ is a vertical asymptote if at least one of the following statements is true:

$$\boxed{1} \qquad \lim_{x \to a^+} f(x) = \infty \qquad \lim_{x \to a^-} f(x) = \infty$$

$$\lim_{x \to a^+} f(x) = -\infty \qquad \lim_{x \to a^-} f(x) = -\infty$$

(For rational functions you can locate the vertical asymptotes by equating the denominator to 0 after canceling any common factors. But for other functions this method does not apply.) Furthermore, in sketching the curve it is very useful to know exactly which of the statements in $\boxed{1}$ is true. If $f(a)$ is not defined but a is an endpoint of the domain of f, then you should compute $\lim_{x \to a^-} f(x)$ or $\lim_{x \to a^+} f(x)$, whether or not this limit is infinite.

   (iii) *Slant Asymptotes.* These are discussed at the end of this section.

E. **Intervals of Increase or Decrease** Use the I/D Test. Compute $f'(x)$ and find the intervals on which $f'(x)$ is positive (f is increasing) and the intervals on which $f'(x)$ is negative (f is decreasing).

F. Local Maximum and Minimum Values   Find the critical numbers of f [the numbers c where $f'(c) = 0$ or $f'(c)$ does not exist]. Then use the First Derivative Test. If f' changes from positive to negative at a critical number c, then $f(c)$ is a local maximum. If f' changes from negative to positive at c, then $f(c)$ is a local minimum. Although it is usually preferable to use the First Derivative Test, you can use the Second Derivative Test if $f'(c) = 0$ and $f''(c) \neq 0$. Then $f''(c) > 0$ implies that $f(c)$ is a local minimum, whereas $f''(c) < 0$ implies that $f(c)$ is a local maximum.

G. Concavity and Points of Inflection   Compute $f''(x)$ and use the Concavity Test. The curve is concave upward where $f''(x) > 0$ and concave downward where $f''(x) < 0$. Inflection points occur where the direction of concavity changes.

H. Sketch the Curve   Using the information in items A–G, draw the graph. Sketch the asymptotes as dashed lines. Plot the intercepts, maximum and minimum points, and inflection points. Then make the curve pass through these points, rising and falling according to E, with concavity according to G, and approaching the asymptotes. If additional accuracy is desired near any point, you can compute the value of the derivative there. The tangent indicates the direction in which the curve proceeds.

V EXAMPLE 1   Use the guidelines to sketch the curve $y = \dfrac{2x^2}{x^2 - 1}$.

A. The domain is

$$\{x \mid x^2 - 1 \neq 0\} = \{x \mid x \neq \pm 1\} = (-\infty, -1) \cup (-1, 1) \cup (1, \infty)$$

B. The x- and y-intercepts are both 0.

C. Since $f(-x) = f(x)$, the function f is even. The curve is symmetric about the y-axis.

D.
$$\lim_{x \to \pm\infty} \frac{2x^2}{x^2 - 1} = \lim_{x \to \pm\infty} \frac{2}{1 - 1/x^2} = 2$$

Therefore the line $y = 2$ is a horizontal asymptote.

Since the denominator is 0 when $x = \pm 1$, we compute the following limits:

$$\lim_{x \to 1^+} \frac{2x^2}{x^2 - 1} = \infty \qquad \lim_{x \to 1^-} \frac{2x^2}{x^2 - 1} = -\infty$$

$$\lim_{x \to -1^+} \frac{2x^2}{x^2 - 1} = -\infty \qquad \lim_{x \to -1^-} \frac{2x^2}{x^2 - 1} = \infty$$

Therefore the lines $x = 1$ and $x = -1$ are vertical asymptotes. This information about limits and asymptotes enables us to draw the preliminary sketch in Figure 5, showing the parts of the curve near the asymptotes.

FIGURE 5
Preliminary sketch

We have shown the curve approaching its horizontal asymptote from above in Figure 5. This is confirmed by the intervals of increase and decrease.

E.
$$f'(x) = \frac{4x(x^2 - 1) - 2x^2 \cdot 2x}{(x^2 - 1)^2} = \frac{-4x}{(x^2 - 1)^2}$$

Since $f'(x) > 0$ when $x < 0$ ($x \neq -1$) and $f'(x) < 0$ when $x > 0$ ($x \neq 1$), f is increasing on $(-\infty, -1)$ and $(-1, 0)$ and decreasing on $(0, 1)$ and $(1, \infty)$.

F. The only critical number is $x = 0$. Since f' changes from positive to negative at 0, $f(0) = 0$ is a local maximum by the First Derivative Test.

G.
$$f''(x) = \frac{-4(x^2 - 1)^2 + 4x \cdot 2(x^2 - 1)2x}{(x^2 - 1)^4} = \frac{12x^2 + 4}{(x^2 - 1)^3}$$

FIGURE 6
Finished sketch of $y = \dfrac{2x^2}{x^2 - 1}$

Since $12x^2 + 4 > 0$ for all x, we have

$$f''(x) > 0 \iff x^2 - 1 > 0 \iff |x| > 1$$

and $f''(x) < 0 \iff |x| < 1$. Thus the curve is concave upward on the intervals $(-\infty, -1)$ and $(1, \infty)$ and concave downward on $(-1, 1)$. It has no point of inflection since 1 and -1 are not in the domain of f.

H. Using the information in E–G, we finish the sketch in Figure 6.

EXAMPLE 2 Sketch the graph of $f(x) = \dfrac{x^2}{\sqrt{x+1}}$.

A. Domain $= \{x \mid x + 1 > 0\} = \{x \mid x > -1\} = (-1, \infty)$

B. The x- and y-intercepts are both 0.

C. Symmetry: None

D. Since

$$\lim_{x \to \infty} \frac{x^2}{\sqrt{x+1}} = \infty$$

there is no horizontal asymptote. Since $\sqrt{x+1} \to 0$ as $x \to -1^+$ and $f(x)$ is always positive, we have

$$\lim_{x \to -1^+} \frac{x^2}{\sqrt{x+1}} = \infty$$

and so the line $x = -1$ is a vertical asymptote.

E.
$$f'(x) = \frac{2x\sqrt{x+1} - x^2 \cdot 1/(2\sqrt{x+1})}{x+1} = \frac{x(3x+4)}{2(x+1)^{3/2}}$$

We see that $f'(x) = 0$ when $x = 0$ $\left(\text{notice that } -\tfrac{4}{3} \text{ is not in the domain of } f\right)$, so the only critical number is 0. Since $f'(x) < 0$ when $-1 < x < 0$ and $f'(x) > 0$ when $x > 0$, f is decreasing on $(-1, 0)$ and increasing on $(0, \infty)$.

F. Since $f'(0) = 0$ and f' changes from negative to positive at 0, $f(0) = 0$ is a local (and absolute) minimum by the First Derivative Test.

G.
$$f''(x) = \frac{2(x+1)^{3/2}(6x+4) - (3x^2 + 4x)3(x+1)^{1/2}}{4(x+1)^3} = \frac{3x^2 + 8x + 8}{4(x+1)^{5/2}}$$

Note that the denominator is always positive. The numerator is the quadratic $3x^2 + 8x + 8$, which is always positive because its discriminant is $b^2 - 4ac = -32$, which is negative, and the coefficient of x^2 is positive. Thus $f''(x) > 0$ for all x in the domain of f, which means that f is concave upward on $(-1, \infty)$ and there is no point of inflection.

H. The curve is sketched in Figure 7.

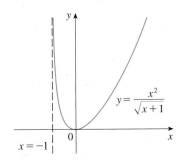

FIGURE 7

EXAMPLE 3 Sketch the graph of $f(x) = \dfrac{\cos x}{2 + \sin x}$.

A. The domain is $\mathbb{R}$.

B. The y-intercept is $f(0) = \tfrac{1}{2}$. The x-intercepts occur when $\cos x = 0$, that is, $x = (2n + 1)\pi/2$, where n is an integer.

C. f is neither even nor odd, but $f(x + 2\pi) = f(x)$ for all x and so f is periodic and has period 2π. Thus, in what follows, we need to consider only $0 \leq x \leq 2\pi$ and then extend the curve by translation in part H.

D. Asymptotes: None

E.
$$f'(x) = \frac{(2 + \sin x)(-\sin x) - \cos x\,(\cos x)}{(2 + \sin x)^2} = -\frac{2 \sin x + 1}{(2 + \sin x)^2}$$

Thus $f'(x) > 0$ when $2 \sin x + 1 < 0 \iff \sin x < -\tfrac{1}{2} \iff$ $7\pi/6 < x < 11\pi/6$. So f is increasing on $(7\pi/6, 11\pi/6)$ and decreasing on $(0, 7\pi/6)$ and $(11\pi/6, 2\pi)$.

F. From part E and the First Derivative Test, we see that the local minimum value is $f(7\pi/6) = -1/\sqrt{3}$ and the local maximum value is $f(11\pi/6) = 1/\sqrt{3}$.

G. If we use the Quotient Rule again and simplify, we get

$$f''(x) = -\frac{2 \cos x\,(1 - \sin x)}{(2 + \sin x)^3}$$

Because $(2 + \sin x)^3 > 0$ and $1 - \sin x \geqslant 0$ for all x, we know that $f''(x) > 0$ when $\cos x < 0$, that is, $\pi/2 < x < 3\pi/2$. So f is concave upward on $(\pi/2, 3\pi/2)$ and concave downward on $(0, \pi/2)$ and $(3\pi/2, 2\pi)$. The inflection points are $(\pi/2, 0)$ and $(3\pi/2, 0)$.

H. The graph of the function restricted to $0 \leqslant x \leqslant 2\pi$ is shown in Figure 8. Then we extend it, using periodicity, to the complete graph in Figure 9.

FIGURE 8

FIGURE 9

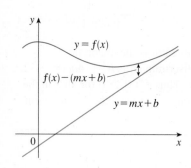

FIGURE 10

Slant Asymptotes

Some curves have asymptotes that are *oblique*, that is, neither horizontal nor vertical. If

$$\lim_{x \to \infty} [f(x) - (mx + b)] = 0$$

then the line $y = mx + b$ is called a **slant asymptote** because the vertical distance between the curve $y = f(x)$ and the line $y = mx + b$ approaches 0, as in Figure 10. (A similar situation exists if we let $x \to -\infty$.) For rational functions, slant asymptotes occur when the degree of the numerator is one more than the degree of the denominator. In such a case the equation of the slant asymptote can be found by long division as in the following example.

V **EXAMPLE 4** Sketch the graph of $f(x) = \dfrac{x^3}{x^2 + 1}$.

A. The domain is $\mathbb{R} = (-\infty, \infty)$.

B. The x- and y-intercepts are both 0.

C. Since $f(-x) = -f(x)$, f is odd and its graph is symmetric about the origin.

D. Since $x^2 + 1$ is never 0, there is no vertical asymptote. Since $f(x) \to \infty$ as $x \to \infty$ and $f(x) \to -\infty$ as $x \to -\infty$, there is no horizontal asymptote. But long division gives

$$f(x) = \frac{x^3}{x^2 + 1} = x - \frac{x}{x^2 + 1}$$

$$f(x) - x = -\frac{x}{x^2 + 1} = -\frac{\dfrac{1}{x}}{1 + \dfrac{1}{x^2}} \to 0 \quad \text{as} \quad x \to \pm\infty$$

So the line $y = x$ is a slant asymptote.

E.
$$f'(x) = \frac{3x^2(x^2 + 1) - x^3 \cdot 2x}{(x^2 + 1)^2} = \frac{x^2(x^2 + 3)}{(x^2 + 1)^2}$$

Since $f'(x) > 0$ for all x (except 0), f is increasing on $(-\infty, \infty)$.

F. Although $f'(0) = 0$, f' does not change sign at 0, so there is no local maximum or minimum.

G.
$$f''(x) = \frac{(4x^3 + 6x)(x^2 + 1)^2 - (x^4 + 3x^2) \cdot 2(x^2 + 1)2x}{(x^2 + 1)^4} = \frac{2x(3 - x^2)}{(x^2 + 1)^3}$$

Since $f''(x) = 0$ when $x = 0$ or $x = \pm\sqrt{3}$, we set up the following chart:

Interval	x	$3 - x^2$	$(x^2 + 1)^3$	$f''(x)$	f
$x < -\sqrt{3}$	$-$	$-$	$+$	$+$	CU on $\left(-\infty, -\sqrt{3}\right)$
$-\sqrt{3} < x < 0$	$-$	$+$	$+$	$-$	CD on $\left(-\sqrt{3}, 0\right)$
$0 < x < \sqrt{3}$	$+$	$+$	$+$	$+$	CU on $\left(0, \sqrt{3}\right)$
$x > \sqrt{3}$	$+$	$-$	$+$	$-$	CD on $\left(\sqrt{3}, \infty\right)$

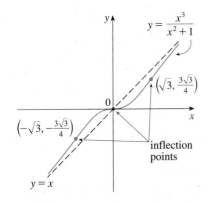

FIGURE 11

The points of inflection are $\left(-\sqrt{3}, -\frac{3}{4}\sqrt{3}\right)$, $(0, 0)$, and $\left(\sqrt{3}, \frac{3}{4}\sqrt{3}\right)$.

H. The graph of f is sketched in Figure 11.

3.6 Graphing with Calculus *and* Calculators

If you have not already read Appendix G, you should do so now. In particular, it explains how to avoid some of the pitfalls of graphing devices by choosing appropriate viewing rectangles.

The method we used to sketch curves in the preceding section was a culmination of much of our study of differential calculus. The graph was the final object that we produced. In this section our point of view is completely different. Here we *start* with a graph produced by a graphing calculator or computer and then we refine it. We use calculus to make sure that we reveal all the important aspects of the curve. And with the use of graphing devices we can tackle curves that would be far too complicated to consider without technology. The theme is the *interaction* between calculus and calculators.

EXAMPLE 1 Graph the polynomial $f(x) = 2x^6 + 3x^5 + 3x^3 - 2x^2$. Use the graphs of f' and f'' to estimate all maximum and minimum points and intervals of concavity.

SOLUTION If we specify a domain but not a range, many graphing devices will deduce a suitable range from the values computed. Figure 1 shows the plot from one such device if we specify that $-5 \le x \le 5$. Although this viewing rectangle is useful for showing that the asymptotic behavior (or end behavior) is the same as for $y = 2x^6$, it is obviously hiding some finer detail. So we change to the viewing rectangle $[-3, 2]$ by $[-50, 100]$ shown in Figure 2.

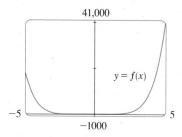

FIGURE 1

FIGURE 2

From this graph it appears that there is an absolute minimum value of about -15.33 when $x \approx -1.62$ (by using the cursor) and f is decreasing on $(-\infty, -1.62)$ and increasing on $(-1.62, \infty)$. Also there appears to be a horizontal tangent at the origin and inflection points when $x = 0$ and when x is somewhere between -2 and -1.

Now let's try to confirm these impressions using calculus. We differentiate and get

$$f'(x) = 12x^5 + 15x^4 + 9x^2 - 4x$$

$$f''(x) = 60x^4 + 60x^3 + 18x - 4$$

When we graph f' in Figure 3 we see that $f'(x)$ changes from negative to positive when $x \approx -1.62$; this confirms (by the First Derivative Test) the minimum value that we found earlier. But, perhaps to our surprise, we also notice that $f'(x)$ changes from positive to negative when $x = 0$ and from negative to positive when $x \approx 0.35$. This means that f has a local maximum at 0 and a local minimum when $x \approx 0.35$, but these were hidden in Figure 2. Indeed, if we now zoom in toward the origin in Figure 4, we see what we missed before: a local maximum value of 0 when $x = 0$ and a local minimum value of about -0.1 when $x \approx 0.35$.

FIGURE 3

FIGURE 4

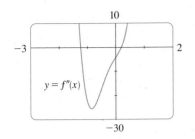

FIGURE 5

What about concavity and inflection points? From Figures 2 and 4 there appear to be inflection points when x is a little to the left of -1 and when x is a little to the right of 0. But it's difficult to determine inflection points from the graph of f, so we graph the second derivative f'' in Figure 5. We see that f'' changes from positive to negative when $x \approx -1.23$ and from negative to positive when $x \approx 0.19$. So, correct to two decimal places, f is concave upward on $(-\infty, -1.23)$ and $(0.19, \infty)$ and concave downward on $(-1.23, 0.19)$. The inflection points are $(-1.23, -10.18)$ and $(0.19, -0.05)$.

We have discovered that no single graph reveals all the important features of this polynomial. But Figures 2 and 4, when taken together, do provide an accurate picture.

V **EXAMPLE 2** Draw the graph of the function

$$f(x) = \frac{x^2 + 7x + 3}{x^2}$$

in a viewing rectangle that contains all the important features of the function. Estimate the maximum and minimum values and the intervals of concavity. Then use calculus to find these quantities exactly.

SOLUTION Figure 6, produced by a computer with automatic scaling, is a disaster. Some graphing calculators use $[-10, 10]$ by $[-10, 10]$ as the default viewing rectangle, so let's try it. We get the graph shown in Figure 7; it's a major improvement.

The y-axis appears to be a vertical asymptote and indeed it is because

$$\lim_{x \to 0} \frac{x^2 + 7x + 3}{x^2} = \infty$$

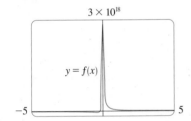

FIGURE 6

Figure 7 also allows us to estimate the x-intercepts: about -0.5 and -6.5. The exact values are obtained by using the quadratic formula to solve the equation $x^2 + 7x + 3 = 0$; we get $x = \left(-7 \pm \sqrt{37}\right)/2$.

FIGURE 7

FIGURE 8

To get a better look at horizontal asymptotes, we change to the viewing rectangle $[-20, 20]$ by $[-5, 10]$ in Figure 8. It appears that $y = 1$ is the horizontal asymptote and this is easily confirmed:

$$\lim_{x \to \pm\infty} \frac{x^2 + 7x + 3}{x^2} = \lim_{x \to \pm\infty} \left(1 + \frac{7}{x} + \frac{3}{x^2}\right) = 1$$

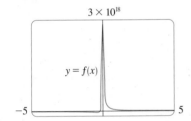

To estimate the minimum value we zoom in to the viewing rectangle $[-3, 0]$ by $[-4, 2]$ in Figure 9. The cursor indicates that the absolute minimum value is about -3.1

FIGURE 9

when $x \approx -0.9$, and we see that the function decreases on $(-\infty, -0.9)$ and $(0, \infty)$ and increases on $(-0.9, 0)$. The exact values are obtained by differentiating:

$$f'(x) = -\frac{7}{x^2} - \frac{6}{x^3} = -\frac{7x + 6}{x^3}$$

This shows that $f'(x) > 0$ when $-\frac{6}{7} < x < 0$ and $f'(x) < 0$ when $x < -\frac{6}{7}$ and when $x > 0$. The exact minimum value is $f\left(-\frac{6}{7}\right) = -\frac{37}{12} \approx -3.08$.

Figure 9 also shows that an inflection point occurs somewhere between $x = -1$ and $x = -2$. We could estimate it much more accurately using the graph of the second derivative, but in this case it's just as easy to find exact values. Since

$$f''(x) = \frac{14}{x^3} + \frac{18}{x^4} = \frac{2(7x + 9)}{x^4}$$

we see that $f''(x) > 0$ when $x > -\frac{9}{7}$ ($x \neq 0$). So f is concave upward on $\left(-\frac{9}{7}, 0\right)$ and $(0, \infty)$ and concave downward on $\left(-\infty, -\frac{9}{7}\right)$. The inflection point is $\left(-\frac{9}{7}, -\frac{71}{27}\right)$.

The analysis using the first two derivatives shows that Figure 8 displays all the major aspects of the curve.

V   EXAMPLE 3 Graph the function $f(x) = \dfrac{x^2(x + 1)^3}{(x - 2)^2(x - 4)^4}$.

SOLUTION Drawing on our experience with a rational function in Example 2, let's start by graphing f in the viewing rectangle $[-10, 10]$ by $[-10, 10]$. From Figure 10 we have the feeling that we are going to have to zoom in to see some finer detail and also zoom out to see the larger picture. But, as a guide to intelligent zooming, let's first take a close look at the expression for $f(x)$. Because of the factors $(x - 2)^2$ and $(x - 4)^4$ in the denominator, we expect $x = 2$ and $x = 4$ to be the vertical asymptotes. Indeed

$$\lim_{x \to 2} \frac{x^2(x + 1)^3}{(x - 2)^2(x - 4)^4} = \infty \quad \text{and} \quad \lim_{x \to 4} \frac{x^2(x + 1)^3}{(x - 2)^2(x - 4)^4} = \infty$$

To find the horizontal asymptotes, we divide numerator and denominator by x^6:

This shows that $f(x) \to 0$ as $x \to \pm\infty$, so the x-axis is a horizontal asymptote.

It is also very useful to consider the behavior of the graph near the x-intercepts using an analysis like that in Example 11 in Section 3.4. Since x^2 is positive, $f(x)$ does not change sign at 0 and so its graph doesn't cross the x-axis at 0. But, because of the factor $(x + 1)^3$, the graph does cross the x-axis at -1 and has a horizontal tangent there. Putting all this information together, but without using derivatives, we see that the curve has to look something like the one in Figure 11.

Now that we know what to look for, we zoom in (several times) to produce the graphs in Figures 12 and 13 and zoom out (several times) to get Figure 14.

FIGURE 10

FIGURE 11

FIGURE 12

FIGURE 13

FIGURE 14

We can read from these graphs that the absolute minimum is about -0.02 and occurs when $x \approx -20$. There is also a local maximum ≈ 0.00002 when $x \approx -0.3$ and a local minimum ≈ 211 when $x \approx 2.5$. These graphs also show three inflection points near -35, -5, and -1 and two between -1 and 0. To estimate the inflection points closely we would need to graph f'', but to compute f'' by hand is an unreasonable chore. If you have a computer algebra system, then it's easy to do (see Exercise 13).

We have seen that, for this particular function, *three* graphs (Figures 12, 13, and 14) are necessary to convey all the useful information. The only way to display all these features of the function on a single graph is to draw it by hand. Despite the exaggerations and distortions, Figure 11 does manage to summarize the essential nature of the function.

The family of functions

$$f(x) = \sin(x + \sin cx)$$

where c is a constant, occurs in applications to frequency modulation (FM) synthesis. A sine wave is modulated by a wave with a different frequency ($\sin cx$). The case where $c = 2$ is studied in Example 4. Exercise 19 explores another special case.

EXAMPLE 4 Graph the function $f(x) = \sin(x + \sin 2x)$. For $0 \le x \le \pi$, estimate all maximum and minimum values, intervals of increase and decrease, and inflection points.

SOLUTION We first note that f is periodic with period 2π. Also, f is odd and $|f(x)| \le 1$ for all x. So the choice of a viewing rectangle is not a problem for this function: We start with $[0, \pi]$ by $[-1.1, 1.1]$. (See Figure 15.)

FIGURE 15

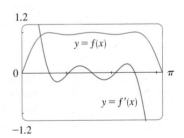

FIGURE 16

It appears that there are three local maximum values and two local minimum values in that window. To confirm this and locate them more accurately, we calculate that

$$f'(x) = \cos(x + \sin 2x) \cdot (1 + 2 \cos 2x)$$

and graph both f and f' in Figure 16.

Using zoom-in and the First Derivative Test, we find the following approximate values:

Intervals of increase:	$(0, 0.6), \ (1.0, 1.6), \ (2.1, 2.5)$
Intervals of decrease:	$(0.6, 1.0), \ (1.6, 2.1), \ (2.5, \pi)$
Local maximum values:	$f(0.6) \approx 1, \ f(1.6) \approx 1, \ f(2.5) \approx 1$
Local minimum values:	$f(1.0) \approx 0.94, \ f(2.1) \approx 0.94$

The second derivative is

$$f''(x) = -(1 + 2\cos 2x)^2 \sin(x + \sin 2x) - 4\sin 2x \cos(x + \sin 2x)$$

Graphing both f and f'' in Figure 17, we obtain the following approximate values:

Concave upward on: (0.8, 1.3), (1.8, 2.3)

Concave downward on: (0, 0.8), (1.3, 1.8), (2.3, π)

Inflection points: (0, 0), (0.8, 0.97), (1.3, 0.97), (1.8, 0.97), (2.3, 0.97)

FIGURE 17

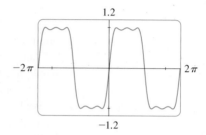

FIGURE 18

Having checked that Figure 15 does indeed represent f accurately for $0 \leqslant x \leqslant \pi$, we can state that the extended graph in Figure 18 represents f accurately for $-2\pi \leqslant x \leqslant 2\pi$.

Our final example is concerned with *families* of functions. As discussed in Appendix G, this means that the functions in the family are related to each other by a formula that contains one or more arbitrary constants. Each value of the constant gives rise to a member of the family and the idea is to see how the graph of the function changes as the constant changes.

V **EXAMPLE 5** How does the graph of $f(x) = 1/(x^2 + 2x + c)$ vary as c varies?

SOLUTION The graphs in Figures 19 and 20 (the special cases $c = 2$ and $c = -2$) show two very different-looking curves. Before drawing any more graphs, let's see what members of this family have in common. Since

$$\lim_{x \to \pm\infty} \frac{1}{x^2 + 2x + c} = 0$$

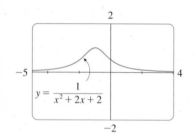

FIGURE 19
$c = 2$

for any value of c, they all have the x-axis as a horizontal asymptote. A vertical asymptote will occur when $x^2 + 2x + c = 0$. Solving this quadratic equation, we get $x = -1 \pm \sqrt{1 - c}$. When $c > 1$, there is no vertical asymptote (as in Figure 19). When $c = 1$, the graph has a single vertical asymptote $x = -1$ because

$$\lim_{x \to -1} \frac{1}{x^2 + 2x + 1} = \lim_{x \to -1} \frac{1}{(x + 1)^2} = \infty$$

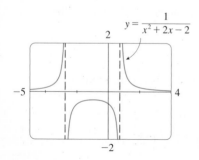

FIGURE 20
$c = -2$

When $c < 1$, there are two vertical asymptotes: $x = -1 \pm \sqrt{1 - c}$ (as in Figure 20). Now we compute the derivative:

$$f'(x) = -\frac{2x + 2}{(x^2 + 2x + c)^2}$$

This shows that $f'(x) = 0$ when $x = -1$ (if $c \neq 1$), $f'(x) > 0$ when $x < -1$, and $f'(x) < 0$ when $x > -1$. For $c \geqslant 1$, this means that f increases on $(-\infty, -1)$ and decreases on $(-1, \infty)$. For $c > 1$, there is an absolute maximum value $f(-1) = 1/(c - 1)$. For $c < 1$, $f(-1) = 1/(c - 1)$ is a local maximum value and the intervals of increase and decrease are interrupted at the vertical asymptotes.

Figure 21 is a "slide show" displaying five members of the family, all graphed in the viewing rectangle $[-5, 4]$ by $[-2, 2]$. As predicted, $c = 1$ is the value at which a transition takes place from two vertical asymptotes to one, and then to none. As c increases from 1, we see that the maximum point becomes lower; this is explained by the fact that $1/(c - 1) \to 0$ as $c \to \infty$. As c decreases from 1, the vertical asymptotes become more widely separated because the distance between them is $2\sqrt{1 - c}$, which becomes large as $c \to -\infty$. Again, the maximum point approaches the x-axis because $1/(c - 1) \to 0$ as $c \to -\infty$.

TEC See an animation of Figure 21 in Visual 3.6.

    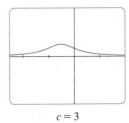

$c = -1$ $\qquad c = 0$ $\qquad c = 1$ $\qquad c = 2$ $\qquad c = 3$

FIGURE 21 The family of functions $f(x) = 1/(x^2 + 2x + c)$

There is clearly no inflection point when $c \leqslant 1$. For $c > 1$ we calculate that

$$f''(x) = \frac{2(3x^2 + 6x + 4 - c)}{(x^2 + 2x + c)^3}$$

and deduce that inflection points occur when $x = -1 \pm \sqrt{3(c - 1)}/3$. So the inflection points become more spread out as c increases and this seems plausible from the last two parts of Figure 21.

3.6 📐 **Exercises**

Access end-of-section exercises and tools for visualizing calculus online at **www.webassign.net**

3.7 **Optimization Problems**

The methods we have learned in this chapter for finding extreme values have practical applications in many areas of life. A businessperson wants to minimize costs and maximize profits. A traveler wants to minimize transportation time. Fermat's Principle in optics states that light follows the path that takes the least time. In this section we solve such problems as maximizing areas, volumes, and profits and minimizing distances, times, and costs.

In solving such practical problems the greatest challenge is often to convert the word problem into a mathematical optimization problem by setting up the function that is to be maximized or minimized. Let's recall the problem-solving principles discussed on page 77 and adapt them to this situation:

PS

> ### Steps in Solving Optimization Problems
>
> 1. **Understand the Problem**   The first step is to read the problem carefully until it is clearly understood. Ask yourself: What is the unknown? What are the given quantities? What are the given conditions?
>
> 2. **Draw a Diagram**   In most problems it is useful to draw a diagram and identify the given and required quantities on the diagram.
>
> 3. **Introduce Notation**   Assign a symbol to the quantity that is to be maximized or minimized (let's call it Q for now). Also select symbols $(a, b, c, \ldots, x, y)$ for other unknown quantities and label the diagram with these symbols. It may help to use initials as suggestive symbols—for example, A for area, h for height, t for time.
>
> 4. Express Q in terms of some of the other symbols from Step 3.
>
> 5. If Q has been expressed as a function of more than one variable in Step 4, use the given information to find relationships (in the form of equations) among these variables. Then use these equations to eliminate all but one of the variables in the expression for Q. Thus Q will be expressed as a function of *one* variable x, say, $Q = f(x)$. Write the domain of this function.
>
> 6. Use the methods of Sections 3.1 and 3.3 to find the *absolute* maximum or minimum value of f. In particular, if the domain of f is a closed interval, then the Closed Interval Method in Section 3.1 can be used.

EXAMPLE 1   A farmer has 2400 ft of fencing and wants to fence off a rectangular field that borders a straight river. He needs no fence along the river. What are the dimensions of the field that has the largest area?

PS Understand the problem
PS Analogy: Try special cases
PS Draw diagrams

SOLUTION   In order to get a feeling for what is happening in this problem, let's experiment with some special cases. Figure 1 (not to scale) shows three possible ways of laying out the 2400 ft of fencing.

Area = 100 · 2200 = 220,000 ft^2

Area = 700 · 1000 = 700,000 ft^2

Area = 1000 · 400 = 400,000 ft^2

FIGURE 1

We see that when we try shallow, wide fields or deep, narrow fields, we get relatively small areas. It seems plausible that there is some intermediate configuration that produces the largest area.

Figure 2 illustrates the general case. We wish to maximize the area A of the rectangle. Let x and y be the depth and width of the rectangle (in feet). Then we express A in terms of x and y:

$$A = xy$$

PS Introduce notation

We want to express A as a function of just one variable, so we eliminate y by expressing it in terms of x. To do this we use the given information that the total length of the fencing is 2400 ft. Thus

$$2x + y = 2400$$

FIGURE 2

From this equation we have $y = 2400 - 2x$, which gives

$$A = x(2400 - 2x) = 2400x - 2x^2$$

Note that $x \geq 0$ and $x \leq 1200$ (otherwise $A < 0$). So the function that we wish to maximize is

$$A(x) = 2400x - 2x^2 \qquad 0 \leq x \leq 1200$$

The derivative is $A'(x) = 2400 - 4x$, so to find the critical numbers we solve the equation

$$2400 - 4x = 0$$

which gives $x = 600$. The maximum value of A must occur either at this critical number or at an endpoint of the interval. Since $A(0) = 0$, $A(600) = 720{,}000$, and $A(1200) = 0$, the Closed Interval Method gives the maximum value as $A(600) = 720{,}000$.

[Alternatively, we could have observed that $A''(x) = -4 < 0$ for all x, so A is always concave downward and the local maximum at $x = 600$ must be an absolute maximum.]

Thus the rectangular field should be 600 ft deep and 1200 ft wide. ∎

V EXAMPLE 2 A cylindrical can is to be made to hold 1 L of oil. Find the dimensions that will minimize the cost of the metal to manufacture the can.

SOLUTION Draw the diagram as in Figure 3, where r is the radius and h the height (both in centimeters). In order to minimize the cost of the metal, we minimize the total surface area of the cylinder (top, bottom, and sides). From Figure 4 we see that the sides are made from a rectangular sheet with dimensions $2\pi r$ and h. So the surface area is

$$A = 2\pi r^2 + 2\pi r h$$

FIGURE 3

Area $2(\pi r^2)$ Area $(2\pi r)h$

FIGURE 4

To eliminate h we use the fact that the volume is given as 1 L, which we take to be 1000 cm³. Thus

$$\pi r^2 h = 1000$$

which gives $h = 1000/(\pi r^2)$. Substitution of this into the expression for A gives

$$A = 2\pi r^2 + 2\pi r\left(\frac{1000}{\pi r^2}\right) = 2\pi r^2 + \frac{2000}{r}$$

Therefore the function that we want to minimize is

$$A(r) = 2\pi r^2 + \frac{2000}{r} \qquad r > 0$$

To find the critical numbers, we differentiate:

$$A'(r) = 4\pi r - \frac{2000}{r^2} = \frac{4(\pi r^3 - 500)}{r^2}$$

Then $A'(r) = 0$ when $\pi r^3 = 500$, so the only critical number is $r = \sqrt[3]{500/\pi}$.

Since the domain of A is $(0, \infty)$, we can't use the argument of Example 1 concerning endpoints. But we can observe that $A'(r) < 0$ for $r < \sqrt[3]{500/\pi}$ and $A'(r) > 0$ for $r > \sqrt[3]{500/\pi}$, so A is decreasing for *all* r to the left of the critical number and increasing for *all* r to the right. Thus $r = \sqrt[3]{500/\pi}$ must give rise to an *absolute* minimum.

In the Applied Project on page 203 we investigate the most economical shape for a can by taking into account other manufacturing costs.

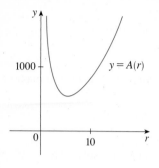

FIGURE 5

[Alternatively, we could argue that $A(r) \to \infty$ as $r \to 0^+$ and $A(r) \to \infty$ as $r \to \infty$, so there must be a minimum value of $A(r)$, which must occur at the critical number. See Figure 5.]

The value of h corresponding to $r = \sqrt[3]{500/\pi}$ is

$$h = \frac{1000}{\pi r^2} = \frac{1000}{\pi (500/\pi)^{2/3}} = 2\sqrt[3]{\frac{500}{\pi}} = 2r$$

Thus, to minimize the cost of the can, the radius should be $\sqrt[3]{500/\pi}$ cm and the height should be equal to twice the radius, namely, the diameter. ▬

NOTE 1 The argument used in Example 2 to justify the absolute minimum is a variant of the First Derivative Test (which applies only to *local* maximum or minimum values) and is stated here for future reference.

TEC Module 3.7 takes you through six additional optimization problems, including animations of the physical situations.

First Derivative Test for Absolute Extreme Values Suppose that c is a critical number of a continuous function f defined on an interval.

(a) If $f'(x) > 0$ for all $x < c$ and $f'(x) < 0$ for all $x > c$, then $f(c)$ is the absolute maximum value of f.

(b) If $f'(x) < 0$ for all $x < c$ and $f'(x) > 0$ for all $x > c$, then $f(c)$ is the absolute minimum value of f.

NOTE 2 An alternative method for solving optimization problems is to use implicit differentiation. Let's look at Example 2 again to illustrate the method. We work with the same equations

$$A = 2\pi r^2 + 2\pi rh \qquad \pi r^2 h = 1000$$

but instead of eliminating h, we differentiate both equations implicitly with respect to r:

$$A' = 4\pi r + 2\pi h + 2\pi rh' \qquad 2\pi rh + \pi r^2 h' = 0$$

The minimum occurs at a critical number, so we set $A' = 0$, simplify, and arrive at the equations

$$2r + h + rh' = 0 \qquad 2h + rh' = 0$$

and subtraction gives $2r - h = 0$, or $h = 2r$.

V **EXAMPLE 3** Find the point on the parabola $y^2 = 2x$ that is closest to the point $(1, 4)$.

SOLUTION The distance between the point $(1, 4)$ and the point (x, y) is

$$d = \sqrt{(x - 1)^2 + (y - 4)^2}$$

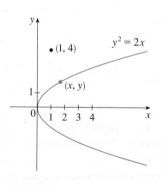

FIGURE 6

(See Figure 6.) But if (x, y) lies on the parabola, then $x = \frac{1}{2}y^2$, so the expression for d becomes

$$d = \sqrt{\left(\tfrac{1}{2}y^2 - 1\right)^2 + (y - 4)^2}$$

(Alternatively, we could have substituted $y = \sqrt{2x}$ to get d in terms of x alone.) Instead of minimizing d, we minimize its square:

$$d^2 = f(y) = \left(\tfrac{1}{2}y^2 - 1\right)^2 + (y - 4)^2$$

(You should convince yourself that the minimum of d occurs at the same point as the minimum of d^2, but d^2 is easier to work with.) Differentiating, we obtain

$$f'(y) = 2\left(\tfrac{1}{2}y^2 - 1\right)y + 2(y - 4) = y^3 - 8$$

so $f'(y) = 0$ when $y = 2$. Observe that $f'(y) < 0$ when $y < 2$ and $f'(y) > 0$ when $y > 2$, so by the First Derivative Test for Absolute Extreme Values, the absolute minimum occurs when $y = 2$. (Or we could simply say that because of the geometric nature of the problem, it's obvious that there is a closest point but not a farthest point.) The corresponding value of x is $x = \tfrac{1}{2}y^2 = 2$. Thus the point on $y^2 = 2x$ closest to $(1, 4)$ is $(2, 2)$.

EXAMPLE 4 A man launches his boat from point A on a bank of a straight river, 3 km wide, and wants to reach point B, 8 km downstream on the opposite bank, as quickly as possible (see Figure 7). He could row his boat directly across the river to point C and then run to B, or he could row directly to B, or he could row to some point D between C and B and then run to B. If he can row 6 km/h and run 8 km/h, where should he land to reach B as soon as possible? (We assume that the speed of the water is negligible compared with the speed at which the man rows.)

SOLUTION If we let x be the distance from C to D, then the running distance is $|DB| = 8 - x$ and the Pythagorean Theorem gives the rowing distance as $|AD| = \sqrt{x^2 + 9}$. We use the equation

$$\text{time} = \frac{\text{distance}}{\text{rate}}$$

Then the rowing time is $\sqrt{x^2 + 9}/6$ and the running time is $(8 - x)/8$, so the total time T as a function of x is

$$T(x) = \frac{\sqrt{x^2 + 9}}{6} + \frac{8 - x}{8}$$

The domain of this function T is $[0, 8]$. Notice that if $x = 0$, he rows to C and if $x = 8$, he rows directly to B. The derivative of T is

$$T'(x) = \frac{x}{6\sqrt{x^2 + 9}} - \frac{1}{8}$$

Thus, using the fact that $x \geqslant 0$, we have

$$T'(x) = 0 \iff \frac{x}{6\sqrt{x^2 + 9}} = \frac{1}{8} \iff 4x = 3\sqrt{x^2 + 9}$$

$$\iff 16x^2 = 9(x^2 + 9) \iff 7x^2 = 81$$

$$\iff x = \frac{9}{\sqrt{7}}$$

The only critical number is $x = 9/\sqrt{7}$. To see whether the minimum occurs at this critical number or at an endpoint of the domain $[0, 8]$, we evaluate T at all three points:

$$T(0) = 1.5 \qquad T\left(\frac{9}{\sqrt{7}}\right) = 1 + \frac{\sqrt{7}}{8} \approx 1.33 \qquad T(8) = \frac{\sqrt{73}}{6} \approx 1.42$$

FIGURE 7

FIGURE 8

FIGURE 9

Since the smallest of these values of T occurs when $x = 9/\sqrt{7}$, the absolute minimum value of T must occur there. Figure 8 illustrates this calculation by showing the graph of T.

Thus the man should land the boat at a point $9/\sqrt{7}$ km (≈ 3.4 km) downstream from his starting point.

V **EXAMPLE 5** Find the area of the largest rectangle that can be inscribed in a semicircle of radius r.

SOLUTION 1 Let's take the semicircle to be the upper half of the circle $x^2 + y^2 = r^2$ with center the origin. Then the word *inscribed* means that the rectangle has two vertices on the semicircle and two vertices on the x-axis as shown in Figure 9.

Let (x, y) be the vertex that lies in the first quadrant. Then the rectangle has sides of lengths $2x$ and y, so its area is

$$A = 2xy$$

To eliminate y we use the fact that (x, y) lies on the circle $x^2 + y^2 = r^2$ and so $y = \sqrt{r^2 - x^2}$. Thus

$$A = 2x\sqrt{r^2 - x^2}$$

The domain of this function is $0 \leq x \leq r$. Its derivative is

$$A' = 2\sqrt{r^2 - x^2} - \frac{2x^2}{\sqrt{r^2 - x^2}} = \frac{2(r^2 - 2x^2)}{\sqrt{r^2 - x^2}}$$

which is 0 when $2x^2 = r^2$, that is, $x = r/\sqrt{2}$ (since $x \geq 0$). This value of x gives a maximum value of A since $A(0) = 0$ and $A(r) = 0$. Therefore the area of the largest inscribed rectangle is

$$A\left(\frac{r}{\sqrt{2}}\right) = 2\frac{r}{\sqrt{2}}\sqrt{r^2 - \frac{r^2}{2}} = r^2$$

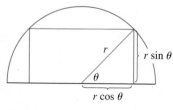

FIGURE 10

SOLUTION 2 A simpler solution is possible if we think of using an angle as a variable. Let θ be the angle shown in Figure 10. Then the area of the rectangle is

$$A(\theta) = (2r \cos \theta)(r \sin \theta) = r^2(2 \sin \theta \, \cos \theta) = r^2 \sin 2\theta$$

We know that $\sin 2\theta$ has a maximum value of 1 and it occurs when $2\theta = \pi/2$. So $A(\theta)$ has a maximum value of r^2 and it occurs when $\theta = \pi/4$.

Notice that this trigonometric solution doesn't involve differentiation. In fact, we didn't need to use calculus at all.

Applications to Business and Economics

In Section 2.7 we introduced the idea of marginal cost. Recall that if $C(x)$, the **cost function**, is the cost of producing x units of a certain product, then the **marginal cost** is the rate of change of C with respect to x. In other words, the marginal cost function is the derivative, $C'(x)$, of the cost function.

Now let's consider marketing. Let $p(x)$ be the price per unit that the company can charge if it sells x units. Then p is called the **demand function** (or **price function**) and we would expect it to be a decreasing function of x. If x units are sold and the price per unit is $p(x)$, then the total revenue is

$$R(x) = xp(x)$$

and R is called the **revenue function**. The derivative R' of the revenue function is called the **marginal revenue function** and is the rate of change of revenue with respect to the number of units sold.

If x units are sold, then the total profit is

$$P(x) = R(x) - C(x)$$

and P is called the **profit function**. The **marginal profit function** is P', the derivative of the profit function. In Exercises 57–62 you are asked to use the marginal cost, revenue, and profit functions to minimize costs and maximize revenues and profits.

V   EXAMPLE 6   A store has been selling 200 Blu-ray disc players a week at \$350 each. A market survey indicates that for each \$10 rebate offered to buyers, the number of units sold will increase by 20 a week. Find the demand function and the revenue function. How large a rebate should the store offer to maximize its revenue?

SOLUTION   If x is the number of Blu-ray players sold per week, then the weekly increase in sales is $x - 200$. For each increase of 20 units sold, the price is decreased by \$10. So for each additional unit sold, the decrease in price will be $\frac{1}{20} \times 10$ and the demand function is

$$p(x) = 350 - \tfrac{10}{20}(x - 200) = 450 - \tfrac{1}{2}x$$

The revenue function is

$$R(x) = xp(x) = 450x - \tfrac{1}{2}x^2$$

Since $R'(x) = 450 - x$, we see that $R'(x) = 0$ when $x = 450$. This value of x gives an absolute maximum by the First Derivative Test (or simply by observing that the graph of R is a parabola that opens downward). The corresponding price is

$$p(450) = 450 - \tfrac{1}{2}(450) = 225$$

and the rebate is $350 - 225 = 125$. Therefore, to maximize revenue, the store should offer a rebate of \$125.

3.7   Exercises

Access end-of-section exercises and tools for visualizing calculus online at **www.webassign.net**

APPLIED PROJECT   **THE SHAPE OF A CAN**

In this project we investigate the most economical shape for a can. We first interpret this to mean that the volume V of a cylindrical can is given and we need to find the height h and radius r that minimize the cost of the metal to make the can (see the figure). If we disregard any waste metal in the manufacturing process, then the problem is to minimize the surface area of the cylinder. We solved this problem in Example 2 in Section 3.7 and we found that $h = 2r$; that is, the height should be the same as the diameter. But if you go to your cupboard or your supermarket with a

⊞   Graphing calculator or computer with graphing software required ruler, you will discover that the height is usually greater than the diameter and the ratio h/r varies from 2 up to about 3.8. Let's see if we can explain this phenomenon.

Discs cut from squares

Discs cut from hexagons

1. The material for the cans is cut from sheets of metal. The cylindrical sides are formed by bending rectangles; these rectangles are cut from the sheet with little or no waste. But if the top and bottom discs are cut from squares of side $2r$ (as in the figure), this leaves considerable waste metal, which may be recycled but has little or no value to the can makers. If this is the case, show that the amount of metal used is minimized when

$$\frac{h}{r} = \frac{8}{\pi} \approx 2.55$$

2. A more efficient packing of the discs is obtained by dividing the metal sheet into hexagons and cutting the circular lids and bases from the hexagons (see the figure). Show that if this strategy is adopted, then

$$\frac{h}{r} = \frac{4\sqrt{3}}{\pi} \approx 2.21$$

3. The values of h/r that we found in Problems 1 and 2 are a little closer to the ones that actually occur on supermarket shelves, but they still don't account for everything. If we look more closely at some real cans, we see that the lid and the base are formed from discs with radius larger than r that are bent over the ends of the can. If we allow for this we would increase h/r. More significantly, in addition to the cost of the metal we need to incorporate the manufacturing of the can into the cost. Let's assume that most of the expense is incurred in joining the sides to the rims of the cans. If we cut the discs from hexagons as in Problem 2, then the total cost is proportional to

$$4\sqrt{3}\, r^2 + 2\pi rh + k(4\pi r + h)$$

where k is the reciprocal of the length that can be joined for the cost of one unit area of metal. Show that this expression is minimized when

$$\frac{\sqrt[3]{V}}{k} = \sqrt{\frac{\pi h}{r}} \cdot \frac{2\pi - h/r}{\pi h/r - 4\sqrt{3}}$$

4. Plot $\sqrt[3]{V}/k$ as a function of $x = h/r$ and use your graph to argue that when a can is large or joining is cheap, we should make h/r approximately 2.21 (as in Problem 2). But when the can is small or joining is costly, h/r should be substantially larger.

5. Our analysis shows that large cans should be almost square but small cans should be tall and thin. Take a look at the relative shapes of the cans in a supermarket. Is our conclusion usually true in practice? Are there exceptions? Can you suggest reasons why small cans are not always tall and thin?

3.8   Newton's Method

Suppose that a car dealer offers to sell you a car for $18,000 or for payments of $375 per month for five years. You would like to know what monthly interest rate the dealer is, in effect, charging you. To find the answer, you have to solve the equation

$$\boxed{1}\qquad\qquad 48x(1+x)^{60} - (1+x)^{60} + 1 = 0$$

(The details are explained in Exercise 39.) How would you solve such an equation?

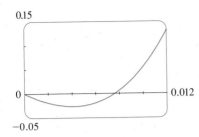

FIGURE 1

Try to solve Equation 1 using the numerical rootfinder on your calculator or computer. Some machines are not able to solve it. Others are successful but require you to specify a starting point for the search.

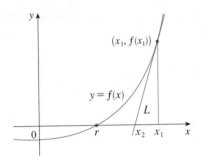

FIGURE 2

For a quadratic equation $ax^2 + bx + c = 0$ there is a well-known formula for the roots. For third- and fourth-degree equations there are also formulas for the roots, but they are extremely complicated. If f is a polynomial of degree 5 or higher, there is no such formula (see the note on page 126). Likewise, there is no formula that will enable us to find the exact roots of a transcendental equation such as $\cos x = x$.

We can find an *approximate* solution to Equation 1 by plotting the left side of the equation. Using a graphing device, and after experimenting with viewing rectangles, we produce the graph in Figure 1.

We see that in addition to the solution $x = 0$, which doesn't interest us, there is a solution between 0.007 and 0.008. Zooming in shows that the root is approximately 0.0076. If we need more accuracy we could zoom in repeatedly, but that becomes tiresome. A faster alternative is to use a numerical rootfinder on a calculator or computer algebra system. If we do so, we find that the root, correct to nine decimal places, is 0.007628603.

How do those numerical rootfinders work? They use a variety of methods, but most of them make some use of **Newton's method**, also called the **Newton-Raphson method**. We will explain how this method works, partly to show what happens inside a calculator or computer, and partly as an application of the idea of linear approximation.

The geometry behind Newton's method is shown in Figure 2, where the root that we are trying to find is labeled r. We start with a first approximation x_1, which is obtained by guessing, or from a rough sketch of the graph of f, or from a computer-generated graph of f. Consider the tangent line L to the curve $y = f(x)$ at the point $(x_1, f(x_1))$ and look at the x-intercept of L, labeled x_2. The idea behind Newton's method is that the tangent line is close to the curve and so its x-intercept, x_2, is close to the x-intercept of the curve (namely, the root r that we are seeking). Because the tangent is a line, we can easily find its x-intercept.

To find a formula for x_2 in terms of x_1 we use the fact that the slope of L is $f'(x_1)$, so its equation is

$$y - f(x_1) = f'(x_1)(x - x_1)$$

Since the x-intercept of L is x_2, we set $y = 0$ and obtain

$$0 - f(x_1) = f'(x_1)(x_2 - x_1)$$

If $f'(x_1) \neq 0$, we can solve this equation for x_2:

$$x_2 = x_1 - \frac{f(x_1)}{f'(x_1)}$$

We use x_2 as a second approximation to r.

Next we repeat this procedure with x_1 replaced by the second approximation x_2, using the tangent line at $(x_2, f(x_2))$. This gives a third approximation:

$$x_3 = x_2 - \frac{f(x_2)}{f'(x_2)}$$

If we keep repeating this process, we obtain a sequence of approximations $x_1, x_2, x_3, x_4, \ldots$ as shown in Figure 3. In general, if the nth approximation is x_n and $f'(x_n) \neq 0$, then the next approximation is given by

$$\boxed{2 \qquad x_{n+1} = x_n - \frac{f(x_n)}{f'(x_n)}}$$

FIGURE 3

If the numbers x_n become closer and closer to r as n becomes large, then we say that the sequence *converges* to r and we write

$$\lim_{n \to \infty} x_n = r$$

Sequences were briefly introduced in *A Preview of Calculus* on page 5. A more thorough discussion starts in Section 11.1.

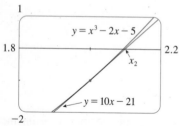

FIGURE 4

⊘ Although the sequence of successive approximations converges to the desired root for functions of the type illustrated in Figure 3, in certain circumstances the sequence may not converge. For example, consider the situation shown in Figure 4. You can see that x_2 is a worse approximation than x_1. This is likely to be the case when $f'(x_1)$ is close to 0. It might even happen that an approximation (such as x_3 in Figure 4) falls outside the domain of f. Then Newton's method fails and a better initial approximation x_1 should be chosen. See Exercises 29–32 for specific examples in which Newton's method works very slowly or does not work at all.

V **EXAMPLE 1** Starting with $x_1 = 2$, find the third approximation x_3 to the root of the equation $x^3 - 2x - 5 = 0$.

SOLUTION We apply Newton's method with

$$f(x) = x^3 - 2x - 5 \quad \text{and} \quad f'(x) = 3x^2 - 2$$

TEC In Module 3.8 you can investigate how Newton's Method works for several functions and what happens when you change x_1.

Newton himself used this equation to illustrate his method and he chose $x_1 = 2$ after some experimentation because $f(1) = -6$, $f(2) = -1$, and $f(3) = 16$. Equation 2 becomes

$$x_{n+1} = x_n - \frac{x_n^3 - 2x_n - 5}{3x_n^2 - 2}$$

With $n = 1$ we have

$$x_2 = x_1 - \frac{x_1^3 - 2x_1 - 5}{3x_1^2 - 2}$$

$$= 2 - \frac{2^3 - 2(2) - 5}{3(2)^2 - 2} = 2.1$$

Figure 5 shows the geometry behind the first step in Newton's method in Example 1. Since $f'(2) = 10$, the tangent line to $y = x^3 - 2x - 5$ at $(2, -1)$ has equation $y = 10x - 21$ so its x-intercept is $x_2 = 2.1$.

Then with $n = 2$ we obtain

$$x_3 = x_2 - \frac{x_2^3 - 2x_2 - 5}{3x_2^2 - 2}$$

$$= 2.1 - \frac{(2.1)^3 - 2(2.1) - 5}{3(2.1)^2 - 2} \approx 2.0946$$

FIGURE 5

It turns out that this third approximation $x_3 \approx 2.0946$ is accurate to four decimal places.

Suppose that we want to achieve a given accuracy, say to eight decimal places, using Newton's method. How do we know when to stop? The rule of thumb that is generally used is that we can stop when successive approximations x_n and x_{n+1} agree to eight decimal places. (A precise statement concerning accuracy in Newton's method will be given in Exercise 39 in Section 11.11.)

Notice that the procedure in going from n to $n + 1$ is the same for all values of n. (It is called an *iterative* process.) This means that Newton's method is particularly convenient for use with a programmable calculator or a computer.

V **EXAMPLE 2** Use Newton's method to find $\sqrt[6]{2}$ correct to eight decimal places.

SOLUTION First we observe that finding $\sqrt[6]{2}$ is equivalent to finding the positive root of the equation

$$x^6 - 2 = 0$$

so we take $f(x) = x^6 - 2$. Then $f'(x) = 6x^5$ and Formula 2 (Newton's method) becomes

$$x_{n+1} = x_n - \frac{x_n^6 - 2}{6x_n^5}$$

If we choose $x_1 = 1$ as the initial approximation, then we obtain

$$x_2 \approx 1.16666667$$
$$x_3 \approx 1.12644368$$
$$x_4 \approx 1.12249707$$
$$x_5 \approx 1.12246205$$
$$x_6 \approx 1.12246205$$

Since x_5 and x_6 agree to eight decimal places, we conclude that

$$\sqrt[6]{2} \approx 1.12246205$$

to eight decimal places.

V **EXAMPLE 3** Find, correct to six decimal places, the root of the equation $\cos x = x$.

SOLUTION We first rewrite the equation in standard form:

$$\cos x - x = 0$$

Therefore we let $f(x) = \cos x - x$. Then $f'(x) = -\sin x - 1$, so Formula 2 becomes

$$x_{n+1} = x_n - \frac{\cos x_n - x_n}{-\sin x_n - 1} = x_n + \frac{\cos x_n - x_n}{\sin x_n + 1}$$

In order to guess a suitable value for x_1 we sketch the graphs of $y = \cos x$ and $y = x$ in Figure 6. It appears that they intersect at a point whose x-coordinate is somewhat less than 1, so let's take $x_1 = 1$ as a convenient first approximation. Then, remembering to put our calculator in radian mode, we get

$$x_2 \approx 0.75036387$$
$$x_3 \approx 0.73911289$$
$$x_4 \approx 0.73908513$$
$$x_5 \approx 0.73908513$$

Since x_4 and x_5 agree to six decimal places (eight, in fact), we conclude that the root of the equation, correct to six decimal places, is 0.739085.

FIGURE 6

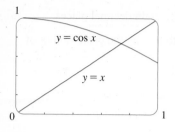

FIGURE 7

Instead of using the rough sketch in Figure 6 to get a starting approximation for Newton's method in Example 3, we could have used the more accurate graph that a calculator or computer provides. Figure 7 suggests that we use $x_1 = 0.75$ as the initial approximation. Then Newton's method gives

$$x_2 \approx 0.73911114$$

$$x_3 \approx 0.73908513$$

$$x_4 \approx 0.73908513$$

and so we obtain the same answer as before, but with one fewer step.

You might wonder why we bother at all with Newton's method if a graphing device is available. Isn't it easier to zoom in repeatedly and find the roots as in Appendix G? If only one or two decimal places of accuracy are required, then indeed Newton's method is inappropriate and a graphing device suffices. But if six or eight decimal places are required, then repeated zooming becomes tiresome. It is usually faster and more efficient to use a computer and Newton's method in tandem—the graphing device to get started and Newton's method to finish.

3.8 Exercises

Access end-of-section exercises and tools for visualizing calculus online at **www.webassign.net**

3.9 Antiderivatives

A physicist who knows the velocity of a particle might wish to know its position at a given time. An engineer who can measure the variable rate at which water is leaking from a tank wants to know the amount leaked over a certain time period. A biologist who knows the rate at which a bacteria population is increasing might want to deduce what the size of the population will be at some future time. In each case, the problem is to find a function F whose derivative is a known function f. If such a function F exists, it is called an *antiderivative* of f.

> **Definition** A function F is called an **antiderivative** of f on an interval I if $F'(x) = f(x)$ for all x in I.

For instance, let $f(x) = x^2$. It isn't difficult to discover an antiderivative of f if we keep the Power Rule in mind. In fact, if $F(x) = \frac{1}{3}x^3$, then $F'(x) = x^2 = f(x)$. But the function $G(x) = \frac{1}{3}x^3 + 100$ also satisfies $G'(x) = x^2$. Therefore both F and G are antiderivatives of f. Indeed, any function of the form $H(x) = \frac{1}{3}x^3 + C$, where C is a constant, is an antiderivative of f. The question arises: Are there any others?

To answer this question, recall that in Section 3.2 we used the Mean Value Theorem to prove that if two functions have identical derivatives on an interval, then they must differ by a constant (Corollary 3.2.7). Thus if F and G are any two antiderivatives of f, then

$$F'(x) = f(x) = G'(x)$$

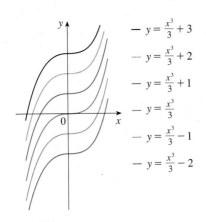

$$— \; y = \frac{x^3}{3} + 3$$
$$— \; y = \frac{x^3}{3} + 2$$
$$— \; y = \frac{x^3}{3} + 1$$
$$— \; y = \frac{x^3}{3}$$
$$— \; y = \frac{x^3}{3} - 1$$
$$— \; y = \frac{x^3}{3} - 2$$

FIGURE 1

Members of the family of
antiderivatives of $f(x) = x^2$

so $G(x) - F(x) = C$, where C is a constant. We can write this as $G(x) = F(x) + C$, so we have the following result.

---

1 Theorem If F is an antiderivative of f on an interval I, then the most general antiderivative of f on I is

$$F(x) + C$$

where C is an arbitrary constant.

Going back to the function $f(x) = x^2$, we see that the general antiderivative of f is $\frac{1}{3}x^3 + C$. By assigning specific values to the constant C, we obtain a family of functions whose graphs are vertical translates of one another (see Figure 1). This makes sense because each curve must have the same slope at any given value of x.

EXAMPLE 1 Find the most general antiderivative of each of the following functions.
(a) $f(x) = \sin x$ (b) $f(x) = x^n$, $n \geq 0$ (c) $f(x) = x^{-3}$

SOLUTION
(a) If $F(x) = -\cos x$, then $F'(x) = \sin x$, so an antiderivative of $\sin x$ is $-\cos x$. By Theorem 1, the most general antiderivative is $G(x) = -\cos x + C$.

(b) We use the Power Rule to discover an antiderivative of x^n:

$$\frac{d}{dx}\left(\frac{x^{n+1}}{n+1}\right) = \frac{(n+1)x^n}{n+1} = x^n$$

Thus the general antiderivative of $f(x) = x^n$ is

$$F(x) = \frac{x^{n+1}}{n+1} + C$$

This is valid for $n \geq 0$ because then $f(x) = x^n$ is defined on an interval.

(c) If we put $n = -3$ in part (b) we get the particular antiderivative $F(x) = x^{-2}/(-2)$ by the same calculation. But notice that $f(x) = x^{-3}$ is not defined at $x = 0$. Thus Theorem 1 tells us only that the general antiderivative of f is $x^{-2}/(-2) + C$ on any interval that does not contain 0. So the general antiderivative of $f(x) = 1/x^3$ is

$$F(x) = \begin{cases} -\dfrac{1}{2x^2} + C_1 & \text{if } x > 0 \\[2mm] -\dfrac{1}{2x^2} + C_2 & \text{if } x < 0 \end{cases}$$

As in Example 1, every differentiation formula, when read from right to left, gives rise to an antidifferentiation formula. In Table 2 we list some particular antiderivatives. Each formula in the table is true because the derivative of the function in the right column appears in the left column. In particular, the first formula says that the antiderivative of a constant times a function is the constant times the antiderivative of the function. The second formula says that the antiderivative of a sum is the sum of the antiderivatives. (We use the notation $F' = f$, $G' = g$.)

2 **Table of Antidifferentiation Formulas**

To obtain the most general antiderivative from the particular ones in Table 2, we have to add a constant (or constants), as in Example 1.

Function	Particular antiderivative	Function	Particular antiderivative
$cf(x)$	$cF(x)$	$\cos x$	$\sin x$
$f(x) + g(x)$	$F(x) + G(x)$	$\sin x$	$-\cos x$
$x^n \; (n \neq -1)$	$\dfrac{x^{n+1}}{n+1}$	$\sec^2 x$	$\tan x$
		$\sec x \tan x$	$\sec x$

EXAMPLE 2 Find all functions g such that

$$g'(x) = 4 \sin x + \frac{2x^5 - \sqrt{x}}{x}$$

SOLUTION We first rewrite the given function as follows:

$$g'(x) = 4 \sin x + \frac{2x^5}{x} - \frac{\sqrt{x}}{x} = 4 \sin x + 2x^4 - \frac{1}{\sqrt{x}}$$

Thus we want to find an antiderivative of

$$g'(x) = 4 \sin x + 2x^4 - x^{-1/2}$$

Using the formulas in Table 2 together with Theorem 1, we obtain

$$g(x) = 4(-\cos x) + 2\frac{x^5}{5} - \frac{x^{1/2}}{\frac{1}{2}} + C$$

$$= -4 \cos x + \tfrac{2}{5}x^5 - 2\sqrt{x} + C \qquad \blacksquare$$

In applications of calculus it is very common to have a situation as in Example 2, where it is required to find a function, given knowledge about its derivatives. An equation that involves the derivatives of a function is called a **differential equation**. Such equations will be studied in some detail in Chapter 9, but for the present we can solve some elementary differential equations. The general solution of a differential equation involves an arbitrary constant (or constants) as in Example 2. However, there may be some extra conditions given that will determine the constants and therefore uniquely specify the solution.

EXAMPLE 3 Find f if $f'(x) = x\sqrt{x}$ and $f(1) = 2$.

SOLUTION The general antiderivative of

$$f'(x) = x^{3/2}$$

is

$$f(x) = \frac{x^{5/2}}{\frac{5}{2}} + C = \tfrac{2}{5}x^{5/2} + C$$

To determine C we use the fact that $f(1) = 2$:

$$f(1) = \tfrac{2}{5} + C = 2$$

Solving for C, we get $C = 2 - \tfrac{2}{5} = \tfrac{8}{5}$, so the particular solution is

$$f(x) = \frac{2x^{5/2} + 8}{5} \qquad \blacksquare$$

V EXAMPLE 4 Find f if $f''(x) = 12x^2 + 6x - 4$, $f(0) = 4$, and $f(1) = 1$.

SOLUTION The general antiderivative of $f''(x) = 12x^2 + 6x - 4$ is

$$f'(x) = 12\frac{x^3}{3} + 6\frac{x^2}{2} - 4x + C = 4x^3 + 3x^2 - 4x + C$$

Using the antidifferentiation rules once more, we find that

$$f(x) = 4\frac{x^4}{4} + 3\frac{x^3}{3} - 4\frac{x^2}{2} + Cx + D = x^4 + x^3 - 2x^2 + Cx + D$$

To determine C and D we use the given conditions that $f(0) = 4$ and $f(1) = 1$. Since $f(0) = 0 + D = 4$, we have $D = 4$. Since

$$f(1) = 1 + 1 - 2 + C + 4 = 1$$

we have $C = -3$. Therefore the required function is

$$f(x) = x^4 + x^3 - 2x^2 - 3x + 4$$

If we are given the graph of a function f, it seems reasonable that we should be able to sketch the graph of an antiderivative F. Suppose, for instance, that we are given that $F(0) = 1$. Then we have a place to start, the point $(0, 1)$, and the direction in which we move our pencil is given at each stage by the derivative $F'(x) = f(x)$. In the next example we use the principles of this chapter to show how to graph F even when we don't have a formula for f. This would be the case, for instance, when $f(x)$ is determined by experimental data.

V EXAMPLE 5 The graph of a function f is given in Figure 2. Make a rough sketch of an antiderivative F, given that $F(0) = 2$.

SOLUTION We are guided by the fact that the slope of $y = F(x)$ is $f(x)$. We start at the point $(0, 2)$ and draw F as an initially decreasing function since $f(x)$ is negative when $0 < x < 1$. Notice that $f(1) = f(3) = 0$, so F has horizontal tangents when $x = 1$ and $x = 3$. For $1 < x < 3$, $f(x)$ is positive and so F is increasing. We see that F has a local minimum when $x = 1$ and a local maximum when $x = 3$. For $x > 3$, $f(x)$ is negative and so F is decreasing on $(3, \infty)$. Since $f(x) \to 0$ as $x \to \infty$, the graph of F becomes flatter as $x \to \infty$. Also notice that $F''(x) = f'(x)$ changes from positive to negative at $x = 2$ and from negative to positive at $x = 4$, so F has inflection points when $x = 2$ and $x = 4$. We use this information to sketch the graph of the antiderivative in Figure 3.

Rectilinear Motion

Antidifferentiation is particularly useful in analyzing the motion of an object moving in a straight line. Recall that if the object has position function $s = f(t)$, then the velocity function is $v(t) = s'(t)$. This means that the position function is an antiderivative of the velocity function. Likewise, the acceleration function is $a(t) = v'(t)$, so the velocity function is an antiderivative of the acceleration. If the acceleration and the initial values $s(0)$ and $v(0)$ are known, then the position function can be found by antidifferentiating twice.

V EXAMPLE 6 A particle moves in a straight line and has acceleration given by $a(t) = 6t + 4$. Its initial velocity is $v(0) = -6$ cm/s and its initial displacement is $s(0) = 9$ cm. Find its position function $s(t)$.

SOLUTION Since $v'(t) = a(t) = 6t + 4$, antidifferentiation gives

$$v(t) = 6\frac{t^2}{2} + 4t + C = 3t^2 + 4t + C$$

FIGURE 2

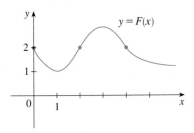

FIGURE 3

Note that $v(0) = C$. But we are given that $v(0) = -6$, so $C = -6$ and

$$v(t) = 3t^2 + 4t - 6$$

Since $v(t) = s'(t)$, s is the antiderivative of v:

$$s(t) = 3\,\frac{t^3}{3} + 4\,\frac{t^2}{2} - 6t + D = t^3 + 2t^2 - 6t + D$$

This gives $s(0) = D$. We are given that $s(0) = 9$, so $D = 9$ and the required position function is

$$s(t) = t^3 + 2t^2 - 6t + 9$$

An object near the surface of the earth is subject to a gravitational force that produces a downward acceleration denoted by g. For motion close to the ground we may assume that g is constant, its value being about 9.8 m/s² (or 32 ft/s²).

EXAMPLE 7   A ball is thrown upward with a speed of 48 ft/s from the edge of a cliff 432 ft above the ground. Find its height above the ground t seconds later. When does it reach its maximum height? When does it hit the ground?

SOLUTION   The motion is vertical and we choose the positive direction to be upward. At time t the distance above the ground is $s(t)$ and the velocity $v(t)$ is decreasing. Therefore the acceleration must be negative and we have

$$a(t) = \frac{dv}{dt} = -32$$

Taking antiderivatives, we have

$$v(t) = -32t + C$$

To determine C we use the given information that $v(0) = 48$. This gives $48 = 0 + C$, so

$$v(t) = -32t + 48$$

The maximum height is reached when $v(t) = 0$, that is, after 1.5 s. Since $s'(t) = v(t)$, we antidifferentiate again and obtain

$$s(t) = -16t^2 + 48t + D$$

Using the fact that $s(0) = 432$, we have $432 = 0 + D$ and so

$$s(t) = -16t^2 + 48t + 432$$

The expression for $s(t)$ is valid until the ball hits the ground. This happens when $s(t) = 0$, that is, when

$$-16t^2 + 48t + 432 = 0$$

or, equivalently,

$$t^2 - 3t - 27 = 0$$

Using the quadratic formula to solve this equation, we get

$$t = \frac{3 \pm 3\sqrt{13}}{2}$$

We reject the solution with the minus sign since it gives a negative value for t. Therefore the ball hits the ground after $3(1 + \sqrt{13})/2 \approx 6.9$ s.

Figure 4 shows the position function of the ball in Example 7. The graph corroborates the conclusions we reached: The ball reaches its maximum height after 1.5 s and hits the ground after 6.9 s.

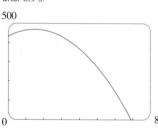

500

0   8

FIGURE 4

3.9   Exercises

Access end-of-section exercises and tools for visualizing calculus online at **www.webassign.net**

3   Review

Concept Check

1. Explain the difference between an absolute maximum and a local maximum. Illustrate with a sketch.

2. (a) What does the Extreme Value Theorem say?
(b) Explain how the Closed Interval Method works.

3. (a) State Fermat's Theorem.
(b) Define a critical number of f.

4. (a) State Rolle's Theorem.
(b) State the Mean Value Theorem and give a geometric interpretation.

5. (a) State the Increasing/Decreasing Test.
(b) What does it mean to say that f is concave upward on an interval I?
(c) State the Concavity Test.
(d) What are inflection points? How do you find them?

6. (a) State the First Derivative Test.
(b) State the Second Derivative Test.
(c) What are the relative advantages and disadvantages of these tests?

7. Explain the meaning of each of the following statements.
(a) $\lim_{x \to \infty} f(x) = L$ (b) $\lim_{x \to -\infty} f(x) = L$ (c) $\lim_{x \to \infty} f(x) = \infty$
(d) The curve $y = f(x)$ has the horizontal asymptote $y = L$.

8. If you have a graphing calculator or computer, why do you need calculus to graph a function?

9. (a) Given an initial approximation x_1 to a root of the equation $f(x) = 0$, explain geometrically, with a diagram, how the second approximation x_2 in Newton's method is obtained.
(b) Write an expression for x_2 in terms of x_1, $f(x_1)$, and $f'(x_1)$.
(c) Write an expression for x_{n+1} in terms of x_n, $f(x_n)$, and $f'(x_n)$.
(d) Under what circumstances is Newton's method likely to fail or to work very slowly?

10. (a) What is an antiderivative of a function f?
(b) Suppose F_1 and F_2 are both antiderivatives of f on an interval I. How are F_1 and F_2 related?

True-False Quiz

Determine whether the statement is true or false. If it is true, explain why. If it is false, explain why or give an example that disproves the statement.

1. If $f'(c) = 0$, then f has a local maximum or minimum at c.

2. If f has an absolute minimum value at c, then $f'(c) = 0$.

3. If f is continuous on (a, b), then f attains an absolute maximum value $f(c)$ and an absolute minimum value $f(d)$ at some numbers c and d in (a, b).

4. If f is differentiable and $f(-1) = f(1)$, then there is a number c such that $|c| < 1$ and $f'(c) = 0$.

5. If $f'(x) < 0$ for $1 < x < 6$, then f is decreasing on $(1, 6)$.

6. If $f''(2) = 0$, then $(2, f(2))$ is an inflection point of the curve $y = f(x)$.

7. If $f'(x) = g'(x)$ for $0 < x < 1$, then $f(x) = g(x)$ for $0 < x < 1$.

8. There exists a function f such that $f(1) = -2$, $f(3) = 0$, and $f'(x) > 1$ for all x.

9. There exists a function f such that $f(x) > 0$, $f'(x) < 0$, and $f''(x) > 0$ for all x.

10. There exists a function f such that $f(x) < 0$, $f'(x) < 0$, and $f''(x) > 0$ for all x.

11. If f and g are increasing on an interval I, then $f + g$ is increasing on I.

12. If f and g are increasing on an interval I, then $f - g$ is increasing on I.

13. If f and g are increasing on an interval I, then fg is increasing on I.

14. If f and g are positive increasing functions on an interval I, then fg is increasing on I.

15. If f is increasing and $f(x) > 0$ on I, then $g(x) = 1/f(x)$ is decreasing on I.

16. If f is even, then f' is even.

17. If f is periodic, then f' is periodic.

18. The most general antiderivative of $f(x) = x^{-2}$ is
$$F(x) = -\frac{1}{x} + C$$

19. If $f'(x)$ exists and is nonzero for all x, then $f(1) \neq f(0)$.

Exercises

1–6 Find the local and absolute extreme values of the function on the given interval.

1. $f(x) = x^3 - 6x^2 + 9x + 1$, $[2, 4]$

2. $f(x) = x\sqrt{1-x}$, $[-1, 1]$

3. $f(x) = \dfrac{3x-4}{x^2+1}$, $[-2, 2]$

4. $f(x) = \sqrt{x^2+x+1}$, $[-2, 1]$

5. $f(x) = x + 2\cos x$, $[-\pi, \pi]$

6. $f(x) = \sin x + \cos^2 x$, $[0, \pi]$

7–12 Find the limit.

7. $\displaystyle\lim_{x\to\infty} \frac{3x^4 + x - 5}{6x^4 - 2x^2 + 1}$

8. $\displaystyle\lim_{t\to\infty} \frac{t^3 - t + 2}{(2t-1)(t^2+t+1)}$

9. $\displaystyle\lim_{x\to-\infty} \frac{\sqrt{4x^2+1}}{3x-1}$

10. $\displaystyle\lim_{x\to-\infty}(x^2 + x^3)$

11. $\displaystyle\lim_{x\to\infty}\left(\sqrt{4x^2+3x} - 2x\right)$

12. $\displaystyle\lim_{x\to\infty} \frac{\sin^4 x}{\sqrt{x}}$

13–15 Sketch the graph of a function that satisfies the given conditions.

13. $f(0) = 0$, $f'(-2) = f'(1) = f'(9) = 0$,
$\displaystyle\lim_{x\to\infty} f(x) = 0$, $\displaystyle\lim_{x\to 6} f(x) = -\infty$,
$f'(x) < 0$ on $(-\infty, -2)$, $(1, 6)$, and $(9, \infty)$,
$f'(x) > 0$ on $(-2, 1)$ and $(6, 9)$,
$f''(x) > 0$ on $(-\infty, 0)$ and $(12, \infty)$,
$f''(x) < 0$ on $(0, 6)$ and $(6, 12)$

14. $f(0) = 0$, f is continuous and even,
$f'(x) = 2x$ if $0 < x < 1$, $f'(x) = -1$ if $1 < x < 3$,
$f'(x) = 1$ if $x > 3$

15. f is odd, $f'(x) < 0$ for $0 < x < 2$,
$f'(x) > 0$ for $x > 2$, $f''(x) > 0$ for $0 < x < 3$,
$f''(x) < 0$ for $x > 3$, $\displaystyle\lim_{x\to\infty} f(x) = -2$

16. The figure shows the graph of the *derivative* f' of a function f.
(a) On what intervals is f increasing or decreasing?
(b) For what values of x does f have a local maximum or minimum?

(c) Sketch the graph of f''.
(d) Sketch a possible graph of f.

17–28 Use the guidelines of Section 3.5 to sketch the curve.

17. $y = 2 - 2x - x^3$

18. $y = x^3 - 6x^2 - 15x + 4$

19. $y = x^4 - 3x^3 + 3x^2 - x$

20. $y = \dfrac{x}{1-x^2}$

21. $y = \dfrac{1}{x(x-3)^2}$

22. $y = \dfrac{1}{x^2} - \dfrac{1}{(x-2)^2}$

23. $y = x^2/(x+8)$

24. $y = \sqrt{1-x} + \sqrt{1+x}$

25. $y = x\sqrt{2+x}$

26. $y = \sqrt[3]{x^2+1}$

27. $y = \sin^2 x - 2\cos x$

28. $y = 4x - \tan x$, $-\pi/2 < x < \pi/2$

 29–32 Produce graphs of f that reveal all the important aspects of the curve. Use graphs of f' and f'' to estimate the intervals of increase and decrease, extreme values, intervals of concavity, and inflection points. In Exercise 29 use calculus to find these quantities exactly.

29. $f(x) = \dfrac{x^2-1}{x^3}$

30. $f(x) = \dfrac{x^3-x}{x^2+x+3}$

31. $f(x) = 3x^6 - 5x^5 + x^4 - 5x^3 - 2x^2 + 2$

32. $f(x) = x^2 + 6.5\sin x$, $-5 \leq x \leq 5$

33. Show that the equation $3x + 2\cos x + 5 = 0$ has exactly one real root.

34. Suppose that f is continuous on $[0, 4]$, $f(0) = 1$, and $2 \leq f'(x) \leq 5$ for all x in $(0, 4)$. Show that $9 \leq f(4) \leq 21$.

35. By applying the Mean Value Theorem to the function $f(x) = x^{1/5}$ on the interval $[32, 33]$, show that

$$2 < \sqrt[5]{33} < 2.0125$$

36. For what values of the constants a and b is $(1, 3)$ a point of inflection of the curve $y = ax^3 + bx^2$?

Graphing calculator or computer required

37. Let $g(x) = f(x^2)$, where f is twice differentiable for all x, $f'(x) > 0$ for all $x \neq 0$, and f is concave downward on $(-\infty, 0)$ and concave upward on $(0, \infty)$.
(a) At what numbers does g have an extreme value?
(b) Discuss the concavity of g.

38. Find two positive integers such that the sum of the first number and four times the second number is 1000 and the product of the numbers is as large as possible.

39. Show that the shortest distance from the point (x_1, y_1) to the straight line $Ax + By + C = 0$ is

$$\frac{|Ax_1 + By_1 + C|}{\sqrt{A^2 + B^2}}$$

40. Find the point on the hyperbola $xy = 8$ that is closest to the point $(3, 0)$.

41. Find the smallest possible area of an isosceles triangle that is circumscribed about a circle of radius r.

42. Find the volume of the largest circular cone that can be inscribed in a sphere of radius r.

43. In $\triangle ABC$, D lies on AB, $CD \perp AB$, $|AD| = |BD| = 4$ cm, and $|CD| = 5$ cm. Where should a point P be chosen on CD so that the sum $|PA| + |PB| + |PC|$ is a minimum?

44. Solve Exercise 43 when $|CD| = 2$ cm.

45. The velocity of a wave of length L in deep water is

$$v = K\sqrt{\frac{L}{C} + \frac{C}{L}}$$

where K and C are known positive constants. What is the length of the wave that gives the minimum velocity?

46. A metal storage tank with volume V is to be constructed in the shape of a right circular cylinder surmounted by a hemisphere. What dimensions will require the least amount of metal?

47. A hockey team plays in an arena with a seating capacity of 15,000 spectators. With the ticket price set at $12, average attendance at a game has been 11,000. A market survey indicates that for each dollar the ticket price is lowered, average attendance will increase by 1000. How should the owners of the team set the ticket price to maximize their revenue from ticket sales?

48. A manufacturer determines that the cost of making x units of a commodity is $C(x) = 1800 + 25x - 0.2x^2 + 0.001x^3$ and the demand function is $p(x) = 48.2 - 0.03x$.
(a) Graph the cost and revenue functions and use the graphs to estimate the production level for maximum profit.
(b) Use calculus to find the production level for maximum profit.
(c) Estimate the production level that minimizes the average cost.

49. Use Newton's method to find the root of the equation $x^5 - x^4 + 3x^2 - 3x - 2 = 0$ in the interval $[1, 2]$ correct to six decimal places.

50. Use Newton's method to find all roots of the equation $\sin x = x^2 - 3x + 1$ correct to six decimal places.

51. Use Newton's method to find the absolute maximum value of the function $f(t) = \cos t + t - t^2$ correct to eight decimal places.

52. Use the guidelines in Section 3.5 to sketch the curve $y = x \sin x$, $0 \le x \le 2\pi$. Use Newton's method when necessary.

53–58 Find f.

53. $f'(x) = \sqrt{x^3} + \sqrt[3]{x^2}$

54. $f'(x) = 8x - 3\sec^2 x$

55. $f'(t) = 2t - 3\sin t$, $f(0) = 5$

56. $f'(u) = \dfrac{u^2 + \sqrt{u}}{u}$, $f(1) = 3$

57. $f''(x) = 1 - 6x + 48x^2$, $f(0) = 1$, $f'(0) = 2$

58. $f''(x) = 2x^3 + 3x^2 - 4x + 5$, $f(0) = 2$, $f(1) = 0$

59–60 A particle is moving with the given data. Find the position of the particle.

59. $v(t) = 2t - \sin t$, $s(0) = 3$

60. $a(t) = \sin t + 3\cos t$, $s(0) = 0$, $v(0) = 2$

61. Use a graphing device to draw a graph of the function $f(x) = x^2 \sin(x^2)$, $0 \le x \le \pi$, and use that graph to sketch the antiderivative F of f that satisfies the initial condition $F(0) = 0$.

62. Investigate the family of curves given by

$$f(x) = x^4 + x^3 + cx^2$$

In particular you should determine the transitional value of c at which the number of critical numbers changes and the transitional value at which the number of inflection points changes. Illustrate the various possible shapes with graphs.

63. A canister is dropped from a helicopter 500 m above the ground. Its parachute does not open, but the canister has been designed to withstand an impact velocity of 100 m/s. Will it burst?

64. In an automobile race along a straight road, car A passed car B twice. Prove that at some time during the race their accelerations were equal. State the assumptions that you make.

65. A rectangular beam will be cut from a cylindrical log of radius 10 inches.
(a) Show that the beam of maximal cross-sectional area is a square.
(b) Four rectangular planks will be cut from the four sections of the log that remain after cutting the square beam. Deter-

mine the dimensions of the planks that will have maximal cross-sectional area.

(c) Suppose that the strength of a rectangular beam is proportional to the product of its width and the square of its depth. Find the dimensions of the strongest beam that can be cut from the cylindrical log.

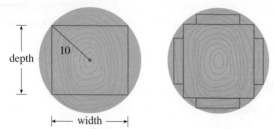

66. If a projectile is fired with an initial velocity v at an angle of inclination θ from the horizontal, then its trajectory, neglecting air resistance, is the parabola

$$y = (\tan \theta)x - \frac{g}{2v^2 \cos^2\theta}x^2 \qquad 0 < \theta < \frac{\pi}{2}$$

(a) Suppose the projectile is fired from the base of a plane that is inclined at an angle α, $\alpha > 0$, from the horizontal, as shown in the figure. Show that the range of the projectile, measured up the slope, is given by

$$R(\theta) = \frac{2v^2 \cos \theta \, \sin(\theta - \alpha)}{g \cos^2\alpha}$$

(b) Determine θ so that R is a maximum.

(c) Suppose the plane is at an angle α *below* the horizontal. Determine the range R in this case, and determine the angle at which the projectile should be fired to maximize R.

Problems Plus

One of the most important principles of problem solving is *analogy* (see page 77). If you are having trouble getting started on a problem, it is sometimes helpful to start by solving a similar, but simpler, problem. The following example illustrates the principle. Cover up the solution and try solving it yourself first.

EXAMPLE 1 If x, y, and z are positive numbers, prove that

$$\frac{(x^2 + 1)(y^2 + 1)(z^2 + 1)}{xyz} \geqslant 8$$

SOLUTION It may be difficult to get started on this problem. (Some students have tackled it by multiplying out the numerator, but that just creates a mess.) Let's try to think of a similar, simpler problem. When several variables are involved, it's often helpful to think of an analogous problem with fewer variables. In the present case we can reduce the number of variables from three to one and prove the analogous inequality

$$\boxed{1} \qquad\qquad \frac{x^2 + 1}{x} \geqslant 2 \qquad \text{for } x > 0$$

In fact, if we are able to prove $\boxed{1}$, then the desired inequality follows because

$$\frac{(x^2 + 1)(y^2 + 1)(z^2 + 1)}{xyz} = \left(\frac{x^2 + 1}{x}\right)\left(\frac{y^2 + 1}{y}\right)\left(\frac{z^2 + 1}{z}\right) \geqslant 2 \cdot 2 \cdot 2 = 8$$

The key to proving $\boxed{1}$ is to recognize that it is a disguised version of a minimum problem. If we let

$$f(x) = \frac{x^2 + 1}{x} = x + \frac{1}{x} \qquad x > 0$$

then $f'(x) = 1 - (1/x^2)$, so $f'(x) = 0$ when $x = 1$. Also, $f'(x) < 0$ for $0 < x < 1$ and $f'(x) > 0$ for $x > 1$. Therefore the absolute minimum value of f is $f(1) = 2$. This means that

$$\frac{x^2 + 1}{x} \geqslant 2 \qquad \text{for all positive values of } x$$

and, as previously mentioned, the given inequality follows by multiplication.

The inequality in $\boxed{1}$ could also be proved without calculus. In fact, if $x > 0$, we have

$$\frac{x^2 + 1}{x} \geqslant 2 \quad\Longleftrightarrow\quad x^2 + 1 \geqslant 2x \quad\Longleftrightarrow\quad x^2 - 2x + 1 \geqslant 0$$

$$\Longleftrightarrow \quad (x - 1)^2 \geqslant 0$$

Because the last inequality is obviously true, the first one is true too. ▪

PS LOOK BACK
What have we learned from the solution to this example?

■ To solve a problem involving several variables, it might help to solve a similar problem with just one variable.

■ When trying to prove an inequality, it might help to think of it as a maximum or minimum problem.

1. Show that $|\sin x - \cos x| \le \sqrt{2}$ for all x.

2. Show that $x^2 y^2 (4 - x^2)(4 - y^2) \le 16$ for all numbers x and y such that $|x| \le 2$ and $|y| \le 2$.

3. Show that the inflection points of the curve $y = (\sin x)/x$ lie on the curve $y^2(x^4 + 4) = 4$.

4. Find the point on the parabola $y = 1 - x^2$ at which the tangent line cuts from the first quadrant the triangle with the smallest area.

5. Find the highest and lowest points on the curve $x^2 + xy + y^2 = 12$.

6. Water is flowing at a constant rate into a spherical tank. Let $V(t)$ be the volume of water in the tank and $H(t)$ be the height of the water in the tank at time t.
   (a) What are the meanings of $V'(t)$ and $H'(t)$? Are these derivatives positive, negative, or zero?
   (b) Is $V''(t)$ positive, negative, or zero? Explain.
   (c) Let t_1, t_2, and t_3 be the times when the tank is one-quarter full, half full, and three-quarters full, respectively. Are the values $H''(t_1)$, $H''(t_2)$, and $H''(t_3)$ positive, negative, or zero? Why?

7. Find the absolute maximum value of the function
$$f(x) = \frac{1}{1 + |x|} + \frac{1}{1 + |x - 2|}$$

8. Find a function f such that $f'(-1) = \tfrac{1}{2}$, $f'(0) = 0$, and $f''(x) > 0$ for all x, or prove that such a function cannot exist.

9. The line $y = mx + b$ intersects the parabola $y = x^2$ in points A and B. (See the figure.) Find the point P on the arc AOB of the parabola that maximizes the area of the triangle PAB.

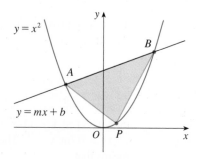

10. Sketch the graph of a function f such that $f'(x) < 0$ for all x, $f''(x) > 0$ for $|x| > 1$, $f''(x) < 0$ for $|x| < 1$, and $\lim_{x \to \pm\infty} [f(x) + x] = 0$.

11. Determine the values of the number a for which the function f has no critical number:
$$f(x) = (a^2 + a - 6) \cos 2x + (a - 2)x + \cos 1$$

12. Sketch the region in the plane consisting of all points (x, y) such that
$$2xy \le |x - y| \le x^2 + y^2$$

13. Let ABC be a triangle with $\angle BAC = 120°$ and $|AB| \cdot |AC| = 1$.
    (a) Express the length of the angle bisector AD in terms of $x = |AB|$.
    (b) Find the largest possible value of $|AD|$.

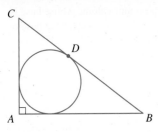

FIGURE FOR PROBLEM 14

14. (a) Let ABC be a triangle with right angle A and hypotenuse $a = |BC|$. (See the figure.) If the inscribed circle touches the hypotenuse at D, show that

$$|CD| = \tfrac{1}{2}(|BC| + |AC| - |AB|)$$

(b) If $\theta = \tfrac{1}{2}\angle C$, express the radius r of the inscribed circle in terms of a and θ.

(c) If a is fixed and θ varies, find the maximum value of r.

15. A triangle with sides a, b, and c varies with time t, but its area never changes. Let θ be the angle opposite the side of length a and suppose θ always remains acute.

(a) Express $d\theta/dt$ in terms of b, c, θ, db/dt, and dc/dt.

(b) Express da/dt in terms of the quantities in part (a).

16. $ABCD$ is a square piece of paper with sides of length 1 m. A quarter-circle is drawn from B to D with center A. The piece of paper is folded along EF, with E on AB and F on AD, so that A falls on the quarter-circle. Determine the maximum and minimum areas that the triangle AEF can have.

17. The speeds of sound c_1 in an upper layer and c_2 in a lower layer of rock and the thickness h of the upper layer can be determined by seismic exploration if the speed of sound in the lower layer is greater than the speed in the upper layer. A dynamite charge is detonated at a point P and the transmitted signals are recorded at a point Q, which is a distance D from P. The first signal to arrive at Q travels along the surface and takes T_1 seconds. The next signal travels from P to a point R, from R to S in the lower layer, and then to Q, taking T_2 seconds. The third signal is reflected off the lower layer at the midpoint O of RS and takes T_3 seconds to reach Q.

(a) Express T_1, T_2, and T_3 in terms of D, h, c_1, c_2, and θ.

(b) Show that T_2 is a minimum when $\sin \theta = c_1/c_2$.

(c) Suppose that $D = 1$ km, $T_1 = 0.26$ s, $T_2 = 0.32$ s, and $T_3 = 0.34$ s. Find c_1, c_2, and h.

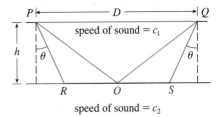

Note: Geophysicists use this technique when studying the structure of the earth's crust, whether searching for oil or examining fault lines.

18. For what values of c is there a straight line that intersects the curve

$$y = x^4 + cx^3 + 12x^2 - 5x + 2$$

in four distinct points?

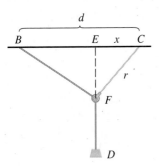

FIGURE FOR PROBLEM 19

19. One of the problems posed by the Marquis de l'Hospital in his calculus textbook *Analyse des Infiniment Petits* concerns a pulley that is attached to the ceiling of a room at a point C by a rope of length r. At another point B on the ceiling, at a distance d from C (where $d > r$), a rope of length ℓ is attached and passed through the pulley at F and connected to a weight W. The weight is released and comes to rest at its equilibrium position D. As l'Hospital argued, this happens when the distance $|ED|$ is maximized. Show that when the system reaches equilibrium, the value of x is

$$\frac{r}{4d}\left(r + \sqrt{r^2 + 8d^2}\right)$$

Notice that this expression is independent of both W and ℓ.

20. Given a sphere with radius r, find the height of a pyramid of minimum volume whose base is a square and whose base and triangular faces are all tangent to the sphere. What if the base of the pyramid is a regular n-gon? (A regular n-gon is a polygon with n equal sides and angles.) (Use the fact that the volume of a pyramid is $\frac{1}{3}Ah$, where A is the area of the base.)

21. Assume that a snowball melts so that its volume decreases at a rate proportional to its surface area. If it takes three hours for the snowball to decrease to half its original volume, how much longer will it take for the snowball to melt completely?

22. A hemispherical bubble is placed on a spherical bubble of radius 1. A smaller hemispherical bubble is then placed on the first one. This process is continued until n chambers, including the sphere, are formed. (The figure shows the case $n = 4$.) Use mathematical induction to prove that the maximum height of any bubble tower with n chambers is $1 + \sqrt{n}$.

4 Integrals

In Example 7 in Section 4.4 you will see how to use power consumption data and an integral to compute the amount of energy used in one day in San Francisco.

© Nathan Jaskowiak / Shutterstock

In Chapter 2 we used the tangent and velocity problems to introduce the derivative, which is the central idea in differential calculus. In much the same way, this chapter starts with the area and distance problems and uses them to formulate the idea of a definite integral, which is the basic concept of integral calculus. We will see in Chapters 5 and 8 how to use the integral to solve problems concerning volumes, lengths of curves, population predictions, cardiac output, forces on a dam, work, consumer surplus, and baseball, among many others.

There is a connection between integral calculus and differential calculus. The Fundamental Theorem of Calculus relates the integral to the derivative, and we will see in this chapter that it greatly simplifies the solution of many problems.

221

4.1   Areas and Distances

Now is a good time to read (or reread) *A Preview of Calculus* (see page 1). It discusses the unifying ideas of calculus and helps put in perspective where we have been and where we are going.

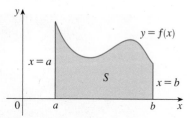

FIGURE 1
$S = \{(x, y) \mid a \le x \le b, 0 \le y \le f(x)\}$

In this section we discover that in trying to find the area under a curve or the distance traveled by a car, we end up with the same special type of limit.

The Area Problem

We begin by attempting to solve the *area problem:* Find the area of the region S that lies under the curve $y = f(x)$ from a to b. This means that S, illustrated in Figure 1, is bounded by the graph of a continuous function f [where $f(x) \ge 0$], the vertical lines $x = a$ and $x = b$, and the x-axis.

In trying to solve the area problem we have to ask ourselves: What is the meaning of the word *area*? This question is easy to answer for regions with straight sides. For a rectangle, the area is defined as the product of the length and the width. The area of a triangle is half the base times the height. The area of a polygon is found by dividing it into triangles (as in Figure 2) and adding the areas of the triangles.

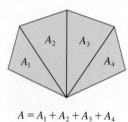

FIGURE 2 $A = lw$ $A = \frac{1}{2}bh$ $A = A_1 + A_2 + A_3 + A_4$

However, it isn't so easy to find the area of a region with curved sides. We all have an intuitive idea of what the area of a region is. But part of the area problem is to make this intuitive idea precise by giving an exact definition of area.

Recall that in defining a tangent we first approximated the slope of the tangent line by slopes of secant lines and then we took the limit of these approximations. We pursue a similar idea for areas. We first approximate the region S by rectangles and then we take the limit of the areas of these rectangles as we increase the number of rectangles. The following example illustrates the procedure.

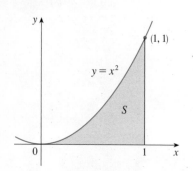

FIGURE 3

V   EXAMPLE 1   Use rectangles to estimate the area under the parabola $y = x^2$ from 0 to 1 (the parabolic region S illustrated in Figure 3).

SOLUTION   We first notice that the area of S must be somewhere between 0 and 1 because S is contained in a square with side length 1, but we can certainly do better than that. Suppose we divide S into four strips S_1, S_2, S_3, and S_4 by drawing the vertical lines $x = \frac{1}{4}$, $x = \frac{1}{2}$, and $x = \frac{3}{4}$ as in Figure 4(a).

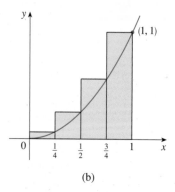

FIGURE 4   (a)   (b)

We can approximate each strip by a rectangle that has the same base as the strip and whose height is the same as the right edge of the strip [see Figure 4(b)]. In other words, the heights of these rectangles are the values of the function $f(x) = x^2$ at the *right* endpoints of the subintervals $\left[0, \frac{1}{4}\right]$, $\left[\frac{1}{4}, \frac{1}{2}\right]$, $\left[\frac{1}{2}, \frac{3}{4}\right]$, and $\left[\frac{3}{4}, 1\right]$.

Each rectangle has width $\frac{1}{4}$ and the heights are $\left(\frac{1}{4}\right)^2$, $\left(\frac{1}{2}\right)^2$, $\left(\frac{3}{4}\right)^2$, and 1^2. If we let R_4 be the sum of the areas of these approximating rectangles, we get

$$R_4 = \frac{1}{4} \cdot \left(\frac{1}{4}\right)^2 + \frac{1}{4} \cdot \left(\frac{1}{2}\right)^2 + \frac{1}{4} \cdot \left(\frac{3}{4}\right)^2 + \frac{1}{4} \cdot 1^2 = \frac{15}{32} = 0.46875$$

From Figure 4(b) we see that the area A of S is less than R_4, so

$$A < 0.46875$$

Instead of using the rectangles in Figure 4(b) we could use the smaller rectangles in Figure 5 whose heights are the values of f at the *left* endpoints of the subintervals. (The leftmost rectangle has collapsed because its height is 0.) The sum of the areas of these approximating rectangles is

$$L_4 = \frac{1}{4} \cdot 0^2 + \frac{1}{4} \cdot \left(\frac{1}{4}\right)^2 + \frac{1}{4} \cdot \left(\frac{1}{2}\right)^2 + \frac{1}{4} \cdot \left(\frac{3}{4}\right)^2 = \frac{7}{32} = 0.21875$$

We see that the area of S is larger than L_4, so we have lower and upper estimates for A:

$$0.21875 < A < 0.46875$$

We can repeat this procedure with a larger number of strips. Figure 6 shows what happens when we divide the region S into eight strips of equal width.

FIGURE 5

(a) Using left endpoints

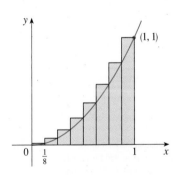

(b) Using right endpoints

FIGURE 6

Approximating S with eight rectangles

By computing the sum of the areas of the smaller rectangles (L_8) and the sum of the areas of the larger rectangles (R_8), we obtain better lower and upper estimates for A:

$$0.2734375 < A < 0.3984375$$

So one possible answer to the question is to say that the true area of S lies somewhere between 0.2734375 and 0.3984375.

We could obtain better estimates by increasing the number of strips. The table at the left shows the results of similar calculations (with a computer) using n rectangles whose heights are found with left endpoints (L_n) or right endpoints (R_n). In particular, we see by using 50 strips that the area lies between 0.3234 and 0.3434. With 1000 strips we narrow it down even more: A lies between 0.3328335 and 0.3338335. A good estimate is obtained by averaging these numbers: $A \approx 0.3333335$.

n	L_n	R_n
10	0.2850000	0.3850000
20	0.3087500	0.3587500
30	0.3168519	0.3501852
50	0.3234000	0.3434000
100	0.3283500	0.3383500
1000	0.3328335	0.3338335

From the values in the table in Example 1, it looks as if R_n is approaching $\frac{1}{3}$ as n increases. We confirm this in the next example.

V **EXAMPLE 2** For the region S in Example 1, show that the sum of the areas of the upper approximating rectangles approaches $\frac{1}{3}$, that is,

$$\lim_{n \to \infty} R_n = \tfrac{1}{3}$$

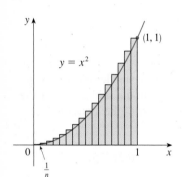

FIGURE 7

SOLUTION R_n is the sum of the areas of the n rectangles in Figure 7. Each rectangle has width $1/n$ and the heights are the values of the function $f(x) = x^2$ at the points $1/n, 2/n, 3/n, \ldots, n/n$; that is, the heights are $(1/n)^2, (2/n)^2, (3/n)^2, \ldots, (n/n)^2$. Thus

$$R_n = \frac{1}{n}\left(\frac{1}{n}\right)^2 + \frac{1}{n}\left(\frac{2}{n}\right)^2 + \frac{1}{n}\left(\frac{3}{n}\right)^2 + \cdots + \frac{1}{n}\left(\frac{n}{n}\right)^2$$

$$= \frac{1}{n} \cdot \frac{1}{n^2}(1^2 + 2^2 + 3^2 + \cdots + n^2)$$

$$= \frac{1}{n^3}(1^2 + 2^2 + 3^2 + \cdots + n^2)$$

Here we need the formula for the sum of the squares of the first n positive integers:

$$\boxed{1} \qquad 1^2 + 2^2 + 3^2 + \cdots + n^2 = \frac{n(n+1)(2n+1)}{6}$$

Perhaps you have seen this formula before. It is proved in Example 5 in Appendix E. Putting Formula 1 into our expression for R_n, we get

$$R_n = \frac{1}{n^3} \cdot \frac{n(n+1)(2n+1)}{6} = \frac{(n+1)(2n+1)}{6n^2}$$

Here we are computing the limit of the sequence $\{R_n\}$. Sequences and their limits were discussed in *A Preview of Calculus* and will be studied in detail in Section 11.1. The idea is very similar to a limit at infinity (Section 3.4) except that in writing $\lim_{n\to\infty}$ we restrict n to be a positive integer. In particular, we know that

$$\lim_{n \to \infty} \frac{1}{n} = 0$$

When we write $\lim_{n\to\infty} R_n = \frac{1}{3}$ we mean that we can make R_n as close to $\frac{1}{3}$ as we like by taking n sufficiently large.

Thus we have

$$\lim_{n \to \infty} R_n = \lim_{n \to \infty} \frac{(n+1)(2n+1)}{6n^2}$$

$$= \lim_{n \to \infty} \frac{1}{6}\left(\frac{n+1}{n}\right)\left(\frac{2n+1}{n}\right)$$

$$= \lim_{n \to \infty} \frac{1}{6}\left(1 + \frac{1}{n}\right)\left(2 + \frac{1}{n}\right)$$

$$= \frac{1}{6} \cdot 1 \cdot 2 = \frac{1}{3}$$

It can be shown that the lower approximating sums also approach $\frac{1}{3}$, that is,

$$\lim_{n \to \infty} L_n = \tfrac{1}{3}$$

From Figures 8 and 9 it appears that, as n increases, both L_n and R_n become better and better approximations to the area of S. Therefore we *define* the area A to be the limit of the sums of the areas of the approximating rectangles, that is,

TEC In Visual 4.1 you can create pictures like those in Figures 8 and 9 for other values of n.

$$A = \lim_{n \to \infty} R_n = \lim_{n \to \infty} L_n = \tfrac{1}{3}$$

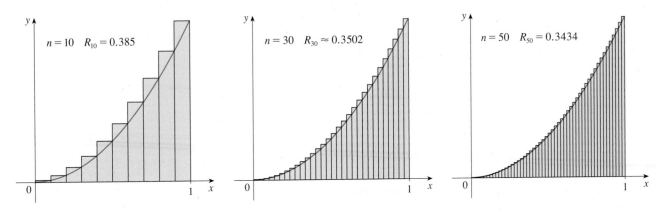

FIGURE 8   Right endpoints produce upper sums because $f(x) = x^2$ is increasing

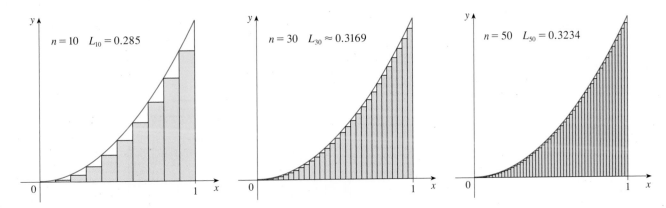

FIGURE 9   Left endpoints produce lower sums because $f(x) = x^2$ is increasing

Let's apply the idea of Examples 1 and 2 to the more general region S of Figure 1. We start by subdividing S into n strips $S_1, S_2, \ldots, S_n$ of equal width as in Figure 10.

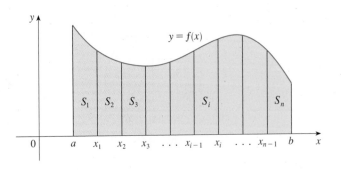

FIGURE 10

The width of the interval $[a, b]$ is $b - a$, so the width of each of the n strips is

$$\Delta x = \frac{b - a}{n}$$

These strips divide the interval $[a, b]$ into n subintervals

$$[x_0, x_1], \quad [x_1, x_2], \quad [x_2, x_3], \quad \ldots, \quad [x_{n-1}, x_n]$$

where $x_0 = a$ and $x_n = b$. The right endpoints of the subintervals are

$$x_1 = a + \Delta x,$$

$$x_2 = a + 2\,\Delta x,$$

$$x_3 = a + 3\,\Delta x,$$

$$\vdots$$

Let's approximate the ith strip S_i by a rectangle with width Δx and height $f(x_i)$, which is the value of f at the right endpoint (see Figure 11). Then the area of the ith rectangle is $f(x_i)\,\Delta x$. What we think of intuitively as the area of S is approximated by the sum of the areas of these rectangles, which is

$$R_n = f(x_1)\,\Delta x + f(x_2)\,\Delta x + \cdots + f(x_n)\,\Delta x$$

FIGURE 11

Figure 12 shows this approximation for $n = 2, 4, 8,$ and 12. Notice that this approximation appears to become better and better as the number of strips increases, that is, as $n \to \infty$. Therefore we define the area A of the region S in the following way.

(a) $n = 2$ (b) $n = 4$ (c) $n = 8$ (d) $n = 12$

FIGURE 12

> **2** **Definition** The **area** A of the region S that lies under the graph of the continuous function f is the limit of the sum of the areas of approximating rectangles:
>
> $$A = \lim_{n \to \infty} R_n = \lim_{n \to \infty} [f(x_1)\,\Delta x + f(x_2)\,\Delta x + \cdots + f(x_n)\,\Delta x]$$

It can be proved that the limit in Definition 2 always exists, since we are assuming that f is continuous. It can also be shown that we get the same value if we use left endpoints:

$$\boxed{3} \qquad A = \lim_{n \to \infty} L_n = \lim_{n \to \infty} [f(x_0)\,\Delta x + f(x_1)\,\Delta x + \cdots + f(x_{n-1})\,\Delta x]$$

In fact, instead of using left endpoints or right endpoints, we could take the height of the ith rectangle to be the value of f at *any* number x_i^* in the ith subinterval $[x_{i-1}, x_i]$. We call the numbers $x_1^*, x_2^*, \ldots, x_n^*$ the **sample points**. Figure 13 shows approximating rectangles when the sample points are not chosen to be endpoints. So a more general expression for the area of S is

$$\boxed{4} \qquad A = \lim_{n \to \infty} [f(x_1^*)\,\Delta x + f(x_2^*)\,\Delta x + \cdots + f(x_n^*)\,\Delta x]$$

FIGURE 13

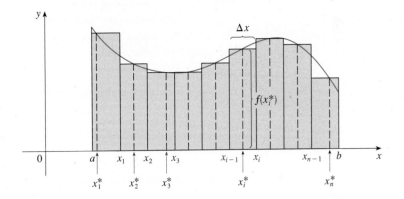

NOTE It can be shown that an equivalent definition of area is the following: *A is the unique number that is smaller than all the upper sums and bigger than all the lower sums.* We saw in Examples 1 and 2, for instance, that the area $\left(A = \frac{1}{3}\right)$ is trapped between all the left approximating sums L_n and all the right approximating sums R_n. The function in those examples, $f(x) = x^2$, happens to be increasing on $[0, 1]$ and so the lower sums arise from left endpoints and the upper sums from right endpoints. (See Figures 8 and 9.) In general, we form **lower** (and **upper**) **sums** by choosing the sample points x_i^* so that $f(x_i^*)$ is the minimum (and maximum) value of f on the ith subinterval. (See Figure 14 and Exercises 7–8).

FIGURE 14
Lower sums (short rectangles)
and upper sums (tall rectangles)

This tells us to
end with $i = n$.

This tells us
to add. $\displaystyle\sum_{i=m}^{n} f(x_i)\,\Delta x$

This tells us to
start with $i = m$.

If you need practice with sigma notation, look at
the examples and try some of the exercises in
Appendix E.

We often use **sigma notation** to write sums with many terms more compactly. For instance,

$$\sum_{i=1}^{n} f(x_i)\,\Delta x = f(x_1)\,\Delta x + f(x_2)\,\Delta x + \cdots + f(x_n)\,\Delta x$$

So the expressions for area in Equations 2, 3, and 4 can be written as follows:

$$A = \lim_{n\to\infty} \sum_{i=1}^{n} f(x_i)\,\Delta x$$

$$A = \lim_{n\to\infty} \sum_{i=1}^{n} f(x_{i-1})\,\Delta x$$

$$A = \lim_{n\to\infty} \sum_{i=1}^{n} f(x_i^*)\,\Delta x$$

We can also rewrite Formula 1 in the following way:

$$\sum_{i=1}^{n} i^2 = \frac{n(n+1)(2n+1)}{6}$$

EXAMPLE 3 Let A be the area of the region that lies under the graph of $f(x) = \cos x$ between $x = 0$ and $x = b$, where $0 \leqslant b \leqslant \pi/2$.
(a) Using right endpoints, find an expression for A as a limit. Do not evaluate the limit.
(b) Estimate the area for the case $b = \pi/2$ by taking the sample points to be midpoints and using four subintervals.

SOLUTION
(a) Since $a = 0$, the width of a subinterval is

$$\Delta x = \frac{b - 0}{n} = \frac{b}{n}$$

So $x_1 = b/n$, $x_2 = 2b/n$, $x_3 = 3b/n$, $x_i = ib/n$, and $x_n = nb/n$. The sum of the areas of the approximating rectangles is

$$\begin{aligned}
R_n &= f(x_1)\,\Delta x + f(x_2)\,\Delta x + \cdots + f(x_n)\,\Delta x \\
&= (\cos x_1)\,\Delta x + (\cos x_2)\,\Delta x + \cdots + (\cos x_n)\,\Delta x \\
&= \left(\cos\frac{b}{n}\right)\frac{b}{n} + \left(\cos\frac{2b}{n}\right)\frac{b}{n} + \cdots + \left(\cos\frac{nb}{n}\right)\frac{b}{n}
\end{aligned}$$

According to Definition 2, the area is

$$A = \lim_{n\to\infty} R_n = \lim_{n\to\infty} \frac{b}{n}\left(\cos\frac{b}{n} + \cos\frac{2b}{n} + \cos\frac{3b}{n} + \cdots + \cos\frac{nb}{n}\right)$$

Using sigma notation we could write

$$A = \lim_{n\to\infty} \frac{b}{n} \sum_{i=1}^{n} \cos\frac{ib}{n}$$

It is very difficult to evaluate this limit directly by hand, but with the aid of a computer algebra system it isn't hard (see Exercise 29). In Section 4.3 we will be able to find A more easily using a different method.

FIGURE 15

(b) With $n = 4$ and $b = \pi/2$ we have $\Delta x = (\pi/2)/4 = \pi/8$, so the subintervals are $[0, \pi/8]$, $[\pi/8, \pi/4]$, $[\pi/4, 3\pi/8]$, and $[3\pi/8, \pi/2]$. The midpoints of these subintervals are

$$x_1^* = \frac{\pi}{16} \qquad x_2^* = \frac{3\pi}{16} \qquad x_3^* = \frac{5\pi}{16} \qquad x_4^* = \frac{7\pi}{16}$$

and the sum of the areas of the four approximating rectangles (see Figure 15) is

$$M_4 = \sum_{i=1}^{4} f(x_i^*)\, \Delta x$$

$$= f(\pi/16)\, \Delta x + f(3\pi/16)\, \Delta x + f(5\pi/16)\, \Delta x + f(7\pi/16)\, \Delta x$$

$$= \left(\cos\frac{\pi}{16}\right)\frac{\pi}{8} + \left(\cos\frac{3\pi}{16}\right)\frac{\pi}{8} + \left(\cos\frac{5\pi}{16}\right)\frac{\pi}{8} + \left(\cos\frac{7\pi}{16}\right)\frac{\pi}{8}$$

$$= \frac{\pi}{8}\left(\cos\frac{\pi}{16} + \cos\frac{3\pi}{16} + \cos\frac{5\pi}{16} + \cos\frac{7\pi}{16}\right) \approx 1.006$$

So an estimate for the area is

$$A \approx 1.006$$

The Distance Problem

Now let's consider the *distance problem:* Find the distance traveled by an object during a certain time period if the velocity of the object is known at all times. (In a sense this is the inverse problem of the velocity problem that we discussed in Section 1.4.) If the velocity remains constant, then the distance problem is easy to solve by means of the formula

$$\text{distance} = \text{velocity} \times \text{time}$$

But if the velocity varies, it's not so easy to find the distance traveled. We investigate the problem in the following example.

V EXAMPLE 4 Suppose the odometer on our car is broken and we want to estimate the distance driven over a 30-second time interval. We take speedometer readings every five seconds and record them in the following table:

Time (s)	0	5	10	15	20	25	30
Velocity (mi/h)	17	21	24	29	32	31	28

In order to have the time and the velocity in consistent units, let's convert the velocity readings to feet per second (1 mi/h = 5280/3600 ft/s):

Time (s)	0	5	10	15	20	25	30
Velocity (ft/s)	25	31	35	43	47	46	41

During the first five seconds the velocity doesn't change very much, so we can estimate the distance traveled during that time by assuming that the velocity is constant. If we take the velocity during that time interval to be the initial velocity (25 ft/s), then we obtain the approximate distance traveled during the first five seconds:

$$25 \text{ ft/s} \times 5 \text{ s} = 125 \text{ ft}$$

Similarly, during the second time interval the velocity is approximately constant and we take it to be the velocity when $t = 5$ s. So our estimate for the distance traveled from $t = 5$ s to $t = 10$ s is

$$31 \text{ ft/s} \times 5 \text{ s} = 155 \text{ ft}$$

If we add similar estimates for the other time intervals, we obtain an estimate for the total distance traveled:

$$(25 \times 5) + (31 \times 5) + (35 \times 5) + (43 \times 5) + (47 \times 5) + (46 \times 5) = 1135 \text{ ft}$$

We could just as well have used the velocity at the *end* of each time period instead of the velocity at the beginning as our assumed constant velocity. Then our estimate becomes

$$(31 \times 5) + (35 \times 5) + (43 \times 5) + (47 \times 5) + (46 \times 5) + (41 \times 5) = 1215 \text{ ft}$$

If we had wanted a more accurate estimate, we could have taken velocity readings every two seconds, or even every second.   ▬

Perhaps the calculations in Example 4 remind you of the sums we used earlier to estimate areas. The similarity is explained when we sketch a graph of the velocity function of the car in Figure 16 and draw rectangles whose heights are the initial velocities for each time interval. The area of the first rectangle is $25 \times 5 = 125$, which is also our estimate for the distance traveled in the first five seconds. In fact, the area of each rectangle can be interpreted as a distance because the height represents velocity and the width represents time. The sum of the areas of the rectangles in Figure 16 is $L_6 = 1135$, which is our initial estimate for the total distance traveled.

FIGURE 16

In general, suppose an object moves with velocity $v = f(t)$, where $a \leqslant t \leqslant b$ and $f(t) \geqslant 0$ (so the object always moves in the positive direction). We take velocity readings at times $t_0 \, (= a), t_1, t_2, \ldots, t_n \, (= b)$ so that the velocity is approximately constant on each subinterval. If these times are equally spaced, then the time between consecutive readings is $\Delta t = (b - a)/n$. During the first time interval the velocity is approximately $f(t_0)$ and so the distance traveled is approximately $f(t_0) \, \Delta t$. Similarly, the distance traveled during the second time interval is about $f(t_1) \, \Delta t$ and the total distance traveled during the time interval $[a, b]$ is approximately

$$f(t_0) \, \Delta t + f(t_1) \, \Delta t + \cdots + f(t_{n-1}) \, \Delta t = \sum_{i=1}^{n} f(t_{i-1}) \, \Delta t$$

If we use the velocity at right endpoints instead of left endpoints, our estimate for the total distance becomes

$$f(t_1) \, \Delta t + f(t_2) \, \Delta t + \cdots + f(t_n) \, \Delta t = \sum_{i=1}^{n} f(t_i) \, \Delta t$$

The more frequently we measure the velocity, the more accurate our estimates become, so it seems plausible that the *exact* distance d traveled is the *limit* of such expressions:

$$\boxed{5} \qquad d = \lim_{n \to \infty} \sum_{i=1}^{n} f(t_{i-1}) \, \Delta t = \lim_{n \to \infty} \sum_{i=1}^{n} f(t_i) \, \Delta t$$

We will see in Section 4.4 that this is indeed true.

Because Equation 5 has the same form as our expressions for area in Equations 2 and 3, it follows that the distance traveled is equal to the area under the graph of the velocity function. In Chapter 5 we will see that other quantities of interest in the natural and social sciences—such as the work done by a variable force or the cardiac output of the heart—can also be interpreted as the area under a curve. So when we compute areas in this chapter, bear in mind that they can be interpreted in a variety of practical ways.

4.1 Exercises

Access end-of-section exercises and tools for visualizing calculus online at **www.webassign.net**

4.2 The Definite Integral

We saw in Section 4.1 that a limit of the form

$$\boxed{1} \qquad \lim_{n \to \infty} \sum_{i=1}^{n} f(x_i^*) \, \Delta x = \lim_{n \to \infty} \left[f(x_1^*) \, \Delta x + f(x_2^*) \, \Delta x + \cdots + f(x_n^*) \, \Delta x \right]$$

arises when we compute an area. We also saw that it arises when we try to find the distance traveled by an object. It turns out that this same type of limit occurs in a wide variety of situations even when f is not necessarily a positive function. In Chapters 5 and 8 we will see that limits of the form $\boxed{1}$ also arise in finding lengths of curves, volumes of solids, centers of mass, force due to water pressure, and work, as well as other quantities. We therefore give this type of limit a special name and notation

---

$\boxed{2}$ **Definition of a Definite Integral** If f is a function defined for $a \le x \le b$, we divide the interval $[a, b]$ into n subintervals of equal width $\Delta x = (b - a)/n$. We let $x_0 \, (= a), x_1, x_2, \ldots, x_n \, (= b)$ be the endpoints of these subintervals and we let $x_1^*, x_2^*, \ldots, x_n^*$ be any **sample points** in these subintervals, so x_i^* lies in the ith subinterval $[x_{i-1}, x_i]$. Then the **definite integral of f from a to b** is

$$\int_a^b f(x) \, dx = \lim_{n \to \infty} \sum_{i=1}^{n} f(x_i^*) \, \Delta x$$

provided that this limit exists and gives the same value for all possible choices of sample points. If it does exist, we say that f is **integrable** on $[a, b]$.

The precise meaning of the limit that defines the integral is as follows:

For every number $\varepsilon > 0$ there is an integer N such that

$$\left| \int_a^b f(x)\, dx - \sum_{i=1}^{n} f(x_i^*)\, \Delta x \right| < \varepsilon$$

for every integer $n > N$ and for every choice of x_i^* in $[x_{i-1}, x_i]$.

NOTE 1 The symbol $\int$ was introduced by Leibniz and is called an **integral sign**. It is an elongated S and was chosen because an integral is a limit of sums. In the notation $\int_a^b f(x)\, dx$, $f(x)$ is called the **integrand** and a and b are called the **limits of integration**; a is the **lower limit** and b is the **upper limit**. For now, the symbol dx has no meaning by itself; $\int_a^b f(x)\, dx$ is all one symbol. The dx simply indicates that the independent variable is x. The procedure of calculating an integral is called **integration**.

NOTE 2 The definite integral $\int_a^b f(x)\, dx$ is a number; it does not depend on x. In fact, we could use any letter in place of x without changing the value of the integral:

$$\int_a^b f(x)\, dx = \int_a^b f(t)\, dt = \int_a^b f(r)\, dr$$

NOTE 3 The sum

$$\sum_{i=1}^{n} f(x_i^*)\, \Delta x$$

that occurs in Definition 2 is called a **Riemann sum** after the German mathematician Bernhard Riemann (1826–1866). So Definition 2 says that the definite integral of an integrable function can be approximated to within any desired degree of accuracy by a Riemann sum.

We know that if f happens to be positive, then the Riemann sum can be interpreted as a sum of areas of approximating rectangles (see Figure 1). By comparing Definition 2 with the definition of area in Section 4.1, we see that the definite integral $\int_a^b f(x)\, dx$ can be interpreted as the area under the curve $y = f(x)$ from a to b. (See Figure 2.)

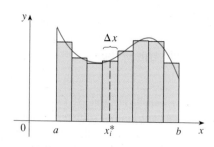

FIGURE 1
If $f(x) \geq 0$, the Riemann sum $\Sigma f(x_i^*)\, \Delta x$ is the sum of areas of rectangles.

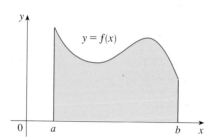

FIGURE 2
If $f(x) \geq 0$, the integral $\int_a^b f(x)\, dx$ is the area under the curve $y = f(x)$ from a to b.

If f takes on both positive and negative values, as in Figure 3, then the Riemann sum is the sum of the areas of the rectangles that lie above the x-axis and the *negatives* of the areas of the rectangles that lie below the x-axis (the areas of the blue rectangles *minus* the areas of the gold rectangles). When we take the limit of such Riemann sums, we get the situation illustrated in Figure 4. A definite integral can be interpreted as a **net area**, that is, a difference of areas:

$$\int_a^b f(x)\, dx = A_1 - A_2$$

where A_1 is the area of the region above the x-axis and below the graph of f, and A_2 is the area of the region below the x-axis and above the graph of f.

FIGURE 3
$\sum f(x_i^*) \Delta x$ is an approximation to the net area.

FIGURE 4
$\int_a^b f(x)\, dx$ is the net area.

NOTE 4 Although we have defined $\int_a^b f(x)\, dx$ by dividing $[a, b]$ into subintervals of equal width, there are situations in which it is advantageous to work with subintervals of unequal width. For instance, in Exercise 16 in Section 4.1 NASA provided velocity data at times that were not equally spaced, but we were still able to estimate the distance traveled. And there are methods for numerical integration that take advantage of unequal subintervals.

If the subinterval widths are $\Delta x_1, \Delta x_2, \ldots, \Delta x_n$, we have to ensure that all these widths approach 0 in the limiting process. This happens if the largest width, max Δx_i, approaches 0. So in this case the definition of a definite integral becomes

$$\int_a^b f(x)\, dx = \lim_{\max \Delta x_i \to 0} \sum_{i=1}^n f(x_i^*) \Delta x_i$$

NOTE 5 We have defined the definite integral for an integrable function, but not all functions are integrable (see Exercises 69–70). The following theorem shows that the most commonly occurring functions are in fact integrable. The theorem is proved in more advanced courses.

---

3 **Theorem** If f is continuous on $[a, b]$, or if f has only a finite number of jump discontinuities, then f is integrable on $[a, b]$; that is, the definite integral $\int_a^b f(x)\, dx$ exists.

If f is integrable on $[a, b]$, then the limit in Definition 2 exists and gives the same value no matter how we choose the sample points x_i^*. To simplify the calculation of the integral we often take the sample points to be right endpoints. Then $x_i^* = x_i$ and the definition of an integral simplifies as follows.

---

4 **Theorem** If f is integrable on $[a, b]$, then

$$\int_a^b f(x)\, dx = \lim_{n \to \infty} \sum_{i=1}^{n} f(x_i)\, \Delta x$$

where $$\Delta x = \frac{b - a}{n} \quad \text{and} \quad x_i = a + i\, \Delta x$$

EXAMPLE 1 Express

$$\lim_{n \to \infty} \sum_{i=1}^{n} (x_i^3 + x_i \sin x_i)\, \Delta x$$

as an integral on the interval $[0, \pi]$.

SOLUTION Comparing the given limit with the limit in Theorem 4, we see that they will be identical if we choose $f(x) = x^3 + x \sin x$. We are given that $a = 0$ and $b = \pi$. Therefore, by Theorem 4, we have

$$\lim_{n \to \infty} \sum_{i=1}^{n} (x_i^3 + x_i \sin x_i)\, \Delta x = \int_0^{\pi} (x^3 + x \sin x)\, dx \qquad \blacksquare$$

Later, when we apply the definite integral to physical situations, it will be important to recognize limits of sums as integrals, as we did in Example 1. When Leibniz chose the notation for an integral, he chose the ingredients as reminders of the limiting process. In general, when we write

$$\lim_{n \to \infty} \sum_{i=1}^{n} f(x_i^*)\, \Delta x = \int_a^b f(x)\, dx$$

we replace $\lim \Sigma$ by $\int$, x_i^* by x, and Δx by dx.

▆ Evaluating Integrals

When we use a limit to evaluate a definite integral, we need to know how to work with sums. The following three equations give formulas for sums of powers of positive integers. Equation 5 may be familiar to you from a course in algebra. Equations 6 and 7 were discussed in Section 4.1 and are proved in Appendix E.

5 $$\sum_{i=1}^{n} i = \frac{n(n + 1)}{2}$$

6 $$\sum_{i=1}^{n} i^2 = \frac{n(n + 1)(2n + 1)}{6}$$

7 $$\sum_{i=1}^{n} i^3 = \left[\frac{n(n + 1)}{2} \right]^2$$

The remaining formulas are simple rules for working with sigma notation:

8	$$\sum_{i=1}^{n} c = nc$$

Formulas 8–11 are proved by writing out each side in expanded form. The left side of Equation 9 is

$$ca_1 + ca_2 + \cdots + ca_n$$

The right side is

$$c(a_1 + a_2 + \cdots + a_n)$$

These are equal by the distributive property. The other formulas are discussed in Appendix E.

9	$$\sum_{i=1}^{n} ca_i = c \sum_{i=1}^{n} a_i$$

10	$$\sum_{i=1}^{n} (a_i + b_i) = \sum_{i=1}^{n} a_i + \sum_{i=1}^{n} b_i$$

11	$$\sum_{i=1}^{n} (a_i - b_i) = \sum_{i=1}^{n} a_i - \sum_{i=1}^{n} b_i$$

EXAMPLE 2

(a) Evaluate the Riemann sum for $f(x) = x^3 - 6x$, taking the sample points to be right endpoints and $a = 0$, $b = 3$, and $n = 6$.

(b) Evaluate $\int_0^3 (x^3 - 6x)\,dx$.

SOLUTION

(a) With $n = 6$ the interval width is

$$\Delta x = \frac{b - a}{n} = \frac{3 - 0}{6} = \frac{1}{2}$$

and the right endpoints are $x_1 = 0.5$, $x_2 = 1.0$, $x_3 = 1.5$, $x_4 = 2.0$, $x_5 = 2.5$, and $x_6 = 3.0$. So the Riemann sum is

$$R_6 = \sum_{i=1}^{6} f(x_i)\,\Delta x$$

$$= f(0.5)\,\Delta x + f(1.0)\,\Delta x + f(1.5)\,\Delta x + f(2.0)\,\Delta x + f(2.5)\,\Delta x + f(3.0)\,\Delta x$$

$$= \tfrac{1}{2}(-2.875 - 5 - 5.625 - 4 + 0.625 + 9)$$

$$= -3.9375$$

Notice that f is not a positive function and so the Riemann sum does not represent a sum of areas of rectangles. But it does represent the sum of the areas of the blue rectangles (above the x-axis) minus the sum of the areas of the gold rectangles (below the x-axis) in Figure 5.

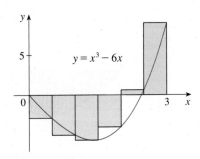

FIGURE 5

(b) With n subintervals we have

$$\Delta x = \frac{b - a}{n} = \frac{3}{n}$$

Thus $x_0 = 0$, $x_1 = 3/n$, $x_2 = 6/n$, $x_3 = 9/n$, and, in general, $x_i = 3i/n$. Since we are using right endpoints, we can use Theorem 4:

$$\int_0^3 (x^3 - 6x)\, dx = \lim_{n \to \infty} \sum_{i=1}^{n} f(x_i)\, \Delta x = \lim_{n \to \infty} \sum_{i=1}^{n} f\left(\frac{3i}{n}\right) \frac{3}{n}$$

In the sum, n is a constant (unlike i), so we can move $3/n$ in front of the Σ sign.

$$= \lim_{n \to \infty} \frac{3}{n} \sum_{i=1}^{n} \left[\left(\frac{3i}{n}\right)^3 - 6\left(\frac{3i}{n}\right) \right] \qquad \text{(Equation 9 with } c = 3/n\text{)}$$

$$= \lim_{n \to \infty} \frac{3}{n} \sum_{i=1}^{n} \left[\frac{27}{n^3} i^3 - \frac{18}{n} i \right]$$

$$= \lim_{n \to \infty} \left[\frac{81}{n^4} \sum_{i=1}^{n} i^3 - \frac{54}{n^2} \sum_{i=1}^{n} i \right] \qquad \text{(Equations 11 and 9)}$$

$$= \lim_{n \to \infty} \left\{ \frac{81}{n^4} \left[\frac{n(n + 1)}{2} \right]^2 - \frac{54}{n^2} \frac{n(n + 1)}{2} \right\} \qquad \text{(Equations 7 and 5)}$$

$$= \lim_{n \to \infty} \left[\frac{81}{4} \left(1 + \frac{1}{n}\right)^2 - 27\left(1 + \frac{1}{n}\right) \right]$$

$$= \frac{81}{4} - 27 = -\frac{27}{4} = -6.75$$

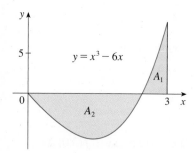

FIGURE 6

$\int_0^3 (x^3 - 6x)\, dx = A_1 - A_2 = -6.75$

This integral can't be interpreted as an area because f takes on both positive and negative values. But it can be interpreted as the difference of areas $A_1 - A_2$, where A_1 and A_2 are shown in Figure 6.

Figure 7 illustrates the calculation by showing the positive and negative terms in the right Riemann sum R_n for $n = 40$. The values in the table show the Riemann sums approaching the exact value of the integral, -6.75, as $n \to \infty$.

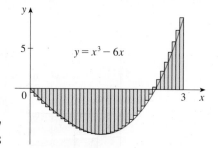

FIGURE 7

$R_{40} \approx -6.3998$

n	R_n
40	-6.3998
100	-6.6130
500	-6.7229
1000	-6.7365
5000	-6.7473

A much simpler method for evaluating the integral in Example 2 will be given in Section 4.4.

Because $f(x) = x^4$ is positive, the integral in Example 3 represents the area shown in Figure 8.

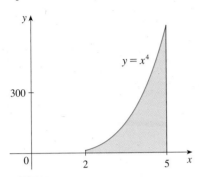

FIGURE 8

EXAMPLE 3
(a) Set up an expression for $\int_2^5 x^4\, dx$ as a limit of sums.
(b) Use a computer algebra system to evaluate the expression.

SOLUTION
(a) Here we have $f(x) = x^4$, $a = 2$, $b = 5$, and

$$\Delta x = \frac{b - a}{n} = \frac{3}{n}$$

So $x_0 = 2$, $x_1 = 2 + 3/n$, $x_2 = 2 + 6/n$, $x_3 = 2 + 9/n$, and

$$x_i = 2 + \frac{3i}{n}$$

From Theorem 4, we get

$$\int_2^5 x^4\, dx = \lim_{n \to \infty} \sum_{i=1}^{n} f(x_i)\, \Delta x = \lim_{n \to \infty} \sum_{i=1}^{n} f\left(2 + \frac{3i}{n}\right) \frac{3}{n}$$

$$= \lim_{n \to \infty} \frac{3}{n} \sum_{i=1}^{n} \left(2 + \frac{3i}{n}\right)^4$$

(b) If we ask a computer algebra system to evaluate the sum and simplify, we obtain

$$\sum_{i=1}^{n} \left(2 + \frac{3i}{n}\right)^4 = \frac{2062n^4 + 3045n^3 + 1170n^2 - 27}{10n^3}$$

Now we ask the computer algebra system to evaluate the limit:

$$\int_2^5 x^4\, dx = \lim_{n \to \infty} \frac{3}{n} \sum_{i=1}^{n} \left(2 + \frac{3i}{n}\right)^4 = \lim_{n \to \infty} \frac{3(2062n^4 + 3045n^3 + 1170n^2 - 27)}{10n^4}$$

$$= \frac{3(2062)}{10} = \frac{3093}{5} = 618.6$$

We will learn a much easier method for the evaluation of integrals in the next section.

V EXAMPLE 4 Evaluate the following integrals by interpreting each in terms of areas.

(a) $\int_0^1 \sqrt{1 - x^2}\, dx$ (b) $\int_0^3 (x - 1)\, dx$

SOLUTION
(a) Since $f(x) = \sqrt{1 - x^2} \geqslant 0$, we can interpret this integral as the area under the curve $y = \sqrt{1 - x^2}$ from 0 to 1. But, since $y^2 = 1 - x^2$, we get $x^2 + y^2 = 1$, which shows that the graph of f is the quarter-circle with radius 1 in Figure 9. Therefore

$$\int_0^1 \sqrt{1 - x^2}\, dx = \tfrac{1}{4}\pi(1)^2 = \frac{\pi}{4}$$

(In Section 7.3 we will be able to *prove* that the area of a circle of radius r is πr^2.)

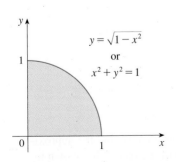

FIGURE 9

(b) The graph of $y = x - 1$ is the line with slope 1 shown in Figure 10. We compute the integral as the difference of the areas of the two triangles:

$$\int_0^3 (x - 1)\, dx = A_1 - A_2 = \tfrac{1}{2}(2 \cdot 2) - \tfrac{1}{2}(1 \cdot 1) = 1.5$$

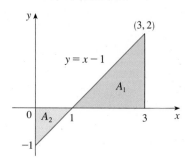

FIGURE 10

The Midpoint Rule

We often choose the sample point x_i^* to be the right endpoint of the ith subinterval because it is convenient for computing the limit. But if the purpose is to find an *approximation* to an integral, it is usually better to choose x_i^* to be the midpoint of the interval, which we denote by $\bar{x}_i$. Any Riemann sum is an approximation to an integral, but if we use midpoints we get the following approximation.

TEC Module 4.2/7.7 shows how the Midpoint Rule estimates improve as n increases.

Midpoint Rule

$$\int_a^b f(x)\, dx \approx \sum_{i=1}^{n} f(\bar{x}_i)\, \Delta x = \Delta x\, [\,f(\bar{x}_1) + \cdots + f(\bar{x}_n)\,]$$

where
$$\Delta x = \frac{b - a}{n}$$

and
$$\bar{x}_i = \tfrac{1}{2}(x_{i-1} + x_i) = \text{midpoint of } [x_{i-1}, x_i]$$

V   EXAMPLE 5 Use the Midpoint Rule with $n = 5$ to approximate $\displaystyle\int_1^2 \frac{1}{x}\, dx$.

SOLUTION The endpoints of the five subintervals are 1, 1.2, 1.4, 1.6, 1.8, and 2.0, so the midpoints are 1.1, 1.3, 1.5, 1.7, and 1.9. The width of the subintervals is $\Delta x = (2 - 1)/5 = \tfrac{1}{5}$, so the Midpoint Rule gives

$$\int_1^2 \frac{1}{x}\, dx \approx \Delta x\, [\,f(1.1) + f(1.3) + f(1.5) + f(1.7) + f(1.9)\,]$$

$$= \frac{1}{5}\left(\frac{1}{1.1} + \frac{1}{1.3} + \frac{1}{1.5} + \frac{1}{1.7} + \frac{1}{1.9} \right)$$

$$\approx 0.691908$$

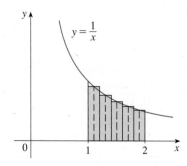

FIGURE 11

Since $f(x) = 1/x > 0$ for $1 \le x \le 2$, the integral represents an area, and the approximation given by the Midpoint Rule is the sum of the areas of the rectangles shown in Figure 11.

At the moment we don't know how accurate the approximation in Example 5 is, but in Section 7.7 we will learn a method for estimating the error involved in using the Midpoint Rule. At that time we will discuss other methods for approximating definite integrals.

If we apply the Midpoint Rule to the integral in Example 2, we get the picture in Figure 12. The approximation $M_{40} \approx -6.7563$ is much closer to the true value -6.75 than the right endpoint approximation, $R_{40} \approx -6.3998$, shown in Figure 7.

TEC In Visual 4.2 you can compare left, right, and midpoint approximations to the integral in Example 2 for different values of n.

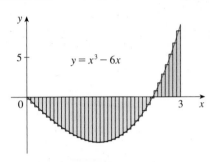

FIGURE 12
$M_{40} \approx -6.7563$

Properties of the Definite Integral

When we defined the definite integral $\int_a^b f(x)\,dx$, we implicitly assumed that $a < b$. But the definition as a limit of Riemann sums makes sense even if $a > b$. Notice that if we reverse a and b, then Δx changes from $(b - a)/n$ to $(a - b)/n$. Therefore

$$\int_b^a f(x)\,dx = -\int_a^b f(x)\,dx$$

If $a = b$, then $\Delta x = 0$ and so

$$\int_a^a f(x)\,dx = 0$$

We now develop some basic properties of integrals that will help us to evaluate integrals in a simple manner. We assume that f and g are continuous functions.

Properties of the Integral
1. $\displaystyle\int_a^b c\,dx = c(b - a)$,   where c is any constant
2. $\displaystyle\int_a^b [f(x) + g(x)]\,dx = \int_a^b f(x)\,dx + \int_a^b g(x)\,dx$
3. $\displaystyle\int_a^b cf(x)\,dx = c\int_a^b f(x)\,dx$,   where c is any constant
4. $\displaystyle\int_a^b [f(x) - g(x)]\,dx = \int_a^b f(x)\,dx - \int_a^b g(x)\,dx$

FIGURE 13
$\int_a^b c\,dx = c(b - a)$

Property 1 says that the integral of a constant function $f(x) = c$ is the constant times the length of the interval. If $c > 0$ and $a < b$, this is to be expected because $c(b - a)$ is the area of the shaded rectangle in Figure 13.

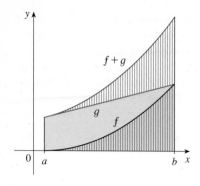

FIGURE 14

$$\int_a^b [f(x) + g(x)] \, dx =$$
$$\int_a^b f(x) \, dx + \int_a^b g(x) \, dx$$

Property 3 seems intuitively reasonable because we know that multiplying a function by a positive number c stretches or shrinks its graph vertically by a factor of c. So it stretches or shrinks each approximating rectangle by a factor c and therefore it has the effect of multiplying the area by c.

Property 2 says that the integral of a sum is the sum of the integrals. For positive functions it says that the area under $f + g$ is the area under f plus the area under g. Figure 14 helps us understand why this is true: In view of how graphical addition works, the corresponding vertical line segments have equal height.

In general, Property 2 follows from Theorem 4 and the fact that the limit of a sum is the sum of the limits:

$$\int_a^b [f(x) + g(x)] \, dx = \lim_{n \to \infty} \sum_{i=1}^n [f(x_i) + g(x_i)] \Delta x$$

$$= \lim_{n \to \infty} \left[\sum_{i=1}^n f(x_i) \Delta x + \sum_{i=1}^n g(x_i) \Delta x \right]$$

$$= \lim_{n \to \infty} \sum_{i=1}^n f(x_i) \Delta x + \lim_{n \to \infty} \sum_{i=1}^n g(x_i) \Delta x$$

$$= \int_a^b f(x) \, dx + \int_a^b g(x) \, dx$$

Property 3 can be proved in a similar manner and says that the integral of a constant times a function is the constant times the integral of the function. In other words, a constant (but *only* a constant) can be taken in front of an integral sign. Property 4 is proved by writing $f - g = f + (-g)$ and using Properties 2 and 3 with $c = -1$.

EXAMPLE 6   Use the properties of integrals to evaluate $\int_0^1 (4 + 3x^2) \, dx$.

SOLUTION   Using Properties 2 and 3 of integrals, we have

$$\int_0^1 (4 + 3x^2) \, dx = \int_0^1 4 \, dx + \int_0^1 3x^2 \, dx = \int_0^1 4 \, dx + 3 \int_0^1 x^2 \, dx$$

We know from Property 1 that

$$\int_0^1 4 \, dx = 4(1 - 0) = 4$$

and we found in Example 2 in Section 4.1 that $\int_0^1 x^2 \, dx = \frac{1}{3}$. So

$$\int_0^1 (4 + 3x^2) \, dx = \int_0^1 4 \, dx + 3 \int_0^1 x^2 \, dx$$

$$= 4 + 3 \cdot \tfrac{1}{3} = 5$$

The next property tells us how to combine integrals of the same function over adjacent intervals:

5.
$$\int_a^c f(x) \, dx + \int_c^b f(x) \, dx = \int_a^b f(x) \, dx$$

FIGURE 15

This is not easy to prove in general, but for the case where $f(x) \geq 0$ and $a < c < b$ Property 5 can be seen from the geometric interpretation in Figure 15: The area under $y = f(x)$ from a to c plus the area from c to b is equal to the total area from a to b.

▼ **EXAMPLE 7** If it is known that $\int_0^{10} f(x)\, dx = 17$ and $\int_0^8 f(x)\, dx = 12$, find $\int_8^{10} f(x)\, dx$.

SOLUTION By Property 5, we have

$$\int_0^8 f(x)\, dx + \int_8^{10} f(x)\, dx = \int_0^{10} f(x)\, dx$$

so $$\int_8^{10} f(x)\, dx = \int_0^{10} f(x)\, dx - \int_0^8 f(x)\, dx = 17 - 12 = 5$$ ▬

Properties 1–5 are true whether $a < b$, $a = b$, or $a > b$. The following properties, in which we compare sizes of functions and sizes of integrals, are true only if $a \le b$.

Comparison Properties of the Integral

6. If $f(x) \ge 0$ for $a \le x \le b$, then $\int_a^b f(x)\, dx \ge 0$.

7. If $f(x) \ge g(x)$ for $a \le x \le b$, then $\int_a^b f(x)\, dx \ge \int_a^b g(x)\, dx$.

8. If $m \le f(x) \le M$ for $a \le x \le b$, then

$$m(b - a) \le \int_a^b f(x)\, dx \le M(b - a)$$

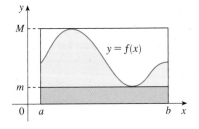

FIGURE 16

If $f(x) \ge 0$, then $\int_a^b f(x)\, dx$ represents the area under the graph of f, so the geometric interpretation of Property 6 is simply that areas are positive. (It also follows directly from the definition because all the quantities involved are positive.) Property 7 says that a bigger function has a bigger integral. It follows from Properties 6 and 4 because $f - g \ge 0$.

Property 8 is illustrated by Figure 16 for the case where $f(x) \ge 0$. If f is continuous we could take m and M to be the absolute minimum and maximum values of f on the interval $[a, b]$. In this case Property 8 says that the area under the graph of f is greater than the area of the rectangle with height m and less than the area of the rectangle with height M.

PROOF OF PROPERTY 8   Since $m \le f(x) \le M$, Property 7 gives

$$\int_a^b m\, dx \le \int_a^b f(x)\, dx \le \int_a^b M\, dx$$

Using Property 1 to evaluate the integrals on the left and right sides, we obtain

$$m(b - a) \le \int_a^b f(x)\, dx \le M(b - a)$$ ▬

Property 8 is useful when all we want is a rough estimate of the size of an integral without going to the bother of using the Midpoint Rule.

EXAMPLE 8   Use Property 8 to estimate $\int_1^4 \sqrt{x}\, dx$.

SOLUTION Since $f(x) = \sqrt{x}$ is an increasing function, its absolute minimum on $[1, 4]$ is $m = f(1) = 1$ and its absolute maximum on $[1, 4]$ is $M = f(4) = \sqrt{4} = 2$. Thus

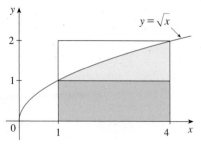

FIGURE 17

Property 8 gives

$$1(4 - 1) \le \int_1^4 \sqrt{x}\, dx \le 2(4 - 1)$$

or

$$3 \le \int_1^4 \sqrt{x}\, dx \le 6$$

The result of Example 8 is illustrated in Figure 17. The area under $y = \sqrt{x}$ from 1 to 4 is greater than the area of the lower rectangle and less than the area of the large rectangle.

4.2   Exercises

Access end-of-section exercises and tools for visualizing calculus online at **www.webassign.net**

DISCOVERY PROJECT   AREA FUNCTIONS

1. (a) Draw the line $y = 2t + 1$ and use geometry to find the area under this line, above the t-axis, and between the vertical lines $t = 1$ and $t = 3$.
   (b) If $x > 1$, let $A(x)$ be the area of the region that lies under the line $y = 2t + 1$ between $t = 1$ and $t = x$. Sketch this region and use geometry to find an expression for $A(x)$.
   (c) Differentiate the area function $A(x)$. What do you notice?

2. (a) If $x \ge -1$, let

$$A(x) = \int_{-1}^x (1 + t^2)\, dt$$

   $A(x)$ represents the area of a region. Sketch that region.
   (b) Use the result of Exercise 28 in Section 4.2 to find an expression for $A(x)$.
   (c) Find $A'(x)$. What do you notice?
   (d) If $x \ge -1$ and h is a small positive number, then $A(x + h) - A(x)$ represents the area of a region. Describe and sketch the region.
   (e) Draw a rectangle that approximates the region in part (d). By comparing the areas of these two regions, show that

$$\frac{A(x + h) - A(x)}{h} \approx 1 + x^2$$

   (f) Use part (e) to give an intuitive explanation for the result of part (c).

3. (a) Draw the graph of the function $f(x) = \cos(x^2)$ in the viewing rectangle $[0, 2]$ by $[-1.25, 1.25]$.
   (b) If we define a new function g by

$$g(x) = \int_0^x \cos(t^2)\, dt$$

   then $g(x)$ is the area under the graph of f from 0 to x [until $f(x)$ becomes negative, at which point $g(x)$ becomes a difference of areas]. Use part (a) to determine the value of

Graphing calculator or computer required

x at which $g(x)$ starts to decrease. [Unlike the integral in Problem 2, it is impossible to evaluate the integral defining g to obtain an explicit expression for $g(x)$.]

(c) Use the integration command on your calculator or computer to estimate $g(0.2)$, $g(0.4)$, $g(0.6)$, ..., $g(1.8)$, $g(2)$. Then use these values to sketch a graph of g.

(d) Use your graph of g from part (c) to sketch the graph of g' using the interpretation of $g'(x)$ as the slope of a tangent line. How does the graph of g' compare with the graph of f?

4. Suppose f is a continuous function on the interval $[a, b]$ and we define a new function g by the equation

$$g(x) = \int_a^x f(t)\,dt$$

Based on your results in Problems 1–3, conjecture an expression for $g'(x)$.

4.3   The Fundamental Theorem of Calculus

The Fundamental Theorem of Calculus is appropriately named because it establishes a connection between the two branches of calculus: differential calculus and integral calculus. Differential calculus arose from the tangent problem, whereas integral calculus arose from a seemingly unrelated problem, the area problem. Newton's mentor at Cambridge, Isaac Barrow (1630–1677), discovered that these two problems are actually closely related. In fact, he realized that differentiation and integration are inverse processes. The Fundamental Theorem of Calculus gives the precise inverse relationship between the derivative and the integral. It was Newton and Leibniz who exploited this relationship and used it to develop calculus into a systematic mathematical method. In particular, they saw that the Fundamental Theorem enabled them to compute areas and integrals very easily without having to compute them as limits of sums as we did in Sections 4.1 and 4.2.

The first part of the Fundamental Theorem deals with functions defined by an equation of the form

$$\boxed{1} \qquad g(x) = \int_a^x f(t)\,dt$$

where f is a continuous function on $[a, b]$ and x varies between a and b. Observe that g depends only on x, which appears as the variable upper limit in the integral. If x is a fixed number, then the integral $\int_a^x f(t)\,dt$ is a definite number. If we then let x vary, the number $\int_a^x f(t)\,dt$ also varies and defines a function of x denoted by $g(x)$.

If f happens to be a positive function, then $g(x)$ can be interpreted as the area under the graph of f from a to x, where x can vary from a to b. (Think of g as the "area so far" function; see Figure 1.)

FIGURE 1

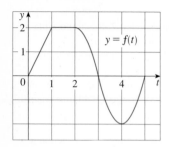

FIGURE 2

V **EXAMPLE 1** If f is the function whose graph is shown in Figure 2 and $g(x) = \int_0^x f(t)\,dt$, find the values of $g(0)$, $g(1)$, $g(2)$, $g(3)$, $g(4)$, and $g(5)$. Then sketch a rough graph of g.

SOLUTION First we notice that $g(0) = \int_0^0 f(t)\,dt = 0$. From Figure 3 we see that $g(1)$ is the area of a triangle:

$$g(1) = \int_0^1 f(t)\,dt = \tfrac{1}{2}(1 \cdot 2) = 1$$

To find $g(2)$ we add to $g(1)$ the area of a rectangle:

$$g(2) = \int_0^2 f(t)\,dt = \int_0^1 f(t)\,dt + \int_1^2 f(t)\,dt = 1 + (1 \cdot 2) = 3$$

We estimate that the area under f from 2 to 3 is about 1.3, so

$$g(3) = g(2) + \int_2^3 f(t)\,dt \approx 3 + 1.3 = 4.3$$

$g(1) = 1$

$g(2) = 3$

$g(3) \approx 4.3$

$g(4) \approx 3$

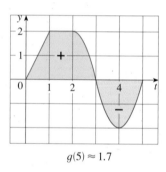

$g(5) \approx 1.7$

FIGURE 3

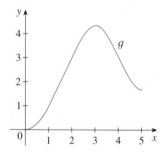

FIGURE 4
$g(x) = \int_a^x f(t)\,dt$

For $t > 3$, $f(t)$ is negative and so we start subtracting areas:

$$g(4) = g(3) + \int_3^4 f(t)\,dt \approx 4.3 + (-1.3) = 3.0$$

$$g(5) = g(4) + \int_4^5 f(t)\,dt \approx 3 + (-1.3) = 1.7$$

We use these values to sketch the graph of g in Figure 4. Notice that, because $f(t)$ is positive for $t < 3$, we keep adding area for $t < 3$ and so g is increasing up to $x = 3$, where it attains a maximum value. For $x > 3$, g decreases because $f(t)$ is negative. ∎

If we take $f(t) = t$ and $a = 0$, then, using Exercise 27 in Section 4.2, we have

$$g(x) = \int_0^x t\,dt = \frac{x^2}{2}$$

Notice that $g'(x) = x$, that is, $g' = f$. In other words, if g is defined as the integral of f by Equation 1, then g turns out to be an antiderivative of f, at least in this case. And if we sketch the derivative of the function g shown in Figure 4 by estimating slopes of tangents, we get a graph like that of f in Figure 2. So we suspect that $g' = f$ in Example 1 too.

To see why this might be generally true we consider any continuous function f with $f(x) \geq 0$. Then $g(x) = \int_a^x f(t)\, dt$ can be interpreted as the area under the graph of f from a to x, as in Figure 1.

In order to compute $g'(x)$ from the definition of a derivative we first observe that, for $h > 0$, $g(x + h) - g(x)$ is obtained by subtracting areas, so it is the area under the graph of f from x to $x + h$ (the blue area in Figure 5). For small h you can see from the figure that this area is approximately equal to the area of the rectangle with height $f(x)$ and width h:

$$g(x + h) - g(x) \approx h f(x)$$

so

$$\frac{g(x + h) - g(x)}{h} \approx f(x)$$

Intuitively, we therefore expect that

$$g'(x) = \lim_{h \to 0} \frac{g(x + h) - g(x)}{h} = f(x)$$

The fact that this is true, even when f is not necessarily positive, is the first part of the Fundamental Theorem of Calculus.

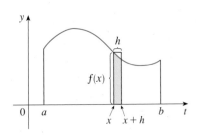

FIGURE 5

We abbreviate the name of this theorem as FTC1. In words, it says that the derivative of a definite integral with respect to its upper limit is the integrand evaluated at the upper limit.

The Fundamental Theorem of Calculus, Part 1 If f is continuous on $[a, b]$, then the function g defined by

$$g(x) = \int_a^x f(t)\, dt \qquad a \leq x \leq b$$

is continuous on $[a, b]$ and differentiable on (a, b), and $g'(x) = f(x)$.

PROOF If x and $x + h$ are in (a, b), then

$$g(x + h) - g(x) = \int_a^{x+h} f(t)\, dt - \int_a^x f(t)\, dt$$

$$= \left(\int_a^x f(t)\, dt + \int_x^{x+h} f(t)\, dt \right) - \int_a^x f(t)\, dt \qquad \text{(by Property 5)}$$

$$= \int_x^{x+h} f(t)\, dt$$

and so, for $h \neq 0$,

$$\boxed{2} \qquad \frac{g(x + h) - g(x)}{h} = \frac{1}{h} \int_x^{x+h} f(t)\, dt$$

For now let's assume that $h > 0$. Since f is continuous on $[x, x + h]$, the Extreme Value Theorem says that there are numbers u and v in $[x, x + h]$ such that $f(u) = m$ and $f(v) = M$, where m and M are the absolute minimum and maximum values of f on $[x, x + h]$. (See Figure 6.)

By Property 8 of integrals, we have

$$mh \leq \int_x^{x+h} f(t)\, dt \leq Mh$$

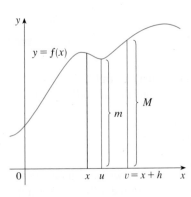

FIGURE 6

that is,

$$f(u)h \leqslant \int_x^{x+h} f(t)\, dt \leqslant f(v)h$$

Since $h > 0$, we can divide this inequality by h:

$$f(u) \leqslant \frac{1}{h} \int_x^{x+h} f(t)\, dt \leqslant f(v)$$

Now we use Equation 2 to replace the middle part of this inequality:

$$\boxed{3} \qquad\qquad f(u) \leqslant \frac{g(x+h) - g(x)}{h} \leqslant f(v)$$

TEC Module 4.3 provides visual
evidence for FTC1.

Inequality 3 can be proved in a similar manner for the case where $h < 0$. (See Exercise 63.)
Now we let $h \to 0$. Then $u \to x$ and $v \to x$, since u and v lie between x and $x + h$.
Therefore

$$\lim_{h \to 0} f(u) = \lim_{u \to x} f(u) = f(x) \qquad \text{and} \qquad \lim_{h \to 0} f(v) = \lim_{v \to x} f(v) = f(x)$$

because f is continuous at x. We conclude, from $\boxed{3}$ and the Squeeze Theorem, that

$$\boxed{4} \qquad\qquad g'(x) = \lim_{h \to 0} \frac{g(x+h) - g(x)}{h} = f(x)$$

If $x = a$ or b, then Equation 4 can be interpreted as a one-sided limit. Then Theorem 2.2.4 (modified for one-sided limits) shows that g is continuous on $[a, b]$.   ■

Using Leibniz notation for derivatives, we can write FTC1 as

$$\boxed{5} \qquad\qquad \frac{d}{dx} \int_a^x f(t)\, dt = f(x)$$

when f is continuous. Roughly speaking, Equation 5 says that if we first integrate f and then differentiate the result, we get back to the original function f.

V EXAMPLE 2 Find the derivative of the function $g(x) = \int_0^x \sqrt{1 + t^2}\, dt$.

SOLUTION Since $f(t) = \sqrt{1 + t^2}$ is continuous, Part 1 of the Fundamental Theorem of Calculus gives

$$g'(x) = \sqrt{1 + x^2} \qquad\qquad ■$$

EXAMPLE 3 Although a formula of the form $g(x) = \int_a^x f(t)\, dt$ may seem like a strange way of defining a function, books on physics, chemistry, and statistics are full of such functions. For instance, the **Fresnel function**

$$S(x) = \int_0^x \sin(\pi t^2 / 2)\, dt$$

is named after the French physicist Augustin Fresnel (1788–1827), who is famous for his works in optics. This function first appeared in Fresnel's theory of the diffraction of light waves, but more recently it has been applied to the design of highways.

Part 1 of the Fundamental Theorem tells us how to differentiate the Fresnel function:

$$S'(x) = \sin(\pi x^2/2)$$

This means that we can apply all the methods of differential calculus to analyze S (see Exercise 57).

Figure 7 shows the graphs of $f(x) = \sin(\pi x^2/2)$ and the Fresnel function $S(x) = \int_0^x f(t)\, dt$. A computer was used to graph S by computing the value of this integral for many values of x. It does indeed look as if $S(x)$ is the area under the graph of f from 0 to x [until $x \approx 1.4$ when $S(x)$ becomes a difference of areas]. Figure 8 shows a larger part of the graph of S.

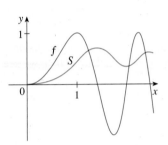

FIGURE 7
$f(x) = \sin(\pi x^2/2)$
$S(x) = \displaystyle\int_0^x \sin(\pi t^2/2)\, dt$

FIGURE 8
The Fresnel function $S(x) = \displaystyle\int_0^x \sin(\pi t^2/2)\, dt$

If we now start with the graph of S in Figure 7 and think about what its derivative should look like, it seems reasonable that $S'(x) = f(x)$. [For instance, S is increasing when $f(x) > 0$ and decreasing when $f(x) < 0$.] So this gives a visual confirmation of Part 1 of the Fundamental Theorem of Calculus.

EXAMPLE 4 Find $\dfrac{d}{dx} \displaystyle\int_1^{x^4} \sec t\, dt$.

SOLUTION Here we have to be careful to use the Chain Rule in conjunction with FTC1. Let $u = x^4$. Then

$$\frac{d}{dx} \int_1^{x^4} \sec t\, dt = \frac{d}{dx} \int_1^{u} \sec t\, dt$$

$$= \frac{d}{du}\left[\int_1^{u} \sec t\, dt \right] \frac{du}{dx} \qquad \text{(by the Chain Rule)}$$

$$= \sec u \, \frac{du}{dx} \qquad \text{(by FTC1)}$$

$$= \sec(x^4) \cdot 4x^3$$

In Section 4.2 we computed integrals from the definition as a limit of Riemann sums and we saw that this procedure is sometimes long and difficult. The second part of the Fundamental Theorem of Calculus, which follows easily from the first part, provides us with a much simpler method for the evaluation of integrals.

We abbreviate this theorem as FTC2.

> **The Fundamental Theorem of Calculus, Part 2**   If f is continuous on $[a, b]$, then
>
> $$\int_a^b f(x)\, dx = F(b) - F(a)$$
>
> where F is any antiderivative of f, that is, a function such that $F' = f$.

PROOF Let $g(x) = \int_a^x f(t)\, dt$. We know from Part 1 that $g'(x) = f(x)$; that is, g is an antiderivative of f. If F is any other antiderivative of f on $[a, b]$, then we know from Corollary 3.2.7 that F and g differ by a constant:

$$\boxed{6} \qquad F(x) = g(x) + C$$

for $a < x < b$. But both F and g are continuous on $[a, b]$ and so, by taking limits of both sides of Equation 6 (as $x \to a^+$ and $x \to b^-$), we see that it also holds when $x = a$ and $x = b$.

If we put $x = a$ in the formula for $g(x)$, we get

$$g(a) = \int_a^a f(t)\, dt = 0$$

So, using Equation 6 with $x = b$ and $x = a$, we have

$$F(b) - F(a) = [g(b) + C] - [g(a) + C]$$
$$= g(b) - g(a) = g(b) = \int_a^b f(t)\, dt \qquad \blacksquare$$

Part 2 of the Fundamental Theorem states that if we know an antiderivative F of f, then we can evaluate $\int_a^b f(x)\, dx$ simply by subtracting the values of F at the endpoints of the interval $[a, b]$. It's very surprising that $\int_a^b f(x)\, dx$, which was defined by a complicated procedure involving all of the values of $f(x)$ for $a \le x \le b$, can be found by knowing the values of $F(x)$ at only two points, a and b.

Although the theorem may be surprising at first glance, it becomes plausible if we interpret it in physical terms. If $v(t)$ is the velocity of an object and $s(t)$ is its position at time t, then $v(t) = s'(t)$, so s is an antiderivative of v. In Section 4.1 we considered an object that always moves in the positive direction and made the guess that the area under the velocity curve is equal to the distance traveled. In symbols:

$$\int_a^b v(t)\, dt = s(b) - s(a)$$

That is exactly what FTC2 says in this context.

V EXAMPLE 5 Evaluate the integral $\int_{-2}^1 x^3\, dx$.

SOLUTION The function $f(x) = x^3$ is continuous on $[-2, 1]$ and we know from Section 3.9 that an antiderivative is $F(x) = \frac{1}{4}x^4$, so Part 2 of the Fundamental Theorem gives

$$\int_{-2}^1 x^3\, dx = F(1) - F(-2) = \tfrac{1}{4}(1)^4 - \tfrac{1}{4}(-2)^4 = -\tfrac{15}{4}$$

Notice that FTC2 says we can use *any* antiderivative F of f. So we may as well use the simplest one, namely $F(x) = \frac{1}{4}x^4$, instead of $\frac{1}{4}x^4 + 7$ or $\frac{1}{4}x^4 + C$. $\qquad \blacksquare$

We often use the notation

$$F(x)\Big]_a^b = F(b) - F(a)$$

So the equation of FTC2 can be written as

$$\int_a^b f(x)\,dx = F(x)\Big]_a^b \qquad \text{where} \qquad F' = f$$

Other common notations are $F(x)\big|_a^b$ and $[F(x)]_a^b$.

EXAMPLE 6 Find the area under the parabola $y = x^2$ from 0 to 1.

SOLUTION An antiderivative of $f(x) = x^2$ is $F(x) = \frac{1}{3}x^3$. The required area A is found using Part 2 of the Fundamental Theorem:

$$A = \int_0^1 x^2\,dx = \frac{x^3}{3}\Bigg]_0^1$$

$$= \frac{1^3}{3} - \frac{0^3}{3} = \frac{1}{3}$$

In applying the Fundamental Theorem we use a particular antiderivative F of f. It is not necessary to use the most general antiderivative.

If you compare the calculation in Example 6 with the one in Example 2 in Section 4.1, you will see that the Fundamental Theorem gives a *much* shorter method.

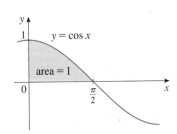

FIGURE 9

EXAMPLE 7 Find the area under the cosine curve from 0 to b, where $0 \le b \le \pi/2$.

SOLUTION Since an antiderivative of $f(x) = \cos x$ is $F(x) = \sin x$, we have

$$A = \int_0^b \cos x\,dx = \sin x\Big]_0^b = \sin b - \sin 0 = \sin b$$

In particular, taking $b = \pi/2$, we have proved that the area under the cosine curve from 0 to $\pi/2$ is $\sin(\pi/2) = 1$. (See Figure 9.)

When the French mathematician Gilles de Roberval first found the area under the sine and cosine curves in 1635, this was a very challenging problem that required a great deal of ingenuity. If we didn't have the benefit of the Fundamental Theorem, we would have to compute a difficult limit of sums using obscure trigonometric identities (or a computer algebra system as in Exercise 29 in Section 4.1). It was even more difficult for Roberval because the apparatus of limits had not been invented in 1635. But in the 1660s and 1670s, when the Fundamental Theorem was discovered by Barrow and exploited by Newton and Leibniz, such problems became very easy, as you can see from Example 7.

EXAMPLE 8 What is wrong with the following calculation?

$$\int_{-1}^{3} \frac{1}{x^2}\,dx = \frac{x^{-1}}{-1}\Bigg]_{-1}^{3} = -\frac{1}{3} - 1 = -\frac{4}{3}$$

SOLUTION To start, we notice that this calculation must be wrong because the answer is negative but $f(x) = 1/x^2 \ge 0$ and Property 6 of integrals says that $\int_a^b f(x)\,dx \ge 0$ when

$f \geqslant 0$. The Fundamental Theorem of Calculus applies to continuous functions. It can't be applied here because $f(x) = 1/x^2$ is not continuous on $[-1, 3]$. In fact, f has an infinite discontinuity at $x = 0$, so

$$\int_{-1}^{3} \frac{1}{x^2} \, dx \qquad \text{does not exist}$$

Differentiation and Integration as Inverse Processes

We end this section by bringing together the two parts of the Fundamental Theorem.

The Fundamental Theorem of Calculus   Suppose f is continuous on $[a, b]$.

1. If $g(x) = \int_{a}^{x} f(t) \, dt$, then $g'(x) = f(x)$.

2. $\int_{a}^{b} f(x) \, dx = F(b) - F(a)$, where F is any antiderivative of f, that is, $F' = f$.

We noted that Part 1 can be rewritten as

$$\frac{d}{dx} \int_{a}^{x} f(t) \, dt = f(x)$$

which says that if f is integrated and then the result is differentiated, we arrive back at the original function f. Since $F'(x) = f(x)$, Part 2 can be rewritten as

$$\int_{a}^{b} F'(x) \, dx = F(b) - F(a)$$

This version says that if we take a function F, first differentiate it, and then integrate the result, we arrive back at the original function F, but in the form $F(b) - F(a)$. Taken together, the two parts of the Fundamental Theorem of Calculus say that differentiation and integration are inverse processes. Each undoes what the other does.

The Fundamental Theorem of Calculus is unquestionably the most important theorem in calculus and, indeed, it ranks as one of the great accomplishments of the human mind. Before it was discovered, from the time of Eudoxus and Archimedes to the time of Galileo and Fermat, problems of finding areas, volumes, and lengths of curves were so difficult that only a genius could meet the challenge. But now, armed with the systematic method that Newton and Leibniz fashioned out of the Fundamental Theorem, we will see in the chapters to come that these challenging problems are accessible to all of us.

4.3 Exercises

Access end-of-section exercises and tools for visualizing calculus online at **www.webassign.net**   ᴱᴺᴴᴬᴺᶜᴱᴰ **WebAssign**

4.4   Indefinite Integrals and the Net Change Theorem

We saw in Section 4.3 that the second part of the Fundamental Theorem of Calculus provides a very powerful method for evaluating the definite integral of a function, assuming that we can find an antiderivative of the function. In this section we introduce a notation for antiderivatives, review the formulas for antiderivatives, and use them to evaluate definite integrals. We also reformulate FTC2 in a way that makes it easier to apply to science and engineering problems.

Indefinite Integrals

Both parts of the Fundamental Theorem establish connections between antiderivatives and definite integrals. Part 1 says that if f is continuous, then $\int_a^x f(t)\,dt$ is an antiderivative of f. Part 2 says that $\int_a^b f(x)\,dx$ can be found by evaluating $F(b) - F(a)$, where F is an antiderivative of f.

We need a convenient notation for antiderivatives that makes them easy to work with. Because of the relation given by the Fundamental Theorem between antiderivatives and integrals, the notation $\int f(x)\,dx$ is traditionally used for an antiderivative of f and is called an **indefinite integral**. Thus

$$\int f(x)\,dx = F(x) \qquad \text{means} \qquad F'(x) = f(x)$$

For example, we can write

$$\int x^2\,dx = \frac{x^3}{3} + C \qquad \text{because} \qquad \frac{d}{dx}\left(\frac{x^3}{3} + C\right) = x^2$$

So we can regard an indefinite integral as representing an entire *family* of functions (one antiderivative for each value of the constant C).

⊘ You should distinguish carefully between definite and indefinite integrals. A definite integral $\int_a^b f(x)\,dx$ is a *number*, whereas an indefinite integral $\int f(x)\,dx$ is a *function* (or family of functions). The connection between them is given by Part 2 of the Fundamental Theorem: If f is continuous on $[a, b]$, then

$$\int_a^b f(x)\,dx = \int f(x)\,dx \Big]_a^b$$

The effectiveness of the Fundamental Theorem depends on having a supply of antiderivatives of functions. We therefore restate the Table of Antidifferentiation Formulas from Section 3.9, together with a few others, in the notation of indefinite integrals. Any formula can be verified by differentiating the function on the right side and obtaining the integrand. For instance,

$$\int \sec^2 x\,dx = \tan x + C \qquad \text{because} \qquad \frac{d}{dx}(\tan x + C) = \sec^2 x$$

---

1   **Table of Indefinite Integrals**

$$\int cf(x)\,dx = c\int f(x)\,dx \qquad \int [f(x) + g(x)]\,dx = \int f(x)\,dx + \int g(x)\,dx$$

$$\int k\,dx = kx + C \qquad \int x^n\,dx = \frac{x^{n+1}}{n+1} + C \quad (n \neq -1)$$

$$\int \sin x\,dx = -\cos x + C \qquad \int \cos x\,dx = \sin x + C$$

$$\int \sec^2 x\,dx = \tan x + C \qquad \int \csc^2 x\,dx = -\cot x + C$$

$$\int \sec x \tan x\,dx = \sec x + C \qquad \int \csc x \cot x\,dx = -\csc x + C$$

Recall from Theorem 3.9.1 that the most general antiderivative *on a given interval* is obtained by adding a constant to a particular antiderivative. **We adopt the convention that when a formula for a general indefinite integral is given, it is valid only on an interval.** Thus we write

$$\int \frac{1}{x^2}\,dx = -\frac{1}{x} + C$$

with the understanding that it is valid on the interval $(0, \infty)$ or on the interval $(-\infty, 0)$. This is true despite the fact that the general antiderivative of the function $f(x) = 1/x^2$, $x \neq 0$, is

$$F(x) = \begin{cases} -\dfrac{1}{x} + C_1 & \text{if } x < 0 \\[2mm] -\dfrac{1}{x} + C_2 & \text{if } x > 0 \end{cases}$$

EXAMPLE 1   Find the general indefinite integral

$$\int (10x^4 - 2\sec^2 x)\,dx$$

SOLUTION   Using our convention and Table 1, we have

$$\int (10x^4 - 2\sec^2 x)\,dx = 10\int x^4\,dx - 2\int \sec^2 x\,dx$$

$$= 10\,\frac{x^5}{5} - 2\tan x + C = 2x^5 - 2\tan x + C$$

You should check this answer by differentiating it.   ▬

V   EXAMPLE 2   Evaluate $\displaystyle\int \frac{\cos\theta}{\sin^2\theta}\,d\theta$.

SOLUTION   This indefinite integral isn't immediately apparent in Table 1, so we use trigonometric identities to rewrite the function before integrating:

$$\int \frac{\cos\theta}{\sin^2\theta}\,d\theta = \int \left(\frac{1}{\sin\theta}\right)\left(\frac{\cos\theta}{\sin\theta}\right)d\theta$$

$$= \int \csc\theta\cot\theta\,d\theta = -\csc\theta + C \quad ▬$$

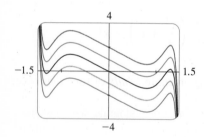

4

−1.5   1.5

−4

FIGURE 1

The indefinite integral in Example 1 is graphed in Figure 1 for several values of C. Here the value of C is the y-intercept.

EXAMPLE 3 Evaluate $\displaystyle\int_0^3 (x^3 - 6x)\,dx$.

SOLUTION Using FTC2 and Table 1, we have

$$\int_0^3 (x^3 - 6x)\,dx = \frac{x^4}{4} - 6\frac{x^2}{2}\Big]_0^3$$

$$= \left(\tfrac{1}{4}\cdot 3^4 - 3\cdot 3^2\right) - \left(\tfrac{1}{4}\cdot 0^4 - 3\cdot 0^2\right)$$

$$= \tfrac{81}{4} - 27 - 0 + 0 = -6.75$$

Compare this calculation with Example 2(b) in Section 4.2.

EXAMPLE 4 Find $\displaystyle\int_0^{12} (x - 12\sin x)\,dx$.

SOLUTION The Fundamental Theorem gives

$$\int_0^{12} (x - 12\sin x)\,dx = \frac{x^2}{2} - 12(-\cos x)\Big]_0^{12}$$

$$= \tfrac{1}{2}(12)^2 + 12(\cos 12 - \cos 0)$$

$$= 72 + 12\cos 12 - 12$$

$$= 60 + 12\cos 12$$

This is the exact value of the integral. If a decimal approximation is desired, we can use a calculator to approximate $\cos 12$. Doing so, we get

$$\int_0^{12} (x - 12\sin x)\,dx \approx 70.1262$$

Figure 2 shows the graph of the integrand in Example 4. We know from Section 4.2 that the value of the integral can be interpreted as the sum of the areas labeled with a plus sign minus the areas labeled with a minus sign.

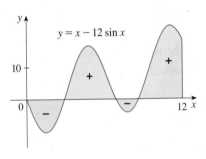

$y = x - 12\sin x$

FIGURE 2

EXAMPLE 5 Evaluate $\displaystyle\int_1^9 \frac{2t^2 + t^2\sqrt{t} - 1}{t^2}\,dt$.

SOLUTION First we need to write the integrand in a simpler form by carrying out the division:

$$\int_1^9 \frac{2t^2 + t^2\sqrt{t} - 1}{t^2}\,dt = \int_1^9 (2 + t^{1/2} - t^{-2})\,dt$$

$$= 2t + \frac{t^{3/2}}{\frac{3}{2}} - \frac{t^{-1}}{-1}\Big]_1^9 = 2t + \tfrac{2}{3}t^{3/2} + \frac{1}{t}\Big]_1^9$$

$$= \left(2\cdot 9 + \tfrac{2}{3}\cdot 9^{3/2} + \tfrac{1}{9}\right) - \left(2\cdot 1 + \tfrac{2}{3}\cdot 1^{3/2} + \tfrac{1}{1}\right)$$

$$= 18 + 18 + \tfrac{1}{9} - 2 - \tfrac{2}{3} - 1 = 32\tfrac{4}{9}$$

▇▇ Applications

Part 2 of the Fundamental Theorem says that if f is continuous on $[a, b]$, then

$$\int_a^b f(x)\,dx = F(b) - F(a)$$

where F is any antiderivative of f. This means that $F' = f$, so the equation can be rewritten as

$$\int_a^b F'(x)\,dx = F(b) - F(a)$$

We know that $F'(x)$ represents the rate of change of $y = F(x)$ with respect to x and $F(b) - F(a)$ is the change in y when x changes from a to b. [Note that y could, for instance, increase, then decrease, then increase again. Although y might change in both directions, $F(b) - F(a)$ represents the *net* change in y.] So we can reformulate FTC2 in words as follows.

Net Change Theorem The integral of a rate of change is the net change:

$$\int_a^b F'(x)\, dx = F(b) - F(a)$$

This principle can be applied to all of the rates of change in the natural and social sciences that we discussed in Section 2.7. Here are a few instances of this idea:

- If $V(t)$ is the volume of water in a reservoir at time t, then its derivative $V'(t)$ is the rate at which water flows into the reservoir at time t. So

$$\int_{t_1}^{t_2} V'(t)\, dt = V(t_2) - V(t_1)$$

  is the change in the amount of water in the reservoir between time t_1 and time t_2.

- If $[C](t)$ is the concentration of the product of a chemical reaction at time t, then the rate of reaction is the derivative $d[C]/dt$. So

$$\int_{t_1}^{t_2} \frac{d[C]}{dt}\, dt = [C](t_2) - [C](t_1)$$

  is the change in the concentration of C from time t_1 to time t_2.

- If the mass of a rod measured from the left end to a point x is $m(x)$, then the linear density is $\rho(x) = m'(x)$. So

$$\int_a^b \rho(x)\, dx = m(b) - m(a)$$

  is the mass of the segment of the rod that lies between $x = a$ and $x = b$.

- If the rate of growth of a population is dn/dt, then

$$\int_{t_1}^{t_2} \frac{dn}{dt}\, dt = n(t_2) - n(t_1)$$

  is the net change in population during the time period from t_1 to t_2. (The population increases when births happen and decreases when deaths occur. The net change takes into account both births and deaths.)

- If $C(x)$ is the cost of producing x units of a commodity, then the marginal cost is the derivative $C'(x)$. So

$$\int_{x_1}^{x_2} C'(x)\, dx = C(x_2) - C(x_1)$$

  is the increase in cost when production is increased from x_1 units to x_2 units.

- If an object moves along a straight line with position function $s(t)$, then its velocity is $v(t) = s'(t)$, so

$$\boxed{2}\qquad \int_{t_1}^{t_2} v(t)\, dt = s(t_2) - s(t_1)$$

is the net change of position, or *displacement,* of the particle during the time period from t_1 to t_2. In Section 4.1 we guessed that this was true for the case where the object moves in the positive direction, but now we have proved that it is always true.

■ If we want to calculate the distance the object travels during the time interval, we have to consider the intervals when $v(t) \geq 0$ (the particle moves to the right) and also the intervals when $v(t) \leq 0$ (the particle moves to the left). In both cases the distance is computed by integrating $|v(t)|$, the speed. Therefore

$$\boxed{3} \qquad \int_{t_1}^{t_2} |v(t)|\, dt = \text{total distance traveled}$$

Figure 3 shows how both displacement and distance traveled can be interpreted in terms of areas under a velocity curve.

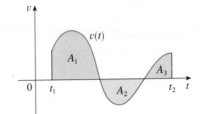

$$\text{displacement} = \int_{t_1}^{t_2} v(t)\, dt = A_1 - A_2 + A_3$$

$$\text{distance} = \int_{t_1}^{t_2} |v(t)|\, dt = A_1 + A_2 + A_3$$

FIGURE 3

■ The acceleration of the object is $a(t) = v'(t)$, so

$$\int_{t_1}^{t_2} a(t)\, dt = v(t_2) - v(t_1)$$

is the change in velocity from time t_1 to time t_2.

V EXAMPLE 6 A particle moves along a line so that its velocity at time t is $v(t) = t^2 - t - 6$ (measured in meters per second).
(a) Find the displacement of the particle during the time period $1 \leq t \leq 4$.
(b) Find the distance traveled during this time period.

SOLUTION
(a) By Equation 2, the displacement is

$$s(4) - s(1) = \int_1^4 v(t)\, dt = \int_1^4 (t^2 - t - 6)\, dt$$

$$= \left[\frac{t^3}{3} - \frac{t^2}{2} - 6t \right]_1^4 = -\frac{9}{2}$$

This means that the particle moved 4.5 m toward the left.

(b) Note that $v(t) = t^2 - t - 6 = (t - 3)(t + 2)$ and so $v(t) \leq 0$ on the interval $[1, 3]$ and $v(t) \geq 0$ on $[3, 4]$. Thus, from Equation 3, the distance traveled is

To integrate the absolute value of $v(t)$, we use Property 5 of integrals from Section 4.2 to split the integral into two parts, one where $v(t) \leq 0$ and one where $v(t) \geq 0$.

$$\int_1^4 |v(t)|\, dt = \int_1^3 [-v(t)]\, dt + \int_3^4 v(t)\, dt$$

$$= \int_1^3 (-t^2 + t + 6)\, dt + \int_3^4 (t^2 - t - 6)\, dt$$

$$= \left[-\frac{t^3}{3} + \frac{t^2}{2} + 6t \right]_1^3 + \left[\frac{t^3}{3} - \frac{t^2}{2} - 6t \right]_3^4$$

$$= \frac{61}{6} \approx 10.17 \text{ m}$$

EXAMPLE 7 Figure 4 shows the power consumption in the city of San Francisco for a day in September (P is measured in megawatts; t is measured in hours starting at midnight). Estimate the energy used on that day.

FIGURE 4

Pacific Gas & Electric

SOLUTION Power is the rate of change of energy: $P(t) = E'(t)$. So, by the Net Change Theorem,

$$\int_0^{24} P(t)\, dt = \int_0^{24} E'(t)\, dt = E(24) - E(0)$$

is the total amount of energy used on that day. We approximate the value of the integral using the Midpoint Rule with 12 subintervals and $\Delta t = 2$:

$$\int_0^{24} P(t)\, dt \approx [P(1) + P(3) + P(5) + \cdots + P(21) + P(23)]\,\Delta t$$

$$\approx (440 + 400 + 420 + 620 + 790 + 840 + 850$$

$$+ 840 + 810 + 690 + 670 + 550)(2)$$

$$= 15{,}840$$

The energy used was approximately 15,840 megawatt-hours. ▬

A note on units

How did we know what units to use for energy in Example 7? The integral $\int_0^{24} P(t)\, dt$ is defined as the limit of sums of terms of the form $P(t_i^*)\,\Delta t$. Now $P(t_i^*)$ is measured in megawatts and Δt is measured in hours, so their product is measured in megawatt-hours. The same is true of the limit. In general, the unit of measurement for $\int_a^b f(x)\, dx$ is the product of the unit for $f(x)$ and the unit for x.

4.4 Exercises

Access end-of-section exercises and tools for visualizing calculus online at **www.webassign.net**   ᴱᴺᴴᴬᴺᴄᴱᴰ **Web**Assign

NEWTON, LEIBNIZ, AND THE INVENTION OF CALCULUS

We sometimes read that the inventors of calculus were Sir Isaac Newton (1642–1727) and Gottfried Wilhelm Leibniz (1646–1716). But we know that the basic ideas behind integration were investigated 2500 years ago by ancient Greeks such as Eudoxus and Archimedes, and methods for finding tangents were pioneered by Pierre Fermat (1601–1665), Isaac Barrow (1630–1677), and others. Barrow—who taught at Cambridge and was a major influence on Newton—was the first to understand the inverse relationship between differentiation and integration. What Newton and Leibniz did was to use this relationship, in the form of the Fundamental Theorem of Calculus, in order to develop calculus into a systematic mathematical discipline. It is in this sense that Newton and Leibniz are credited with the invention of calculus.

Read about the contributions of these men in one or more of the given references and write a report on one of the following three topics. You can include biographical details, but the main thrust of your report should be a description, in some detail, of their methods and notations. In particular, you should consult one of the sourcebooks, which give excerpts from the original publications of Newton and Leibniz, translated from Latin to English.

- The Role of Newton in the Development of Calculus

- The Role of Leibniz in the Development of Calculus

- The Controversy between the Followers of Newton and Leibniz over Priority in the Invention of Calculus

References

1. Carl Boyer and Uta Merzbach, *A History of Mathematics* (New York: Wiley, 1987), Chapter 19.

2. Carl Boyer, *The History of the Calculus and Its Conceptual Development* (New York: Dover, 1959), Chapter V.

3. C. H. Edwards, *The Historical Development of the Calculus* (New York: Springer-Verlag, 1979), Chapters 8 and 9.

4. Howard Eves, *An Introduction to the History of Mathematics,* 6th ed. (New York: Saunders, 1990), Chapter 11.

5. C. C. Gillispie, ed., *Dictionary of Scientific Biography* (New York: Scribner's, 1974). See the article on Leibniz by Joseph Hofmann in Volume VIII and the article on Newton by I. B. Cohen in Volume X.

6. Victor Katz, *A History of Mathematics: An Introduction* (New York: HarperCollins, 1993), Chapter 12.

7. Morris Kline, *Mathematical Thought from Ancient to Modern Times* (New York: Oxford University Press, 1972), Chapter 17.

Sourcebooks

1. John Fauvel and Jeremy Gray, eds., *The History of Mathematics: A Reader* (London: MacMillan Press, 1987), Chapters 12 and 13.

2. D. E. Smith, ed., *A Sourcebook in Mathematics* (New York: Dover, 1959), Chapter V.

3. D. J. Struik, ed., *A Sourcebook in Mathematics, 1200–1800* (Princeton, NJ: Princeton University Press, 1969), Chapter V.

4.5 The Substitution Rule

Because of the Fundamental Theorem, it's important to be able to find antiderivatives. But our antidifferentiation formulas don't tell us how to evaluate integrals such as

$$\boxed{1} \qquad \int 2x\sqrt{1 + x^2}\, dx$$

To find this integral we use the problem-solving strategy of *introducing something extra*. Here the "something extra" is a new variable; we change from the variable x to a new variable u. Suppose that we let u be the quantity under the root sign in $\boxed{1}$, $u = 1 + x^2$. Then the differential of u is $du = 2x\, dx$. Notice that if the dx in the notation for an integral were to be interpreted as a differential, then the differential $2x\, dx$ would occur in $\boxed{1}$ and so, formally, without justifying our calculation, we could write

Differentials were defined in Section 2.9. If $u = f(x)$, then

$$du = f'(x)\, dx$$

$$\boxed{2} \qquad \int 2x\sqrt{1 + x^2}\, dx = \int \sqrt{1 + x^2}\ 2x\, dx = \int \sqrt{u}\ du$$

$$= \tfrac{2}{3}u^{3/2} + C = \tfrac{2}{3}(x^2 + 1)^{3/2} + C$$

But now we can check that we have the correct answer by using the Chain Rule to differentiate the final function of Equation 2:

$$\frac{d}{dx}\left[\tfrac{2}{3}(x^2 + 1)^{3/2} + C\right] = \tfrac{2}{3}\cdot\tfrac{3}{2}(x^2 + 1)^{1/2}\cdot 2x = 2x\sqrt{x^2 + 1}$$

In general, this method works whenever we have an integral that we can write in the form $\int f(g(x))\, g'(x)\, dx$. Observe that if $F' = f$, then

$$\boxed{3} \qquad \int F'(g(x))\, g'(x)\, dx = F(g(x)) + C$$

because, by the Chain Rule,

$$\frac{d}{dx}[F(g(x))] = F'(g(x))\, g'(x)$$

If we make the "change of variable" or "substitution" $u = g(x)$, then from Equation 3 we have

$$\int F'(g(x))\, g'(x)\, dx = F(g(x)) + C = F(u) + C = \int F'(u)\, du$$

or, writing $F' = f$, we get

$$\int f(g(x))\, g'(x)\, dx = \int f(u)\, du$$

Thus we have proved the following rule.

$\boxed{4}$ **The Substitution Rule** If $u = g(x)$ is a differentiable function whose range is an interval I and f is continuous on I, then

$$\int f(g(x))\, g'(x)\, dx = \int f(u)\, du$$

Notice that the Substitution Rule for integration was proved using the Chain Rule for differentiation. Notice also that if $u = g(x)$, then $du = g'(x)\,dx$, so a way to remember the Substitution Rule is to think of dx and du in $\boxed{4}$ as differentials.

Thus the Substitution Rule says: **It is permissible to operate with dx and du after integral signs as if they were differentials.**

EXAMPLE 1 Find $\int x^3 \cos(x^4 + 2)\,dx$.

SOLUTION We make the substitution $u = x^4 + 2$ because its differential is $du = 4x^3\,dx$, which, apart from the constant factor 4, occurs in the integral. Thus, using $x^3\,dx = \tfrac{1}{4}\,du$ and the Substitution Rule, we have

$$\int x^3 \cos(x^4 + 2)\,dx = \int \cos u \cdot \tfrac{1}{4}\,du = \tfrac{1}{4}\int \cos u\,du$$
$$= \tfrac{1}{4}\sin u + C$$
$$= \tfrac{1}{4}\sin(x^4 + 2) + C$$

Check the answer by differentiating it.

Notice that at the final stage we had to return to the original variable x.

The idea behind the Substitution Rule is to replace a relatively complicated integral by a simpler integral. This is accomplished by changing from the original variable x to a new variable u that is a function of x. Thus in Example 1 we replaced the integral $\int x^3 \cos(x^4 + 2)\,dx$ by the simpler integral $\tfrac{1}{4}\int \cos u\,du$.

The main challenge in using the Substitution Rule is to think of an appropriate substitution. You should try to choose u to be some function in the integrand whose differential also occurs (except for a constant factor). This was the case in Example 1. If that is not possible, try choosing u to be some complicated part of the integrand (perhaps the inner function in a composite function). Finding the right substitution is a bit of an art. It's not unusual to guess wrong; if your first guess doesn't work, try another substitution.

EXAMPLE 2 Evaluate $\int \sqrt{2x + 1}\,dx$.

SOLUTION 1 Let $u = 2x + 1$. Then $du = 2\,dx$, so $dx = \tfrac{1}{2}\,du$. Thus the Substitution Rule gives

$$\int \sqrt{2x + 1}\,dx = \int \sqrt{u}\cdot\tfrac{1}{2}\,du = \tfrac{1}{2}\int u^{1/2}\,du$$
$$= \frac{1}{2}\cdot\frac{u^{3/2}}{3/2} + C = \tfrac{1}{3}u^{3/2} + C$$
$$= \tfrac{1}{3}(2x + 1)^{3/2} + C$$

SOLUTION 2 Another possible substitution is $u = \sqrt{2x + 1}$. Then

$$du = \frac{dx}{\sqrt{2x + 1}} \qquad \text{so} \qquad dx = \sqrt{2x + 1}\,du = u\,du$$

(Or observe that $u^2 = 2x + 1$, so $2u\,du = 2\,dx$.) Therefore

$$\int \sqrt{2x + 1}\,dx = \int u\cdot u\,du = \int u^2\,du$$
$$= \frac{u^3}{3} + C = \tfrac{1}{3}(2x + 1)^{3/2} + C$$

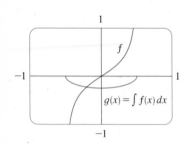

FIGURE 1

$f(x) = \dfrac{x}{\sqrt{1 - 4x^2}}$

$g(x) = \displaystyle\int f(x)\,dx = -\tfrac{1}{4}\sqrt{1 - 4x^2}$

V   EXAMPLE 3 Find $\displaystyle\int \frac{x}{\sqrt{1 - 4x^2}}\,dx$.

SOLUTION Let $u = 1 - 4x^2$. Then $du = -8x\,dx$, so $x\,dx = -\tfrac{1}{8}\,du$ and

$$\int \frac{x}{\sqrt{1 - 4x^2}}\,dx = -\tfrac{1}{8}\int \frac{1}{\sqrt{u}}\,du = -\tfrac{1}{8}\int u^{-1/2}\,du$$

$$= -\tfrac{1}{8}\left(2\sqrt{u}\right) + C = -\tfrac{1}{4}\sqrt{1 - 4x^2} + C \qquad \blacksquare$$

The answer to Example 3 could be checked by differentiation, but instead let's check it with a graph. In Figure 1 we have used a computer to graph both the integrand $f(x) = x/\sqrt{1 - 4x^2}$ and its indefinite integral $g(x) = -\tfrac{1}{4}\sqrt{1 - 4x^2}$ (we take the case $C = 0$). Notice that $g(x)$ decreases when $f(x)$ is negative, increases when $f(x)$ is positive, and has its minimum value when $f(x) = 0$. So it seems reasonable, from the graphical evidence, that g is an antiderivative of f.

EXAMPLE 4 Calculate $\displaystyle\int \cos 5x\,dx$.

SOLUTION If we let $u = 5x$, then $du = 5\,dx$, so $dx = \tfrac{1}{5}\,du$. Therefore

$$\int \cos 5x\,dx = \tfrac{1}{5}\int \cos u\,du = \tfrac{1}{5}\sin u + C = \tfrac{1}{5}\sin 5x + C \qquad \blacksquare$$

NOTE With some experience, you might be able to evaluate integrals like those in Examples 1–4 without going to the trouble of making an explicit substitution. By recognizing the pattern in Equation 3, where the integrand on the left side is the product of the derivative of an outer function and the derivative of the inner function, we could work Example 1 as follows:

$$\int x^3 \cos(x^4 + 2)\,dx = \int \cos(x^4 + 2) \cdot x^3\,dx = \tfrac{1}{4}\int \cos(x^4 + 2) \cdot (4x^3)\,dx$$

$$= \tfrac{1}{4}\int \cos(x^4 + 2) \cdot \frac{d}{dx}(x^4 + 2)\,dx = \tfrac{1}{4}\sin(x^4 + 2) + C$$

Similarly, the solution to Example 4 could be written like this:

$$\int \cos 5x\,dx = \tfrac{1}{5}\int 5\cos 5x\,dx = \tfrac{1}{5}\int \frac{d}{dx}(\sin 5x)\,dx = \tfrac{1}{5}\sin 5x + C$$

The following example, however, is more complicated and so an explicit substitution is advisable.

EXAMPLE 5 Find $\displaystyle\int \sqrt{1 + x^2}\,x^5\,dx$.

SOLUTION An appropriate substitution becomes more obvious if we factor x^5 as $x^4 \cdot x$. Let $u = 1 + x^2$. Then $du = 2x\,dx$, so $x\,dx = \frac{1}{2}\,du$. Also $x^2 = u - 1$, so $x^4 = (u-1)^2$:

$$\int \sqrt{1+x^2}\,x^5\,dx = \int \sqrt{1+x^2}\,x^4 \cdot x\,dx$$

$$= \int \sqrt{u}\,(u-1)^2 \cdot \tfrac{1}{2}\,du = \tfrac{1}{2}\int \sqrt{u}\,(u^2 - 2u + 1)\,du$$

$$= \tfrac{1}{2}\int (u^{5/2} - 2u^{3/2} + u^{1/2})\,du$$

$$= \tfrac{1}{2}\big(\tfrac{2}{7}u^{7/2} - 2\cdot\tfrac{2}{5}u^{5/2} + \tfrac{2}{3}u^{3/2}\big) + C$$

$$= \tfrac{1}{7}(1+x^2)^{7/2} - \tfrac{2}{5}(1+x^2)^{5/2} + \tfrac{1}{3}(1+x^2)^{3/2} + C$$

Definite Integrals

When evaluating a *definite* integral by substitution, two methods are possible. One method is to evaluate the indefinite integral first and then use the Fundamental Theorem. For instance, using the result of Example 2, we have

$$\int_0^4 \sqrt{2x+1}\,dx = \int \sqrt{2x+1}\,dx\Big]_0^4$$

$$= \tfrac{1}{3}(2x+1)^{3/2}\Big]_0^4 = \tfrac{1}{3}(9)^{3/2} - \tfrac{1}{3}(1)^{3/2}$$

$$= \tfrac{1}{3}(27 - 1) = \tfrac{26}{3}$$

Another method, which is usually preferable, is to change the limits of integration when the variable is changed.

This rule says that when using a substitution in a definite integral, we must put everything in terms of the new variable u, not only x and dx but also the limits of integration. The new limits of integration are the values of u that correspond to $x = a$ and $x = b$.

$\boxed{5}$ **The Substitution Rule for Definite Integrals** If g' is continuous on $[a, b]$ and f is continuous on the range of $u = g(x)$, then

$$\int_a^b f(g(x))g'(x)\,dx = \int_{g(a)}^{g(b)} f(u)\,du$$

PROOF Let F be an antiderivative of f. Then, by $\boxed{3}$, $F(g(x))$ is an antiderivative of $f(g(x))g'(x)$, so by Part 2 of the Fundamental Theorem, we have

$$\int_a^b f(g(x))g'(x)\,dx = F(g(x))\Big]_a^b = F(g(b)) - F(g(a))$$

But, applying FTC2 a second time, we also have

$$\int_{g(a)}^{g(b)} f(u)\,du = F(u)\Big]_{g(a)}^{g(b)} = F(g(b)) - F(g(a))$$

V EXAMPLE 6 Evaluate $\int_0^4 \sqrt{2x+1}\,dx$ using $\boxed{5}$.

SOLUTION Using the substitution from Solution 1 of Example 2, we have $u = 2x + 1$ and $dx = \frac{1}{2}\,du$. To find the new limits of integration we note that

$$\text{when } x = 0,\ u = 2(0) + 1 = 1 \qquad \text{and} \qquad \text{when } x = 4,\ u = 2(4) + 1 = 9$$

Therefore

$$\int_0^4 \sqrt{2x + 1}\,dx = \int_1^9 \tfrac{1}{2}\sqrt{u}\,du$$

$$= \tfrac{1}{2} \cdot \tfrac{2}{3} u^{3/2}\Big]_1^9$$

$$= \tfrac{1}{3}(9^{3/2} - 1^{3/2}) = \tfrac{26}{3}$$

Observe that when using ⑤ we do *not* return to the variable x after integrating. We simply evaluate the expression in u between the appropriate values of u. ▬

EXAMPLE 7 Evaluate $\displaystyle\int_1^2 \frac{dx}{(3 - 5x)^2}$.

The integral given in Example 7 is an abbreviation for

$$\int_1^2 \frac{1}{(3 - 5x)^2}\,dx$$

SOLUTION Let $u = 3 - 5x$. Then $du = -5\,dx$, so $dx = -\frac{1}{5}\,du$. When $x = 1$, $u = -2$ and when $x = 2$, $u = -7$. Thus

$$\int_1^2 \frac{dx}{(3 - 5x)^2} = -\frac{1}{5}\int_{-2}^{-7} \frac{du}{u^2}$$

$$= -\frac{1}{5}\left[-\frac{1}{u}\right]_{-2}^{-7} = \frac{1}{5u}\Big]_{-2}^{-7}$$

$$= \frac{1}{5}\left(-\frac{1}{7} + \frac{1}{2}\right) = \frac{1}{14}$$ ▬

▬ Symmetry

The next theorem uses the Substitution Rule for Definite Integrals ⑤ to simplify the calculation of integrals of functions that possess symmetry properties.

> ⑥ **Integrals of Symmetric Functions** Suppose f is continuous on $[-a, a]$.
> (a) If f is even $[f(-x) = f(x)]$, then $\int_{-a}^a f(x)\,dx = 2\int_0^a f(x)\,dx$.
> (b) If f is odd $[f(-x) = -f(x)]$, then $\int_{-a}^a f(x)\,dx = 0$.

PROOF We split the integral in two:

$$\boxed{7} \qquad \int_{-a}^a f(x)\,dx = \int_{-a}^0 f(x)\,dx + \int_0^a f(x)\,dx = -\int_0^{-a} f(x)\,dx + \int_0^a f(x)\,dx$$

In the first integral on the far right side we make the substitution $u = -x$. Then $du = -dx$ and when $x = -a$, $u = a$. Therefore

$$-\int_0^{-a} f(x)\,dx = -\int_0^a f(-u)\,(-du) = \int_0^a f(-u)\,du$$

and so Equation 7 becomes

$$\boxed{8} \qquad \int_{-a}^{a} f(x)\, dx = \int_{0}^{a} f(-u)\, du + \int_{0}^{a} f(x)\, dx$$

(a) f even, $\displaystyle\int_{-a}^{a} f(x)\, dx = 2 \int_{0}^{a} f(x)\, dx$

(a) If f is even, then $f(-u) = f(u)$ so Equation 8 gives

$$\int_{-a}^{a} f(x)\, dx = \int_{0}^{a} f(u)\, du + \int_{0}^{a} f(x)\, dx = 2 \int_{0}^{a} f(x)\, dx$$

(b) If f is odd, then $f(-u) = -f(u)$ and so Equation 8 gives

$$\int_{-a}^{a} f(x)\, dx = -\int_{0}^{a} f(u)\, du + \int_{0}^{a} f(x)\, dx = 0 \qquad \blacksquare$$

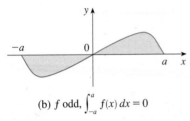

(b) f odd, $\displaystyle\int_{-a}^{a} f(x)\, dx = 0$

FIGURE 2

Theorem 6 is illustrated by Figure 2. For the case where f is positive and even, part (a) says that the area under $y = f(x)$ from $-a$ to a is twice the area from 0 to a because of symmetry. Recall that an integral $\int_{a}^{b} f(x)\, dx$ can be expressed as the area above the x-axis and below $y = f(x)$ minus the area below the axis and above the curve. Thus part (b) says the integral is 0 because the areas cancel.

V EXAMPLE 8 Since $f(x) = x^6 + 1$ satisfies $f(-x) = f(x)$, it is even and so

$$\int_{-2}^{2} (x^6 + 1)\, dx = 2 \int_{0}^{2} (x^6 + 1)\, dx$$

$$= 2\big[\tfrac{1}{7}x^7 + x\big]_{0}^{2} = 2\big(\tfrac{128}{7} + 2\big) = \tfrac{284}{7} \qquad \blacksquare$$

EXAMPLE 9 Since $f(x) = (\tan x)/(1 + x^2 + x^4)$ satisfies $f(-x) = -f(x)$, it is odd and so

$$\int_{-1}^{1} \frac{\tan x}{1 + x^2 + x^4}\, dx = 0 \qquad \blacksquare$$

4.5 Exercises

Access end-of-section exercises and tools for visualizing calculus online at **www.webassign.net**

---

4 Review

Concept Check

1. (a) Write an expression for a Riemann sum of a function f. Explain the meaning of the notation that you use.
   (b) If $f(x) \geqslant 0$, what is the geometric interpretation of a Riemann sum? Illustrate with a diagram.
   (c) If $f(x)$ takes on both positive and negative values, what is the geometric interpretation of a Riemann sum? Illustrate with a diagram.

2. (a) Write the definition of the definite integral of a continuous function from a to b.
   (b) What is the geometric interpretation of $\int_{a}^{b} f(x)\, dx$ if $f(x) \geqslant 0$?
   (c) What is the geometric interpretation of $\int_{a}^{b} f(x)\, dx$ if $f(x)$ takes on both positive and negative values? Illustrate with a diagram.

3. State both parts of the Fundamental Theorem of Calculus.

4. (a) State the Net Change Theorem.
   (b) If $r(t)$ is the rate at which water flows into a reservoir, what does $\int_{t_1}^{t_2} r(t)\,dt$ represent?

5. Suppose a particle moves back and forth along a straight line with velocity $v(t)$, measured in feet per second, and acceleration $a(t)$.
   (a) What is the meaning of $\int_{60}^{120} v(t)\,dt$?

(b) What is the meaning of $\int_{60}^{120} |v(t)|\,dt$?

(c) What is the meaning of $\int_{60}^{120} a(t)\,dt$?

6. (a) Explain the meaning of the indefinite integral $\int f(x)\,dx$.
   (b) What is the connection between the definite integral $\int_a^b f(x)\,dx$ and the indefinite integral $\int f(x)\,dx$?

7. Explain exactly what is meant by the statement that "differentiation and integration are inverse processes."

8. State the Substitution Rule. In practice, how do you use it?

True-False Quiz

Determine whether the statement is true or false. If it is true, explain why. If it is false, explain why or give an example that disproves the statement.

1. If f and g are continuous on $[a, b]$, then
$$\int_a^b [f(x) + g(x)]\,dx = \int_a^b f(x)\,dx + \int_a^b g(x)\,dx$$

2. If f and g are continuous on $[a, b]$, then
$$\int_a^b [f(x)g(x)]\,dx = \left(\int_a^b f(x)\,dx\right)\left(\int_a^b g(x)\,dx\right)$$

3. If f is continuous on $[a, b]$, then
$$\int_a^b 5f(x)\,dx = 5\int_a^b f(x)\,dx$$

4. If f is continuous on $[a, b]$, then
$$\int_a^b xf(x)\,dx = x\int_a^b f(x)\,dx$$

5. If f is continuous on $[a, b]$ and $f(x) \geq 0$, then
$$\int_a^b \sqrt{f(x)}\,dx = \sqrt{\int_a^b f(x)\,dx}$$

6. If f' is continuous on $[1, 3]$, then $\int_1^3 f'(v)\,dv = f(3) - f(1)$.

7. If f and g are continuous and $f(x) \geq g(x)$ for $a \leq x \leq b$, then
$$\int_a^b f(x)\,dx \geq \int_a^b g(x)\,dx$$

8. If f and g are differentiable and $f(x) \geq g(x)$ for $a < x < b$, then $f'(x) \geq g'(x)$ for $a < x < b$.

9. $\int_{-1}^1 \left(x^5 - 6x^9 + \frac{\sin x}{(1+x^4)^2}\right)dx = 0$

10. $\int_{-5}^5 (ax^2 + bx + c)\,dx = 2\int_0^5 (ax^2 + c)\,dx$

11. All continuous functions have derivatives.

12. All continuous functions have antiderivatives.

13. $\int_\pi^{2\pi} \frac{\sin x}{x}\,dx = \int_\pi^{3\pi} \frac{\sin x}{x}\,dx + \int_{3\pi}^{2\pi} \frac{\sin x}{x}\,dx$

14. If $\int_0^1 f(x)\,dx = 0$, then $f(x) = 0$ for $0 \leq x \leq 1$.

15. If f is continuous on $[a, b]$, then
$$\frac{d}{dx}\left(\int_a^b f(x)\,dx\right) = f(x)$$

16. $\int_0^2 (x - x^3)\,dx$ represents the area under the curve $y = x - x^3$ from 0 to 2.

17. $\int_{-2}^1 \frac{1}{x^4}\,dx = -\frac{3}{8}$

18. If f has a discontinuity at 0, then $\int_{-1}^1 f(x)\,dx$ does not exist.

Exercises

1. Use the given graph of f to find the Riemann sum with six subintervals. Take the sample points to be (a) left endpoints and (b) midpoints. In each case draw a diagram and explain what the Riemann sum represents.

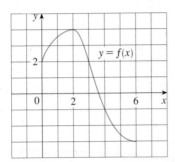

2. (a) Evaluate the Riemann sum for

$$f(x) = x^2 - x \qquad 0 \le x \le 2$$

with four subintervals, taking the sample points to be right endpoints. Explain, with the aid of a diagram, what the Riemann sum represents.

(b) Use the definition of a definite integral (with right endpoints) to calculate the value of the integral

$$\int_0^2 (x^2 - x)\, dx$$

(c) Use the Fundamental Theorem to check your answer to part (b).

(d) Draw a diagram to explain the geometric meaning of the integral in part (b).

3. Evaluate

$$\int_0^1 \left(x + \sqrt{1 - x^2}\right) dx$$

by interpreting it in terms of areas.

4. Express

$$\lim_{n \to \infty} \sum_{i=1}^{n} \sin x_i\, \Delta x$$

as a definite integral on the interval $[0, \pi]$ and then evaluate the integral.

5. If $\int_0^6 f(x)\, dx = 10$ and $\int_0^4 f(x)\, dx = 7$, find $\int_4^6 f(x)\, dx$.

CAS **6.** (a) Write $\int_1^5 (x + 2x^5)\, dx$ as a limit of Riemann sums, taking the sample points to be right endpoints. Use a computer algebra system to evaluate the sum and to compute the limit.

(b) Use the Fundamental Theorem to check your answer to part (a).

7. The following figure shows the graphs of f, f', and $\int_0^x f(t)\, dt$. Identify each graph, and explain your choices.

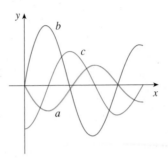

8. Evaluate:

(a) $\displaystyle \int_0^{\pi/2} \frac{d}{dx} \left(\sin \frac{x}{2} \cos \frac{x}{3} \right) dx$

(b) $\displaystyle \frac{d}{dx} \int_0^{\pi/2} \sin \frac{x}{2} \cos \frac{x}{3}\, dx$

(c) $\displaystyle \frac{d}{dx} \int_x^{\pi/2} \sin \frac{t}{2} \cos \frac{t}{3}\, dt$

9–28 Evaluate the integral.

9. $\displaystyle \int_1^2 (8x^3 + 3x^2)\, dx$

10. $\displaystyle \int_0^T (x^4 - 8x + 7)\, dx$

11. $\displaystyle \int_0^1 (1 - x^9)\, dx$

12. $\displaystyle \int_0^1 (1 - x)^9\, dx$

13. $\displaystyle \int_1^9 \frac{\sqrt{u} - 2u^2}{u}\, du$

14. $\displaystyle \int_0^1 \left(\sqrt[4]{u} + 1\right)^2 du$

15. $\displaystyle \int_0^1 y(y^2 + 1)^5\, dy$

16. $\displaystyle \int_0^2 y^2 \sqrt{1 + y^3}\, dy$

17. $\displaystyle \int_1^5 \frac{dt}{(t - 4)^2}$

18. $\displaystyle \int_0^1 \sin(3\pi t)\, dt$

19. $\displaystyle \int_0^1 v^2 \cos(v^3)\, dv$

20. $\displaystyle \int_{-1}^1 \frac{\sin x}{1 + x^2}\, dx$

21. $\displaystyle \int_{-\pi/4}^{\pi/4} \frac{t^4 \tan t}{2 + \cos t}\, dt$

22. $\displaystyle \int \frac{x + 2}{\sqrt{x^2 + 4x}}\, dx$

23. $\displaystyle \int \sin \pi t \cos \pi t\, dt$

24. $\displaystyle \int \sin x \cos(\cos x)\, dx$

25. $\displaystyle \int_0^{\pi/8} \sec 2\theta \tan 2\theta\, d\theta$

26. $\displaystyle \int_0^{\pi/4} (1 + \tan t)^3 \sec^2 t\, dt$

27. $\displaystyle \int_0^3 |x^2 - 4|\, dx$

28. $\displaystyle \int_0^4 |\sqrt{x} - 1|\, dx$

⊞ Graphing calculator or computer required   CAS Computer algebra system required

29–30 Evaluate the indefinite integral. Illustrate and check that your answer is reasonable by graphing both the function and its antiderivative (take $C = 0$).

29. $\displaystyle\int \frac{\cos x}{\sqrt{1 + \sin x}}\, dx$

30. $\displaystyle\int \frac{x^3}{\sqrt{x^2 + 1}}\, dx$

31. Use a graph to give a rough estimate of the area of the region that lies under the curve $y = x\sqrt{x}$, $0 \leqslant x \leqslant 4$. Then find the exact area.

32. Graph the function $f(x) = \cos^2 x \sin x$ and use the graph to guess the value of the integral $\int_0^{2\pi} f(x)\, dx$. Then evaluate the integral to confirm your guess.

33–38 Find the derivative of the function.

33. $\displaystyle F(x) = \int_0^x \frac{t^2}{1 + t^3}\, dt$

34. $\displaystyle F(x) = \int_x^1 \sqrt{t + \sin t}\, dt$

35. $\displaystyle g(x) = \int_0^{x^4} \cos(t^2)\, dt$

36. $\displaystyle g(x) = \int_1^{\sin x} \frac{1 - t^2}{1 + t^4}\, dt$

37. $\displaystyle y = \int_{\sqrt{x}}^x \frac{\cos \theta}{\theta}\, d\theta$

38. $\displaystyle y = \int_{2x}^{3x+1} \sin(t^4)\, dt$

39–40 Use Property 8 of integrals to estimate the value of the integral.

39. $\displaystyle\int_1^3 \sqrt{x^2 + 3}\, dx$

40. $\displaystyle\int_3^5 \frac{1}{x + 1}\, dx$

41–42 Use the properties of integrals to verify the inequality.

41. $\displaystyle\int_0^1 x^2 \cos x\, dx \leqslant \frac{1}{3}$

42. $\displaystyle\int_{\pi/4}^{\pi/2} \frac{\sin x}{x}\, dx \leqslant \frac{\sqrt{2}}{2}$

43. Use the Midpoint Rule with $n = 6$ to approximate $\int_0^3 \sin(x^3)\, dx$.

44. A particle moves along a line with velocity function $v(t) = t^2 - t$, where v is measured in meters per second. Find (a) the displacement and (b) the distance traveled by the particle during the time interval $[0, 5]$.

45. Let $r(t)$ be the rate at which the world's oil is consumed, where t is measured in years starting at $t = 0$ on January 1, 2000, and $r(t)$ is measured in barrels per year. What does $\int_0^8 r(t)\, dt$ represent?

46. A radar gun was used to record the speed of a runner at the times given in the table. Use the Midpoint Rule to estimate the distance the runner covered during those 5 seconds.

t (s)	v (m/s)	t (s)	v (m/s)
0	0	3.0	10.51
0.5	4.67	3.5	10.67
1.0	7.34	4.0	10.76
1.5	8.86	4.5	10.81
2.0	9.73	5.0	10.81
2.5	10.22		

47. A population of honeybees increased at a rate of $r(t)$ bees per week, where the graph of r is as shown. Use the Midpoint Rule with six subintervals to estimate the increase in the bee population during the first 24 weeks.

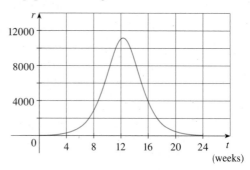

48. Let

$$f(x) = \begin{cases} -x - 1 & \text{if } -3 \leqslant x \leqslant 0 \\ -\sqrt{1 - x^2} & \text{if } 0 \leqslant x \leqslant 1 \end{cases}$$

Evaluate $\int_{-3}^1 f(x)\, dx$ by interpreting the integral as a difference of areas.

49. If f is continuous and $\int_0^2 f(x)\, dx = 6$, evaluate $\int_0^{\pi/2} f(2 \sin \theta) \cos \theta\, d\theta$.

50. The Fresnel function $S(x) = \int_0^x \sin(\frac{1}{2}\pi t^2)\, dt$ was introduced in Section 4.3. Fresnel also used the function

$$C(x) = \int_0^x \cos(\tfrac{1}{2}\pi t^2)\, dt$$

in his theory of the diffraction of light waves.
(a) On what intervals is C increasing?
(b) On what intervals is C concave upward?
(c) Use a graph to solve the following equation correct to two decimal places:

$$\int_0^x \cos(\tfrac{1}{2}\pi t^2)\, dt = 0.7$$

(d) Plot the graphs of C and S on the same screen. How are these graphs related?

51. If f is a continuous function such that

$$\int_0^x f(t)\, dt = x \sin x + \int_0^x \frac{f(t)}{1 + t^2}\, dt$$

for all x, find an explicit formula for $f(x)$.

52. Find a function f and a value of the constant a such that

$$2 \int_a^x f(t)\, dt = 2 \sin x - 1$$

53. If f' is continuous on $[a, b]$, show that

$$2 \int_a^b f(x) f'(x)\, dx = [f(b)]^2 - [f(a)]^2$$

54. Find $\displaystyle \lim_{h \to 0} \frac{1}{h} \int_2^{2+h} \sqrt{1 + t^3}\, dt$.

55. If f is continuous on $[0, 1]$, prove that

$$\int_0^1 f(x)\, dx = \int_0^1 f(1 - x)\, dx$$

56. Evaluate

$$\lim_{n \to \infty} \frac{1}{n} \left[\left(\frac{1}{n} \right)^9 + \left(\frac{2}{n} \right)^9 + \left(\frac{3}{n} \right)^9 + \cdots + \left(\frac{n}{n} \right)^9 \right]$$

Problems Plus

Before you look at the solution of the following example, cover it up and first try to solve the problem yourself.

EXAMPLE 1 Evaluate $\lim\limits_{x \to 3} \left(\dfrac{x}{x-3} \displaystyle\int_3^x \dfrac{\sin t}{t}\, dt \right)$.

SOLUTION Let's start by having a preliminary look at the ingredients of the function. What happens to the first factor, $x/(x-3)$, when x approaches 3? The numerator approaches 3 and the denominator approaches 0, so we have

$$\frac{x}{x-3} \to \infty \quad \text{as} \quad x \to 3^+ \qquad \text{and} \qquad \frac{x}{x-3} \to -\infty \quad \text{as} \quad x \to 3^-$$

PS The principles of problem solving are discussed on page 77.

The second factor approaches $\int_3^3 (\sin t)/t\, dt$, which is 0. It's not clear what happens to the function as a whole. (One factor is becoming large while the other is becoming small.) So how do we proceed?

One of the principles of problem solving is *recognizing something familiar*. Is there a part of the function that reminds us of something we've seen before? Well, the integral

$$\int_3^x \frac{\sin t}{t}\, dt$$

has x as its upper limit of integration and that type of integral occurs in Part 1 of the Fundamental Theorem of Calculus:

$$\frac{d}{dx} \int_a^x f(t)\, dt = f(x)$$

This suggests that differentiation might be involved.

Once we start thinking about differentiation, the denominator $(x-3)$ reminds us of something else that should be familiar: One of the forms of the definition of the derivative in Chapter 2 is

$$F'(a) = \lim_{x \to a} \frac{F(x) - F(a)}{x - a}$$

and with $a = 3$ this becomes

$$F'(3) = \lim_{x \to 3} \frac{F(x) - F(3)}{x - 3}$$

So what is the function F in our situation? Notice that if we define

$$F(x) = \int_3^x \frac{\sin t}{t}\, dt$$

then $F(3) = 0$. What about the factor x in the numerator? That's just a red herring, so let's factor it out and put together the calculation:

$$\lim_{x \to 3} \left(\frac{x}{x-3} \int_3^x \frac{\sin t}{t}\, dt \right) = \lim_{x \to 3} x \cdot \lim_{x \to 3} \frac{\displaystyle\int_3^x \frac{\sin t}{t}\, dt}{x - 3}$$

$$= 3 \lim_{x \to 3} \frac{F(x) - F(3)}{x - 3}$$

$$= 3F'(3) = 3\, \frac{\sin 3}{3} \qquad \text{(FTC1)}$$

$$= \sin 3$$

1. If $x \sin \pi x = \int_0^{x^2} f(t)\, dt$, where f is a continuous function, find $f(4)$.

2. Find the maximum value of the area of the region under the curve $y = 4x - x^3$ from $x = a$ to $x = a + 1$, for all $a > 0$.

3. If f is a differentiable function such that $f(x)$ is never 0 and $\int_0^x f(t)\, dt = [f(x)]^2$ for all x, find f.

4. (a) Graph several members of the family of functions $f(x) = (2cx - x^2)/c^3$ for $c > 0$ and look at the regions enclosed by these curves and the x-axis. Make a conjecture about how the areas of these regions are related.
   (b) Prove your conjecture in part (a).
   (c) Take another look at the graphs in part (a) and use them to sketch the curve traced out by the vertices (highest points) of the family of functions. Can you guess what kind of curve this is?
   (d) Find an equation of the curve you sketched in part (c).

5. If $f(x) = \int_0^{g(x)} \frac{1}{\sqrt{1 + t^3}}\, dt$, where $g(x) = \int_0^{\cos x} [1 + \sin(t^2)]\, dt$, find $f'(\pi/2)$.

6. If $f(x) = \int_0^x x^2 \sin(t^2)\, dt$, find $f'(x)$.

7. Find the interval $[a, b]$ for which the value of the integral $\int_a^b (2 + x - x^2)\, dx$ is a maximum.

8. Use an integral to estimate the sum $\sum_{i=1}^{10000} \sqrt{i}$.

9. (a) Evaluate $\int_0^n [\![x]\!]\, dx$, where n is a positive integer.
   (b) Evaluate $\int_a^b [\![x]\!]\, dx$, where a and b are real numbers with $0 \le a < b$.

10. Find $\dfrac{d^2}{dx^2} \int_0^x \left(\int_1^{\sin t} \sqrt{1 + u^4}\, du \right) dt$.

11. Suppose the coefficients of the cubic polynomial $P(x) = a + bx + cx^2 + dx^3$ satisfy the equation
$$ a + \frac{b}{2} + \frac{c}{3} + \frac{d}{4} = 0 $$
Show that the equation $P(x) = 0$ has a root between 0 and 1. Can you generalize this result for an nth-degree polynomial?

12. A circular disk of radius r is used in an evaporator and is rotated in a vertical plane. If it is to be partially submerged in the liquid so as to maximize the exposed wetted area of the disk, show that the center of the disk should be positioned at a height $r/\sqrt{1 + \pi^2}$ above the surface of the liquid.

13. Prove that if f is continuous, then $\int_0^x f(u)(x - u)\, du = \int_0^x \left(\int_0^u f(t)\, dt \right) du$.

14. The figure shows a region consisting of all points inside a square that are closer to the center than to the sides of the square. Find the area of the region.

15. Evaluate $\displaystyle\lim_{n \to \infty} \left(\frac{1}{\sqrt{n}\,\sqrt{n+1}} + \frac{1}{\sqrt{n}\,\sqrt{n+2}} + \cdots + \frac{1}{\sqrt{n}\,\sqrt{n+n}} \right)$.

16. For any number c, we let $f_c(x)$ be the smaller of the two numbers $(x - c)^2$ and $(x - c - 2)^2$. Then we define $g(c) = \int_0^1 f_c(x)\, dx$. Find the maximum and minimum values of $g(c)$ if $-2 \le c \le 2$.

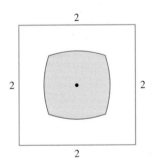

2

2　　　　　　2

2

FIGURE FOR PROBLEM 14

Applications of Integration

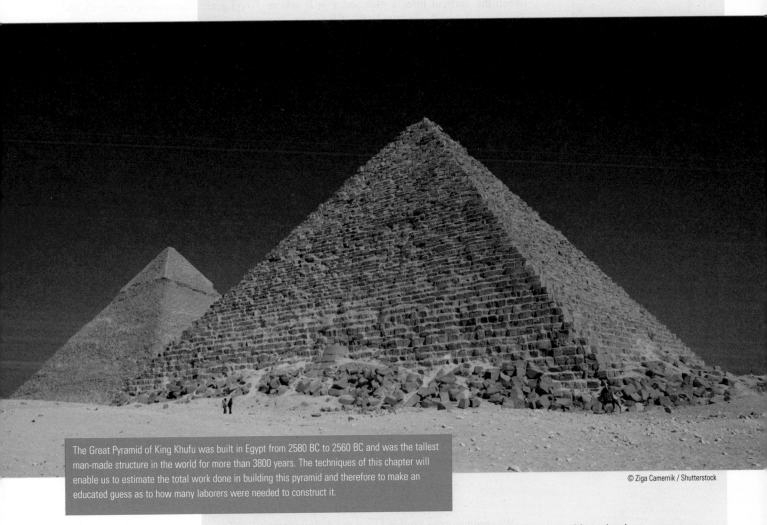

The Great Pyramid of King Khufu was built in Egypt from 2580 BC to 2560 BC and was the tallest man-made structure in the world for more than 3800 years. The techniques of this chapter will enable us to estimate the total work done in building this pyramid and therefore to make an educated guess as to how many laborers were needed to construct it.

© Ziga Camernik / Shutterstock

xpIn this chapter we explore some of the applications of the definite integral by using it to compute areas between curves, volumes of solids, and the work done by a varying force. The common theme is the following general method, which is similar to the one we used to find areas under curves: We break up a quantity Q into a large number of small parts. We next approximate each small part by a quantity of the form $f(x_i^*) \, \Delta x$ and thus approximate Q by a Riemann sum. Then we take the limit and express Q as an integral. Finally we evaluate the integral using the Fundamental Theorem of Calculus or the Midpoint Rule.

5.1   Areas Between Curves

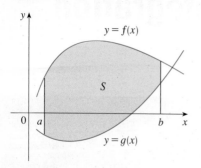

FIGURE 1
$S = \{(x, y) \mid a \leqslant x \leqslant b, g(x) \leqslant y \leqslant f(x)\}$

In Chapter 4 we defined and calculated areas of regions that lie under the graphs of functions. Here we use integrals to find areas of regions that lie between the graphs of two functions.

Consider the region S that lies between two curves $y = f(x)$ and $y = g(x)$ and between the vertical lines $x = a$ and $x = b$, where f and g are continuous functions and $f(x) \geqslant g(x)$ for all x in $[a, b]$. (See Figure 1.)

Just as we did for areas under curves in Section 4.1, we divide S into n strips of equal width and then we approximate the ith strip by a rectangle with base Δx and height $f(x_i^*) - g(x_i^*)$. (See Figure 2. If we liked, we could take all of the sample points to be right endpoints, in which case $x_i^* = x_i$.) The Riemann sum

$$\sum_{i=1}^{n} [f(x_i^*) - g(x_i^*)] \, \Delta x$$

is therefore an approximation to what we intuitively think of as the area of S.

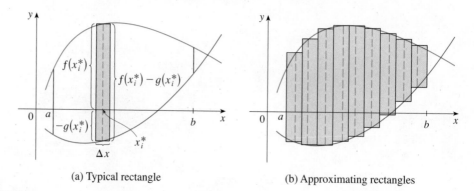

FIGURE 2   (a) Typical rectangle   (b) Approximating rectangles

This approximation appears to become better and better as $n \to \infty$. Therefore we define the **area** A of the region S as the limiting value of the sum of the areas of these approximating rectangles.

$$\boxed{1} \qquad A = \lim_{n \to \infty} \sum_{i=1}^{n} [f(x_i^*) - g(x_i^*)] \, \Delta x$$

We recognize the limit in $\boxed{1}$ as the definite integral of $f - g$. Therefore we have the following formula for area.

> $\boxed{2}$ The area A of the region bounded by the curves $y = f(x)$, $y = g(x)$, and the lines $x = a$, $x = b$, where f and g are continuous and $f(x) \geqslant g(x)$ for all x in $[a, b]$, is
>
> $$A = \int_a^b [f(x) - g(x)] \, dx$$

Notice that in the special case where $g(x) = 0$, S is the region under the graph of f and our general definition of area $\boxed{1}$ reduces to our previous definition (Definition 2 in Section 4.1).

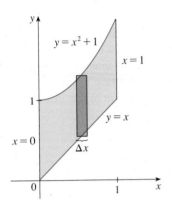

FIGURE 3

$$A = \int_a^b f(x)\,dx - \int_a^b g(x)\,dx$$

FIGURE 4

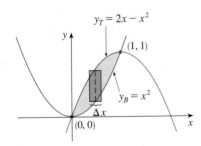

FIGURE 5

In the case where both f and g are positive, you can see from Figure 3 why $\boxed{2}$ is true:

$$A = [\text{area under } y = f(x)] - [\text{area under } y = g(x)]$$

$$= \int_a^b f(x)\,dx - \int_a^b g(x)\,dx = \int_a^b [f(x) - g(x)]\,dx$$

EXAMPLE 1   Find the area of the region bounded above by $y = x^2 + 1$, bounded below by $y = x$, and bounded on the sides by $x = 0$ and $x = 1$.

SOLUTION   The region is shown in Figure 4. The upper boundary curve is $y = x^2 + 1$ and the lower boundary curve is $y = x$. So we use the area formula $\boxed{2}$ with $f(x) = x^2 + 1$, $g(x) = x$, $a = 0$, and $b = 1$:

$$A = \int_0^1 [(x^2 + 1) - x]\,dx = \int_0^1 (x^2 - x + 1)\,dx$$

$$= \frac{x^3}{3} - \frac{x^2}{2} + x \bigg]_0^1 = \frac{1}{3} - \frac{1}{2} + 1 = \frac{5}{6}$$

In Figure 4 we drew a typical approximating rectangle with width Δx as a reminder of the procedure by which the area is defined in $\boxed{1}$. In general, when we set up an integral for an area, it's helpful to sketch the region to identify the top curve y_T, the bottom curve y_B, and a typical approximating rectangle as in Figure 5. Then the area of a typical rectangle is $(y_T - y_B)\,\Delta x$ and the equation

$$A = \lim_{n \to \infty} \sum_{i=1}^n (y_T - y_B)\,\Delta x = \int_a^b (y_T - y_B)\,dx$$

summarizes the procedure of adding (in a limiting sense) the areas of all the typical rectangles.

Notice that in Figure 5 the left-hand boundary reduces to a point, whereas in Figure 3 the right-hand boundary reduces to a point. In the next example both of the side boundaries reduce to a point, so the first step is to find a and b.

V EXAMPLE 2   Find the area of the region enclosed by the parabolas $y = x^2$ and $y = 2x - x^2$.

SOLUTION   We first find the points of intersection of the parabolas by solving their equations simultaneously. This gives $x^2 = 2x - x^2$, or $2x^2 - 2x = 0$. Thus $2x(x - 1) = 0$, so $x = 0$ or 1. The points of intersection are $(0, 0)$ and $(1, 1)$.

We see from Figure 6 that the top and bottom boundaries are

$$y_T = 2x - x^2 \qquad \text{and} \qquad y_B = x^2$$

The area of a typical rectangle is

$$(y_T - y_B)\,\Delta x = (2x - x^2 - x^2)\,\Delta x$$

and the region lies between $x = 0$ and $x = 1$. So the total area is

$$A = \int_0^1 (2x - 2x^2)\,dx = 2\int_0^1 (x - x^2)\,dx$$

$$= 2\left[\frac{x^2}{2} - \frac{x^3}{3}\right]_0^1 = 2\left(\frac{1}{2} - \frac{1}{3}\right) = \frac{1}{3}$$

FIGURE 6

Sometimes it's difficult, or even impossible, to find the points of intersection of two curves exactly. As shown in the following example, we can use a graphing calculator or computer to find approximate values for the intersection points and then proceed as before.

EXAMPLE 3 Find the approximate area of the region bounded by the curves $y = x/\sqrt{x^2+1}$ and $y = x^4 - x$.

SOLUTION If we were to try to find the exact intersection points, we would have to solve the equation

$$\frac{x}{\sqrt{x^2+1}} = x^4 - x$$

This looks like a very difficult equation to solve exactly (in fact, it's impossible), so instead we use a graphing device to draw the graphs of the two curves in Figure 7. One intersection point is the origin. We zoom in toward the other point of intersection and find that $x \approx 1.18$. (If greater accuracy is required, we could use Newton's method or a rootfinder, if available on our graphing device.) Thus an approximation to the area between the curves is

$$A \approx \int_0^{1.18} \left[\frac{x}{\sqrt{x^2+1}} - (x^4 - x) \right] dx$$

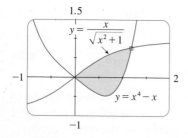

FIGURE 7

To integrate the first term we use the substitution $u = x^2 + 1$. Then $du = 2x\,dx$, and when $x = 1.18$, we have $u \approx 2.39$. So

$$A \approx \frac{1}{2} \int_1^{2.39} \frac{du}{\sqrt{u}} - \int_0^{1.18} (x^4 - x)\,dx$$

$$= \sqrt{u}\,\Big]_1^{2.39} - \left[\frac{x^5}{5} - \frac{x^2}{2} \right]_0^{1.18}$$

$$= \sqrt{2.39} - 1 - \frac{(1.18)^5}{5} + \frac{(1.18)^2}{2}$$

$$\approx 0.785$$

EXAMPLE 4 Figure 8 shows velocity curves for two cars, A and B, that start side by side and move along the same road. What does the area between the curves represent? Use the Midpoint Rule to estimate it.

SOLUTION We know from Section 4.4 that the area under the velocity curve A represents the distance traveled by car A during the first 16 seconds. Similarly, the area under curve B is the distance traveled by car B during that time period. So the area between these curves, which is the difference of the areas under the curves, is the distance between the cars after 16 seconds. We read the velocities from the graph and convert them to feet per second (1 mi/h = $\frac{5280}{3600}$ ft/s).

FIGURE 8

t	0	2	4	6	8	10	12	14	16
v_A	0	34	54	67	76	84	89	92	95
v_B	0	21	34	44	51	56	60	63	65
$v_A - v_B$	0	13	20	23	25	28	29	29	30

We use the Midpoint Rule with $n = 4$ intervals, so that $\Delta t = 4$. The midpoints of the intervals are $\bar{t}_1 = 2$, $\bar{t}_2 = 6$, $\bar{t}_3 = 10$, and $\bar{t}_4 = 14$. We estimate the distance between the cars after 16 seconds as follows:

$$\int_0^{16} (v_A - v_B)\, dt \approx \Delta t \,[13 + 23 + 28 + 29]$$

$$= 4(93) = 372 \text{ ft}$$

■

If we are asked to find the area between the curves $y = f(x)$ and $y = g(x)$ where $f(x) \geqslant g(x)$ for some values of x but $g(x) \geqslant f(x)$ for other values of x, then we split the given region S into several regions $S_1, S_2, \ldots$ with areas $A_1, A_2, \ldots$ as shown in Figure 9. We then define the area of the region S to be the sum of the areas of the smaller regions $S_1, S_2, \ldots$, that is, $A = A_1 + A_2 + \cdots$. Since

$$|f(x) - g(x)| = \begin{cases} f(x) - g(x) & \text{when } f(x) \geqslant g(x) \\ g(x) - f(x) & \text{when } g(x) \geqslant f(x) \end{cases}$$

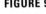

FIGURE 9

we have the following expression for A.

> **3** The area between the curves $y = f(x)$ and $y = g(x)$ and between $x = a$ and $x = b$ is
>
> $$A = \int_a^b |f(x) - g(x)|\, dx$$

When evaluating the integral in $\boxed{3}$, however, we must still split it into integrals corresponding to $A_1, A_2, \ldots$.

V EXAMPLE 5 Find the area of the region bounded by the curves $y = \sin x$, $y = \cos x$, $x = 0$, and $x = \pi/2$.

SOLUTION The points of intersection occur when $\sin x = \cos x$, that is, when $x = \pi/4$ (since $0 \leqslant x \leqslant \pi/2$). The region is sketched in Figure 10. Observe that $\cos x \geqslant \sin x$ when $0 \leqslant x \leqslant \pi/4$ but $\sin x \geqslant \cos x$ when $\pi/4 \leqslant x \leqslant \pi/2$. Therefore the required area is

$$A = \int_0^{\pi/2} |\cos x - \sin x|\, dx = A_1 + A_2$$

$$= \int_0^{\pi/4} (\cos x - \sin x)\, dx + \int_{\pi/4}^{\pi/2} (\sin x - \cos x)\, dx$$

$$= \Big[\sin x + \cos x\Big]_0^{\pi/4} + \Big[-\cos x - \sin x\Big]_{\pi/4}^{\pi/2}$$

$$= \left(\frac{1}{\sqrt{2}} + \frac{1}{\sqrt{2}} - 0 - 1\right) + \left(-0 - 1 + \frac{1}{\sqrt{2}} + \frac{1}{\sqrt{2}}\right)$$

$$= 2\sqrt{2} - 2$$

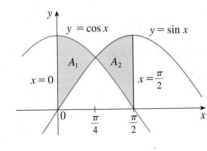

FIGURE 10

In this particular example we could have saved some work by noticing that the region is symmetric about $x = \pi/4$ and so

$$A = 2A_1 = 2 \int_0^{\pi/4} (\cos x - \sin x)\, dx$$

■

Some regions are best treated by regarding x as a function of y. If a region is bounded by curves with equations $x = f(y)$, $x = g(y)$, $y = c$, and $y = d$, where f and g are continuous and $f(y) \geqslant g(y)$ for $c \leqslant y \leqslant d$ (see Figure 11), then its area is

$$A = \int_c^d [f(y) - g(y)]\, dy$$

FIGURE 11

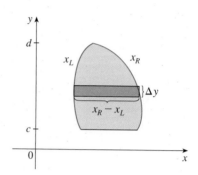

FIGURE 12

If we write x_R for the right boundary and x_L for the left boundary, then, as Figure 12 illustrates, we have

$$A = \int_c^d (x_R - x_L)\, dy$$

Here a typical approximating rectangle has dimensions $x_R - x_L$ and Δy.

V EXAMPLE 6 Find the area enclosed by the line $y = x - 1$ and the parabola $y^2 = 2x + 6$.

SOLUTION By solving the two equations we find that the points of intersection are $(-1, -2)$ and $(5, 4)$. We solve the equation of the parabola for x and notice from Figure 13 that the left and right boundary curves are

$$x_L = \tfrac{1}{2}y^2 - 3 \qquad \text{and} \qquad x_R = y + 1$$

We must integrate between the appropriate y-values, $y = -2$ and $y = 4$. Thus

$$A = \int_{-2}^{4} (x_R - x_L)\, dy = \int_{-2}^{4} \left[(y + 1) - \left(\tfrac{1}{2}y^2 - 3\right)\right] dy$$

$$= \int_{-2}^{4} \left(-\tfrac{1}{2}y^2 + y + 4\right) dy$$

$$= -\frac{1}{2}\left(\frac{y^3}{3}\right) + \frac{y^2}{2} + 4y \Bigg]_{-2}^{4}$$

$$= -\tfrac{1}{6}(64) + 8 + 16 - \left(\tfrac{4}{3} + 2 - 8\right) = 18$$

NOTE We could have found the area in Example 6 by integrating with respect to x instead of y, but the calculation is much more involved. It would have meant splitting the region in two and computing the areas labeled A_1 and A_2 in Figure 14. The method we used in Example 6 is *much* easier.

FIGURE 13

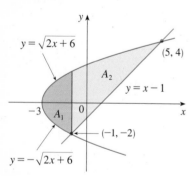

FIGURE 14

| 5.1 | **Exercises** |

Access end-of-section exercises and tools for visualizing calculus online at **www.webassign.net**

APPLIED PROJECT

THE GINI INDEX

How is it possible to measure the distribution of income among the inhabitants of a given country? One such measure is the *Gini index*, named after the Italian economist Corrado Gini who first devised it in 1912.

We first rank all households in a country by income and then we compute the percentage of households whose income is at most a given percentage of the country's total income. We define a **Lorenz curve** $y = L(x)$ on the interval $[0, 1]$ by plotting the point $(a/100, b/100)$ on the curve if the bottom $a\%$ of households receive at most $b\%$ of the total income. For instance, in Figure 1 the point $(0.4, 0.12)$ is on the Lorenz curve for the United States in 2008 because the poorest 40% of the population received just 12% of the total income. Likewise, the bottom 80% of the population received 50% of the total income, so the point $(0.8, 0.5)$ lies on the Lorenz curve. (The Lorenz curve is named after the American economist Max Lorenz.)

Figure 2 shows some typical Lorenz curves. They all pass through the points $(0, 0)$ and $(1, 1)$ and are concave upward. In the extreme case $L(x) = x$, society is perfectly egalitarian: The poorest $a\%$ of the population receives $a\%$ of the total income and so everybody receives the same income. The area between a Lorenz curve $y = L(x)$ and the line $y = x$ measures how much the income distribution differs from absolute equality. The **Gini index** (sometimes called the **Gini coefficient** or the **coefficient of inequality**) is the area between the Lorenz curve and the line $y = x$ (shaded in Figure 3) divided by the area under $y = x$.

FIGURE 1
Lorenz curve for the US in 2008

FIGURE 2

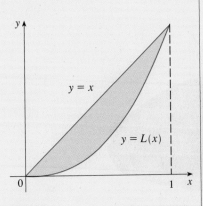

FIGURE 3

1. (a) Show that the Gini index G is twice the area between the Lorenz curve and the line $y = x$, that is,
$$G = 2\int_0^1 [x - L(x)]\, dx$$

   (b) What is the value of G for a perfectly egalitarian society (everybody has the same income)? What is the value of G for a perfectly totalitarian society (a single person receives all the income)?

2. The following table (derived from data supplied by the US Census Bureau) shows values of the Lorenz function for income distribution in the United States for the year 2008.

x	0.0	0.2	0.4	0.6	0.8	1.0
$L(x)$	0.000	0.034	0.120	0.267	0.500	1.000

   (a) What percentage of the total US income was received by the richest 20% of the population in 2008?
   (b) Use a calculator or computer to fit a quadratic function to the data in the table. Graph the data points and the quadratic function. Is the quadratic model a reasonable fit?
   (c) Use the quadratic model for the Lorenz function to estimate the Gini index for the United States in 2008.

3. The following table gives values for the Lorenz function in the years 1970, 1980, 1990, and 2000. Use the method of Problem 2 to estimate the Gini index for the United States for those years and compare with your answer to Problem 2(c). Do you notice a trend?

x	0.0	0.2	0.4	0.6	0.8	1.0
1970	0.000	0.041	0.149	0.323	0.568	1.000
1980	0.000	0.042	0.144	0.312	0.559	1.000
1990	0.000	0.038	0.134	0.293	0.530	1.000
2000	0.000	0.036	0.125	0.273	0.503	1.000

4. A power model often provides a more accurate fit than a quadratic model for a Lorenz function. If you have a computer with Maple or Mathematica, fit a power function $(y = ax^k)$ to the data in Problem 2 and use it to estimate the Gini index for the United States in 2008. Compare with your answer to parts (b) and (c) of Problem 2.

5.2 Volumes

In trying to find the volume of a solid we face the same type of problem as in finding areas. We have an intuitive idea of what volume means, but we must make this idea precise by using calculus to give an exact definition of volume.

We start with a simple type of solid called a **cylinder** (or, more precisely, a *right cylinder*). As illustrated in Figure 1(a), a cylinder is bounded by a plane region B_1, called the **base**, and a congruent region B_2 in a parallel plane. The cylinder consists of all points on line segments that are perpendicular to the base and join B_1 to B_2. If the area of the base is A and the height of the cylinder (the distance from B_1 to B_2) is h, then the volume V of the cylinder is defined as

$$V = Ah$$

In particular, if the base is a circle with radius r, then the cylinder is a circular cylinder with volume $V = \pi r^2 h$ [see Figure 1(b)], and if the base is a rectangle with length l and width w, then the cylinder is a rectangular box (also called a *rectangular parallelepiped*) with volume $V = lwh$ [see Figure 1(c)].

FIGURE 1   (a) Cylinder $V = Ah$ (b) Circular cylinder $V = \pi r^2 h$ (c) Rectangular box $V = lwh$

For a solid S that isn't a cylinder we first "cut" S into pieces and approximate each piece by a cylinder. We estimate the volume of S by adding the volumes of the cylinders. We arrive at the exact volume of S through a limiting process in which the number of pieces becomes large.

We start by intersecting S with a plane and obtaining a plane region that is called a **cross-section** of S. Let $A(x)$ be the area of the cross-section of S in a plane P_x perpendicular to the x-axis and passing through the point x, where $a \leqslant x \leqslant b$. (See Figure 2. Think of slicing S with a knife through x and computing the area of this slice.) The cross-sectional area $A(x)$ will vary as x increases from a to b.

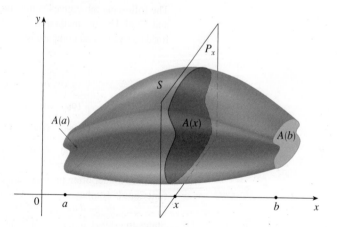

FIGURE 2

Let's divide S into n "slabs" of equal width Δx by using the planes $P_{x_1}, P_{x_2}, \ldots$ to slice the solid. (Think of slicing a loaf of bread.) If we choose sample points x_i^* in $[x_{i-1}, x_i]$, we can approximate the ith slab S_i (the part of S that lies between the planes $P_{x_{i-1}}$ and P_{x_i}) by a cylinder with base area $A(x_i^*)$ and "height" Δx. (See Figure 3.)

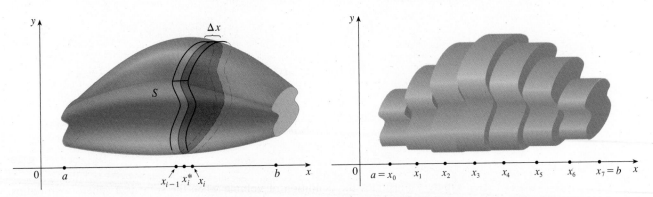

FIGURE 3

The volume of this cylinder is $A(x_i^*)\,\Delta x$, so an approximation to our intuitive conception of the volume of the ith slab S_i is

$$V(S_i) \approx A(x_i^*)\,\Delta x$$

Adding the volumes of these slabs, we get an approximation to the total volume (that is, what we think of intuitively as the volume):

$$V \approx \sum_{i=1}^{n} A(x_i^*)\,\Delta x$$

This approximation appears to become better and better as $n \to \infty$. (Think of the slices as becoming thinner and thinner.) Therefore we *define* the volume as the limit of these sums as $n \to \infty$. But we recognize the limit of Riemann sums as a definite integral and so we have the following definition.

It can be proved that this definition is independent of how S is situated with respect to the x-axis. In other words, no matter how we slice S with parallel planes, we always get the same answer for V.

> **Definition of Volume** Let S be a solid that lies between $x = a$ and $x = b$. If the cross-sectional area of S in the plane P_x, through x and perpendicular to the x-axis, is $A(x)$, where A is a continuous function, then the **volume** of S is
>
> $$V = \lim_{n \to \infty} \sum_{i=1}^{n} A(x_i^*)\,\Delta x = \int_a^b A(x)\,dx$$

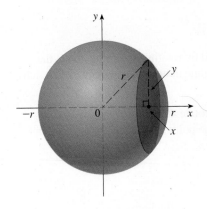

FIGURE 4

When we use the volume formula $V = \int_a^b A(x)\,dx$, it is important to remember that $A(x)$ is the area of a moving cross-section obtained by slicing through x perpendicular to the x-axis.

Notice that, for a cylinder, the cross-sectional area is constant: $A(x) = A$ for all x. So our definition of volume gives $V = \int_a^b A\,dx = A(b - a)$; this agrees with the formula $V = Ah$.

EXAMPLE 1 Show that the volume of a sphere of radius r is $V = \frac{4}{3}\pi r^3$.

SOLUTION If we place the sphere so that its center is at the origin (see Figure 4), then the plane P_x intersects the sphere in a circle whose radius (from the Pythagorean Theorem) is $y = \sqrt{r^2 - x^2}$. So the cross-sectional area is

$$A(x) = \pi y^2 = \pi(r^2 - x^2)$$

Using the definition of volume with $a = -r$ and $b = r$, we have

$$V = \int_{-r}^{r} A(x)\, dx = \int_{-r}^{r} \pi(r^2 - x^2)\, dx$$

$$= 2\pi \int_{0}^{r} (r^2 - x^2)\, dx \qquad \text{(The integrand is even.)}$$

$$= 2\pi \left[r^2 x - \frac{x^3}{3} \right]_{0}^{r} = 2\pi \left(r^3 - \frac{r^3}{3} \right)$$

$$= \tfrac{4}{3}\pi r^3$$

Figure 5 illustrates the definition of volume when the solid is a sphere with radius $r = 1$. From the result of Example 1, we know that the volume of the sphere is $\frac{4}{3}\pi$, which is approximately 4.18879. Here the slabs are circular cylinders, or *disks*, and the three parts of Figure 5 show the geometric interpretations of the Riemann sums

$$\sum_{i=1}^{n} A(\bar{x}_i)\, \Delta x = \sum_{i=1}^{n} \pi(1^2 - \bar{x}_i^2)\, \Delta x$$

when $n = 5$, 10, and 20 if we choose the sample points x_i^* to be the midpoints $\bar{x}_i$. Notice that as we increase the number of approximating cylinders, the corresponding Riemann sums become closer to the true volume.

TEC Visual 5.2A shows an animation of Figure 5.

(a) Using 5 disks, $V \approx 4.2726$

(b) Using 10 disks, $V \approx 4.2097$

(c) Using 20 disks, $V \approx 4.1940$

FIGURE 5 Approximating the volume of a sphere with radius 1

V **EXAMPLE 2** Find the volume of the solid obtained by rotating about the x-axis the region under the curve $y = \sqrt{x}$ from 0 to 1. Illustrate the definition of volume by sketching a typical approximating cylinder.

SOLUTION The region is shown in Figure 6(a). If we rotate about the x-axis, we get the solid shown in Figure 6(b). When we slice through the point x, we get a disk with radius $\sqrt{x}$. The area of this cross-section is

$$A(x) = \pi\left(\sqrt{x}\right)^2 = \pi x$$

and the volume of the approximating cylinder (a disk with thickness Δx) is

$$A(x)\, \Delta x = \pi x\, \Delta x$$

The solid lies between $x = 0$ and $x = 1$, so its volume is

$$V = \int_0^1 A(x)\,dx = \int_0^1 \pi x\,dx = \pi \frac{x^2}{2}\Big]_0^1 = \frac{\pi}{2}$$

Did we get a reasonable answer in Example 2? As a check on our work, let's replace the given region by a square with base $[0, 1]$ and height 1. If we rotate this square, we get a cylinder with radius 1, height 1, and volume $\pi \cdot 1^2 \cdot 1 = \pi$. We computed that the given solid has half this volume. That seems about right.

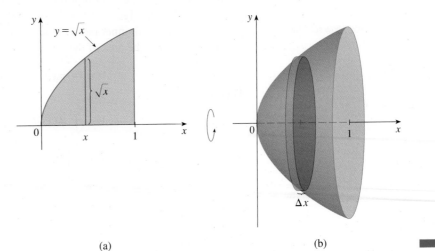

FIGURE 6

(a)   (b)

V **EXAMPLE 3** Find the volume of the solid obtained by rotating the region bounded by $y = x^3$, $y = 8$, and $x = 0$ about the y-axis.

SOLUTION The region is shown in Figure 7(a) and the resulting solid is shown in Figure 7(b). Because the region is rotated about the y-axis, it makes sense to slice the solid perpendicular to the y-axis and therefore to integrate with respect to y. If we slice at height y, we get a circular disk with radius x, where $x = \sqrt[3]{y}$. So the area of a cross-section through y is

$$A(y) = \pi x^2 = \pi\left(\sqrt[3]{y}\,\right)^2 = \pi y^{2/3}$$

and the volume of the approximating cylinder pictured in Figure 7(b) is

$$A(y)\,\Delta y = \pi y^{2/3}\,\Delta y$$

Since the solid lies between $y = 0$ and $y = 8$, its volume is

$$V = \int_0^8 A(y)\,dy = \int_0^8 \pi y^{2/3}\,dy = \pi\left[\tfrac{3}{5}y^{5/3}\right]_0^8 = \frac{96\pi}{5}$$

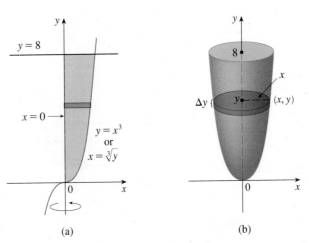

FIGURE 7

(a)   (b)

EXAMPLE 4   The region $\mathcal{R}$ enclosed by the curves $y = x$ and $y = x^2$ is rotated about the x-axis. Find the volume of the resulting solid.

SOLUTION   The curves $y = x$ and $y = x^2$ intersect at the points $(0, 0)$ and $(1, 1)$. The region between them, the solid of rotation, and a cross-section perpendicular to the x-axis are shown in Figure 8. A cross-section in the plane P_x has the shape of a *washer* (an annular ring) with inner radius x^2 and outer radius x, so we find the cross-sectional area by subtracting the area of the inner circle from the area of the outer circle:

$$A(x) = \pi x^2 - \pi(x^2)^2 = \pi(x^2 - x^4)$$

Therefore we have

$$V = \int_0^1 A(x)\, dx = \int_0^1 \pi(x^2 - x^4)\, dx$$

$$= \pi \left[\frac{x^3}{3} - \frac{x^5}{5} \right]_0^1 = \frac{2\pi}{15}$$

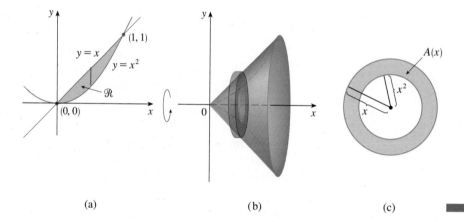

FIGURE 8   (a)   (b)   (c)

EXAMPLE 5   Find the volume of the solid obtained by rotating the region in Example 4 about the line $y = 2$.

SOLUTION   The solid and a cross-section are shown in Figure 9. Again the cross-section is a washer, but this time the inner radius is $2 - x$ and the outer radius is $2 - x^2$.

FIGURE 9

The cross-sectional area is

$$A(x) = \pi(2 - x^2)^2 - \pi(2 - x)^2$$

and so the volume of S is

$$V = \int_0^1 A(x)\,dx$$

$$= \pi \int_0^1 \left[(2 - x^2)^2 - (2 - x)^2\right] dx$$

$$= \pi \int_0^1 (x^4 - 5x^2 + 4x)\,dx$$

$$= \pi \left[\frac{x^5}{5} - 5\frac{x^3}{3} + 4\frac{x^2}{2}\right]_0^1$$

$$= \frac{8\pi}{15}$$

The solids in Examples 1–5 are all called **solids of revolution** because they are obtained by revolving a region about a line. In general, we calculate the volume of a solid of revolution by using the basic defining formula

$$V = \int_a^b A(x)\,dx \qquad \text{or} \qquad V = \int_c^d A(y)\,dy$$

and we find the cross-sectional area $A(x)$ or $A(y)$ in one of the following ways:

- If the cross-section is a disk (as in Examples 1–3), we find the radius of the disk (in terms of x or y) and use

$$A = \pi(\text{radius})^2$$

- If the cross-section is a washer (as in Examples 4 and 5), we find the inner radius r_{in} and outer radius r_{out} from a sketch (as in Figures 8, 9, and 10) and compute the area of the washer by subtracting the area of the inner disk from the area of the outer disk:

$$A = \pi(\text{outer radius})^2 - \pi(\text{inner radius})^2$$

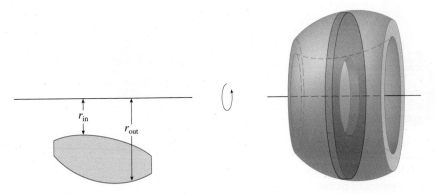

FIGURE 10

The next example gives a further illustration of the procedure.

EXAMPLE 6 Find the volume of the solid obtained by rotating the region in Example 4 about the line $x = -1$.

SOLUTION Figure 11 shows a horizontal cross-section. It is a washer with inner radius $1 + y$ and outer radius $1 + \sqrt{y}$, so the cross-sectional area is

$$A(y) = \pi(\text{outer radius})^2 - \pi(\text{inner radius})^2$$
$$= \pi\left(1 + \sqrt{y}\right)^2 - \pi(1 + y)^2$$

The volume is

$$V = \int_0^1 A(y)\, dy = \pi \int_0^1 \left[\left(1 + \sqrt{y}\right)^2 - (1 + y)^2\right] dy$$

$$= \pi \int_0^1 \left(2\sqrt{y} - y - y^2\right) dy = \pi \left[\frac{4y^{3/2}}{3} - \frac{y^2}{2} - \frac{y^3}{3}\right]_0^1 = \frac{\pi}{2}$$

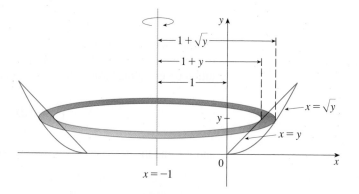

FIGURE 11

We now find the volumes of two solids that are *not* solids of revolution.

EXAMPLE 7 Figure 12 shows a solid with a circular base of radius 1. Parallel cross-sections perpendicular to the base are equilateral triangles. Find the volume of the solid.

SOLUTION Let's take the circle to be $x^2 + y^2 = 1$. The solid, its base, and a typical cross-section at a distance x from the origin are shown in Figure 13.

FIGURE 12
Computer-generated picture
of the solid in Example 7

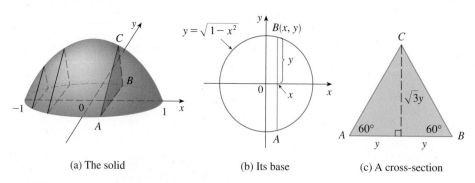

(a) The solid   (b) Its base   (c) A cross-section

FIGURE 13

Since B lies on the circle, we have $y = \sqrt{1 - x^2}$ and so the base of the triangle ABC is $|AB| = 2\sqrt{1 - x^2}$. Since the triangle is equilateral, we see from Figure 13(c) that its height is $\sqrt{3}\, y = \sqrt{3}\sqrt{1 - x^2}$. The cross-sectional area is therefore

$$A(x) = \tfrac{1}{2} \cdot 2\sqrt{1 - x^2} \cdot \sqrt{3}\sqrt{1 - x^2} = \sqrt{3}\,(1 - x^2)$$

and the volume of the solid is

$$V = \int_{-1}^{1} A(x)\,dx = \int_{-1}^{1} \sqrt{3}\,(1 - x^2)\,dx$$

$$= 2\int_{0}^{1} \sqrt{3}\,(1 - x^2)\,dx = 2\sqrt{3}\left[x - \frac{x^3}{3}\right]_0^1 = \frac{4\sqrt{3}}{3}$$ ▬

V **EXAMPLE 8** Find the volume of a pyramid whose base is a square with side L and whose height is h.

SOLUTION We place the origin O at the vertex of the pyramid and the x-axis along its central axis as in Figure 14. Any plane P_x that passes through x and is perpendicular to the x-axis intersects the pyramid in a square with side of length s, say. We can express s in terms of x by observing from the similar triangles in Figure 15 that

$$\frac{x}{h} = \frac{s/2}{L/2} = \frac{s}{L}$$

and so $s = Lx/h$. [Another method is to observe that the line OP has slope $L/(2h)$ and so its equation is $y = Lx/(2h)$.] Thus the cross-sectional area is

$$A(x) = s^2 = \frac{L^2}{h^2}x^2$$

FIGURE 14

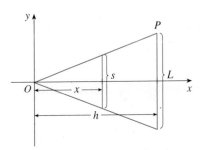

FIGURE 15

The pyramid lies between $x = 0$ and $x = h$, so its volume is

$$V = \int_0^h A(x)\,dx = \int_0^h \frac{L^2}{h^2}x^2\,dx = \frac{L^2}{h^2}\frac{x^3}{3}\bigg]_0^h = \frac{L^2 h}{3}$$ ▬

NOTE We didn't need to place the vertex of the pyramid at the origin in Example 8. We did so merely to make the equations simple. If, instead, we had placed the center of the base at the origin and the vertex on the positive y-axis, as in Figure 16, you can verify that we would have obtained the integral

$$V = \int_0^h \frac{L^2}{h^2}(h - y)^2\,dy = \frac{L^2 h}{3}$$

FIGURE 16

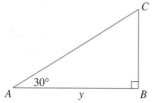

FIGURE 17

EXAMPLE 9 A wedge is cut out of a circular cylinder of radius 4 by two planes. One plane is perpendicular to the axis of the cylinder. The other intersects the first at an angle of 30° along a diameter of the cylinder. Find the volume of the wedge.

SOLUTION If we place the x-axis along the diameter where the planes meet, then the base of the solid is a semicircle with equation $y = \sqrt{16 - x^2}$, $-4 \leqslant x \leqslant 4$. A cross-section perpendicular to the x-axis at a distance x from the origin is a triangle ABC, as shown in Figure 17, whose base is $y = \sqrt{16 - x^2}$ and whose height is $|BC| = y \tan 30° = \sqrt{16 - x^2}/\sqrt{3}$. Thus the cross-sectional area is

$$A(x) = \tfrac{1}{2}\sqrt{16 - x^2} \cdot \frac{1}{\sqrt{3}} \sqrt{16 - x^2} = \frac{16 - x^2}{2\sqrt{3}}$$

and the volume is

$$V = \int_{-4}^{4} A(x)\, dx = \int_{-4}^{4} \frac{16 - x^2}{2\sqrt{3}}\, dx$$

$$= \frac{1}{\sqrt{3}} \int_{0}^{4} (16 - x^2)\, dx = \frac{1}{\sqrt{3}} \left[16x - \frac{x^3}{3} \right]_{0}^{4}$$

$$= \frac{128}{3\sqrt{3}}$$

For another method see Exercise 62.

5.2 Exercises

Access end-of-section exercises and tools for visualizing calculus online at **www.webassign.net**

5.3 Volumes by Cylindrical Shells

Some volume problems are very difficult to handle by the methods of the preceding section. For instance, let's consider the problem of finding the volume of the solid obtained by rotating about the y-axis the region bounded by $y = 2x^2 - x^3$ and $y = 0$. (See Figure 1.) If we slice perpendicular to the y-axis, we get a washer. But to compute the inner radius and the outer radius of the washer, we'd have to solve the cubic equation $y = 2x^2 - x^3$ for x in terms of y; that's not easy.

FIGURE 1

FIGURE 2

Fortunately, there is a method, called the **method of cylindrical shells**, that is easier to use in such a case. Figure 2 shows a cylindrical shell with inner radius r_1, outer radius r_2, and height h. Its volume V is calculated by subtracting the volume V_1 of the inner cylinder from the volume V_2 of the outer cylinder:

$$V = V_2 - V_1$$

$$= \pi r_2^2 h - \pi r_1^2 h = \pi(r_2^2 - r_1^2)h$$

$$= \pi(r_2 + r_1)(r_2 - r_1)h$$

$$= 2\pi \frac{r_2 + r_1}{2} h(r_2 - r_1)$$

If we let $\Delta r = r_2 - r_1$ (the thickness of the shell) and $r = \frac{1}{2}(r_2 + r_1)$ (the average radius of the shell), then this formula for the volume of a cylindrical shell becomes

$$\boxed{1} \qquad \boxed{V = 2\pi r h\, \Delta r}$$

and it can be remembered as

$$V = [\text{circumference}][\text{height}][\text{thickness}]$$

Now let S be the solid obtained by rotating about the y-axis the region bounded by $y = f(x)$ [where $f(x) \geq 0$], $y = 0$, $x = a$, and $x = b$, where $b > a \geq 0$. (See Figure 3.)

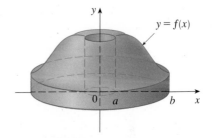

FIGURE 3

We divide the interval $[a, b]$ into n subintervals $[x_{i-1}, x_i]$ of equal width Δx and let $\bar{x}_i$ be the midpoint of the ith subinterval. If the rectangle with base $[x_{i-1}, x_i]$ and height $f(\bar{x}_i)$ is rotated about the y-axis, then the result is a cylindrical shell with average radius $\bar{x}_i$, height $f(\bar{x}_i)$, and thickness Δx (see Figure 4), so by Formula 1 its volume is

$$V_i = (2\pi \bar{x}_i)[f(\bar{x}_i)]\,\Delta x$$

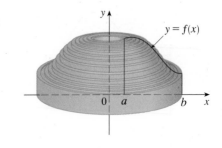

FIGURE 4

Therefore an approximation to the volume V of S is given by the sum of the volumes of these shells:

$$V \approx \sum_{i=1}^{n} V_i = \sum_{i=1}^{n} 2\pi \bar{x}_i f(\bar{x}_i) \, \Delta x$$

This approximation appears to become better as $n \rightarrow \infty$. But, from the definition of an integral, we know that

$$\lim_{n \to \infty} \sum_{i=1}^{n} 2\pi \bar{x}_i f(\bar{x}_i) \, \Delta x = \int_a^b 2\pi x f(x) \, dx$$

Thus the following appears plausible:

> **2**   The volume of the solid in Figure 3, obtained by rotating about the y-axis the region under the curve $y = f(x)$ from a to b, is
>
> $$V = \int_a^b 2\pi x f(x) \, dx \qquad \text{where } 0 \leqslant a < b$$

The argument using cylindrical shells makes Formula 2 seem reasonable, but later we will be able to prove it (see Exercise 71 in Section 7.1).

The best way to remember Formula 2 is to think of a typical shell, cut and flattened as in Figure 5, with radius x, circumference $2\pi x$, height $f(x)$, and thickness Δx or dx:

$$\int_a^b \underbrace{(2\pi x)}_{\text{circumference}} \underbrace{[f(x)]}_{\text{height}} \underbrace{dx}_{\text{thickness}}$$

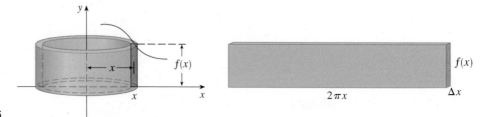

FIGURE 5

This type of reasoning will be helpful in other situations, such as when we rotate about lines other than the y-axis.

EXAMPLE 1 Find the volume of the solid obtained by rotating about the y-axis the region bounded by $y = 2x^2 - x^3$ and $y = 0$.

SOLUTION From the sketch in Figure 6 we see that a typical shell has radius x, circumference $2\pi x$, and height $f(x) = 2x^2 - x^3$. So, by the shell method, the volume is

$$V = \int_0^2 (2\pi x)(2x^2 - x^3) \, dx = 2\pi \int_0^2 (2x^3 - x^4) \, dx$$

$$= 2\pi \left[\tfrac{1}{2}x^4 - \tfrac{1}{5}x^5 \right]_0^2 = 2\pi \left(8 - \tfrac{32}{5} \right) = \tfrac{16}{5}\pi$$

FIGURE 6

It can be verified that the shell method gives the same answer as slicing.

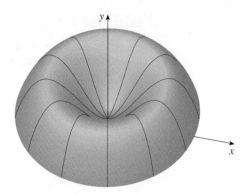

Figure 7 shows a computer-generated picture of the solid whose volume we computed in Example 1.

FIGURE 7

NOTE Comparing the solution of Example 1 with the remarks at the beginning of this section, we see that the method of cylindrical shells is much easier than the washer method for this problem. We did not have to find the coordinates of the local maximum and we did not have to solve the equation of the curve for x in terms of y. However, in other examples the methods of the preceding section may be easier.

V **EXAMPLE 2** Find the volume of the solid obtained by rotating about the y-axis the region between $y = x$ and $y = x^2$.

SOLUTION The region and a typical shell are shown in Figure 8. We see that the shell has radius x, circumference $2\pi x$, and height $x - x^2$. So the volume is

$$V = \int_0^1 (2\pi x)(x - x^2)\, dx = 2\pi \int_0^1 (x^2 - x^3)\, dx$$

$$= 2\pi \left[\frac{x^3}{3} - \frac{x^4}{4} \right]_0^1 = \frac{\pi}{6}$$

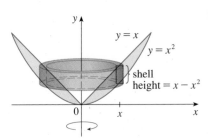

FIGURE 8

As the following example shows, the shell method works just as well if we rotate about the x-axis. We simply have to draw a diagram to identify the radius and height of a shell.

V **EXAMPLE 3** Use cylindrical shells to find the volume of the solid obtained by rotating about the x-axis the region under the curve $y = \sqrt{x}$ from 0 to 1.

SOLUTION This problem was solved using disks in Example 2 in Section 5.2. To use shells we relabel the curve $y = \sqrt{x}$ (in the figure in that example) as $x = y^2$ in Figure 9. For rotation about the x-axis we see that a typical shell has radius y, circumference $2\pi y$, and height $1 - y^2$. So the volume is

$$V = \int_0^1 (2\pi y)(1 - y^2)\, dy = 2\pi \int_0^1 (y - y^3)\, dy$$

$$= 2\pi \left[\frac{y^2}{2} - \frac{y^4}{4} \right]_0^1 = \frac{\pi}{2}$$

FIGURE 9

In this problem the disk method was simpler.

V **EXAMPLE 4** Find the volume of the solid obtained by rotating the region bounded by $y = x - x^2$ and $y = 0$ about the line $x = 2$.

SOLUTION Figure 10 shows the region and a cylindrical shell formed by rotation about the line $x = 2$. It has radius $2 - x$, circumference $2\pi(2 - x)$, and height $x - x^2$.

FIGURE 10

The volume of the given solid is

$$V = \int_0^1 2\pi(2 - x)(x - x^2)\,dx$$

$$= 2\pi \int_0^1 (x^3 - 3x^2 + 2x)\,dx$$

$$= 2\pi\left[\frac{x^4}{4} - x^3 + x^2\right]_0^1 = \frac{\pi}{2}$$

5.3 Exercises

Access end-of-section exercises and tools for visualizing calculus online at **www.webassign.net**   ENHANCED WebAssign

---

5.4 Work

The term *work* is used in everyday language to mean the total amount of effort required to perform a task. In physics it has a technical meaning that depends on the idea of a *force*. Intuitively, you can think of a force as describing a push or pull on an object—for example, a horizontal push of a book across a table or the downward pull of the earth's gravity on a ball. In general, if an object moves along a straight line with position function $s(t)$, then the **force** F on the object (in the same direction) is given by Newton's Second Law of Motion as the product of its mass m and its acceleration:

$$\boxed{1} \qquad F = m\frac{d^2s}{dt^2}$$

In the SI metric system, the mass is measured in kilograms (kg), the displacement in meters (m), the time in seconds (s), and the force in newtons (N = kg·m/s²). Thus a force of 1 N acting on a mass of 1 kg produces an acceleration of 1 m/s². In the US Customary system the fundamental unit is chosen to be the unit of force, which is the pound.

In the case of constant acceleration, the force F is also constant and the work done is defined to be the product of the force F and the distance d that the object moves:

$$\boxed{2} \qquad W = Fd \qquad \text{work} = \text{force} \times \text{distance}$$

If F is measured in newtons and d in meters, then the unit for W is a newton-meter, which is called a joule (J). If F is measured in pounds and d in feet, then the unit for W is a foot-pound (ft-lb), which is about 1.36 J.

V EXAMPLE 1

(a) How much work is done in lifting a 1.2-kg book off the floor to put it on a desk that is 0.7 m high? Use the fact that the acceleration due to gravity is $g = 9.8$ m/s^2.

(b) How much work is done in lifting a 20-lb weight 6 ft off the ground?

SOLUTION

(a) The force exerted is equal and opposite to that exerted by gravity, so Equation 1 gives

$$F = mg = (1.2)(9.8) = 11.76 \text{ N}$$

and then Equation 2 gives the work done as

$$W = Fd = (11.76)(0.7) \approx 8.2 \text{ J}$$

(b) Here the force is given as $F = 20$ lb, so the work done is

$$W = Fd = 20 \cdot 6 = 120 \text{ ft-lb}$$

Notice that in part (b), unlike part (a), we did not have to multiply by g because we were given the *weight* (which is a force) and not the mass of the object. ▬

Equation 2 defines work as long as the force is constant, but what happens if the force is variable? Let's suppose that the object moves along the x-axis in the positive direction, from $x = a$ to $x = b$, and at each point x between a and b a force $f(x)$ acts on the object, where f is a continuous function. We divide the interval $[a, b]$ into n subintervals with endpoints $x_0, x_1, \ldots, x_n$ and equal width Δx. We choose a sample point x_i^* in the ith subinterval $[x_{i-1}, x_i]$. Then the force at that point is $f(x_i^*)$. If n is large, then Δx is small, and since f is continuous, the values of f don't change very much over the interval $[x_{i-1}, x_i]$. In other words, f is almost constant on the interval and so the work W_i that is done in moving the particle from x_{i-1} to x_i is approximately given by Equation 2:

$$W_i \approx f(x_i^*)\, \Delta x$$

Thus we can approximate the total work by

$$\boxed{3} \qquad W \approx \sum_{i=1}^{n} f(x_i^*)\, \Delta x$$

It seems that this approximation becomes better as we make n larger. Therefore we define the **work done in moving the object from a to b** as the limit of this quantity as $n \to \infty$. Since the right side of $\boxed{3}$ is a Riemann sum, we recognize its limit as being a definite integral and so

$$\boxed{4} \qquad W = \lim_{n \to \infty} \sum_{i=1}^{n} f(x_i^*)\, \Delta x = \int_a^b f(x)\, dx$$

frictionless   0 x
surface

(a) Natural position of spring

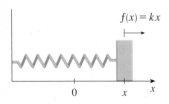

$f(x) = kx$

0   x x

(b) Stretched position of spring

FIGURE 1
Hooke's Law

EXAMPLE 2   When a particle is located a distance x feet from the origin, a force of $x^2 + 2x$ pounds acts on it. How much work is done in moving it from $x = 1$ to $x = 3$?

SOLUTION
$$W = \int_1^3 (x^2 + 2x)\, dx = \frac{x^3}{3} + x^2 \Big]_1^3 = \frac{50}{3}$$

The work done is $16\frac{2}{3}$ ft-lb.

In the next example we use a law from physics: **Hooke's Law** states that the force required to maintain a spring stretched x units beyond its natural length is proportional to x:

$$f(x) = kx$$

where k is a positive constant (called the **spring constant**). Hooke's Law holds provided that x is not too large (see Figure 1).

EXAMPLE 3   A force of 40 N is required to hold a spring that has been stretched from its natural length of 10 cm to a length of 15 cm. How much work is done in stretching the spring from 15 cm to 18 cm?

SOLUTION   According to Hooke's Law, the force required to hold the spring stretched x meters beyond its natural length is $f(x) = kx$. When the spring is stretched from 10 cm to 15 cm, the amount stretched is 5 cm $= 0.05$ m. This means that $f(0.05) = 40$, so

$$0.05k = 40 \qquad k = \frac{40}{0.05} = 800$$

Thus $f(x) = 800x$ and the work done in stretching the spring from 15 cm to 18 cm is

$$W = \int_{0.05}^{0.08} 800x\, dx = 800\,\frac{x^2}{2}\Big]_{0.05}^{0.08}$$
$$= 400[(0.08)^2 - (0.05)^2] = 1.56 \text{ J}$$

EXAMPLE 4   A 200-lb cable is 100 ft long and hangs vertically from the top of a tall building. How much work is required to lift the cable to the top of the building?

SOLUTION   Here we don't have a formula for the force function, but we can use an argument similar to the one that led to Definition 4.

Let's place the origin at the top of the building and the x-axis pointing downward as in Figure 2. We divide the cable into small parts with length Δx. If x_i^* is a point in the ith such interval, then all points in the interval are lifted by approximately the same amount, namely x_i^*. The cable weighs 2 pounds per foot, so the weight of the ith part is $2\Delta x$. Thus the work done on the ith part, in foot-pounds, is

$$\underbrace{(2\Delta x)}_{\text{force}} \cdot \underbrace{x_i^*}_{\text{distance}} = 2x_i^* \,\Delta x$$

x_i^* Δx

0

100

x

FIGURE 2

If we had placed the origin at the bottom of the cable and the x-axis upward, we would have gotten

$$W = \int_0^{100} 2(100 - x)\, dx$$

which gives the same answer.

We get the total work done by adding all these approximations and letting the number of parts become large (so $\Delta x \to 0$):

$$W = \lim_{n \to \infty} \sum_{i=1}^n 2x_i^* \,\Delta x = \int_0^{100} 2x\, dx$$
$$= x^2 \Big]_0^{100} = 10{,}000 \text{ ft-lb}$$

EXAMPLE 5 A tank has the shape of an inverted circular cone with height 10 m and base radius 4 m. It is filled with water to a height of 8 m. Find the work required to empty the tank by pumping all of the water to the top of the tank. (The density of water is 1000 kg/m³.)

SOLUTION Let's measure depths from the top of the tank by introducing a vertical coordinate line as in Figure 3. The water extends from a depth of 2 m to a depth of 10 m and so we divide the interval [2, 10] into n subintervals with endpoints $x_0, x_1, \ldots, x_n$ and choose x_i^* in the ith subinterval. This divides the water into n layers. The ith layer is approximated by a circular cylinder with radius r_i and height Δx. We can compute r_i from similar triangles, using Figure 4, as follows:

$$\frac{r_i}{10 - x_i^*} = \frac{4}{10} \qquad r_i = \tfrac{2}{5}(10 - x_i^*)$$

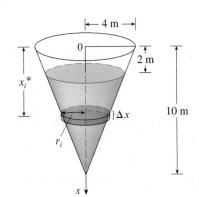

FIGURE 3

Thus an approximation to the volume of the ith layer of water is

$$V_i \approx \pi r_i^2 \, \Delta x = \frac{4\pi}{25}(10 - x_i^*)^2 \, \Delta x$$

and so its mass is

$$m_i = \text{density} \times \text{volume}$$

$$\approx 1000 \cdot \frac{4\pi}{25}(10 - x_i^*)^2 \, \Delta x = 160\pi(10 - x_i^*)^2 \, \Delta x$$

The force required to raise this layer must overcome the force of gravity and so

$$F_i = m_i g \approx (9.8)160\pi(10 - x_i^*)^2 \, \Delta x$$

$$= 1568\pi(10 - x_i^*)^2 \, \Delta x$$

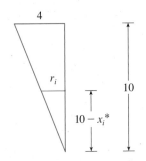

FIGURE 4

Each particle in the layer must travel a distance of approximately x_i^*. The work W_i done to raise this layer to the top is approximately the product of the force F_i and the distance x_i^*:

$$W_i \approx F_i x_i^* \approx 1568\pi x_i^*(10 - x_i^*)^2 \, \Delta x$$

To find the total work done in emptying the entire tank, we add the contributions of each of the n layers and then take the limit as $n \to \infty$:

$$W = \lim_{n \to \infty} \sum_{i=1}^{n} 1568\pi x_i^*(10 - x_i^*)^2 \, \Delta x = \int_2^{10} 1568\pi x(10 - x)^2 \, dx$$

$$= 1568\pi \int_2^{10} (100x - 20x^2 + x^3) \, dx = 1568\pi \left[50x^2 - \frac{20x^3}{3} + \frac{x^4}{4} \right]_2^{10}$$

$$= 1568\pi \left(\tfrac{2048}{3} \right) \approx 3.4 \times 10^6 \, \text{J}$$

5.4   Exercises

Access end-of-section exercises and tools for visualizing calculus online at **www.webassign.net**

ENHANCED
WebAssign

5.5 Average Value of a Function

It is easy to calculate the average value of finitely many numbers $y_1, y_2, \ldots, y_n$:

$$y_{\text{ave}} = \frac{y_1 + y_2 + \cdots + y_n}{n}$$

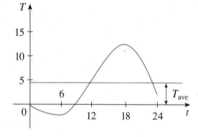

FIGURE 1

But how do we compute the average temperature during a day if infinitely many temperature readings are possible? Figure 1 shows the graph of a temperature function $T(t)$, where t is measured in hours and T in °C, and a guess at the average temperature, T_{ave}.

In general, let's try to compute the average value of a function $y = f(x)$, $a \leqslant x \leqslant b$. We start by dividing the interval $[a, b]$ into n equal subintervals, each with length $\Delta x = (b - a)/n$. Then we choose points $x_1^*, \ldots, x_n^*$ in successive subintervals and calculate the average of the numbers $f(x_1^*), \ldots, f(x_n^*)$:

$$\frac{f(x_1^*) + \cdots + f(x_n^*)}{n}$$

(For example, if f represents a temperature function and $n = 24$, this means that we take temperature readings every hour and then average them.) Since $\Delta x = (b - a)/n$, we can write $n = (b - a)/\Delta x$ and the average value becomes

$$\frac{f(x_1^*) + \cdots + f(x_n^*)}{\dfrac{b - a}{\Delta x}} = \frac{1}{b - a} [f(x_1^*) \Delta x + \cdots + f(x_n^*) \Delta x]$$

$$= \frac{1}{b - a} \sum_{i=1}^{n} f(x_i^*) \Delta x$$

If we let n increase, we would be computing the average value of a large number of closely spaced values. (For example, we would be averaging temperature readings taken every minute or even every second.) The limiting value is

$$\lim_{n \to \infty} \frac{1}{b - a} \sum_{i=1}^{n} f(x_i^*) \Delta x = \frac{1}{b - a} \int_a^b f(x)\,dx$$

by the definition of a definite integral.

Therefore we define the **average value of f** on the interval $[a, b]$ as

$$\boxed{f_{\text{ave}} = \frac{1}{b - a} \int_a^b f(x)\,dx}$$

For a positive function, we can think of this definition as saying

$$\frac{\text{area}}{\text{width}} = \text{average height}$$

V EXAMPLE 1 Find the average value of the function $f(x) = 1 + x^2$ on the interval $[-1, 2]$.

SOLUTION With $a = -1$ and $b = 2$ we have

$$f_{\text{ave}} = \frac{1}{b - a} \int_a^b f(x)\,dx = \frac{1}{2 - (-1)} \int_{-1}^{2} (1 + x^2)\,dx$$

$$= \frac{1}{3} \left[x + \frac{x^3}{3} \right]_{-1}^{2} = 2$$

If $T(t)$ is the temperature at time t, we might wonder if there is a specific time when the temperature is the same as the average temperature. For the temperature function graphed in Figure 1, we see that there are two such times—just before noon and just before midnight. In general, is there a number c at which the value of a function f is exactly equal to the average value of the function, that is, $f(c) = f_{\text{ave}}$? The following theorem says that this is true for continuous functions.

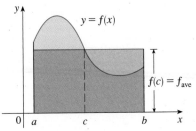

The Mean Value Theorem for Integrals If f is continuous on $[a, b]$, then there exists a number c in $[a, b]$ such that

$$f(c) = f_{\text{ave}} = \frac{1}{b - a} \int_a^b f(x)\, dx$$

that is,

$$\int_a^b f(x)\, dx = f(c)(b - a)$$

The Mean Value Theorem for Integrals is a consequence of the Mean Value Theorem for derivatives and the Fundamental Theorem of Calculus. The proof is outlined in Exercise 23.

The geometric interpretation of the Mean Value Theorem for Integrals is that, for *positive* functions f, there is a number c such that the rectangle with base $[a, b]$ and height $f(c)$ has the same area as the region under the graph of f from a to b. (See Figure 2 and the more picturesque interpretation in the margin note.)

FIGURE 2

You can always chop off the top of a (two-dimensional) mountain at a certain height and use it to fill in the valleys so that the mountain becomes completely flat.

V EXAMPLE 2 Since $f(x) = 1 + x^2$ is continuous on the interval $[-1, 2]$, the Mean Value Theorem for Integrals says there is a number c in $[-1, 2]$ such that

$$\int_{-1}^{2} (1 + x^2)\, dx = f(c)[2 - (-1)]$$

In this particular case we can find c explicitly. From Example 1 we know that $f_{\text{ave}} = 2$, so the value of c satisfies

$$f(c) = f_{\text{ave}} = 2$$

Therefore $\qquad\qquad\qquad 1 + c^2 = 2 \qquad$ so $\qquad c^2 = 1$

So in this case there happen to be two numbers $c = \pm 1$ in the interval $[-1, 2]$ that work in the Mean Value Theorem for Integrals.

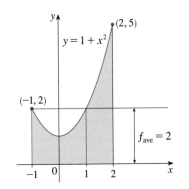

FIGURE 3

Examples 1 and 2 are illustrated by Figure 3.

V EXAMPLE 3 Show that the average velocity of a car over a time interval $[t_1, t_2]$ is the same as the average of its velocities during the trip.

SOLUTION If $s(t)$ is the displacement of the car at time t, then, by definition, the average velocity of the car over the interval is

$$\frac{\Delta s}{\Delta t} = \frac{s(t_2) - s(t_1)}{t_2 - t_1}$$

On the other hand, the average value of the velocity function on the interval is

$$v_{\text{ave}} = \frac{1}{t_2 - t_1} \int_{t_1}^{t_2} v(t)\, dt = \frac{1}{t_2 - t_1} \int_{t_1}^{t_2} s'(t)\, dt$$

$$= \frac{1}{t_2 - t_1} [s(t_2) - s(t_1)] \qquad \text{(by the Net Change Theorem)}$$

$$= \frac{s(t_2) - s(t_1)}{t_2 - t_1} = \text{average velocity}$$

5.5 Exercises

Access end-of-section exercises and tools for visualizing calculus online at **www.webassign.net**

APPLIED PROJECT

CALCULUS AND BASEBALL

In this project we explore two of the many applications of calculus to baseball. The physical interactions of the game, especially the collision of ball and bat, are quite complex and their models are discussed in detail in a book by Robert Adair, *The Physics of Baseball,* 3d ed. (New York, 2002).

An overhead view of the position of a baseball bat, shown every fiftieth of a second during a typical swing. (Adapted from *The Physics of Baseball*)

1. It may surprise you to learn that the collision of baseball and bat lasts only about a thousandth of a second. Here we calculate the average force on the bat during this collision by first computing the change in the ball's momentum.

    The *momentum p* of an object is the product of its mass m and its velocity v, that is, $p = mv$. Suppose an object, moving along a straight line, is acted on by a force $F = F(t)$ that is a continuous function of time.

    (a) Show that the change in momentum over a time interval $[t_0, t_1]$ is equal to the integral of F from t_0 to t_1; that is, show that

    $$p(t_1) - p(t_0) = \int_{t_0}^{t_1} F(t)\, dt$$

    This integral is called the *impulse* of the force over the time interval.

    (b) A pitcher throws a 90-mi/h fastball to a batter, who hits a line drive directly back to the pitcher. The ball is in contact with the bat for 0.001 s and leaves the bat with velocity 110 mi/h. A baseball weighs 5 oz and, in US Customary units, its mass is measured in slugs: $m = w/g$ where $g = 32$ ft/s^2.

      (i) Find the change in the ball's momentum.
      (ii) Find the average force on the bat.

2. In this problem we calculate the work required for a pitcher to throw a 90-mi/h fastball by first considering kinetic energy.

    The *kinetic energy K* of an object of mass m and velocity v is given by $K = \frac{1}{2}mv^2$. Suppose an object of mass m, moving in a straight line, is acted on by a force $F = F(s)$ that depends on its position s. According to Newton's Second Law

    $$F(s) = ma = m\frac{dv}{dt}$$

    where a and v denote the acceleration and velocity of the object.

(a) Show that the work done in moving the object from a position s_0 to a position s_1 is equal to the change in the object's kinetic energy; that is, show that

$$W = \int_{s_0}^{s_1} F(s)\, ds = \tfrac{1}{2}mv_1^2 - \tfrac{1}{2}mv_0^2$$

where $v_0 = v(s_0)$ and $v_1 = v(s_1)$ are the velocities of the object at the positions s_0 and s_1. *Hint:* By the Chain Rule,

$$m\,\frac{dv}{dt} = m\,\frac{dv}{ds}\,\frac{ds}{dt} = mv\,\frac{dv}{ds}$$

(b) How many foot-pounds of work does it take to throw a baseball at a speed of 90 mi/h?

NOTE: Another application of calculus to baseball can be found in Problem 16 on page 521.

5 Review

Concept Check

1. (a) Draw two typical curves $y = f(x)$ and $y = g(x)$, where $f(x) \geqslant g(x)$ for $a \leqslant x \leqslant b$. Show how to approximate the area between these curves by a Riemann sum and sketch the corresponding approximating rectangles. Then write an expression for the exact area.

   (b) Explain how the situation changes if the curves have equations $x = f(y)$ and $x = g(y)$, where $f(y) \geqslant g(y)$ for $c \leqslant y \leqslant d$.

2. Suppose that Sue runs faster than Kathy throughout a 1500-meter race. What is the physical meaning of the area between their velocity curves for the first minute of the race?

3. (a) Suppose S is a solid with known cross-sectional areas. Explain how to approximate the volume of S by a Riemann sum. Then write an expression for the exact volume.

   (b) If S is a solid of revolution, how do you find the cross-sectional areas?

4. (a) What is the volume of a cylindrical shell?

   (b) Explain how to use cylindrical shells to find the volume of a solid of revolution.

   (c) Why might you want to use the shell method instead of slicing?

5. Suppose that you push a book across a 6-meter-long table by exerting a force $f(x)$ at each point from $x = 0$ to $x = 6$. What does $\int_0^6 f(x)\, dx$ represent? If $f(x)$ is measured in newtons, what are the units for the integral?

6. (a) What is the average value of a function f on an interval $[a, b]$?

   (b) What does the Mean Value Theorem for Integrals say? What is its geometric interpretation?

Exercises

1–6 Find the area of the region bounded by the given curves.

1. $y = x^2$, $\quad y = 4x - x^2$

2. $y = 20 - x^2$, $\quad y = x^2 - 12$

3. $y = 1 - 2x^2$, $\quad y = |x|$

4. $x + y = 0$, $\quad x = y^2 + 3y$

5. $y = \sin(\pi x/2)$, $\quad y = x^2 - 2x$

6. $y = \sqrt{x}$, $\quad y = x^2$, $\quad x = 2$

7–11 Find the volume of the solid obtained by rotating the region bounded by the given curves about the specified axis.

7. $y = 2x$, $y = x^2$; $\quad$ about the x-axis

8. $x = 1 + y^2$, $y = x - 3$; $\quad$ about the y-axis

9. $x = 0$, $x = 9 - y^2$; $\quad$ about $x = -1$

10. $y = x^2 + 1$, $y = 9 - x^2$; $\quad$ about $y = -1$

11. $x^2 - y^2 = a^2$, $x = a + h$ (where $a > 0$, $h > 0$); about the y-axis

Graphing calculator or computer required

12–14 Set up, but do not evaluate, an integral for the volume of the solid obtained by rotating the region bounded by the given curves about the specified axis.

12. $y = \tan x$, $y = x$, $x = \pi/3$;   about the y-axis

13. $y = \cos^2 x$, $|x| \leq \pi/2$, $y = \frac{1}{4}$;   about $x = \pi/2$

14. $y = \sqrt{x}$, $y = x^2$;   about $y = 2$

15. Find the volumes of the solids obtained by rotating the region bounded by the curves $y = x$ and $y = x^2$ about the following lines.
(a) The x-axis   (b) The y-axis   (c) $y = 2$

16. Let $\mathcal{R}$ be the region in the first quadrant bounded by the curves $y = x^3$ and $y = 2x - x^2$. Calculate the following quantities.
(a) The area of $\mathcal{R}$
(b) The volume obtained by rotating $\mathcal{R}$ about the x-axis
(c) The volume obtained by rotating $\mathcal{R}$ about the y-axis

17. Let $\mathcal{R}$ be the region bounded by the curves $y = \tan(x^2)$, $x = 1$, and $y = 0$. Use the Midpoint Rule with $n = 4$ to estimate the following quantities.
(a) The area of $\mathcal{R}$
(b) The volume obtained by rotating $\mathcal{R}$ about the x-axis

18. Let $\mathcal{R}$ be the region bounded by the curves $y = 1 - x^2$ and $y = x^6 - x + 1$. Estimate the following quantities.
(a) The x-coordinates of the points of intersection of the curves
(b) The area of $\mathcal{R}$
(c) The volume generated when $\mathcal{R}$ is rotated about the x-axis
(d) The volume generated when $\mathcal{R}$ is rotated about the y-axis

19–22 Each integral represents the volume of a solid. Describe the solid.

19. $\displaystyle\int_0^{\pi/2} 2\pi x \cos x \, dx$ **20.** $\displaystyle\int_0^{\pi/2} 2\pi \cos^2 x \, dx$

21. $\displaystyle\int_0^{\pi} \pi(2 - \sin x)^2 \, dx$ **22.** $\displaystyle\int_0^4 2\pi(6 - y)(4y - y^2) \, dy$

23. The base of a solid is a circular disk with radius 3. Find the volume of the solid if parallel cross-sections perpendicular too the base are isosceles right triangles with hypotenuse lying along the base.

24. The base of a solid is the region bounded by the parabolas $y = x^2$ and $y = 2 - x^2$. Find the volume of the solid if the cross-sections perpendicular to the x-axis are squares with one side lying along the base.

25. The height of a monument is 20 m. A horizontal cross-section at a distance x meters from the top is an equilateral triangle with side $\frac{1}{4}x$ meters. Find the volume of the monument.

26. (a) The base of a solid is a square with vertices located at $(1, 0)$, $(0, 1)$, $(-1, 0)$, and $(0, -1)$. Each cross-section perpendicular to the x-axis is a semicircle. Find the volume of the solid.
(b) Show that by cutting the solid of part (a), we can rearrange it to form a cone. Thus compute its volume more simply.

27. A force of 30 N is required to maintain a spring stretched from its natural length of 12 cm to a length of 15 cm. How much work is done in stretching the spring from 12 cm to 20 cm?

28. A 1600-lb elevator is suspended by a 200-ft cable that weighs 10 lb/ft. How much work is required to raise the elevator from the basement to the third floor, a distance of 30 ft?

29. A tank full of water has the shape of a paraboloid of revolution as shown in the figure; that is, its shape is obtained by rotating a parabola about a vertical axis.
(a) If its height is 4 ft and the radius at the top is 4 ft, find the work required to pump the water out of the tank.
(b) After 4000 ft-lb of work has been done, what is the depth of the water remaining in the tank?

30. Find the average value of the function $f(t) = t \sin(t^2)$ on the interval $[0, 10]$.

31. If f is a continuous function, what is the limit as $h \to 0$ of the average value of f on the interval $[x, x + h]$?

32. Let $\mathcal{R}_1$ be the region bounded by $y = x^2$, $y = 0$, and $x = b$, where $b > 0$. Let $\mathcal{R}_2$ be the region bounded by $y = x^2$, $x = 0$, and $y = b^2$.
(a) Is there a value of b such that $\mathcal{R}_1$ and $\mathcal{R}_2$ have the same area?
(b) Is there a value of b such that $\mathcal{R}_1$ sweeps out the same volume when rotated about the x-axis and the y-axis?
(c) Is there a value of b such that $\mathcal{R}_1$ and $\mathcal{R}_2$ sweep out the same volume when rotated about the x-axis?
(d) Is there a value of b such that $\mathcal{R}_1$ and $\mathcal{R}_2$ sweep out the same volume when rotated about the y-axis?

Problems Plus

1. (a) Find a positive continuous function f such that the area under the graph of f from 0 to t is $A(t) = t^3$ for all $t > 0$.
   (b) A solid is generated by rotating bout the x-axis the region under the curve $y = f(x)$, where f is a positive function and $x \geqslant 0$. The volume generated by the part of the curve from $x = 0$ to $x = b$ is b^2 for all $b > 0$. Find the function f.

2. There is a line through the origin that divides the region bounded by the parabola $y = x - x^2$ and the x-axis into two regions with equal area. What is the slope of that line?

3. The figure shows a horizontal line $y = c$ intersecting the curve $y = 8x - 27x^3$. Find the number c such that the areas of the shaded regions are equal.

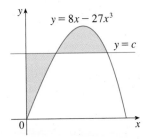

$y = 8x - 27x^3$

$y = c$

FIGURE FOR PROBLEM 3

4. A cylindrical glass of radius r and height L is filled with water and then tilted until the water remaining in the glass exactly covers its base.
   (a) Determine a way to "slice" the water into parallel rectangular cross-sections and then *set up* a definite integral for the volume of the water in the glass.
   (b) Determine a way to "slice" the water into parallel cross-sections that are trapezoids and then *set up* a definite integral for the volume of the water.
   (c) Find the volume of water in the glass by evaluating one of the integrals in part (a) or part (b).
   (d) Find the volume of the water in the glass from purely geometric considerations.
   (e) Suppose the glass is tilted until the water exactly covers half the base. In what direction can you "slice" the water into triangular cross-sections? Rectangular cross-sections? Cross-sections that are segments of circles? Find the volume of water in the glass.

5. (a) Show that the volume of a segment of height h of a sphere of radius r is

$$V = \tfrac{1}{3}\pi h^2(3r - h)$$

   (See the figure.)
   (b) Show that if a sphere of radius 1 is sliced by a plane at a distance x from the center in such a way that the volume of one segment is twice the volume of the other, then x is a solution of the equation

$$3x^3 - 9x + 2 = 0$$

   where $0 < x < 1$. Use Newton's method to find x accurate to four decimal places.
   (c) Using the formula for the volume of a segment of a sphere, it can be shown that the depth x to which a floating sphere of radius r sinks in water is a root of the equation

$$x^3 - 3rx^2 + 4r^3s = 0$$

   where s is the specific gravity of the sphere. Suppose a wooden sphere of radius 0.5 m has specific gravity 0.75. Calculate, to four-decimal-place accuracy, the depth to which the sphere will sink.

FIGURE FOR PROBLEM 5

$y = L - h$

$y = 0$

L

h

$y = -h$

FIGURE FOR PROBLEM 6

(d) A hemispherical bowl has radius 5 inches and water is running into the bowl at the rate of $0.2 \text{ in}^3/\text{s}$.
  (i) How fast is the water level in the bowl rising at the instant the water is 3 inches deep?
  (ii) At a certain instant, the water is 4 inches deep. How long will it take to fill the bowl?

6. Archimedes' Principle states that the buoyant force on an object partially or fully submerged in a fluid is equal to the weight of the fluid that the object displaces. Thus, for an object of density ρ_0 floating partly submerged in a fluid of density ρ_f, the buoyant force is given by $F = \rho_f g \int_{-h}^{0} A(y) \, dy$, where g is the acceleration due to gravity and $A(y)$ is the area of a typical cross-section of the object (see the figure). The weight of the object is given by

$$W = \rho_0 g \int_{-h}^{L-h} A(y) \, dy$$

(a) Show that the percentage of the volume of the object above the surface of the liquid is

$$100 \frac{\rho_f - \rho_0}{\rho_f}$$

(b) The density of ice is 917 kg/m^3 and the density of seawater is 1030 kg/m^3. What percentage of the volume of an iceberg is above water?
(c) An ice cube floats in a glass filled to the brim with water. Does the water overflow when the ice melts?
(d) A sphere of radius 0.4 m and having negligible weight is floating in a large freshwater lake. How much work is required to completely submerge the sphere? The density of the water is 1000 kg/m^3.

7. Water in an open bowl evaporates at a rate proportional to the area of the surface of the water. (This means that the rate of decrease of the volume is proportional to the area of the surface.) Show that the depth of the water decreases at a constant rate, regardless of the shape of the bowl.

8. A sphere of radius 1 overlaps a smaller sphere of radius r in such a way that their intersection is a circle of radius r. (In other words, they intersect in a great circle of the small sphere.) Find r so that the volume inside the small sphere and outside the large sphere is as large as possible.

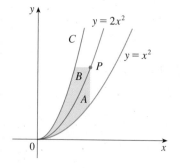

y

$y = 2x^2$

C

$y = x^2$

P

B

A

0

x

FIGURE FOR PROBLEM 9

9. The figure shows a curve C with the property that, for every point P on the middle curve $y = 2x^2$, the areas A and B are equal. Find an equation for C.

10. A paper drinking cup filled with water has the shape of a cone with height h and semivertical angle θ. (See the figure.) A ball is placed carefully in the cup, thereby displacing some of the water and making it overflow. What is the radius of the ball that causes the greatest volume of water to spill out of the cup?

h

θ

11. A *clepsydra*, or water clock, is a glass container with a small hole in the bottom through which water can flow. The "clock" is calibrated for measuring time by placing markings on the container corresponding to water levels at equally spaced times. Let $x = f(y)$ be continuous on the interval $[0, b]$ and assume that the container is formed by rotating the graph of f about the y-axis. Let V denote the volume of water and h the height of the water level at time t.

(a) Determine V as a function of h.

(b) Show that

$$\frac{dV}{dt} = \pi[f(h)]^2 \frac{dh}{dt}$$

(c) Suppose that A is the area of the hole in the bottom of the container. It follows from Torricelli's Law that the rate of change of the volume of the water is given by

$$\frac{dV}{dt} = kA\sqrt{h}$$

where k is a negative constant. Determine a formula for the function f such that dh/dt is a constant C. What is the advantage in having $dh/dt = C$?

12. A cylindrical container of radius r and height L is partially filled with a liquid whose volume is V. If the container is rotated about its axis of symmetry with constant angular speed ω, then the container will induce a rotational motion in the liquid around the same axis. Eventually, the liquid will be rotating at the same angular speed as the container. The surface of the liquid will be convex, as indicated in the figure, because the centrifugal force on the liquid particles increases with the distance from the axis of the container. It can be shown that the surface of the liquid is a paraboloid of revolution generated by rotating the parabola

$$y = h + \frac{\omega^2 x^2}{2g}$$

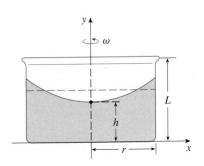

FIGURE FOR PROBLEM 12

about the y-axis, where g is the acceleration due to gravity.

(a) Determine h as a function of ω.

(b) At what angular speed will the surface of the liquid touch the bottom? At what speed will it spill over the top?

(c) Suppose the radius of the container is 2 ft, the height is 7 ft, and the container and liquid are rotating at the same constant angular speed. The surface of the liquid is 5 ft below the top of the tank at the central axis and 4 ft below the top of the tank 1 ft out from the central axis.

(i) Determine the angular speed of the container and the volume of the fluid.

(ii) How far below the top of the tank is the liquid at the wall of the container?

13. Suppose the graph of a cubic polynomial intersects the parabola $y = x^2$ when $x = 0$, $x = a$, and $x = b$, where $0 < a < b$. If the two regions between the curves have the same area, how is b related to a?

6

Inverse Functions:
Exponential, Logarithmic, and Inverse Trigonometric Functions

Exponential functions are used to describe the rapid growth of populations, including the bacteria pictured here.

The common theme that links the functions of this chapter is that they occur as pairs of inverse functions. In particular, two of the most important functions that occur in mathematics and its applications are the exponential function $f(x) = a^x$ and its inverse function, the logarithmic function $g(x) = \log_a x$. In this chapter we investigate their properties, compute their derivatives, and use them to describe exponential growth and decay in biology, physics, chemistry, and other sciences. We also study the inverses of trigonometric and hyperbolic functions. Finally, we look at a method (l'Hospital's Rule) for computing difficult limits and apply it to sketching curves.

There are two possible ways of defining the exponential and logarithmic functions and developing their properties and derivatives. One is to start with the exponential function (defined as in algebra or precalculus courses) and then define the logarithm as its inverse. That is the approach taken in Sections 6.2, 6.3, and 6.4 and is probably the most intuitive method. The other way is to start by defining the logarithm as an integral and then define the exponential function as its inverse. This approach is followed in Sections 6.2*, 6.3*, and 6.4* and, although it is less intuitive, many instructors prefer it because it is more rigorous and the properties follow more easily. You need only read one of these two approaches (whichever your instructor recommends).

6.1   Inverse Functions

Table 1 gives data from an experiment in which a bacteria culture started with 100 bacteria in a limited nutrient medium; the size of the bacteria population was recorded at hourly intervals. The number of bacteria N is a function of the time t: $N = f(t)$.

Suppose, however, that the biologist changes her point of view and becomes interested in the time required for the population to reach various levels. In other words, she is thinking of t as a function of N. This function is called the *inverse function* of f, denoted by f^{-1}, and read "f inverse." Thus $t = f^{-1}(N)$ is the time required for the population level to reach N. The values of f^{-1} can be found by reading Table 1 from right to left or by consulting Table 2. For instance, $f^{-1}(550) = 6$ because $f(6) = 550$.

TABLE 1 N as a function of t

t (hours)	$N = f(t)$ = population at time t
0	100
1	168
2	259
3	358
4	445
5	509
6	550
7	573
8	586

TABLE 2 t as a function of N

N	$t = f^{-1}(N)$ = time to reach N bacteria
100	0
168	1
259	2
358	3
445	4
509	5
550	6
573	7
586	8

Not all functions possess inverses. Let's compare the functions f and g whose arrow diagrams are shown in Figure 1. Note that f never takes on the same value twice (any two inputs in A have different outputs), whereas g does take on the same value twice (both 2 and 3 have the same output, 4). In symbols,

$$g(2) = g(3)$$

but
$$f(x_1) \neq f(x_2) \qquad \text{whenever } x_1 \neq x_2$$

Functions that share this property with f are called *one-to-one functions*.

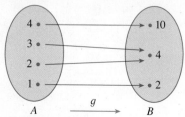

FIGURE 1

f is one-to-one; g is not

> **1**   **Definition**   A function f is called a **one-to-one function** if it never takes on the same value twice; that is,
>
> $$f(x_1) \neq f(x_2) \qquad \text{whenever } x_1 \neq x_2$$

In the language of inputs and outputs, this definition says that f is one-to-one if each output corresponds to only one input.

If a horizontal line intersects the graph of f in more than one point, then we see from Figure 2 that there are numbers x_1 and x_2 such that $f(x_1) = f(x_2)$. This means that f is not one-to-one. Therefore we have the following geometric method for determining whether a function is one-to-one.

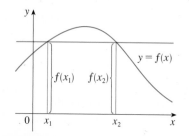

FIGURE 2

This function is not one-to-one because $f(x_1) = f(x_2)$.

> **Horizontal Line Test**   A function is one-to-one if and only if no horizontal line intersects its graph more than once.

FIGURE 3

$f(x) = x^3$ is one-to-one.

FIGURE 4

$g(x) = x^2$ is not one-to-one.

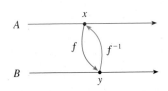

FIGURE 5

V EXAMPLE 1 Is the function $f(x) = x^3$ one-to-one?

SOLUTION 1 If $x_1 \neq x_2$, then $x_1^3 \neq x_2^3$ (two different numbers can't have the same cube). Therefore, by Definition 1, $f(x) = x^3$ is one-to-one.

SOLUTION 2 From Figure 3 we see that no horizontal line intersects the graph of $f(x) = x^3$ more than once. Therefore, by the Horizontal Line Test, f is one-to-one. ▪

V EXAMPLE 2 Is the function $g(x) = x^2$ one-to-one?

SOLUTION 1 This function is not one-to-one because, for instance,

$$g(1) = 1 = g(-1)$$

and so 1 and -1 have the same output.

SOLUTION 2 From Figure 4 we see that there are horizontal lines that intersect the graph of g more than once. Therefore, by the Horizontal Line Test, g is not one-to-one. ▪

One-to-one functions are important because they are precisely the functions that possess inverse functions according to the following definition.

---

2 **Definition** Let f be a one-to-one function with domain A and range B. Then its **inverse function** f^{-1} has domain B and range A and is defined by

$$f^{-1}(y) = x \iff f(x) = y$$

for any y in B.

This definition says that if f maps x into y, then f^{-1} maps y back into x. (If f were not one-to-one, then f^{-1} would not be uniquely defined.) The arrow diagram in Figure 5 indicates that f^{-1} reverses the effect of f. Note that

$$\text{domain of } f^{-1} = \text{range of } f$$
$$\text{range of } f^{-1} = \text{domain of } f$$

For example, the inverse function of $f(x) = x^3$ is $f^{-1}(x) = x^{1/3}$ because if $y = x^3$, then

$$f^{-1}(y) = f^{-1}(x^3) = (x^3)^{1/3} = x$$

⊘ **CAUTION** Do not mistake the -1 in f^{-1} for an exponent. Thus

$$f^{-1}(x) \quad \text{does } not \text{ mean} \quad \frac{1}{f(x)}$$

The reciprocal $1/f(x)$ could, however, be written as $[f(x)]^{-1}$.

☑ EXAMPLE 3 If $f(1) = 5$, $f(3) = 7$, and $f(8) = -10$, find $f^{-1}(7)$, $f^{-1}(5)$, and $f^{-1}(-10)$.

SOLUTION From the definition of f^{-1} we have

$$f^{-1}(7) = 3 \qquad \text{because} \qquad f(3) = 7$$

$$f^{-1}(5) = 1 \qquad \text{because} \qquad f(1) = 5$$

$$f^{-1}(-10) = 8 \qquad \text{because} \qquad f(8) = -10$$

The diagram in Figure 6 makes it clear how f^{-1} reverses the effect of f in this case.

 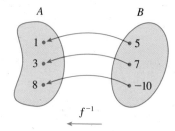

FIGURE 6
The inverse function reverses
inputs and outputs.

The letter x is traditionally used as the independent variable, so when we concentrate on f^{-1} rather than on f, we usually reverse the roles of x and y in Definition 2 and write

3
$$f^{-1}(x) = y \quad \Longleftrightarrow \quad f(y) = x$$

By substituting for y in Definition 2 and substituting for x in **3**, we get the following **cancellation equations**:

4
$$f^{-1}(f(x)) = x \quad \text{for every } x \text{ in } A$$
$$f(f^{-1}(x)) = x \quad \text{for every } x \text{ in } B$$

The first cancellation equation says that if we start with x, apply f, and then apply f^{-1}, we arrive back at x, where we started (see the machine diagram in Figure 7). Thus f^{-1} undoes what f does. The second equation says that f undoes what f^{-1} does.

FIGURE 7

$$x \longrightarrow \boxed{f} \longrightarrow f(x) \longrightarrow \boxed{f^{-1}} \longrightarrow x$$

For example, if $f(x) = x^3$, then $f^{-1}(x) = x^{1/3}$ and so the cancellation equations become

$$f^{-1}(f(x)) = (x^3)^{1/3} = x$$
$$f(f^{-1}(x)) = (x^{1/3})^3 = x$$

These equations simply say that the cube function and the cube root function cancel each other when applied in succession.

Now let's see how to compute inverse functions. If we have a function $y = f(x)$ and are able to solve this equation for x in terms of y, then according to Definition 2 we must have $x = f^{-1}(y)$. If we want to call the independent variable x, we then interchange x and y and arrive at the equation $y = f^{-1}(x)$.

> **5** **How to Find the Inverse Function of a One-to-One Function f**
>
> **Step 1** Write $y = f(x)$.
> **Step 2** Solve this equation for x in terms of y (if possible).
> **Step 3** To express f^{-1} as a function of x, interchange x and y.
> The resulting equation is $y = f^{-1}(x)$.

V **EXAMPLE 4** Find the inverse function of $f(x) = x^3 + 2$.

SOLUTION According to **5** we first write

$$y = x^3 + 2$$

Then we solve this equation for x:

$$x^3 = y - 2$$

$$x = \sqrt[3]{y - 2}$$

Finally, we interchange x and y:

$$y = \sqrt[3]{x - 2}$$

Therefore the inverse function is $f^{-1}(x) = \sqrt[3]{x - 2}$.

In Example 4, notice how f^{-1} reverses the effect of f. The function f is the rule "Cube, then add 2"; f^{-1} is the rule "Subtract 2, then take the cube root."

The principle of interchanging x and y to find the inverse function also gives us the method for obtaining the graph of f^{-1} from the graph of f. Since $f(a) = b$ if and only if $f^{-1}(b) = a$, the point (a, b) is on the graph of f if and only if the point (b, a) is on the graph of f^{-1}. But we get the point (b, a) from (a, b) by reflecting about the line $y = x$. (See Figure 8.)

FIGURE 8

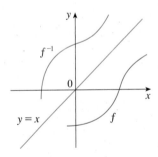

FIGURE 9

Therefore, as illustrated by Figure 9:

> The graph of f^{-1} is obtained by reflecting the graph of f about the line $y = x$.

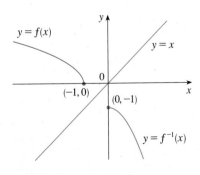

FIGURE 10

EXAMPLE 5 Sketch the graphs of $f(x) = \sqrt{-1 - x}$ and its inverse function using the same coordinate axes.

SOLUTION First we sketch the curve $y = \sqrt{-1 - x}$ (the top half of the parabola $y^2 = -1 - x$, or $x = -y^2 - 1$) and then we reflect about the line $y = x$ to get the graph of f^{-1}. (See Figure 10.) As a check on our graph, notice that the expression for f^{-1} is $f^{-1}(x) = -x^2 - 1$, $x \geq 0$. So the graph of f^{-1} is the right half of the parabola $y = -x^2 - 1$ and this seems reasonable from Figure 10.

The Calculus of Inverse Functions

Now let's look at inverse functions from the point of view of calculus. Suppose that f is both one-to-one and continuous. We think of a continuous function as one whose graph has no break in it. (It consists of just one piece.) Since the graph of f^{-1} is obtained from the graph of f by reflecting about the line $y = x$, the graph of f^{-1} has no break in it either (see Figure 9). Thus we might expect that f^{-1} is also a continuous function.

This geometrical argument does not prove the following theorem but at least it makes the theorem plausible. A proof can be found in Appendix F.

---

6 **Theorem** If f is a one-to-one continuous function defined on an interval, then its inverse function f^{-1} is also continuous.

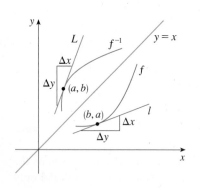

FIGURE 11

Now suppose that f is a one-to-one differentiable function. Geometrically we can think of a differentiable function as one whose graph has no corner or kink in it. We get the graph of f^{-1} by reflecting the graph of f about the line $y = x$, so the graph of f^{-1} has no corner or kink in it either. We therefore expect that f^{-1} is also differentiable (except where its tangents are vertical). In fact, we can predict the value of the derivative of f^{-1} at a given point by a geometric argument. In Figure 11 the graphs of f and its inverse f^{-1} are shown. If $f(b) = a$, then $f^{-1}(a) = b$ and $(f^{-1})'(a)$ is the slope of the tangent line L to the graph of f^{-1} at (a, b), which is $\Delta y/\Delta x$. Reflecting in the line $y = x$ has the effect of interchanging the x- and y-coordinates. So the slope of the reflected line ℓ [the tangent to the graph of f at (b, a)] is $\Delta x/\Delta y$. Thus the slope of L is the reciprocal of the slope of ℓ, that is,

$$(f^{-1})'(a) = \frac{\Delta y}{\Delta x} = \frac{1}{\Delta x/\Delta y} = \frac{1}{f'(b)}$$

---

7 **Theorem** If f is a one-to-one differentiable function with inverse function f^{-1} and $f'(f^{-1}(a)) \neq 0$, then the inverse function is differentiable at a and

$$(f^{-1})'(a) = \frac{1}{f'(f^{-1}(a))}$$

PROOF   Write the definition of derivative as in Equation 2.1.5:

$$(f^{-1})'(a) = \lim_{x \to a} \frac{f^{-1}(x) - f^{-1}(a)}{x - a}$$

If $f(b) = a$, then $f^{-1}(a) = b$. And if we let $y = f^{-1}(x)$, then $f(y) = x$. Since f is differentiable, it is continuous, so f^{-1} is continuous by Theorem 6. Thus if $x \to a$, then $f^{-1}(x) \to f^{-1}(a)$, that is, $y \to b$. Therefore

Note that $x \neq a \;\Rightarrow\; f(y) \neq f(b)$ because f is one-to-one.

$$(f^{-1})'(a) = \lim_{x \to a} \frac{f^{-1}(x) - f^{-1}(a)}{x - a} = \lim_{y \to b} \frac{y - b}{f(y) - f(b)}$$

$$= \lim_{y \to b} \frac{1}{\dfrac{f(y) - f(b)}{y - b}} = \frac{1}{\displaystyle\lim_{y \to b} \frac{f(y) - f(b)}{y - b}}$$

$$= \frac{1}{f'(b)} = \frac{1}{f'(f^{-1}(a))}$$

NOTE 1 Replacing a by the general number x in the formula of Theorem 7, we get

$$\boxed{8} \qquad (f^{-1})'(x) = \frac{1}{f'(f^{-1}(x))}$$

If we write $y = f^{-1}(x)$, then $f(y) = x$, so Equation 8, when expressed in Leibniz notation, becomes

$$\frac{dy}{dx} = \frac{1}{\dfrac{dx}{dy}}$$

NOTE 2 If it is known in advance that f^{-1} is differentiable, then its derivative can be computed more easily than in the proof of Theorem 7 by using implicit differentiation. If $y = f^{-1}(x)$, then $f(y) = x$. Differentiating the equation $f(y) = x$ implicitly with respect to x, remembering that y is a function of x, and using the Chain Rule, we get

$$f'(y)\frac{dy}{dx} = 1$$

Therefore
$$\frac{dy}{dx} = \frac{1}{f'(y)} = \frac{1}{\dfrac{dx}{dy}}$$

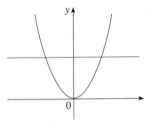

(a) $y = x^2, \quad x \in \mathbb{R}$

(b) $f(x) = x^2, \quad 0 \le x \le 2$

FIGURE 12

EXAMPLE 6 Although the function $y = x^2$, $x \in \mathbb{R}$, is not one-to-one and therefore does not have an inverse function, we can turn it into a one-to-one function by restricting its domain. For instance, the function $f(x) = x^2$, $0 \le x \le 2$, is one-to-one (by the Horizontal Line Test) and has domain $[0, 2]$ and range $[0, 4]$. (See Figure 12.) Thus f has an inverse function f^{-1} with domain $[0, 4]$ and range $[0, 2]$.

Without computing a formula for $(f^{-1})'$ we can still calculate $(f^{-1})'(1)$. Since $f(1) = 1$, we have $f^{-1}(1) = 1$. Also $f'(x) = 2x$. So by Theorem 7 we have

$$(f^{-1})'(1) = \frac{1}{f'(f^{-1}(1))} = \frac{1}{f'(1)} = \frac{1}{2}$$

In this case it is easy to find f^{-1} explicitly. In fact, $f^{-1}(x) = \sqrt{x}$, $0 \le x \le 4$. [In general, we could use the method given by $\boxed{5}$.] Then $(f^{-1})'(x) = 1/(2\sqrt{x})$, so $(f^{-1})'(1) = \frac{1}{2}$, which agrees with the preceding computation. The functions f and f^{-1} are graphed in Figure 13.

V EXAMPLE 7 If $f(x) = 2x + \cos x$, find $(f^{-1})'(1)$.

FIGURE 13

SOLUTION Notice that f is one-to-one because

$$f'(x) = 2 - \sin x > 0$$

and so f is increasing. To use Theorem 7 we need to know $f^{-1}(1)$ and we can find it by inspection:

$$f(0) = 1 \quad \Rightarrow \quad f^{-1}(1) = 0$$

Therefore

$$(f^{-1})'(1) = \frac{1}{f'(f^{-1}(1))} = \frac{1}{f'(0)} = \frac{1}{2 - \sin 0} = \frac{1}{2}$$

6.1   Exercises

Access end-of-section exercises and tools for visualizing calculus online at **www.webassign.net**   **WebAssign**

6.2   Exponential Functions and Their Derivatives

If your instructor has assigned Sections 6.2*, 6.3*, and 6.4*, you don't need to read Sections 6.2–6.4 (pp. 310–331).

The function $f(x) = 2^x$ is called an *exponential function* because the variable, x, is the exponent. It should not be confused with the power function $g(x) = x^2$, in which the variable is the base.

In general, an **exponential function** is a function of the form

$$f(x) = a^x$$

where a is a positive constant. Let's recall what this means.

If $x = n$, a positive integer, then

$$a^n = \underbrace{a \cdot a \cdot \cdots \cdot a}_{n \text{ factors}}$$

If $x = 0$, then $a^0 = 1$, and if $x = -n$, where n is a positive integer, then

$$a^{-n} = \frac{1}{a^n}$$

If x is a rational number, $x = p/q$, where p and q are integers and $q > 0$, then

$$a^x = a^{p/q} = \sqrt[q]{a^p} = \left(\sqrt[q]{a}\right)^p$$

But what is the meaning of a^x if x is an irrational number? For instance, what is meant by $2^{\sqrt{3}}$ or 5^π?

To help us answer this question we first look at the graph of the function $y = 2^x$, where x is rational. A representation of this graph is shown in Figure 1. We want to enlarge the domain of $y = 2^x$ to include both rational and irrational numbers.

There are holes in the graph in Figure 1 corresponding to irrational values of x. We want to fill in the holes by defining $f(x) = 2^x$, where $x \in \mathbb{R}$, so that f is an increasing continuous function. In particular, since the irrational number $\sqrt{3}$ satisfies

$$1.7 < \sqrt{3} < 1.8$$

we must have

$$2^{1.7} < 2^{\sqrt{3}} < 2^{1.8}$$

and we know what $2^{1.7}$ and $2^{1.8}$ mean because 1.7 and 1.8 are rational numbers. Similarly, if we use better approximations for $\sqrt{3}$, we obtain better approximations for $2^{\sqrt{3}}$:

$$1.73 < \sqrt{3} < 1.74 \quad \Rightarrow \quad 2^{1.73} < 2^{\sqrt{3}} < 2^{1.74}$$
$$1.732 < \sqrt{3} < 1.733 \quad \Rightarrow \quad 2^{1.732} < 2^{\sqrt{3}} < 2^{1.733}$$
$$1.7320 < \sqrt{3} < 1.7321 \quad \Rightarrow \quad 2^{1.7320} < 2^{\sqrt{3}} < 2^{1.7321}$$
$$1.73205 < \sqrt{3} < 1.73206 \quad \Rightarrow \quad 2^{1.73205} < 2^{\sqrt{3}} < 2^{1.73206}$$

$$\vdots \qquad\qquad \vdots \qquad\qquad \vdots \qquad\qquad \vdots$$

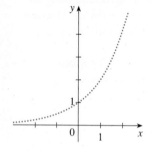

FIGURE 1
Representation of $y = 2^x$, x rational

A proof of this fact is given in J. Marsden and A. Weinstein, *Calculus Unlimited* (Menlo Park, CA, 1981). For an online version, see caltechbook.library.caltech.edu/197/

It can be shown that there is exactly one number that is greater than all of the numbers

$$2^{1.7}, \quad 2^{1.73}, \quad 2^{1.732}, \quad 2^{1.7320}, \quad 2^{1.73205}, \quad \ldots$$

and less than all of the numbers

$$2^{1.8}, \quad 2^{1.74}, \quad 2^{1.733}, \quad 2^{1.7321}, \quad 2^{1.73206}, \quad \ldots$$

We define $2^{\sqrt{3}}$ to be this number. Using the preceding approximation process we can compute it correct to six decimal places:

$$2^{\sqrt{3}} \approx 3.321997$$

Similarly, we can define 2^x (or a^x, if $a > 0$) where x is any irrational number. Figure 2 shows how all the holes in Figure 1 have been filled to complete the graph of the function $f(x) = 2^x, x \in \mathbb{R}$.

In general, if a is any positive number, we define

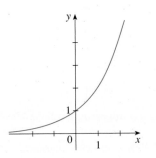

FIGURE 2
$y = 2^x$, x real

$$\boxed{1} \qquad \boxed{a^x = \lim_{r \to x} a^r \qquad r \text{ rational}}$$

This definition makes sense because any irrational number can be approximated as closely as we like by a rational number. For instance, because $\sqrt{3}$ has the decimal representation $\sqrt{3} = 1.7320508\ldots$, Definition 1 says that $2^{\sqrt{3}}$ is the limit of the sequence of numbers

$$2^{1.7}, \quad 2^{1.73}, \quad 2^{1.732}, \quad 2^{1.7320}, \quad 2^{1.73205}, \quad 2^{1.732050}, \quad 2^{1.7320508}, \quad \ldots$$

Similarly, 5^π is the limit of the sequence of numbers

$$5^{3.1}, \quad 5^{3.14}, \quad 5^{3.141}, \quad 5^{3.1415}, \quad 5^{3.14159}, \quad 5^{3.141592}, \quad 5^{3.1415926}, \quad \ldots$$

It can be shown that Definition 1 uniquely specifies a^x and makes the function $f(x) = a^x$ continuous.

The graphs of members of the family of functions $y = a^x$ are shown in Figure 3 for various values of the base a. Notice that all of these graphs pass through the same point $(0, 1)$ because $a^0 = 1$ for $a \neq 0$. Notice also that as the base a gets larger, the exponential function grows more rapidly (for $x > 0$).

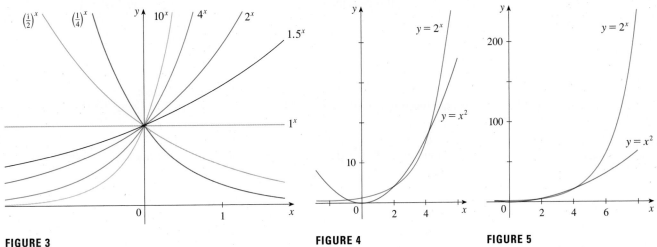

FIGURE 3
Members of the family of exponential functions

FIGURE 4

FIGURE 5

Figure 4 shows how the exponential function $y = 2^x$ compares with the power function $y = x^2$. The graphs intersect three times, but ultimately the exponential curve $y = 2^x$ grows far more rapidly than the parabola $y = x^2$. (See also Figure 5.)

You can see from Figure 3 that there are basically three kinds of exponential functions $y = a^x$. If $0 < a < 1$, the exponential function decreases; if $a = 1$, it is a constant; and if $a > 1$, it increases. These three cases are illustrated in Figure 6. Because $(1/a)^x = 1/a^x = a^{-x}$, the graph of $y = (1/a)^x$ is just the reflection of the graph of $y = a^x$ about the y-axis.

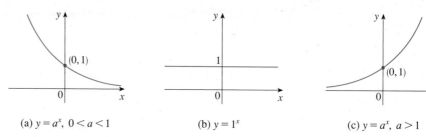

FIGURE 6   (a) $y = a^x$, $0 < a < 1$ (b) $y = 1^x$ (c) $y = a^x$, $a > 1$

The properties of the exponential function are summarized in the following theorem.

---

2   Theorem If $a > 0$ and $a \neq 1$, then $f(x) = a^x$ is a continuous function with domain $\mathbb{R}$ and range $(0, \infty)$. In particular, $a^x > 0$ for all x. If $0 < a < 1$, $f(x) = a^x$ is a decreasing function; if $a > 1$, f is an increasing function. If $a, b > 0$ and $x, y \in \mathbb{R}$, then

1. $a^{x+y} = a^x a^y$ **2.** $a^{x-y} = \dfrac{a^x}{a^y}$ **3.** $(a^x)^y = a^{xy}$ **4.** $(ab)^x = a^x b^x$

---

www.stewartcalculus.com

For review and practice using the Laws of Exponents, click on *Review of Algebra*.

The reason for the importance of the exponential function lies in properties 1–4, which are called the **Laws of Exponents**. If x and y are rational numbers, then these laws are well known from elementary algebra. For arbitrary real numbers x and y these laws can be deduced from the special case where the exponents are rational by using Equation 1.

The following limits can be read from the graphs shown in Figure 6 or proved from the definition of a limit at infinity. (See Exercise 71 in Section 6.3.)

---

3   If $a > 1$, then $\lim\limits_{x \to \infty} a^x = \infty$ and $\lim\limits_{x \to -\infty} a^x = 0$

        If $0 < a < 1$, then $\lim\limits_{x \to \infty} a^x = 0$ and $\lim\limits_{x \to -\infty} a^x = \infty$

In particular, if $a \neq 1$, then the x-axis is a horizontal asymptote of the graph of the exponential function $y = a^x$.

EXAMPLE 1
(a) Find $\lim_{x \to \infty} (2^{-x} - 1)$.
(b) Sketch the graph of the function $y = 2^{-x} - 1$.

SOLUTION

(a) $\lim\limits_{x \to \infty} (2^{-x} - 1) = \lim\limits_{x \to \infty} \left[\left(\tfrac{1}{2} \right)^x - 1 \right]$

                        $= 0 - 1$ $\left[\text{by } \boxed{3} \text{ with } a = \tfrac{1}{2} < 1 \right]$

                        $= -1$

(b) We write $y = \left(\frac{1}{2}\right)^x - 1$ as in part (a). The graph of $y = \left(\frac{1}{2}\right)^x$ is shown in Figure 3, so we shift it down one unit to obtain the graph of $y = \left(\frac{1}{2}\right)^x - 1$ shown in Figure 7. (For a review of shifting graphs, see Section 1.3.) Part (a) shows that the line $y = -1$ is a horizontal asymptote.

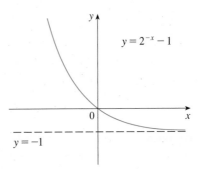

FIGURE 7

Applications of Exponential Functions

The exponential function occurs very frequently in mathematical models of nature and society. Here we indicate briefly how it arises in the description of population growth. In Section 6.5 we will pursue these and other applications in greater detail.

In Section 2.7 we considered a bacteria population that doubles every hour and saw that if the initial population is n_0, then the population after t hours is given by the function $f(t) = n_0 2^t$. This population function is a constant multiple of the exponential function $y = 2^t$, so it exhibits the rapid growth that we observed in Figures 2 and 5. Under ideal conditions (unlimited space and nutrition and freedom from disease), this exponential growth is typical of what actually occurs in nature.

What about the human population? Table 1 shows data for the population of the world in the 20th century, where $t = 0$ corresponds to 1900. Figure 8 shows the corresponding scatter plot.

TABLE 1

t	Population (millions)
0	1650
10	1750
20	1860
30	2070
40	2300
50	2560
60	3040
70	3710
80	4450
90	5280
100	6080
110	6870

FIGURE 8 Scatter plot for world population growth

The pattern of the data points in Figure 8 suggests exponential growth, so we use a graphing calculator with exponential regression capability to apply the method of least squares and obtain the exponential model

$$P = (1436.53) \cdot (1.01395)^t$$

Figure 9 shows the graph of this exponential function together with the original data points. We see that the exponential curve fits the data reasonably well. The period of relatively slow population growth is explained by the two world wars and the Great Depression of the 1930s.

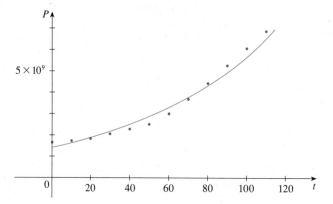

FIGURE 9

Exponential model for population growth

Derivatives of Exponential Functions

Let's try to compute the derivative of the exponential function $f(x) = a^x$ using the definition of a derivative:

$$f'(x) = \lim_{h \to 0} \frac{f(x + h) - f(x)}{h} = \lim_{h \to 0} \frac{a^{x+h} - a^x}{h}$$

$$= \lim_{h \to 0} \frac{a^x a^h - a^x}{h} = \lim_{h \to 0} \frac{a^x(a^h - 1)}{h}$$

The factor a^x doesn't depend on h, so we can take it in front of the limit:

$$f'(x) = a^x \lim_{h \to 0} \frac{a^h - 1}{h}$$

Notice that the limit is the value of the derivative of f at 0, that is,

$$\lim_{h \to 0} \frac{a^h - 1}{h} = f'(0)$$

Therefore we have shown that if the exponential function $f(x) = a^x$ is differentiable at 0, then it is differentiable everywhere and

$$\boxed{4} \qquad\qquad f'(x) = f'(0)a^x$$

This equation says that *the rate of change of any exponential function is proportional to the function itself.* (The slope is proportional to the height.)

Numerical evidence for the existence of $f'(0)$ is given in the table at the left for the cases $a = 2$ and $a = 3$. (Values are stated correct to four decimal places.) It appears that the limits exist and

h	$\dfrac{2^h - 1}{h}$	$\dfrac{3^h - 1}{h}$
0.1	0.7177	1.1612
0.01	0.6956	1.1047
0.001	0.6934	1.0992
0.0001	0.6932	1.0987

$$\text{for } a = 2, \quad f'(0) = \lim_{h \to 0} \frac{2^h - 1}{h} \approx 0.69$$

$$\text{for } a = 3, \quad f'(0) = \lim_{h \to 0} \frac{3^h - 1}{h} \approx 1.10$$

In fact, it can be proved that these limits exist and, correct to six decimal places, the values are

$$\boxed{5} \qquad \frac{d}{dx}(2^x)\bigg|_{x=0} \approx 0.693147 \qquad \frac{d}{dx}(3^x)\bigg|_{x=0} \approx 1.098612$$

Thus, from Equation 4, we have

$$\boxed{6} \qquad \frac{d}{dx}(2^x) \approx (0.69)2^x \qquad \frac{d}{dx}(3^x) \approx (1.10)3^x$$

Of all possible choices for the base a in Equation 4, the simplest differentiation formula occurs when $f'(0) = 1$. In view of the estimates of $f'(0)$ for $a = 2$ and $a = 3$, it seems reasonable that there is a number a between 2 and 3 for which $f'(0) = 1$. It is traditional to denote this value by the letter e. Thus we have the following definition.

$\boxed{7}$ **Definition of the Number e**

e is the number such that $\quad \displaystyle\lim_{h \to 0} \frac{e^h - 1}{h} = 1$

Geometrically this means that of all the possible exponential functions $y = a^x$, the function $f(x) = e^x$ is the one whose tangent line at $(0, 1)$ has a slope $f'(0)$ that is exactly 1. (See Figures 10 and 11.) We call the function $f(x) = e^x$ the *natural exponential function*.

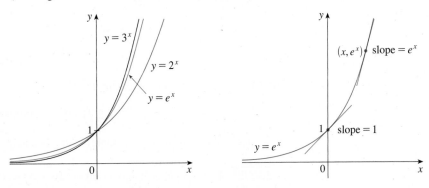

FIGURE 10   **FIGURE 11**

If we put $a = e$ and, therefore, $f'(0) = 1$ in Equation 4, it becomes the following important differentiation formula.

TEC Visual 6.2/6.3* uses the slope-a-scope to illustrate this formula.

$\boxed{8}$ **Derivative of the Natural Exponential Function**

$$\frac{d}{dx}(e^x) = e^x$$

Thus the exponential function $f(x) = e^x$ has the property that it is its own derivative. The geometrical significance of this fact is that the slope of a tangent line to the curve $y = e^x$ at any point is equal to the y-coordinate of the point (see Figure 11).

V **EXAMPLE 2** Differentiate the function $y = e^{\tan x}$.

SOLUTION To use the Chain Rule, we let $u = \tan x$. Then we have $y = e^u$, so

$$\frac{dy}{dx} = \frac{dy}{du}\frac{du}{dx} = e^u\frac{du}{dx} = e^{\tan x}\sec^2 x \qquad \blacksquare$$

In general if we combine Formula 8 with the Chain Rule, as in Example 2, we get

9

$$\frac{d}{dx}(e^u) = e^u\frac{du}{dx}$$

EXAMPLE 3 Find y' if $y = e^{-4x}\sin 5x$.

SOLUTION Using Formula 9 and the Product Rule, we have

$$y' = e^{-4x}(\cos 5x)(5) + (\sin 5x)e^{-4x}(-4) = e^{-4x}(5\cos 5x - 4\sin 5x) \qquad \blacksquare$$

We have seen that e is a number that lies somewhere between 2 and 3, but we can use Equation 4 to estimate the numerical value of e more accurately. Let $e = 2^c$. Then $e^x = 2^{cx}$. If $f(x) = 2^x$, then from Equation 4 we have $f'(x) = k2^x$, where the value of k is $f'(0) \approx 0.693147$. Thus, by the Chain Rule,

$$e^x = \frac{d}{dx}(e^x) = \frac{d}{dx}(2^{cx}) = k2^{cx}\frac{d}{dx}(cx) = ck2^{cx}$$

Putting $x = 0$, we have $1 = ck$, so $c = 1/k$ and

$$e = 2^{1/k} \approx 2^{1/0.693147} \approx 2.71828$$

It can be shown that the approximate value to 20 decimal places is

$$e \approx 2.71828182845904523536$$

The decimal expansion of e is nonrepeating because e is an irrational number.

EXAMPLE 4 In Example 6 in Section 2.7 we considered a population of bacteria cells in a homogeneous nutrient medium. We showed that if the population doubles every hour, then the population after t hours is

$$n = n_0 2^t$$

where n_0 is the initial population. Now we can use $\boxed{4}$ and $\boxed{5}$ to compute the growth rate:

The rate of growth is proportional to the size of the population.

$$\frac{dn}{dt} \approx n_0(0.693147)2^t$$

For instance, if the initial population is $n_0 = 1000$ cells, then the growth rate after two hours is

$$\frac{dn}{dt}\bigg|_{t=2} \approx (1000)(0.693147)2^t\big|_{t=2}$$

$$= (4000)(0.693147) \approx 2773 \text{ cells/h} \qquad \blacksquare$$

V **EXAMPLE 5** Find the absolute maximum value of the function $f(x) = xe^{-x}$.

SOLUTION We differentiate to find any critical numbers:

$$f'(x) = xe^{-x}(-1) + e^{-x}(1) = e^{-x}(1 - x)$$

Since exponential functions are always positive, we see that $f'(x) > 0$ when $1 - x > 0$, that is, when $x < 1$. Similarly, $f'(x) < 0$ when $x > 1$. By the First Derivative Test for Absolute Extreme Values, f has an absolute maximum value when $x = 1$ and the value is

$$f(1) = (1)e^{-1} = \frac{1}{e} \approx 0.37$$

Exponential Graphs

The exponential function $f(x) = e^x$ is one of the most frequently occurring functions in calculus and its applications, so it is important to be familiar with its graph (Figure 12) and properties. We summarize these properties as follows, using the fact that this function is just a special case of the exponential functions considered in Theorem 2 but with base $a = e > 1$.

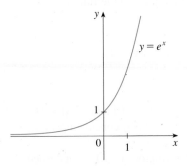

FIGURE 12
The natural exponential function

---

10 **Properties of the Natural Exponential Function** The exponential function $f(x) = e^x$ is an increasing continuous function with domain $\mathbb{R}$ and range $(0, \infty)$. Thus $e^x > 0$ for all x. Also

$$\lim_{x \to -\infty} e^x = 0 \qquad \lim_{x \to \infty} e^x = \infty$$

So the x-axis is a horizontal asymptote of $f(x) = e^x$.

EXAMPLE 6 Find $\displaystyle\lim_{x \to \infty} \frac{e^{2x}}{e^{2x} + 1}$.

SOLUTION We divide numerator and denominator by e^{2x}:

$$\lim_{x \to \infty} \frac{e^{2x}}{e^{2x} + 1} = \lim_{x \to \infty} \frac{1}{1 + e^{-2x}} = \frac{1}{1 + \displaystyle\lim_{x \to \infty} e^{-2x}}$$

$$= \frac{1}{1 + 0} = 1$$

We have used the fact that $t = -2x \to -\infty$ as $x \to \infty$ and so

$$\lim_{x \to \infty} e^{-2x} = \lim_{t \to -\infty} e^t = 0$$

EXAMPLE 7 Use the first and second derivatives of $f(x) = e^{1/x}$, together with asymptotes, to sketch its graph.

SOLUTION Notice that the domain of f is $\{x \mid x \neq 0\}$, so we check for vertical asymptotes by computing the left and right limits as $x \to 0$. As $x \to 0^+$, we know that $t = 1/x \to \infty$, so

$$\lim_{x \to 0^+} e^{1/x} = \lim_{t \to \infty} e^t = \infty$$

and this shows that $x = 0$ is a vertical asymptote. As $x \to 0^-$, we have $t = 1/x \to -\infty$, so

$$\lim_{x \to 0^-} e^{1/x} = \lim_{t \to -\infty} e^t = 0$$

As $x \to \pm\infty$, we have $1/x \to 0$ and so

$$\lim_{x \to \pm\infty} e^{1/x} = e^0 = 1$$

This shows that $y = 1$ is a horizontal asymptote.

Now let's compute the derivative. The Chain Rule gives

$$f'(x) = -\frac{e^{1/x}}{x^2}$$

Since $e^{1/x} > 0$ and $x^2 > 0$ for all $x \neq 0$, we have $f'(x) < 0$ for all $x \neq 0$. Thus f is decreasing on $(-\infty, 0)$ and on $(0, \infty)$. There is no critical number, so the function has no maximum or minimum. The second derivative is

$$f''(x) = -\frac{x^2 e^{1/x}(-1/x^2) - e^{1/x}(2x)}{x^4} = \frac{e^{1/x}(2x + 1)}{x^4}$$

Since $e^{1/x} > 0$ and $x^4 > 0$, we have $f''(x) > 0$ when $x > -\frac{1}{2}$ ($x \neq 0$) and $f''(x) < 0$ when $x < -\frac{1}{2}$. So the curve is concave downward on $\left(-\infty, -\frac{1}{2}\right)$ and concave upward on $\left(-\frac{1}{2}, 0\right)$ and on $(0, \infty)$. The inflection point is $\left(-\frac{1}{2}, e^{-2}\right)$.

To sketch the graph of f we first draw the horizontal asymptote $y = 1$ (as a dashed line), together with the parts of the curve near the asymptotes in a preliminary sketch [Figure 13(a)]. These parts reflect the information concerning limits and the fact that f is decreasing on both $(-\infty, 0)$ and $(0, \infty)$. Notice that we have indicated that $f(x) \to 0$ as $x \to 0^-$ even though $f(0)$ does not exist. In Figure 13(b) we finish the sketch by incorporating the information concerning concavity and the inflection point. In Figure 13(c) we check our work with a graphing device.

(a) Preliminary sketch

(b) Finished sketch

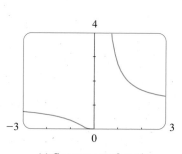

(c) Computer confirmation

FIGURE 13

Integration

Because the exponential function $y = e^x$ has a simple derivative, its integral is also simple:

11
$$\int e^x \, dx = e^x + C$$

V **EXAMPLE 8** Evaluate $\int x^2 e^{x^3}\, dx$.

SOLUTION We substitute $u = x^3$. Then $du = 3x^2\, dx$, so $x^2\, dx = \frac{1}{3}\, du$ and

$$\int x^2 e^{x^3}\, dx = \frac{1}{3}\int e^u\, du = \frac{1}{3}e^u + C = \frac{1}{3}e^{x^3} + C$$

EXAMPLE 9 Find the area under the curve $y = e^{-3x}$ from 0 to 1.

SOLUTION The area is

$$A = \int_0^1 e^{-3x}\, dx = -\frac{1}{3}e^{-3x}\Big]_0^1 = \frac{1}{3}(1 - e^{-3})$$

6.2 Exercises

Access end-of-section exercises and tools for visualizing calculus online at **www.webassign.net**

6.3 Logarithmic Functions

If $a > 0$ and $a \neq 1$, the exponential function $f(x) = a^x$ is either increasing or decreasing and so it is one-to-one. It therefore has an inverse function f^{-1}, which is called the **logarithmic function with base a** and is denoted by $\log_a$. If we use the formulation of an inverse function given by (6.1.3),

$$f^{-1}(x) = y \iff f(y) = x$$

then we have

1

$$\log_a x = y \iff a^y = x$$

Thus, if $x > 0$, then $\log_a x$ is the exponent to which the base a must be raised to give x.

EXAMPLE 1 Evaluate (a) $\log_3 81$, (b) $\log_{25} 5$, and (c) $\log_{10} 0.001$.

SOLUTION
(a) $\log_3 81 = 4$ because $3^4 = 81$
(b) $\log_{25} 5 = \frac{1}{2}$ because $25^{1/2} = 5$
(c) $\log_{10} 0.001 = -3$ because $10^{-3} = 0.001$

The cancellation equations (6.1.4), when applied to $f(x) = a^x$ and $f^{-1}(x) = \log_a x$, become

2

$$\log_a(a^x) = x \quad \text{for every } x \in \mathbb{R}$$

$$a^{\log_a x} = x \quad \text{for every } x > 0$$

FIGURE 1

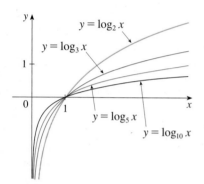

FIGURE 2

The logarithmic function $\log_a$ has domain $(0, \infty)$ and range $\mathbb{R}$ and is continuous since it is the inverse of a continuous function, namely, the exponential function. Its graph is the reflection of the graph of $y = a^x$ about the line $y = x$.

Figure 1 shows the case where $a > 1$. (The most important logarithmic functions have base $a > 1$.) The fact that $y = a^x$ is a very rapidly increasing function for $x > 0$ is reflected in the fact that $y = \log_a x$ is a very slowly increasing function for $x > 1$.

Figure 2 shows the graphs of $y = \log_a x$ with various values of the base a. Since $\log_a 1 = 0$, the graphs of all logarithmic functions pass through the point $(1, 0)$.

The following theorem summarizes the properties of logarithmic functions.

---

3   **Theorem**   If $a > 1$, the function $f(x) = \log_a x$ is a one-to-one, continuous, increasing function with domain $(0, \infty)$ and range $\mathbb{R}$. If $x, y > 0$ and r is any real number, then

1. $\log_a(xy) = \log_a x + \log_a y$

2. $\log_a\left(\dfrac{x}{y}\right) = \log_a x - \log_a y$

3. $\log_a(x^r) = r \log_a x$

Properties 1, 2, and 3 follow from the corresponding properties of exponential functions given in Section 6.2.

EXAMPLE 2   Use the properties of logarithms in Theorem 3 to evaluate the following.
(a) $\log_4 2 + \log_4 32$ (b) $\log_2 80 - \log_2 5$

SOLUTION
(a) Using Property 1 in Theorem 3, we have

$$\log_4 2 + \log_4 32 = \log_4(2 \cdot 32) = \log_4 64 = 3$$

since $4^3 = 64$.

(b) Using Property 2 we have

$$\log_2 80 - \log_2 5 = \log_2\left(\tfrac{80}{5}\right) = \log_2 16 = 4$$

since $2^4 = 16$.

The limits of exponential functions given in Section 6.2 are reflected in the following limits of logarithmic functions. (Compare with Figure 1.)

---

4   If $a > 1$, then

$$\lim_{x \to \infty} \log_a x = \infty \qquad \text{and} \qquad \lim_{x \to 0^+} \log_a x = -\infty$$

In particular, the y-axis is a vertical asymptote of the curve $y = \log_a x$.

EXAMPLE 3   Find $\lim_{x \to 0} \log_{10}(\tan^2 x)$.

SOLUTION As $x \to 0$, we know that $t = \tan^2 x \to \tan^2 0 = 0$ and the values of t are positive. So by $\boxed{4}$ with $a = 10 > 1$, we have

$$\lim_{x \to 0} \log_{10}(\tan^2 x) = \lim_{t \to 0^+} \log_{10} t = -\infty \qquad \blacksquare$$

▨ Natural Logarithms

Notation for Logarithms

Most textbooks in calculus and the sciences, as well as calculators, use the notation $\ln x$ for the natural logarithm and $\log x$ for the "common logarithm," $\log_{10} x$. In the more advanced mathematical and scientific literature and in computer languages, however, the notation $\log x$ usually denotes the natural logarithm.

Of all possible bases a for logarithms, we will see in the next section that the most convenient choice of a base is the number e, which was defined in Section 6.2. The logarithm with base e is called the **natural logarithm** and has a special notation:

$$\log_e x = \ln x$$

If we put $a = e$ and replace $\log_e$ with "ln" in $\boxed{1}$ and $\boxed{2}$, then the defining properties of the natural logarithm function become

$\boxed{5}$

$$\ln x = y \quad \Longleftrightarrow \quad e^y = x$$

$\boxed{6}$

$$\ln(e^x) = x \qquad x \in \mathbb{R}$$
$$e^{\ln x} = x \qquad x > 0$$

In particular, if we set $x = 1$, we get

$$\ln e = 1$$

EXAMPLE 4 Find x if $\ln x = 5$.

SOLUTION 1 From $\boxed{5}$ we see that

$$\ln x = 5 \qquad \text{means} \qquad e^5 = x$$

Therefore $x = e^5$.

(If you have trouble working with the "ln" notation, just replace it by $\log_e$. Then the equation becomes $\log_e x = 5$; so, by the definition of logarithm, $e^5 = x$.)

SOLUTION 2 Start with the equation

$$\ln x = 5$$

and apply the exponential function to both sides of the equation:

$$e^{\ln x} = e^5$$

But the second cancellation equation in $\boxed{6}$ says that $e^{\ln x} = x$. Therefore $x = e^5$. $\qquad \blacksquare$

V **EXAMPLE 5** Solve the equation $e^{5-3x} = 10$.

SOLUTION We take natural logarithms of both sides of the equation and use $\boxed{6}$:

$$\ln(e^{5-3x}) = \ln 10$$

$$5 - 3x = \ln 10$$

$$3x = 5 - \ln 10$$

$$x = \tfrac{1}{3}(5 - \ln 10)$$

Since the natural logarithm is found on scientific calculators, we can approximate the solution: to four decimal places, $x \approx 0.8991$. ▬

V **EXAMPLE 6** Express $\ln a + \tfrac{1}{2} \ln b$ as a single logarithm.

SOLUTION Using Properties 3 and 1 of logarithms, we have

$$\ln a + \tfrac{1}{2} \ln b = \ln a + \ln b^{1/2}$$

$$= \ln a + \ln \sqrt{b}$$

$$= \ln\left(a\sqrt{b}\right)$$ ▬

The following formula shows that logarithms with any base can be expressed in terms of the natural logarithm.

$\boxed{7}$ **Change of Base Formula** For any positive number a $(a \neq 1)$, we have

$$\log_a x = \frac{\ln x}{\ln a}$$

PROOF Let $y = \log_a x$. Then, from $\boxed{1}$, we have $a^y = x$. Taking natural logarithms of both sides of this equation, we get $y \ln a = \ln x$. Therefore

$$y = \frac{\ln x}{\ln a}$$ ▬

Scientific calculators have a key for natural logarithms, so Formula 7 enables us to use a calculator to compute a logarithm with any base (as shown in the following example). Similarly, Formula 7 allows us to graph any logarithmic function on a graphing calculator or computer (see Exercises 20–22).

EXAMPLE 7 Evaluate $\log_8 5$ correct to six decimal places.

SOLUTION Formula 7 gives

$$\log_8 5 = \frac{\ln 5}{\ln 8} \approx 0.773976$$ ▬

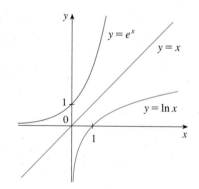

FIGURE 3
The graph of $y = \ln x$ is the reflection of the graph of $y = e^x$ about the line $y = x$.

Graph and Growth of the Natural Logarithm

The graphs of the exponential function $y = e^x$ and its inverse function, the natural logarithm function, are shown in Figure 3. Because the curve $y = e^x$ crosses the y-axis with a slope of 1, it follows that the reflected curve $y = \ln x$ crosses the x-axis with a slope of 1.

In common with all other logarithmic functions with base greater than 1, the natural logarithm is a continuous, increasing function defined on $(0, \infty)$ and the y-axis is a vertical asymptote.

If we put $a = e$ in $\boxed{4}$, then we have the following limits:

$\boxed{8}$

$$\lim_{x \to \infty} \ln x = \infty \qquad \lim_{x \to 0^+} \ln x = -\infty$$

V **EXAMPLE 8** Sketch the graph of the function $y = \ln(x - 2) - 1$.

SOLUTION We start with the graph of $y = \ln x$ as given in Figure 3. Using the transformations of Section 1.3, we shift it 2 units to the right to get the graph of $y = \ln(x - 2)$ and then we shift it 1 unit downward to get the graph of $y = \ln(x - 2) - 1$. (See Figure 4.)

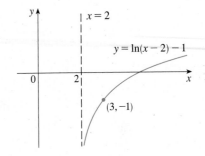

FIGURE 4

Notice that the line $x = 2$ is a vertical asymptote since

$$\lim_{x \to 2^+} \left[\ln(x - 2) - 1 \right] = -\infty$$

We have seen that $\ln x \to \infty$ as $x \to \infty$. But this happens *very* slowly. In fact, $\ln x$ grows more slowly than any positive power of x. To illustrate this fact, we compare approximate values of the functions $y = \ln x$ and $y = x^{1/2} = \sqrt{x}$ in the following table and we graph them in Figures 5 and 6.

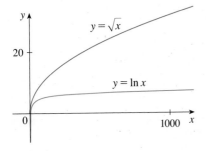

FIGURE 5

x	1	2	5	10	50	100	500	1000	10,000	100,000
$\ln x$	0	0.69	1.61	2.30	3.91	4.6	6.2	6.9	9.2	11.5
$\sqrt{x}$	1	1.41	2.24	3.16	7.07	10.0	22.4	31.6	100	316
$\dfrac{\ln x}{\sqrt{x}}$	0	0.49	0.72	0.73	0.55	0.46	0.28	0.22	0.09	0.04

You can see that initially the graphs of $y = \sqrt{x}$ and $y = \ln x$ grow at comparable rates, but eventually the root function far surpasses the logarithm. In fact, we will be able to show in Section 6.8 that

$$\lim_{x \to \infty} \frac{\ln x}{x^p} = 0$$

for any positive power p. So for large x, the values of $\ln x$ are very small compared with x^p. (See Exercise 72.)

FIGURE 6

6.3 Exercises

Access end-of-section exercises and tools for visualizing calculus online at **www.webassign.net**

6.4 Derivatives of Logarithmic Functions

In this section we find the derivatives of the logarithmic functions $y = \log_a x$ and the exponential functions $y = a^x$. We start with the natural logarithmic function $y = \ln x$. We know that it is differentiable because it is the inverse of the differentiable function $y = e^x$.

$\boxed{1}$

$$\frac{d}{dx}(\ln x) = \frac{1}{x}$$

PROOF Let $y = \ln x$. Then

$$e^y = x$$

Differentiating this equation implicitly with respect to x, we get

$$e^y \frac{dy}{dx} = 1$$

and so

$$\frac{dy}{dx} = \frac{1}{e^y} = \frac{1}{x}$$

V **EXAMPLE 1** Differentiate $y = \ln(x^3 + 1)$.

SOLUTION To use the Chain Rule, we let $u = x^3 + 1$. Then $y = \ln u$, so

$$\frac{dy}{dx} = \frac{dy}{du} \frac{du}{dx} = \frac{1}{u} \frac{du}{dx} = \frac{1}{x^3 + 1}(3x^2) = \frac{3x^2}{x^3 + 1}$$

In general, if we combine Formula 1 with the Chain Rule as in Example 1, we get

$\boxed{2}$

$$\frac{d}{dx}(\ln u) = \frac{1}{u}\frac{du}{dx}$$ or $$\frac{d}{dx}[\ln g(x)] = \frac{g'(x)}{g(x)}$$

V **EXAMPLE 2** Find $\dfrac{d}{dx}\ln(\sin x)$.

SOLUTION Using $\boxed{2}$, we have

$$\frac{d}{dx}\ln(\sin x) = \frac{1}{\sin x}\frac{d}{dx}(\sin x) = \frac{1}{\sin x}\cos x = \cot x$$

EXAMPLE 3 Differentiate $f(x) = \sqrt{\ln x}$.

SOLUTION This time the logarithm is the inner function, so the Chain Rule gives

$$f'(x) = \tfrac{1}{2}(\ln x)^{-1/2} \frac{d}{dx}(\ln x) = \frac{1}{2\sqrt{\ln x}} \cdot \frac{1}{x} = \frac{1}{2x\sqrt{\ln x}}$$

EXAMPLE 4 Find $\dfrac{d}{dx} \ln \dfrac{x + 1}{\sqrt{x - 2}}$.

SOLUTION 1

$$\frac{d}{dx} \ln \frac{x + 1}{\sqrt{x - 2}} = \frac{1}{\dfrac{x + 1}{\sqrt{x - 2}}} \frac{d}{dx} \frac{x + 1}{\sqrt{x - 2}}$$

$$= \frac{\sqrt{x - 2}}{x + 1} \frac{\sqrt{x - 2} \cdot 1 - (x + 1)(\tfrac{1}{2})(x - 2)^{-1/2}}{x - 2}$$

$$= \frac{x - 2 - \tfrac{1}{2}(x + 1)}{(x + 1)(x - 2)} = \frac{x - 5}{2(x + 1)(x - 2)}$$

Figure 1 shows the graph of the function f of Example 4 together with the graph of its derivative. It gives a visual check on our calculation. Notice that $f'(x)$ is large negative when f is rapidly decreasing and $f'(x) = 0$ when f has a minimum.

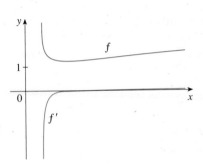

FIGURE 1

SOLUTION 2 If we first simplify the given function using the laws of logarithms, then the differentiation becomes easier:

$$\frac{d}{dx} \ln \frac{x + 1}{\sqrt{x - 2}} = \frac{d}{dx}\left[\ln(x + 1) - \tfrac{1}{2}\ln(x - 2)\right] = \frac{1}{x + 1} - \frac{1}{2}\left(\frac{1}{x - 2}\right)$$

(This answer can be left as written, but if we used a common denominator we would see that it gives the same answer as in Solution 1.)

EXAMPLE 5 Find the absolute minimum value of $f(x) = x^2 \ln x$.

SOLUTION The domain is $(0, \infty)$ and the Product Rule gives

$$f'(x) = x^2 \cdot \frac{1}{x} + 2x \ln x = x(1 + 2 \ln x)$$

Therefore $f'(x) = 0$ when $2 \ln x = -1$, that is, $\ln x = -\tfrac{1}{2}$, or $x = e^{-1/2}$. Also, $f'(x) > 0$ when $x > e^{-1/2}$ and $f'(x) < 0$ for $0 < x < e^{-1/2}$. So, by the First Derivative Test for Absolute Extreme Values, $f(1/\sqrt{e}) = -1/(2e)$ is the absolute minimum.

EXAMPLE 6 Discuss the curve $y = \ln(4 - x^2)$ using the guidelines of Section 3.5.

SOLUTION
A. The domain is

$$\{x \mid 4 - x^2 > 0\} = \{x \mid x^2 < 4\} = \{x \mid |x| < 2\} = (-2, 2)$$

B. The y-intercept is $f(0) = \ln 4$. To find the x-intercept we set

$$y = \ln(4 - x^2) = 0$$

We know that $\ln 1 = \log_e 1 = 0$ (since $e^0 = 1$), so we have $4 - x^2 = 1 \Rightarrow x^2 = 3$ and therefore the x-intercepts are $\pm\sqrt{3}$.
C. Since $f(-x) = f(x)$, f is even and the curve is symmetric about the y-axis.
D. We look for vertical asymptotes at the endpoints of the domain. Since $4 - x^2 \to 0^+$ as $x \to 2^-$ and also as $x \to -2^+$, we have

$$\lim_{x \to 2^-} \ln(4 - x^2) = -\infty \quad \text{and} \quad \lim_{x \to -2^+} \ln(4 - x^2) = -\infty$$

by (6.3.8). Thus the lines $x = 2$ and $x = -2$ are vertical asymptotes.

FIGURE 2
$y = \ln(4 - x^2)$

E.

$$f'(x) = \frac{-2x}{4 - x^2}$$

Since $f'(x) > 0$ when $-2 < x < 0$ and $f'(x) < 0$ when $0 < x < 2$, f is increasing on $(-2, 0)$ and decreasing on $(0, 2)$.

F. The only critical number is $x = 0$. Since f' changes from positive to negative at 0, $f(0) = \ln 4$ is a local maximum by the First Derivative Test.

G.

$$f''(x) = \frac{(4 - x^2)(-2) + 2x(-2x)}{(4 - x^2)^2} = \frac{-8 - 2x^2}{(4 - x^2)^2}$$

Since $f''(x) < 0$ for all x, the curve is concave downward on $(-2, 2)$ and has no inflection point.

H. Using this information, we sketch the curve in Figure 2.

V **EXAMPLE 7** Find $f'(x)$ if $f(x) = \ln|x|$.

SOLUTION Since

$$f(x) = \begin{cases} \ln x & \text{if } x > 0 \\ \ln(-x) & \text{if } x < 0 \end{cases}$$

it follows that

$$f'(x) = \begin{cases} \dfrac{1}{x} & \text{if } x > 0 \\ \dfrac{1}{-x}(-1) = \dfrac{1}{x} & \text{if } x < 0 \end{cases}$$

Thus $f'(x) = 1/x$ for all $x \neq 0$.

The result of Example 7 is worth remembering:

3

$$\frac{d}{dx}\left(\ln|x|\right) = \frac{1}{x}$$

The corresponding integration formula is

4

$$\int \frac{1}{x}\,dx = \ln|x| + C$$

Notice that this fills the gap in the rule for integrating power functions:

$$\int x^n\,dx = \frac{x^{n+1}}{n + 1} + C \qquad \text{if } n \neq -1$$

The missing case $(n = -1)$ is supplied by Formula 4.

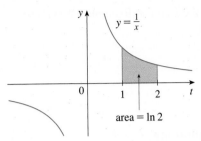

FIGURE 3

area = ln 2

EXAMPLE 8 Find, correct to three decimal places, the area of the region under the hyperbola $xy = 1$ from $x = 1$ to $x = 2$.

SOLUTION The given region is shown in Figure 3. Using Formula 4 (without the absolute value sign, since $x > 0$), we see that the area is

$$A = \int_1^2 \frac{1}{x}\,dx = \ln x \Big]_1^2$$

$$= \ln 2 - \ln 1 = \ln 2 \approx 0.693$$

V EXAMPLE 9 Evaluate $\int \dfrac{x}{x^2 + 1}\,dx$.

SOLUTION We make the substitution $u = x^2 + 1$ because the differential $du = 2x\,dx$ occurs (except for the constant factor 2). Thus $x\,dx = \frac{1}{2}\,du$ and

$$\int \frac{x}{x^2 + 1}\,dx = \frac{1}{2}\int \frac{du}{u} = \frac{1}{2}\ln|u| + C$$

$$= \frac{1}{2}\ln|x^2 + 1| + C = \frac{1}{2}\ln(x^2 + 1) + C$$

Notice that we removed the absolute value signs because $x^2 + 1 > 0$ for all x. We could use the properties of logarithms to write the answer as

$$\ln \sqrt{x^2 + 1} + C$$

but this isn't necessary.

Since the function $f(x) = (\ln x)/x$ in Example 10 is positive for $x > 1$, the integral represents the area of the shaded region in Figure 4.

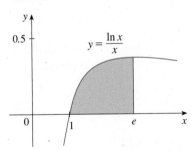

FIGURE 4

V EXAMPLE 10 Calculate $\int_1^e \dfrac{\ln x}{x}\,dx$.

SOLUTION We let $u = \ln x$ because its differential $du = dx/x$ occurs in the integral. When $x = 1$, $u = \ln 1 = 0$; when $x = e$, $u = \ln e = 1$. Thus

$$\int_1^e \frac{\ln x}{x}\,dx = \int_0^1 u\,du = \frac{u^2}{2}\bigg]_0^1 = \frac{1}{2}$$

V EXAMPLE 11 Calculate $\int \tan x\,dx$.

SOLUTION First we write tangent in terms of sine and cosine:

$$\int \tan x\,dx = \int \frac{\sin x}{\cos x}\,dx$$

This suggests that we should substitute $u = \cos x$ since then $du = -\sin x\,dx$ and so $\sin x\,dx = -du$:

$$\int \tan x\,dx = \int \frac{\sin x}{\cos x}\,dx = -\int \frac{du}{u}$$

$$= -\ln|u| + C = -\ln|\cos x| + C$$

Since $-\ln|\cos x| = \ln(1/|\cos x|) = \ln|\sec x|$, the result of Example 11 can also be written as

5
$$\int \tan x \, dx = \ln|\sec x| + C$$

▇ General Logarithmic and Exponential Functions

Formula 7 in Section 6.3 expresses a logarithmic function with base a in terms of the natural logarithmic function:

$$\log_a x = \frac{\ln x}{\ln a}$$

Since $\ln a$ is a constant, we can differentiate as follows:

$$\frac{d}{dx}(\log_a x) = \frac{d}{dx}\frac{\ln x}{\ln a} = \frac{1}{\ln a}\frac{d}{dx}(\ln x) = \frac{1}{x \ln a}$$

6
$$\frac{d}{dx}(\log_a x) = \frac{1}{x \ln a}$$

EXAMPLE 12 Using Formula 6 and the Chain Rule, we get

$$\frac{d}{dx}\log_{10}(2 + \sin x) = \frac{1}{(2 + \sin x)\ln 10}\frac{d}{dx}(2 + \sin x) = \frac{\cos x}{(2 + \sin x)\ln 10}$$ ▇

From Formula 6 we see one of the main reasons that natural logarithms (logarithms with base e) are used in calculus: The differentiation formula is simplest when $a = e$ because $\ln e = 1$.

EXPONENTIAL FUNCTIONS WITH BASE a In Section 6.2 we showed that the derivative of the general exponential function $f(x) = a^x$, $a > 0$, is a constant multiple of itself:

$$f'(x) = f'(0)a^x \qquad \text{where} \qquad f'(0) = \lim_{h \to 0}\frac{a^h - 1}{h}$$

We are now in a position to show that the value of the constant is $f'(0) = \ln a$.

7
$$\frac{d}{dx}(a^x) = a^x \ln a$$

PROOF We use the fact that $e^{\ln a} = a$:

$$\frac{d}{dx}(a^x) = \frac{d}{dx}(e^{\ln a})^x = \frac{d}{dx}e^{(\ln a)x} = e^{(\ln a)x}\frac{d}{dx}(\ln a)x$$

$$= (e^{\ln a})^x(\ln a) = a^x \ln a \qquad ▇$$

In Example 6 in Section 2.7 we considered a population of bacteria cells that doubles every hour and we saw that the population after t hours is $n = n_0 2^t$, where n_0 is the initial population. Formula 7 enables us to find the growth rate:

$$\frac{dn}{dt} = n_0 2^t \ln 2$$

EXAMPLE 13 Combining Formula 7 with the Chain Rule, we have

$$\frac{d}{dx}\left(10^{x^2}\right) = 10^{x^2}(\ln 10)\frac{d}{dx}\left(x^2\right) = (2\ln 10)x10^{x^2}$$

The integration formula that follows from Formula 7 is

$$\int a^x\,dx = \frac{a^x}{\ln a} + C \qquad a \neq 1$$

EXAMPLE 14

$$\int_0^5 2^x\,dx = \frac{2^x}{\ln 2}\bigg]_0^5 = \frac{2^5}{\ln 2} - \frac{2^0}{\ln 2} = \frac{31}{\ln 2}$$

Logarithmic Differentiation

The calculation of derivatives of complicated functions involving products, quotients, or powers can often be simplified by taking logarithms. The method used in the following example is called **logarithmic differentiation**.

EXAMPLE 15 Differentiate $y = \dfrac{x^{3/4}\sqrt{x^2 + 1}}{(3x + 2)^5}$.

SOLUTION We take logarithms of both sides of the equation and use the properties of logarithms to simplify:

$$\ln y = \tfrac{3}{4}\ln x + \tfrac{1}{2}\ln(x^2 + 1) - 5\ln(3x + 2)$$

Differentiating implicitly with respect to x gives

$$\frac{1}{y}\frac{dy}{dx} = \frac{3}{4}\cdot\frac{1}{x} + \frac{1}{2}\cdot\frac{2x}{x^2 + 1} - 5\cdot\frac{3}{3x + 2}$$

Solving for dy/dx, we get

$$\frac{dy}{dx} = y\left(\frac{3}{4x} + \frac{x}{x^2 + 1} - \frac{15}{3x + 2}\right)$$

Because we have an explicit expression for y, we can substitute and write

$$\frac{dy}{dx} = \frac{x^{3/4}\sqrt{x^2 + 1}}{(3x + 2)^5}\left(\frac{3}{4x} + \frac{x}{x^2 + 1} - \frac{15}{3x + 2}\right)$$

If we hadn't used logarithmic differentiation in Example 15, we would have had to use both the Quotient Rule and the Product Rule. The resulting calculation would have been horrendous.

> **Steps in Logarithmic Differentiation**
>
> **1.** Take natural logarithms of both sides of an equation $y = f(x)$ and use the properties of logarithms to simplify.
> **2.** Differentiate implicitly with respect to x.
> **3.** Solve the resulting equation for y'.

If $f(x) < 0$ for some values of x, then $\ln f(x)$ is not defined, but we can write $|y| = |f(x)|$ and use Equation 3. We illustrate this procedure by proving the general version of the Power Rule, as promised in Section 2.3.

> **The Power Rule** If n is any real number and $f(x) = x^n$, then
>
> $$f'(x) = nx^{n-1}$$

If $x = 0$, we can show that $f'(0) = 0$ for $n > 1$ directly from the definition of a derivative.

PROOF Let $y = x^n$ and use logarithmic differentiation:

$$\ln|y| = \ln|x|^n = n\ln|x| \qquad x \neq 0$$

Therefore
$$\frac{y'}{y} = \frac{n}{x}$$

Hence
$$y' = n\frac{y}{x} = n\frac{x^n}{x} = nx^{n-1}$$

⊘   You should distinguish carefully between the Power Rule $[(d/dx)\,x^n = nx^{n-1}]$, where the base is variable and the exponent is constant, and the rule for differentiating exponential functions $[(d/dx)\,a^x = a^x \ln a]$, where the base is constant and the exponent is variable.
In general there are four cases for exponents and bases:

Constant base, constant exponent
1. $\dfrac{d}{dx}(a^b) = 0$ (a and b are constants)

Variable base, constant exponent
2. $\dfrac{d}{dx}[f(x)]^b = b[f(x)]^{b-1}f'(x)$

Constant base, variable exponent
3. $\dfrac{d}{dx}[a^{g(x)}] = a^{g(x)}(\ln a)g'(x)$

Variable base, variable exponent
4. To find $(d/dx)[f(x)]^{g(x)}$, logarithmic differentiation can be used, as in the next example.

V   **EXAMPLE 16**   Differentiate $y = x^{\sqrt{x}}$.

SOLUTION 1   Since both the base and the exponent are variable, we use logarithmic differentiation:

$$\ln y = \ln x^{\sqrt{x}} = \sqrt{x}\,\ln x$$

$$\frac{y'}{y} = \sqrt{x} \cdot \frac{1}{x} + (\ln x)\frac{1}{2\sqrt{x}}$$

$$y' = y\left(\frac{1}{\sqrt{x}} + \frac{\ln x}{2\sqrt{x}}\right) = x^{\sqrt{x}}\left(\frac{2 + \ln x}{2\sqrt{x}}\right)$$

Figure 5 illustrates Example 16 by showing the graphs of $f(x) = x^{\sqrt{x}}$ and its derivative.

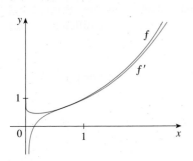

FIGURE 5

SOLUTION 2 Another method is to write $x^{\sqrt{x}} = \left(e^{\ln x}\right)^{\sqrt{x}}$:

$$\frac{d}{dx}\left(x^{\sqrt{x}}\right) = \frac{d}{dx}\left(e^{\sqrt{x}\,\ln x}\right) = e^{\sqrt{x}\,\ln x}\,\frac{d}{dx}\left(\sqrt{x}\,\ln x\right)$$

$$= x^{\sqrt{x}}\left(\frac{2 + \ln x}{2\sqrt{x}}\right) \qquad \text{(as in Solution 1)}$$

The Number e as a Limit

We have shown that if $f(x) = \ln x$, then $f'(x) = 1/x$. Thus $f'(1) = 1$. We now use this fact to express the number e as a limit.

From the definition of a derivative as a limit, we have

$$f'(1) = \lim_{h\to 0}\frac{f(1+h) - f(1)}{h} = \lim_{x\to 0}\frac{f(1+x) - f(1)}{x}$$

$$= \lim_{x\to 0}\frac{\ln(1+x) - \ln 1}{x} = \lim_{x\to 0}\frac{1}{x}\ln(1+x)$$

$$= \lim_{x\to 0}\ln(1+x)^{1/x}$$

Because $f'(1) = 1$, we have

$$\lim_{x\to 0}\ln(1+x)^{1/x} = 1$$

Then, by Theorem 1.8.8 and the continuity of the exponential function, we have

$$e = e^1 = e^{\lim_{x\to 0}\ln(1+x)^{1/x}} = \lim_{x\to 0}e^{\ln(1+x)^{1/x}} = \lim_{x\to 0}(1+x)^{1/x}$$

FIGURE 6

x	$(1+x)^{1/x}$
0.1	2.59374246
0.01	2.70481383
0.001	2.71692393
0.0001	2.71814593
0.00001	2.71826824
0.000001	2.71828047
0.0000001	2.71828169
0.00000001	2.71828181

8
$$e = \lim_{x\to 0}(1+x)^{1/x}$$

Formula 8 is illustrated by the graph of the function $y = (1+x)^{1/x}$ in Figure 6 and a table of values for small values of x.

If we put $n = 1/x$ in Formula 8, then $n \to \infty$ as $x \to 0^+$ and so an alternative expression for e is

9
$$e = \lim_{n\to\infty}\left(1 + \frac{1}{n}\right)^n$$

  Exercises

Access end-of-section exercises and tools for visualizing calculus online at **www.webassign.net**

6.2* The Natural Logarithmic Function

If your instructor has assigned Sections 6.2–6.4, you need not read Sections 6.2*, 6.3*, and 6.4* (pp. 332–350).

In this section we define the natural logarithm as an integral and then show that it obeys the usual laws of logarithms. The Fundamental Theorem makes it easy to differentiate this function.

> **1 Definition** The **natural logarithmic function** is the function defined by
>
> $$\ln x = \int_1^x \frac{1}{t}\, dt \qquad x > 0$$

The existence of this function depends on the fact that the integral of a continuous function always exists. If $x > 1$, then $\ln x$ can be interpreted geometrically as the area under the hyperbola $y = 1/t$ from $t = 1$ to $t = x$. (See Figure 1.) For $x = 1$, we have

$$\ln 1 = \int_1^1 \frac{1}{t}\, dt = 0$$

FIGURE 1

For $0 < x < 1$,

$$\ln x = \int_1^x \frac{1}{t}\, dt = -\int_x^1 \frac{1}{t}\, dt < 0$$

and so $\ln x$ is the negative of the area shaded in Figure 2.

FIGURE 2

V EXAMPLE 1

(a) By comparing areas, show that $\frac{1}{2} < \ln 2 < \frac{3}{4}$.

(b) Use the Midpoint Rule with $n = 10$ to estimate the value of $\ln 2$.

SOLUTION

(a) We can interpret $\ln 2$ as the area under the curve $y = 1/t$ from 1 to 2. From Figure 3 we see that this area is larger than the area of rectangle *BCDE* and smaller than the area of trapezoid *ABCD*. Thus we have

$$\frac{1}{2} \cdot 1 < \ln 2 < 1 \cdot \frac{1}{2}\left(1 + \frac{1}{2}\right)$$

$$\frac{1}{2} < \ln 2 < \frac{3}{4}$$

(b) If we use the Midpoint Rule with $f(t) = 1/t$, $n = 10$, and $\Delta t = 0.1$, we get

$$\ln 2 = \int_1^2 \frac{1}{t}\, dt \approx (0.1)[f(1.05) + f(1.15) + \cdots + f(1.95)]$$

$$= (0.1)\left(\frac{1}{1.05} + \frac{1}{1.15} + \cdots + \frac{1}{1.95}\right) \approx 0.693 \quad \blacksquare$$

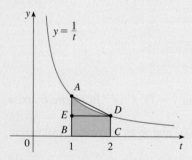

FIGURE 3

Notice that the integral that defines $\ln x$ is exactly the type of integral discussed in Part 1 of the Fundamental Theorem of Calculus (see Section 4.3). In fact, using that theorem, we have

$$\frac{d}{dx}\int_1^x \frac{1}{t}\, dt = \frac{1}{x}$$

and so

> **2**
> $$\frac{d}{dx}(\ln x) = \frac{1}{x}$$

We now use this differentiation rule to prove the following properties of the logarithm function.

> **3** **Laws of Logarithms** If x and y are positive numbers and r is a rational number, then
>
> **1.** $\ln(xy) = \ln x + \ln y$ **2.** $\ln\left(\dfrac{x}{y}\right) = \ln x - \ln y$ **3.** $\ln(x^r) = r \ln x$

PROOF

1. Let $f(x) = \ln(ax)$, where a is a positive constant. Then, using Equation 2 and the Chain Rule, we have

$$f'(x) = \frac{1}{ax} \frac{d}{dx}(ax) = \frac{1}{ax} \cdot a = \frac{1}{x}$$

Therefore $f(x)$ and $\ln x$ have the same derivative and so they must differ by a constant:

$$\ln(ax) = \ln x + C$$

Putting $x = 1$ in this equation, we get $\ln a = \ln 1 + C = 0 + C = C$. Thus

$$\ln(ax) = \ln x + \ln a$$

If we now replace the constant a by any number y, we have

$$\ln(xy) = \ln x + \ln y$$

2. Using Law 1 with $x = 1/y$, we have

$$\ln \frac{1}{y} + \ln y = \ln\left(\frac{1}{y} \cdot y\right) = \ln 1 = 0$$

and so

$$\ln \frac{1}{y} = -\ln y$$

Using Law 1 again, we have

$$\ln\left(\frac{x}{y}\right) = \ln\left(x \cdot \frac{1}{y}\right) = \ln x + \ln \frac{1}{y} = \ln x - \ln y$$

The proof of Law 3 is left as an exercise. ■

EXAMPLE 2 Expand the expression $\ln \dfrac{(x^2 + 5)^4 \sin x}{x^3 + 1}$.

SOLUTION Using Laws 1, 2, and 3, we get

$$\ln \frac{(x^2 + 5)^4 \sin x}{x^3 + 1} = \ln(x^2 + 5)^4 + \ln \sin x - \ln(x^3 + 1)$$

$$= 4 \ln(x^2 + 5) + \ln \sin x - \ln(x^3 + 1)$$ ■

V EXAMPLE 3 Express $\ln a + \frac{1}{2} \ln b$ as a single logarithm.

SOLUTION Using Laws 3 and 1 of logarithms, we have

$$\ln a + \tfrac{1}{2} \ln b = \ln a + \ln b^{1/2} = \ln a + \ln \sqrt{b} = \ln\!\left(a\sqrt{b}\right)$$

In order to graph $y = \ln x$, we first determine its limits:

4
$$\text{(a) } \lim_{x \to \infty} \ln x = \infty \qquad \text{(b) } \lim_{x \to 0^+} \ln x = -\infty$$

PROOF

(a) Using Law 3 with $x = 2$ and $r = n$ (where n is any positive integer), we have $\ln(2^n) = n \ln 2$. Now $\ln 2 > 0$, so this shows that $\ln(2^n) \to \infty$ as $n \to \infty$. But $\ln x$ is an increasing function since its derivative $1/x$ is positive. Therefore $\ln x \to \infty$ as $x \to \infty$.

(b) If we let $t = 1/x$, then $t \to \infty$ as $x \to 0^+$. Thus, using (a), we have

$$\lim_{x \to 0^+} \ln x = \lim_{t \to \infty} \ln\!\left(\frac{1}{t}\right) = \lim_{t \to \infty} (-\ln t) = -\infty$$

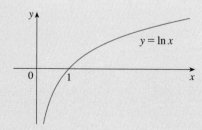

FIGURE 4

If $y = \ln x, \ x > 0$, then

$$\frac{dy}{dx} = \frac{1}{x} > 0 \qquad \text{and} \qquad \frac{d^2y}{dx^2} = -\frac{1}{x^2} < 0$$

which shows that $\ln x$ is increasing and concave downward on $(0, \infty)$. Putting this information together with **4**, we draw the graph of $y = \ln x$ in Figure 4.

Since $\ln 1 = 0$ and $\ln x$ is an increasing continuous function that takes on arbitrarily large values, the Intermediate Value Theorem shows that there is a number where $\ln x$ takes on the value 1. (See Figure 5.) This important number is denoted by e.

FIGURE 5

5 Definition e is the number such that $\ln e = 1$.

EXAMPLE 4 Use a graphing calculator or computer to estimate the value of e.

SOLUTION According to Definition 5, we estimate the value of e by graphing the curves $y = \ln x$ and $y = 1$ and determining the x-coordinate of the point of intersection. By zooming in repeatedly, as in Figure 6, we find that

$$e \approx 2.718$$

With more sophisticated methods, it can be shown that the approximate value of e, to 20 decimal places, is

$$e \approx 2.71828182845904523536$$

FIGURE 6

The decimal expansion of e is nonrepeating because e is an irrational number.

Now let's use Formula 2 to differentiate functions that involve the natural logarithmic function.

V EXAMPLE 5 Differentiate $y = \ln(x^3 + 1)$.

SOLUTION To use the Chain Rule, we let $u = x^3 + 1$. Then $y = \ln u$, so

$$\frac{dy}{dx} = \frac{dy}{du} \frac{du}{dx} = \frac{1}{u} \frac{du}{dx} = \frac{1}{x^3 + 1} (3x^2) = \frac{3x^2}{x^3 + 1}$$

In general, if we combine Formula 2 with the Chain Rule as in Example 5, we get

$$\boxed{6} \qquad \boxed{\frac{d}{dx} (\ln u) = \frac{1}{u} \frac{du}{dx}} \quad \text{or} \quad \boxed{\frac{d}{dx} [\ln g(x)] = \frac{g'(x)}{g(x)}}$$

V **EXAMPLE 6** Find $\dfrac{d}{dx} \ln(\sin x)$.

SOLUTION Using $\boxed{6}$, we have

$$\frac{d}{dx} \ln(\sin x) = \frac{1}{\sin x} \frac{d}{dx} (\sin x) = \frac{1}{\sin x} \cos x = \cot x$$

EXAMPLE 7 Differentiate $f(x) = \sqrt{\ln x}$.

SOLUTION This time the logarithm is the inner function, so the Chain Rule gives

$$f'(x) = \tfrac{1}{2}(\ln x)^{-1/2} \frac{d}{dx} (\ln x) = \frac{1}{2\sqrt{\ln x}} \cdot \frac{1}{x} = \frac{1}{2x\sqrt{\ln x}}$$

EXAMPLE 8 Find $\dfrac{d}{dx} \ln \dfrac{x + 1}{\sqrt{x - 2}}$.

SOLUTION 1

$$\frac{d}{dx} \ln \frac{x + 1}{\sqrt{x - 2}} = \frac{1}{\dfrac{x + 1}{\sqrt{x - 2}}} \frac{d}{dx} \frac{x + 1}{\sqrt{x - 2}}$$

$$= \frac{\sqrt{x - 2}}{x + 1} \frac{\sqrt{x - 2} \cdot 1 - (x + 1)(\tfrac{1}{2})(x - 2)^{-1/2}}{x - 2}$$

$$= \frac{x - 2 - \tfrac{1}{2}(x + 1)}{(x + 1)(x - 2)}$$

$$= \frac{x - 5}{2(x + 1)(x - 2)}$$

Figure 7 shows the graph of the function f of Example 8 together with the graph of its derivative. It gives a visual check on our calculation. Notice that $f'(x)$ is large negative when f is rapidly decreasing and $f'(x) = 0$ when f has a minimum.

FIGURE 7

SOLUTION 2 If we first simplify the given function using the laws of logarithms, then the differentiation becomes easier:

$$\frac{d}{dx} \ln \frac{x + 1}{\sqrt{x - 2}} = \frac{d}{dx} \left[\ln(x + 1) - \tfrac{1}{2} \ln(x - 2) \right]$$

$$= \frac{1}{x + 1} - \frac{1}{2} \left(\frac{1}{x - 2} \right)$$

(This answer can be left as written, but if we used a common denominator we would see that it gives the same answer as in Solution 1.)

EXAMPLE 9 Discuss the curve $y = \ln(4 - x^2)$ using the guidelines of Section 3.5.

SOLUTION

A. The domain is

$$\{x \mid 4 - x^2 > 0\} = \{x \mid x^2 < 4\} = \{x \mid |x| < 2\} = (-2, 2)$$

B. The y-intercept is $f(0) = \ln 4$. To find the x-intercept we set

$$y = \ln(4 - x^2) = 0$$

We know that $\ln 1 = 0$, so we have $4 - x^2 = 1 \Rightarrow x^2 = 3$ and therefore the x-intercepts are $\pm\sqrt{3}$.

C. Since $f(-x) = f(x)$, f is even and the curve is symmetric about the y-axis.

D. We look for vertical asymptotes at the endpoints of the domain. Since $4 - x^2 \to 0^+$ as $x \to 2^-$ and also as $x \to -2^+$, we have

$$\lim_{x \to 2^-} \ln(4 - x^2) = -\infty \quad \text{and} \quad \lim_{x \to -2^+} \ln(4 - x^2) = -\infty$$

by $\boxed{4}$. Thus the lines $x = 2$ and $x = -2$ are vertical asymptotes.

E.
$$f'(x) = \frac{-2x}{4 - x^2}$$

$x = -2$ $(0, \ln 4)$ $x = 2$

$(-\sqrt{3}, 0)$ $(\sqrt{3}, 0)$

FIGURE 8

$y = \ln(4 - x^2)$

Since $f'(x) > 0$ when $-2 < x < 0$ and $f'(x) < 0$ when $0 < x < 2$, f is increasing on $(-2, 0)$ and decreasing on $(0, 2)$.

F. The only critical number is $x = 0$. Since f' changes from positive to negative at 0, $f(0) = \ln 4$ is a local maximum by the First Derivative Test.

G.
$$f''(x) = \frac{(4 - x^2)(-2) + 2x(-2x)}{(4 - x^2)^2} = \frac{-8 - 2x^2}{(4 - x^2)^2}$$

Since $f''(x) < 0$ for all x, the curve is concave downward on $(-2, 2)$ and has no inflection point.

H. Using this information, we sketch the curve in Figure 8. ▬

V **EXAMPLE 10** Find $f'(x)$ if $f(x) = \ln|x|$.

SOLUTION Since

$$f(x) = \begin{cases} \ln x & \text{if } x > 0 \\ \ln(-x) & \text{if } x < 0 \end{cases}$$

it follows that

$$f'(x) = \begin{cases} \dfrac{1}{x} & \text{if } x > 0 \\ \dfrac{1}{-x}(-1) = \dfrac{1}{x} & \text{if } x < 0 \end{cases}$$

Thus $f'(x) = 1/x$ for all $x \neq 0$. ▬

The result of Example 10 is worth remembering:

$\boxed{7}$

$$\frac{d}{dx}(\ln|x|) = \frac{1}{x}$$

The corresponding integration formula is

8

$$\int \frac{1}{x}\, dx = \ln|x| + C$$

Notice that this fills the gap in the rule for integrating power functions:

$$\int x^n\, dx = \frac{x^{n+1}}{n+1} + C \qquad \text{if } n \neq -1$$

The missing case ($n = -1$) is supplied by Formula 8.

V **EXAMPLE 11** Evaluate $\displaystyle\int \frac{x}{x^2 + 1}\, dx$.

SOLUTION We make the substitution $u = x^2 + 1$ because the differential $du = 2x\,dx$ occurs (except for the constant factor 2). Thus $x\,dx = \frac{1}{2}\,du$ and

$$\int \frac{x}{x^2 + 1}\, dx = \tfrac{1}{2} \int \frac{du}{u} = \tfrac{1}{2} \ln|u| + C$$

$$= \tfrac{1}{2} \ln|x^2 + 1| + C = \tfrac{1}{2} \ln(x^2 + 1) + C$$

Notice that we removed the absolute value signs because $x^2 + 1 > 0$ for all x. We could use the properties of logarithms to write the answer as

$$\ln \sqrt{x^2 + 1} + C$$

but this isn't necessary.

V **EXAMPLE 12** Calculate $\displaystyle\int_1^e \frac{\ln x}{x}\, dx$.

SOLUTION We let $u = \ln x$ because its differential $du = dx/x$ occurs in the integral. When $x = 1$, $u = \ln 1 = 0$; when $x = e$, $u = \ln e = 1$. Thus

$$\int_1^e \frac{\ln x}{x}\, dx = \int_0^1 u\, du = \frac{u^2}{2} \Bigg]_0^1 = \frac{1}{2}$$

V **EXAMPLE 13** Calculate $\displaystyle\int \tan x\, dx$.

SOLUTION First we write tangent in terms of sine and cosine:

$$\int \tan x\, dx = \int \frac{\sin x}{\cos x}\, dx$$

This suggests that we should substitute $u = \cos x$ since then $du = -\sin x\,dx$ and so $\sin x\,dx = -du$:

$$\int \tan x\, dx = \int \frac{\sin x}{\cos x}\, dx = -\int \frac{du}{u}$$

$$= -\ln|u| + C = -\ln|\cos x| + C$$

Since the function $f(x) = (\ln x)/x$ in Example 12 is positive for $x > 1$, the integral represents the area of the shaded region in Figure 9.

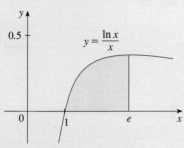

FIGURE 9

Since $-\ln|\cos x| = \ln(1/|\cos x|) = \ln|\sec x|$, the result of Example 13 can also be written as

$$\boxed{\quad \int \tan x\, dx = \ln|\sec x| + C \quad}$$

⑨

Logarithmic Differentiation

The calculation of derivatives of complicated functions involving products, quotients, or powers can often be simplified by taking logarithms. The method used in the following example is called **logarithmic differentiation**.

V EXAMPLE 14 Differentiate $y = \dfrac{x^{3/4}\sqrt{x^2 + 1}}{(3x + 2)^5}$.

SOLUTION We take logarithms of both sides of the equation and use the Laws of Logarithms to simplify:

$$\ln y = \tfrac{3}{4}\ln x + \tfrac{1}{2}\ln(x^2 + 1) - 5\ln(3x + 2)$$

Differentiating implicitly with respect to x gives

$$\frac{1}{y}\frac{dy}{dx} = \frac{3}{4}\cdot\frac{1}{x} + \frac{1}{2}\cdot\frac{2x}{x^2 + 1} - 5\cdot\frac{3}{3x + 2}$$

Solving for dy/dx, we get

$$\frac{dy}{dx} = y\left(\frac{3}{4x} + \frac{x}{x^2 + 1} - \frac{15}{3x + 2}\right)$$

If we hadn't used logarithmic differentiation in Example 14, we would have had to use both the Quotient Rule and the Product Rule. The resulting calculation would have been horrendous.

Because we have an explicit expression for y, we can substitute and write

$$\frac{dy}{dx} = \frac{x^{3/4}\sqrt{x^2 + 1}}{(3x + 2)^5}\left(\frac{3}{4x} + \frac{x}{x^2 + 1} - \frac{15}{3x + 2}\right)$$

━━

Steps in Logarithmic Differentiation

1. Take natural logarithms of both sides of an equation $y = f(x)$ and use the Laws of Logarithms to simplify.
2. Differentiate implicitly with respect to x.
3. Solve the resulting equation for y'.

If $f(x) < 0$ for some values of x, then $\ln f(x)$ is not defined, but we can write $|y| = |f(x)|$ and use Equation 7.

6.3* The Natural Exponential Function

Since ln is an increasing function, it is one-to-one and therefore has an inverse function, which we denote by exp. Thus, according to the definition of an inverse function,

$f^{-1}(x) = y \iff f(y) = x$

1 $\exp(x) = y \iff \ln y = x$

and the cancellation equations are

$f^{-1}(f(x)) = x$
$f(f^{-1}(x)) = x$

2 $\exp(\ln x) = x \quad \text{and} \quad \ln(\exp x) = x$

In particular, we have

$$\exp(0) = 1 \quad \text{since} \quad \ln 1 = 0$$

$$\exp(1) = e \quad \text{since} \quad \ln e = 1$$

We obtain the graph of $y = \exp x$ by reflecting the graph of $y = \ln x$ about the line $y = x$. (See Figure 1.) The domain of exp is the range of ln, that is, $(-\infty, \infty)$; the range of exp is the domain of ln, that is, $(0, \infty)$.

If r is any rational number, then the third law of logarithms gives

$$\ln(e^r) = r \ln e = r$$

Therefore, by **1**, $\exp(r) = e^r$

FIGURE 1

Thus $\exp(x) = e^x$ whenever x is a rational number. This leads us to define e^x, even for irrational values of x, by the equation

$$e^x = \exp(x)$$

In other words, for the reasons given, we define e^x to be the inverse of the function $\ln x$. In this notation **1** becomes

3 $e^x = y \iff \ln y = x$

and the cancellation equations **2** become

4 $e^{\ln x} = x \quad x > 0$

5 $\ln(e^x) = x \quad \text{for all } x$

EXAMPLE 1 Find x if $\ln x = 5$.

SOLUTION 1 From $\boxed{3}$ we see that

$$\ln x = 5 \qquad \text{means} \qquad e^5 = x$$

Therefore $x = e^5$.

SOLUTION 2 Start with the equation

$$\ln x = 5$$

and apply the exponential function to both sides of the equation:

$$e^{\ln x} = e^5$$

But $\boxed{4}$ says that $e^{\ln x} = x$. Therefore $x = e^5$.

V **EXAMPLE 2** Solve the equation $e^{5-3x} = 10$.

SOLUTION We take natural logarithms of both sides of the equation and use $\boxed{5}$:

$$\ln(e^{5-3x}) = \ln 10$$

$$5 - 3x = \ln 10$$

$$3x = 5 - \ln 10$$

$$x = \tfrac{1}{3}(5 - \ln 10)$$

Since the natural logarithm is found on scientific calculators, we can approximate the solution to four decimal places: $x \approx 0.8991$.

The exponential function $f(x) = e^x$ is one of the most frequently occurring functions in calculus and its applications, so it is important to be familiar with its graph (Figure 2) and its properties (which follow from the fact that it is the inverse of the natural logarithmic function).

FIGURE 2
The natural exponential function

$\boxed{6}$ **Properties of the Natural Exponential Function** The exponential function $f(x) = e^x$ is an increasing continuous function with domain $\mathbb{R}$ and range $(0, \infty)$. Thus $e^x > 0$ for all x. Also

$$\lim_{x \to -\infty} e^x = 0 \qquad \lim_{x \to \infty} e^x = \infty$$

So the x-axis is a horizontal asymptote of $f(x) = e^x$.

EXAMPLE 3 Find $\displaystyle\lim_{x \to \infty} \frac{e^{2x}}{e^{2x} + 1}$.

SOLUTION We divide numerator and denominator by e^{2x}:

$$\lim_{x \to \infty} \frac{e^{2x}}{e^{2x} + 1} = \lim_{x \to \infty} \frac{1}{1 + e^{-2x}} = \frac{1}{1 + \lim_{x \to \infty} e^{-2x}}$$

$$= \frac{1}{1 + 0} = 1$$

We have used the fact that $t = -2x \to -\infty$ as $x \to \infty$ and so

$$\lim_{x \to \infty} e^{-2x} = \lim_{t \to -\infty} e^t = 0$$

We now verify that $f(x) = e^x$ has the properties expected of an exponential function.

---

7   **Laws of Exponents**   If x and y are real numbers and r is rational, then

1. $e^{x+y} = e^x e^y$ **2.** $e^{x-y} = \dfrac{e^x}{e^y}$ **3.** $(e^x)^r = e^{rx}$

PROOF OF LAW 1   Using the first law of logarithms and Equation 5, we have

$$\ln(e^x e^y) = \ln(e^x) + \ln(e^y) = x + y = \ln(e^{x+y})$$

Since ln is a one-to-one function, it follows that $e^x e^y = e^{x+y}$.

Laws 2 and 3 are proved similarly (see Exercises 105 and 106). As we will see in the next section, Law 3 actually holds when r is any real number. ∎

Differentiation

The natural exponential function has the remarkable property that *it is its own derivative.*

TEC Visual 6.2/6.3* uses the slope-a-scope to illustrate this formula.

8
$$\frac{d}{dx}(e^x) = e^x$$

FIGURE 3

PROOF   The function $y = e^x$ is differentiable because it is the inverse function of $y = \ln x$, which we know is differentiable with nonzero derivative. To find its derivative, we use the inverse function method. Let $y = e^x$. Then $\ln y = x$ and, differentiating this latter equation implicitly with respect to x, we get

$$\frac{1}{y}\frac{dy}{dx} = 1$$

$$\frac{dy}{dx} = y = e^x$$ ∎

The geometric interpretation of Formula 8 is that the slope of a tangent line to the curve $y = e^x$ at any point is equal to the y-coordinate of the point (see Figure 3). This property implies that the exponential curve $y = e^x$ grows very rapidly (see Exercise 110).

V   **EXAMPLE 4**   Differentiate the function $y = e^{\tan x}$.

SOLUTION   To use the Chain Rule, we let $u = \tan x$. Then we have $y = e^u$, so

$$\frac{dy}{dx} = \frac{dy}{du}\frac{du}{dx} = e^u \frac{du}{dx} = e^{\tan x}\sec^2 x$$

In general, if we combine Formula 8 with the Chain Rule, as in Example 4, we get

9
$$\frac{d}{dx}(e^u) = e^u \frac{du}{dx}$$

EXAMPLE 5 Find y' if $y = e^{-4x} \sin 5x$.

SOLUTION Using Formula 9 and the Product Rule, we have

$$y' = e^{-4x}(\cos 5x)(5) + (\sin 5x)e^{-4x}(-4) = e^{-4x}(5\cos 5x - 4\sin 5x)$$

▼ EXAMPLE 6 Find the absolute maximum value of the function $f(x) = xe^{-x}$.

SOLUTION We differentiate to find any critical numbers:

$$f'(x) = xe^{-x}(-1) + e^{-x}(1) = e^{-x}(1 - x)$$

Since exponential functions are always positive, we see that $f'(x) > 0$ when $1 - x > 0$, that is, when $x < 1$. Similarly, $f'(x) < 0$ when $x > 1$. By the First Derivative Test for Absolute Extreme Values, f has an absolute maximum value when $x = 1$ and the value is

$$f(1) = (1)e^{-1} = \frac{1}{e} \approx 0.37$$

EXAMPLE 7 Use the first and second derivatives of $f(x) = e^{1/x}$, together with asymptotes, to sketch its graph.

SOLUTION Notice that the domain of f is $\{x \mid x \neq 0\}$, so we check for vertical asymptotes by computing the left and right limits as $x \to 0$. As $x \to 0^+$, we know that $t = 1/x \to \infty$, so

$$\lim_{x \to 0^+} e^{1/x} = \lim_{t \to \infty} e^t = \infty$$

and this shows that $x = 0$ is a vertical asymptote. As $x \to 0^-$, we have $t = 1/x \to -\infty$, so

$$\lim_{x \to 0^-} e^{1/x} = \lim_{t \to -\infty} e^t = 0$$

As $x \to \pm\infty$, we have $1/x \to 0$ and so

$$\lim_{x \to \pm\infty} e^{1/x} = e^0 = 1$$

This shows that $y = 1$ is a horizontal asymptote.

Now let's compute the derivative. The Chain Rule gives

$$f'(x) = -\frac{e^{1/x}}{x^2}$$

Since $e^{1/x} > 0$ and $x^2 > 0$ for all $x \neq 0$, we have $f'(x) < 0$ for all $x \neq 0$. Thus f is decreasing on $(-\infty, 0)$ and on $(0, \infty)$. There is no critical number, so the function has no maximum or minimum. The second derivative is

$$f''(x) = -\frac{x^2 e^{1/x}(-1/x^2) - e^{1/x}(2x)}{x^4} = \frac{e^{1/x}(2x + 1)}{x^4}$$

Since $e^{1/x} > 0$ and $x^4 > 0$, we have $f''(x) > 0$ when $x > -\frac{1}{2}$ $(x \neq 0)$ and $f''(x) < 0$ when $x < -\frac{1}{2}$. So the curve is concave downward on $\left(-\infty, -\frac{1}{2}\right)$ and concave upward on $\left(-\frac{1}{2}, 0\right)$ and on $(0, \infty)$. The inflection point is $\left(-\frac{1}{2}, e^{-2}\right)$.

To sketch the graph of f we first draw the horizontal asymptote $y = 1$ (as a dashed line), together with the parts of the curve near the asymptotes in a preliminary sketch [Figure 4(a)]. These parts reflect the information concerning limits and the fact that f is decreasing on both $(-\infty, 0)$ and $(0, \infty)$. Notice that we have indicated that $f(x) \to 0$ as $x \to 0^-$ even though $f(0)$ does not exist. In Figure 4(b) we finish the sketch by incorporating the information concerning concavity and the inflection point. In Figure 4(c) we check our work with a graphing device.

(a) Preliminary sketch

(b) Finished sketch

(c) Computer confirmation

FIGURE 4

Integration

Because the exponential function $y = e^x$ has a simple derivative, its integral is also simple:

$$\boxed{\int e^x \, dx = e^x + C}$$

10

V EXAMPLE 8 Evaluate $\int x^2 e^{x^3} \, dx$.

SOLUTION We substitute $u = x^3$. Then $du = 3x^2 \, dx$, so $x^2 \, dx = \frac{1}{3} \, du$ and

$$\int x^2 e^{x^3} \, dx = \frac{1}{3} \int e^u \, du = \frac{1}{3} e^u + C = \frac{1}{3} e^{x^3} + C$$

EXAMPLE 9 Find the area under the curve $y = e^{-3x}$ from 0 to 1.

SOLUTION The area is

$$A = \int_0^1 e^{-3x} \, dx = -\frac{1}{3} e^{-3x} \Big]_0^1 = \frac{1}{3}(1 - e^{-3})$$

6.3* Exercises

Access end-of-section exercises and tools for visualizing calculus online at **www.webassign.net**

6.4* General Logarithmic and Exponential Functions

In this section we use the natural exponential and logarithmic functions to study exponential and logarithmic functions with base $a > 0$.

General Exponential Functions

If $a > 0$ and r is any rational number, then by $\boxed{4}$ and $\boxed{7}$ in Section 6.3*,

$$a^r = (e^{\ln a})^r = e^{r \ln a}$$

Therefore, even for irrational numbers x, we *define*

$\boxed{1}$

$$a^x = e^{x \ln a}$$

Thus, for instance,

$$2^{\sqrt{3}} = e^{\sqrt{3} \ln 2} \approx e^{1.20} \approx 3.32$$

The function $f(x) = a^x$ is called the **exponential function with base a**. Notice that a^x is positive for all x because e^x is positive for all x.

Definition 1 allows us to extend one of the laws of logarithms. We already know that $\ln(a^r) = r \ln a$ when r is rational. But if we now let r be *any* real number we have, from Definition 1,

$$\ln a^r = \ln(e^{r \ln a}) = r \ln a$$

Thus

$\boxed{2}$ $$\ln a^r = r \ln a \qquad \text{for any real number } r$$

The general laws of exponents follow from Definition 1 together with the laws of exponents for e^x.

$\boxed{3}$ **Laws of Exponents** If x and y are real numbers and $a, b > 0$, then

1. $a^{x+y} = a^x a^y$ **2.** $a^{x-y} = a^x/a^y$ **3.** $(a^x)^y = a^{xy}$ **4.** $(ab)^x = a^x b^x$

PROOF

1. Using Definition 1 and the laws of exponents for e^x, we have

$$a^{x+y} = e^{(x+y) \ln a} = e^{x \ln a + y \ln a}$$

$$= e^{x \ln a} e^{y \ln a} = a^x a^y$$

3. Using Equation 2 we obtain

$$(a^x)^y = e^{y \ln(a^x)} = e^{yx \ln a}$$

$$= e^{xy \ln a} = a^{xy}$$

The remaining proofs are left as exercises.   ▬

The differentiation formula for exponential functions is also a consequence of Definition 1:

$\boxed{4}$

$$\frac{d}{dx}(a^x) = a^x \ln a$$

PROOF

$$\frac{d}{dx}(a^x) = \frac{d}{dx}(e^{x \ln a}) = e^{x \ln a}\frac{d}{dx}(x \ln a)$$

$$= a^x \ln a$$

Notice that if $a = e$, then $\ln e = 1$ and Formula 4 simplifies to a formula that we already know: $(d/dx)\, e^x = e^x$. In fact, the reason that the natural exponential function is used more often than other exponential functions is that its differentiation formula is simpler.

EXAMPLE 1 In Example 6 in Section 2.7 we considered a population of bacteria cells in a homogeneous nutrient medium. We showed that if the population doubles every hour, then the population after t hours is

$$n = n_0 2^t$$

where n_0 is the initial population. Now we can use $\boxed{4}$ to compute the growth rate:

$$\frac{dn}{dt} = n_0 2^t \ln 2$$

For instance, if the initial population is $n_0 = 1000$ cells, then the growth rate after two hours is

$$\frac{dn}{dt}\bigg|_{t=2} = (1000)2^t \ln 2 \,\big|_{t=2}$$

$$= 4000 \ln 2 \approx 2773 \text{ cells/h}$$

EXAMPLE 2 Combining Formula 4 with the Chain Rule, we have

$$\frac{d}{dx}\left(10^{x^2}\right) = 10^{x^2}(\ln 10)\frac{d}{dx}(x^2) = (2 \ln 10)x10^{x^2}$$

Exponential Graphs

If $a > 1$, then $\ln a > 0$, so $(d/dx)\, a^x = a^x \ln a > 0$, which shows that $y = a^x$ is increasing (see Figure 1). If $0 < a < 1$, then $\ln a < 0$ and so $y = a^x$ is decreasing (see Figure 2).

$$\lim_{x \to -\infty} a^x = 0, \ \lim_{x \to \infty} a^x = \infty$$

FIGURE 1 $y = a^x, \ a > 1$

$$\lim_{x \to -\infty} a^x = \infty, \ \lim_{x \to \infty} a^x = 0$$

FIGURE 2 $y = a^x, \ 0 < a < 1$

Notice from Figure 3 that as the base a gets larger, the exponential function grows more rapidly (for $x > 0$).

FIGURE 3

FIGURE 4

FIGURE 5

Figure 4 shows how the exponential function $y = 2^x$ compares with the power function $y = x^2$. The graphs intersect three times, but ultimately the exponential curve $y = 2^x$ grows far more rapidly than the parabola $y = x^2$. (See also Figure 5.)

In Section 6.5 we will show how exponential functions occur in the description of population growth and radioactive decay. Let's look at human population growth. Table 1 shows data for the population of the world in the 20th century, where $t = 0$ corresponds to 1900. Figure 6 shows the corresponding scatter plot.

TABLE 1

t	Population (millions)
0	1650
10	1750
20	1860
30	2070
40	2300
50	2560
60	3040
70	3710
80	4450
90	5280
100	6080
110	6870

FIGURE 6 Scatter plot for world population growth

The pattern of the data points in Figure 6 suggests exponential growth, so we use a graphing calculator with exponential regression capability to apply the method of least squares and obtain the exponential model

$$P = (1436.53) \cdot (1.01395)^t$$

Figure 7 shows the graph of this exponential function together with the original data points. We see that the exponential curve fits the data reasonably well. The period of relatively slow population growth is explained by the two world wars and the Great Depression of the 1930s.

FIGURE 7
Exponential model for
population growth

Exponential Integrals

The integration formula that follows from Formula 4 is

$$\int a^x \, dx = \frac{a^x}{\ln a} + C \qquad a \neq 1$$

V EXAMPLE 3
$$\int_0^5 2^x \, dx = \frac{2^x}{\ln 2} \Big]_0^5 = \frac{2^5}{\ln 2} - \frac{2^0}{\ln 2} = \frac{31}{\ln 2}$$

The Power Rule Versus the Exponential Rule

Now that we have defined arbitrary powers of numbers, we are in a position to prove the general version of the Power Rule, as promised in Section 2.3.

> **The Power Rule** If n is any real number and $f(x) = x^n$, then
>
> $$f'(x) = nx^{n-1}$$

If $x = 0$, we can show that $f'(0) = 0$ for $n > 1$ directly from the definition of a derivative.

PROOF Let $y = x^n$ and use logarithmic differentiation:

$$\ln|y| = \ln|x|^n = n \ln|x| \qquad x \neq 0$$

Therefore
$$\frac{y'}{y} = \frac{n}{x}$$

Hence
$$y' = n\frac{y}{x} = n\frac{x^n}{x} = nx^{n-1}$$

∅   You should distinguish carefully between the Power Rule [$(d/dx)\, x^n = nx^{n-1}$], where the base is variable and the exponent is constant, and the rule for differentiating exponential functions [$(d/dx)\, a^x = a^x \ln a$], where the base is constant and the exponent is variable.

In general there are four cases for exponents and bases:

Constant base, constant exponent

1. $\dfrac{d}{dx}(a^b) = 0$ (a and b are constants)

Variable base, constant exponent

2. $\dfrac{d}{dx}[f(x)]^b = b[f(x)]^{b-1}f'(x)$

Constant base, variable exponent

3. $\dfrac{d}{dx}[a^{g(x)}] = a^{g(x)}(\ln a)g'(x)$

Variable base, variable exponent

4. To find $(d/dx)[f(x)]^{g(x)}$, logarithmic differentiation can be used, as in the next example.

V EXAMPLE 4 Differentiate $y = x^{\sqrt{x}}$.

SOLUTION 1 Since both the base and the exponent are variable, we use logarithmic differentiation:

$$\ln y = \ln x^{\sqrt{x}} = \sqrt{x}\,\ln x$$

$$\frac{y'}{y} = \sqrt{x}\cdot\frac{1}{x} + (\ln x)\frac{1}{2\sqrt{x}}$$

$$y' = y\left(\frac{1}{\sqrt{x}} + \frac{\ln x}{2\sqrt{x}}\right) = x^{\sqrt{x}}\left(\frac{2 + \ln x}{2\sqrt{x}}\right)$$

SOLUTION 2 Another method is to write $x^{\sqrt{x}} = (e^{\ln x})^{\sqrt{x}}$:

$$\frac{d}{dx}\left(x^{\sqrt{x}}\right) = \frac{d}{dx}\left(e^{\sqrt{x}\,\ln x}\right) = e^{\sqrt{x}\,\ln x}\frac{d}{dx}\left(\sqrt{x}\,\ln x\right)$$

$$= x^{\sqrt{x}}\left(\frac{2 + \ln x}{2\sqrt{x}}\right) \qquad \text{(as in Solution 1)}$$

Figure 8 illustrates Example 4 by showing the graphs of $f(x) = x^{\sqrt{x}}$ and its derivative.

FIGURE 8

General Logarithmic Functions

If $a > 0$ and $a \neq 1$, then $f(x) = a^x$ is a one-to-one function. Its inverse function is called the **logarithmic function with base a** and is denoted by $\log_a$. Thus

5
$$\log_a x = y \iff a^y = x$$

In particular, we see that

$$\log_e x = \ln x$$

The cancellation equations for the inverse functions $\log_a x$ and a^x are

$$a^{\log_a x} = x \qquad \text{and} \qquad \log_a(a^x) = x$$

FIGURE 9

Figure 9 shows the case where $a > 1$. (The most important logarithmic functions have base $a > 1$.) The fact that $y = a^x$ is a very rapidly increasing function for $x > 0$ is reflected in the fact that $y = \log_a x$ is a very slowly increasing function for $x > 1$.

FIGURE 10

Figure 10 shows the graphs of $y = \log_a x$ with various values of the base a. Since $\log_a 1 = 0$, the graphs of all logarithmic functions pass through the point $(1, 0)$.

The laws of logarithms are similar to those for the natural logarithm and can be deduced from the laws of exponents (see Exercise 65).

The following formula shows that logarithms with any base can be expressed in terms of the natural logarithm.

---

6 **Change of Base Formula** For any positive number a $(a \neq 1)$, we have

$$\log_a x = \frac{\ln x}{\ln a}$$

PROOF Let $y = \log_a x$. Then, from $\boxed{5}$, we have $a^y = x$. Taking natural logarithms of both sides of this equation, we get $y \ln a = \ln x$. Therefore

$$y = \frac{\ln x}{\ln a}$$

Scientific calculators have a key for natural logarithms, so Formula 6 enables us to use a calculator to compute a logarithm with any base (as shown in the following example). Similarly, Formula 6 allows us to graph any logarithmic function on a graphing calculator or computer (see Exercises 14–16).

EXAMPLE 5 Evaluate $\log_8 5$ correct to six decimal places.

SOLUTION Formula 6 gives

$$\log_8 5 = \frac{\ln 5}{\ln 8} \approx 0.773976$$

Notation for Logarithms

Most textbooks in calculus and the sciences, as well as calculators, use the notation $\ln x$ for the natural logarithm and $\log x$ for the "common logarithm," $\log_{10} x$. In the more advanced mathematical and scientific literature and in computer languages, however, the notation $\log x$ usually denotes the natural logarithm.

Formula 6 enables us to differentiate any logarithmic function. Since $\ln a$ is a constant, we can differentiate as follows:

$$\frac{d}{dx}(\log_a x) = \frac{d}{dx}\left(\frac{\ln x}{\ln a}\right) = \frac{1}{\ln a}\frac{d}{dx}(\ln x) = \frac{1}{x \ln a}$$

7
$$\frac{d}{dx}(\log_a x) = \frac{1}{x \ln a}$$

V **EXAMPLE 6** Using Formula 7 and the Chain Rule, we get

$$\frac{d}{dx}\log_{10}(2 + \sin x) = \frac{1}{(2 + \sin x)\ln 10}\frac{d}{dx}(2 + \sin x)$$

$$= \frac{\cos x}{(2 + \sin x)\ln 10}$$

From Formula 7 we see one of the main reasons that natural logarithms (logarithms with base e) are used in calculus: The differentiation formula is simplest when $a = e$ because $\ln e = 1$.

The Number e as a Limit

We have shown that if $f(x) = \ln x$, then $f'(x) = 1/x$. Thus $f'(1) = 1$. We now use this fact to express the number e as a limit.

From the definition of a derivative as a limit, we have

$$f'(1) = \lim_{h \to 0} \frac{f(1+h) - f(1)}{h} = \lim_{x \to 0} \frac{f(1+x) - f(1)}{x}$$

$$= \lim_{x \to 0} \frac{\ln(1+x) - \ln 1}{x} = \lim_{x \to 0} \frac{1}{x} \ln(1+x)$$

$$= \lim_{x \to 0} \ln(1+x)^{1/x}$$

Because $f'(1) = 1$, we have

$$\lim_{x \to 0} \ln(1+x)^{1/x} = 1$$

Then, by Theorem 1.8.8 and the continuity of the exponential function, we have

$$e = e^1 = e^{\lim_{x \to 0} \ln(1+x)^{1/x}} = \lim_{x \to 0} e^{\ln(1+x)^{1/x}} = \lim_{x \to 0} (1+x)^{1/x}$$

$$\boxed{8} \qquad \boxed{e = \lim_{x \to 0} (1+x)^{1/x}}$$

Formula 8 is illustrated by the graph of the function $y = (1+x)^{1/x}$ in Figure 11 and a table of values for small values of x.

x	$(1 + x)^{1/x}$
0.1	2.59374246
0.01	2.70481383
0.001	2.71692393
0.0001	2.71814593
0.00001	2.71826824
0.000001	2.71828047
0.0000001	2.71828169
0.00000001	2.71828181

FIGURE 11

If we put $n = 1/x$ in Formula 8, then $n \to \infty$ as $x \to 0^+$ and so an alternative expression for e is

$$\boxed{9} \qquad \boxed{e = \lim_{n \to \infty} \left(1 + \frac{1}{n}\right)^n}$$

6.4* **Exercises**

Access end-of-section exercises and tools for visualizing calculus online at **www.webassign.net**

ENHANCED
WebAssign

In many natural phenomena, quantities grow or decay at a rate proportional to their size. For instance, if $y = f(t)$ is the number of individuals in a population of animals or bacteria at time t, then it seems reasonable to expect that the rate of growth $f'(t)$ is proportional to the population $f(t)$; that is, $f'(t) = kf(t)$ for some constant k. Indeed, under ideal conditions (unlimited environment, adequate nutrition, immunity to disease) the mathematical model given by the equation $f'(t) = kf(t)$ predicts what actually happens fairly accurately. Another example occurs in nuclear physics where the mass of a radioactive substance decays at a rate proportional to the mass. In chemistry, the rate of a unimolecular first-order reaction is proportional to the concentration of the substance. In finance, the value of a savings account with continuously compounded interest increases at a rate proportional to that value.

In general, if $y(t)$ is the value of a quantity y at time t and if the rate of change of y with respect to t is proportional to its size $y(t)$ at any time, then

| 1 |

$$\frac{dy}{dt} = ky$$

where k is a constant. Equation 1 is sometimes called the **law of natural growth** (if $k > 0$) or the **law of natural decay** (if $k < 0$). It is called a **differential equation** because it involves an unknown function y and its derivative dy/dt.

It's not hard to think of a solution of Equation 1. This equation asks us to find a function whose derivative is a constant multiple of itself. We have met such functions in this chapter. Any exponential function of the form $y(t) = Ce^{kt}$, where C is a constant, satisfies

$$y'(t) = C(ke^{kt}) = k(Ce^{kt}) = ky(t)$$

We will see in Section 9.4 that *any* function that satisfies $dy/dt = ky$ must be of the form $y = Ce^{kt}$. To see the significance of the constant C, we observe that

$$y(0) = Ce^{k \cdot 0} = C$$

Therefore C is the initial value of the function.

| 2 | **Theorem**   The only solutions of the differential equation $dy/dt = ky$ are the exponential functions
$$y(t) = y(0)e^{kt}$$

Population Growth

What is the significance of the proportionality constant k? In the context of population growth, where $P(t)$ is the size of a population at time t, we can write

| 3 | $$\frac{dP}{dt} = kP \qquad \text{or} \qquad \frac{1}{P}\frac{dP}{dt} = k$$

The quantity

$$\frac{1}{P}\frac{dP}{dt}$$

is the growth rate divided by the population size; it is called the **relative growth rate**. According to 3, instead of saying "the growth rate is proportional to population size" we could say "the relative growth rate is constant." Then ☐2 says that a population with constant relative growth rate must grow exponentially. Notice that the relative growth rate k appears as the coefficient of t in the exponential function Ce^{kt}. For instance, if

$$\frac{dP}{dt} = 0.02P$$

and t is measured in years, then the relative growth rate is $k = 0.02$ and the population grows at a relative rate of 2% per year. If the population at time 0 is P_0, then the expression for the population is

$$P(t) = P_0 e^{0.02t}$$

V EXAMPLE 1 Use the fact that the world population was 2560 million in 1950 and 3040 million in 1960 to model the population of the world in the second half of the 20th century. (Assume that the growth rate is proportional to the population size.) What is the relative growth rate? Use the model to estimate the world population in 1993 and to predict the population in the year 2020.

SOLUTION We measure the time t in years and let $t = 0$ in the year 1950. We measure the population $P(t)$ in millions of people. Then $P(0) = 2560$ and $P(10) = 3040$. Since we are assuming that $dP/dt = kP$, Theorem 2 gives

$$P(t) = P(0)e^{kt} = 2560e^{kt}$$

$$P(10) = 2560e^{10k} = 3040$$

$$k = \frac{1}{10}\ln\frac{3040}{2560} \approx 0.017185$$

The relative growth rate is about 1.7% per year and the model is

$$P(t) = 2560e^{0.017185t}$$

We estimate that the world population in 1993 was

$$P(43) = 2560e^{0.017185(43)} \approx 5360 \text{ million}$$

The model predicts that the population in 2020 will be

$$P(70) = 2560e^{0.017185(70)} \approx 8524 \text{ million}$$

The graph in Figure 1 shows that the model is fairly accurate to the end of the 20th century (the dots represent the actual population), so the estimate for 1993 is quite reliable. But the prediction for 2020 is riskier.

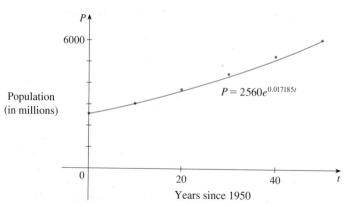

FIGURE 1
A model for world population growth in the second half of the 20th century

▪ Radioactive Decay

Radioactive substances decay by spontaneously emitting radiation. If $m(t)$ is the mass remaining from an initial mass m_0 of the substance after time t, then the relative decay rate

$$-\frac{1}{m}\frac{dm}{dt}$$

has been found experimentally to be constant. (Since dm/dt is negative, the relative decay rate is positive.) It follows that

$$\frac{dm}{dt} = km$$

where k is a negative constant. In other words, radioactive substances decay at a rate proportional to the remaining mass. This means that we can use $\boxed{2}$ to show that the mass decays exponentially:

$$m(t) = m_0 e^{kt}$$

Physicists express the rate of decay in terms of **half-life**, the time required for half of any given quantity to decay.

▽ EXAMPLE 2 The half-life of radium-226 is 1590 years.
(a) A sample of radium-226 has a mass of 100 mg. Find a formula for the mass of the sample that remains after t years.
(b) Find the mass after 1000 years correct to the nearest milligram.
(c) When will the mass be reduced to 30 mg?

SOLUTION
(a) Let $m(t)$ be the mass of radium-226 (in milligrams) that remains after t years. Then $dm/dt = km$ and $y(0) = 100$, so $\boxed{2}$ gives

$$m(t) = m(0)e^{kt} = 100e^{kt}$$

In order to determine the value of k, we use the fact that $y(1590) = \frac{1}{2}(100)$. Thus

$$100e^{1590k} = 50 \qquad \text{so} \qquad e^{1590k} = \tfrac{1}{2}$$

and

$$1590k = \ln\tfrac{1}{2} = -\ln 2$$

$$k = -\frac{\ln 2}{1590}$$

Therefore

$$m(t) = 100e^{-(\ln 2)t/1590}$$

We could use the fact that $e^{\ln 2} = 2$ to write the expression for $m(t)$ in the alternative form

$$m(t) = 100 \times 2^{-t/1590}$$

(b) The mass after 1000 years is

$$m(1000) = 100e^{-(\ln 2)1000/1590} \approx 65 \text{ mg}$$

(c) We want to find the value of t such that $m(t) = 30$, that is,

$$100e^{-(\ln 2)t/1590} = 30 \qquad \text{or} \qquad e^{-(\ln 2)t/1590} = 0.3$$

We solve this equation for t by taking the natural logarithm of both sides:

$$-\frac{\ln 2}{1590} t = \ln 0.3$$

Thus

$$t = -1590 \frac{\ln 0.3}{\ln 2} \approx 2762 \text{ years}$$

150

$m = 100e^{-(\ln 2)t/1590}$

$m = 30$

0   4000

FIGURE 2

As a check on our work in Example 2, we use a graphing device to draw the graph of $m(t)$ in Figure 2 together with the horizontal line $m = 30$. These curves intersect when $t \approx 2800$, and this agrees with the answer to part (c).

Newton's Law of Cooling

Newton's Law of Cooling states that the rate of cooling of an object is proportional to the temperature difference between the object and its surroundings, provided that this difference is not too large. (This law also applies to warming.) If we let $T(t)$ be the temperature of the object at time t and T_s be the temperature of the surroundings, then we can formulate Newton's Law of Cooling as a differential equation:

$$\frac{dT}{dt} = k(T - T_s)$$

where k is a constant. This equation is not quite the same as Equation 1, so we make the change of variable $y(t) = T(t) - T_s$. Because T_s is constant, we have $y'(t) = T'(t)$ and so the equation becomes

$$\frac{dy}{dt} = ky$$

We can then use $\boxed{2}$ to find an expression for y, from which we can find T.

EXAMPLE 3 A bottle of soda pop at room temperature (72°F) is placed in a refrigerator where the temperature is 44°F. After half an hour the soda pop has cooled to 61°F.
(a) What is the temperature of the soda pop after another half hour?
(b) How long does it take for the soda pop to cool to 50°F?

SOLUTION
(a) Let $T(t)$ be the temperature of the soda after t minutes. The surrounding temperature is $T_s = 44°F$, so Newton's Law of Cooling states that

$$\frac{dT}{dt} = k(T - 44)$$

If we let $y = T - 44$, then $y(0) = T(0) - 44 = 72 - 44 = 28$, so y satisfies

$$\frac{dy}{dt} = ky \qquad y(0) = 28$$

and by $\boxed{2}$ we have

$$y(t) = y(0)e^{kt} = 28e^{kt}$$

We are given that $T(30) = 61$, so $y(30) = 61 - 44 = 17$ and

$$28e^{30k} = 17 \qquad e^{30k} = \tfrac{17}{28}$$

Taking logarithms, we have

$$k = \frac{\ln\left(\frac{17}{28}\right)}{30} \approx -0.01663$$

Thus

$$y(t) = 28e^{-0.01663t}$$

$$T(t) = 44 + 28e^{-0.01663t}$$

$$T(60) = 44 + 28e^{-0.01663(60)} \approx 54.3$$

So after another half hour the pop has cooled to about 54°F.

(b) We have $T(t) = 50$ when

$$44 + 28e^{-0.01663t} = 50$$

$$e^{-0.01663t} = \tfrac{6}{28}$$

$$t = \frac{\ln\left(\frac{6}{28}\right)}{-0.01663} \approx 92.6$$

The pop cools to 50°F after about 1 hour 33 minutes.

Notice that in Example 3, we have

$$\lim_{t \to \infty} T(t) = \lim_{t \to \infty} (44 + 28e^{-0.01663t}) = 44 + 28 \cdot 0 = 44$$

which is to be expected. The graph of the temperature function is shown in Figure 3.

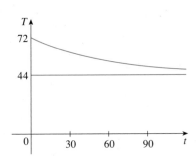

FIGURE 3

Continuously Compounded Interest

EXAMPLE 4 If $1000 is invested at 6% interest, compounded annually, then after 1 year the investment is worth $1000(1.06) = $1060, after 2 years it's worth $[1000(1.06)]1.06 = $1123.60, and after t years it's worth $1000(1.06)^t$. In general, if an amount A_0 is invested at an interest rate r ($r = 0.06$ in this example), then after t years it's worth $A_0(1 + r)^t$. Usually, however, interest is compounded more frequently, say, n times a year. Then in each compounding period the interest rate is r/n and there are nt compounding periods in t years, so the value of the investment is

$$A_0\left(1 + \frac{r}{n}\right)^{nt}$$

For instance, after 3 years at 6% interest a $1000 investment will be worth

$$\$1000(1.06)^3 = \$1191.02 \quad \text{with annual compounding}$$

$$\$1000(1.03)^6 = \$1194.05 \quad \text{with semiannual compounding}$$

$$\$1000(1.015)^{12} = \$1195.62 \quad \text{with quarterly compounding}$$

$$\$1000(1.005)^{36} = \$1196.68 \quad \text{with monthly compounding}$$

$$\$1000\left(1 + \frac{0.06}{365}\right)^{365 \cdot 3} = \$1197.20 \quad \text{with daily compounding}$$

You can see that the interest paid increases as the number of compounding periods (n) increases. If we let $n \to \infty$, then we will be compounding the interest **continuously** and the value of the investment will be

$$A(t) = \lim_{n \to \infty} A_0 \left(1 + \frac{r}{n} \right)^{nt}$$

$$= \lim_{n \to \infty} A_0 \left[\left(1 + \frac{r}{n} \right)^{n/r} \right]^{rt}$$

$$= A_0 \left[\lim_{n \to \infty} \left(1 + \frac{r}{n} \right)^{n/r} \right]^{rt}$$

$$= A_0 \left[\lim_{m \to \infty} \left(1 + \frac{1}{m} \right)^{m} \right]^{rt} \qquad \text{(where } m = n/r\text{)}$$

But the limit in this expression is equal to the number e (see Equation 6.4.9 or 6.4*.9). So with continuous compounding of interest at interest rate r, the amount after t years is

$$A(t) = A_0 e^{rt}$$

If we differentiate this equation, we get

$$\frac{dA}{dt} = r A_0 e^{rt} = r A(t)$$

which says that, with continuous compounding of interest, the rate of increase of an investment is proportional to its size.

Returning to the example of $1000 invested for 3 years at 6% interest, we see that with continuous compounding of interest the value of the investment will be

$$A(3) = \$1000 e^{(0.06)3} = \$1197.22$$

Notice how close this is to the amount we calculated for daily compounding, $1197.20. But the amount is easier to compute if we use continuous compounding. ▬

6.5 Exercises

Access end-of-section exercises and tools for visualizing calculus online at **www.webassign.net**

6.6 Inverse Trigonometric Functions

is section we apply the ideas of Section 6.1 to find the derivatives of the so-called inverse trigonometric functions. We have a slight difficulty in this task: Because the trigonometric functions are not one-to-one, they do not have inverse functions. The difficulty is overcome by restricting the domains of these functions so that they become one-to-one.

You can see from Figure 1 that the sine function $y = \sin x$ is not one-to-one (use the Horizontal Line Test). But the function $f(x) = \sin x$, $-\pi/2 \leqslant x \leqslant \pi/2$, *is* one-to-one (see

Figure 2). The inverse function of this restricted sine function f exists and is denoted by $\sin^{-1}$ or arcsin. It is called the **inverse sine function** or the **arcsine function**.

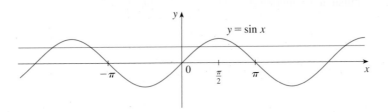

FIGURE 1

FIGURE 2 $y = \sin x, -\frac{\pi}{2} \leq x \leq \frac{\pi}{2}$

Since the definition of an inverse function says that

$$f^{-1}(x) = y \iff f(y) = x$$

we have

$$\boxed{1} \qquad \sin^{-1}x = y \iff \sin y = x \quad \text{and} \quad -\frac{\pi}{2} \leq y \leq \frac{\pi}{2}$$

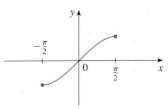

Thus, if $-1 \leq x \leq 1$, $\sin^{-1}x$ is the number between $-\pi/2$ and $\pi/2$ whose sine is x.

EXAMPLE 1 Evaluate (a) $\sin^{-1}\left(\frac{1}{2}\right)$ and (b) $\tan\left(\arcsin\frac{1}{3}\right)$.

SOLUTION
(a) We have

$$\sin^{-1}\left(\tfrac{1}{2}\right) = \frac{\pi}{6}$$

because $\sin(\pi/6) = \frac{1}{2}$ and $\pi/6$ lies between $-\pi/2$ and $\pi/2$.

(b) Let $\theta = \arcsin\frac{1}{3}$, so $\sin\theta = \frac{1}{3}$. Then we can draw a right triangle with angle θ as in Figure 3 and deduce from the Pythagorean Theorem that the third side has length $\sqrt{9-1} = 2\sqrt{2}$. This enables us to read from the triangle that

$$\tan\left(\arcsin\tfrac{1}{3}\right) = \tan\theta = \frac{1}{2\sqrt{2}}$$

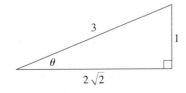

FIGURE 3

The cancellation equations for inverse functions become, in this case,

$$
\boxed{2} \qquad
\begin{aligned}
\sin^{-1}(\sin x) &= x \quad \text{for } -\frac{\pi}{2} \leq x \leq \frac{\pi}{2} \\[2mm]
\sin(\sin^{-1}x) &= x \quad \text{for } -1 \leq x \leq 1
\end{aligned}
$$

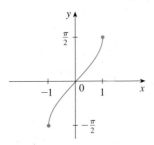

FIGURE 4
$y = \sin^{-1}x = \arcsin x$

The inverse sine function, $\sin^{-1}$, has domain $[-1, 1]$ and range $[-\pi/2, \pi/2]$, and its graph, shown in Figure 4, is obtained from that of the restricted sine function (Figure 2) by reflection about the line $y = x$.

We know that the sine function f is continuous, so the inverse sine function is also continuous. We also know from Section 2.4 that the sine function is differentiable, so the inverse sine function is also differentiable. We could calculate the derivative of $\sin^{-1}$ by the formula in Theorem 6.1.7, but since we know that $\sin^{-1}$ is differentiable, we can just as easily calculate it by implicit differentiation as follows.

Let $y = \sin^{-1}x$. Then $\sin y = x$ and $-\pi/2 \leqslant y \leqslant \pi/2$. Differentiating $\sin y = x$ implicitly with respect to x, we obtain

$$\cos y \, \frac{dy}{dx} = 1$$

and

$$\frac{dy}{dx} = \frac{1}{\cos y}$$

Now $\cos y \geqslant 0$ since $-\pi/2 \leqslant y \leqslant \pi/2$, so

$$\cos y = \sqrt{1 - \sin^2 y} = \sqrt{1 - x^2}$$

Therefore

$$\frac{dy}{dx} = \frac{1}{\cos y} = \frac{1}{\sqrt{1 - x^2}}$$

3
$$\frac{d}{dx}(\sin^{-1}x) = \frac{1}{\sqrt{1 - x^2}} \qquad -1 < x < 1$$

V **EXAMPLE 2** If $f(x) = \sin^{-1}(x^2 - 1)$, find (a) the domain of f, (b) $f'(x)$, and (c) the domain of f'.

SOLUTION

(a) Since the domain of the inverse sine function is $[-1, 1]$, the domain of f is

$$\{x \mid -1 \leqslant x^2 - 1 \leqslant 1\} = \{x \mid 0 \leqslant x^2 \leqslant 2\}$$
$$= \{x \mid |x| \leqslant \sqrt{2}\} = \left[-\sqrt{2}, \sqrt{2}\right]$$

(b) Combining Formula 3 with the Chain Rule, we have

$$f'(x) = \frac{1}{\sqrt{1 - (x^2 - 1)^2}} \, \frac{d}{dx}(x^2 - 1)$$

$$= \frac{1}{\sqrt{1 - (x^4 - 2x^2 + 1)}} \, 2x = \frac{2x}{\sqrt{2x^2 - x^4}}$$

(c) The domain of f' is

$$\{x \mid -1 < x^2 - 1 < 1\} = \{x \mid 0 < x^2 < 2\}$$
$$= \{x \mid 0 < |x| < \sqrt{2}\} = \left(-\sqrt{2}, 0\right) \cup \left(0, \sqrt{2}\right) \qquad \blacksquare$$

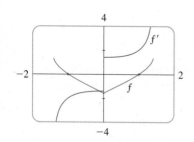

FIGURE 5

The graphs of the function f of Example 2 and its derivative are shown in Figure 5. Notice that f is not differentiable at 0 and this is consistent with the fact that the graph of f' makes a sudden jump at $x = 0$.

The **inverse cosine function** is handled similarly. The restricted cosine function $f(x) = \cos x$, $0 \leqslant x \leqslant \pi$, is one-to-one (see Figure 6) and so it has an inverse function denoted by $\cos^{-1}$ or arccos.

4
$$\cos^{-1}x = y \quad \Longleftrightarrow \quad \cos y = x \quad \text{and} \quad 0 \leqslant y \leqslant \pi$$

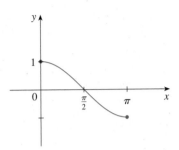

FIGURE 6

$y = \cos x, 0 \leqslant x \leqslant \pi$

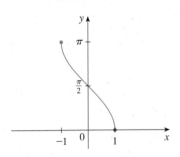

FIGURE 7

$y = \cos^{-1} x = \arccos x$

The cancellation equations are

$$\boxed{5} \qquad \begin{aligned} &\cos^{-1}(\cos x) = x \quad \text{for } 0 \leqslant x \leqslant \pi \\ &\cos(\cos^{-1} x) = x \quad \text{for } -1 \leqslant x \leqslant 1 \end{aligned}$$

The inverse cosine function, $\cos^{-1}$, has domain $[-1, 1]$ and range $[0, \pi]$ and is a continuous function whose graph is shown in Figure 7. Its derivative is given by

$$\boxed{6} \qquad \frac{d}{dx}(\cos^{-1} x) = -\frac{1}{\sqrt{1 - x^2}} \qquad -1 < x < 1$$

Formula 6 can be proved by the same method as for Formula 3 and is left as Exercise 17.

The tangent function can be made one-to-one by restricting it to the interval $(-\pi/2, \pi/2)$. Thus the **inverse tangent function** is defined as the inverse of the function $f(x) = \tan x$, $-\pi/2 < x < \pi/2$. (See Figure 8.) It is denoted by $\tan^{-1}$ or arctan.

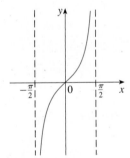

FIGURE 8

$y = \tan x, -\frac{\pi}{2} < x < \frac{\pi}{2}$

$$\boxed{7} \qquad \tan^{-1} x = y \iff \tan y = x \quad \text{and} \quad -\frac{\pi}{2} < y < \frac{\pi}{2}$$

EXAMPLE 3 Simplify the expression $\cos(\tan^{-1} x)$.

SOLUTION 1 Let $y = \tan^{-1} x$. Then $\tan y = x$ and $-\pi/2 < y < \pi/2$. We want to find $\cos y$ but, since $\tan y$ is known, it is easier to find $\sec y$ first:

$$\sec^2 y = 1 + \tan^2 y = 1 + x^2$$

$$\sec y = \sqrt{1 + x^2} \qquad \text{(since } \sec y > 0 \text{ for } -\pi/2 < y < \pi/2\text{)}$$

Thus

$$\cos(\tan^{-1} x) = \cos y = \frac{1}{\sec y} = \frac{1}{\sqrt{1 + x^2}}$$

FIGURE 9

SOLUTION 2 Instead of using trigonometric identities as in Solution 1, it is perhaps easier to use a diagram. If $y = \tan^{-1}x$, then $\tan y = x$, and we can read from Figure 9 (which illustrates the case $y > 0$) that

$$\cos(\tan^{-1}x) = \cos y = \frac{1}{\sqrt{1 + x^2}}$$

The inverse tangent function, $\tan^{-1} = \arctan$, has domain $\mathbb{R}$ and range $(-\pi/2, \pi/2)$. Its graph is shown in Figure 10.

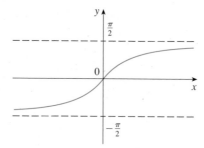

FIGURE 10
$y = \tan^{-1}x = \arctan x$

We know that

$$\lim_{x \to (\pi/2)^-} \tan x = \infty \qquad \text{and} \qquad \lim_{x \to -(\pi/2)^+} \tan x = -\infty$$

and so the lines $x = \pm\pi/2$ are vertical asymptotes of the graph of tan. Since the graph of $\tan^{-1}$ is obtained by reflecting the graph of the restricted tangent function about the line $y = x$, it follows that the lines $y = \pi/2$ and $y = -\pi/2$ are horizontal asymptotes of the graph of $\tan^{-1}$. This fact is expressed by the following limits:

8

$$\lim_{x \to \infty} \tan^{-1}x = \frac{\pi}{2} \qquad \lim_{x \to -\infty} \tan^{-1}x = -\frac{\pi}{2}$$

EXAMPLE 4 Evaluate $\displaystyle\lim_{x \to 2^+} \arctan\left(\frac{1}{x - 2}\right)$.

SOLUTION If we let $t = 1/(x - 2)$, we know that $t \to \infty$ as $x \to 2^+$. Therefore, by the first equation in $\boxed{8}$, we have

$$\lim_{x \to 2^+} \arctan\left(\frac{1}{x - 2}\right) = \lim_{t \to \infty} \arctan t = \frac{\pi}{2}$$

Because tan is differentiable, $\tan^{-1}$ is also differentiable. To find its derivative, we let $y = \tan^{-1}x$. Then $\tan y = x$. Differentiating this latter equation implicitly with respect to x, we have

$$\sec^2 y \, \frac{dy}{dx} = 1$$

and so

$$\frac{dy}{dx} = \frac{1}{\sec^2 y} = \frac{1}{1 + \tan^2 y} = \frac{1}{1 + x^2}$$

$$\boxed{9} \qquad \frac{d}{dx}(\tan^{-1}x) = \frac{1}{1 + x^2}$$

The remaining inverse trigonometric functions are not used as frequently and are summarized here.

$$\boxed{10} \quad y = \csc^{-1}x \ (|x| \geqslant 1) \iff \csc y = x \quad \text{and} \quad y \in (0, \pi/2] \cup (\pi, 3\pi/2]$$

$$y = \sec^{-1}x \ (|x| \geqslant 1) \iff \sec y = x \quad \text{and} \quad y \in [0, \pi/2) \cup [\pi, 3\pi/2)$$

$$y = \cot^{-1}x \ (x \in \mathbb{R}) \iff \cot y = x \quad \text{and} \quad y \in (0, \pi)$$

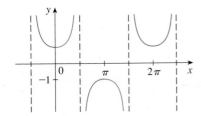

FIGURE 11

$y = \sec x$

The choice of intervals for y in the definitions of $\csc^{-1}$ and $\sec^{-1}$ is not universally agreed upon. For instance, some authors use $y \in [0, \pi/2) \cup (\pi/2, \pi]$ in the definition of $\sec^{-1}$. [You can see from the graph of the secant function in Figure 11 that both this choice and the one in $\boxed{10}$ will work.] The reason for the choice in $\boxed{10}$ is that the differentiation formulas are simpler (see Exercise 79).

We collect in Table 11 the differentiation formulas for all of the inverse trigonometric functions. The proofs of the formulas for the derivatives of $\csc^{-1}$, $\sec^{-1}$, and $\cot^{-1}$ are left as Exercises 19–21.

$\boxed{11}$ **Table of Derivatives of Inverse Trigonometric Functions**

$$\frac{d}{dx}(\sin^{-1}x) = \frac{1}{\sqrt{1 - x^2}} \qquad\qquad \frac{d}{dx}(\csc^{-1}x) = -\frac{1}{x\sqrt{x^2 - 1}}$$

$$\frac{d}{dx}(\cos^{-1}x) = -\frac{1}{\sqrt{1 - x^2}} \qquad\qquad \frac{d}{dx}(\sec^{-1}x) = \frac{1}{x\sqrt{x^2 - 1}}$$

$$\frac{d}{dx}(\tan^{-1}x) = \frac{1}{1 + x^2} \qquad\qquad \frac{d}{dx}(\cot^{-1}x) = -\frac{1}{1 + x^2}$$

Each of these formulas can be combined with the Chain Rule. For instance, if u is a differentiable function of x, then

$$\frac{d}{dx}(\sin^{-1}u) = \frac{1}{\sqrt{1 - u^2}}\frac{du}{dx} \quad \text{and} \quad \frac{d}{dx}(\tan^{-1}u) = \frac{1}{1 + u^2}\frac{du}{dx}$$

V **EXAMPLE 5**   Differentiate (a) $y = \dfrac{1}{\sin^{-1}x}$ and (b) $f(x) = x \arctan\sqrt{x}$.

SOLUTION

(a)
$$\frac{dy}{dx} = \frac{d}{dx}(\sin^{-1}x)^{-1} = -(\sin^{-1}x)^{-2}\frac{d}{dx}(\sin^{-1}x)$$

$$= -\frac{1}{(\sin^{-1}x)^2\sqrt{1 - x^2}}$$

Recall that arctan x is an alternative
notation for $\tan^{-1}x$.

(b)
$$f'(x) = x\,\frac{1}{1 + (\sqrt{x})^2}\left(\tfrac{1}{2}x^{-1/2}\right) + \arctan\sqrt{x}$$

$$= \frac{\sqrt{x}}{2(1 + x)} + \arctan\sqrt{x}$$

EXAMPLE 6 Prove the identity $\tan^{-1}x + \cot^{-1}x = \pi/2$.

SOLUTION Although calculus is not needed to prove this identity, the proof using calculus is quite simple. If $f(x) = \tan^{-1}x + \cot^{-1}x$, then

$$f'(x) = \frac{1}{1 + x^2} - \frac{1}{1 + x^2} = 0$$

for all values of x. Therefore $f(x) = C$, a constant. To determine the value of C, we put $x = 1$. Then

$$C = f(1) = \tan^{-1}1 + \cot^{-1}1 = \frac{\pi}{4} + \frac{\pi}{4} = \frac{\pi}{2}$$

Thus $\tan^{-1}x + \cot^{-1}x = \pi/2$.

Each of the formulas in Table 11 gives rise to an integration formula. The two most useful of these are the following:

12
$$\int \frac{1}{\sqrt{1 - x^2}}\,dx = \sin^{-1}x + C$$

13
$$\int \frac{1}{x^2 + 1}\,dx = \tan^{-1}x + C$$

EXAMPLE 7 Find $\displaystyle\int_0^{1/4} \frac{1}{\sqrt{1 - 4x^2}}\,dx$.

SOLUTION If we write

$$\int_0^{1/4} \frac{1}{\sqrt{1 - 4x^2}}\,dx = \int_0^{1/4} \frac{1}{\sqrt{1 - (2x)^2}}\,dx$$

then the integral resembles Equation 12 and the substitution $u = 2x$ is suggested. This gives $du = 2\,dx$, so $dx = du/2$. When $x = 0$, $u = 0$; when $x = \frac{1}{4}$, $u = \frac{1}{2}$. So

$$\int_0^{1/4} \frac{1}{\sqrt{1 - 4x^2}}\,dx = \tfrac{1}{2}\int_0^{1/2} \frac{du}{\sqrt{1 - u^2}} = \tfrac{1}{2}\sin^{-1}u\Big]_0^{1/2}$$

$$= \tfrac{1}{2}\left[\sin^{-1}\left(\tfrac{1}{2}\right) - \sin^{-1}0\right] = \frac{1}{2}\cdot\frac{\pi}{6} = \frac{\pi}{12}$$

EXAMPLE 8 Evaluate $\displaystyle\int \frac{1}{x^2 + a^2}\,dx$.

SOLUTION To make the given integral more like Equation 13 we write

$$\int \frac{dx}{x^2 + a^2} = \int \frac{dx}{a^2\left(\dfrac{x^2}{a^2} + 1\right)} = \frac{1}{a^2} \int \frac{dx}{\left(\dfrac{x}{a}\right)^2 + 1}$$

This suggests that we substitute $u = x/a$. Then $du = dx/a$, $dx = a\,du$, and

$$\int \frac{dx}{x^2 + a^2} = \frac{1}{a^2} \int \frac{a\,du}{u^2 + 1} = \frac{1}{a} \int \frac{du}{u^2 + 1} = \frac{1}{a} \tan^{-1}u + C$$

Thus we have the formula

One of the main uses of inverse trigonometric functions is that they often arise when we integrate rational functions.

14
$$\int \frac{1}{x^2 + a^2}\,dx = \frac{1}{a}\tan^{-1}\left(\frac{x}{a}\right) + C$$

EXAMPLE 9 Find $\displaystyle\int \frac{x}{x^4 + 9}\,dx$.

SOLUTION We substitute $u = x^2$ because then $du = 2x\,dx$ and we can use Equation 14 with $a = 3$:

$$\int \frac{x}{x^4 + 9}\,dx = \frac{1}{2}\int \frac{du}{u^2 + 9} = \frac{1}{2} \cdot \frac{1}{3}\tan^{-1}\left(\frac{u}{3}\right) + C = \frac{1}{6}\tan^{-1}\left(\frac{x^2}{3}\right) + C$$

6.6   Exercises

Access end-of-section exercises and tools for visualizing calculus online at **www.webassign.net**   WebAssign

---

APPLIED PROJECT   [CAS] **WHERE TO SIT AT THE MOVIES**

A movie theater has a screen that is positioned 10 ft off the floor and is 25 ft high. The first row of seats is placed 9 ft from the screen and the rows are set 3 ft apart. The floor of the seating area is inclined at an angle of $\alpha = 20°$ above the horizontal and the distance up the incline that you sit is x. The theater has 21 rows of seats, so $0 \le x \le 60$. Suppose you decide that the best place to sit is in the row where the angle θ subtended by the screen at your eyes is a maximum. Let's also suppose that your eyes are 4 ft above the floor, as shown in the figure. (In Exercise 48 in Section 6.6 we looked at a simpler version of this problem, where the floor is horizontal, but this project involves a more complicated situation and requires technology.)

1. Show that

$$\theta = \arccos\left(\frac{a^2 + b^2 - 625}{2ab}\right)$$

[CAS] Computer algebra system required

where

$$a^2 = (9 + x \cos \alpha)^2 + (31 - x \sin \alpha)^2$$

and

$$b^2 = (9 + x \cos \alpha)^2 + (x \sin \alpha - 6)^2$$

2. Use a graph of θ as a function of x to estimate the value of x that maximizes θ. In which row should you sit? What is the viewing angle θ in this row?

3. Use your computer algebra system to differentiate θ and find a numerical value for the root of the equation $d\theta/dx = 0$. Does this value confirm your result in Problem 2?

4. Use the graph of θ to estimate the average value of θ on the interval $0 \leqslant x \leqslant 60$. Then use your CAS to compute the average value. Compare with the maximum and minimum values of θ.

6.7   Hyperbolic Functions

Certain even and odd combinations of the exponential functions e^x and e^{-x} arise so frequently in mathematics and its applications that they deserve to be given special names. In many ways they are analogous to the trigonometric functions, and they have the same relationship to the hyperbola that the trigonometric functions have to the circle. For this reason they are collectively called **hyperbolic functions** and individually called **hyperbolic sine**, **hyperbolic cosine**, and so on.

Definition of the Hyperbolic Functions

$$\sinh x = \frac{e^x - e^{-x}}{2} \qquad\qquad \operatorname{csch} x = \frac{1}{\sinh x}$$

$$\cosh x = \frac{e^x + e^{-x}}{2} \qquad\qquad \operatorname{sech} x = \frac{1}{\cosh x}$$

$$\tanh x = \frac{\sinh x}{\cosh x} \qquad\qquad \coth x = \frac{\cosh x}{\sinh x}$$

The graphs of hyperbolic sine and cosine can be sketched using graphical addition as in Figures 1 and 2.

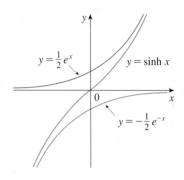

FIGURE 1
$y = \sinh x = \frac{1}{2}e^x - \frac{1}{2}e^{-x}$

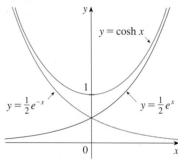

FIGURE 2
$y = \cosh x = \frac{1}{2}e^x + \frac{1}{2}e^{-x}$

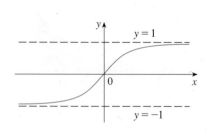

FIGURE 3
$y = \tanh x$

Note that sinh has domain $\mathbb{R}$ and range $\mathbb{R}$, while cosh has domain $\mathbb{R}$ and range $[1, \infty)$. The graph of tanh is shown in Figure 3. It has the horizontal asymptotes $y = \pm 1$. (See Exercise 23.)

Some of the mathematical uses of hyperbolic functions will be seen in Chapter 7. Applications to science and engineering occur whenever an entity such as light, velocity, electricity, or radioactivity is gradually absorbed or extinguished, for the decay can be represented by hyperbolic functions. The most famous application is the use of hyperbolic cosine to describe the shape of a hanging wire. It can be proved that if a heavy flexible cable (such as a telephone or power line) is suspended between two points at the same height, then it takes the shape of a curve with equation $y = c + a \cosh(x/a)$ called a *catenary* (see Figure 4). (The Latin word *catena* means "chain.")

Another application of hyperbolic functions occurs in the description of ocean waves: The velocity of a water wave with length L moving across a body of water with depth d is modeled by the function

$$v = \sqrt{\frac{gL}{2\pi} \tanh\left(\frac{2\pi d}{L}\right)}$$

where g is the acceleration due to gravity. (See Figure 5 and Exercise 49.)

The hyperbolic functions satisfy a number of identities that are similar to well-known trigonometric identities. We list some of them here and leave most of the proofs to the exercises.

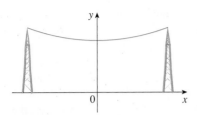

FIGURE 4
A catenary $y = c + a \cosh(x/a)$

FIGURE 5
Idealized ocean wave

Hyperbolic Identities

$$\sinh(-x) = -\sinh x \qquad \cosh(-x) = \cosh x$$

$$\cosh^2 x - \sinh^2 x = 1 \qquad 1 - \tanh^2 x = \operatorname{sech}^2 x$$

$$\sinh(x + y) = \sinh x \cosh y + \cosh x \sinh y$$

$$\cosh(x + y) = \cosh x \cosh y + \sinh x \sinh y$$

▽ EXAMPLE 1 Prove (a) $\cosh^2 x - \sinh^2 x = 1$ and (b) $1 - \tanh^2 x = \operatorname{sech}^2 x$.

SOLUTION

(a)
$$\cosh^2 x - \sinh^2 x = \left(\frac{e^x + e^{-x}}{2}\right)^2 - \left(\frac{e^x - e^{-x}}{2}\right)^2$$

$$= \frac{e^{2x} + 2 + e^{-2x}}{4} - \frac{e^{2x} - 2 + e^{-2x}}{4} = \frac{4}{4} = 1$$

(b) We start with the identity proved in part (a):

$$\cosh^2 x - \sinh^2 x = 1$$

If we divide both sides by $\cosh^2 x$, we get

$$1 - \frac{\sinh^2 x}{\cosh^2 x} = \frac{1}{\cosh^2 x}$$

or
$$1 - \tanh^2 x = \operatorname{sech}^2 x$$

The Gateway Arch in St. Louis was designed using a hyperbolic cosine function (Exercise 48).

FIGURE 6

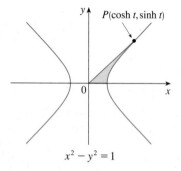

FIGURE 7

The identity proved in Example 1(a) gives a clue to the reason for the name "hyperbolic" functions:

If t is any real number, then the point $P(\cos t, \sin t)$ lies on the unit circle $x^2 + y^2 = 1$ because $\cos^2 t + \sin^2 t = 1$. In fact, t can be interpreted as the radian measure of $\angle POQ$ in Figure 6. For this reason the trigonometric functions are sometimes called *circular* functions.

Likewise, if t is any real number, then the point $P(\cosh t, \sinh t)$ lies on the right branch of the hyperbola $x^2 - y^2 = 1$ because $\cosh^2 t - \sinh^2 t = 1$ and $\cosh t \geq 1$. This time, t does not represent the measure of an angle. However, it turns out that t represents twice the area of the shaded hyperbolic sector in Figure 7, just as in the trigonometric case t represents twice the area of the shaded circular sector in Figure 6.

The derivatives of the hyperbolic functions are easily computed. For example,

$$\frac{d}{dx}(\sinh x) = \frac{d}{dx}\left(\frac{e^x - e^{-x}}{2}\right) = \frac{e^x + e^{-x}}{2} = \cosh x$$

We list the differentiation formulas for the hyperbolic functions as Table 1. The remaining proofs are left as exercises. Note the analogy with the differentiation formulas for trigonometric functions, but beware that the signs are different in some cases.

---

1 **Derivatives of Hyperbolic Functions**

$$\frac{d}{dx}(\sinh x) = \cosh x \qquad\qquad \frac{d}{dx}(\operatorname{csch} x) = -\operatorname{csch} x \coth x$$

$$\frac{d}{dx}(\cosh x) = \sinh x \qquad\qquad \frac{d}{dx}(\operatorname{sech} x) = -\operatorname{sech} x \tanh x$$

$$\frac{d}{dx}(\tanh x) = \operatorname{sech}^2 x \qquad\qquad \frac{d}{dx}(\coth x) = -\operatorname{csch}^2 x$$

V EXAMPLE 2 Any of these differentiation rules can be combined with the Chain Rule. For instance,

$$\frac{d}{dx}\left(\cosh \sqrt{x}\right) = \sinh \sqrt{x} \cdot \frac{d}{dx}\sqrt{x} = \frac{\sinh \sqrt{x}}{2\sqrt{x}}$$

Inverse Hyperbolic Functions

You can see from Figures 1 and 3 that sinh and tanh are one-to-one functions and so they have inverse functions denoted by $\sinh^{-1}$ and $\tanh^{-1}$. Figure 2 shows that cosh is not one-to-one, but when restricted to the domain $[0, \infty)$ it becomes one-to-one. The inverse hyperbolic cosine function is defined as the inverse of this restricted function.

2

$$y = \sinh^{-1}x \iff \sinh y = x$$

$$y = \cosh^{-1}x \iff \cosh y = x \quad \text{and} \quad y \geq 0$$

$$y = \tanh^{-1}x \iff \tanh y = x$$

The remaining inverse hyperbolic functions are defined similarly (see Exercise 28).

We can sketch the graphs of $\sinh^{-1}$, $\cosh^{-1}$, and $\tanh^{-1}$ in Figures 8, 9, and 10 by using Figures 1, 2, and 3.

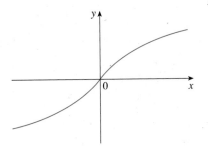

FIGURE 8 $y = \sinh^{-1} x$
domain $= \mathbb{R}$ range $= \mathbb{R}$

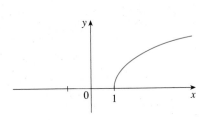

FIGURE 9 $y = \cosh^{-1} x$
domain $= [1, \infty)$ range $= [0, \infty)$

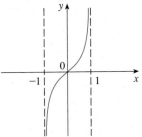

FIGURE 10 $y = \tanh^{-1} x$
domain $= (-1, 1)$ range $= \mathbb{R}$

Since the hyperbolic functions are defined in terms of exponential functions, it's not surprising to learn that the inverse hyperbolic functions can be expressed in terms of logarithms. In particular, we have:

Formula 3 is proved in Example 3. The proofs of Formulas 4 and 5 are requested in Exercises 26 and 27.

$$
\boxed{3} \qquad \sinh^{-1}x = \ln\left(x + \sqrt{x^2 + 1}\right) \qquad x \in \mathbb{R}
$$

$$
\boxed{4} \qquad \cosh^{-1}x = \ln\left(x + \sqrt{x^2 - 1}\right) \qquad x \geqslant 1
$$

$$
\boxed{5} \qquad \tanh^{-1}x = \tfrac{1}{2} \ln\left(\frac{1 + x}{1 - x}\right) \qquad -1 < x < 1
$$

EXAMPLE 3 Show that $\sinh^{-1}x = \ln\left(x + \sqrt{x^2 + 1}\right)$.

SOLUTION Let $y = \sinh^{-1}x$. Then

$$
x = \sinh y = \frac{e^y - e^{-y}}{2}
$$

so

$$
e^y - 2x - e^{-y} = 0
$$

or, multiplying by e^y,

$$
e^{2y} - 2xe^y - 1 = 0
$$

This is really a quadratic equation in e^y:

$$
(e^y)^2 - 2x(e^y) - 1 = 0
$$

Solving by the quadratic formula, we get

$$
e^y = \frac{2x \pm \sqrt{4x^2 + 4}}{2} = x \pm \sqrt{x^2 + 1}
$$

Note that $e^y > 0$, but $x - \sqrt{x^2 + 1} < 0$ (because $x < \sqrt{x^2 + 1}$). Thus the minus sign is inadmissible and we have

$$
e^y = x + \sqrt{x^2 + 1}
$$

Therefore

$$
y = \ln(e^y) = \ln\left(x + \sqrt{x^2 + 1}\right)
$$

(See Exercise 25 for another method.)

6 | Derivatives of Inverse Hyperbolic Functions

$$\frac{d}{dx}(\sinh^{-1}x) = \frac{1}{\sqrt{1+x^2}} \qquad \frac{d}{dx}(\text{csch}^{-1}x) = -\frac{1}{|x|\sqrt{x^2+1}}$$

$$\frac{d}{dx}(\cosh^{-1}x) = \frac{1}{\sqrt{x^2-1}} \qquad \frac{d}{dx}(\text{sech}^{-1}x) = -\frac{1}{x\sqrt{1-x^2}}$$

$$\frac{d}{dx}(\tanh^{-1}x) = \frac{1}{1-x^2} \qquad \frac{d}{dx}(\coth^{-1}x) = \frac{1}{1-x^2}$$

Notice that the formulas for the derivatives of $\tanh^{-1}x$ and $\coth^{-1}x$ appear to be identical. But the domains of these functions have no numbers in common: $\tanh^{-1}x$ is defined for $|x| < 1$, whereas $\coth^{-1}x$ is defined for $|x| > 1$.

The inverse hyperbolic functions are all differentiable because the hyperbolic functions are differentiable. The formulas in Table 6 can be proved either by the method for inverse functions or by differentiating Formulas 3, 4, and 5.

V EXAMPLE 4 Prove that $\dfrac{d}{dx}(\sinh^{-1}x) = \dfrac{1}{\sqrt{1+x^2}}$.

SOLUTION 1 Let $y = \sinh^{-1}x$. Then $\sinh y = x$. If we differentiate this equation implicitly with respect to x, we get

$$\cosh y \frac{dy}{dx} = 1$$

Since $\cosh^2 y - \sinh^2 y = 1$ and $\cosh y \geq 0$, we have $\cosh y = \sqrt{1 + \sinh^2 y}$, so

$$\frac{dy}{dx} = \frac{1}{\cosh y} = \frac{1}{\sqrt{1 + \sinh^2 y}} = \frac{1}{\sqrt{1+x^2}}$$

SOLUTION 2 From Equation 3 (proved in Example 3), we have

$$\frac{d}{dx}(\sinh^{-1}x) = \frac{d}{dx}\ln\left(x + \sqrt{x^2+1}\right)$$

$$= \frac{1}{x + \sqrt{x^2+1}}\frac{d}{dx}\left(x + \sqrt{x^2+1}\right)$$

$$= \frac{1}{x + \sqrt{x^2+1}}\left(1 + \frac{x}{\sqrt{x^2+1}}\right)$$

$$= \frac{\sqrt{x^2+1} + x}{\left(x + \sqrt{x^2+1}\right)\sqrt{x^2+1}}$$

$$= \frac{1}{\sqrt{x^2+1}}$$

V EXAMPLE 5 Find $\dfrac{d}{dx}[\tanh^{-1}(\sin x)]$.

SOLUTION Using Table 6 and the Chain Rule, we have

$$\frac{d}{dx}[\tanh^{-1}(\sin x)] = \frac{1}{1 - (\sin x)^2}\frac{d}{dx}(\sin x)$$

$$= \frac{1}{1 - \sin^2 x}\cos x = \frac{\cos x}{\cos^2 x} = \sec x$$

V **EXAMPLE 6** Evaluate $\int_0^1 \dfrac{dx}{\sqrt{1+x^2}}$.

SOLUTION Using Table 6 (or Example 4) we know that an antiderivative of $1/\sqrt{1+x^2}$ is $\sinh^{-1}x$. Therefore

$$\int_0^1 \frac{dx}{\sqrt{1+x^2}} = \sinh^{-1}x\Big]_0^1$$

$$= \sinh^{-1} 1$$

$$= \ln\left(1 + \sqrt{2}\,\right) \qquad \text{(from Equation 3)}$$

6.7 Exercises

Access end-of-section exercises and tools for visualizing calculus online at **www.webassign.net**   WebAssign ᴱᴺᴴᴬᴺᶜᴱᴰ

6.8 Indeterminate Forms and l'Hospital's Rule

Suppose we are trying to analyze the behavior of the function

$$F(x) = \frac{\ln x}{x - 1}$$

Although F is not defined when $x = 1$, we need to know how F behaves *near* 1. In particular, we would like to know the value of the limit

1 $$\lim_{x \to 1} \frac{\ln x}{x - 1}$$

In computing this limit we can't apply Law 5 of limits (the limit of a quotient is the quotient of the limits, see Section 1.6) because the limit of the denominator is 0. In fact, although the limit in 1 exists, its value is not obvious because both numerator and denominator approach 0 and $\frac{0}{0}$ is not defined.

In general, if we have a limit of the form

$$\lim_{x \to a} \frac{f(x)}{g(x)}$$

where both $f(x) \to 0$ and $g(x) \to 0$ as $x \to a$, then this limit may or may not exist and is called an **indeterminate form of type $\frac{0}{0}$**. We met some limits of this type in Chapter 1. For rational functions, we can cancel common factors:

$$\lim_{x \to 1} \frac{x^2 - x}{x^2 - 1} = \lim_{x \to 1} \frac{x(x - 1)}{(x + 1)(x - 1)} = \lim_{x \to 1} \frac{x}{x + 1} = \frac{1}{2}$$

We used a geometric argument to show that

$$\lim_{x \to 0} \frac{\sin x}{x} = 1$$

But these methods do not work for limits such as 1, so in this section we introduce a systematic method, known as *l'Hospital's Rule*, for the evaluation of indeterminate forms.

Another situation in which a limit is not obvious occurs when we look for a horizontal asymptote of F and need to evaluate its limit at infinity:

$$\boxed{2} \qquad \lim_{x \to \infty} \frac{\ln x}{x - 1}$$

It isn't obvious how to evaluate this limit because both numerator and denominator become large as $x \to \infty$. There is a struggle between numerator and denominator. If the numerator wins, the limit will be ∞; if the denominator wins, the answer will be 0. Or there may be some compromise, in which case the answer will be some finite positive number.

In general, if we have a limit of the form

$$\lim_{x \to a} \frac{f(x)}{g(x)}$$

where both $f(x) \to \infty$ (or $-\infty$) and $g(x) \to \infty$ (or $-\infty$), then the limit may or may not exist and is called an **indeterminate form of type ∞/∞**. We saw in Section 3.4 that this type of limit can be evaluated for certain functions, including rational functions, by dividing numerator and denominator by the highest power of x that occurs in the denominator. For instance,

$$\lim_{x \to \infty} \frac{x^2 - 1}{2x^2 + 1} = \lim_{x \to \infty} \frac{1 - \dfrac{1}{x^2}}{2 + \dfrac{1}{x^2}} = \frac{1 - 0}{2 + 0} = \frac{1}{2}$$

This method does not work for limits such as $\boxed{2}$, but l'Hospital's Rule also applies to this type of indeterminate form.

L'Hospital

L'Hospital's Rule is named after a French nobleman, the Marquis de l'Hospital (1661–1704), but was discovered by a Swiss mathematician, John Bernoulli (1667–1748). You might sometimes see l'Hospital spelled as l'Hôpital, but he spelled his own name l'Hospital, as was common in the 17th century. See Exercise 93 for the example that the Marquis used to illustrate his rule. See the project on page 377 for further historical details.

> **L'Hospital's Rule** Suppose f and g are differentiable and $g'(x) \neq 0$ on an open interval I that contains a (except possibly at a). Suppose that
>
> $$\lim_{x \to a} f(x) = 0 \qquad \text{and} \qquad \lim_{x \to a} g(x) = 0$$
>
> or that
>
> $$\lim_{x \to a} f(x) = \pm\infty \qquad \text{and} \qquad \lim_{x \to a} g(x) = \pm\infty$$
>
> (In other words, we have an indeterminate form of type $\frac{0}{0}$ or ∞/∞.) Then
>
> $$\lim_{x \to a} \frac{f(x)}{g(x)} = \lim_{x \to a} \frac{f'(x)}{g'(x)}$$
>
> if the limit on the right side exists (or is ∞ or $-\infty$).

NOTE 1 L'Hospital's Rule says that the limit of a quotient of functions is equal to the limit of the quotient of their derivatives, provided that the given conditions are satisfied. It is especially important to verify the conditions regarding the limits of f and g before using l'Hospital's Rule.

FIGURE 1

Figure 1 suggests visually why l'Hospital's Rule might be true. The first graph shows two differentiable functions f and g, each of which approaches 0 as $x \to a$. If we were to zoom in toward the point $(a, 0)$, the graphs would start to look almost linear. But if the functions actually *were* linear, as in the second graph, then their ratio would be

$$\frac{m_1(x - a)}{m_2(x - a)} = \frac{m_1}{m_2}$$

which is the ratio of their derivatives. This suggests that

$$\lim_{x \to a} \frac{f(x)}{g(x)} = \lim_{x \to a} \frac{f'(x)}{g'(x)}$$

⊘ Notice that when using l'Hospital's Rule we differentiate the numerator and denominator *separately*. We do *not* use the Quotient Rule.

The graph of the function of Example 2 is shown in Figure 2. We have noticed previously that exponential functions grow far more rapidly than power functions, so the result of Example 2 is not unexpected. See also Exercise 71.

FIGURE 2

NOTE 2 L'Hospital's Rule is also valid for one-sided limits and for limits at infinity or negative infinity; that is, "$x \to a$" can be replaced by any of the symbols $x \to a^+$, $x \to a^-$, $x \to \infty$, or $x \to -\infty$.

NOTE 3 For the special case in which $f(a) = g(a) = 0$, f' and g' are continuous, and $g'(a) \neq 0$, it is easy to see why l'Hospital's Rule is true. In fact, using the alternative form of the definition of a derivative, we have

$$\lim_{x \to a} \frac{f'(x)}{g'(x)} = \frac{f'(a)}{g'(a)} = \frac{\displaystyle\lim_{x \to a} \frac{f(x) - f(a)}{x - a}}{\displaystyle\lim_{x \to a} \frac{g(x) - g(a)}{x - a}} = \lim_{x \to a} \frac{\dfrac{f(x) - f(a)}{x - a}}{\dfrac{g(x) - g(a)}{x - a}}$$

$$= \lim_{x \to a} \frac{f(x) - f(a)}{g(x) - g(a)} = \lim_{x \to a} \frac{f(x)}{g(x)}$$

The general version of l'Hospital's Rule for the indeterminate form $\frac{0}{0}$ is somewhat more difficult and its proof is deferred to the end of this section. The proof for the indeterminate form ∞/∞ can be found in more advanced books.

V **EXAMPLE 1** Find $\displaystyle\lim_{x \to 1} \frac{\ln x}{x - 1}$.

SOLUTION Since

$$\lim_{x \to 1} \ln x = \ln 1 = 0 \qquad \text{and} \qquad \lim_{x \to 1} (x - 1) = 0$$

we can apply l'Hospital's Rule:

$$\lim_{x \to 1} \frac{\ln x}{x - 1} = \lim_{x \to 1} \frac{\dfrac{d}{dx} (\ln x)}{\dfrac{d}{dx} (x - 1)} = \lim_{x \to 1} \frac{1/x}{1}$$

$$= \lim_{x \to 1} \frac{1}{x} = 1$$

V **EXAMPLE 2** Calculate $\displaystyle\lim_{x \to \infty} \frac{e^x}{x^2}$.

SOLUTION We have $\lim_{x \to \infty} e^x = \infty$ and $\lim_{x \to \infty} x^2 = \infty$, so l'Hospital's Rule gives

$$\lim_{x \to \infty} \frac{e^x}{x^2} = \lim_{x \to \infty} \frac{\dfrac{d}{dx} (e^x)}{\dfrac{d}{dx} (x^2)} = \lim_{x \to \infty} \frac{e^x}{2x}$$

Since $e^x \to \infty$ and $2x \to \infty$ as $x \to \infty$, the limit on the right side is also indeterminate, but a second application of l'Hospital's Rule gives

$$\lim_{x \to \infty} \frac{e^x}{x^2} = \lim_{x \to \infty} \frac{e^x}{2x} = \lim_{x \to \infty} \frac{e^x}{2} = \infty$$

The graph of the function of Example 3 is shown in Figure 3. We have discussed previously the slow growth of logarithms, so it isn't surprising that this ratio approaches 0 as $x \to \infty$. See also Exercise 72.

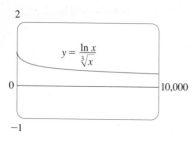

FIGURE 3

V **EXAMPLE 3** Calculate $\lim_{x \to \infty} \dfrac{\ln x}{\sqrt[3]{x}}$.

SOLUTION Since $\ln x \to \infty$ and $\sqrt[3]{x} \to \infty$ as $x \to \infty$, l'Hospital's Rule applies:

$$\lim_{x \to \infty} \frac{\ln x}{\sqrt[3]{x}} = \lim_{x \to \infty} \frac{1/x}{\frac{1}{3} x^{-2/3}}$$

Notice that the limit on the right side is now indeterminate of type $\frac{0}{0}$. But instead of applying l'Hospital's Rule a second time as we did in Example 2, we simplify the expression and see that a second application is unnecessary:

$$\lim_{x \to \infty} \frac{\ln x}{\sqrt[3]{x}} = \lim_{x \to \infty} \frac{1/x}{\frac{1}{3} x^{-2/3}} = \lim_{x \to \infty} \frac{3}{\sqrt[3]{x}} = 0$$

EXAMPLE 4 Find $\lim_{x \to 0} \dfrac{\tan x - x}{x^3}$. (See Exercise 42 in Section 1.5.)

SOLUTION Noting that both $\tan x - x \to 0$ and $x^3 \to 0$ as $x \to 0$, we use l'Hospital's Rule:

$$\lim_{x \to 0} \frac{\tan x - x}{x^3} = \lim_{x \to 0} \frac{\sec^2 x - 1}{3x^2}$$

Since the limit on the right side is still indeterminate of type $\frac{0}{0}$, we apply l'Hospital's Rule again:

$$\lim_{x \to 0} \frac{\sec^2 x - 1}{3x^2} = \lim_{x \to 0} \frac{2 \sec^2 x \tan x}{6x}$$

Because $\lim_{x \to 0} \sec^2 x = 1$, we simplify the calculation by writing

$$\lim_{x \to 0} \frac{2 \sec^2 x \tan x}{6x} = \frac{1}{3} \lim_{x \to 0} \sec^2 x \cdot \lim_{x \to 0} \frac{\tan x}{x} = \frac{1}{3} \lim_{x \to 0} \frac{\tan x}{x}$$

The graph in Figure 4 gives visual confirmation of the result of Example 4. If we were to zoom in too far, however, we would get an inaccurate graph because $\tan x$ is close to x when x is small. See Exercise 42(d) in Section 1.5.

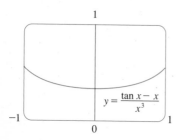

FIGURE 4

We can evaluate this last limit either by using l'Hospital's Rule a third time or by writing $\tan x$ as $(\sin x)/(\cos x)$ and making use of our knowledge of trigonometric limits. Putting together all the steps, we get

$$\lim_{x \to 0} \frac{\tan x - x}{x^3} = \lim_{x \to 0} \frac{\sec^2 x - 1}{3x^2} = \lim_{x \to 0} \frac{2 \sec^2 x \tan x}{6x}$$

$$= \frac{1}{3} \lim_{x \to 0} \frac{\tan x}{x} = \frac{1}{3} \lim_{x \to 0} \frac{\sec^2 x}{1} = \frac{1}{3}$$

V **EXAMPLE 5** Find $\lim_{x \to \pi^-} \dfrac{\sin x}{1 - \cos x}$.

SOLUTION If we blindly attempted to use l'Hospital's Rule, we would get

$$\lim_{x \to \pi^-} \frac{\sin x}{1 - \cos x} = \lim_{x \to \pi^-} \frac{\cos x}{\sin x} = -\infty$$

This is **wrong!** Although the numerator $\sin x \to 0$ as $x \to \pi^-$, notice that the denominator $(1 - \cos x)$ does not approach 0, so l'Hospital's Rule can't be applied here.

The required limit is, in fact, easy to find because the function is continuous at π and the denominator is nonzero there:

$$\lim_{x \to \pi^-} \frac{\sin x}{1 - \cos x} = \frac{\sin \pi}{1 - \cos \pi} = \frac{0}{1 - (-1)} = 0$$

Example 5 shows what can go wrong if you use l'Hospital's Rule without thinking. Other limits *can* be found using l'Hospital's Rule but are more easily found by other methods. (See Examples 3 and 5 in Section 1.6, Example 3 in Section 3.4, and the discussion at the beginning of this section.) So when evaluating any limit, you should consider other methods before using l'Hospital's Rule.

Indeterminate Products

If $\lim_{x \to a} f(x) = 0$ and $\lim_{x \to a} g(x) = \infty$ (or $-\infty$), then it isn't clear what the value of $\lim_{x \to a} [f(x) g(x)]$, if any, will be. There is a struggle between f and g. If f wins, the answer will be 0; if g wins, the answer will be ∞ (or $-\infty$). Or there may be a compromise where the answer is a finite nonzero number. This kind of limit is called an **indeterminate form of type $0 \cdot \infty$**. We can deal with it by writing the product fg as a quotient:

$$fg = \frac{f}{1/g} \qquad \text{or} \qquad fg = \frac{g}{1/f}$$

This converts the given limit into an indeterminate form of type $\frac{0}{0}$ or ∞/∞ so that we can use l'Hospital's Rule.

V **EXAMPLE 6** Evaluate $\lim_{x \to 0^+} x \ln x$.

SOLUTION The given limit is indeterminate because, as $x \to 0^+$, the first factor (x) approaches 0 while the second factor $(\ln x)$ approaches $-\infty$. Writing $x = 1/(1/x)$, we have $1/x \to \infty$ as $x \to 0^+$, so l'Hospital's Rule gives

$$\lim_{x \to 0^+} x \ln x = \lim_{x \to 0^+} \frac{\ln x}{1/x} = \lim_{x \to 0^+} \frac{1/x}{-1/x^2} = \lim_{x \to 0^+} (-x) = 0$$

NOTE In solving Example 6 another possible option would have been to write

$$\lim_{x \to 0^+} x \ln x = \lim_{x \to 0^+} \frac{x}{1/\ln x}$$

This gives an indeterminate form of the type 0/0, but if we apply l'Hospital's Rule we get a more complicated expression than the one we started with. In general, when we rewrite an indeterminate product, we try to choose the option that leads to the simpler limit.

EXAMPLE 7 Use l'Hospital's Rule to help sketch the graph of $f(x) = xe^x$.

SOLUTION Because both x and e^x become large as $x \to \infty$, we have $\lim_{x \to \infty} xe^x = \infty$. As $x \to -\infty$, however, $e^x \to 0$ and so we have an indeterminate product that requires the use of l'Hospital's Rule:

$$\lim_{x \to -\infty} xe^x = \lim_{x \to -\infty} \frac{x}{e^{-x}} = \lim_{x \to -\infty} \frac{1}{-e^{-x}} = \lim_{x \to -\infty} (-e^x) = 0$$

Thus the x-axis is a horizontal asymptote.

Figure 5 shows the graph of the function in Example 6. Notice that the function is undefined at $x = 0$; the graph approaches the origin but never quite reaches it.

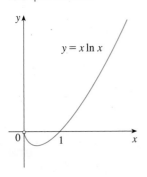

$y = x \ln x$

FIGURE 5

We use the methods of Chapter 3 to gather other information concerning the graph. The derivative is

$$f'(x) = xe^x + e^x = (x + 1)e^x$$

Since e^x is always positive, we see that $f'(x) > 0$ when $x + 1 > 0$, and $f'(x) < 0$ when $x + 1 < 0$. So f is increasing on $(-1, \infty)$ and decreasing on $(-\infty, -1)$. Because $f'(-1) = 0$ and f' changes from negative to positive at $x = -1$, $f(-1) = -e^{-1}$ is a local (and absolute) minimum. The second derivative is

$$f''(x) = (x + 1)e^x + e^x = (x + 2)e^x$$

Since $f''(x) > 0$ if $x > -2$ and $f''(x) < 0$ if $x < -2$, f is concave upward on $(-2, \infty)$ and concave downward on $(-\infty, -2)$. The inflection point is $(-2, -2e^{-2})$.

We use this information to sketch the curve in Figure 6.

FIGURE 6

Indeterminate Differences

If $\lim_{x \to a} f(x) = \infty$ and $\lim_{x \to a} g(x) = \infty$, then the limit

$$\lim_{x \to a} [f(x) - g(x)]$$

is called an **indeterminate form of type $\infty - \infty$**. Again there is a contest between f and g. Will the answer be ∞ (f wins) or will it be $-\infty$ (g wins) or will they compromise on a finite number? To find out, we try to convert the difference into a quotient (for instance, by using a common denominator, or rationalization, or factoring out a common factor) so that we have an indeterminate form of type $\frac{0}{0}$ or ∞/∞.

EXAMPLE 8 Compute $\displaystyle\lim_{x \to (\pi/2)^-} (\sec x - \tan x)$.

SOLUTION First notice that $\sec x \to \infty$ and $\tan x \to \infty$ as $x \to (\pi/2)^-$, so the limit is indeterminate. Here we use a common denominator:

$$\lim_{x \to (\pi/2)^-} (\sec x - \tan x) = \lim_{x \to (\pi/2)^-} \left(\frac{1}{\cos x} - \frac{\sin x}{\cos x} \right)$$

$$= \lim_{x \to (\pi/2)^-} \frac{1 - \sin x}{\cos x} = \lim_{x \to (\pi/2)^-} \frac{-\cos x}{-\sin x} = 0$$

Note that the use of l'Hospital's Rule is justified because $1 - \sin x \to 0$ and $\cos x \to 0$ as $x \to (\pi/2)^-$.

Indeterminate Powers

Several indeterminate forms arise from the limit

$$\lim_{x \to a} [f(x)]^{g(x)}$$

1. $\displaystyle\lim_{x \to a} f(x) = 0$ and $\displaystyle\lim_{x \to a} g(x) = 0$ type 0^0

2. $\displaystyle\lim_{x \to a} f(x) = \infty$ and $\displaystyle\lim_{x \to a} g(x) = 0$ type ∞^0

3. $\displaystyle\lim_{x \to a} f(x) = 1$ and $\displaystyle\lim_{x \to a} g(x) = \pm\infty$ type 1^∞

Although forms of the type 0^0, ∞^0, and 1^∞ are indeterminate, the form 0^∞ is not indeterminate. (See Exercise 96.)

Each of these three cases can be treated either by taking the natural logarithm:

$$\text{let }\quad y = [f(x)]^{g(x)}, \quad\text{then}\quad \ln y = g(x)\ln f(x)$$

or by writing the function as an exponential:

$$[f(x)]^{g(x)} = e^{g(x)\ln f(x)}$$

(Recall that both of these methods were used in differentiating such functions.) In either method we are led to the indeterminate product $g(x)\ln f(x)$, which is of type $0 \cdot \infty$.

EXAMPLE 9   Calculate $\displaystyle\lim_{x\to 0^+}(1 + \sin 4x)^{\cot x}$.

SOLUTION   First notice that as $x \to 0^+$, we have $1 + \sin 4x \to 1$ and $\cot x \to \infty$, so the given limit is indeterminate. Let

$$y = (1 + \sin 4x)^{\cot x}$$

Then

$$\ln y = \ln[(1 + \sin 4x)^{\cot x}] = \cot x \ln(1 + \sin 4x)$$

so l'Hospital's Rule gives

$$\lim_{x\to 0^+}\ln y = \lim_{x\to 0^+}\frac{\ln(1 + \sin 4x)}{\tan x}$$

$$= \lim_{x\to 0^+}\frac{\dfrac{4\cos 4x}{1 + \sin 4x}}{\sec^2 x} = 4$$

So far we have computed the limit of $\ln y$, but what we want is the limit of y. To find this we use the fact that $y = e^{\ln y}$:

$$\lim_{x\to 0^+}(1 + \sin 4x)^{\cot x} = \lim_{x\to 0^+}y = \lim_{x\to 0^+}e^{\ln y} = e^4 \qquad\blacksquare$$

The graph of the function $y = x^x$, $x > 0$, is shown in Figure 7. Notice that although 0^0 is not defined, the values of the function approach 1 as $x \to 0^+$. This confirms the result of Example 10.

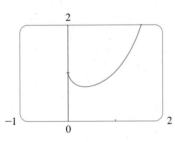

FIGURE 7

V   EXAMPLE 10   Find $\displaystyle\lim_{x\to 0^+}x^x$.

SOLUTION   Notice that this limit is indeterminate since $0^x = 0$ for any $x > 0$ but $x^0 = 1$ for any $x \neq 0$. We could proceed as in Example 9 or by writing the function as an exponential:

$$x^x = (e^{\ln x})^x = e^{x\ln x}$$

In Example 6 we used l'Hospital's Rule to show that

$$\lim_{x\to 0^+}x\ln x = 0$$

Therefore

$$\lim_{x\to 0^+}x^x = \lim_{x\to 0^+}e^{x\ln x} = e^0 = 1 \qquad\blacksquare$$

In order to give the promised proof of l'Hospital's Rule, we first need a generalization of the Mean Value Theorem. The following theorem is named after another French mathematician, Augustin-Louis Cauchy (1789–1857).

See the biographical sketch of Cauchy
on page 60.

> **3** **Cauchy's Mean Value Theorem** Suppose that the functions f and g are continuous on $[a, b]$ and differentiable on (a, b), and $g'(x) \neq 0$ for all x in (a, b). Then there is a number c in (a, b) such that
>
> $$\frac{f'(c)}{g'(c)} = \frac{f(b) - f(a)}{g(b) - g(a)}$$

Notice that if we take the special case in which $g(x) = x$, then $g'(c) = 1$ and Theorem 3 is just the ordinary Mean Value Theorem. Furthermore, Theorem 3 can be proved in a similar manner. You can verify that all we have to do is change the function h given by Equation 3.2.4 to the function

$$h(x) = f(x) - f(a) - \frac{f(b) - f(a)}{g(b) - g(a)} [g(x) - g(a)]$$

and apply Rolle's Theorem as before.

PROOF OF L'HOSPITAL'S RULE We are assuming that $\lim_{x \to a} f(x) = 0$ and $\lim_{x \to a} g(x) = 0$. Let

$$L = \lim_{x \to a} \frac{f'(x)}{g'(x)}$$

We must show that $\lim_{x \to a} [f(x)/g(x)] = L$. Define

$$F(x) = \begin{cases} f(x) & \text{if } x \neq a \\ 0 & \text{if } x = a \end{cases} \qquad G(x) = \begin{cases} g(x) & \text{if } x \neq a \\ 0 & \text{if } x = a \end{cases}$$

Then F is continuous on I since f is continuous on $\{x \in I \mid x \neq a\}$ and

$$\lim_{x \to a} F(x) = \lim_{x \to a} f(x) = 0 = F(a)$$

Likewise, G is continuous on I. Let $x \in I$ and $x > a$. Then F and G are continuous on $[a, x]$ and differentiable on (a, x) and $G' \neq 0$ there (since $F' = f'$ and $G' = g'$). Therefore, by Cauchy's Mean Value Theorem, there is a number y such that $a < y < x$ and

$$\frac{F'(y)}{G'(y)} = \frac{F(x) - F(a)}{G(x) - G(a)} = \frac{F(x)}{G(x)}$$

Here we have used the fact that, by definition, $F(a) = 0$ and $G(a) = 0$. Now, if we let $x \to a^+$, then $y \to a^+$ (since $a < y < x$), so

$$\lim_{x \to a^+} \frac{f(x)}{g(x)} = \lim_{x \to a^+} \frac{F(x)}{G(x)} = \lim_{y \to a^+} \frac{F'(y)}{G'(y)} = \lim_{y \to a^+} \frac{f'(y)}{g'(y)} = L$$

A similar argument shows that the left-hand limit is also L. Therefore

$$\lim_{x \to a} \frac{f(x)}{g(x)} = L$$

This proves l'Hospital's Rule for the case where a is finite.

If a is infinite, we let $t = 1/x$. Then $t \to 0^+$ as $x \to \infty$, so we have

$$\lim_{x \to \infty} \frac{f(x)}{g(x)} = \lim_{t \to 0^+} \frac{f(1/t)}{g(1/t)}$$

$$= \lim_{t \to 0^+} \frac{f'(1/t)(-1/t^2)}{g'(1/t)(-1/t^2)} \quad \text{(by l'Hospital's Rule for finite } a\text{)}$$

$$= \lim_{t \to 0^+} \frac{f'(1/t)}{g'(1/t)} = \lim_{x \to \infty} \frac{f'(x)}{g'(x)}$$

6.8 Exercises

Access end-of-section exercises and tools for visualizing calculus online at **www.webassign.net**

WRITING PROJECT | THE ORIGINS OF L'HOSPITAL'S RULE

www.stewartcalculus.com

The Internet is another source of information for this project. Click on *History of Mathematics* for a list of reliable websites.

L'Hospital's Rule was first published in 1696 in the Marquis de l'Hospital's calculus textbook *Analyse des Infiniment Petits,* but the rule was discovered in 1694 by the Swiss mathematician John (Johann) Bernoulli. The explanation is that these two mathematicians had entered into a curious business arrangement whereby the Marquis de l'Hospital bought the rights to Bernoulli's mathematical discoveries. The details, including a translation of l'Hospital's letter to Bernoulli proposing the arrangement, can be found in the book by Eves [1].

Write a report on the historical and mathematical origins of l'Hospital's Rule. Start by providing brief biographical details of both men (the dictionary edited by Gillispie [2] is a good source) and outline the business deal between them. Then give l'Hospital's statement of his rule, which is found in Struik's sourcebook [4] and more briefly in the book of Katz [3]. Notice that l'Hospital and Bernoulli formulated the rule geometrically and gave the answer in terms of differentials. Compare their statement with the version of l'Hospital's Rule given in Section 6.8 and show that the two statements are essentially the same.

1. Howard Eves, *In Mathematical Circles (Volume 2: Quadrants III and IV)* (Boston: Prindle, Weber and Schmidt, 1969), pp. 20–22.

2. C. C. Gillispie, ed., *Dictionary of Scientific Biography* (New York: Scribner's, 1974). See the article on Johann Bernoulli by E. A. Fellmann and J. O. Fleckenstein in Volume II and the article on the Marquis de l'Hospital by Abraham Robinson in Volume VIII.

3. Victor Katz, *A History of Mathematics: An Introduction* (New York: HarperCollins, 1993), p. 484.

4. D. J. Struik, ed., *A Sourcebook in Mathematics, 1200–1800* (Princeton, NJ: Princeton University Press, 1969), pp. 315–316.

6   Review

Concept Check

1. (a) What is a one-to-one function? How can you tell if a function is one-to-one by looking at its graph?
   (b) If f is a one-to-one function, how is its inverse function f^{-1} defined? How do you obtain the graph of f^{-1} from the graph of f?
   (c) If f is a one-to-one function and $f'(f^{-1}(a)) \neq 0$, write a formula for $(f^{-1})'(a)$.

2. (a) What are the domain and range of the natural exponential function $f(x) = e^x$?
   (b) What are the domain and range of the natural logarithmic function $f(x) = \ln x$?
   (c) How are the graphs of these functions related? Sketch these graphs by hand, using the same axes.
   (d) If a is a positive number, $a \neq 1$, write an equation that expresses $\log_a x$ in terms of $\ln x$.

3. (a) How is the inverse sine function $f(x) = \sin^{-1}x$ defined? What are its domain and range?
   (b) How is the inverse cosine function $f(x) = \cos^{-1}x$ defined? What are its domain and range?
   (c) How is the inverse tangent function $f(x) = \tan^{-1}x$ defined? What are its domain and range? Sketch its graph.

4. Write the definitions of the hyperbolic functions $\sinh x$, $\cosh x$, and $\tanh x$.

5. State the derivative of each function.
   (a) $y = e^x$ (b) $y = a^x$ (c) $y = \ln x$

(d) $y = \log_a x$ (e) $y = \sin^{-1}x$ (f) $y = \cos^{-1}x$
(g) $y = \tan^{-1}x$ (h) $y = \sinh x$ (i) $y = \cosh x$
(j) $y = \tanh x$ (k) $y = \sinh^{-1}x$ (l) $y = \cosh^{-1}x$
(m) $y = \tanh^{-1}x$

6. (a) How is the number e defined?
   (b) Express e as a limit.
   (c) Why is the natural exponential function $y = e^x$ used more often in calculus than the other exponential functions $y = a^x$?
   (d) Why is the natural logarithmic function $y = \ln x$ used more often in calculus than the other logarithmic functions $y = \log_a x$?

7. (a) Write a differential equation that expresses the law of natural growth.
   (b) Under what circumstances is this an appropriate model for population growth?
   (c) What are the solutions of this equation?

8. (a) What does l'Hospital's Rule say?
   (b) How can you use l'Hospital's Rule if you have a product $f(x) g(x)$ where $f(x) \to 0$ and $g(x) \to \infty$ as $x \to a$?
   (c) How can you use l'Hospital's Rule if you have a difference $f(x) - g(x)$ where $f(x) \to \infty$ and $g(x) \to \infty$ as $x \to a$?
   (d) How can you use l'Hospital's Rule if you have a power $[f(x)]^{g(x)}$ where $f(x) \to 0$ and $g(x) \to 0$ as $x \to a$?

True-False Quiz

Determine whether the statement is true or false. If it is true, explain why. If it is false, explain why or give an example that disproves the statement.

1. If f is one-to-one, with domain $\mathbb{R}$, then $f^{-1}(f(6)) = 6$.

2. If f is one-to-one and differentiable, with domain $\mathbb{R}$, then $(f^{-1})'(6) = 1/f'(6)$.

3. The function $f(x) = \cos x$, $-\pi/2 \leqslant x \leqslant \pi/2$, is one-to-one.

4. $\tan^{-1}(-1) = 3\pi/4$

5. If $0 < a < b$, then $\ln a < \ln b$.

6. $\pi^{\sqrt{5}} = e^{\sqrt{5} \ln \pi}$

7. You can always divide by e^x.

8. If $a > 0$ and $b > 0$, then $\ln(a + b) = \ln a + \ln b$.

9. If $x > 0$, then $(\ln x)^6 = 6 \ln x$.

10. $\dfrac{d}{dx}(10^x) = x10^{x-1}$

11. $\dfrac{d}{dx}(\ln 10) = \dfrac{1}{10}$

12. The inverse function of $y = e^{3x}$ is $y = \frac{1}{3} \ln x$.

13. $\cos^{-1}x = \dfrac{1}{\cos x}$

14. $\tan^{-1}x = \dfrac{\sin^{-1}x}{\cos^{-1}x}$

15. $\cosh x \geqslant 1$ for all x

16. $\ln \dfrac{1}{10} = -\displaystyle\int_1^{10} \dfrac{dx}{x}$

17. $\displaystyle\int_2^{16} \dfrac{dx}{x} = 3 \ln 2$

18. $\displaystyle\lim_{x \to \pi^-} \dfrac{\tan x}{1 - \cos x} = \lim_{x \to \pi^-} \dfrac{\sec^2 x}{\sin x} = \infty$

Exercises

1. The graph of f is shown. Is f one-to-one? Explain.

2. The graph of g is given.
   (a) Why is g one-to-one?
   (b) Estimate the value of $g^{-1}(2)$.
   (c) Estimate the domain of g^{-1}.
   (d) Sketch the graph of g^{-1}.

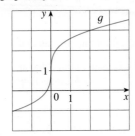

3. Suppose f is one-to-one, $f(7) = 3$, and $f'(7) = 8$. Find
   (a) $f^{-1}(3)$ and (b) $(f^{-1})'(3)$.

4. Find the inverse function of $f(x) = \dfrac{x+1}{2x+1}$.

5–9 Sketch a rough graph of the function without using a calculator.

5. $y = 5^x - 1$

6. $y = -e^{-x}$

7. $y = -\ln x$

8. $y = \ln(x - 1)$

9. $y = 2\arctan x$

10. Let $a > 1$. For large values of x, which of the functions $y = x^a$, $y = a^x$, and $y = \log_a x$ has the largest values and which has the smallest values?

11–12 Find the exact value of each expression.

11. (a) $e^{2\ln 3}$ (b) $\log_{10} 25 + \log_{10} 4$

12. (a) $\ln e^{\pi}$ (b) $\tan\!\left(\arcsin \tfrac{1}{2}\right)$

13–20 Solve the equation for x.

13. $\ln x = \tfrac{1}{3}$

14. $e^x = \tfrac{1}{3}$

15. $e^{e^x} = 17$

16. $\ln(1 + e^{-x}) = 3$

17. $\ln(x+1) + \ln(x-1) = 1$

18. $\log_5(c^x) = d$

19. $\tan^{-1}x = 1$

20. $\sin x = 0.3$

21–47 Differentiate.

21. $f(t) = t^2 \ln t$

22. $g(t) = \dfrac{e^t}{1 + e^t}$

23. $h(\theta) = e^{\tan 2\theta}$

24. $h(u) = 10^{\sqrt{u}}$

25. $y = \ln|\sec 5x + \tan 5x|$

26. $y = x\cos^{-1}x$

27. $y = x\tan^{-1}(4x)$

28. $y = e^{mx}\cos nx$

29. $y = \ln(\sec^2 x)$

30. $y = \sqrt{t}\,\ln(t^4)$

31. $y = \dfrac{e^{1/x}}{x^2}$

32. $y = (\arcsin 2x)^2$

33. $y = 3^{x\ln x}$

34. $y = e^{\cos x} + \cos(e^x)$

35. $H(v) = v\tan^{-1}v$

36. $F(z) = \log_{10}(1 + z^2)$

37. $y = x\sinh(x^2)$

38. $y = (\cos x)^x$

39. $y = \ln\sin x - \tfrac{1}{2}\sin^2 x$

40. $y = \arctan(\arcsin\sqrt{x}\,)$

41. $y = \ln\!\left(\dfrac{1}{x}\right) + \dfrac{1}{\ln x}$

42. $xe^y = y - 1$

43. $y = \ln(\cosh 3x)$

44. $y = \dfrac{(x^2+1)^4}{(2x+1)^3(3x-1)^5}$

45. $y = \cosh^{-1}(\sinh x)$

46. $y = x\tanh^{-1}\sqrt{x}$

47. $y = \cos\!\left(e^{\sqrt{\tan 3x}}\right)$

48. Show that
$$\frac{d}{dx}\left(\tfrac{1}{2}\tan^{-1}x + \tfrac{1}{4}\ln\frac{(x+1)^2}{x^2+1}\right) = \frac{1}{(1+x)(1+x^2)}$$

49–52 Find f' in terms of g'.

49. $f(x) = e^{g(x)}$

50. $f(x) = g(e^x)$

51. $f(x) = \ln|g(x)|$

52. $f(x) = g(\ln x)$

53–54 Find $f^{(n)}(x)$.

53. $f(x) = 2^x$

54. $f(x) = \ln(2x)$

55. Use mathematical induction to show that if $f(x) = xe^x$, then $f^{(n)}(x) = (x+n)e^x$.

56. Find y' if $y = x + \arctan y$.

57–58 Find an equation of the tangent to the curve at the given point.

57. $y = (2+x)e^{-x}$, $(0, 2)$

58. $y = x\ln x$, (e, e)

59. At what point on the curve $y = [\ln(x+4)]^2$ is the tangent horizontal?

60. If $f(x) = xe^{\sin x}$, find $f'(x)$. Graph f and f' on the same screen and comment.

61. (a) Find an equation of the tangent to the curve $y = e^x$ that is parallel to the line $x - 4y = 1$.
(b) Find an equation of the tangent to the curve $y = e^x$ that passes through the origin.

62. The function $C(t) = K(e^{-at} - e^{-bt})$, where a, b, and K are positive constants and $b > a$, is used to model the concentration at time t of a drug injected into the bloodstream.
(a) Show that $\lim_{t \to \infty} C(t) = 0$.
(b) Find $C'(t)$, the rate at which the drug is cleared from circulation.
(c) When is this rate equal to 0?

63–78 Evaluate the limit.

63. $\lim_{x \to \infty} e^{-3x}$

64. $\lim_{x \to 10^-} \ln(100 - x^2)$

65. $\lim_{x \to 3^-} e^{2/(x-3)}$

66. $\lim_{x \to \infty} \arctan(x^3 - x)$

67. $\lim_{x \to 0^+} \ln(\sinh x)$

68. $\lim_{x \to \infty} e^{-x} \sin x$

69. $\lim_{x \to \infty} \dfrac{1 + 2^x}{1 - 2^x}$

70. $\lim_{x \to \infty} \left(1 + \dfrac{4}{x}\right)^x$

71. $\lim_{x \to 0} \dfrac{e^x - 1}{\tan x}$

72. $\lim_{x \to 0} \dfrac{1 - \cos x}{x^2 + x}$

73. $\lim_{x \to 0} \dfrac{e^{4x} - 1 - 4x}{x^2}$

74. $\lim_{x \to \infty} \dfrac{e^{4x} - 1 - 4x}{x^2}$

75. $\lim_{x \to -\infty} (x^2 - x^3)e^{2x}$

76. $\lim_{x \to 0^+} x^2 \ln x$

77. $\lim_{x \to 1^+} \left(\dfrac{x}{x-1} - \dfrac{1}{\ln x}\right)$

78. $\lim_{x \to (\pi/2)^-} (\tan x)^{\cos x}$

79–84 Sketch the curve using the guidelines of Section 3.5.

79. $y = e^x \sin x$, $-\pi \le x \le \pi$
80. $y = \sin^{-1}(1/x)$
81. $y = x \ln x$
82. $y = e^{2x - x^2}$
83. $y = (x - 2)e^{-x}$
84. $y = x + \ln(x^2 + 1)$

85. Investigate the family of curves given by $f(x) = xe^{-cx}$, where c is a real number. Start by computing the limits as $x \to \pm\infty$. Identify any transitional values of c where the basic shape changes. What happens to the maximum or minimum points and inflection points as c changes? Illustrate by graphing several members of the family.

86. Investigate the family of functions $f(x) = cxe^{-cx^2}$. What happens to the maximum and minimum points and the inflection points as c changes? Illustrate your conclusions by graphing several members of the family.

87. An equation of motion of the form $s = Ae^{-ct}\cos(\omega t + \delta)$ represents damped oscillation of an object. Find the velocity and acceleration of the object.

88. (a) Show that there is exactly one root of the equation $\ln x = 3 - x$ and that it lies between 2 and e.
(b) Find the root of the equation in part (a) correct to four decimal places.

89. A bacteria culture contains 200 cells initially and grows at a rate proportional to its size. After half an hour the population has increased to 360 cells.
(a) Find the number of bacteria after t hours.
(b) Find the number of bacteria after 4 hours.
(c) Find the rate of growth after 4 hours.
(d) When will the population reach 10,000?

90. Cobalt-60 has a half-life of 5.24 years.
(a) Find the mass that remains from a 100-mg sample after 20 years.
(b) How long would it take for the mass to decay to 1 mg?

91. The biologist G. F. Gause conducted an experiment in the 1930s with the protozoan *Paramecium* and used the population function

$$P(t) = \dfrac{64}{1 + 31e^{-0.7944t}}$$

to model his data, where t was measured in days. Use this model to determine when the population was increasing most rapidly.

92–105 Evaluate the integral.

92. $\displaystyle\int_0^4 \dfrac{1}{16 + t^2}\, dt$

93. $\displaystyle\int_0^1 ye^{-2y^2}\, dy$

94. $\displaystyle\int_2^5 \dfrac{dr}{1 + 2r}$

95. $\displaystyle\int_0^1 \dfrac{e^x}{1 + e^{2x}}\, dx$

96. $\displaystyle\int_0^{\pi/2} \dfrac{\cos x}{1 + \sin^2 x}\, dx$

97. $\displaystyle\int \dfrac{e^{\sqrt{x}}}{\sqrt{x}}\, dx$

98. $\displaystyle\int \dfrac{\cos(\ln x)}{x}\, dx$

99. $\displaystyle\int \dfrac{x + 1}{x^2 + 2x}\, dx$

100. $\displaystyle\int \dfrac{\csc^2 x}{1 + \cot x}\, dx$

101. $\displaystyle\int \tan x \ln(\cos x)\, dx$

102. $\displaystyle\int \dfrac{x}{\sqrt{1 - x^4}}\, dx$

103. $\displaystyle\int 2^{\tan\theta} \sec^2\theta\, d\theta$

104. $\displaystyle\int \sinh au\, du$

105. $\displaystyle\int \left(\dfrac{1 - x}{x}\right)^2 dx$

106–108 Use properties of integrals to prove the inequality.

106. $\displaystyle\int_0^1 \sqrt{1 + e^{2x}}\, dx \ge e - 1$

107. $\int_0^1 e^x \cos x \, dx \leqslant e - 1$ **108.** $\int_0^1 x \sin^{-1}x \, dx \leqslant \pi/4$

109–110 Find $f'(x)$.

109. $f(x) = \int_1^{\sqrt{x}} \dfrac{e^s}{s} \, ds$ **110.** $f(x) = \int_{\ln x}^{2x} e^{-t^2} \, dt$

---

111. Find the average value of the function $f(x) = 1/x$ on the interval $[1, 4]$.

112. Find the area of the region bounded by the curves $y = e^x$, $y = e^{-x}$, $x = -2$, and $x = 1$.

113. Find the volume of the solid obtained by rotating about the y-axis the region under the curve $y = 1/(1 + x^4)$ from $x = 0$ to $x = 1$.

114. If $f(x) = x + x^2 + e^x$, find $(f^{-1})'(1)$.

115. If $f(x) = \ln x + \tan^{-1}x$, find $(f^{-1})'(\pi/4)$.

116. What is the area of the largest rectangle in the first quadrant with two sides on the axes and one vertex on the curve $y = e^{-x}$?

117. What is the area of the largest triangle in the first quadrant with two sides on the axes and the third side tangent to the curve $y = e^{-x}$?

118. Evaluate $\int_0^1 e^x \, dx$ without using the Fundamental Theorem of Calculus. [*Hint:* Use the definition of a definite integral with right endpoints, sum a geometric series, and then use l'Hospital's Rule.]

119. If $F(x) = \int_a^b t^x \, dt$, where $a, b > 0$, then, by the Fundamental Theorem,

$$F(x) = \frac{b^{x+1} - a^{x+1}}{x + 1} \qquad x \neq -1$$

$$F(-1) = \ln b - \ln a$$

Use l'Hospital's Rule to show that F is continuous at -1.

120. Show that

$$\cos\{\arctan[\sin(\operatorname{arccot} x)]\} = \sqrt{\frac{x^2 + 1}{x^2 + 2}}$$

121. If f is a continuous function such that

$$\int_0^x f(t) \, dt = xe^{2x} + \int_0^x e^{-t}f(t) \, dt$$

for all x, find an explicit formula for $f(x)$.

122. The figure shows two regions in the first quadrant: $A(t)$ is the area under the curve $y = \sin(x^2)$ from 0 to t, and $B(t)$ is the area of the triangle with vertices O, P, and $(t, 0)$. Find $\displaystyle\lim_{t \to 0^+} [A(t)/B(t)]$.

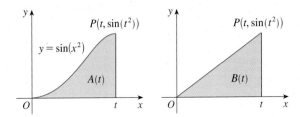

Problems Plus

Cover up the solution to the example and try it yourself.

FIGURE 1

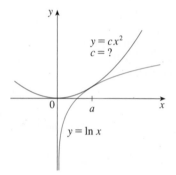

FIGURE 2

EXAMPLE 1 For what values of c does the equation $\ln x = cx^2$ have exactly one solution?

SOLUTION One of the most important principles of problem solving is to draw a diagram, even if the problem as stated doesn't explicitly mention a geometric situation. Our present problem can be reformulated geometrically as follows: For what values of c does the curve $y = \ln x$ intersect the curve $y = cx^2$ in exactly one point?

Let's start by graphing $y = \ln x$ and $y = cx^2$ for various values of c. We know that, for $c \neq 0$, $y = cx^2$ is a parabola that opens upward if $c > 0$ and downward if $c < 0$. Figure 1 shows the parabolas $y = cx^2$ for several positive values of c. Most of them don't intersect $y = \ln x$ at all and one intersects twice. We have the feeling that there must be a value of c (somewhere between 0.1 and 0.3) for which the curves intersect exactly once, as in Figure 2.

To find that particular value of c, we let a be the x-coordinate of the single point of intersection. In other words, $\ln a = ca^2$, so a is the unique solution of the given equation. We see from Figure 2 that the curves just touch, so they have a common tangent line when $x = a$. That means the curves $y = \ln x$ and $y = cx^2$ have the same slope when $x = a$. Therefore

$$\frac{1}{a} = 2ca$$

Solving the equations $\ln a = ca^2$ and $1/a = 2ca$, we get

$$\ln a = ca^2 = c \cdot \frac{1}{2c} = \frac{1}{2}$$

Thus $a = e^{1/2}$ and

$$c = \frac{\ln a}{a^2} = \frac{\ln e^{1/2}}{e} = \frac{1}{2e}$$

For negative values of c we have the situation illustrated in Figure 3: All parabolas $y = cx^2$ with negative values of c intersect $y = \ln x$ exactly once. And let's not forget about $c = 0$: The curve $y = 0x^2 = 0$ is just the x-axis, which intersects $y = \ln x$ exactly once.

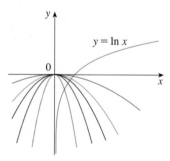

FIGURE 3

To summarize, the required values of c are $c = 1/(2e)$ and $c \leq 0$.

1. If a rectangle has its base on the x-axis and two vertices on the curve $y = e^{-x^2}$, show that the rectangle has the largest possible area when the two vertices are at the points of inflection of the curve.

2. Prove that $\log_2 5$ is an irrational number.

3. Does the function $f(x) = e^{10|x-2|-x^2}$ have an absolute maximum? If so, find it. What about an absolute minimum?

4. If $\int_0^4 e^{(x-2)^4}\, dx = k$, find the value of $\int_0^4 xe^{(x-2)^4}\, dx$.

5. Show that

$$\frac{d^n}{dx^n}(e^{ax}\sin bx) = r^n e^{ax}\sin(bx + n\theta)$$

where a and b are positive numbers, $r^2 = a^2 + b^2$, and $\theta = \tan^{-1}(b/a)$.

6. Show that $\sin^{-1}(\tanh x) = \tan^{-1}(\sinh x)$.

7. Show that, for $x > 0$,

$$\frac{x}{1 + x^2} < \tan^{-1}x < x$$

8. Suppose f is continuous, $f(0) = 0$, $f(1) = 1$, $f'(x) > 0$, and $\int_0^1 f(x)\, dx = \frac{1}{3}$. Find the value of the integral $\int_0^1 f^{-1}(y)\, dy$.

9. Show that $f(x) = \int_1^x \sqrt{1 + t^3}\, dt$ is one-to-one and find $(f^{-1})'(0)$.

10. If

$$y = \frac{x}{\sqrt{a^2 - 1}} - \frac{2}{\sqrt{a^2 - 1}}\arctan\frac{\sin x}{a + \sqrt{a^2 - 1} + \cos x}$$

show that $y' = \dfrac{1}{a + \cos x}$.

11. For what value of a is the following equation true?

$$\lim_{x \to \infty}\left(\frac{x + a}{x - a}\right)^x = e$$

12. Sketch the set of all points (x, y) such that $|x + y| \le e^x$.

13. Prove that $\cosh(\sinh x) < \sinh(\cosh x)$ for all x.

14. Show that, for all positive values of x and y,

$$\frac{e^{x+y}}{xy} \ge e^2$$

15. For what value of k does the equation $e^{2x} = k\sqrt{x}$ have exactly one solution?

16. For which positive numbers a is it true that $a^x \ge 1 + x$ for all x?

17. For which positive numbers a does the curve $y = a^x$ intersect the line $y = x$?

18. For what values of c does the curve $y = cx^3 + e^x$ have inflection points?

Techniques of Integration

Shown is a photograph of Omega Centauri, which contains several million stars and is the largest globular cluster in our galaxy. Astronomers use stellar stereography to determine the actual density of stars in a star cluster from the (two-dimensional) density that can be analyzed from a photograph. In Section 7.8 you are asked to evaluate an integral to calculate the perceived density from the actual density.

Because of the Fundamental Theorem of Calculus, we can integrate a function if we know an antiderivative, that is, an indefinite integral. We summarize here the most important integrals that we have learned so far.

$$\int x^n \, dx = \frac{x^{n+1}}{n+1} + C \quad (n \neq -1) \qquad \int \frac{1}{x} \, dx = \ln|x| + C$$

$$\int e^x \, dx = e^x + C \qquad \int a^x \, dx = \frac{a^x}{\ln a} + C$$

$$\int \sin x \, dx = -\cos x + C \qquad \int \cos x \, dx = \sin x + C$$

$$\int \sec^2 x \, dx = \tan x + C \qquad \int \csc^2 x \, dx = -\cot x + C$$

$$\int \sec x \tan x \, dx = \sec x + C \qquad \int \csc x \cot x \, dx = -\csc x + C$$

$$\int \sinh x \, dx = \cosh x + C \qquad \int \cosh x \, dx = \sinh x + C$$

$$\int \tan x \, dx = \ln|\sec x| + C \qquad \int \cot x \, dx = \ln|\sin x| + C$$

$$\int \frac{1}{x^2 + a^2} \, dx = \frac{1}{a} \tan^{-1}\left(\frac{x}{a}\right) + C \qquad \int \frac{1}{\sqrt{a^2 - x^2}} \, dx = \sin^{-1}\left(\frac{x}{a}\right) + C, \quad a > 0$$

In this chapter we develop techniques for using these basic integration formulas to obtain indefinite integrals of more complicated functions. We learned the most important method of integration, the Substitution Rule, in Section 4.5. The other general technique, integration by parts, is presented in Section 7.1. Then we learn methods that are special to particular classes of functions, such as trigonometric functions and rational functions.

Integration is not as straightforward as differentiation; there are no rules that absolutely guarantee obtaining an indefinite integral of a function. Therefore we discuss a strategy for integration in Section 7.5.

7.1 Integration by Parts

Every differentiation rule has a corresponding integration rule. For instance, the Substitution Rule for integration corresponds to the Chain Rule for differentiation. The rule that corresponds to the Product Rule for differentiation is called the rule for *integration by parts*.

The Product Rule states that if f and g are differentiable functions, then

$$\frac{d}{dx}[f(x)g(x)] = f(x)g'(x) + g(x)f'(x)$$

In the notation for indefinite integrals this equation becomes

$$\int [f(x)g'(x) + g(x)f'(x)]\, dx = f(x)g(x)$$

or

$$\int f(x)g'(x)\, dx + \int g(x)f'(x)\, dx = f(x)g(x)$$

We can rearrange this equation as

$$\boxed{1} \qquad \int f(x)g'(x)\, dx = f(x)g(x) - \int g(x)f'(x)\, dx$$

Formula 1 is called the **formula for integration by parts**. It is perhaps easier to remember in the following notation. Let $u = f(x)$ and $v = g(x)$. Then the differentials are $du = f'(x)\, dx$ and $dv = g'(x)\, dx$, so, by the Substitution Rule, the formula for integration by parts becomes

$$\boxed{2} \qquad \int u\, dv = uv - \int v\, du$$

EXAMPLE 1 Find $\int x \sin x\, dx$.

SOLUTION USING FORMULA 1 Suppose we choose $f(x) = x$ and $g'(x) = \sin x$. Then $f'(x) = 1$ and $g(x) = -\cos x$. (For g we can choose *any* antiderivative of g'.) Thus, using Formula 1, we have

$$\int x \sin x\, dx = f(x)g(x) - \int g(x)f'(x)\, dx$$

$$= x(-\cos x) - \int (-\cos x)\, dx$$

$$= -x \cos x + \int \cos x\, dx$$

$$= -x \cos x + \sin x + C$$

It's wise to check the answer by differentiating it. If we do so, we get $x \sin x$, as expected.

SOLUTION USING FORMULA 2 Let

$$u = x \qquad\qquad dv = \sin x\, dx$$

Then

$$du = dx \qquad\qquad v = -\cos x$$

and so

It is helpful to use the pattern:
$$u = \square \qquad dv = \square$$
$$du = \square \qquad v = \square$$

$$\int x \sin x\, dx = \int \overset{u}{x}\ \overset{dv}{\overbrace{\sin x\, dx}} = \overset{u}{x}\ \overset{v}{\overbrace{(-\cos x)}} - \int \overset{v}{\overbrace{(-\cos x)}}\ \overset{du}{dx}$$

$$= -x \cos x + \int \cos x\, dx$$

$$= -x \cos x + \sin x + C$$

NOTE Our aim in using integration by parts is to obtain a simpler integral than the one we started with. Thus in Example 1 we started with $\int x \sin x\, dx$ and expressed it in terms of the simpler integral $\int \cos x\, dx$. If we had instead chosen $u = \sin x$ and $dv = x\, dx$, then $du = \cos x\, dx$ and $v = x^2/2$, so integration by parts gives

$$\int x \sin x\, dx = (\sin x)\frac{x^2}{2} - \frac{1}{2}\int x^2 \cos x\, dx$$

Although this is true, $\int x^2 \cos x\, dx$ is a more difficult integral than the one we started with. In general, when deciding on a choice for u and dv, we usually try to choose $u = f(x)$ to be a function that becomes simpler when differentiated (or at least not more complicated) as long as $dv = g'(x)\, dx$ can be readily integrated to give v.

V EXAMPLE 2 Evaluate $\int \ln x\, dx$.

SOLUTION Here we don't have much choice for u and dv. Let

$$u = \ln x \qquad dv = dx$$

Then

$$du = \frac{1}{x}\, dx \qquad v = x$$

Integrating by parts, we get

$$\int \ln x\, dx = x \ln x - \int x\,\frac{dx}{x}$$

It's customary to write $\int 1\, dx$ as $\int dx$.

$$= x \ln x - \int dx$$

Check the answer by differentiating it.

$$= x \ln x - x + C$$

Integration by parts is effective in this example because the derivative of the function $f(x) = \ln x$ is simpler than f.

V EXAMPLE 3 Find $\int t^2 e^t\, dt$.

SOLUTION Notice that t^2 becomes simpler when differentiated (whereas e^t is unchanged when differentiated or integrated), so we choose

$$u = t^2 \qquad dv = e^t\, dt$$

Then

$$du = 2t\, dt \qquad v = e^t$$

Integration by parts gives

$$\boxed{3} \qquad\qquad \int t^2 e^t \, dt = t^2 e^t - 2 \int t e^t \, dt$$

The integral that we obtained, $\int t e^t \, dt$, is simpler than the original integral but is still not obvious. Therefore we use integration by parts a second time, this time with $u = t$ and $dv = e^t \, dt$. Then $du = dt$, $v = e^t$, and

$$\int t e^t \, dt = t e^t - \int e^t \, dt$$

$$= t e^t - e^t + C$$

Putting this in Equation 3, we get

$$\int t^2 e^t \, dt = t^2 e^t - 2 \int t e^t \, dt$$

$$= t^2 e^t - 2(t e^t - e^t + C)$$

$$= t^2 e^t - 2 t e^t + 2 e^t + C_1 \qquad \text{where } C_1 = -2C$$ ▬

V   **EXAMPLE 4**   Evaluate $\int e^x \sin x \, dx$.

An easier method, using complex numbers, is given in Exercise 50 in Appendix H.

SOLUTION Neither e^x nor $\sin x$ becomes simpler when differentiated, but we try choosing $u = e^x$ and $dv = \sin x \, dx$ anyway. Then $du = e^x \, dx$ and $v = -\cos x$, so integration by parts gives

$$\boxed{4} \qquad\qquad \int e^x \sin x \, dx = -e^x \cos x + \int e^x \cos x \, dx$$

The integral that we have obtained, $\int e^x \cos x \, dx$, is no simpler than the original one, but at least it's no more difficult. Having had success in the preceding example integrating by parts twice, we persevere and integrate by parts again. This time we use $u = e^x$ and $dv = \cos x \, dx$. Then $du = e^x \, dx$, $v = \sin x$, and

$$\boxed{5} \qquad\qquad \int e^x \cos x \, dx = e^x \sin x - \int e^x \sin x \, dx$$

At first glance, it appears as if we have accomplished nothing because we have arrived at $\int e^x \sin x \, dx$, which is where we started. However, if we put the expression for $\int e^x \cos x \, dx$ from Equation 5 into Equation 4 we get

$$\int e^x \sin x \, dx = -e^x \cos x + e^x \sin x - \int e^x \sin x \, dx$$

This can be regarded as an equation to be solved for the unknown integral. Adding $\int e^x \sin x \, dx$ to both sides, we obtain

$$2 \int e^x \sin x \, dx = -e^x \cos x + e^x \sin x$$

Dividing by 2 and adding the constant of integration, we get

Figure 1 illustrates Example 4 by showing the graphs of $f(x) = e^x \sin x$ and $F(x) = \frac{1}{2} e^x (\sin x - \cos x)$. As a visual check on our work, notice that $f(x) = 0$ when F has a maximum or minimum.

FIGURE 1

$$\int e^x \sin x \, dx = \tfrac{1}{2} e^x (\sin x - \cos x) + C$$ ▬

If we combine the formula for integration by parts with Part 2 of the Fundamental Theorem of Calculus, we can evaluate definite integrals by parts. Evaluating both sides of Formula 1 between a and b, assuming f' and g' are continuous, and using the Fundamental Theorem, we obtain

$$\boxed{6 \qquad \int_a^b f(x)g'(x)\,dx = f(x)g(x)\Big]_a^b - \int_a^b g(x)f'(x)\,dx}$$

EXAMPLE 5 Calculate $\displaystyle\int_0^1 \tan^{-1}x\,dx$.

SOLUTION Let

$$u = \tan^{-1}x \qquad\qquad dv = dx$$

Then

$$du = \frac{dx}{1+x^2} \qquad\qquad v = x$$

So Formula 6 gives

$$\int_0^1 \tan^{-1}x\,dx = x\tan^{-1}x\Big]_0^1 - \int_0^1 \frac{x}{1+x^2}\,dx$$

$$= 1 \cdot \tan^{-1}1 - 0 \cdot \tan^{-1}0 - \int_0^1 \frac{x}{1+x^2}\,dx$$

$$= \frac{\pi}{4} - \int_0^1 \frac{x}{1+x^2}\,dx$$

To evaluate this integral we use the substitution $t = 1 + x^2$ (since u has another meaning in this example). Then $dt = 2x\,dx$, so $x\,dx = \frac{1}{2}\,dt$. When $x = 0$, $t = 1$; when $x = 1$, $t = 2$; so

$$\int_0^1 \frac{x}{1+x^2}\,dx = \frac{1}{2}\int_1^2 \frac{dt}{t} = \frac{1}{2}\ln|t|\,\Big]_1^2$$

$$= \frac{1}{2}(\ln 2 - \ln 1) = \frac{1}{2}\ln 2$$

Therefore

$$\int_0^1 \tan^{-1}x\,dx = \frac{\pi}{4} - \int_0^1 \frac{x}{1+x^2}\,dx = \frac{\pi}{4} - \frac{\ln 2}{2}$$

Since $\tan^{-1}x \geqslant 0$ for $x \geqslant 0$, the integral in Example 5 can be interpreted as the area of the region shown in Figure 2.

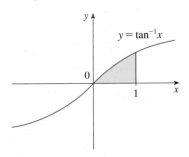

FIGURE 2

EXAMPLE 6 Prove the reduction formula

Equation 7 is called a *reduction formula* because the exponent n has been *reduced* to $n - 1$ and $n - 2$.

$$\boxed{7 \qquad \int \sin^n x\,dx = -\frac{1}{n}\cos x\,\sin^{n-1}x + \frac{n-1}{n}\int \sin^{n-2}x\,dx}$$

where $n \geqslant 2$ is an integer.

SOLUTION Let

$$u = \sin^{n-1}x \qquad\qquad dv = \sin x\,dx$$

Then

$$du = (n-1)\sin^{n-2}x\,\cos x\,dx \qquad\qquad v = -\cos x$$

so integration by parts gives

$$\int \sin^n x \, dx = -\cos x \sin^{n-1} x + (n-1) \int \sin^{n-2} x \cos^2 x \, dx$$

Since $\cos^2 x = 1 - \sin^2 x$, we have

$$\int \sin^n x \, dx = -\cos x \sin^{n-1} x + (n-1) \int \sin^{n-2} x \, dx - (n-1) \int \sin^n x \, dx$$

As in Example 4, we solve this equation for the desired integral by taking the last term on the right side to the left side. Thus we have

$$n \int \sin^n x \, dx = -\cos x \sin^{n-1} x + (n-1) \int \sin^{n-2} x \, dx$$

or

$$\int \sin^n x \, dx = -\frac{1}{n} \cos x \sin^{n-1} x + \frac{n-1}{n} \int \sin^{n-2} x \, dx$$

The reduction formula $\boxed{7}$ is useful because by using it repeatedly we could eventually express $\int \sin^n x \, dx$ in terms of $\int \sin x \, dx$ (if n is odd) or $\int (\sin x)^0 \, dx = \int dx$ (if n is even).

7.1 Exercises

Access end-of-section exercises and tools for visualizing calculus online at **www.webassign.net**

ENHANCED
WebAssign

7.2 Trigonometric Integrals

In this section we use trigonometric identities to integrate certain combinations of trigonometric functions. We start with powers of sine and cosine.

EXAMPLE 1 Evaluate $\int \cos^3 x \, dx$.

SOLUTION Simply substituting $u = \cos x$ isn't helpful, since then $du = -\sin x \, dx$. In order to integrate powers of cosine, we would need an extra $\sin x$ factor. Similarly, a power of sine would require an extra $\cos x$ factor. Thus here we can separate one cosine factor and convert the remaining $\cos^2 x$ factor to an expression involving sine using the identity $\sin^2 x + \cos^2 x = 1$:

$$\cos^3 x = \cos^2 x \cdot \cos x = (1 - \sin^2 x) \cos x$$

We can then evaluate the integral by substituting $u = \sin x$, so $du = \cos x \, dx$ and

$$\int \cos^3 x \, dx = \int \cos^2 x \cdot \cos x \, dx = \int (1 - \sin^2 x) \cos x \, dx$$

$$= \int (1 - u^2) \, du = u - \tfrac{1}{3} u^3 + C$$

$$= \sin x - \tfrac{1}{3} \sin^3 x + C$$

In general, we try to write an integrand involving powers of sine and cosine in a form where we have only one sine factor (and the remainder of the expression in terms of cosine) or only one cosine factor (and the remainder of the expression in terms of sine). The identity $\sin^2 x + \cos^2 x = 1$ enables us to convert back and forth between even powers of sine and cosine.

V **EXAMPLE 2** Find $\int \sin^5 x \cos^2 x \, dx$.

SOLUTION We could convert $\cos^2 x$ to $1 - \sin^2 x$, but we would be left with an expression in terms of $\sin x$ with no extra $\cos x$ factor. Instead, we separate a single sine factor and rewrite the remaining $\sin^4 x$ factor in terms of $\cos x$:

$$\sin^5 x \cos^2 x = (\sin^2 x)^2 \cos^2 x \sin x = (1 - \cos^2 x)^2 \cos^2 x \sin x$$

Substituting $u = \cos x$, we have $du = -\sin x \, dx$ and so

$$\int \sin^5 x \cos^2 x \, dx = \int (\sin^2 x)^2 \cos^2 x \sin x \, dx$$

$$= \int (1 - \cos^2 x)^2 \cos^2 x \sin x \, dx$$

$$= \int (1 - u^2)^2 u^2 (-du) = -\int (u^2 - 2u^4 + u^6) \, du$$

$$= -\left(\frac{u^3}{3} - 2\frac{u^5}{5} + \frac{u^7}{7}\right) + C$$

$$= -\tfrac{1}{3}\cos^3 x + \tfrac{2}{5}\cos^5 x - \tfrac{1}{7}\cos^7 x + C \qquad \blacksquare$$

Figure 1 shows the graphs of the integrand $\sin^5 x \cos^2 x$ in Example 2 and its indefinite integral (with $C = 0$). Which is which?

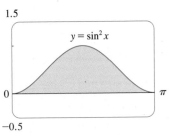

FIGURE 1

In the preceding examples, an odd power of sine or cosine enabled us to separate a single factor and convert the remaining even power. If the integrand contains even powers of both sine and cosine, this strategy fails. In this case, we can take advantage of the following half-angle identities (see Equations 17b and 17a in Appendix D):

$$\sin^2 x = \tfrac{1}{2}(1 - \cos 2x) \qquad \text{and} \qquad \cos^2 x = \tfrac{1}{2}(1 + \cos 2x)$$

V **EXAMPLE 3** Evaluate $\int_0^\pi \sin^2 x \, dx$.

SOLUTION If we write $\sin^2 x = 1 - \cos^2 x$, the integral is no simpler to evaluate. Using the half-angle formula for $\sin^2 x$, however, we have

$$\int_0^\pi \sin^2 x \, dx = \tfrac{1}{2}\int_0^\pi (1 - \cos 2x) \, dx$$

$$= \left[\tfrac{1}{2}\left(x - \tfrac{1}{2}\sin 2x\right)\right]_0^\pi$$

$$= \tfrac{1}{2}\left(\pi - \tfrac{1}{2}\sin 2\pi\right) - \tfrac{1}{2}\left(0 - \tfrac{1}{2}\sin 0\right) = \tfrac{1}{2}\pi$$

Example 3 shows that the area of the region shown in Figure 2 is $\pi/2$.

1.5

$y = \sin^2 x$

0   π

−0.5

FIGURE 2

Notice that we mentally made the substitution $u = 2x$ when integrating $\cos 2x$. Another method for evaluating this integral was given in Exercise 47 in Section 7.1. $\qquad \blacksquare$

EXAMPLE 4 Find $\int \sin^4 x \, dx$.

SOLUTION We could evaluate this integral using the reduction formula for $\int \sin^n x \, dx$ (Equation 7.1.7) together with Example 3 (as in Exercise 47 in Section 7.1), but a better method is to write $\sin^4 x = (\sin^2 x)^2$ and use a half-angle formula:

$$\int \sin^4 x \, dx = \int (\sin^2 x)^2 \, dx$$

$$= \int \left(\frac{1 - \cos 2x}{2} \right)^2 dx$$

$$= \tfrac{1}{4} \int (1 - 2\cos 2x + \cos^2 2x) \, dx$$

Since $\cos^2 2x$ occurs, we must use another half-angle formula

$$\cos^2 2x = \tfrac{1}{2}(1 + \cos 4x)$$

This gives

$$\int \sin^4 x \, dx = \tfrac{1}{4} \int \left[1 - 2\cos 2x + \tfrac{1}{2}(1 + \cos 4x) \right] dx$$

$$= \tfrac{1}{4} \int \left(\tfrac{3}{2} - 2\cos 2x + \tfrac{1}{2}\cos 4x \right) dx$$

$$= \tfrac{1}{4} \left(\tfrac{3}{2}x - \sin 2x + \tfrac{1}{8}\sin 4x \right) + C$$

To summarize, we list guidelines to follow when evaluating integrals of the form $\int \sin^m x \cos^n x \, dx$, where $m \geq 0$ and $n \geq 0$ are integers.

Strategy for Evaluating $\int \sin^m x \cos^n x \, dx$

(a) If the power of cosine is odd ($n = 2k + 1$), save one cosine factor and use $\cos^2 x = 1 - \sin^2 x$ to express the remaining factors in terms of sine:

$$\int \sin^m x \cos^{2k+1} x \, dx = \int \sin^m x \, (\cos^2 x)^k \cos x \, dx$$

$$= \int \sin^m x \, (1 - \sin^2 x)^k \cos x \, dx$$

Then substitute $u = \sin x$.

(b) If the power of sine is odd ($m = 2k + 1$), save one sine factor and use $\sin^2 x = 1 - \cos^2 x$ to express the remaining factors in terms of cosine:

$$\int \sin^{2k+1} x \cos^n x \, dx = \int (\sin^2 x)^k \cos^n x \sin x \, dx$$

$$= \int (1 - \cos^2 x)^k \cos^n x \sin x \, dx$$

Then substitute $u = \cos x$. [Note that if the powers of both sine and cosine are odd, either (a) or (b) can be used.]

(c) If the powers of both sine and cosine are even, use the half-angle identities

$$\sin^2 x = \tfrac{1}{2}(1 - \cos 2x) \qquad \cos^2 x = \tfrac{1}{2}(1 + \cos 2x)$$

It is sometimes helpful to use the identity

$$\sin x \cos x = \tfrac{1}{2}\sin 2x$$

We can use a similar strategy to evaluate integrals of the form $\int \tan^m x \sec^n x \, dx$. Since $(d/dx) \tan x = \sec^2 x$, we can separate a $\sec^2 x$ factor and convert the remaining (even) power of secant to an expression involving tangent using the identity $\sec^2 x = 1 + \tan^2 x$. Or, since $(d/dx) \sec x = \sec x \tan x$, we can separate a $\sec x \tan x$ factor and convert the remaining (even) power of tangent to secant.

V EXAMPLE 5 Evaluate $\int \tan^6 x \sec^4 x \, dx$.

SOLUTION If we separate one $\sec^2 x$ factor, we can express the remaining $\sec^2 x$ factor in terms of tangent using the identity $\sec^2 x = 1 + \tan^2 x$. We can then evaluate the integral by substituting $u = \tan x$ so that $du = \sec^2 x \, dx$:

$$\int \tan^6 x \sec^4 x \, dx = \int \tan^6 x \sec^2 x \sec^2 x \, dx$$

$$= \int \tan^6 x \,(1 + \tan^2 x) \sec^2 x \, dx$$

$$= \int u^6 (1 + u^2) \, du = \int (u^6 + u^8) \, du$$

$$= \frac{u^7}{7} + \frac{u^9}{9} + C$$

$$= \tfrac{1}{7} \tan^7 x + \tfrac{1}{9} \tan^9 x + C$$

EXAMPLE 6 Find $\int \tan^5 \theta \sec^7 \theta \, d\theta$.

SOLUTION If we separate a $\sec^2 \theta$ factor, as in the preceding example, we are left with a $\sec^5 \theta$ factor, which isn't easily converted to tangent. However, if we separate a $\sec \theta \tan \theta$ factor, we can convert the remaining power of tangent to an expression involving only secant using the identity $\tan^2 \theta = \sec^2 \theta - 1$. We can then evaluate the integral by substituting $u = \sec \theta$, so $du = \sec \theta \tan \theta \, d\theta$:

$$\int \tan^5 \theta \sec^7 \theta \, d\theta = \int \tan^4 \theta \sec^6 \theta \sec \theta \tan \theta \, d\theta$$

$$= \int (\sec^2 \theta - 1)^2 \sec^6 \theta \sec \theta \tan \theta \, d\theta$$

$$= \int (u^2 - 1)^2 u^6 \, du$$

$$= \int (u^{10} - 2u^8 + u^6) \, du$$

$$= \frac{u^{11}}{11} - 2\frac{u^9}{9} + \frac{u^7}{7} + C$$

$$= \tfrac{1}{11} \sec^{11} \theta - \tfrac{2}{9} \sec^9 \theta + \tfrac{1}{7} \sec^7 \theta + C$$

The preceding examples demonstrate strategies for evaluating integrals of the form $\int \tan^m x \sec^n x \, dx$ for two cases, which we summarize here.

Strategy for Evaluating $\int \tan^m x \, \sec^n x \, dx$

(a) If the power of secant is even ($n = 2k$, $k \geq 2$), save a factor of $\sec^2 x$ and use $\sec^2 x = 1 + \tan^2 x$ to express the remaining factors in terms of $\tan x$:

$$\int \tan^m x \, \sec^{2k} x \, dx = \int \tan^m x \, (\sec^2 x)^{k-1} \sec^2 x \, dx$$

$$= \int \tan^m x \, (1 + \tan^2 x)^{k-1} \sec^2 x \, dx$$

Then substitute $u = \tan x$.

(b) If the power of tangent is odd ($m = 2k + 1$), save a factor of $\sec x \tan x$ and use $\tan^2 x = \sec^2 x - 1$ to express the remaining factors in terms of $\sec x$:

$$\int \tan^{2k+1} x \, \sec^n x \, dx = \int (\tan^2 x)^k \sec^{n-1} x \, \sec x \tan x \, dx$$

$$= \int (\sec^2 x - 1)^k \sec^{n-1} x \, \sec x \tan x \, dx$$

Then substitute $u = \sec x$.

For other cases, the guidelines are not as clear-cut. We may need to use identities, integration by parts, and occasionally a little ingenuity. We will sometimes need to be able to integrate $\tan x$ by using the formula established in Chapter 6:

$$\int \tan x \, dx = \ln |\sec x| + C$$

We will also need the indefinite integral of secant:

Formula 1 was discovered by James Gregory in 1668. (See his biography on page 117.) Gregory used this formula to solve a problem in constructing nautical tables.

$$\boxed{1} \qquad \int \sec x \, dx = \ln |\sec x + \tan x| + C$$

We could verify Formula 1 by differentiating the right side, or as follows. First we multiply numerator and denominator by $\sec x + \tan x$:

$$\int \sec x \, dx = \int \sec x \, \frac{\sec x + \tan x}{\sec x + \tan x} \, dx$$

$$= \int \frac{\sec^2 x + \sec x \tan x}{\sec x + \tan x} \, dx$$

If we substitute $u = \sec x + \tan x$, then $du = (\sec x \tan x + \sec^2 x) \, dx$, so the integral becomes $\int (1/u) \, du = \ln |u| + C$. Thus we have

$$\int \sec x \, dx = \ln |\sec x + \tan x| + C$$

EXAMPLE 7 Find $\int \tan^3 x \, dx$.

SOLUTION Here only $\tan x$ occurs, so we use $\tan^2 x = \sec^2 x - 1$ to rewrite a $\tan^2 x$ factor in terms of $\sec^2 x$:

$$\int \tan^3 x \, dx = \int \tan x \, \tan^2 x \, dx$$

$$= \int \tan x \, (\sec^2 x - 1) \, dx$$

$$= \int \tan x \, \sec^2 x \, dx - \int \tan x \, dx$$

$$= \frac{\tan^2 x}{2} - \ln |\sec x| + C$$

In the first integral we mentally substituted $u = \tan x$ so that $du = \sec^2 x \, dx$.   ■

If an even power of tangent appears with an odd power of secant, it is helpful to express the integrand completely in terms of $\sec x$. Powers of $\sec x$ may require integration by parts, as shown in the following example.

EXAMPLE 8 Find $\int \sec^3 x \, dx$.

SOLUTION Here we integrate by parts with

$$u = \sec x \qquad\qquad dv = \sec^2 x \, dx$$
$$du = \sec x \tan x \, dx \qquad v = \tan x$$

Then

$$\int \sec^3 x \, dx = \sec x \tan x - \int \sec x \, \tan^2 x \, dx$$

$$= \sec x \tan x - \int \sec x \, (\sec^2 x - 1) \, dx$$

$$= \sec x \tan x - \int \sec^3 x \, dx + \int \sec x \, dx$$

Using Formula 1 and solving for the required integral, we get

$$\int \sec^3 x \, dx = \tfrac{1}{2}\left(\sec x \tan x + \ln |\sec x + \tan x|\right) + C$$
   ■

Integrals such as the one in the preceding example may seem very special but they occur frequently in applications of integration, as we will see in Chapter 8. Integrals of the form $\int \cot^m x \, \csc^n x \, dx$ can be found by similar methods because of the identity $1 + \cot^2 x = \csc^2 x$.

Finally, we can make use of another set of trigonometric identities:

---

2   To evaluate the integrals (a) $\int \sin mx \cos nx \, dx$, (b) $\int \sin mx \sin nx \, dx$, or (c) $\int \cos mx \cos nx \, dx$, use the corresponding identity:

(a) $\sin A \cos B = \tfrac{1}{2}[\sin(A - B) + \sin(A + B)]$

(b) $\sin A \sin B = \tfrac{1}{2}[\cos(A - B) - \cos(A + B)]$

(c) $\cos A \cos B = \tfrac{1}{2}[\cos(A - B) + \cos(A + B)]$

These product identities are discussed in Appendix D.

EXAMPLE 9 Evaluate $\int \sin 4x \cos 5x \, dx$.

SOLUTION This integral could be evaluated using integration by parts, but it's easier to use the identity in Equation 2(a) as follows:

$$\int \sin 4x \cos 5x \, dx = \int \tfrac{1}{2}[\sin(-x) + \sin 9x] \, dx$$

$$= \tfrac{1}{2} \int (-\sin x + \sin 9x) \, dx$$

$$= \tfrac{1}{2}\left(\cos x - \tfrac{1}{9}\cos 9x\right) + C$$

7.2   Exercises

Access end-of-section exercises and tools for visualizing calculus online at **www.webassign.net**

7.3   Trigonometric Substitution

In finding the area of a circle or an ellipse, an integral of the form $\int \sqrt{a^2 - x^2} \, dx$ arises, where $a > 0$. If it were $\int x\sqrt{a^2 - x^2} \, dx$, the substitution $u = a^2 - x^2$ would be effective but, as it stands, $\int \sqrt{a^2 - x^2} \, dx$ is more difficult. If we change the variable from x to θ by the substitution $x = a \sin \theta$, then the identity $1 - \sin^2\theta = \cos^2\theta$ allows us to get rid of the root sign because

$$\sqrt{a^2 - x^2} = \sqrt{a^2 - a^2\sin^2\theta} = \sqrt{a^2(1 - \sin^2\theta)} = \sqrt{a^2\cos^2\theta} = a\,|\cos\theta|$$

Notice the difference between the substitution $u = a^2 - x^2$ (in which the new variable is a function of the old one) and the substitution $x = a \sin \theta$ (the old variable is a function of the new one).

In general, we can make a substitution of the form $x = g(t)$ by using the Substitution Rule in reverse. To make our calculations simpler, we assume that g has an inverse function; that is, g is one-to-one. In this case, if we replace u by x and x by t in the Substitution Rule (Equation 4.5.4), we obtain

$$\int f(x) \, dx = \int f(g(t))g'(t) \, dt$$

This kind of substitution is called *inverse substitution*.

We can make the inverse substitution $x = a \sin \theta$ provided that it defines a one-to-one function. This can be accomplished by restricting θ to lie in the interval $[-\pi/2, \pi/2]$.

In the following table we list trigonometric substitutions that are effective for the given radical expressions because of the specified trigonometric identities. In each case the restriction on θ is imposed to ensure that the function that defines the substitution is one-to-one. (These are the same intervals used in Section 6.6 in defining the inverse functions.)

Table of Trigonometric Substitutions

Expression	Substitution	Identity
$\sqrt{a^2 - x^2}$	$x = a \sin\theta, \quad -\dfrac{\pi}{2} \le \theta \le \dfrac{\pi}{2}$	$1 - \sin^2\theta = \cos^2\theta$
$\sqrt{a^2 + x^2}$	$x = a \tan\theta, \quad -\dfrac{\pi}{2} < \theta < \dfrac{\pi}{2}$	$1 + \tan^2\theta = \sec^2\theta$
$\sqrt{x^2 - a^2}$	$x = a \sec\theta, \quad 0 \le \theta < \dfrac{\pi}{2} \text{ or } \pi \le \theta < \dfrac{3\pi}{2}$	$\sec^2\theta - 1 = \tan^2\theta$

V **EXAMPLE 1** Evaluate $\displaystyle\int \frac{\sqrt{9 - x^2}}{x^2}\, dx$.

SOLUTION Let $x = 3 \sin\theta$, where $-\pi/2 \le \theta \le \pi/2$. Then $dx = 3 \cos\theta\, d\theta$ and

$$\sqrt{9 - x^2} = \sqrt{9 - 9\sin^2\theta} = \sqrt{9\cos^2\theta} = 3\,|\cos\theta| = 3\cos\theta$$

(Note that $\cos\theta \ge 0$ because $-\pi/2 \le \theta \le \pi/2$.) Thus the Inverse Substitution Rule gives

$$\int \frac{\sqrt{9 - x^2}}{x^2}\, dx = \int \frac{3\cos\theta}{9\sin^2\theta}\, 3\cos\theta\, d\theta$$

$$= \int \frac{\cos^2\theta}{\sin^2\theta}\, d\theta = \int \cot^2\theta\, d\theta$$

$$= \int (\csc^2\theta - 1)\, d\theta$$

$$= -\cot\theta - \theta + C$$

Since this is an indefinite integral, we must return to the original variable x. This can be done either by using trigonometric identities to express $\cot\theta$ in terms of $\sin\theta = x/3$ or by drawing a diagram, as in Figure 1, where θ is interpreted as an angle of a right triangle. Since $\sin\theta = x/3$, we label the opposite side and the hypotenuse as having lengths x and 3. Then the Pythagorean Theorem gives the length of the adjacent side as $\sqrt{9 - x^2}$, so we can simply read the value of $\cot\theta$ from the figure:

$$\cot\theta = \frac{\sqrt{9 - x^2}}{x}$$

(Although $\theta > 0$ in the diagram, this expression for $\cot\theta$ is valid even when $\theta < 0$.) Since $\sin\theta = x/3$, we have $\theta = \sin^{-1}(x/3)$ and so

$$\int \frac{\sqrt{9 - x^2}}{x^2}\, dx = -\frac{\sqrt{9 - x^2}}{x} - \sin^{-1}\!\left(\frac{x}{3}\right) + C$$

FIGURE 1

$\sin\theta = \dfrac{x}{3}$

V **EXAMPLE 2** Find the area enclosed by the ellipse

$$\frac{x^2}{a^2} + \frac{y^2}{b^2} = 1$$

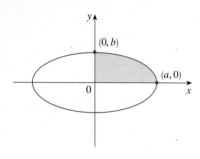

FIGURE 2

$$\frac{x^2}{a^2} + \frac{y^2}{b^2} = 1$$

SOLUTION Solving the equation of the ellipse for y, we get

$$\frac{y^2}{b^2} = 1 - \frac{x^2}{a^2} = \frac{a^2 - x^2}{a^2} \qquad \text{or} \qquad y = \pm\frac{b}{a}\sqrt{a^2 - x^2}$$

Because the ellipse is symmetric with respect to both axes, the total area A is four times the area in the first quadrant (see Figure 2). The part of the ellipse in the first quadrant is given by the function

$$y = \frac{b}{a}\sqrt{a^2 - x^2} \qquad 0 \le x \le a$$

and so

$$\tfrac{1}{4}A = \int_0^a \frac{b}{a}\sqrt{a^2 - x^2}\,dx$$

To evaluate this integral we substitute $x = a\sin\theta$. Then $dx = a\cos\theta\,d\theta$. To change the limits of integration we note that when $x = 0$, $\sin\theta = 0$, so $\theta = 0$; when $x = a$, $\sin\theta = 1$, so $\theta = \pi/2$. Also

$$\sqrt{a^2 - x^2} = \sqrt{a^2 - a^2\sin^2\theta} = \sqrt{a^2\cos^2\theta} = a|\cos\theta| = a\cos\theta$$

since $0 \le \theta \le \pi/2$. Therefore

$$A = 4\frac{b}{a}\int_0^a \sqrt{a^2 - x^2}\,dx = 4\frac{b}{a}\int_0^{\pi/2} a\cos\theta \cdot a\cos\theta\,d\theta$$

$$= 4ab\int_0^{\pi/2}\cos^2\theta\,d\theta = 4ab\int_0^{\pi/2}\tfrac{1}{2}(1 + \cos 2\theta)\,d\theta$$

$$= 2ab\big[\theta + \tfrac{1}{2}\sin 2\theta\big]_0^{\pi/2} = 2ab\left(\frac{\pi}{2} + 0 - 0\right) = \pi ab$$

We have shown that the area of an ellipse with semiaxes a and b is πab. In particular, taking $a = b = r$, we have proved the famous formula that the area of a circle with radius r is πr^2.

NOTE Since the integral in Example 2 was a definite integral, we changed the limits of integration and did not have to convert back to the original variable x.

V EXAMPLE 3 Find $\displaystyle\int \frac{1}{x^2\sqrt{x^2 + 4}}\,dx$.

SOLUTION Let $x = 2\tan\theta$, $-\pi/2 < \theta < \pi/2$. Then $dx = 2\sec^2\theta\,d\theta$ and

$$\sqrt{x^2 + 4} = \sqrt{4(\tan^2\theta + 1)} = \sqrt{4\sec^2\theta} = 2|\sec\theta| = 2\sec\theta$$

Thus we have

$$\int \frac{dx}{x^2\sqrt{x^2 + 4}} = \int \frac{2\sec^2\theta\,d\theta}{4\tan^2\theta \cdot 2\sec\theta} = \frac{1}{4}\int \frac{\sec\theta}{\tan^2\theta}\,d\theta$$

To evaluate this trigonometric integral we put everything in terms of $\sin\theta$ and $\cos\theta$:

$$\frac{\sec\theta}{\tan^2\theta} = \frac{1}{\cos\theta} \cdot \frac{\cos^2\theta}{\sin^2\theta} = \frac{\cos\theta}{\sin^2\theta}$$

Therefore, making the substitution $u = \sin \theta$, we have

$$\int \frac{dx}{x^2\sqrt{x^2+4}} = \frac{1}{4} \int \frac{\cos \theta}{\sin^2\theta} \, d\theta = \frac{1}{4} \int \frac{du}{u^2}$$

$$= \frac{1}{4}\left(-\frac{1}{u}\right) + C = -\frac{1}{4 \sin \theta} + C$$

$$= -\frac{\csc \theta}{4} + C$$

We use Figure 3 to determine that $\csc \theta = \sqrt{x^2+4}/x$ and so

$$\int \frac{dx}{x^2\sqrt{x^2+4}} = -\frac{\sqrt{x^2+4}}{4x} + C$$

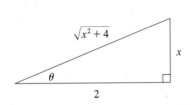

FIGURE 3

$\tan \theta = \dfrac{x}{2}$

EXAMPLE 4 Find $\displaystyle\int \frac{x}{\sqrt{x^2+4}} \, dx$.

SOLUTION It would be possible to use the trigonometric substitution $x = 2 \tan \theta$ here (as in Example 3). But the direct substitution $u = x^2 + 4$ is simpler, because then $du = 2x \, dx$ and

$$\int \frac{x}{\sqrt{x^2+4}} \, dx = \frac{1}{2} \int \frac{du}{\sqrt{u}} = \sqrt{u} + C = \sqrt{x^2+4} + C$$

NOTE Example 4 illustrates the fact that even when trigonometric substitutions are possible, they may not give the easiest solution. You should look for a simpler method first.

EXAMPLE 5 Evaluate $\displaystyle\int \frac{dx}{\sqrt{x^2-a^2}}$, where $a > 0$.

SOLUTION 1 We let $x = a \sec \theta$, where $0 < \theta < \pi/2$ or $\pi < \theta < 3\pi/2$. Then $dx = a \sec \theta \tan \theta \, d\theta$ and

$$\sqrt{x^2-a^2} = \sqrt{a^2(\sec^2\theta - 1)} = \sqrt{a^2\tan^2\theta} = a|\tan \theta| = a \tan \theta$$

Therefore

$$\int \frac{dx}{\sqrt{x^2-a^2}} = \int \frac{a \sec \theta \tan \theta}{a \tan \theta} \, d\theta = \int \sec \theta \, d\theta = \ln|\sec \theta + \tan \theta| + C$$

The triangle in Figure 4 gives $\tan \theta = \sqrt{x^2-a^2}/a$, so we have

$$\int \frac{dx}{\sqrt{x^2-a^2}} = \ln\left|\frac{x}{a} + \frac{\sqrt{x^2-a^2}}{a}\right| + C$$

$$= \ln|x + \sqrt{x^2-a^2}| - \ln a + C$$

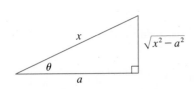

FIGURE 4

$\sec \theta = \dfrac{x}{a}$

Writing $C_1 = C - \ln a$, we have

$$\boxed{1} \qquad \int \frac{dx}{\sqrt{x^2-a^2}} = \ln|x + \sqrt{x^2-a^2}| + C_1$$

SOLUTION 2 For $x > 0$ the hyperbolic substitution $x = a \cosh t$ can also be used. Using the identity $\cosh^2 y - \sinh^2 y = 1$, we have

$$\sqrt{x^2 - a^2} = \sqrt{a^2(\cosh^2 t - 1)} = \sqrt{a^2 \sinh^2 t} = a \sinh t$$

Since $dx = a \sinh t \, dt$, we obtain

$$\int \frac{dx}{\sqrt{x^2 - a^2}} = \int \frac{a \sinh t \, dt}{a \sinh t} = \int dt = t + C$$

Since $\cosh t = x/a$, we have $t = \cosh^{-1}(x/a)$ and

2

$$\int \frac{dx}{\sqrt{x^2 - a^2}} = \cosh^{-1}\left(\frac{x}{a}\right) + C$$

Although Formulas 1 and 2 look quite different, they are actually equivalent by Formula 6.7.4.

NOTE As Example 5 illustrates, hyperbolic substitutions can be used in place of trigonometric substitutions and sometimes they lead to simpler answers. But we usually use trigonometric substitutions because trigonometric identities are more familiar than hyperbolic identities.

As Example 6 shows, trigonometric substitution is sometimes a good idea when $(x^2 + a^2)^{n/2}$ occurs in an integral, where n is any integer. The same is true when $(a^2 - x^2)^{n/2}$ or $(x^2 - a^2)^{n/2}$ occur.

EXAMPLE 6 Find $\displaystyle\int_0^{3\sqrt{3}/2} \frac{x^3}{(4x^2 + 9)^{3/2}} \, dx$.

SOLUTION First we note that $(4x^2 + 9)^{3/2} = (\sqrt{4x^2 + 9})^3$ so trigonometric substitution is appropriate. Although $\sqrt{4x^2 + 9}$ is not quite one of the expressions in the table of trigonometric substitutions, it becomes one of them if we make the preliminary substitution $u = 2x$. When we combine this with the tangent substitution, we have $x = \frac{3}{2} \tan \theta$, which gives $dx = \frac{3}{2} \sec^2\theta \, d\theta$ and

$$\sqrt{4x^2 + 9} = \sqrt{9 \tan^2\theta + 9} = 3 \sec \theta$$

When $x = 0$, $\tan \theta = 0$, so $\theta = 0$; when $x = 3\sqrt{3}/2$, $\tan \theta = \sqrt{3}$, so $\theta = \pi/3$.

$$\int_0^{3\sqrt{3}/2} \frac{x^3}{(4x^2 + 9)^{3/2}} \, dx = \int_0^{\pi/3} \frac{\frac{27}{8} \tan^3\theta}{27 \sec^3\theta} \, \tfrac{3}{2} \sec^2\theta \, d\theta$$

$$= \tfrac{3}{16} \int_0^{\pi/3} \frac{\tan^3\theta}{\sec \theta} \, d\theta = \tfrac{3}{16} \int_0^{\pi/3} \frac{\sin^3\theta}{\cos^2\theta} \, d\theta$$

$$= \tfrac{3}{16} \int_0^{\pi/3} \frac{1 - \cos^2\theta}{\cos^2\theta} \sin \theta \, d\theta$$

Now we substitute $u = \cos \theta$ so that $du = -\sin \theta \, d\theta$. When $\theta = 0$, $u = 1$; when

$\theta = \pi/3$, $u = \frac{1}{2}$. Therefore

$$\int_0^{3\sqrt{3}/2} \frac{x^3}{(4x^2 + 9)^{3/2}} \, dx = -\frac{3}{16} \int_1^{1/2} \frac{1 - u^2}{u^2} \, du$$

$$= \frac{3}{16} \int_1^{1/2} (1 - u^{-2}) \, du = \frac{3}{16} \left[u + \frac{1}{u} \right]_1^{1/2}$$

$$= \frac{3}{16} \left[\left(\frac{1}{2} + 2 \right) - (1 + 1) \right] = \frac{3}{32}$$

EXAMPLE 7 Evaluate $\displaystyle\int \frac{x}{\sqrt{3 - 2x - x^2}} \, dx$.

SOLUTION We can transform the integrand into a function for which trigonometric substitution is appropriate by first completing the square under the root sign:

$$3 - 2x - x^2 = 3 - (x^2 + 2x) = 3 + 1 - (x^2 + 2x + 1)$$

$$= 4 - (x + 1)^2$$

This suggests that we make the substitution $u = x + 1$. Then $du = dx$ and $x = u - 1$, so

$$\int \frac{x}{\sqrt{3 - 2x - x^2}} \, dx = \int \frac{u - 1}{\sqrt{4 - u^2}} \, du$$

We now substitute $u = 2 \sin \theta$, giving $du = 2 \cos \theta \, d\theta$ and $\sqrt{4 - u^2} = 2 \cos \theta$, so

$$\int \frac{x}{\sqrt{3 - 2x - x^2}} \, dx = \int \frac{2 \sin \theta - 1}{2 \cos \theta} 2 \cos \theta \, d\theta$$

$$= \int (2 \sin \theta - 1) \, d\theta$$

$$= -2 \cos \theta - \theta + C$$

$$= -\sqrt{4 - u^2} - \sin^{-1}\left(\frac{u}{2} \right) + C$$

$$= -\sqrt{3 - 2x - x^2} - \sin^{-1}\left(\frac{x + 1}{2} \right) + C$$

Figure 5 shows the graphs of the integrand in Example 7 and its indefinite integral (with $C = 0$). Which is which?

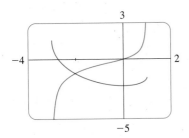

FIGURE 5

7.3 Exercises

Access end-of-section exercises and tools for visualizing calculus online at **www.webassign.net** ᴱᴺᴴᴬᴺᶜᴱᴰ WebAssign

7.4   Integration of Rational Functions by Partial Fractions

In this section we show how to integrate any rational function (a ratio of polynomials) by expressing it as a sum of simpler fractions, called *partial fractions,* that we already know how to integrate. To illustrate the method, observe that by taking the fractions $2/(x - 1)$ and $1/(x + 2)$ to a common denominator we obtain

$$\frac{2}{x - 1} - \frac{1}{x + 2} = \frac{2(x + 2) - (x - 1)}{(x - 1)(x + 2)} = \frac{x + 5}{x^2 + x - 2}$$

If we now reverse the procedure, we see how to integrate the function on the right side of this equation:

$$\int \frac{x + 5}{x^2 + x - 2}\, dx = \int \left(\frac{2}{x - 1} - \frac{1}{x + 2} \right) dx$$

$$= 2 \ln |x - 1| - \ln |x + 2| + C$$

To see how the method of partial fractions works in general, let's consider a rational function

$$f(x) = \frac{P(x)}{Q(x)}$$

where P and Q are polynomials. It's possible to express f as a sum of simpler fractions provided that the degree of P is less than the degree of Q. Such a rational function is called *proper.* Recall that if

$$P(x) = a_n x^n + a_{n-1} x^{n-1} + \cdots + a_1 x + a_0$$

where $a_n \neq 0$, then the degree of P is n and we write $\deg(P) = n$.

If f is *improper,* that is, $\deg(P) \geq \deg(Q)$, then we must take the preliminary step of dividing Q into P (by long division) until a remainder $R(x)$ is obtained such that $\deg(R) < \deg(Q)$. The division statement is

$$\boxed{1} \qquad\qquad f(x) = \frac{P(x)}{Q(x)} = S(x) + \frac{R(x)}{Q(x)}$$

where S and R are also polynomials.

As the following example illustrates, sometimes this preliminary step is all that is required.

V   EXAMPLE 1   Find $\displaystyle\int \frac{x^3 + x}{x - 1}\, dx$.

$$\begin{array}{r}
x^2 + x + 2 \\
x - 1 \overline{)\, x^3 \qquad + x } \\
\underline{x^3 - x^2} \\
x^2 + x \\
\underline{x^2 - x} \\
2x \\
\underline{2x - 2} \\
2
\end{array}$$

SOLUTION   Since the degree of the numerator is greater than the degree of the denominator, we first perform the long division. This enables us to write

$$\int \frac{x^3 + x}{x - 1}\, dx = \int \left(x^2 + x + 2 + \frac{2}{x - 1} \right) dx$$

$$= \frac{x^3}{3} + \frac{x^2}{2} + 2x + 2 \ln |x - 1| + C \qquad\blacksquare$$

The next step is to factor the denominator $Q(x)$ as far as possible. It can be shown that any polynomial Q can be factored as a product of linear factors (of the form $ax + b$) and irreducible quadratic factors (of the form $ax^2 + bx + c$, where $b^2 - 4ac < 0$). For instance, if $Q(x) = x^4 - 16$, we could factor it as

$$Q(x) = (x^2 - 4)(x^2 + 4) = (x - 2)(x + 2)(x^2 + 4)$$

The third step is to express the proper rational function $R(x)/Q(x)$ (from Equation 1) as a sum of **partial fractions** of the form

$$\frac{A}{(ax + b)^i} \quad \text{or} \quad \frac{Ax + B}{(ax^2 + bx + c)^j}$$

A theorem in algebra guarantees that it is always possible to do this. We explain the details for the four cases that occur.

CASE I The denominator $Q(x)$ is a product of distinct linear factors.
This means that we can write

$$Q(x) = (a_1 x + b_1)(a_2 x + b_2) \cdots (a_k x + b_k)$$

where no factor is repeated (and no factor is a constant multiple of another). In this case the partial fraction theorem states that there exist constants $A_1, A_2, \ldots, A_k$ such that

$$\boxed{2} \qquad \frac{R(x)}{Q(x)} = \frac{A_1}{a_1 x + b_1} + \frac{A_2}{a_2 x + b_2} + \cdots + \frac{A_k}{a_k x + b_k}$$

These constants can be determined as in the following example.

▼ EXAMPLE 2 Evaluate $\displaystyle\int \frac{x^2 + 2x - 1}{2x^3 + 3x^2 - 2x}\, dx$.

SOLUTION Since the degree of the numerator is less than the degree of the denominator, we don't need to divide. We factor the denominator as

$$2x^3 + 3x^2 - 2x = x(2x^2 + 3x - 2) = x(2x - 1)(x + 2)$$

Since the denominator has three distinct linear factors, the partial fraction decomposition of the integrand $\boxed{2}$ has the form

$$\boxed{3} \qquad \frac{x^2 + 2x - 1}{x(2x - 1)(x + 2)} = \frac{A}{x} + \frac{B}{2x - 1} + \frac{C}{x + 2}$$

Another method for finding A, B, and C is given in the note after this example.

To determine the values of A, B, and C, we multiply both sides of this equation by the product of the denominators, $x(2x - 1)(x + 2)$, obtaining

$$\boxed{4} \qquad x^2 + 2x - 1 = A(2x - 1)(x + 2) + Bx(x + 2) + Cx(2x - 1)$$

Expanding the right side of Equation 4 and writing it in the standard form for polynomials, we get

$$\boxed{5} \qquad x^2 + 2x - 1 = (2A + B + 2C)x^2 + (3A + 2B - C)x - 2A$$

The polynomials in Equation 5 are identical, so their coefficients must be equal. The coefficient of x^2 on the right side, $2A + B + 2C$, must equal the coefficient of x^2 on the left side—namely, 1. Likewise, the coefficients of x are equal and the constant terms are equal. This gives the following system of equations for A, B, and C:

$$2A + B + 2C = 1$$
$$3A + 2B - C = 2$$
$$-2A = -1$$

Solving, we get $A = \frac{1}{2}$, $B = \frac{1}{5}$, and $C = -\frac{1}{10}$, and so

We could check our work by taking the terms to a common denominator and adding them.

$$\int \frac{x^2 + 2x - 1}{2x^3 + 3x^2 - 2x}\, dx = \int \left[\frac{1}{2}\frac{1}{x} + \frac{1}{5}\frac{1}{2x-1} - \frac{1}{10}\frac{1}{x+2} \right] dx$$

$$= \tfrac{1}{2}\ln|x| + \tfrac{1}{10}\ln|2x-1| - \tfrac{1}{10}\ln|x+2| + K$$

Figure 1 shows the graphs of the integrand in Example 2 and its indefinite integral (with $K = 0$). Which is which?

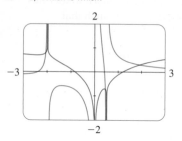

FIGURE 1

In integrating the middle term we have made the mental substitution $u = 2x - 1$, which gives $du = 2\,dx$ and $dx = \frac{1}{2}\,du$.

NOTE We can use an alternative method to find the coefficients A, B, and C in Example 2. Equation 4 is an identity; it is true for every value of x. Let's choose values of x that simplify the equation. If we put $x = 0$ in Equation 4, then the second and third terms on the right side vanish and the equation then becomes $-2A = -1$, or $A = \frac{1}{2}$. Likewise, $x = \frac{1}{2}$ gives $5B/4 = \frac{1}{4}$ and $x = -2$ gives $10C = -1$, so $B = \frac{1}{5}$ and $C = -\frac{1}{10}$. (You may object that Equation 3 is not valid for $x = 0, \frac{1}{2}$, or -2, so why should Equation 4 be valid for those values? In fact, Equation 4 is true for all values of x, even $x = 0, \frac{1}{2}$, and -2. See Exercise 71 for the reason.)

EXAMPLE 3 Find $\int \dfrac{dx}{x^2 - a^2}$, where $a \neq 0$.

SOLUTION The method of partial fractions gives

$$\frac{1}{x^2 - a^2} = \frac{1}{(x-a)(x+a)} = \frac{A}{x-a} + \frac{B}{x+a}$$

and therefore

$$A(x+a) + B(x-a) = 1$$

Using the method of the preceding note, we put $x = a$ in this equation and get $A(2a) = 1$, so $A = 1/(2a)$. If we put $x = -a$, we get $B(-2a) = 1$, so $B = -1/(2a)$. Thus

$$\int \frac{dx}{x^2 - a^2} = \frac{1}{2a} \int \left(\frac{1}{x-a} - \frac{1}{x+a} \right) dx$$

$$= \frac{1}{2a} \left(\ln|x-a| - \ln|x+a| \right) + C$$

Since $\ln x - \ln y = \ln(x/y)$, we can write the integral as

$$\boxed{6} \qquad \int \frac{dx}{x^2 - a^2} = \frac{1}{2a} \ln \left| \frac{x - a}{x + a} \right| + C$$

See Exercises 57–58 for ways of using Formula 6.

CASE II $Q(x)$ **is a product of linear factors, some of which are repeated.**
Suppose the first linear factor $(a_1 x + b_1)$ is repeated r times; that is, $(a_1 x + b_1)^r$ occurs in the factorization of $Q(x)$. Then instead of the single term $A_1/(a_1 x + b_1)$ in Equation 2, we would use

$$\boxed{7} \qquad \frac{A_1}{a_1 x + b_1} + \frac{A_2}{(a_1 x + b_1)^2} + \cdots + \frac{A_r}{(a_1 x + b_1)^r}$$

By way of illustration, we could write

$$\frac{x^3 - x + 1}{x^2(x - 1)^3} = \frac{A}{x} + \frac{B}{x^2} + \frac{C}{x - 1} + \frac{D}{(x - 1)^2} + \frac{E}{(x - 1)^3}$$

but we prefer to work out in detail a simpler example.

EXAMPLE 4 Find $\displaystyle\int \frac{x^4 - 2x^2 + 4x + 1}{x^3 - x^2 - x + 1} \, dx$.

SOLUTION The first step is to divide. The result of long division is

$$\frac{x^4 - 2x^2 + 4x + 1}{x^3 - x^2 - x + 1} = x + 1 + \frac{4x}{x^3 - x^2 - x + 1}$$

The second step is to factor the denominator $Q(x) = x^3 - x^2 - x + 1$. Since $Q(1) = 0$, we know that $x - 1$ is a factor and we obtain

$$x^3 - x^2 - x + 1 = (x - 1)(x^2 - 1) = (x - 1)(x - 1)(x + 1)$$

$$= (x - 1)^2(x + 1)$$

Since the linear factor $x - 1$ occurs twice, the partial fraction decomposition is

$$\frac{4x}{(x - 1)^2(x + 1)} = \frac{A}{x - 1} + \frac{B}{(x - 1)^2} + \frac{C}{x + 1}$$

Multiplying by the least common denominator, $(x - 1)^2(x + 1)$, we get

$$\boxed{8} \qquad 4x = A(x - 1)(x + 1) + B(x + 1) + C(x - 1)^2$$

$$= (A + C)x^2 + (B - 2C)x + (-A + B + C)$$

Another method for finding the coefficients:
Put $x = 1$ in $\boxed{8}$: $B = 2$.
Put $x = -1$: $C = -1$.
Put $x = 0$: $A = B + C = 1$.

Now we equate coefficients:

$$A \qquad + \quad C = 0$$

$$B - 2C = 4$$

$$-A + B + \quad C = 0$$

Solving, we obtain $A = 1$, $B = 2$, and $C = -1$, so

$$\int \frac{x^4 - 2x^2 + 4x + 1}{x^3 - x^2 - x + 1}\,dx = \int \left[x + 1 + \frac{1}{x - 1} + \frac{2}{(x - 1)^2} - \frac{1}{x + 1} \right] dx$$

$$= \frac{x^2}{2} + x + \ln|x - 1| - \frac{2}{x - 1} - \ln|x + 1| + K$$

$$= \frac{x^2}{2} + x - \frac{2}{x - 1} + \ln\left|\frac{x - 1}{x + 1}\right| + K \qquad \blacksquare$$

CASE III $Q(x)$ **contains irreducible quadratic factors, none of which is repeated.**

If $Q(x)$ has the factor $ax^2 + bx + c$, where $b^2 - 4ac < 0$, then, in addition to the partial fractions in Equations 2 and 7, the expression for $R(x)/Q(x)$ will have a term of the form

$$\boxed{9} \qquad\qquad\qquad \frac{Ax + B}{ax^2 + bx + c}$$

where A and B are constants to be determined. For instance, the function given by $f(x) = x/[(x - 2)(x^2 + 1)(x^2 + 4)]$ has a partial fraction decomposition of the form

$$\frac{x}{(x - 2)(x^2 + 1)(x^2 + 4)} = \frac{A}{x - 2} + \frac{Bx + C}{x^2 + 1} + \frac{Dx + E}{x^2 + 4}$$

The term given in $\boxed{9}$ can be integrated by completing the square (if necessary) and using the formula

$$\boxed{10} \qquad\qquad \boxed{\;\int \frac{dx}{x^2 + a^2} = \frac{1}{a} \tan^{-1}\left(\frac{x}{a}\right) + C\;}$$

V **EXAMPLE 5** Evaluate $\displaystyle\int \frac{2x^2 - x + 4}{x^3 + 4x}\,dx$.

SOLUTION Since $x^3 + 4x = x(x^2 + 4)$ can't be factored further, we write

$$\frac{2x^2 - x + 4}{x(x^2 + 4)} = \frac{A}{x} + \frac{Bx + C}{x^2 + 4}$$

Multiplying by $x(x^2 + 4)$, we have

$$2x^2 - x + 4 = A(x^2 + 4) + (Bx + C)x$$

$$= (A + B)x^2 + Cx + 4A$$

Equating coefficients, we obtain

$$A + B = 2 \qquad\qquad C = -1 \qquad\qquad 4A = 4$$

Thus $A = 1$, $B = 1$, and $C = -1$ and so

$$\int \frac{2x^2 - x + 4}{x^3 + 4x}\,dx = \int \left(\frac{1}{x} + \frac{x - 1}{x^2 + 4} \right) dx$$

In order to integrate the second term we split it into two parts:

$$\int \frac{x - 1}{x^2 + 4}\, dx = \int \frac{x}{x^2 + 4}\, dx - \int \frac{1}{x^2 + 4}\, dx$$

We make the substitution $u = x^2 + 4$ in the first of these integrals so that $du = 2x\, dx$. We evaluate the second integral by means of Formula 10 with $a = 2$:

$$\int \frac{2x^2 - x + 4}{x(x^2 + 4)}\, dx = \int \frac{1}{x}\, dx + \int \frac{x}{x^2 + 4}\, dx - \int \frac{1}{x^2 + 4}\, dx$$

$$= \ln |x| + \tfrac{1}{2}\ln(x^2 + 4) - \tfrac{1}{2}\tan^{-1}(x/2) + K \qquad \blacksquare$$

EXAMPLE 6 Evaluate $\displaystyle\int \frac{4x^2 - 3x + 2}{4x^2 - 4x + 3}\, dx$.

SOLUTION Since the degree of the numerator is *not less than* the degree of the denominator, we first divide and obtain

$$\frac{4x^2 - 3x + 2}{4x^2 - 4x + 3} = 1 + \frac{x - 1}{4x^2 - 4x + 3}$$

Notice that the quadratic $4x^2 - 4x + 3$ is irreducible because its discriminant is $b^2 - 4ac = -32 < 0$. This means it can't be factored, so we don't need to use the partial fraction technique.

To integrate the given function we complete the square in the denominator:

$$4x^2 - 4x + 3 = (2x - 1)^2 + 2$$

This suggests that we make the substitution $u = 2x - 1$. Then $du = 2\, dx$ and $x = \tfrac{1}{2}(u + 1)$, so

$$\int \frac{4x^2 - 3x + 2}{4x^2 - 4x + 3}\, dx = \int \left(1 + \frac{x - 1}{4x^2 - 4x + 3} \right) dx$$

$$= x + \tfrac{1}{2}\int \frac{\tfrac{1}{2}(u + 1) - 1}{u^2 + 2}\, du = x + \tfrac{1}{4}\int \frac{u - 1}{u^2 + 2}\, du$$

$$= x + \tfrac{1}{4}\int \frac{u}{u^2 + 2}\, du - \tfrac{1}{4}\int \frac{1}{u^2 + 2}\, du$$

$$= x + \tfrac{1}{8}\ln(u^2 + 2) - \frac{1}{4} \cdot \frac{1}{\sqrt{2}}\tan^{-1}\left(\frac{u}{\sqrt{2}} \right) + C$$

$$= x + \tfrac{1}{8}\ln(4x^2 - 4x + 3) - \frac{1}{4\sqrt{2}}\tan^{-1}\left(\frac{2x - 1}{\sqrt{2}} \right) + C \qquad \blacksquare$$

NOTE Example 6 illustrates the general procedure for integrating a partial fraction of the form

$$\frac{Ax + B}{ax^2 + bx + c} \qquad \text{where } b^2 - 4ac < 0$$

We complete the square in the denominator and then make a substitution that brings the integral into the form

$$\int \frac{Cu + D}{u^2 + a^2}\, du = C \int \frac{u}{u^2 + a^2}\, du + D \int \frac{1}{u^2 + a^2}\, du$$

Then the first integral is a logarithm and the second is expressed in terms of $\tan^{-1}$.

CASE IV $Q(x)$ **contains a repeated irreducible quadratic factor.**
If $Q(x)$ has the factor $(ax^2 + bx + c)^r$, where $b^2 - 4ac < 0$, then instead of the single partial fraction $\boxed{9}$, the sum

$$\boxed{11} \qquad \frac{A_1 x + B_1}{ax^2 + bx + c} + \frac{A_2 x + B_2}{(ax^2 + bx + c)^2} + \cdots + \frac{A_r x + B_r}{(ax^2 + bx + c)^r}$$

occurs in the partial fraction decomposition of $R(x)/Q(x)$. Each of the terms in $\boxed{11}$ can be integrated by using a substitution or by first completing the square if necessary.

It would be extremely tedious to work out by hand the numerical values of the coefficients in Example 7. Most computer algebra systems, however, can find the numerical values very quickly. For instance, the Maple command

$$\texttt{convert(f, parfrac, x)}$$

or the Mathematica command

$$\texttt{Apart[f]}$$

gives the following values:

$$A = -1, \quad B = \tfrac{1}{8}, \quad C = D = -1,$$

$$E = \tfrac{15}{8}, \quad F = -\tfrac{1}{8}, \quad G = H = \tfrac{3}{4},$$

$$I = -\tfrac{1}{2}, \quad J = \tfrac{1}{2}$$

EXAMPLE 7 Write out the form of the partial fraction decomposition of the function

$$\frac{x^3 + x^2 + 1}{x(x - 1)(x^2 + x + 1)(x^2 + 1)^3}$$

SOLUTION

$$\frac{x^3 + x^2 + 1}{x(x - 1)(x^2 + x + 1)(x^2 + 1)^3}$$

$$= \frac{A}{x} + \frac{B}{x - 1} + \frac{Cx + D}{x^2 + x + 1} + \frac{Ex + F}{x^2 + 1} + \frac{Gx + H}{(x^2 + 1)^2} + \frac{Ix + J}{(x^2 + 1)^3} \qquad \blacksquare$$

EXAMPLE 8 Evaluate $\displaystyle \int \frac{1 - x + 2x^2 - x^3}{x(x^2 + 1)^2}\, dx$.

SOLUTION The form of the partial fraction decomposition is

$$\frac{1 - x + 2x^2 - x^3}{x(x^2 + 1)^2} = \frac{A}{x} + \frac{Bx + C}{x^2 + 1} + \frac{Dx + E}{(x^2 + 1)^2}$$

Multiplying by $x(x^2 + 1)^2$, we have

$$-x^3 + 2x^2 - x + 1 = A(x^2 + 1)^2 + (Bx + C)x(x^2 + 1) + (Dx + E)x$$

$$= A(x^4 + 2x^2 + 1) + B(x^4 + x^2) + C(x^3 + x) + Dx^2 + Ex$$

$$= (A + B)x^4 + Cx^3 + (2A + B + D)x^2 + (C + E)x + A$$

If we equate coefficients, we get the system

$$A + B = 0 \qquad C = -1 \qquad 2A + B + D = 2 \qquad C + E = -1 \qquad A = 1$$

which has the solution $A = 1$, $B = -1$, $C = -1$, $D = 1$, and $E = 0$. Thus

$$\int \frac{1 - x + 2x^2 - x^3}{x(x^2 + 1)^2} \, dx = \int \left(\frac{1}{x} - \frac{x + 1}{x^2 + 1} + \frac{x}{(x^2 + 1)^2} \right) dx$$

$$= \int \frac{dx}{x} - \int \frac{x}{x^2 + 1} \, dx - \int \frac{dx}{x^2 + 1} + \int \frac{x \, dx}{(x^2 + 1)^2}$$

In the second and fourth terms we made the mental substitution $u = x^2 + 1$.

$$= \ln|x| - \tfrac{1}{2}\ln(x^2 + 1) - \tan^{-1}x - \frac{1}{2(x^2 + 1)} + K$$

We note that sometimes partial fractions can be avoided when integrating a rational function. For instance, although the integral

$$\int \frac{x^2 + 1}{x(x^2 + 3)} \, dx$$

could be evaluated by the method of Case III, it's much easier to observe that if $u = x(x^2 + 3) = x^3 + 3x$, then $du = (3x^2 + 3) \, dx$ and so

$$\int \frac{x^2 + 1}{x(x^2 + 3)} \, dx = \tfrac{1}{3}\ln|x^3 + 3x| + C$$

■ Rationalizing Substitutions

Some nonrational functions can be changed into rational functions by means of appropriate substitutions. In particular, when an integrand contains an expression of the form $\sqrt[n]{g(x)}$, then the substitution $u = \sqrt[n]{g(x)}$ may be effective. Other instances appear in the exercises.

EXAMPLE 9 Evaluate $\displaystyle\int \frac{\sqrt{x + 4}}{x} \, dx$.

SOLUTION Let $u = \sqrt{x + 4}$. Then $u^2 = x + 4$, so $x = u^2 - 4$ and $dx = 2u \, du$. Therefore

$$\int \frac{\sqrt{x + 4}}{x} \, dx = \int \frac{u}{u^2 - 4} \, 2u \, du = 2 \int \frac{u^2}{u^2 - 4} \, du$$

$$= 2 \int \left(1 + \frac{4}{u^2 - 4} \right) du$$

We can evaluate this integral either by factoring $u^2 - 4$ as $(u - 2)(u + 2)$ and using partial fractions or by using Formula 6 with $a = 2$:

$$\int \frac{\sqrt{x + 4}}{x} \, dx = 2 \int du + 8 \int \frac{du}{u^2 - 4} = 2u + 8 \cdot \frac{1}{2 \cdot 2} \ln\left| \frac{u - 2}{u + 2} \right| + C$$

$$= 2\sqrt{x + 4} + 2\ln\left| \frac{\sqrt{x + 4} - 2}{\sqrt{x + 4} + 2} \right| + C$$

7.4   Exercises

Access end-of-section exercises and tools for visualizing calculus online at **www.webassign.net**

7.5   Strategy for Integration

As we have seen, integration is more challenging than differentiation. In finding the derivative of a function it is obvious which differentiation formula we should apply. But it may not be obvious which technique we should use to integrate a given function.

Until now individual techniques have been applied in each section. For instance, we usually used substitution in Exercises 4.5, integration by parts in Exercises 7.1, and partial fractions in Exercises 7.4. But in this section we present a collection of miscellaneous integrals in random order and the main challenge is to recognize which technique or formula to use. No hard and fast rules can be given as to which method applies in a given situation, but we give some advice on strategy that you may find useful.

A prerequisite for applying a strategy is a knowledge of the basic integration formulas. In the following table we have collected the integrals from our previous list together with several additional formulas that we have learned in this chapter. Most of them should be memorized. It is useful to know them all, but the ones marked with an asterisk need not be memorized since they are easily derived. Formula 19 can be avoided by using partial fractions, and trigonometric substitutions can be used in place of Formula 20.

Table of Integration Formulas   Constants of integration have been omitted.

1. $\int x^n \, dx = \dfrac{x^{n+1}}{n+1} \quad (n \neq -1)$ 2. $\int \dfrac{1}{x} \, dx = \ln|x|$

3. $\int e^x \, dx = e^x$ 4. $\int a^x \, dx = \dfrac{a^x}{\ln a}$

5. $\int \sin x \, dx = -\cos x$ 6. $\int \cos x \, dx = \sin x$

7. $\int \sec^2 x \, dx = \tan x$ 8. $\int \csc^2 x \, dx = -\cot x$

9. $\int \sec x \tan x \, dx = \sec x$ 10. $\int \csc x \cot x \, dx = -\csc x$

11. $\int \sec x \, dx = \ln|\sec x + \tan x|$ 12. $\int \csc x \, dx = \ln|\csc x - \cot x|$

13. $\int \tan x \, dx = \ln|\sec x|$ 14. $\int \cot x \, dx = \ln|\sin x|$

15. $\int \sinh x \, dx = \cosh x$ 16. $\int \cosh x \, dx = \sinh x$

17. $\int \dfrac{dx}{x^2 + a^2} = \dfrac{1}{a} \tan^{-1}\left(\dfrac{x}{a}\right)$ 18. $\int \dfrac{dx}{\sqrt{a^2 - x^2}} = \sin^{-1}\left(\dfrac{x}{a}\right), \quad a > 0$

*19. $\int \dfrac{dx}{x^2 - a^2} = \dfrac{1}{2a} \ln\left|\dfrac{x-a}{x+a}\right|$ *20. $\int \dfrac{dx}{\sqrt{x^2 \pm a^2}} = \ln\left|x + \sqrt{x^2 \pm a^2}\right|$

Once you are armed with these basic integration formulas, if you don't immediately see how to attack a given integral, you might try the following four-step strategy.

1. Simplify the Integrand if Possible Sometimes the use of algebraic manipulation or trigonometric identities will simplify the integrand and make the method of integration obvious. Here are some examples:

$$\int \sqrt{x}\,(1 + \sqrt{x}\,)\,dx = \int (\sqrt{x} + x)\,dx$$

$$\int \frac{\tan \theta}{\sec^2\theta}\,d\theta = \int \frac{\sin \theta}{\cos \theta}\cos^2\theta\,d\theta$$

$$= \int \sin \theta \cos \theta\,d\theta = \tfrac{1}{2}\int \sin 2\theta\,d\theta$$

$$\int (\sin x + \cos x)^2\,dx = \int (\sin^2 x + 2 \sin x \cos x + \cos^2 x)\,dx$$

$$= \int (1 + 2 \sin x \cos x)\,dx$$

2. Look for an Obvious Substitution Try to find some function $u = g(x)$ in the integrand whose differential $du = g'(x)\,dx$ also occurs, apart from a constant factor. For instance, in the integral

$$\int \frac{x}{x^2 - 1}\,dx$$

we notice that if $u = x^2 - 1$, then $du = 2x\,dx$. Therefore we use the substitution $u = x^2 - 1$ instead of the method of partial fractions.

3. Classify the Integrand According to Its Form If Steps 1 and 2 have not led to the solution, then we take a look at the form of the integrand $f(x)$.

(a) *Trigonometric functions.* If $f(x)$ is a product of powers of $\sin x$ and $\cos x$, of $\tan x$ and $\sec x$, or of $\cot x$ and $\csc x$, then we use the substitutions recommended in Section 7.2.

(b) *Rational functions.* If f is a rational function, we use the procedure of Section 7.4 involving partial fractions.

(c) *Integration by parts.* If $f(x)$ is a product of a power of x (or a polynomial) and a transcendental function (such as a trigonometric, exponential, or logarithmic function), then we try integration by parts, choosing u and dv according to the advice given in Section 7.1. If you look at the functions in Exercises 7.1, you will see that most of them are the type just described.

(d) *Radicals.* Particular kinds of substitutions are recommended when certain radicals appear.

  (i) If $\sqrt{\pm x^2 \pm a^2}$ occurs, we use a trigonometric substitution according to the table in Section 7.3.

  (ii) If $\sqrt[n]{ax + b}$ occurs, we use the rationalizing substitution $u = \sqrt[n]{ax + b}$. More generally, this sometimes works for $\sqrt[n]{g(x)}$.

4. Try Again If the first three steps have not produced the answer, remember that there are basically only two methods of integration: substitution and parts.

(a) *Try substitution.* Even if no substitution is obvious (Step 2), some inspiration or ingenuity (or even desperation) may suggest an appropriate substitution.

(b) *Try parts.* Although integration by parts is used most of the time on products of the form described in Step 3(c), it is sometimes effective on single functions. Looking at Section 7.1, we see that it works on $\tan^{-1}x$, $\sin^{-1}x$, and $\ln x$, and these are all inverse functions.

(c) *Manipulate the integrand.* Algebraic manipulations (perhaps rationalizing the denominator or using trigonometric identities) may be useful in transforming the integral into an easier form. These manipulations may be more substantial than in Step 1 and may involve some ingenuity. Here is an example:

$$\int \frac{dx}{1 - \cos x} = \int \frac{1}{1 - \cos x} \cdot \frac{1 + \cos x}{1 + \cos x}\,dx = \int \frac{1 + \cos x}{1 - \cos^2 x}\,dx$$

$$= \int \frac{1 + \cos x}{\sin^2 x}\,dx = \int \left(\csc^2 x + \frac{\cos x}{\sin^2 x}\right) dx$$

(d) *Relate the problem to previous problems.* When you have built up some experience in integration, you may be able to use a method on a given integral that is similar to a method you have already used on a previous integral. Or you may even be able to express the given integral in terms of a previous one. For instance, $\int \tan^2 x \sec x\,dx$ is a challenging integral, but if we make use of the identity $\tan^2 x = \sec^2 x - 1$, we can write

$$\int \tan^2 x \sec x\,dx = \int \sec^3 x\,dx - \int \sec x\,dx$$

and if $\int \sec^3 x\,dx$ has previously been evaluated (see Example 8 in Section 7.2), then that calculation can be used in the present problem.

(e) *Use several methods.* Sometimes two or three methods are required to evaluate an integral. The evaluation could involve several successive substitutions of different types, or it might combine integration by parts with one or more substitutions.

In the following examples we indicate a method of attack but do not fully work out the integral.

EXAMPLE 1 $\int \dfrac{\tan^3 x}{\cos^3 x}\,dx$

In Step 1 we rewrite the integral:

$$\int \frac{\tan^3 x}{\cos^3 x}\,dx = \int \tan^3 x \sec^3 x\,dx$$

The integral is now of the form $\int \tan^m x \sec^n x\,dx$ with m odd, so we can use the advice in Section 7.2.

Alternatively, if in Step 1 we had written

$$\int \frac{\tan^3 x}{\cos^3 x}\,dx = \int \frac{\sin^3 x}{\cos^3 x}\frac{1}{\cos^3 x}\,dx = \int \frac{\sin^3 x}{\cos^6 x}\,dx$$

then we could have continued as follows with the substitution $u = \cos x$:

$$\int \frac{\sin^3 x}{\cos^6 x}\,dx = \int \frac{1 - \cos^2 x}{\cos^6 x}\sin x\,dx = \int \frac{1 - u^2}{u^6}(-du)$$

$$= \int \frac{u^2 - 1}{u^6}\,du = \int (u^{-4} - u^{-6})\,du$$

V EXAMPLE 2 $\displaystyle\int e^{\sqrt{x}}\,dx$

According to (ii) in Step 3(d), we substitute $u = \sqrt{x}$. Then $x = u^2$, so $dx = 2u\,du$ and

$$\int e^{\sqrt{x}}\,dx = 2 \int u e^{u}\,du$$

The integrand is now a product of u and the transcendental function e^u so it can be integrated by parts. ▬

EXAMPLE 3 $\displaystyle\int \frac{x^5 + 1}{x^3 - 3x^2 - 10x}\,dx$

No algebraic simplification or substitution is obvious, so Steps 1 and 2 don't apply here. The integrand is a rational function so we apply the procedure of Section 7.4, remembering that the first step is to divide. ▬

V EXAMPLE 4 $\displaystyle\int \frac{dx}{x\sqrt{\ln x}}$

Here Step 2 is all that is needed. We substitute $u = \ln x$ because its differential is $du = dx/x$, which occurs in the integral. ▬

V EXAMPLE 5 $\displaystyle\int \sqrt{\frac{1 - x}{1 + x}}\,dx$

Although the rationalizing substitution

$$u = \sqrt{\frac{1 - x}{1 + x}}$$

works here [(ii) in Step 3(d)], it leads to a very complicated rational function. An easier method is to do some algebraic manipulation [either as Step 1 or as Step 4(c)]. Multiplying numerator and denominator by $\sqrt{1 - x}$, we have

$$\int \sqrt{\frac{1 - x}{1 + x}}\,dx = \int \frac{1 - x}{\sqrt{1 - x^2}}\,dx$$

$$= \int \frac{1}{\sqrt{1 - x^2}}\,dx - \int \frac{x}{\sqrt{1 - x^2}}\,dx$$

$$= \sin^{-1}x + \sqrt{1 - x^2} + C \qquad ▬$$

▬ **Can We Integrate All Continuous Functions?**

The question arises: Will our strategy for integration enable us to find the integral of every continuous function? For example, can we use it to evaluate $\int e^{x^2}\,dx$? The answer is No, at least not in terms of the functions that we are familiar with.

The functions that we have been dealing with in this book are called **elementary functions**. These are the polynomials, rational functions, power functions (x^a), exponential functions (a^x), logarithmic functions, trigonometric and inverse trigonometric functions, hyperbolic and inverse hyperbolic functions, and all functions that can be obtained from these by the five operations of addition, subtraction, multiplication, division, and composition. For instance, the function

$$f(x) = \sqrt{\frac{x^2 - 1}{x^3 + 2x - 1}} + \ln(\cosh x) - xe^{\sin 2x}$$

is an elementary function.

If f is an elementary function, then f' is an elementary function but $\int f(x)\,dx$ need not be an elementary function. Consider $f(x) = e^{x^2}$. Since f is continuous, its integral exists, and if we define the function F by

$$F(x) = \int_0^x e^{t^2}\,dt$$

then we know from Part 1 of the Fundamental Theorem of Calculus that

$$F'(x) = e^{x^2}$$

Thus $f(x) = e^{x^2}$ has an antiderivative F, but it has been proved that F is not an elementary function. This means that no matter how hard we try, we will never succeed in evaluating $\int e^{x^2}\,dx$ in terms of the functions we know. (In Chapter 11, however, we will see how to express $\int e^{x^2}\,dx$ as an infinite series.) The same can be said of the following integrals:

$$\int \frac{e^x}{x}\,dx \qquad \int \sin(x^2)\,dx \qquad \int \cos(e^x)\,dx$$

$$\int \sqrt{x^3 + 1}\,dx \qquad \int \frac{1}{\ln x}\,dx \qquad \int \frac{\sin x}{x}\,dx$$

In fact, the majority of elementary functions don't have elementary antiderivatives. You may be assured, though, that the integrals in the following exercises are all elementary functions.

7.5 Exercises

Access end-of-section exercises and tools for visualizing calculus online at **www.webassign.net**

7.6 Integration Using Tables and Computer Algebra Systems

In this section we describe how to use tables and computer algebra systems to integrate functions that have elementary antiderivatives. You should bear in mind, though, that even the most powerful computer algebra systems can't find explicit formulas for the antiderivatives of functions like e^{x^2} or the other functions described at the end of Section 7.5.

▮ Tables of Integrals

Tables of indefinite integrals are very useful when we are confronted by an integral that is difficult to evaluate by hand and we don't have access to a computer algebra system. A relatively brief table of 120 integrals, categorized by form, is provided on the Reference Pages at the back of the book. More extensive tables are available in the CRC Standard Mathematical Tables and Formulae, 31st ed. by Daniel Zwillinger (Boca Raton, FL, 2002) (709 entries) or in Gradshteyn and Ryzhik's Table of Integrals, Series, and Products, 7e (San Diego, 2007), which contains hundreds of pages of integrals. It should be remembered, however, that integrals do not often occur in exactly the form listed in a table. Usually we need to use the Substitution Rule or algebraic manipulation to transform a given integral into one of the forms in the table.

EXAMPLE 1 The region bounded by the curves $y = \arctan x$, $y = 0$, and $x = 1$ is rotated about the y-axis. Find the volume of the resulting solid.

SOLUTION Using the method of cylindrical shells, we see that the volume is

$$V = \int_0^1 2\pi x \arctan x \, dx$$

The Table of Integrals appears on Reference Pages 6–10 at the back of the book.

In the section of the Table of Integrals titled *Inverse Trigonometric Forms* we locate Formula 92:

$$\int u \tan^{-1}u \, du = \frac{u^2 + 1}{2} \tan^{-1}u - \frac{u}{2} + C$$

Thus the volume is

$$V = 2\pi \int_0^1 x \tan^{-1}x \, dx = 2\pi \left[\frac{x^2 + 1}{2} \tan^{-1}x - \frac{x}{2} \right]_0^1$$

$$= \pi \left[(x^2 + 1) \tan^{-1}x - x \right]_0^1 = \pi(2 \tan^{-1}1 - 1)$$

$$= \pi[2(\pi/4) - 1] = \tfrac{1}{2}\pi^2 - \pi$$

V **EXAMPLE 2** Use the Table of Integrals to find $\displaystyle\int \frac{x^2}{\sqrt{5 - 4x^2}} \, dx$.

SOLUTION If we look at the section of the table titled *Forms Involving $\sqrt{a^2 - u^2}$*, we see that the closest entry is number 34:

$$\int \frac{u^2}{\sqrt{a^2 - u^2}} \, du = -\frac{u}{2} \sqrt{a^2 - u^2} + \frac{a^2}{2} \sin^{-1}\left(\frac{u}{a}\right) + C$$

This is not exactly what we have, but we will be able to use it if we first make the substitution $u = 2x$:

$$\int \frac{x^2}{\sqrt{5 - 4x^2}} \, dx = \int \frac{(u/2)^2}{\sqrt{5 - u^2}} \frac{du}{2} = \frac{1}{8} \int \frac{u^2}{\sqrt{5 - u^2}} \, du$$

Then we use Formula 34 with $a^2 = 5$ (so $a = \sqrt{5}$):

$$\int \frac{x^2}{\sqrt{5 - 4x^2}} \, dx = \frac{1}{8} \int \frac{u^2}{\sqrt{5 - u^2}} \, du = \frac{1}{8}\left(-\frac{u}{2} \sqrt{5 - u^2} + \frac{5}{2} \sin^{-1}\frac{u}{\sqrt{5}} \right) + C$$

$$= -\frac{x}{8} \sqrt{5 - 4x^2} + \frac{5}{16} \sin^{-1}\left(\frac{2x}{\sqrt{5}}\right) + C$$

EXAMPLE 3 Use the Table of Integrals to evaluate $\displaystyle\int x^3 \sin x \, dx$.

SOLUTION If we look in the section called *Trigonometric Forms*, we see that none of the entries explicitly includes a u^3 factor. However, we can use the reduction formula in entry 84 with $n = 3$:

$$\int x^3 \sin x \, dx = -x^3 \cos x + 3 \int x^2 \cos x \, dx$$

85. $\displaystyle\int u^n \cos u \, du$

$\displaystyle = u^n \sin u - n \int u^{n-1} \sin u \, du$

We now need to evaluate $\int x^2 \cos x \, dx$. We can use the reduction formula in entry 85 with $n = 2$, followed by entry 82:

$$\int x^2 \cos x \, dx = x^2 \sin x - 2 \int x \sin x \, dx$$

$$= x^2 \sin x - 2(\sin x - x \cos x) + K$$

Combining these calculations, we get

$$\int x^3 \sin x \, dx = -x^3 \cos x + 3x^2 \sin x + 6x \cos x - 6 \sin x + C$$

where $C = 3K$.

◤ EXAMPLE 4   Use the Table of Integrals to find $\int x\sqrt{x^2 + 2x + 4} \, dx$.

SOLUTION   Since the table gives forms involving $\sqrt{a^2 + x^2}$, $\sqrt{a^2 - x^2}$, and $\sqrt{x^2 - a^2}$, but not $\sqrt{ax^2 + bx + c}$, we first complete the square:

$$x^2 + 2x + 4 = (x + 1)^2 + 3$$

If we make the substitution $u = x + 1$ (so $x = u - 1$), the integrand will involve the pattern $\sqrt{a^2 + u^2}$:

$$\int x\sqrt{x^2 + 2x + 4} \, dx = \int (u - 1)\sqrt{u^2 + 3} \, du$$

$$= \int u\sqrt{u^2 + 3} \, du - \int \sqrt{u^2 + 3} \, du$$

The first integral is evaluated using the substitution $t = u^2 + 3$:

$$\int u\sqrt{u^2 + 3} \, du = \tfrac{1}{2}\int \sqrt{t} \, dt = \tfrac{1}{2} \cdot \tfrac{2}{3}t^{3/2} = \tfrac{1}{3}(u^2 + 3)^{3/2}$$

21. $\displaystyle\int \sqrt{a^2 + u^2} \, du = \frac{u}{2}\sqrt{a^2 + u^2}$

$\displaystyle + \frac{a^2}{2}\ln\!\left(u + \sqrt{a^2 + u^2}\right) + C$

For the second integral we use Formula 21 with $a = \sqrt{3}$:

$$\int \sqrt{u^2 + 3} \, du = \frac{u}{2}\sqrt{u^2 + 3} + \tfrac{3}{2}\ln\!\left(u + \sqrt{u^2 + 3}\right)$$

Thus

$$\int x\sqrt{x^2 + 2x + 4} \, dx$$

$$= \tfrac{1}{3}(x^2 + 2x + 4)^{3/2} - \frac{x + 1}{2}\sqrt{x^2 + 2x + 4} - \tfrac{3}{2}\ln\!\left(x + 1 + \sqrt{x^2 + 2x + 4}\right) + C$$

▬ Computer Algebra Systems

We have seen that the use of tables involves matching the form of the given integrand with the forms of the integrands in the tables. Computers are particularly good at matching patterns. And just as we used substitutions in conjunction with tables, a CAS can perform substitutions that transform a given integral into one that occurs in its stored formulas. So it isn't surprising that computer algebra systems excel at integration. That doesn't mean that integration by hand is an obsolete skill. We will see that a hand computation sometimes produces an indefinite integral in a form that is more convenient than a machine answer.

To begin, let's see what happens when we ask a machine to integrate the relatively simple function $y = 1/(3x - 2)$. Using the substitution $u = 3x - 2$, an easy calculation by hand gives

$$\int \frac{1}{3x - 2} \, dx = \tfrac{1}{3} \ln|3x - 2| + C$$

whereas Derive, Mathematica, and Maple all return the answer

$$\tfrac{1}{3} \ln(3x - 2)$$

The first thing to notice is that computer algebra systems omit the constant of integration. In other words, they produce a *particular* antiderivative, not the most general one. Therefore, when making use of a machine integration, we might have to add a constant. Second, the absolute value signs are omitted in the machine answer. That is fine if our problem is concerned only with values of x greater than $\tfrac{2}{3}$. But if we are interested in other values of x, then we need to insert the absolute value symbol.

In the next example we reconsider the integral of Example 4, but this time we ask a machine for the answer.

EXAMPLE 5 Use a computer algebra system to find $\int x\sqrt{x^2 + 2x + 4} \, dx$.

SOLUTION Maple responds with the answer

$$\tfrac{1}{3}(x^2 + 2x + 4)^{3/2} - \tfrac{1}{4}(2x + 2)\sqrt{x^2 + 2x + 4} - \frac{3}{2} \operatorname{arcsinh} \frac{\sqrt{3}}{3}(1 + x)$$

This looks different from the answer we found in Example 4, but it is equivalent because the third term can be rewritten using the identity

This is Equation 6.7.3.

$$\operatorname{arcsinh} x = \ln\left(x + \sqrt{x^2 + 1}\right)$$

Thus

$$\operatorname{arcsinh} \frac{\sqrt{3}}{3}(1 + x) = \ln\left[\frac{\sqrt{3}}{3}(1 + x) + \sqrt{\tfrac{1}{3}(1 + x)^2 + 1}\right]$$

$$= \ln \frac{1}{\sqrt{3}}\left[1 + x + \sqrt{(1 + x)^2 + 3}\right]$$

$$= \ln \frac{1}{\sqrt{3}} + \ln\left(x + 1 + \sqrt{x^2 + 2x + 4}\right)$$

The resulting extra term $-\tfrac{3}{2}\ln(1/\sqrt{3})$ can be absorbed into the constant of integration.

Mathematica gives the answer

$$\left(\frac{5}{6} + \frac{x}{6} + \frac{x^2}{3}\right)\sqrt{x^2 + 2x + 4} - \frac{3}{2} \operatorname{arcsinh}\left(\frac{1 + x}{\sqrt{3}}\right)$$

Mathematica combined the first two terms of Example 4 (and the Maple result) into a single term by factoring.

Derive gives the answer

$$\tfrac{1}{6}\sqrt{x^2 + 2x + 4}\,(2x^2 + x + 5) - \tfrac{3}{2}\ln\left(\sqrt{x^2 + 2x + 4} + x + 1\right)$$

The first term is like the first term in the Mathematica answer, and the second term is identical to the last term in Example 4.

EXAMPLE 6 Use a CAS to evaluate $\int x(x^2 + 5)^8 \, dx$.

SOLUTION Maple and Mathematica give the same answer:

$$\tfrac{1}{18} x^{18} + \tfrac{5}{2} x^{16} + 50x^{14} + \tfrac{1750}{3} x^{12} + 4375x^{10} + 21875x^8 + \tfrac{218750}{3} x^6 + 156250x^4 + \tfrac{390625}{2} x^2$$

It's clear that both systems must have expanded $(x^2 + 5)^8$ by the Binomial Theorem and then integrated each term.

Derive and the TI-89 and TI-92 also give this answer.

If we integrate by hand instead, using the substitution $u = x^2 + 5$, we get

$$\int x(x^2 + 5)^8 \, dx = \tfrac{1}{18}(x^2 + 5)^9 + C$$

For most purposes, this is a more convenient form of the answer.

EXAMPLE 7 Use a CAS to find $\int \sin^5 x \, \cos^2 x \, dx$.

SOLUTION In Example 2 in Section 7.2 we found that

$$\boxed{1} \qquad \int \sin^5 x \, \cos^2 x \, dx = -\tfrac{1}{3} \cos^3 x + \tfrac{2}{5} \cos^5 x - \tfrac{1}{7} \cos^7 x + C$$

Derive and Maple report the answer

$$-\tfrac{1}{7} \sin^4 x \, \cos^3 x - \tfrac{4}{35} \sin^2 x \, \cos^3 x - \tfrac{8}{105} \cos^3 x$$

whereas Mathematica produces

$$-\tfrac{5}{64} \cos x - \tfrac{1}{192} \cos 3x + \tfrac{3}{320} \cos 5x - \tfrac{1}{448} \cos 7x$$

We suspect that there are trigonometric identities which show that these three answers are equivalent. Indeed, if we ask Derive, Maple, and Mathematica to simplify their expressions using trigonometric identities, they ultimately produce the same form of the answer as in Equation 1.

7.6   **Exercises**

Access end-of-section exercises and tools for visualizing calculus online at **www.webassign.net**   ENHANCED WebAssign

DISCOVERY PROJECT   CAS **PATTERNS IN INTEGRALS**

In this project a computer algebra system is used to investigate indefinite integrals of families of functions. By observing the patterns that occur in the integrals of several members of the family, you will first guess, and then prove, a general formula for the integral of any member of the family.

1. (a) Use a computer algebra system to evaluate the following integrals.

(i) $\displaystyle\int \frac{1}{(x+2)(x+3)}\, dx$ (ii) $\displaystyle\int \frac{1}{(x+1)(x+5)}\, dx$

(iii) $\displaystyle\int \frac{1}{(x+2)(x-5)}\, dx$ (iv) $\displaystyle\int \frac{1}{(x+2)^2}\, dx$

(b) Based on the pattern of your responses in part (a), guess the value of the integral

$$\int \frac{1}{(x+a)(x+b)}\, dx$$

if $a \neq b$. What if $a = b$?

(c) Check your guess by asking your CAS to evaluate the integral in part (b). Then prove it using partial fractions.

2. (a) Use a computer algebra system to evaluate the following integrals.

(i) $\displaystyle\int \sin x \cos 2x\, dx$ (ii) $\displaystyle\int \sin 3x \cos 7x\, dx$ (iii) $\displaystyle\int \sin 8x \cos 3x\, dx$

(b) Based on the pattern of your responses in part (a), guess the value of the integral

$$\int \sin ax \cos bx\, dx$$

(c) Check your guess with a CAS. Then prove it using the techniques of Section 7.2. For what values of a and b is it valid?

3. (a) Use a computer algebra system to evaluate the following integrals.

(i) $\displaystyle\int \ln x\, dx$ (ii) $\displaystyle\int x \ln x\, dx$ (iii) $\displaystyle\int x^2 \ln x\, dx$

(iv) $\displaystyle\int x^3 \ln x\, dx$ (v) $\displaystyle\int x^7 \ln x\, dx$

(b) Based on the pattern of your responses in part (a), guess the value of

$$\int x^n \ln x\, dx$$

(c) Use integration by parts to prove the conjecture that you made in part (b). For what values of n is it valid?

4. (a) Use a computer algebra system to evaluate the following integrals.

(i) $\displaystyle\int xe^x\, dx$ (ii) $\displaystyle\int x^2 e^x\, dx$ (iii) $\displaystyle\int x^3 e^x\, dx$

(iv) $\displaystyle\int x^4 e^x\, dx$ (v) $\displaystyle\int x^5 e^x\, dx$

CAS Computer algebra system required

(b) Based on the pattern of your responses in part (a), guess the value of $\int x^6 e^x\,dx$. Then use your CAS to check your guess.

(c) Based on the patterns in parts (a) and (b), make a conjecture as to the value of the integral

$$\int x^n e^x\,dx$$

when n is a positive integer.

(d) Use mathematical induction to prove the conjecture you made in part (c).

7.7 Approximate Integration

There are two situations in which it is impossible to find the exact value of a definite integral.

The first situation arises from the fact that in order to evaluate $\int_a^b f(x)\,dx$ using the Fundamental Theorem of Calculus we need to know an antiderivative of f. Sometimes, however, it is difficult, or even impossible, to find an antiderivative (see Section 7.5). For example, it is impossible to evaluate the following integrals exactly:

$$\int_0^1 e^{x^2}\,dx \qquad \int_{-1}^1 \sqrt{1+x^3}\,dx$$

The second situation arises when the function is determined from a scientific experiment through instrument readings or collected data. There may be no formula for the function (see Example 5).

In both cases we need to find approximate values of definite integrals. We already know one such method. Recall that the definite integral is defined as a limit of Riemann sums, so any Riemann sum could be used as an approximation to the integral: If we divide $[a,b]$ into n subintervals of equal length $\Delta x = (b-a)/n$, then we have

$$\int_a^b f(x)\,dx \approx \sum_{i=1}^n f(x_i^*)\,\Delta x$$

where x_i^* is any point in the ith subinterval $[x_{i-1},x_i]$. If x_i^* is chosen to be the left endpoint of the interval, then $x_i^* = x_{i-1}$ and we have

$$\boxed{1} \qquad \int_a^b f(x)\,dx \approx L_n = \sum_{i=1}^n f(x_{i-1})\,\Delta x$$

If $f(x) \geq 0$, then the integral represents an area and $\boxed{1}$ represents an approximation of this area by the rectangles shown in Figure 1(a). If we choose x_i^* to be the right endpoint, then $x_i^* = x_i$ and we have

$$\boxed{2} \qquad \int_a^b f(x)\,dx \approx R_n = \sum_{i=1}^n f(x_i)\,\Delta x$$

[See Figure 1(b).] The approximations L_n and R_n defined by Equations 1 and 2 are called the **left endpoint approximation** and **right endpoint approximation**, respectively.

In Section 4.2 we also considered the case where x_i^* is chosen to be the midpoint $\bar{x}_i$ of the subinterval $[x_{i-1},x_i]$. Figure 1(c) shows the midpoint approximation M_n, which appears to be better than either L_n or R_n.

(a) Left endpoint approximation

(b) Right endpoint approximation

(c) Midpoint approximation

FIGURE 1

Midpoint Rule

$$\int_a^b f(x)\,dx \approx M_n = \Delta x\,[f(\bar{x}_1) + f(\bar{x}_2) + \cdots + f(\bar{x}_n)]$$

where

$$\Delta x = \frac{b - a}{n}$$

and

$$\bar{x}_i = \tfrac{1}{2}(x_{i-1} + x_i) = \text{midpoint of } [x_{i-1}, x_i]$$

Another approximation, called the Trapezoidal Rule, results from averaging the approximations in Equations 1 and 2:

$$\int_a^b f(x)\,dx \approx \frac{1}{2}\left[\sum_{i=1}^n f(x_{i-1})\,\Delta x + \sum_{i=1}^n f(x_i)\,\Delta x\right] = \frac{\Delta x}{2}\left[\sum_{i=1}^n \big(f(x_{i-1}) + f(x_i)\big)\right]$$

$$= \frac{\Delta x}{2}\left[\big(f(x_0) + f(x_1)\big) + \big(f(x_1) + f(x_2)\big) + \cdots + \big(f(x_{n-1}) + f(x_n)\big)\right]$$

$$= \frac{\Delta x}{2}\left[f(x_0) + 2f(x_1) + 2f(x_2) + \cdots + 2f(x_{n-1}) + f(x_n)\right]$$

Trapezoidal Rule

$$\int_a^b f(x)\,dx \approx T_n = \frac{\Delta x}{2}\left[f(x_0) + 2f(x_1) + 2f(x_2) + \cdots + 2f(x_{n-1}) + f(x_n)\right]$$

where $\Delta x = (b - a)/n$ and $x_i = a + i\,\Delta x$.

The reason for the name Trapezoidal Rule can be seen from Figure 2, which illustrates the case with $f(x) \geq 0$ and $n = 4$. The area of the trapezoid that lies above the ith subinterval is

$$\Delta x\left(\frac{f(x_{i-1}) + f(x_i)}{2}\right) = \frac{\Delta x}{2}\left[f(x_{i-1}) + f(x_i)\right]$$

and if we add the areas of all these trapezoids, we get the right side of the Trapezoidal Rule.

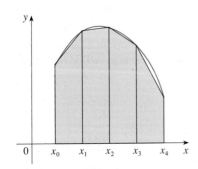

FIGURE 2

Trapezoidal approximation

EXAMPLE 1 Use (a) the Trapezoidal Rule and (b) the Midpoint Rule with $n = 5$ to approximate the integral $\int_1^2 (1/x)\,dx$.

SOLUTION
(a) With $n = 5$, $a = 1$, and $b = 2$, we have $\Delta x = (2 - 1)/5 = 0.2$, and so the Trapezoidal Rule gives

$$\int_1^2 \frac{1}{x}\,dx \approx T_5 = \frac{0.2}{2}\left[f(1) + 2f(1.2) + 2f(1.4) + 2f(1.6) + 2f(1.8) + f(2)\right]$$

$$= 0.1\left(\frac{1}{1} + \frac{2}{1.2} + \frac{2}{1.4} + \frac{2}{1.6} + \frac{2}{1.8} + \frac{1}{2}\right)$$

$$\approx 0.695635$$

This approximation is illustrated in Figure 3.

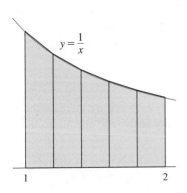

$y = \dfrac{1}{x}$

1   2

FIGURE 3

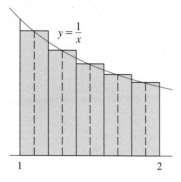

FIGURE 4

$$\int_a^b f(x)\, dx = \text{approximation} + \text{error}$$

Approximations to $\int_1^2 \dfrac{1}{x}\, dx$

Corresponding errors

(b) The midpoints of the five subintervals are 1.1, 1.3, 1.5, 1.7, and 1.9, so the Midpoint Rule gives

$$\int_1^2 \frac{1}{x}\, dx \approx \Delta x \left[f(1.1) + f(1.3) + f(1.5) + f(1.7) + f(1.9) \right]$$

$$= \frac{1}{5}\left(\frac{1}{1.1} + \frac{1}{1.3} + \frac{1}{1.5} + \frac{1}{1.7} + \frac{1}{1.9} \right)$$

$$\approx 0.691908$$

This approximation is illustrated in Figure 4.

In Example 1 we deliberately chose an integral whose value can be computed explicitly so that we can see how accurate the Trapezoidal and Midpoint Rules are. By the Fundamental Theorem of Calculus,

$$\int_1^2 \frac{1}{x}\, dx = \ln x \Big]_1^2 = \ln 2 = 0.693147\ldots$$

The **error** in using an approximation is defined to be the amount that needs to be added to the approximation to make it exact. From the values in Example 1 we see that the errors in the Trapezoidal and Midpoint Rule approximations for $n = 5$ are

$$E_T \approx -0.002488 \qquad \text{and} \qquad E_M \approx 0.001239$$

In general, we have

$$E_T = \int_a^b f(x)\, dx - T_n \qquad \text{and} \qquad E_M = \int_a^b f(x)\, dx - M_n$$

The following tables show the results of calculations similar to those in Example 1, but for $n = 5$, 10, and 20 and for the left and right endpoint approximations as well as the Trapezoidal and Midpoint Rules.

n	L_n	R_n	T_n	M_n
5	0.745635	0.645635	0.695635	0.691908
10	0.718771	0.668771	0.693771	0.692835
20	0.705803	0.680803	0.693303	0.693069

n	E_L	E_R	E_T	E_M
5	−0.052488	0.047512	−0.002488	0.001239
10	−0.025624	0.024376	−0.000624	0.000312
20	−0.012656	0.012344	−0.000156	0.000078

We can make several observations from these tables:

1. In all of the methods we get more accurate approximations when we increase the value of n. (But very large values of n result in so many arithmetic operations that we have to beware of accumulated round-off error.)

2. The errors in the left and right endpoint approximations are opposite in sign and appear to decrease by a factor of about 2 when we double the value of n.

3. The Trapezoidal and Midpoint Rules are much more accurate than the endpoint approximations.

It turns out that these observations are true in most cases.

4. The errors in the Trapezoidal and Midpoint Rules are opposite in sign and appear to decrease by a factor of about 4 when we double the value of n.

5. The size of the error in the Midpoint Rule is about half the size of the error in the Trapezoidal Rule.

Figure 5 shows why we can usually expect the Midpoint Rule to be more accurate than the Trapezoidal Rule. The area of a typical rectangle in the Midpoint Rule is the same as the area of the trapezoid $ABCD$ whose upper side is tangent to the graph at P. The area of this trapezoid is closer to the area under the graph than is the area of the trapezoid $AQRD$ used in the Trapezoidal Rule. [The midpoint error (shaded red) is smaller than the trapezoidal error (shaded blue).]

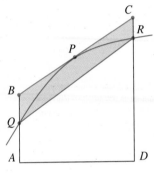

FIGURE 5

These observations are corroborated in the following error estimates, which are proved in books on numerical analysis. Notice that Observation 4 corresponds to the n^2 in each denominator because $(2n)^2 = 4n^2$. The fact that the estimates depend on the size of the second derivative is not surprising if you look at Figure 5, because $f''(x)$ measures how much the graph is curved. [Recall that $f''(x)$ measures how fast the slope of $y = f(x)$ changes.]

---

3 **Error Bounds** Suppose $|f''(x)| \leq K$ for $a \leq x \leq b$. If E_T and E_M are the errors in the Trapezoidal and Midpoint Rules, then

$$|E_T| \leq \frac{K(b-a)^3}{12n^2} \quad \text{and} \quad |E_M| \leq \frac{K(b-a)^3}{24n^2}$$

Let's apply this error estimate to the Trapezoidal Rule approximation in Example 1. If $f(x) = 1/x$, then $f'(x) = -1/x^2$ and $f''(x) = 2/x^3$. Since $1 \leq x \leq 2$, we have $1/x \leq 1$, so

$$|f''(x)| = \left|\frac{2}{x^3}\right| \leq \frac{2}{1^3} = 2$$

Therefore, taking $K = 2$, $a = 1$, $b = 2$, and $n = 5$ in the error estimate $\boxed{3}$, we see that

K can be any number larger than all the values of $|f''(x)|$, but smaller values of K give better error bounds.

$$|E_T| \leq \frac{2(2-1)^3}{12(5)^2} = \frac{1}{150} \approx 0.006667$$

Comparing this error estimate of 0.006667 with the actual error of about 0.002488, we see that it can happen that the actual error is substantially less than the upper bound for the error given by $\boxed{3}$.

V **EXAMPLE 2** How large should we take n in order to guarantee that the Trapezoidal and Midpoint Rule approximations for $\int_1^2 (1/x)\, dx$ are accurate to within 0.0001?

SOLUTION We saw in the preceding calculation that $|f''(x)| \leq 2$ for $1 \leq x \leq 2$, so we can take $K = 2$, $a = 1$, and $b = 2$ in $\boxed{3}$. Accuracy to within 0.0001 means that the size of the error should be less than 0.0001. Therefore we choose n so that

$$\frac{2(1)^3}{12n^2} < 0.0001$$

Solving the inequality for n, we get

$$n^2 > \frac{2}{12(0.0001)}$$

It's quite possible that a lower value for n would suffice, but 41 is the smallest value for which the error bound formula can *guarantee* us accuracy to within 0.0001.

or

$$n > \frac{1}{\sqrt{0.0006}} \approx 40.8$$

Thus $n = 41$ will ensure the desired accuracy.

For the same accuracy with the Midpoint Rule we choose n so that

$$\frac{2(1)^3}{24n^2} < 0.0001 \qquad \text{and so} \qquad n > \frac{1}{\sqrt{0.0012}} \approx 29$$

V **EXAMPLE 3**
(a) Use the Midpoint Rule with $n = 10$ to approximate the integral $\int_0^1 e^{x^2}\, dx$.
(b) Give an upper bound for the error involved in this approximation.

SOLUTION
(a) Since $a = 0$, $b = 1$, and $n = 10$, the Midpoint Rule gives

$$\int_0^1 e^{x^2}\, dx \approx \Delta x \left[f(0.05) + f(0.15) + \cdots + f(0.85) + f(0.95) \right]$$

$$= 0.1\left[e^{0.0025} + e^{0.0225} + e^{0.0625} + e^{0.1225} + e^{0.2025} + e^{0.3025} \right.$$
$$\left. + e^{0.4225} + e^{0.5625} + e^{0.7225} + e^{0.9025} \right]$$

$$\approx 1.460393$$

$y = e^{x^2}$

FIGURE 6

Figure 6 illustrates this approximation.

(b) Since $f(x) = e^{x^2}$, we have $f'(x) = 2xe^{x^2}$ and $f''(x) = (2 + 4x^2)e^{x^2}$. Also, since $0 \leq x \leq 1$, we have $x^2 \leq 1$ and so

$$0 \leq f''(x) = (2 + 4x^2)e^{x^2} \leq 6e$$

Error estimates give upper bounds for the error. They are theoretical, worst-case scenarios. The actual error in this case turns out to be about 0.0023.

Taking $K = 6e$, $a = 0$, $b = 1$, and $n = 10$ in the error estimate $\boxed{3}$, we see that an upper bound for the error is

$$\frac{6e(1)^3}{24(10)^2} = \frac{e}{400} \approx 0.007$$

Simpson's Rule

Another rule for approximate integration results from using parabolas instead of straight line segments to approximate a curve. As before, we divide $[a, b]$ into n subintervals of equal length $h = \Delta x = (b - a)/n$, but this time we assume that n is an *even* number. Then on each consecutive pair of intervals we approximate the curve $y = f(x) \geq 0$ by a parabola as shown in Figure 7. If $y_i = f(x_i)$, then $P_i(x_i, y_i)$ is the point on the curve lying above x_i. A typical parabola passes through three consecutive points P_i, P_{i+1}, and P_{i+2}.

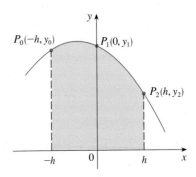

FIGURE 7

FIGURE 8

To simplify our calculations, we first consider the case where $x_0 = -h$, $x_1 = 0$, and $x_2 = h$. (See Figure 8.) We know that the equation of the parabola through P_0, P_1, and P_2 is of the form $y = Ax^2 + Bx + C$ and so the area under the parabola from $x = -h$ to $x = h$ is

$$\int_{-h}^{h} (Ax^2 + Bx + C)\, dx = 2 \int_0^h (Ax^2 + C)\, dx$$

Here we have used Theorem 4.5.6.
Notice that $Ax^2 + C$ is even and Bx is odd.

$$= 2\left[A\frac{x^3}{3} + Cx \right]_0^h$$

$$= 2\left(A\frac{h^3}{3} + Ch \right) = \frac{h}{3}(2Ah^2 + 6C)$$

But, since the parabola passes through $P_0(-h, y_0)$, $P_1(0, y_1)$, and $P_2(h, y_2)$, we have

$$y_0 = A(-h)^2 + B(-h) + C = Ah^2 - Bh + C$$

$$y_1 = C$$

$$y_2 = Ah^2 + Bh + C$$

and therefore $\quad y_0 + 4y_1 + y_2 = 2Ah^2 + 6C$

Thus we can rewrite the area under the parabola as

$$\frac{h}{3}(y_0 + 4y_1 + y_2)$$

Now by shifting this parabola horizontally we do not change the area under it. This means that the area under the parabola through P_0, P_1, and P_2 from $x = x_0$ to $x = x_2$ in Figure 7 is still

$$\frac{h}{3}(y_0 + 4y_1 + y_2)$$

Similarly, the area under the parabola through P_2, P_3, and P_4 from $x = x_2$ to $x = x_4$ is

$$\frac{h}{3}(y_2 + 4y_3 + y_4)$$

If we compute the areas under all the parabolas in this manner and add the results, we get

$$\int_a^b f(x)\, dx \approx \frac{h}{3}(y_0 + 4y_1 + y_2) + \frac{h}{3}(y_2 + 4y_3 + y_4)$$

$$+ \cdots + \frac{h}{3}(y_{n-2} + 4y_{n-1} + y_n)$$

$$= \frac{h}{3}(y_0 + 4y_1 + 2y_2 + 4y_3 + 2y_4 + \cdots + 2y_{n-2} + 4y_{n-1} + y_n)$$

Although we have derived this approximation for the case in which $f(x) \geq 0$, it is a reasonable approximation for any continuous function f and is called Simpson's Rule after the English mathematician Thomas Simpson (1710–1761). Note the pattern of coefficients: $1, 4, 2, 4, 2, 4, 2, \ldots, 4, 2, 4, 1$.

Simpson

Thomas Simpson was a weaver who taught himself mathematics and went on to become one of the best English mathematicians of the 18th century. What we call Simpson's Rule was actually known to Cavalieri and Gregory in the 17th century, but Simpson popularized it in his best-selling calculus textbook, *A New Treatise of Fluxions*.

Simpson's Rule

$$\int_a^b f(x)\, dx \approx S_n = \frac{\Delta x}{3}[f(x_0) + 4f(x_1) + 2f(x_2) + 4f(x_3) + \cdots$$

$$+ 2f(x_{n-2}) + 4f(x_{n-1}) + f(x_n)]$$

where n is even and $\Delta x = (b - a)/n$.

EXAMPLE 4   Use Simpson's Rule with $n = 10$ to approximate $\int_1^2 (1/x)\, dx$.

SOLUTION   Putting $f(x) = 1/x$, $n = 10$, and $\Delta x = 0.1$ in Simpson's Rule, we obtain

$$\int_1^2 \frac{1}{x}\, dx \approx S_{10}$$

$$= \frac{\Delta x}{3}[f(1) + 4f(1.1) + 2f(1.2) + 4f(1.3) + \cdots + 2f(1.8) + 4f(1.9) + f(2)]$$

$$= \frac{0.1}{3}\left(\frac{1}{1} + \frac{4}{1.1} + \frac{2}{1.2} + \frac{4}{1.3} + \frac{2}{1.4} + \frac{4}{1.5} + \frac{2}{1.6} + \frac{4}{1.7} + \frac{2}{1.8} + \frac{4}{1.9} + \frac{1}{2}\right)$$

$$\approx 0.693150$$

Notice that, in Example 4, Simpson's Rule gives us a *much* better approximation ($S_{10} \approx 0.693150$) to the true value of the integral ($\ln 2 \approx 0.693147\ldots$) than does the Trapezoidal Rule ($T_{10} \approx 0.693771$) or the Midpoint Rule ($M_{10} \approx 0.692835$). It turns out (see Exercise 50) that the approximations in Simpson's Rule are weighted averages of those in the Trapezoidal and Midpoint Rules:

$$S_{2n} = \tfrac{1}{3}T_n + \tfrac{2}{3}M_n$$

(Recall that E_T and E_M usually have opposite signs and $|E_M|$ is about half the size of $|E_T|$.)

In many applications of calculus we need to evaluate an integral even if no explicit formula is known for y as a function of x. A function may be given graphically or as a table of values of collected data. If there is evidence that the values are not changing rapidly, then the Trapezoidal Rule or Simpson's Rule can still be used to find an approximate value for $\int_a^b y \, dx$, the integral of y with respect to x.

V **EXAMPLE 5** Figure 9 shows data traffic on the link from the United States to SWITCH, the Swiss academic and research network, on February 10, 1998. $D(t)$ is the data through-put, measured in megabits per second (Mb/s). Use Simpson's Rule to estimate the total amount of data transmitted on the link from midnight to noon on that day.

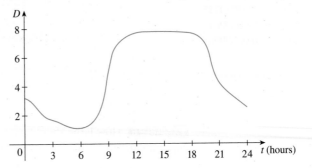

FIGURE 9

SOLUTION Because we want the units to be consistent and $D(t)$ is measured in megabits per second, we convert the units for t from hours to seconds. If we let $A(t)$ be the amount of data (in megabits) transmitted by time t, where t is measured in seconds, then $A'(t) = D(t)$. So, by the Net Change Theorem (see Section 4.4), the total amount of data transmitted by noon (when $t = 12 \times 60^2 = 43{,}200$) is

$$A(43{,}200) = \int_0^{43{,}200} D(t) \, dt$$

We estimate the values of $D(t)$ at hourly intervals from the graph and compile them in the table.

t (hours)	t (seconds)	$D(t)$	t (hours)	t (seconds)	$D(t)$
0	0	3.2	7	25,200	1.3
1	3,600	2.7	8	28,800	2.8
2	7,200	1.9	9	32,400	5.7
3	10,800	1.7	10	36,000	7.1
4	14,400	1.3	11	39,600	7.7
5	18,000	1.0	12	43,200	7.9
6	21,600	1.1			

Then we use Simpson's Rule with $n = 12$ and $\Delta t = 3600$ to estimate the integral:

$$\int_0^{43{,}200} A(t) \, dt \approx \frac{\Delta t}{3} [D(0) + 4D(3600) + 2D(7200) + \cdots + 4D(39{,}600) + D(43{,}200)]$$

$$\approx \frac{3600}{3} [3.2 + 4(2.7) + 2(1.9) + 4(1.7) + 2(1.3) + 4(1.0)$$
$$+ 2(1.1) + 4(1.3) + 2(2.8) + 4(5.7) + 2(7.1) + 4(7.7) + 7.9]$$

$$= 143{,}880$$

Thus the total amount of data transmitted from midnight to noon is about 144,000 mega-bits, or 144 gigabits.

n	M_n	S_n
4	0.69121989	0.69315453
8	0.69266055	0.69314765
16	0.69302521	0.69314721

n	E_M	E_S
4	0.00192729	−0.00000735
8	0.00048663	−0.00000047
16	0.00012197	−0.00000003

The table in the margin shows how Simpson's Rule compares with the Midpoint Rule for the integral $\int_1^2 (1/x)\,dx$, whose value is about 0.69314718. The second table shows how the error E_s in Simpson's Rule decreases by a factor of about 16 when n is doubled. (In Exercises 27 and 28 you are asked to verify this for two additional integrals.) That is consistent with the appearance of n^4 in the denominator of the following error estimate for Simpson's Rule. It is similar to the estimates given in $\boxed{3}$ for the Trapezoidal and Midpoint Rules, but it uses the fourth derivative of f.

$\boxed{4}$ **Error Bound for Simpson's Rule** Suppose that $|f^{(4)}(x)| \leq K$ for $a \leq x \leq b$. If E_S is the error involved in using Simpson's Rule, then

$$|E_S| \leq \frac{K(b-a)^5}{180n^4}$$

EXAMPLE 6 How large should we take n in order to guarantee that the Simpson's Rule approximation for $\int_1^2 (1/x)\,dx$ is accurate to within 0.0001?

SOLUTION If $f(x) = 1/x$, then $f^{(4)}(x) = 24/x^5$. Since $x \geq 1$, we have $1/x \leq 1$ and so

$$|f^{(4)}(x)| = \left|\frac{24}{x^5}\right| \leq 24$$

Many calculators and computer algebra systems have a built-in algorithm that computes an approximation of a definite integral. Some of these machines use Simpson's Rule; others use more sophisticated techniques such as adaptive numerical integration. This means that if a function fluctuates much more on a certain part of the interval than it does elsewhere, then that part gets divided into more subintervals. This strategy reduces the number of calculations required to achieve a prescribed accuracy.

Therefore we can take $K = 24$ in $\boxed{4}$. Thus, for an error less than 0.0001, we should choose n so that

$$\frac{24(1)^5}{180n^4} < 0.0001$$

This gives

$$n^4 > \frac{24}{180(0.0001)}$$

or

$$n > \frac{1}{\sqrt[4]{0.00075}} \approx 6.04$$

Therefore $n = 8$ (n must be even) gives the desired accuracy. (Compare this with Example 2, where we obtained $n = 41$ for the Trapezoidal Rule and $n = 29$ for the Midpoint Rule.)

Figure 10 illustrates the calculation in Example 7. Notice that the parabolic arcs are so close to the graph of $y = e^{x^2}$ that they are practically indistinguishable from it.

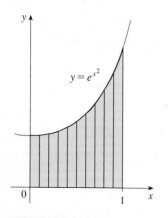

FIGURE 10

EXAMPLE 7
(a) Use Simpson's Rule with $n = 10$ to approximate the integral $\int_0^1 e^{x^2}\,dx$.
(b) Estimate the error involved in this approximation.

SOLUTION
(a) If $n = 10$, then $\Delta x = 0.1$ and Simpson's Rule gives

$$\int_0^1 e^{x^2}\,dx \approx \frac{\Delta x}{3}[f(0) + 4f(0.1) + 2f(0.2) + \cdots + 2f(0.8) + 4f(0.9) + f(1)]$$

$$= \frac{0.1}{3}[e^0 + 4e^{0.01} + 2e^{0.04} + 4e^{0.09} + 2e^{0.16} + 4e^{0.25} + 2e^{0.36}$$

$$+ 4e^{0.49} + 2e^{0.64} + 4e^{0.81} + e^1]$$

$$\approx 1.462681$$

(b) The fourth derivative of $f(x) = e^{x^2}$ is

$$f^{(4)}(x) = (12 + 48x^2 + 16x^4)e^{x^2}$$

and so, since $0 \leq x \leq 1$, we have

$$0 \leq f^{(4)}(x) \leq (12 + 48 + 16)e^1 = 76e$$

Therefore, putting $K = 76e$, $a = 0$, $b = 1$, and $n = 10$ in $\boxed{4}$, we see that the error is at most

$$\frac{76e(1)^5}{180(10)^4} \approx 0.000115$$

(Compare this with Example 3.) Thus, correct to three decimal places, we have

$$\int_0^1 e^{x^2}\, dx \approx 1.463$$

7.7 Exercises

Access end-of-section exercises and tools for visualizing calculus online at **www.webassign.net**

7.8 Improper Integrals

In defining a definite integral $\int_a^b f(x)\, dx$ we dealt with a function f defined on a finite interval $[a, b]$ and we assumed that f does not have an infinite discontinuity (see Section 4.2). In this section we extend the concept of a definite integral to the case where the interval is infinite and also to the case where f has an infinite discontinuity in $[a, b]$. In either case the integral is called an *improper* integral. One of the most important applications of this idea, probability distributions, will be studied in Section 8.5.

■ Type 1: Infinite Intervals

Consider the infinite region S that lies under the curve $y = 1/x^2$, above the x-axis, and to the right of the line $x = 1$. You might think that, since S is infinite in extent, its area must be infinite, but let's take a closer look. The area of the part of S that lies to the left of the line $x = t$ (shaded in Figure 1) is

$$A(t) = \int_1^t \frac{1}{x^2}\, dx = -\frac{1}{x}\Big]_1^t = 1 - \frac{1}{t}$$

Notice that $A(t) < 1$ no matter how large t is chosen.

We also observe that

$$\lim_{t \to \infty} A(t) = \lim_{t \to \infty}\left(1 - \frac{1}{t}\right) = 1$$

The area of the shaded region approaches 1 as $t \to \infty$ (see Figure 2 on page 430), so we say that the area of the infinite region S is equal to 1 and we write

$$\int_1^\infty \frac{1}{x^2}\, dx = \lim_{t \to \infty}\int_1^t \frac{1}{x^2}\, dx = 1$$

FIGURE 1

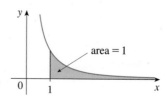

FIGURE 2

Using this example as a guide, we define the integral of f (not necessarily a positive function) over an infinite interval as the limit of integrals over finite intervals.

---

1 **Definition of an Improper Integral of Type 1**

(a) If $\int_a^t f(x)\,dx$ exists for every number $t \geqslant a$, then

$$\int_a^\infty f(x)\,dx = \lim_{t \to \infty} \int_a^t f(x)\,dx$$

provided this limit exists (as a finite number).

(b) If $\int_t^b f(x)\,dx$ exists for every number $t \leqslant b$, then

$$\int_{-\infty}^b f(x)\,dx = \lim_{t \to -\infty} \int_t^b f(x)\,dx$$

provided this limit exists (as a finite number).

The improper integrals $\int_a^\infty f(x)\,dx$ and $\int_{-\infty}^b f(x)\,dx$ are called **convergent** if the corresponding limit exists and **divergent** if the limit does not exist.

(c) If both $\int_a^\infty f(x)\,dx$ and $\int_{-\infty}^a f(x)\,dx$ are convergent, then we define

$$\int_{-\infty}^\infty f(x)\,dx = \int_{-\infty}^a f(x)\,dx + \int_a^\infty f(x)\,dx$$

In part (c) any real number a can be used (see Exercise 74).

Any of the improper integrals in Definition 1 can be interpreted as an area provided that f is a positive function. For instance, in case (a) if $f(x) \geqslant 0$ and the integral $\int_a^\infty f(x)\,dx$ is convergent, then we define the area of the region $S = \{(x, y) \mid x \geqslant a, 0 \leqslant y \leqslant f(x)\}$ in Figure 3 to be

$$A(S) = \int_a^\infty f(x)\,dx$$

This is appropriate because $\int_a^\infty f(x)\,dx$ is the limit as $t \to \infty$ of the area under the graph of f from a to t.

FIGURE 3

V **EXAMPLE 1** Determine whether the integral $\int_1^\infty (1/x)\,dx$ is convergent or divergent.

SOLUTION According to part (a) of Definition 1, we have

$$\int_1^\infty \frac{1}{x}\,dx = \lim_{t\to\infty} \int_1^t \frac{1}{x}\,dx = \lim_{t\to\infty} \ln|x|\Big]_1^t$$

$$= \lim_{t\to\infty}(\ln t - \ln 1) = \lim_{t\to\infty} \ln t = \infty$$

The limit does not exist as a finite number and so the improper integral $\int_1^\infty (1/x)\,dx$ is divergent.

Let's compare the result of Example 1 with the example given at the beginning of this section:

$$\int_1^\infty \frac{1}{x^2}\,dx \text{ converges} \qquad \int_1^\infty \frac{1}{x}\,dx \text{ diverges}$$

Geometrically, this says that although the curves $y = 1/x^2$ and $y = 1/x$ look very similar for $x > 0$, the region under $y = 1/x^2$ to the right of $x = 1$ (the shaded region in Figure 4) has finite area whereas the corresponding region under $y = 1/x$ (in Figure 5) has infinite area. Note that both $1/x^2$ and $1/x$ approach 0 as $x \to \infty$ but $1/x^2$ approaches 0 faster than $1/x$. The values of $1/x$ don't decrease fast enough for its integral to have a finite value.

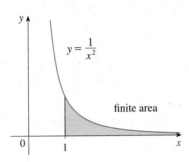

FIGURE 4 $\int_1^\infty (1/x^2)\,dx$ converges

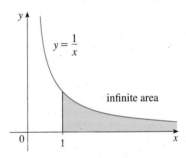

FIGURE 5 $\int_1^\infty (1/x)\,dx$ diverges

EXAMPLE 2 Evaluate $\int_{-\infty}^0 xe^x\,dx$.

SOLUTION Using part (b) of Definition 1, we have

$$\int_{-\infty}^0 xe^x\,dx = \lim_{t\to-\infty} \int_t^0 xe^x\,dx$$

We integrate by parts with $u = x$, $dv = e^x\,dx$ so that $du = dx$, $v = e^x$:

$$\int_t^0 xe^x\,dx = xe^x\Big]_t^0 - \int_t^0 e^x\,dx$$

$$= -te^t - 1 + e^t$$

We know that $e^t \to 0$ as $t \to -\infty$, and by l'Hospital's Rule we have

$$\lim_{t\to-\infty} te^t = \lim_{t\to-\infty} \frac{t}{e^{-t}} = \lim_{t\to-\infty} \frac{1}{-e^{-t}}$$

$$= \lim_{t\to-\infty}(-e^t) = 0$$

TEC In Module 7.8 you can investigate visually and numerically whether several improper integrals are convergent or divergent.

Therefore

$$\int_{-\infty}^0 xe^x\,dx = \lim_{t\to-\infty}\ (-te^t - 1 + e^t)$$

$$= -0 - 1 + 0 = -1 \qquad \blacksquare$$

EXAMPLE 3 Evaluate $\displaystyle\int_{-\infty}^{\infty} \frac{1}{1+x^2}\,dx$.

SOLUTION It's convenient to choose $a = 0$ in Definition 1(c):

$$\int_{-\infty}^{\infty} \frac{1}{1+x^2}\,dx = \int_{-\infty}^0 \frac{1}{1+x^2}\,dx + \int_0^{\infty} \frac{1}{1+x^2}\,dx$$

We must now evaluate the integrals on the right side separately:

$$\int_0^{\infty} \frac{1}{1+x^2}\,dx = \lim_{t\to\infty}\int_0^t \frac{dx}{1+x^2} = \lim_{t\to\infty} \tan^{-1}x\Big]_0^t$$

$$= \lim_{t\to\infty}\ (\tan^{-1}t - \tan^{-1}0) = \lim_{t\to\infty}\tan^{-1}t = \frac{\pi}{2}$$

$$\int_{-\infty}^0 \frac{1}{1+x^2}\,dx = \lim_{t\to-\infty}\int_t^0 \frac{dx}{1+x^2} = \lim_{t\to-\infty} \tan^{-1}x\Big]_t^0$$

$$= \lim_{t\to-\infty}\ (\tan^{-1}0 - \tan^{-1}t)$$

$$= 0 - \left(-\frac{\pi}{2}\right) = \frac{\pi}{2}$$

Since both of these integrals are convergent, the given integral is convergent and

$$\int_{-\infty}^{\infty} \frac{1}{1+x^2}\,dx = \frac{\pi}{2} + \frac{\pi}{2} = \pi$$

Since $1/(1+x^2) > 0$, the given improper integral can be interpreted as the area of the infinite region that lies under the curve $y = 1/(1+x^2)$ and above the x-axis (see Figure 6). $\qquad \blacksquare$

$y = \dfrac{1}{1+x^2}$ area $= \pi$

FIGURE 6

EXAMPLE 4 For what values of p is the integral $\displaystyle\int_1^{\infty} \frac{1}{x^p}\,dx$ convergent?

SOLUTION We know from Example 1 that if $p = 1$, then the integral is divergent, so let's assume that $p \neq 1$. Then

$$\int_1^{\infty} \frac{1}{x^p}\,dx = \lim_{t\to\infty}\int_1^t x^{-p}\,dx$$

$$= \lim_{t\to\infty} \frac{x^{-p+1}}{-p+1}\bigg]_{x=1}^{x=t}$$

$$= \lim_{t\to\infty} \frac{1}{1-p}\left[\frac{1}{t^{p-1}} - 1\right]$$

If $p > 1$, then $p - 1 > 0$, so as $t \to \infty$, $t^{p-1} \to \infty$ and $1/t^{p-1} \to 0$. Therefore

$$\int_1^{\infty} \frac{1}{x^p}\,dx = \frac{1}{p-1} \qquad \text{if } p > 1$$

and so the integral converges. But if $p < 1$, then $p - 1 < 0$ and so

$$\frac{1}{t^{p-1}} = t^{1-p} \to \infty \qquad \text{as } t \to \infty$$

and the integral diverges.

We summarize the result of Example 4 for future reference:

$$\boxed{2} \qquad \int_1^\infty \frac{1}{x^p}\, dx \quad \text{is convergent if } p > 1 \text{ and divergent if } p \le 1.$$

$y = f(x)$

$x = b$

$a \qquad t\ b$

FIGURE 7

Type 2: Discontinuous Integrands

Suppose that f is a positive continuous function defined on a finite interval $[a, b)$ but has a vertical asymptote at b. Let S be the unbounded region under the graph of f and above the x-axis between a and b. (For Type 1 integrals, the regions extended indefinitely in a horizontal direction. Here the region is infinite in a vertical direction.) The area of the part of S between a and t (the shaded region in Figure 7) is

$$A(t) = \int_a^t f(x)\, dx$$

If it happens that $A(t)$ approaches a definite number A as $t \to b^-$, then we say that the area of the region S is A and we write

$$\int_a^b f(x)\, dx = \lim_{t \to b^-} \int_a^t f(x)\, dx$$

Parts (b) and (c) of Definition 3 are illustrated in Figures 8 and 9 for the case where $f(x) \ge 0$ and f has vertical asymptotes at a and c, respectively.

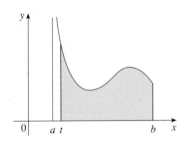

$a\ t \qquad\qquad b$

FIGURE 8

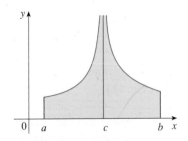

$a \qquad c \qquad b$

FIGURE 9

We use this equation to define an improper integral of Type 2 even when f is not a positive function, no matter what type of discontinuity f has at b.

> **3**   **Definition of an Improper Integral of Type 2**
>
> (a) If f is continuous on $[a, b)$ and is discontinuous at b, then
>
> $$\int_a^b f(x)\, dx = \lim_{t \to b^-} \int_a^t f(x)\, dx$$
>
> if this limit exists (as a finite number).
>
> (b) If f is continuous on $(a, b]$ and is discontinuous at a, then
>
> $$\int_a^b f(x)\, dx = \lim_{t \to a^+} \int_t^b f(x)\, dx$$
>
> if this limit exists (as a finite number).
>
> The improper integral $\int_a^b f(x)\, dx$ is called **convergent** if the corresponding limit exists and **divergent** if the limit does not exist.
>
> (c) If f has a discontinuity at c, where $a < c < b$, and both $\int_a^c f(x)\, dx$ and $\int_c^b f(x)\, dx$ are convergent, then we define
>
> $$\int_a^b f(x)\, dx = \int_a^c f(x)\, dx + \int_c^b f(x)\, dx$$

EXAMPLE 5 Find $\int_2^5 \dfrac{1}{\sqrt{x-2}}\,dx$.

SOLUTION We note first that the given integral is improper because $f(x) = 1/\sqrt{x-2}$ has the vertical asymptote $x = 2$. Since the infinite discontinuity occurs at the left end-point of $[2, 5]$, we use part (b) of Definition 3:

$$\int_2^5 \frac{dx}{\sqrt{x-2}} = \lim_{t \to 2^+} \int_t^5 \frac{dx}{\sqrt{x-2}}$$

$$= \lim_{t \to 2^+} 2\sqrt{x-2}\,\Big]_t^5$$

$$= \lim_{t \to 2^+} 2(\sqrt{3} - \sqrt{t-2})$$

$$= 2\sqrt{3}$$

area $= 2\sqrt{3}$

FIGURE 10

Thus the given improper integral is convergent and, since the integrand is positive, we can interpret the value of the integral as the area of the shaded region in Figure 10.

V EXAMPLE 6 Determine whether $\int_0^{\pi/2} \sec x\,dx$ converges or diverges.

SOLUTION Note that the given integral is improper because $\lim_{x \to (\pi/2)^-} \sec x = \infty$. Using part (a) of Definition 3 and Formula 14 from the Table of Integrals, we have

$$\int_0^{\pi/2} \sec x\,dx = \lim_{t \to (\pi/2)^-} \int_0^t \sec x\,dx = \lim_{t \to (\pi/2)^-} \ln|\sec x + \tan x|\,\Big]_0^t$$

$$= \lim_{t \to (\pi/2)^-} [\ln(\sec t + \tan t) - \ln 1] = \infty$$

because $\sec t \to \infty$ and $\tan t \to \infty$ as $t \to (\pi/2)^-$. Thus the given improper integral is divergent.

EXAMPLE 7 Evaluate $\int_0^3 \dfrac{dx}{x-1}$ if possible.

SOLUTION Observe that the line $x = 1$ is a vertical asymptote of the integrand. Since it occurs in the middle of the interval $[0, 3]$, we must use part (c) of Definition 3 with $c = 1$:

$$\int_0^3 \frac{dx}{x-1} = \int_0^1 \frac{dx}{x-1} + \int_1^3 \frac{dx}{x-1}$$

where

$$\int_0^1 \frac{dx}{x-1} = \lim_{t \to 1^-} \int_0^t \frac{dx}{x-1} = \lim_{t \to 1^-} \ln|x-1|\,\Big]_0^t$$

$$= \lim_{t \to 1^-} (\ln|t-1| - \ln|-1|)$$

$$= \lim_{t \to 1^-} \ln(1-t) = -\infty$$

because $1 - t \to 0^+$ as $t \to 1^-$. Thus $\int_0^1 dx/(x-1)$ is divergent. This implies that $\int_0^3 dx/(x-1)$ is divergent. [We do not need to evaluate $\int_1^3 dx/(x-1)$.]

⊘   **WARNING** If we had not noticed the asymptote $x = 1$ in Example 7 and had instead confused the integral with an ordinary integral, then we might have made the following erroneous calculation:

$$\int_0^3 \frac{dx}{x-1} = \ln|x-1|\Big]_0^3 = \ln 2 - \ln 1 = \ln 2$$

This is wrong because the integral is improper and must be calculated in terms of limits.

From now on, whenever you meet the symbol $\int_a^b f(x)\,dx$ you must decide, by looking at the function f on $[a, b]$, whether it is an ordinary definite integral or an improper integral.

EXAMPLE 8 Evaluate $\int_0^1 \ln x\,dx$.

SOLUTION We know that the function $f(x) = \ln x$ has a vertical asymptote at 0 since $\lim_{x\to 0^+} \ln x = -\infty$. Thus the given integral is improper and we have

$$\int_0^1 \ln x\,dx = \lim_{t\to 0^+} \int_t^1 \ln x\,dx$$

Now we integrate by parts with $u = \ln x$, $dv = dx$, $du = dx/x$, and $v = x$:

$$\int_t^1 \ln x\,dx = x \ln x\Big]_t^1 - \int_t^1 dx$$
$$= 1 \ln 1 - t \ln t - (1 - t)$$
$$= -t \ln t - 1 + t$$

To find the limit of the first term we use l'Hospital's Rule:

$$\lim_{t\to 0^+} t \ln t = \lim_{t\to 0^+} \frac{\ln t}{1/t} = \lim_{t\to 0^+} \frac{1/t}{-1/t^2} = \lim_{t\to 0^+} (-t) = 0$$

Therefore $\displaystyle\int_0^1 \ln x\,dx = \lim_{t\to 0^+} (-t \ln t - 1 + t) = -0 - 1 + 0 = -1$

Figure 11 shows the geometric interpretation of this result. The area of the shaded region above $y = \ln x$ and below the x-axis is 1.

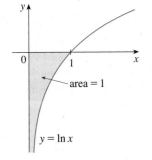

FIGURE 11

area = 1

$y = \ln x$

A Comparison Test for Improper Integrals

Sometimes it is impossible to find the exact value of an improper integral and yet it is important to know whether it is convergent or divergent. In such cases the following theorem is useful. Although we state it for Type 1 integrals, a similar theorem is true for Type 2 integrals.

Comparison Theorem Suppose that f and g are continuous functions with $f(x) \geqslant g(x) \geqslant 0$ for $x \geqslant a$.

(a) If $\int_a^\infty f(x)\,dx$ is convergent, then $\int_a^\infty g(x)\,dx$ is convergent.

(b) If $\int_a^\infty g(x)\,dx$ is divergent, then $\int_a^\infty f(x)\,dx$ is divergent.

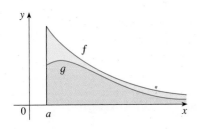

FIGURE 12

We omit the proof of the Comparison Theorem, but Figure 12 makes it seem plausible. If the area under the top curve $y = f(x)$ is finite, then so is the area under the bottom curve $y = g(x)$. And if the area under $y = g(x)$ is infinite, then so is the area under $y = f(x)$. [Note that the reverse is not necessarily true: If $\int_a^\infty g(x)\,dx$ is convergent, $\int_a^\infty f(x)\,dx$ may or may not be convergent, and if $\int_a^\infty f(x)\,dx$ is divergent, $\int_a^\infty g(x)\,dx$ may or may not be divergent.]

V EXAMPLE 9 Show that $\int_0^\infty e^{-x^2}dx$ is convergent.

SOLUTION We can't evaluate the integral directly because the antiderivative of e^{-x^2} is not an elementary function (as explained in Section 7.5). We write

$$\int_0^\infty e^{-x^2}dx = \int_0^1 e^{-x^2}dx + \int_1^\infty e^{-x^2}dx$$

and observe that the first integral on the right-hand side is just an ordinary definite integral. In the second integral we use the fact that for $x \geqslant 1$ we have $x^2 \geqslant x$, so $-x^2 \leqslant -x$ and therefore $e^{-x^2} \leqslant e^{-x}$. (See Figure 13.) The integral of e^{-x} is easy to evaluate:

$$\int_1^\infty e^{-x}dx = \lim_{t\to\infty}\int_1^t e^{-x}dx = \lim_{t\to\infty}(e^{-1} - e^{-t}) = e^{-1}$$

Thus, taking $f(x) = e^{-x}$ and $g(x) = e^{-x^2}$ in the Comparison Theorem, we see that $\int_1^\infty e^{-x^2}dx$ is convergent. It follows that $\int_0^\infty e^{-x^2}dx$ is convergent. ▄

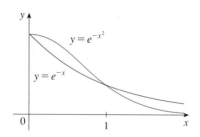

FIGURE 13

In Example 9 we showed that $\int_0^\infty e^{-x^2}dx$ is convergent without computing its value. In Exercise 70 we indicate how to show that its value is approximately 0.8862. In probability theory it is important to know the exact value of this improper integral, as we will see in Section 8.5; using the methods of multivariable calculus it can be shown that the exact value is $\sqrt{\pi}/2$. Table 1 illustrates the definition of an improper integral by showing how the (computer-generated) values of $\int_0^t e^{-x^2}dx$ approach $\sqrt{\pi}/2$ as t becomes large. In fact, these values converge quite quickly because $e^{-x^2} \to 0$ very rapidly as $x \to \infty$.

TABLE 1

t	$\int_0^t e^{-x^2}\,dx$
1	0.7468241328
2	0.8820813908
3	0.8862073483
4	0.8862269118
5	0.8862269255
6	0.8862269255

TABLE 2

t	$\int_1^t [(1 + e^{-x})/x]\,dx$
2	0.8636306042
5	1.8276735512
10	2.5219648704
100	4.8245541204
1000	7.1271392134
10000	9.4297243064

EXAMPLE 10 The integral $\int_1^\infty \dfrac{1 + e^{-x}}{x}\,dx$ is divergent by the Comparison Theorem because

$$\frac{1 + e^{-x}}{x} > \frac{1}{x}$$

and $\int_1^\infty (1/x)\,dx$ is divergent by Example 1 [or by ⎰2⎱ with $p = 1$]. ▄

Table 2 illustrates the divergence of the integral in Example 10. It appears that the values are not approaching any fixed number.

7.8 Exercises

Access end-of-section exercises and tools for visualizing calculus online at **www.webassign.net** ᴱᴺᴴᴬᴺᶜᴱᴰ **WebAssign**

7. Review

Concept Check

1. State the rule for integration by parts. In practice, how do you use it?

2. How do you evaluate $\int \sin^m x \cos^n x \, dx$ if m is odd? What if n is odd? What if m and n are both even?

3. If the expression $\sqrt{a^2 - x^2}$ occurs in an integral, what substitution might you try? What if $\sqrt{a^2 + x^2}$ occurs? What if $\sqrt{x^2 - a^2}$ occurs?

4. What is the form of the partial fraction decomposition of a rational function $P(x)/Q(x)$ if the degree of P is less than the degree of Q and $Q(x)$ has only distinct linear factors? What if a linear factor is repeated? What if $Q(x)$ has an irreducible quadratic factor (not repeated)? What if the quadratic factor is repeated?

5. State the rules for approximating the definite integral $\int_a^b f(x) \, dx$ with the Midpoint Rule, the Trapezoidal Rule, and Simpson's Rule. Which would you expect to give the best estimate? How do you approximate the error for each rule?

6. Define the following improper integrals.

   (a) $\int_a^\infty f(x) \, dx$ (b) $\int_{-\infty}^b f(x) \, dx$ (c) $\int_{-\infty}^\infty f(x) \, dx$

7. Define the improper integral $\int_a^b f(x) \, dx$ for each of the following cases.
   (a) f has an infinite discontinuity at a.
   (b) f has an infinite discontinuity at b.
   (c) f has an infinite discontinuity at c, where $a < c < b$.

8. State the Comparison Theorem for improper integrals.

True-False Quiz

Determine whether the statement is true or false. If it is true, explain why. If it is false, explain why or give an example that disproves the statement.

1. $\dfrac{x(x^2 + 4)}{x^2 - 4}$ can be put in the form $\dfrac{A}{x + 2} + \dfrac{B}{x - 2}$.

2. $\dfrac{x^2 + 4}{x(x^2 - 4)}$ can be put in the form $\dfrac{A}{x} + \dfrac{B}{x + 2} + \dfrac{C}{x - 2}$.

3. $\dfrac{x^2 + 4}{x^2(x - 4)}$ can be put in the form $\dfrac{A}{x^2} + \dfrac{B}{x - 4}$.

4. $\dfrac{x^2 - 4}{x(x^2 + 4)}$ can be put in the form $\dfrac{A}{x} + \dfrac{B}{x^2 + 4}$.

5. $\displaystyle\int_0^4 \dfrac{x}{x^2 - 1} \, dx = \tfrac{1}{2} \ln 15$

6. $\displaystyle\int_1^\infty \dfrac{1}{x^{\sqrt{2}}} \, dx$ is convergent.

7. If f is continuous, then $\int_{-\infty}^\infty f(x) \, dx = \lim_{t \to \infty} \int_{-t}^t f(x) \, dx$.

8. The Midpoint Rule is always more accurate than the Trapezoidal Rule.

9. (a) Every elementary function has an elementary derivative.
   (b) Every elementary function has an elementary antiderivative.

10. If f is continuous on $[0, \infty)$ and $\int_1^\infty f(x) \, dx$ is convergent, then $\int_0^\infty f(x) \, dx$ is convergent.

11. If f is a continuous, decreasing function on $[1, \infty)$ and $\lim_{x \to \infty} f(x) = 0$, then $\int_1^\infty f(x) \, dx$ is convergent.

12. If $\int_a^\infty f(x) \, dx$ and $\int_a^\infty g(x) \, dx$ are both convergent, then $\int_a^\infty [f(x) + g(x)] \, dx$ is convergent.

13. If $\int_a^\infty f(x) \, dx$ and $\int_a^\infty g(x) \, dx$ are both divergent, then $\int_a^\infty [f(x) + g(x)] \, dx$ is divergent.

14. If $f(x) \le g(x)$ and $\int_0^\infty g(x) \, dx$ diverges, then $\int_0^\infty f(x) \, dx$ also diverges.

Exercises

Note: Additional practice in techniques of integration is provided in Exercises 7.5.

1–40 Evaluate the integral.

1. $\displaystyle\int_1^2 \dfrac{(x + 1)^2}{x} \, dx$

2. $\displaystyle\int_1^2 \dfrac{x}{(x + 1)^2} \, dx$

3. $\displaystyle\int_0^{\pi/2} \sin \theta \, e^{\cos \theta} \, d\theta$

4. $\displaystyle\int_0^{\pi/6} t \sin 2t \, dt$

5. $\displaystyle\int \dfrac{dt}{2t^2 + 3t + 1}$

6. $\displaystyle\int_1^2 x^5 \ln x \, dx$

7. $\displaystyle\int_0^{\pi/2} \sin^3 \theta \cos^2 \theta \, d\theta$

8. $\displaystyle\int \dfrac{dx}{\sqrt{e^x - 1}}$

9. $\displaystyle\int \dfrac{\sin(\ln t)}{t} \, dt$

10. $\displaystyle\int_0^1 \dfrac{\sqrt{\arctan x}}{1 + x^2} \, dx$

▦ Graphing calculator or computer required   CAS Computer algebra system required

11. $\int_1^2 \frac{\sqrt{x^2-1}}{x}\,dx$

12. $\int \frac{e^{2x}}{1+e^{4x}}\,dx$

13. $\int e^{\sqrt[3]{x}}\,dx$

14. $\int \frac{x^2+2}{x+2}\,dx$

15. $\int \frac{x-1}{x^2+2x}\,dx$

16. $\int \frac{\sec^6\theta}{\tan^2\theta}\,d\theta$

17. $\int x\sec x\tan x\,dx$

18. $\int \frac{x^2+8x-3}{x^3+3x^2}\,dx$

19. $\int \frac{x+1}{9x^2+6x+5}\,dx$

20. $\int \tan^5\theta\sec^3\theta\,d\theta$

21. $\int \frac{dx}{\sqrt{x^2-4x}}$

22. $\int te^{\sqrt{t}}\,dt$

23. $\int \frac{dx}{x\sqrt{x^2+1}}$

24. $\int e^x\cos x\,dx$

25. $\int \frac{3x^3-x^2+6x-4}{(x^2+1)(x^2+2)}\,dx$

26. $\int x\sin x\cos x\,dx$

27. $\int_0^{\pi/2} \cos^3 x\sin 2x\,dx$

28. $\int \frac{\sqrt[3]{x}+1}{\sqrt[3]{x}-1}\,dx$

29. $\int_{-3}^3 \frac{x}{1+|x|}\,dx$

30. $\int \frac{dx}{e^x\sqrt{1-e^{-2x}}}$

31. $\int_0^{\ln 10} \frac{e^x\sqrt{e^x-1}}{e^x+8}\,dx$

32. $\int_0^{\pi/4} \frac{x\sin x}{\cos^3 x}\,dx$

33. $\int \frac{x^2}{(4-x^2)^{3/2}}\,dx$

34. $\int (\arcsin x)^2\,dx$

35. $\int \frac{1}{\sqrt{x+x^{3/2}}}\,dx$

36. $\int \frac{1-\tan\theta}{1+\tan\theta}\,d\theta$

37. $\int (\cos x+\sin x)^2\cos 2x\,dx$

38. $\int \frac{2^{\sqrt{x}}}{\sqrt{x}}\,dx$

39. $\int_0^{1/2} \frac{xe^{2x}}{(1+2x)^2}\,dx$

40. $\int_{\pi/4}^{\pi/3} \frac{\sqrt{\tan\theta}}{\sin 2\theta}\,d\theta$

41–50 Evaluate the integral or show that it is divergent.

41. $\int_1^\infty \frac{1}{(2x+1)^3}\,dx$

42. $\int_1^\infty \frac{\ln x}{x^4}\,dx$

43. $\int_2^\infty \frac{dx}{x\ln x}$

44. $\int_2^6 \frac{y}{\sqrt{y-2}}\,dy$

45. $\int_0^4 \frac{\ln x}{\sqrt{x}}\,dx$

46. $\int_0^1 \frac{1}{2-3x}\,dx$

47. $\int_0^1 \frac{x-1}{\sqrt{x}}\,dx$

48. $\int_{-1}^1 \frac{dx}{x^2-2x}$

49. $\int_{-\infty}^\infty \frac{dx}{4x^2+4x+5}$

50. $\int_1^\infty \frac{\tan^{-1}x}{x^2}\,dx$

51–52 Evaluate the indefinite integral. Illustrate and check that your answer is reasonable by graphing both the function and its antiderivative (take $C=0$).

51. $\int \ln(x^2+2x+2)\,dx$

52. $\int \frac{x^3}{\sqrt{x^2+1}}\,dx$

53. Graph the function $f(x)=\cos^2 x\sin^3 x$ and use the graph to guess the value of the integral $\int_0^{2\pi} f(x)\,dx$. Then evaluate the integral to confirm your guess.

CAS 54. (a) How would you evaluate $\int x^5 e^{-2x}\,dx$ by hand? (Don't actually carry out the integration.)
(b) How would you evaluate $\int x^5 e^{-2x}\,dx$ using tables? (Don't actually do it.)
(c) Use a CAS to evaluate $\int x^5 e^{-2x}\,dx$.
(d) Graph the integrand and the indefinite integral on the same screen.

55–58 Use the Table of Integrals on the Reference Pages to evaluate the integral.

55. $\int \sqrt{4x^2-4x-3}\,dx$

56. $\int \csc^5 t\,dt$

57. $\int \cos x\sqrt{4+\sin^2 x}\,dx$

58. $\int \frac{\cot x}{\sqrt{1+2\sin x}}\,dx$

59. Verify Formula 33 in the Table of Integrals (a) by differentiation and (b) by using a trigonometric substitution.

60. Verify Formula 62 in the Table of Integrals.

61. Is it possible to find a number n such that $\int_0^\infty x^n\,dx$ is convergent?

62. For what values of a is $\int_0^\infty e^{ax}\cos x\,dx$ convergent? Evaluate the integral for those values of a.

63–64 Use (a) the Trapezoidal Rule, (b) the Midpoint Rule, and (c) Simpson's Rule with $n=10$ to approximate the given integral. Round your answers to six decimal places.

63. $\int_2^4 \frac{1}{\ln x}\,dx$

64. $\int_1^4 \sqrt{x}\cos x\,dx$

65. Estimate the errors involved in Exercise 63, parts (a) and (b). How large should n be in each case to guarantee an error of less than 0.00001?

66. Use Simpson's Rule with $n=6$ to estimate the area under the curve $y=e^x/x$ from $x=1$ to $x=4$.

67. The speedometer reading (v) on a car was observed at 1-minute intervals and recorded in the chart. Use Simpson's Rule to estimate the distance traveled by the car.

t (min)	v (mi/h)	t (min)	v (mi/h)
0	40	6	56
1	42	7	57
2	45	8	57
3	49	9	55
4	52	10	56
5	54		

68. A population of honeybees increased at a rate of $r(t)$ bees per week, where the graph of r is as shown. Use Simpson's Rule with six subintervals to estimate the increase in the bee population during the first 24 weeks.

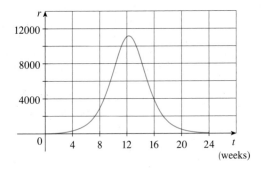

CAS 69. (a) If $f(x) = \sin(\sin x)$, use a graph to find an upper bound for $|f^{(4)}(x)|$.
(b) Use Simpson's Rule with $n = 10$ to approximate $\int_0^\pi f(x)\, dx$ and use part (a) to estimate the error.
(c) How large should n be to guarantee that the size of the error in using S_n is less than 0.00001?

70. Suppose you are asked to estimate the volume of a football. You measure and find that a football is 28 cm long. You use a piece of string and measure the circumference at its widest point to be 53 cm. The circumference 7 cm from each end is 45 cm. Use Simpson's Rule to make your estimate.

|← 28 cm →|

71. Use the Comparison Theorem to determine whether the integral is convergent or divergent.
(a) $\int_1^\infty \dfrac{2 + \sin x}{\sqrt{x}}\, dx$ (b) $\int_1^\infty \dfrac{1}{\sqrt{1 + x^4}}\, dx$

72. Find the area of the region bounded by the hyperbola $y^2 - x^2 = 1$ and the line $y = 3$.

73. Find the area bounded by the curves $y = \cos x$ and $y = \cos^2 x$ between $x = 0$ and $x = \pi$.

74. Find the area of the region bounded by the curves $y = 1/(2 + \sqrt{x})$, $y = 1/(2 - \sqrt{x})$, and $x = 1$.

75. The region under the curve $y = \cos^2 x$, $0 \le x \le \pi/2$, is rotated about the x-axis. Find the volume of the resulting solid.

76. The region in Exercise 75 is rotated about the y-axis. Find the volume of the resulting solid.

77. If f' is continuous on $[0, \infty)$ and $\lim_{x \to \infty} f(x) = 0$, show that
$$\int_0^\infty f'(x)\, dx = -f(0)$$

78. We can extend our definition of average value of a continuous function to an infinite interval by defining the average value of f on the interval $[a, \infty)$ to be
$$\lim_{t \to \infty} \frac{1}{t - a} \int_a^t f(x)\, dx$$
(a) Find the average value of $y = \tan^{-1} x$ on the interval $[0, \infty)$
(b) If $f(x) \ge 0$ and $\int_a^\infty f(x)\, dx$ is divergent, show that the average value of f on the interval $[a, \infty)$ is $\lim_{x \to \infty} f(x)$, if this limit exists.
(c) If $\int_a^\infty f(x)\, dx$ is convergent, what is the average value of f on the interval $[a, \infty)$?
(d) Find the average value of $y = \sin x$ on the interval $[0, \infty)$.

79. Use the substitution $u = 1/x$ to show that
$$\int_0^\infty \frac{\ln x}{1 + x^2}\, dx = 0$$

80. The magnitude of the repulsive force between two point charges with the same sign, one of size 1 and the other of size q, is
$$F = \frac{q}{4\pi\varepsilon_0 r^2}$$
where r is the distance between the charges and ε_0 is a constant. The *potential* V at a point P due to the charge q is defined to be the work expended in bringing a unit charge to P from infinity along the straight line that joins q and P. Find a formula for V.

Problems Plus

Cover up the solution to the example and try it yourself first.

EXAMPLE 1

(a) Prove that if f is a continuous function, then

$$\int_0^a f(x)\, dx = \int_0^a f(a - x)\, dx$$

(b) Use part (a) to show that

$$\int_0^{\pi/2} \frac{\sin^n x}{\sin^n x + \cos^n x}\, dx = \frac{\pi}{4}$$

for all positive numbers n.

SOLUTION

(a) At first sight, the given equation may appear somewhat baffling. How is it possible to connect the left side to the right side? Connections can often be made through one of the principles of problem solving: *introduce something extra*. Here the extra ingredient is a new variable. We often think of introducing a new variable when we use the Substitution Rule to integrate a specific function. But that technique is still useful in the present circumstance in which we have a general function f.

Once we think of making a substitution, the form of the right side suggests that it should be $u = a - x$. Then $du = -dx$. When $x = 0$, $u = a$; when $x = a$, $u = 0$. So

$$\int_0^a f(a - x)\, dx = -\int_a^0 f(u)\, du = \int_0^a f(u)\, du$$

But this integral on the right side is just another way of writing $\int_0^a f(x)\, dx$. So the given equation is proved.

(b) If we let the given integral be I and apply part (a) with $a = \pi/2$, we get

$$I = \int_0^{\pi/2} \frac{\sin^n x}{\sin^n x + \cos^n x}\, dx = \int_0^{\pi/2} \frac{\sin^n(\pi/2 - x)}{\sin^n(\pi/2 - x) + \cos^n(\pi/2 - x)}\, dx$$

A well-known trigonometric identity tells us that $\sin(\pi/2 - x) = \cos x$ and $\cos(\pi/2 - x) = \sin x$, so we get

$$I = \int_0^{\pi/2} \frac{\cos^n x}{\cos^n x + \sin^n x}\, dx$$

Notice that the two expressions for I are very similar. In fact, the integrands have the same denominator. This suggests that we should add the two expressions. If we do so, we get

$$2I = \int_0^{\pi/2} \frac{\sin^n x + \cos^n x}{\sin^n x + \cos^n x}\, dx = \int_0^{\pi/2} 1\, dx = \frac{\pi}{2}$$

Therefore $I = \pi/4$.

PS The principles of problem solving are discussed on page 77.

The computer graphs in Figure 1 make it seem plausible that all of the integrals in the example have the same value. The graph of each integrand is labeled with the corresponding value of n.

FIGURE 1

|←——— 14 in ———→|

FIGURE FOR PROBLEM 1

FIGURE FOR PROBLEM 6

 1. Three mathematics students have ordered a 14-inch pizza. Instead of slicing it in the traditional way, they decide to slice it by parallel cuts, as shown in the figure. Being mathematics majors, they are able to determine where to slice so that each gets the same amount of pizza. Where are the cuts made?

2. Evaluate

$$\int \frac{1}{x^7 - x}\, dx$$

The straightforward approach would be to start with partial fractions, but that would be brutal. Try a substitution.

3. Evaluate $\int_0^1 \left(\sqrt[3]{1 - x^7} - \sqrt[7]{1 - x^3} \right) dx$.

4. The centers of two disks with radius 1 are one unit apart. Find the area of the union of the two disks.

5. An ellipse is cut out of a circle with radius a. The major axis of the ellipse coincides with a diameter of the circle and the minor axis has length $2b$. Prove that the area of the remaining part of the circle is the same as the area of an ellipse with semiaxes a and $a - b$.

6. A man initially standing at the point O walks along a pier pulling a rowboat by a rope of length L. The man keeps the rope straight and taut. The path followed by the boat is a curve called a *tractrix* and it has the property that the rope is always tangent to the curve (see the figure).
(a) Show that if the path followed by the boat is the graph of the function $y = f(x)$, then

$$f'(x) = \frac{dy}{dx} = \frac{-\sqrt{L^2 - x^2}}{x}$$

(b) Determine the function $y = f(x)$.

7. A function f is defined by

$$f(x) = \int_0^\pi \cos t \, \cos(x - t)\, dt \qquad 0 \le x \le 2\pi$$

Find the minimum value of f.

8. If n is a positive integer, prove that

$$\int_0^1 (\ln x)^n\, dx = (-1)^n n!$$

9. Show that

$$\int_0^1 (1 - x^2)^n\, dx = \frac{2^{2n}(n!)^2}{(2n + 1)!}$$

Hint: Start by showing that if I_n denotes the integral, then

$$I_{k+1} = \frac{2k + 2}{2k + 3} I_k$$

 Graphing calculator or computer required

10. Suppose that f is a positive function such that f' is continuous.
   (a) How is the graph of $y = f(x) \sin nx$ related to the graph of $y = f(x)$? What happens as $n \to \infty$?
   (b) Make a guess as to the value of the limit

$$\lim_{n \to \infty} \int_0^1 f(x) \sin nx \, dx$$

   based on graphs of the integrand.
   (c) Using integration by parts, confirm the guess that you made in part (b). [Use the fact that, since f' is continuous, there is a constant M such that $|f'(x)| \le M$ for $0 \le x \le 1$.]

11. If $0 < a < b$, find $\displaystyle \lim_{t \to 0} \left\{ \int_0^1 [bx + a(1 - x)]^t \, dx \right\}^{1/t}$.

12. Graph $f(x) = \sin(e^x)$ and use the graph to estimate the value of t such that $\int_t^{t+1} f(x) \, dx$ is a maximum. Then find the exact value of t that maximizes this integral.

13. Evaluate $\displaystyle \int_{-1}^{\infty} \left(\frac{x^4}{1 + x^6} \right)^2 \, dx$.

14. Evaluate $\displaystyle \int \sqrt{\tan x} \, dx$.

15. The circle with radius 1 shown in the figure touches the curve $y = |2x|$ twice. Find the area of the region that lies between the two curves.

16. A rocket is fired straight up, burning fuel at the constant rate of b kilograms per second. Let $v = v(t)$ be the velocity of the rocket at time t and suppose that the velocity u of the exhaust gas is constant. Let $M = M(t)$ be the mass of the rocket at time t and note that M decreases as the fuel burns. If we neglect air resistance, it follows from Newton's Second Law that

$$F = M \frac{dv}{dt} - ub$$

where the force $F = -Mg$. Thus

$$\boxed{1} \qquad M \frac{dv}{dt} - ub = -Mg$$

Let M_1 be the mass of the rocket without fuel, M_2 the initial mass of the fuel, and $M_0 = M_1 + M_2$. Then, until the fuel runs out at time $t = M_2/b$, the mass is $M = M_0 - bt$.
   (a) Substitute $M = M_0 - bt$ into Equation 1 and solve the resulting equation for v. Use the initial condition $v(0) = 0$ to evaluate the constant.
   (b) Determine the velocity of the rocket at time $t = M_2/b$. This is called the *burnout velocity*.
   (c) Determine the height of the rocket $y = y(t)$ at the burnout time.
   (d) Find the height of the rocket at any time t.

FIGURE FOR PROBLEM 15

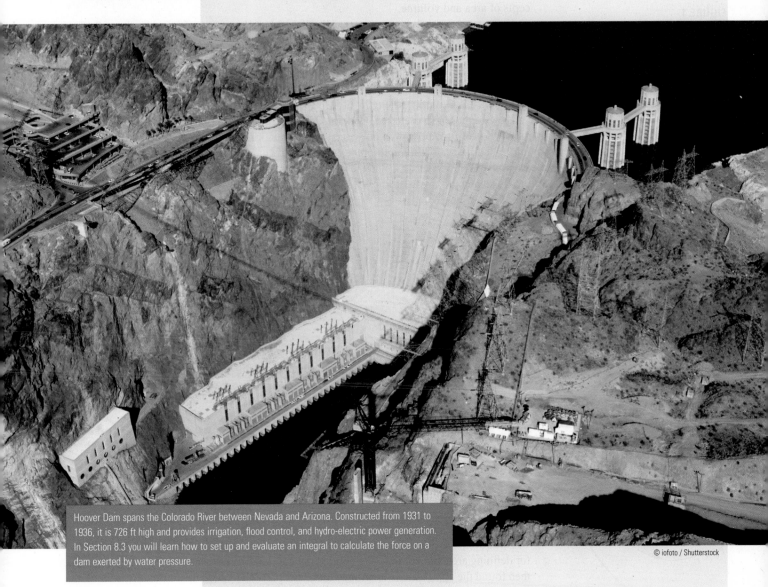

8 Further Applications of Integration

Hoover Dam spans the Colorado River between Nevada and Arizona. Constructed from 1931 to 1936, it is 726 ft high and provides irrigation, flood control, and hydro-electric power generation. In Section 8.3 you will learn how to set up and evaluate an integral to calculate the force on a dam exerted by water pressure.

© iofoto / Shutterstock

We looked at some applications of integrals in Chapter 5: areas, volumes, work, and average values. Here we explore some of the many other geometric applications of integration—the length of a curve, the area of a surface—as well as quantities of interest in physics, engineering, biology, economics, and statistics. For instance, we will investigate the center of gravity of a plate, the force exerted by water pressure on a dam, the flow of blood from the human heart, and the average time spent on hold during a customer support telephone call.

8.1   Arc Length

TEC Visual 8.1 shows an animation of Figure 2.

FIGURE 1

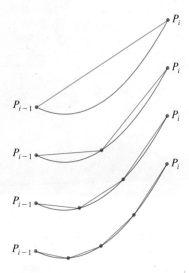

FIGURE 2

What do we mean by the length of a curve? We might think of fitting a piece of string to the curve in Figure 1 and then measuring the string against a ruler. But that might be difficult to do with much accuracy if we have a complicated curve. We need a precise definition for the length of an arc of a curve, in the same spirit as the definitions we developed for the concepts of area and volume.

If the curve is a polygon, we can easily find its length; we just add the lengths of the line segments that form the polygon. (We can use the distance formula to find the distance between the endpoints of each segment.) We are going to define the length of a general curve by first approximating it by a polygon and then taking a limit as the number of segments of the polygon is increased. This process is familiar for the case of a circle, where the circumference is the limit of lengths of inscribed polygons (see Figure 2).

Now suppose that a curve C is defined by the equation $y = f(x)$, where f is continuous and $a \le x \le b$. We obtain a polygonal approximation to C by dividing the interval $[a, b]$ into n subintervals with endpoints $x_0, x_1, \ldots, x_n$ and equal width Δx. If $y_i = f(x_i)$, then the point $P_i(x_i, y_i)$ lies on C and the polygon with vertices $P_0, P_1, \ldots, P_n$, illustrated in Figure 3, is an approximation to C.

FIGURE 3

FIGURE 4

The length L of C is approximately the length of this polygon and the approximation gets better as we let n increase. (See Figure 4, where the arc of the curve between P_{i-1} and P_i has been magnified and approximations with successively smaller values of Δx are shown.) Therefore we define the **length** L of the curve C with equation $y = f(x)$, $a \le x \le b$, as the limit of the lengths of these inscribed polygons (if the limit exists):

$$\boxed{1} \qquad L = \lim_{n \to \infty} \sum_{i=1}^{n} |P_{i-1}P_i|$$

Notice that the procedure for defining arc length is very similar to the procedure we used for defining area and volume: We divided the curve into a large number of small parts. We then found the approximate lengths of the small parts and added them. Finally, we took the limit as $n \to \infty$.

The definition of arc length given by Equation 1 is not very convenient for computational purposes, but we can derive an integral formula for L in the case where f has a continuous derivative. [Such a function f is called **smooth** because a small change in x produces a small change in $f'(x)$.]

If we let $\Delta y_i = y_i - y_{i-1}$, then

$$|P_{i-1}P_i| = \sqrt{(x_i - x_{i-1})^2 + (y_i - y_{i-1})^2} = \sqrt{(\Delta x)^2 + (\Delta y_i)^2}$$

By applying the Mean Value Theorem to f on the interval $[x_{i-1}, x_i]$, we find that there is a number x_i^* between x_{i-1} and x_i such that

$$f(x_i) - f(x_{i-1}) = f'(x_i^*)(x_i - x_{i-1})$$

that is,

$$\Delta y_i = f'(x_i^*)\, \Delta x$$

Thus we have

$$|P_{i-1}P_i| = \sqrt{(\Delta x)^2 + (\Delta y_i)^2} = \sqrt{(\Delta x)^2 + [f'(x_i^*)\, \Delta x]^2}$$

$$= \sqrt{1 + [f'(x_i^*)]^2}\,\sqrt{(\Delta x)^2} = \sqrt{1 + [f'(x_i^*)]^2}\;\Delta x \qquad \text{(since } \Delta x > 0)$$

Therefore, by Definition 1,

$$L = \lim_{n \to \infty} \sum_{i=1}^{n} |P_{i-1}P_i| = \lim_{n \to \infty} \sum_{i=1}^{n} \sqrt{1 + [f'(x_i^*)]^2}\;\Delta x$$

We recognize this expression as being equal to

$$\int_a^b \sqrt{1 + [f'(x)]^2}\;dx$$

by the definition of a definite integral. This integral exists because the function $g(x) = \sqrt{1 + [f'(x)]^2}$ is continuous. Thus we have proved the following theorem:

---

2 **The Arc Length Formula** If f' is continuous on $[a, b]$, then the length of the curve $y = f(x)$, $a \leqslant x \leqslant b$, is

$$L = \int_a^b \sqrt{1 + [f'(x)]^2}\;dx$$

If we use Leibniz notation for derivatives, we can write the arc length formula as follows:

3

$$L = \int_a^b \sqrt{1 + \left(\frac{dy}{dx}\right)^2}\;dx$$

EXAMPLE 1 Find the length of the arc of the semicubical parabola $y^2 = x^3$ between the points $(1, 1)$ and $(4, 8)$. (See Figure 5.)

SOLUTION For the top half of the curve we have

$$y = x^{3/2} \qquad \frac{dy}{dx} = \tfrac{3}{2}x^{1/2}$$

and so the arc length formula gives

$$L = \int_1^4 \sqrt{1 + \left(\frac{dy}{dx}\right)^2}\;dx = \int_1^4 \sqrt{1 + \tfrac{9}{4}x}\;dx$$

If we substitute $u = 1 + \tfrac{9}{4}x$, then $du = \tfrac{9}{4}\,dx$. When $x = 1$, $u = \tfrac{13}{4}$; when $x = 4$, $u = 10$.

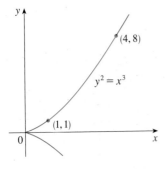

FIGURE 5

As a check on our answer to Example 1, notice from Figure 5 that the arc length ought to be slightly larger than the distance from $(1, 1)$ to $(4, 8)$, which is

$$\sqrt{58} \approx 7.615773$$

According to our calculation in Example 1, we have

$$L = \tfrac{1}{27}(80\sqrt{10} - 13\sqrt{13}) \approx 7.633705$$

Sure enough, this is a bit greater than the length of the line segment.

Therefore

$$L = \tfrac{4}{9} \int_{13/4}^{10} \sqrt{u}\; du = \tfrac{4}{9} \cdot \tfrac{2}{3} u^{3/2} \Big]_{13/4}^{10}$$

$$= \tfrac{8}{27}\Big[10^{3/2} - \big(\tfrac{13}{4}\big)^{3/2}\Big] = \tfrac{1}{27}\big(80\sqrt{10} - 13\sqrt{13}\big)$$

If a curve has the equation $x = g(y)$, $c \leq y \leq d$, and $g'(y)$ is continuous, then by interchanging the roles of x and y in Formula 2 or Equation 3, we obtain the following formula for its length:

$$\boxed{4 \qquad L = \int_c^d \sqrt{1 + [g'(y)]^2}\; dy = \int_c^d \sqrt{1 + \left(\frac{dx}{dy}\right)^2}\; dy}$$

V EXAMPLE 2 Find the length of the arc of the parabola $y^2 = x$ from $(0, 0)$ to $(1, 1)$.

SOLUTION Since $x = y^2$, we have $dx/dy = 2y$, and Formula 4 gives

$$L = \int_0^1 \sqrt{1 + \left(\frac{dx}{dy}\right)^2}\; dy = \int_0^1 \sqrt{1 + 4y^2}\; dy$$

We make the trigonometric substitution $y = \tfrac{1}{2} \tan \theta$, which gives $dy = \tfrac{1}{2} \sec^2\theta\; d\theta$ and $\sqrt{1 + 4y^2} = \sqrt{1 + \tan^2\theta} = \sec \theta$. When $y = 0$, $\tan \theta = 0$, so $\theta = 0$; when $y = 1$, $\tan \theta = 2$, so $\theta = \tan^{-1} 2 = \alpha$, say. Thus

$$L = \int_0^\alpha \sec \theta \cdot \tfrac{1}{2} \sec^2\theta\; d\theta = \tfrac{1}{2} \int_0^\alpha \sec^3\theta\; d\theta$$

$$= \tfrac{1}{2} \cdot \tfrac{1}{2}\Big[\sec \theta \tan \theta + \ln |\sec \theta + \tan \theta|\Big]_0^\alpha \qquad \text{(from Example 8 in Section 7.2)}$$

$$= \tfrac{1}{4}\big(\sec \alpha \tan \alpha + \ln |\sec \alpha + \tan \alpha|\big)$$

(We could have used Formula 21 in the Table of Integrals.) Since $\tan \alpha = 2$, we have $\sec^2\alpha = 1 + \tan^2\alpha = 5$, so $\sec \alpha = \sqrt{5}$ and

$$L = \frac{\sqrt{5}}{2} + \frac{\ln(\sqrt{5} + 2)}{4}$$

Figure 6 shows the arc of the parabola whose length is computed in Example 2, together with polygonal approximations having $n = 1$ and $n = 2$ line segments, respectively. For $n = 1$ the approximate length is $L_1 = \sqrt{2}$, the diagonal of a square. The table shows the approximations L_n that we get by dividing $[0, 1]$ into n equal subintervals. Notice that each time we double the number of sides of the polygon, we get closer to the exact length, which is

$$L = \frac{\sqrt{5}}{2} + \frac{\ln(\sqrt{5} + 2)}{4} \approx 1.478943$$

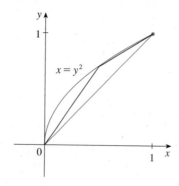

FIGURE 6

n	L_n
1	1.414
2	1.445
4	1.464
8	1.472
16	1.476
32	1.478
64	1.479

Because of the presence of the square root sign in Formulas 2 and 4, the calculation of an arc length often leads to an integral that is very difficult or even impossible to evaluate explicitly. Thus we sometimes have to be content with finding an approximation to the length of a curve, as in the following example.

V EXAMPLE 3

(a) Set up an integral for the length of the arc of the hyperbola $xy = 1$ from the point $(1, 1)$ to the point $\left(2, \frac{1}{2}\right)$.

(b) Use Simpson's Rule with $n = 10$ to estimate the arc length.

SOLUTION

(a) We have

$$y = \frac{1}{x} \qquad \frac{dy}{dx} = -\frac{1}{x^2}$$

and so the arc length is

$$L = \int_1^2 \sqrt{1 + \left(\frac{dy}{dx}\right)^2}\, dx = \int_1^2 \sqrt{1 + \frac{1}{x^4}}\, dx = \int_1^2 \frac{\sqrt{x^4 + 1}}{x^2}\, dx$$

(b) Using Simpson's Rule (see Section 7.7) with $a = 1$, $b = 2$, $n = 10$, $\Delta x = 0.1$, and $f(x) = \sqrt{1 + 1/x^4}$, we have

$$L = \int_1^2 \sqrt{1 + \frac{1}{x^4}}\, dx$$

Checking the value of the definite integral with a more accurate approximation produced by a computer algebra system, we see that the approximation using Simpson's Rule is accurate to four decimal places.

$$\approx \frac{\Delta x}{3}\left[f(1) + 4f(1.1) + 2f(1.2) + 4f(1.3) + \cdots + 2f(1.8) + 4f(1.9) + f(2)\right]$$

$$\approx 1.1321$$

The Arc Length Function

We will find it useful to have a function that measures the arc length of a curve from a particular starting point to any other point on the curve. Thus if a smooth curve C has the equation $y = f(x)$, $a \le x \le b$, let $s(x)$ be the distance along C from the initial point $P_0(a, f(a))$ to the point $Q(x, f(x))$. Then s is a function, called the **arc length function**, and, by Formula 2,

$$\boxed{5} \qquad s(x) = \int_a^x \sqrt{1 + [f'(t)]^2}\, dt$$

(We have replaced the variable of integration by t so that x does not have two meanings.) We can use Part 1 of the Fundamental Theorem of Calculus to differentiate Equation 5 (since the integrand is continuous):

$$\boxed{6} \qquad \frac{ds}{dx} = \sqrt{1 + [f'(x)]^2} = \sqrt{1 + \left(\frac{dy}{dx}\right)^2}$$

Equation 6 shows that the rate of change of s with respect to x is always at least 1 and is equal to 1 when $f'(x)$, the slope of the curve, is 0. The differential of arc length is

$$\boxed{7} \qquad ds = \sqrt{1 + \left(\frac{dy}{dx}\right)^2}\, dx$$

and this equation is sometimes written in the symmetric form

$$\boxed{8} \qquad (ds)^2 = (dx)^2 + (dy)^2$$

The geometric interpretation of Equation 8 is shown in Figure 7. It can be used as a mnemonic device for remembering both of the Formulas 3 and 4. If we write $L = \int ds$, then from Equation 8 either we can solve to get $\boxed{7}$, which gives $\boxed{3}$, or we can solve to get

$$ds = \sqrt{1 + \left(\frac{dx}{dy}\right)^2}\, dy$$

which gives $\boxed{4}$.

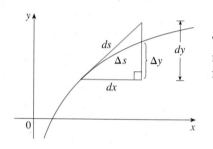

FIGURE 7

V EXAMPLE 4 Find the arc length function for the curve $y = x^2 - \frac{1}{8}\ln x$ taking $P_0(1, 1)$ as the starting point.

SOLUTION If $f(x) = x^2 - \frac{1}{8}\ln x$, then

$$f'(x) = 2x - \frac{1}{8x}$$

$$1 + [f'(x)]^2 = 1 + \left(2x - \frac{1}{8x}\right)^2 = 1 + 4x^2 - \frac{1}{2} + \frac{1}{64x^2}$$

$$= 4x^2 + \frac{1}{2} + \frac{1}{64x^2} = \left(2x + \frac{1}{8x}\right)^2$$

$$\sqrt{1 + [f'(x)]^2} = 2x + \frac{1}{8x}$$

Thus the arc length function is given by

$$s(x) = \int_1^x \sqrt{1 + [f'(t)]^2}\, dt$$

$$= \int_1^x \left(2t + \frac{1}{8t}\right) dt = t^2 + \tfrac{1}{8}\ln t \Big]_1^x$$

$$= x^2 + \tfrac{1}{8}\ln x - 1$$

For instance, the arc length along the curve from $(1, 1)$ to $(3, f(3))$ is

$$s(3) = 3^2 + \tfrac{1}{8}\ln 3 - 1 = 8 + \frac{\ln 3}{8} \approx 8.1373$$

Figure 8 shows the interpretation of the arc length function in Example 4. Figure 9 shows the graph of this arc length function. Why is $s(x)$ negative when x is less than 1?

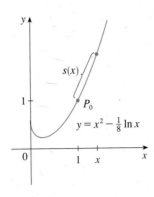

$y = x^2 - \frac{1}{8} \ln x$

FIGURE 8

$s(x) = x^2 + \frac{1}{8} \ln x - 1$

FIGURE 9

<table><tr><td>8.1</td><td>**Exercises**</td></tr></table>

Access end-of-section exercises and tools for visualizing calculus online at **www.webassign.net**   WebAssign

---

DISCOVERY PROJECT   **ARC LENGTH CONTEST**

The curves shown are all examples of graphs of continuous functions f that have the following properties.

1. $f(0) = 0$ and $f(1) = 0$

2. $f(x) \geq 0$ for $0 \leq x \leq 1$

3. The area under the graph of f from 0 to 1 is equal to 1.

The lengths L of these curves, however, are different.

$L \approx 3.249$

$L \approx 2.919$

$L \approx 3.152$

$L \approx 3.213$

Try to discover formulas for two functions that satisfy the given conditions 1, 2, and 3. (Your graphs might be similar to the ones shown or could look quite different.) Then calculate the arc length of each graph. The winning entry will be the one with the smallest arc length.

8.2   Area of a Surface of Revolution

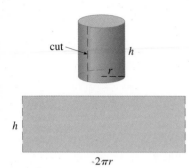

FIGURE 1

A surface of revolution is formed when a curve is rotated about a line. Such a surface is the lateral boundary of a solid of revolution of the type discussed in Sections 5.2 and 5.3.

We want to define the area of a surface of revolution in such a way that it corresponds to our intuition. If the surface area is A, we can imagine that painting the surface would require the same amount of paint as does a flat region with area A.

Let's start with some simple surfaces. The lateral surface area of a circular cylinder with radius r and height h is taken to be $A = 2\pi r h$ because we can imagine cutting the cylinder and unrolling it (as in Figure 1) to obtain a rectangle with dimensions $2\pi r$ and h.

Likewise, we can take a circular cone with base radius r and slant height l, cut it along the dashed line in Figure 2, and flatten it to form a sector of a circle with radius l and central angle $\theta = 2\pi r/l$. We know that, in general, the area of a sector of a circle with radius l and angle θ is $\frac{1}{2}l^2\theta$ (see Exercise 35 in Section 7.3) and so in this case the area is

$$A = \tfrac{1}{2}l^2\theta = \tfrac{1}{2}l^2\left(\frac{2\pi r}{l}\right) = \pi r l$$

Therefore we define the lateral surface area of a cone to be $A = \pi r l$.

FIGURE 2

What about more complicated surfaces of revolution? If we follow the strategy we used with arc length, we can approximate the original curve by a polygon. When this polygon is rotated about an axis, it creates a simpler surface whose surface area approximates the actual surface area. By taking a limit, we can determine the exact surface area.

The approximating surface, then, consists of a number of *bands*, each formed by rotating a line segment about an axis. To find the surface area, each of these bands can be considered a portion of a circular cone, as shown in Figure 3. The area of the band (or frustum of a cone) with slant height l and upper and lower radii r_1 and r_2 is found by subtracting the areas of two cones:

$$\boxed{1} \qquad A = \pi r_2(l_1 + l) - \pi r_1 l_1 = \pi[(r_2 - r_1)l_1 + r_2 l]$$

From similar triangles we have

$$\frac{l_1}{r_1} = \frac{l_1 + l}{r_2}$$

FIGURE 3

which gives

$$r_2 l_1 = r_1 l_1 + r_1 l \qquad \text{or} \qquad (r_2 - r_1)l_1 = r_1 l$$

Putting this in Equation 1, we get

$$A = \pi(r_1 l + r_2 l)$$

or

$$\boxed{2} \qquad \boxed{A = 2\pi r l}$$

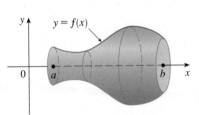

$y \uparrow$ $y = f(x)$

(a) Surface of revolution

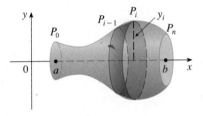

$y \uparrow$ P_0 P_{i-1} P_i y_i P_n

(b) Approximating band

FIGURE 4

where $r = \frac{1}{2}(r_1 + r_2)$ is the average radius of the band.

Now we apply this formula to our strategy. Consider the surface shown in Figure 4, which is obtained by rotating the curve $y = f(x)$, $a \leqslant x \leqslant b$, about the x-axis, where f is positive and has a continuous derivative. In order to define its surface area, we divide the interval $[a, b]$ into n subintervals with endpoints $x_0, x_1, \ldots, x_n$ and equal width Δx, as we did in determining arc length. If $y_i = f(x_i)$, then the point $P_i(x_i, y_i)$ lies on the curve. The part of the surface between x_{i-1} and x_i is approximated by taking the line segment $P_{i-1}P_i$ and rotating it about the x-axis. The result is a band with slant height $l = |P_{i-1}P_i|$ and average radius $r = \frac{1}{2}(y_{i-1} + y_i)$ so, by Formula 2, its surface area is

$$2\pi \frac{y_{i-1} + y_i}{2} |P_{i-1}P_i|$$

As in the proof of Theorem 8.1.2, we have

$$|P_{i-1}P_i| = \sqrt{1 + [f'(x_i^*)]^2}\, \Delta x$$

where x_i^* is some number in $[x_{i-1}, x_i]$. When Δx is small, we have $y_i = f(x_i) \approx f(x_i^*)$ and also $y_{i-1} = f(x_{i-1}) \approx f(x_i^*)$, since f is continuous. Therefore

$$2\pi \frac{y_{i-1} + y_i}{2} |P_{i-1}P_i| \approx 2\pi f(x_i^*) \sqrt{1 + [f'(x_i^*)]^2}\, \Delta x$$

and so an approximation to what we think of as the area of the complete surface of revolution is

$$\boxed{3} \qquad \sum_{i=1}^{n} 2\pi f(x_i^*) \sqrt{1 + [f'(x_i^*)]^2}\, \Delta x$$

This approximation appears to become better as $n \to \infty$ and, recognizing $\boxed{3}$ as a Riemann sum for the function $g(x) = 2\pi f(x) \sqrt{1 + [f'(x)]^2}$, we have

$$\lim_{n \to \infty} \sum_{i=1}^{n} 2\pi f(x_i^*) \sqrt{1 + [f'(x_i^*)]^2}\, \Delta x = \int_a^b 2\pi f(x) \sqrt{1 + [f'(x)]^2}\, dx$$

Therefore, in the case where f is positive and has a continuous derivative, we define the **surface area** of the surface obtained by rotating the curve $y = f(x)$, $a \leqslant x \leqslant b$, about the x-axis as

$$\boxed{4} \qquad S = \int_a^b 2\pi f(x) \sqrt{1 + [f'(x)]^2}\, dx$$

With the Leibniz notation for derivatives, this formula becomes

$$\boxed{5} \qquad S = \int_a^b 2\pi y \sqrt{1 + \left(\frac{dy}{dx}\right)^2}\, dx$$

If the curve is described as $x = g(y)$, $c \leqslant y \leqslant d$, then the formula for surface area becomes

$$\boxed{6} \qquad S = \int_c^d 2\pi y \sqrt{1 + \left(\frac{dx}{dy}\right)^2}\, dy$$

and both Formulas 5 and 6 can be summarized symbolically, using the notation for arc length given in Section 8.1, as

$$\boxed{7} \qquad S = \int 2\pi y\, ds$$

For rotation about the y-axis, the surface area formula becomes

$$\boxed{8} \qquad S = \int 2\pi x\, ds$$

where, as before, we can use either

$$ds = \sqrt{1 + \left(\frac{dy}{dx}\right)^2}\, dx \qquad \text{or} \qquad ds = \sqrt{1 + \left(\frac{dx}{dy}\right)^2}\, dy$$

These formulas can be remembered by thinking of $2\pi y$ or $2\pi x$ as the circumference of a circle traced out by the point (x, y) on the curve as it is rotated about the x-axis or y-axis, respectively (see Figure 5).

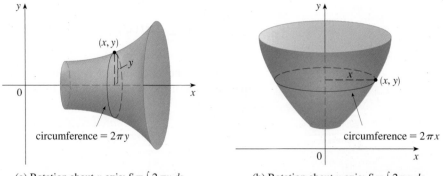

FIGURE 5   (a) Rotation about x-axis: $S = \int 2\pi y\, ds$ (b) Rotation about y-axis: $S = \int 2\pi x\, ds$

FIGURE 6

Figure 6 shows the portion of the sphere whose surface area is computed in Example 1.

V EXAMPLE 1 The curve $y = \sqrt{4 - x^2}$, $-1 \le x \le 1$, is an arc of the circle $x^2 + y^2 = 4$. Find the area of the surface obtained by rotating this arc about the x-axis. (The surface is a portion of a sphere of radius 2. See Figure 6.)

SOLUTION We have

$$\frac{dy}{dx} = \frac{1}{2}(4 - x^2)^{-1/2}(-2x) = \frac{-x}{\sqrt{4 - x^2}}$$

and so, by Formula 5, the surface area is

$$S = \int_{-1}^{1} 2\pi y \sqrt{1 + \left(\frac{dy}{dx}\right)^2} \, dx$$

$$= 2\pi \int_{-1}^{1} \sqrt{4 - x^2} \sqrt{1 + \frac{x^2}{4 - x^2}} \, dx$$

$$= 2\pi \int_{-1}^{1} \sqrt{4 - x^2} \frac{2}{\sqrt{4 - x^2}} \, dx$$

$$= 4\pi \int_{-1}^{1} 1 \, dx = 4\pi(2) = 8\pi$$

Figure 7 shows the surface of revolution whose area is computed in Example 2.

FIGURE 7

As a check on our answer to Example 2, notice from Figure 7 that the surface area should be close to that of a circular cylinder with the same height and radius halfway between the upper and lower radius of the surface: $2\pi(1.5)(3) \approx 28.27$. We computed that the surface area was

$$\frac{\pi}{6}\left(17\sqrt{17} - 5\sqrt{5}\right) \approx 30.85$$

which seems reasonable. Alternatively, the surface area should be slightly larger than the area of a frustum of a cone with the same top and bottom edges. From Equation 2, this is $2\pi(1.5)(\sqrt{10}) \approx 29.80$.

V EXAMPLE 2 The arc of the parabola $y = x^2$ from $(1, 1)$ to $(2, 4)$ is rotated about the y-axis. Find the area of the resulting surface.

SOLUTION 1 Using

$$y = x^2 \quad \text{and} \quad \frac{dy}{dx} = 2x$$

we have, from Formula 8,

$$S = \int 2\pi x \, ds$$

$$= \int_{1}^{2} 2\pi x \sqrt{1 + \left(\frac{dy}{dx}\right)^2} \, dx$$

$$= 2\pi \int_{1}^{2} x \sqrt{1 + 4x^2} \, dx$$

Substituting $u = 1 + 4x^2$, we have $du = 8x \, dx$. Remembering to change the limits of integration, we have

$$S = \frac{\pi}{4} \int_{5}^{17} \sqrt{u} \, du = \frac{\pi}{4} \left[\frac{2}{3} u^{3/2}\right]_{5}^{17}$$

$$= \frac{\pi}{6}\left(17\sqrt{17} - 5\sqrt{5}\right)$$

SOLUTION 2 Using

$$x = \sqrt{y} \quad \text{and} \quad \frac{dx}{dy} = \frac{1}{2\sqrt{y}}$$

we have

$$S = \int 2\pi x \, ds = \int_1^4 2\pi x \sqrt{1 + \left(\frac{dx}{dy}\right)^2} \, dy$$

$$= 2\pi \int_1^4 \sqrt{y} \sqrt{1 + \frac{1}{4y}} \, dy = \pi \int_1^4 \sqrt{4y + 1} \, dy$$

$$= \frac{\pi}{4} \int_5^{17} \sqrt{u} \, du \qquad \text{(where } u = 1 + 4y)$$

$$= \frac{\pi}{6} \left(17\sqrt{17} - 5\sqrt{5}\right) \qquad \text{(as in Solution 1)}$$

V EXAMPLE 3   Find the area of the surface generated by rotating the curve $y = e^x$, $0 \leqslant x \leqslant 1$, about the x-axis.

Another method: Use Formula 6 with $x = \ln y$.

SOLUTION Using Formula 5 with

$$y = e^x \qquad \text{and} \qquad \frac{dy}{dx} = e^x$$

we have

$$S = \int_0^1 2\pi y \sqrt{1 + \left(\frac{dy}{dx}\right)^2} \, dx = 2\pi \int_0^1 e^x \sqrt{1 + e^{2x}} \, dx$$

$$= 2\pi \int_1^e \sqrt{1 + u^2} \, du \qquad \text{(where } u = e^x)$$

$$= 2\pi \int_{\pi/4}^{\alpha} \sec^3\theta \, d\theta \qquad \text{(where } u = \tan\theta \text{ and } \alpha = \tan^{-1}e)$$

Or use Formula 21 in the Table of Integrals.

$$= 2\pi \cdot \tfrac{1}{2}\left[\sec\theta \tan\theta + \ln|\sec\theta + \tan\theta|\right]_{\pi/4}^{\alpha} \qquad \text{(by Example 8 in Section 7.2)}$$

$$= \pi\left[\sec\alpha \tan\alpha + \ln(\sec\alpha + \tan\alpha) - \sqrt{2} - \ln(\sqrt{2} + 1)\right]$$

Since $\tan\alpha = e$, we have $\sec^2\alpha = 1 + \tan^2\alpha = 1 + e^2$ and

$$S = \pi\left[e\sqrt{1 + e^2} + \ln\left(e + \sqrt{1 + e^2}\right) - \sqrt{2} - \ln(\sqrt{2} + 1)\right]$$

8.2   Exercises

Access end-of-section exercises and tools for visualizing calculus online at **www.webassign.net**

DISCOVERY PROJECT   ROTATING ON A SLANT

We know how to find the volume of a solid of revolution obtained by rotating a region about a horizontal or vertical line (see Section 5.2). We also know how to find the surface area of a surface of revolution if we rotate a curve about a horizontal or vertical line (see Section 8.2). But what if we rotate about a slanted line, that is, a line that is neither horizontal nor vertical? In this project you are asked to discover formulas for the volume of a solid of revolution and for the area of a surface of revolution when the axis of rotation is a slanted line.

Let C be the arc of the curve $y = f(x)$ between the points $P(p, f(p))$ and $Q(q, f(q))$ and let $\mathcal{R}$ be the region bounded by C, by the line $y = mx + b$ (which lies entirely below C), and by the perpendiculars to the line from P and Q.

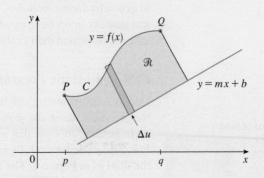

1. Show that the area of $\mathcal{R}$ is

$$\frac{1}{1 + m^2} \int_p^q [f(x) - mx - b][1 + mf'(x)] \, dx$$

[*Hint:* This formula can be verified by subtracting areas, but it will be helpful throughout the project to derive it by first approximating the area using rectangles perpendicular to the line, as shown in the following figure. Use the figure to help express Δu in terms of Δx.]

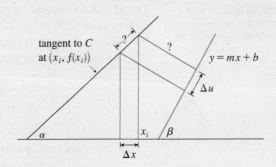

2. Find the area of the region shown in the figure at the left.

3. Find a formula (similar to the one in Problem 1) for the volume of the solid obtained by rotating $\mathcal{R}$ about the line $y = mx + b$.

4. Find the volume of the solid obtained by rotating the region of Problem 2 about the line $y = x - 2$.

5. Find a formula for the area of the surface obtained by rotating C about the line $y = mx + b$.

CAS **6.** Use a computer algebra system to find the exact area of the surface obtained by rotating the curve $y = \sqrt{x}$, $0 \le x \le 4$, about the line $y = \frac{1}{2}x$. Then approximate your result to three decimal places.

_____

CAS Computer algebra system required

8.3 Applications to Physics and Engineering

Among the many applications of integral calculus to physics and engineering, we consider two here: force due to water pressure and centers of mass. As with our previous applications to geometry (areas, volumes, and lengths) and to work, our strategy is to break up the physical quantity into a large number of small parts, approximate each small part, add the results, take the limit, and then evaluate the resulting integral.

▬ Hydrostatic Pressure and Force

Deep-sea divers realize that water pressure increases as they dive deeper. This is because the weight of the water above them increases.

In general, suppose that a thin horizontal plate with area A square meters is submerged in a fluid of density ρ kilograms per cubic meter at a depth d meters below the surface of the fluid as in Figure 1. The fluid directly above the plate has volume $V = Ad$, so its mass is $m = \rho V = \rho Ad$. The force exerted by the fluid on the plate is therefore

$$F = mg = \rho gAd$$

where g is the acceleration due to gravity. The **pressure** P on the plate is defined to be the force per unit area:

$$P = \frac{F}{A} = \rho gd$$

surface of fluid

FIGURE 1

When using US Customary units, we write $P = \rho gd = \delta d$, where $\delta = \rho g$ is the *weight density* (as opposed to ρ, which is the *mass density*). For instance, the weight density of water is $\delta = 62.5 \text{ lb/ft}^3$.

The SI unit for measuring pressure is newtons per square meter, which is called a pascal (abbreviation: $1 \text{ N/m}^2 = 1$ Pa). Since this is a small unit, the kilopascal (kPa) is often used. For instance, because the density of water is $\rho = 1000 \text{ kg/m}^3$, the pressure at the bottom of a swimming pool 2 m deep is

$$P = \rho gd = 1000 \text{ kg/m}^3 \times 9.8 \text{ m/s}^2 \times 2 \text{ m}$$

$$= 19{,}600 \text{ Pa} = 19.6 \text{ kPa}$$

An important principle of fluid pressure is the experimentally verified fact that *at any point in a liquid the pressure is the same in all directions.* (A diver feels the same pressure on nose and both ears.) Thus the pressure in *any* direction at a depth d in a fluid with mass density ρ is given by

$$\boxed{1} \qquad\qquad P = \rho gd = \delta d$$

This helps us determine the hydrostatic force against a *vertical* plate or wall or dam in a fluid. This is not a straightforward problem because the pressure is not constant but increases as the depth increases.

50 m

20 m

30 m

FIGURE 2

V **EXAMPLE 1** A dam has the shape of the trapezoid shown in Figure 2. The height is 20 m and the width is 50 m at the top and 30 m at the bottom. Find the force on the dam due to hydrostatic pressure if the water level is 4 m from the top of the dam.

SOLUTION We choose a vertical x-axis with origin at the surface of the water and directed downward as in Figure 3(a). The depth of the water is 16 m, so we divide the interval $[0, 16]$ into subintervals of equal length with endpoints x_i and we choose

FIGURE 3

$x_i^* \in [x_{i-1}, x_i]$. The ith horizontal strip of the dam is approximated by a rectangle with height Δx and width w_i, where, from similar triangles in Figure 3(b),

$$\frac{a}{16 - x_i^*} = \frac{10}{20} \qquad \text{or} \qquad a = \frac{16 - x_i^*}{2} = 8 - \frac{x_i^*}{2}$$

and so

$$w_i = 2(15 + a) = 2\left(15 + 8 - \tfrac{1}{2}x_i^*\right) = 46 - x_i^*$$

If A_i is the area of the ith strip, then

$$A_i \approx w_i \, \Delta x = (46 - x_i^*) \, \Delta x$$

If Δx is small, then the pressure P_i on the ith strip is almost constant and we can use Equation 1 to write

$$P_i \approx 1000 g x_i^*$$

The hydrostatic force F_i acting on the ith strip is the product of the pressure and the area:

$$F_i = P_i A_i \approx 1000 g x_i^* (46 - x_i^*) \, \Delta x$$

Adding these forces and taking the limit as $n \to \infty$, we obtain the total hydrostatic force on the dam:

$$F = \lim_{n \to \infty} \sum_{i=1}^{n} 1000 g x_i^* (46 - x_i^*) \, \Delta x$$

$$= \int_0^{16} 1000 g x (46 - x) \, dx$$

$$= 1000(9.8) \int_0^{16} (46x - x^2) \, dx$$

$$= 9800 \left[23x^2 - \frac{x^3}{3} \right]_0^{16}$$

$$\approx 4.43 \times 10^7 \text{ N}$$

EXAMPLE 2 Find the hydrostatic force on one end of a cylindrical drum with radius 3 ft if the drum is submerged in water 10 ft deep.

SOLUTION In this example it is convenient to choose the axes as in Figure 4 so that the origin is placed at the center of the drum. Then the circle has a simple equation, $x^2 + y^2 = 9$. As in Example 1 we divide the circular region into horizontal strips of equal width. From the equation of the circle, we see that the length of the ith strip is $2\sqrt{9 - (y_i^*)^2}$ and so its area is

$$A_i = 2\sqrt{9 - (y_i^*)^2} \, \Delta y$$

The pressure on this strip is approximately

$$\delta d_i = 62.5(7 - y_i^*)$$

and so the force on the strip is approximately

$$\delta d_i A_i = 62.5(7 - y_i^*) 2\sqrt{9 - (y_i^*)^2} \, \Delta y$$

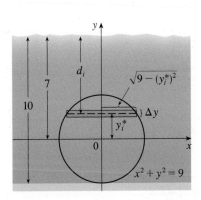

FIGURE 4

The total force is obtained by adding the forces on all the strips and taking the limit:

$$F = \lim_{n \to \infty} \sum_{i=1}^{n} 62.5(7 - y_i^*)2\sqrt{9 - (y_i^*)^2}\, \Delta y$$

$$= 125 \int_{-3}^{3} (7 - y)\sqrt{9 - y^2}\, dy$$

$$= 125 \cdot 7 \int_{-3}^{3} \sqrt{9 - y^2}\, dy - 125 \int_{-3}^{3} y\sqrt{9 - y^2}\, dy$$

The second integral is 0 because the integrand is an odd function (see Theorem 4.5.6). The first integral can be evaluated using the trigonometric substitution $y = 3\sin\theta$, but it's simpler to observe that it is the area of a semicircular disk with radius 3. Thus

$$F = 875 \int_{-3}^{3} \sqrt{9 - y^2}\, dy = 875 \cdot \tfrac{1}{2}\pi(3)^2$$

$$= \frac{7875\pi}{2} \approx 12{,}370 \text{ lb}$$

Moments and Centers of Mass

Our main objective here is to find the point P on which a thin plate of any given shape balances horizontally as in Figure 5. This point is called the **center of mass** (or center of gravity) of the plate.

    We first consider the simpler situation illustrated in Figure 6, where two masses m_1 and m_2 are attached to a rod of negligible mass on opposite sides of a fulcrum and at distances d_1 and d_2 from the fulcrum. The rod will balance if

FIGURE 5

$$\boxed{2} \qquad\qquad m_1 d_1 = m_2 d_2$$

This is an experimental fact discovered by Archimedes and called the Law of the Lever. (Think of a lighter person balancing a heavier one on a seesaw by sitting farther away from the center.)

    Now suppose that the rod lies along the x-axis with m_1 at x_1 and m_2 at x_2 and the center of mass at $\bar{x}$. If we compare Figures 6 and 7, we see that $d_1 = \bar{x} - x_1$ and $d_2 = x_2 - \bar{x}$ and so Equation 2 gives

FIGURE 6

$$m_1(\bar{x} - x_1) = m_2(x_2 - \bar{x})$$

$$m_1\bar{x} + m_2\bar{x} = m_1 x_1 + m_2 x_2$$

$$\boxed{3} \qquad\qquad \bar{x} = \frac{m_1 x_1 + m_2 x_2}{m_1 + m_2}$$

The numbers $m_1 x_1$ and $m_2 x_2$ are called the **moments** of the masses m_1 and m_2 (with respect to the origin), and Equation 3 says that the center of mass $\bar{x}$ is obtained by adding the moments of the masses and dividing by the total mass $m = m_1 + m_2$.

FIGURE 7

In general, if we have a system of n particles with masses $m_1, m_2, \ldots, m_n$ located at the points $x_1, x_2, \ldots, x_n$ on the x-axis, it can be shown similarly that the center of mass of the system is located at

$$\boxed{4} \qquad \bar{x} = \frac{\sum_{i=1}^{n} m_i x_i}{\sum_{i=1}^{n} m_i} = \frac{\sum_{i=1}^{n} m_i x_i}{m}$$

where $m = \Sigma\, m_i$ is the total mass of the system, and the sum of the individual moments

$$M = \sum_{i=1}^{n} m_i x_i$$

is called the **moment of the system about the origin**. Then Equation 4 could be rewritten as $m\bar{x} = M$, which says that if the total mass were considered as being concentrated at the center of mass $\bar{x}$, then its moment would be the same as the moment of the system.

Now we consider a system of n particles with masses $m_1, m_2, \ldots, m_n$ located at the points $(x_1, y_1), (x_2, y_2), \ldots, (x_n, y_n)$ in the xy-plane as shown in Figure 8. By analogy with the one-dimensional case, we define the **moment of the system about the y-axis** to be

$$\boxed{5} \qquad M_y = \sum_{i=1}^{n} m_i x_i$$

and the **moment of the system about the x-axis** as

$$\boxed{6} \qquad M_x = \sum_{i=1}^{n} m_i y_i$$

Then M_y measures the tendency of the system to rotate about the y-axis and M_x measures the tendency to rotate about the x-axis.

As in the one-dimensional case, the coordinates $(\bar{x}, \bar{y})$ of the center of mass are given in terms of the moments by the formulas

$$\boxed{7} \qquad \bar{x} = \frac{M_y}{m} \qquad \bar{y} = \frac{M_x}{m}$$

where $m = \Sigma\, m_i$ is the total mass. Since $m\bar{x} = M_y$ and $m\bar{y} = M_x$, the center of mass $(\bar{x}, \bar{y})$ is the point where a single particle of mass m would have the same moments as the system.

V **EXAMPLE 3** Find the moments and center of mass of the system of objects that have masses 3, 4, and 8 at the points $(-1, 1)$, $(2, -1)$, and $(3, 2)$, respectively.

SOLUTION We use Equations 5 and 6 to compute the moments:

$$M_y = 3(-1) + 4(2) + 8(3) = 29$$
$$M_x = 3(1) + 4(-1) + 8(2) = 15$$

Since $m = 3 + 4 + 8 = 15$, we use Equations 7 to obtain

$$\bar{x} = \frac{M_y}{m} = \frac{29}{15} \qquad \bar{y} = \frac{M_x}{m} = \frac{15}{15} = 1$$

Thus the center of mass is $\left(1\frac{14}{15}, 1\right)$. (See Figure 9.)

FIGURE 8

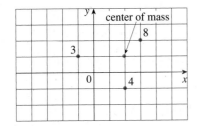

FIGURE 9

Next we consider a flat plate (called a *lamina*) with uniform density ρ that occupies a region $\mathcal{R}$ of the plane. We wish to locate the center of mass of the plate, which is called the **centroid** of $\mathcal{R}$. In doing so we use the following physical principles: The **symmetry principle** says that if $\mathcal{R}$ is symmetric about a line l, then the centroid of $\mathcal{R}$ lies on l. (If $\mathcal{R}$ is reflected about l, then $\mathcal{R}$ remains the same so its centroid remains fixed. But the only fixed points lie on l.) Thus the centroid of a rectangle is its center. Moments should be defined so that if the entire mass of a region is concentrated at the center of mass, then its moments remain unchanged. Also, the moment of the union of two nonoverlapping regions should be the sum of the moments of the individual regions.

Suppose that the region $\mathcal{R}$ is of the type shown in Figure 10(a); that is, $\mathcal{R}$ lies between the lines $x = a$ and $x = b$, above the x-axis, and beneath the graph of f, where f is a continuous function. We divide the interval $[a, b]$ into n subintervals with endpoints x_0, $x_1, \ldots, x_n$ and equal width Δx. We choose the sample point x_i^* to be the midpoint $\overline{x}_i$ of the ith subinterval, that is, $\overline{x}_i = (x_{i-1} + x_i)/2$. This determines the polygonal approximation to $\mathcal{R}$ shown in Figure 10(b).

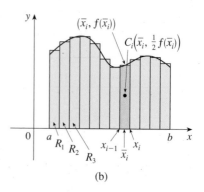

FIGURE 10

(a)

(b)

The centroid of the ith approximating rectangle R_i is its center $C_i\left(\overline{x}_i, \frac{1}{2} f(\overline{x}_i)\right)$. Its area is $f(\overline{x}_i) \, \Delta x$, so its mass is

$$\rho f(\overline{x}_i) \, \Delta x$$

The moment of R_i about the y-axis is the product of its mass and the distance from C_i to the y-axis, which is $\overline{x}_i$. Thus

$$M_y(R_i) = [\rho f(\overline{x}_i) \, \Delta x] \, \overline{x}_i = \rho \overline{x}_i f(\overline{x}_i) \, \Delta x$$

Adding these moments, we obtain the moment of the polygonal approximation to $\mathcal{R}$, and then by taking the limit as $n \to \infty$ we obtain the moment of $\mathcal{R}$ itself about the y-axis:

$$M_y = \lim_{n \to \infty} \sum_{i=1}^{n} \rho \overline{x}_i f(\overline{x}_i) \, \Delta x = \rho \int_a^b x f(x) \, dx$$

In a similar fashion we compute the moment of R_i about the x-axis as the product of its mass and the distance from C_i to the x-axis:

$$M_x(R_i) = [\rho f(\overline{x}_i) \, \Delta x] \tfrac{1}{2} f(\overline{x}_i) = \rho \cdot \tfrac{1}{2} [f(\overline{x}_i)]^2 \, \Delta x$$

Again we add these moments and take the limit to obtain the moment of $\mathcal{R}$ about the x-axis:

$$M_x = \lim_{n \to \infty} \sum_{i=1}^{n} \rho \cdot \tfrac{1}{2}[f(\bar{x}_i)]^2 \, \Delta x = \rho \int_a^b \tfrac{1}{2}[f(x)]^2 \, dx$$

Just as for systems of particles, the center of mass of the plate is defined so that $m\bar{x} = M_y$ and $m\bar{y} = M_x$. But the mass of the plate is the product of its density and its area:

$$m = \rho A = \rho \int_a^b f(x) \, dx$$

and so

$$\bar{x} = \frac{M_y}{m} = \frac{\rho \int_a^b x f(x) \, dx}{\rho \int_a^b f(x) \, dx} = \frac{\int_a^b x f(x) \, dx}{\int_a^b f(x) \, dx}$$

$$\bar{y} = \frac{M_x}{m} = \frac{\rho \int_a^b \tfrac{1}{2}[f(x)]^2 \, dx}{\rho \int_a^b f(x) \, dx} = \frac{\int_a^b \tfrac{1}{2}[f(x)]^2 \, dx}{\int_a^b f(x) \, dx}$$

Notice the cancellation of the ρ's. The location of the center of mass is independent of the density.

In summary, the center of mass of the plate (or the centroid of $\mathcal{R}$) is located at the point $(\bar{x}, \bar{y})$, where

$$\boxed{8} \qquad \bar{x} = \frac{1}{A} \int_a^b x f(x) \, dx \qquad \bar{y} = \frac{1}{A} \int_a^b \tfrac{1}{2}[f(x)]^2 \, dx$$

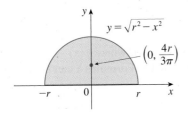

FIGURE 11

EXAMPLE 4 Find the center of mass of a semicircular plate of radius r.

SOLUTION In order to use $\boxed{8}$ we place the semicircle as in Figure 11 so that $f(x) = \sqrt{r^2 - x^2}$ and $a = -r$, $b = r$. Here there is no need to use the formula to calculate $\bar{x}$ because, by the symmetry principle, the center of mass must lie on the y-axis, so $\bar{x} = 0$. The area of the semicircle is $A = \tfrac{1}{2}\pi r^2$, so

$$\bar{y} = \frac{1}{A} \int_{-r}^{r} \tfrac{1}{2}[f(x)]^2 \, dx$$

$$= \frac{1}{\frac{1}{2}\pi r^2} \cdot \frac{1}{2} \int_{-r}^{r} \left(\sqrt{r^2 - x^2} \right)^2 dx$$

$$= \frac{2}{\pi r^2} \int_0^r (r^2 - x^2) \, dx = \frac{2}{\pi r^2} \left[r^2 x - \frac{x^3}{3} \right]_0^r$$

$$= \frac{2}{\pi r^2} \frac{2r^3}{3} = \frac{4r}{3\pi}$$

The center of mass is located at the point $(0, 4r/(3\pi))$.

EXAMPLE 5   Find the centroid of the region bounded by the curves $y = \cos x$, $y = 0$, $x = 0$, and $x = \pi/2$.

SOLUTION The area of the region is

$$A = \int_0^{\pi/2} \cos x \, dx = \sin x \Big]_0^{\pi/2} = 1$$

so Formulas 8 give

$$\bar{x} = \frac{1}{A} \int_0^{\pi/2} x f(x) \, dx = \int_0^{\pi/2} x \cos x \, dx$$

$$= x \sin x \Big]_0^{\pi/2} - \int_0^{\pi/2} \sin x \, dx \qquad \text{(by integration by parts)}$$

$$= \frac{\pi}{2} - 1$$

$$\bar{y} = \frac{1}{A} \int_0^{\pi/2} \tfrac{1}{2} [f(x)]^2 \, dx = \tfrac{1}{2} \int_0^{\pi/2} \cos^2 x \, dx$$

$$= \tfrac{1}{4} \int_0^{\pi/2} (1 + \cos 2x) \, dx = \tfrac{1}{4} \Big[x + \tfrac{1}{2} \sin 2x \Big]_0^{\pi/2}$$

$$= \frac{\pi}{8}$$

The centroid is $\left(\tfrac{1}{2}\pi - 1, \tfrac{1}{8}\pi \right)$ and is shown in Figure 12.

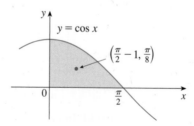

FIGURE 12

If the region $\mathcal{R}$ lies between two curves $y = f(x)$ and $y = g(x)$, where $f(x) \ge g(x)$, as illustrated in Figure 13, then the same sort of argument that led to Formulas 8 can be used to show that the centroid of $\mathcal{R}$ is $(\bar{x}, \bar{y})$, where

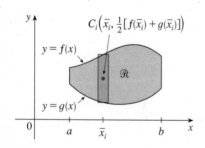

FIGURE 13

9

$$\bar{x} = \frac{1}{A} \int_a^b x [f(x) - g(x)] \, dx$$

$$\bar{y} = \frac{1}{A} \int_a^b \tfrac{1}{2} \{ [f(x)]^2 - [g(x)]^2 \} \, dx$$

(See Exercise 47.)

EXAMPLE 6   Find the centroid of the region bounded by the line $y = x$ and the parabola $y = x^2$.

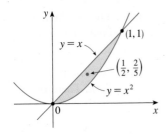

FIGURE 14

SOLUTION The region is sketched in Figure 14. We take $f(x) = x$, $g(x) = x^2$, $a = 0$, and $b = 1$ in Formulas 9. First we note that the area of the region is

$$A = \int_0^1 (x - x^2)\,dx = \frac{x^2}{2} - \frac{x^3}{3}\,\Bigg]_0^1 = \frac{1}{6}$$

Therefore

$$\bar{x} = \frac{1}{A}\int_0^1 x[f(x) - g(x)]\,dx = \frac{1}{\frac{1}{6}}\int_0^1 x(x - x^2)\,dx$$

$$= 6\int_0^1 (x^2 - x^3)\,dx = 6\left[\frac{x^3}{3} - \frac{x^4}{4}\right]_0^1 = \frac{1}{2}$$

$$\bar{y} = \frac{1}{A}\int_0^1 \tfrac{1}{2}\{[f(x)]^2 - [g(x)]^2\}\,dx = \frac{1}{\frac{1}{6}}\int_0^1 \tfrac{1}{2}(x^2 - x^4)\,dx$$

$$= 3\left[\frac{x^3}{3} - \frac{x^5}{5}\right]_0^1 = \frac{2}{5}$$

The centroid is $\left(\frac{1}{2}, \frac{2}{5}\right)$.

We end this section by showing a surprising connection between centroids and volumes of revolution.

This theorem is named after the Greek mathematician Pappus of Alexandria, who lived in the fourth century AD.

> **Theorem of Pappus** Let $\mathcal{R}$ be a plane region that lies entirely on one side of a line l in the plane. If $\mathcal{R}$ is rotated about l, then the volume of the resulting solid is the product of the area A of $\mathcal{R}$ and the distance d traveled by the centroid of $\mathcal{R}$.

PROOF We give the proof for the special case in which the region lies between $y = f(x)$ and $y = g(x)$ as in Figure 13 and the line l is the y-axis. Using the method of cylindrical shells (see Section 5.3), we have

$$V = \int_a^b 2\pi x[f(x) - g(x)]\,dx$$

$$= 2\pi \int_a^b x[f(x) - g(x)]\,dx$$

$$= 2\pi(\bar{x}A) \qquad \text{(by Formulas 9)}$$

$$= (2\pi\bar{x})A = Ad$$

where $d = 2\pi\bar{x}$ is the distance traveled by the centroid during one rotation about the y-axis.

V EXAMPLE 7 A torus is formed by rotating a circle of radius r about a line in the plane of the circle that is a distance R ($> r$) from the center of the circle. Find the volume of the torus.

SOLUTION The circle has area $A = \pi r^2$. By the symmetry principle, its centroid is its center and so the distance traveled by the centroid during a rotation is $d = 2\pi R$. Therefore, by the Theorem of Pappus, the volume of the torus is

$$V = Ad = (2\pi R)(\pi r^2) = 2\pi^2 r^2 R$$

The method of Example 7 should be compared with the method of Exercise 61 in Section 5.2.

8.3 Exercises

Access end-of-section exercises and tools for visualizing calculus online at **www.webassign.net** ENHANCED WebAssign

---

DISCOVERY PROJECT **COMPLEMENTARY COFFEE CUPS**

Suppose you have a choice of two coffee cups of the type shown, one that bends outward and one inward, and you notice that they have the same height and their shapes fit together snugly. You wonder which cup holds more coffee. Of course you could fill one cup with water and pour it into the other one but, being a calculus student, you decide on a more mathematical approach. Ignoring the handles, you observe that both cups are surfaces of revolution, so you can think of the coffee as a volume of revolution.

Cup A   Cup B

1. Suppose the cups have height h, cup A is formed by rotating the curve $x = f(y)$ about the y-axis, and cup B is formed by rotating the same curve about the line $x = k$. Find the value of k such that the two cups hold the same amount of coffee.

2. What does your result from Problem 1 say about the areas A_1 and A_2 shown in the figure?

3. Use Pappus's Theorem to explain your result in Problems 1 and 2.

4. Based on your own measurements and observations, suggest a value for h and an equation for $x = f(y)$ and calculate the amount of coffee that each cup holds.

8.4 Applications to Economics and Biology

In this section we consider some applications of integration to economics (consumer surplus) and biology (blood flow, cardiac output). Others are described in the exercises.

Consumer Surplus

Recall from Section 3.7 that the demand function $p(x)$ is the price that a company has to charge in order to sell x units of a commodity. Usually, selling larger quantities requires lowering prices, so the demand function is a decreasing function. The graph of a typical demand function, called a **demand curve**, is shown in Figure 1. If X is the amount of the commodity that is currently available, then $P = p(X)$ is the current selling price.

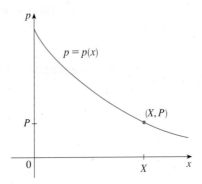

FIGURE 1

A typical demand curve

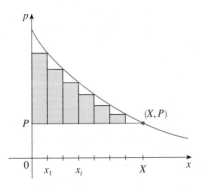

FIGURE 2

We divide the interval $[0, X]$ into n subintervals, each of length $\Delta x = X/n$, and let $x_i^* = x_i$ be the right endpoint of the ith subinterval, as in Figure 2. If, after the first x_{i-1} units were sold, a total of only x_i units had been available and the price per unit had been set at $p(x_i)$ dollars, then the additional Δx units could have been sold (but no more). The consumers who would have paid $p(x_i)$ dollars placed a high value on the product; they would have paid what it was worth to them. So in paying only P dollars they have saved an amount of

$$(\text{savings per unit})(\text{number of units}) = [p(x_i) - P]\Delta x$$

Considering similar groups of willing consumers for each of the subintervals and adding the savings, we get the total savings:

$$\sum_{i=1}^{n} [p(x_i) - P]\Delta x$$

(This sum corresponds to the area enclosed by the rectangles in Figure 2.) If we let $n \to \infty$, this Riemann sum approaches the integral

$$\boxed{1} \qquad \int_0^X [p(x) - P]\,dx$$

which economists call the **consumer surplus** for the commodity.

The consumer surplus represents the amount of money saved by consumers in purchasing the commodity at price P, corresponding to an amount demanded of X. Figure 3 shows the interpretation of the consumer surplus as the area under the demand curve and above the line $p = P$.

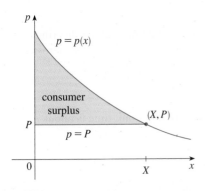

FIGURE 3

V EXAMPLE 1 The demand for a product, in dollars, is

$$p = 1200 - 0.2x - 0.0001x^2$$

Find the consumer surplus when the sales level is 500.

SOLUTION Since the number of products sold is $X = 500$, the corresponding price is

$$P = 1200 - (0.2)(500) - (0.0001)(500)^2 = 1075$$

Therefore, from Definition 1, the consumer surplus is

$$\int_0^{500} [p(x) - P]\, dx = \int_0^{500} (1200 - 0.2x - 0.0001x^2 - 1075)\, dx$$

$$= \int_0^{500} (125 - 0.2x - 0.0001x^2)\, dx$$

$$= 125x - 0.1x^2 - (0.0001)\left(\frac{x^3}{3}\right)\Bigg]_0^{500}$$

$$= (125)(500) - (0.1)(500)^2 - \frac{(0.0001)(500)^3}{3}$$

$$= \$33{,}333.33$$

Blood Flow

In Example 7 in Section 2.7 we discussed the law of laminar flow:

$$v(r) = \frac{P}{4\eta l}(R^2 - r^2)$$

which gives the velocity v of blood that flows along a blood vessel with radius R and length l at a distance r from the central axis, where P is the pressure difference between the ends of the vessel and η is the viscosity of the blood. Now, in order to compute the rate of blood flow, or *flux* (volume per unit time), we consider smaller, equally spaced radii $r_1, r_2, \ldots$. The approximate area of the ring (or washer) with inner radius r_{i-1} and outer radius r_i is

$$2\pi r_i\, \Delta r \qquad \text{where} \quad \Delta r = r_i - r_{i-1}$$

FIGURE 4

(See Figure 4.) If Δr is small, then the velocity is almost constant throughout this ring and can be approximated by $v(r_i)$. Thus the volume of blood per unit time that flows across the ring is approximately

$$(2\pi r_i\, \Delta r)\, v(r_i) = 2\pi r_i v(r_i)\, \Delta r$$

and the total volume of blood that flows across a cross-section per unit time is about

$$\sum_{i=1}^{n} 2\pi r_i\, v(r_i)\, \Delta r$$

FIGURE 5

This approximation is illustrated in Figure 5. Notice that the velocity (and hence the volume per unit time) increases toward the center of the blood vessel. The approximation gets better as n increases. When we take the limit we get the exact value of the **flux** (or *discharge*), which is the volume of blood that passes a cross-section per unit time:

$$F = \lim_{n \to \infty} \sum_{i=1}^{n} 2\pi r_i\, v(r_i)\, \Delta r = \int_0^R 2\pi r\, v(r)\, dr$$

$$= \int_0^R 2\pi r\, \frac{P}{4\eta l}\, (R^2 - r^2)\, dr$$

$$= \frac{\pi P}{2\eta l} \int_0^R (R^2 r - r^3)\, dr = \frac{\pi P}{2\eta l} \left[R^2 \frac{r^2}{2} - \frac{r^4}{4} \right]_{r=0}^{r=R}$$

$$= \frac{\pi P}{2\eta l} \left[\frac{R^4}{2} - \frac{R^4}{4} \right] = \frac{\pi P R^4}{8\eta l}$$

The resulting equation

$$\boxed{2} \qquad\qquad F = \frac{\pi P R^4}{8\eta l}$$

is called **Poiseuille's Law**; it shows that the flux is proportional to the fourth power of the radius of the blood vessel.

Cardiac Output

Figure 6 shows the human cardiovascular system. Blood returns from the body through the veins, enters the right atrium of the heart, and is pumped to the lungs through the pulmonary arteries for oxygenation. It then flows back into the left atrium through the pulmonary veins and then out to the rest of the body through the aorta. The **cardiac output** of the heart is the volume of blood pumped by the heart per unit time, that is, the rate of flow into the aorta.

The *dye dilution method* is used to measure the cardiac output. Dye is injected into the right atrium and flows through the heart into the aorta. A probe inserted into the aorta measures the concentration of the dye leaving the heart at equally spaced times over a time interval $[0, T]$ until the dye has cleared. Let $c(t)$ be the concentration of the dye at time t. If we divide $[0, T]$ into subintervals of equal length Δt, then the amount of dye that flows past the measuring point during the subinterval from $t = t_{i-1}$ to $t = t_i$ is approximately

$$(\text{concentration})(\text{volume}) = c(t_i)(F\, \Delta t)$$

where F is the rate of flow that we are trying to determine. Thus the total amount of dye is approximately

$$\sum_{i=1}^{n} c(t_i) F\, \Delta t = F \sum_{i=1}^{n} c(t_i)\, \Delta t$$

and, letting $n \to \infty$, we find that the amount of dye is

$$A = F \int_0^T c(t)\, dt$$

Thus the cardiac output is given by

$$\boxed{3} \qquad\qquad F = \frac{A}{\displaystyle\int_0^T c(t)\, dt}$$

where the amount of dye A is known and the integral can be approximated from the concentration readings.

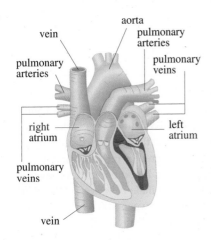

FIGURE 6

vein
aorta
pulmonary arteries
pulmonary arteries
pulmonary veins
right atrium
left atrium
pulmonary veins
vein

t	$c(t)$	t	$c(t)$
0	0	6	6.1
1	0.4	7	4.0
2	2.8	8	2.3
3	6.5	9	1.1
4	9.8	10	0
5	8.9		

V **EXAMPLE 2** A 5-mg bolus of dye is injected into a right atrium. The concentration of the dye (in milligrams per liter) is measured in the aorta at one-second intervals as shown in the chart. Estimate the cardiac output.

SOLUTION Here $A = 5$, $\Delta t = 1$, and $T = 10$. We use Simpson's Rule to approximate the integral of the concentration:

$$\int_0^{10} c(t)\, dt \approx \tfrac{1}{3}[0 + 4(0.4) + 2(2.8) + 4(6.5) + 2(9.8) + 4(8.9)$$
$$+ 2(6.1) + 4(4.0) + 2(2.3) + 4(1.1) + 0]$$
$$\approx 41.87$$

Thus Formula 3 gives the cardiac output to be

$$F = \frac{A}{\displaystyle\int_0^{10} c(t)\, dt} \approx \frac{5}{41.87} \approx 0.12 \text{ L/s} = 7.2 \text{ L/min}$$

8.4 Exercises

Access end-of-section exercises and tools for visualizing calculus online at **www.webassign.net**   **ENHANCED** **WebAssign**

8.5 Probability

Calculus plays a role in the analysis of random behavior. Suppose we consider the cholesterol level of a person chosen at random from a certain age group, or the height of an adult female chosen at random, or the lifetime of a randomly chosen battery of a certain type. Such quantities are called **continuous random variables** because their values actually range over an interval of real numbers, although they might be measured or recorded only to the nearest integer. We might want to know the probability that a blood cholesterol level is greater than 250, or the probability that the height of an adult female is between 60 and 70 inches, or the probability that the battery we are buying lasts between 100 and 200 hours. If X represents the lifetime of that type of battery, we denote this last probability as follows:

$$P(100 \le X \le 200)$$

According to the frequency interpretation of probability, this number is the long-run proportion of all batteries of the specified type whose lifetimes are between 100 and 200 hours. Since it represents a proportion, the probability naturally falls between 0 and 1.

Every continuous random variable X has a **probability density function** f. This means that the probability that X lies between a and b is found by integrating f from a to b:

1
$$P(a \le X \le b) = \int_a^b f(x)\, dx$$

For example, Figure 1 shows the graph of a model for the probability density function f for a random variable X defined to be the height in inches of an adult female in the United States (according to data from the National Health Survey). The probability that the height of a woman chosen at random from this population is between 60 and 70 inches is equal to the area under the graph of f from 60 to 70.

FIGURE 1

Probability density function
for the height of an adult female

In general, the probability density function f of a random variable X satisfies the condition $f(x) \geq 0$ for all x. Because probabilities are measured on a scale from 0 to 1, it follows that

$$\boxed{2} \qquad \int_{-\infty}^{\infty} f(x)\, dx = 1$$

EXAMPLE 1 Let $f(x) = 0.006x(10 - x)$ for $0 \leq x \leq 10$ and $f(x) = 0$ for all other values of x.
(a) Verify that f is a probability density function.
(b) Find $P(4 \leq X \leq 8)$.

SOLUTION
(a) For $0 \leq x \leq 10$ we have $0.006x(10 - x) \geq 0$, so $f(x) \geq 0$ for all x. We also need to check that Equation 2 is satisfied:

$$\int_{-\infty}^{\infty} f(x)\, dx = \int_{0}^{10} 0.006x(10 - x)\, dx = 0.006 \int_{0}^{10} (10x - x^2)\, dx$$

$$= 0.006\left[5x^2 - \tfrac{1}{3}x^3\right]_{0}^{10} = 0.006\left(500 - \tfrac{1000}{3}\right) = 1$$

Therefore f is a probability density function.

(b) The probability that X lies between 4 and 8 is

$$P(4 \leq X \leq 8) = \int_{4}^{8} f(x)\, dx = 0.006 \int_{4}^{8} (10x - x^2)\, dx$$

$$= 0.006\left[5x^2 - \tfrac{1}{3}x^3\right]_{4}^{8} = 0.544$$

V EXAMPLE 2 Phenomena such as waiting times and equipment failure times are commonly modeled by exponentially decreasing probability density functions. Find the exact form of such a function.

SOLUTION Think of the random variable as being the time you wait on hold before an agent of a company you're telephoning answers your call. So instead of x, let's use t to represent time, in minutes. If f is the probability density function and you call at time $t = 0$, then, from Definition 1, $\int_{0}^{2} f(t)\, dt$ represents the probability that an agent answers within the first two minutes and $\int_{4}^{5} f(t)\, dt$ is the probability that your call is answered during the fifth minute.

It's clear that $f(t) = 0$ for $t < 0$ (the agent can't answer before you place the call). For $t > 0$ we are told to use an exponentially decreasing function, that is, a function of the form $f(t) = Ae^{-ct}$, where A and c are positive constants. Thus

$$f(t) = \begin{cases} 0 & \text{if } t < 0 \\ Ae^{-ct} & \text{if } t \geq 0 \end{cases}$$

We use Equation 2 to determine the value of A:

$$1 = \int_{-\infty}^{\infty} f(t)\, dt = \int_{-\infty}^{0} f(t)\, dt + \int_{0}^{\infty} f(t)\, dt$$

$$= \int_{0}^{\infty} Ae^{-ct}\, dt = \lim_{x \to \infty} \int_{0}^{x} Ae^{-ct}\, dt$$

$$= \lim_{x \to \infty} \left[-\frac{A}{c} e^{-ct} \right]_{0}^{x} = \lim_{x \to \infty} \frac{A}{c}(1 - e^{-cx})$$

$$= \frac{A}{c}$$

Therefore $A/c = 1$ and so $A = c$. Thus every exponential density function has the form

$$f(t) = \begin{cases} 0 & \text{if } t < 0 \\ ce^{-ct} & \text{if } t \geqslant 0 \end{cases}$$

A typical graph is shown in Figure 2.

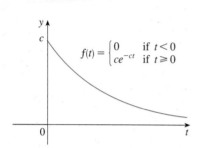

FIGURE 2

An exponential density function

$$f(t) = \begin{cases} 0 & \text{if } t < 0 \\ ce^{-ct} & \text{if } t \geqslant 0 \end{cases}$$

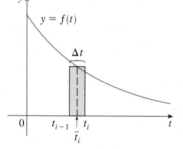

FIGURE 3

■ Average Values

Suppose you're waiting for a company to answer your phone call and you wonder how long, on average, you can expect to wait. Let $f(t)$ be the corresponding density function, where t is measured in minutes, and think of a sample of N people who have called this company. Most likely, none of them had to wait more than an hour, so let's restrict our attention to the interval $0 \leqslant t \leqslant 60$. Let's divide that interval into n intervals of length Δt and endpoints $0, t_1, t_2, \ldots, t_{60}$. (Think of Δt as lasting a minute, or half a minute, or 10 seconds, or even a second.) The probability that somebody's call gets answered during the time period from t_{i-1} to t_i is the area under the curve $y = f(t)$ from t_{i-1} to t_i, which is approximately equal to $f(\bar{t}_i)\, \Delta t$. (This is the area of the approximating rectangle in Figure 3, where $\bar{t}_i$ is the midpoint of the interval.)

Since the long-run proportion of calls that get answered in the time period from t_{i-1} to t_i is $f(\bar{t}_i)\, \Delta t$, we expect that, out of our sample of N callers, the number whose call was answered in that time period is approximately $Nf(\bar{t}_i)\, \Delta t$ and the time that each waited is about $\bar{t}_i$. Therefore the total time they waited is the product of these numbers: approximately $\bar{t}_i[Nf(\bar{t}_i)\, \Delta t]$. Adding over all such intervals, we get the approximate total of everybody's waiting times:

$$\sum_{i=1}^{n} N\bar{t}_i\, f(\bar{t}_i)\, \Delta t$$

If we now divide by the number of callers N, we get the approximate *average* waiting time:

$$\sum_{i=1}^{n} \bar{t}_i\, f(\bar{t}_i)\, \Delta t$$

We recognize this as a Riemann sum for the function $t f(t)$. As the time interval shrinks (that is, $\Delta t \to 0$ and $n \to \infty$), this Riemann sum approaches the integral

$$\int_{0}^{60} t f(t)\, dt$$

This integral is called the *mean waiting time*.

In general, the **mean** of any probability density function f is defined to be

It is traditional to denote the mean by the Greek letter μ (mu).

$$\mu = \int_{-\infty}^{\infty} x f(x)\, dx$$

The mean can be interpreted as the long-run average value of the random variable X. It can also be interpreted as a measure of centrality of the probability density function.

The expression for the mean resembles an integral we have seen before. If $\mathcal{R}$ is the region that lies under the graph of f, we know from Formula 8.3.8 that the x-coordinate of the centroid of $\mathcal{R}$ is

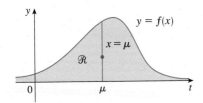

$$\bar{x} = \frac{\displaystyle\int_{-\infty}^{\infty} x f(x)\, dx}{\displaystyle\int_{-\infty}^{\infty} f(x)\, dx} = \int_{-\infty}^{\infty} x f(x)\, dx = \mu$$

FIGURE 4
$\mathcal{R}$ balances at a point on the line $x = \mu$

because of Equation 2. So a thin plate in the shape of $\mathcal{R}$ balances at a point on the vertical line $x = \mu$. (See Figure 4.)

EXAMPLE 3 Find the mean of the exponential distribution of Example 2:

$$f(t) = \begin{cases} 0 & \text{if } t < 0 \\ ce^{-ct} & \text{if } t \geqslant 0 \end{cases}$$

SOLUTION According to the definition of a mean, we have

$$\mu = \int_{-\infty}^{\infty} t f(t)\, dt = \int_{0}^{\infty} tce^{-ct}\, dt$$

To evaluate this integral we use integration by parts, with $u = t$ and $dv = ce^{-ct}\, dt$:

$$\int_{0}^{\infty} tce^{-ct}\, dt = \lim_{x \to \infty} \int_{0}^{x} tce^{-ct}\, dt = \lim_{x \to \infty} \left(-te^{-ct}\Big|_{0}^{x} + \int_{0}^{x} e^{-ct}\, dt \right)$$

The limit of the first term is 0 by l'Hospital's Rule.

$$= \lim_{x \to \infty} \left(-xe^{-cx} + \frac{1}{c} - \frac{e^{-cx}}{c} \right) = \frac{1}{c}$$

The mean is $\mu = 1/c$, so we can rewrite the probability density function as

$$f(t) = \begin{cases} 0 & \text{if } t < 0 \\ \mu^{-1}e^{-t/\mu} & \text{if } t \geqslant 0 \end{cases}$$

V EXAMPLE 4 Suppose the average waiting time for a customer's call to be answered by a company representative is five minutes.
(a) Find the probability that a call is answered during the first minute.
(b) Find the probability that a customer waits more than five minutes to be answered.

SOLUTION
(a) We are given that the mean of the exponential distribution is $\mu = 5$ min and so, from the result of Example 3, we know that the probability density function is

$$f(t) = \begin{cases} 0 & \text{if } t < 0 \\ 0.2e^{-t/5} & \text{if } t \geqslant 0 \end{cases}$$

Thus the probability that a call is answered during the first minute is

$$P(0 \leqslant T \leqslant 1) = \int_0^1 f(t) \, dt$$

$$= \int_0^1 0.2 e^{-t/5} \, dt = 0.2(-5) e^{-t/5} \Big]_0^1$$

$$= 1 - e^{-1/5} \approx 0.1813$$

So about 18% of customers' calls are answered during the first minute.

(b) The probability that a customer waits more than five minutes is

$$P(T > 5) = \int_5^\infty f(t) \, dt = \int_5^\infty 0.2 e^{-t/5} \, dt$$

$$= \lim_{x \to \infty} \int_5^x 0.2 e^{-t/5} \, dt = \lim_{x \to \infty} (e^{-1} - e^{-x/5})$$

$$= \frac{1}{e} \approx 0.368$$

About 37% of customers wait more than five minutes before their calls are answered.

Notice the result of Example 4(b): Even though the mean waiting time is 5 minutes, only 37% of callers wait more than 5 minutes. The reason is that some callers have to wait much longer (maybe 10 or 15 minutes), and this brings up the average.

Another measure of centrality of a probability density function is the *median*. That is a number m such that half the callers have a waiting time less than m and the other callers have a waiting time longer than m. In general, the **median** of a probability density function is the number m such that

$$\int_m^\infty f(x) \, dx = \tfrac{1}{2}$$

This means that half the area under the graph of f lies to the right of m. In Exercise 9 you are asked to show that the median waiting time for the company described in Example 4 is approximately 3.5 minutes.

Normal Distributions

Many important random phenomena—such as test scores on aptitude tests, heights and weights of individuals from a homogeneous population, annual rainfall in a given location—are modeled by a **normal distribution**. This means that the probability density function of the random variable X is a member of the family of functions

$$\boxed{3} \qquad\qquad f(x) = \frac{1}{\sigma \sqrt{2\pi}} \, e^{-(x-\mu)^2/(2\sigma^2)}$$

The standard deviation is denoted by the lowercase Greek letter σ (sigma).

You can verify that the mean for this function is μ. The positive constant σ is called the **standard deviation**; it measures how spread out the values of X are. From the bell-shaped graphs of members of the family in Figure 5, we see that for small values of σ the values of X are clustered about the mean, whereas for larger values of σ the values of X are more spread out. Statisticians have methods for using sets of data to estimate μ and σ.

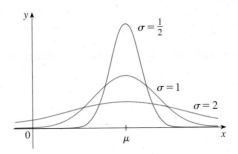

FIGURE 5
Normal distributions

The factor $1/(\sigma\sqrt{2\pi})$ is needed to make f a probability density function. In fact, it can be verified using the methods of multivariable calculus that

$$\int_{-\infty}^{\infty} \frac{1}{\sigma\sqrt{2\pi}} e^{-(x-\mu)^2/(2\sigma^2)}\, dx = 1$$

FIGURE 6
Distribution of IQ scores

V **EXAMPLE 5** Intelligence Quotient (IQ) scores are distributed normally with mean 100 and standard deviation 15. (Figure 6 shows the corresponding probability density function.)
(a) What percentage of the population has an IQ score between 85 and 115?
(b) What percentage of the population has an IQ above 140?

SOLUTION
(a) Since IQ scores are normally distributed, we use the probability density function given by Equation 3 with $\mu = 100$ and $\sigma = 15$:

$$P(85 \leqslant X \leqslant 115) = \int_{85}^{115} \frac{1}{15\sqrt{2\pi}} e^{-(x-100)^2/(2\cdot 15^2)}\, dx$$

Recall from Section 7.5 that the function $y = e^{-x^2}$ doesn't have an elementary antiderivative, so we can't evaluate the integral exactly. But we can use the numerical integration capability of a calculator or computer (or the Midpoint Rule or Simpson's Rule) to estimate the integral. Doing so, we find that

$$P(85 \leqslant X \leqslant 115) \approx 0.68$$

So about 68% of the population has an IQ between 85 and 115, that is, within one standard deviation of the mean.

(b) The probability that the IQ score of a person chosen at random is more than 140 is

$$P(X > 140) = \int_{140}^{\infty} \frac{1}{15\sqrt{2\pi}} e^{-(x-100)^2/450}\, dx$$

To avoid the improper integral we could approximate it by the integral from 140 to 200. (It's quite safe to say that people with an IQ over 200 are extremely rare.) Then

$$P(X > 140) \approx \int_{140}^{200} \frac{1}{15\sqrt{2\pi}} e^{-(x-100)^2/450}\, dx \approx 0.0038$$

Therefore about 0.4% of the population has an IQ over 140. ∎

8.5   Exercises

Access end-of-section exercises and tools for visualizing calculus online at **www.webassign.net**

8   Review

Concept Check

1. (a) How is the length of a curve defined?
   (b) Write an expression for the length of a smooth curve given by $y = f(x)$, $a \leqslant x \leqslant b$.
   (c) What if x is given as a function of y?

2. (a) Write an expression for the surface area of the surface obtained by rotating the curve $y = f(x)$, $a \leqslant x \leqslant b$, about the x-axis.
   (b) What if x is given as a function of y?
   (c) What if the curve is rotated about the y-axis?

3. Describe how we can find the hydrostatic force against a vertical wall submersed in a fluid.

4. (a) What is the physical significance of the center of mass of a thin plate?
   (b) If the plate lies between $y = f(x)$ and $y = 0$, where $a \leqslant x \leqslant b$, write expressions for the coordinates of the center of mass.

5. What does the Theorem of Pappus say?

6. Given a demand function $p(x)$, explain what is meant by the consumer surplus when the amount of a commodity currently available is X and the current selling price is P. Illustrate with a sketch.

7. (a) What is the cardiac output of the heart?
   (b) Explain how the cardiac output can be measured by the dye dilution method.

8. What is a probability density function? What properties does such a function have?

9. Suppose $f(x)$ is the probability density function for the weight of a female college student, where x is measured in pounds.
   (a) What is the meaning of the integral $\int_0^{130} f(x)\, dx$?
   (b) Write an expression for the mean of this density function.
   (c) How can we find the median of this density function?

10. What is a normal distribution? What is the significance of the standard deviation?

Exercises

1–2 Find the length of the curve.

1. $y = \frac{1}{6}(x^2 + 4)^{3/2}$, $\quad 0 \leqslant x \leqslant 3$

2. $y = 2 \ln\left(\sin \frac{1}{2}x\right)$, $\quad \pi/3 \leqslant x \leqslant \pi$

3. (a) Find the length of the curve
$$y = \frac{x^4}{16} + \frac{1}{2x^2} \quad 1 \leqslant x \leqslant 2$$
   (b) Find the area of the surface obtained by rotating the curve in part (a) about the y-axis.

4. (a) The curve $y = x^2$, $0 \leqslant x \leqslant 1$, is rotated about the y-axis. Find the area of the resulting surface.
   (b) Find the area of the surface obtained by rotating the curve in part (a) about the x-axis.

5. Use Simpson's Rule with $n = 10$ to estimate the length of the sine curve $y = \sin x$, $0 \leqslant x \leqslant \pi$.

6. Use Simpson's Rule with $n = 10$ to estimate the area of the surface obtained by rotating the sine curve in Exercise 5 about the x-axis.

7. Find the length of the curve
$$y = \int_1^x \sqrt{\sqrt{t} - 1}\ dt \qquad 1 \leqslant x \leqslant 16$$

8. Find the area of the surface obtained by rotating the curve in Exercise 7 about the y-axis.

9. A gate in an irrigation canal is constructed in the form of a trapezoid 3 ft wide at the bottom, 5 ft wide at the top, and 2 ft high. It is placed vertically in the canal so that the water just covers the gate. Find the hydrostatic force on one side of the gate.

10. A trough is filled with water and its vertical ends have the shape of the parabolic region in the figure. Find the hydrostatic force on one end of the trough.

11–12 Find the centroid of the region bounded by the given curves.

11. $y = \frac{1}{2}x, \quad y = \sqrt{x}$

12. $y = \sin x, \quad y = 0, \quad x = \pi/4, \quad x = 3\pi/4$

13–14 Find the centroid of the region shown

13.

14.

---

15. Find the volume obtained when the circle of radius 1 with center $(1, 0)$ is rotated about the y-axis.

16. Use the Theorem of Pappus and the fact that the volume of a sphere of radius r is $\frac{4}{3}\pi r^3$ to find the centroid of the semicircular region bounded by the curve $y = \sqrt{r^2 - x^2}$ and the x-axis.

17. The demand function for a commodity is given by

$$p = 2000 - 0.1x - 0.01x^2$$

Find the consumer surplus when the sales level is 100.

18. After a 6-mg injection of dye into a heart, the readings of dye concentration at two-second intervals are as shown in the table. Use Simpson's Rule to estimate the cardiac output.

t	$c(t)$	t	$c(t)$
0	0	14	4.7
2	1.9	16	3.3
4	3.3	18	2.1
6	5.1	20	1.1
8	7.6	22	0.5
10	7.1	24	0
12	5.8		

19. (a) Explain why the function

$$f(x) = \begin{cases} \dfrac{\pi}{20} \sin\left(\dfrac{\pi x}{10}\right) & \text{if } 0 \le x \le 10 \\ 0 & \text{if } x < 0 \text{ or } x > 10 \end{cases}$$

is a probability density function.
(b) Find $P(X < 4)$.
(c) Calculate the mean. Is the value what you would expect?

20. Lengths of human pregnancies are normally distributed with mean 268 days and standard deviation 15 days. What percentage of pregnancies last between 250 days and 280 days?

21. The length of time spent waiting in line at a certain bank is modeled by an exponential density function with mean 8 minutes.
(a) What is the probability that a customer is served in the first 3 minutes?
(b) What is the probability that a customer has to wait more than 10 minutes?
(c) What is the median waiting time?

1. Find the area of the region $S = \{(x, y) \mid x \geq 0, \ y \leq 1, \ x^2 + y^2 \leq 4y\}$.

2. Find the centroid of the region enclosed by the loop of the curve $y^2 = x^3 - x^4$.

3. If a sphere of radius r is sliced by a plane whose distance from the center of the sphere is d, then the sphere is divided into two pieces called segments of one base. The corresponding surfaces are called *spherical zones of one base*.
   (a) Determine the surface areas of the two spherical zones indicated in the figure.
   (b) Determine the approximate area of the Arctic Ocean by assuming that it is approximately circular in shape, with center at the North Pole and "circumference" at 75° north latitude. Use $r = 3960$ mi for the radius of the earth.
   (c) A sphere of radius r is inscribed in a right circular cylinder of radius r. Two planes perpendicular to the central axis of the cylinder and a distance h apart cut off a *spherical zone of two bases* on the sphere. Show that the surface area of the spherical zone equals the surface area of the region that the two planes cut off on the cylinder.
   (d) The *Torrid Zone* is the region on the surface of the earth that is between the Tropic of Cancer (23.45° north latitude) and the Tropic of Capricorn (23.45° south latitude). What is the area of the Torrid Zone?

4. (a) Show that an observer at height H above the north pole of a sphere of radius r can see a part of the sphere that has area

$$\frac{2\pi r^2 H}{r + H}$$

   (b) Two spheres with radii r and R are placed so that the distance between their centers is d, where $d > r + R$. Where should a light be placed on the line joining the centers of the spheres in order to illuminate the largest total surface?

5. Suppose that the density of seawater, $\rho = \rho(z)$, varies with the depth z below the surface.
   (a) Show that the hydrostatic pressure is governed by the differential equation

$$\frac{dP}{dz} = \rho(z)g$$

   where g is the acceleration due to gravity. Let P_0 and ρ_0 be the pressure and density at $z = 0$. Express the pressure at depth z as an integral.
   (b) Suppose the density of seawater at depth z is given by $\rho = \rho_0 e^{z/H}$, where H is a positive constant. Find the total force, expressed as an integral, exerted on a vertical circular porthole of radius r whose center is located at a distance $L > r$ below the surface.

6. The figure shows a semicircle with radius 1, horizontal diameter PQ, and tangent lines at P and Q. At what height above the diameter should the horizontal line be placed so as to minimize the shaded area?

P   Q

FIGURE FOR PROBLEM 6

7. Let P be a pyramid with a square base of side $2b$ and suppose that S is a sphere with its center on the base of P and S is tangent to all eight edges of P. Find the height of P. Then find the volume of the intersection of S and P.

8. Consider a flat metal plate to be placed vertically under water with its top 2 m below the surface of the water. Determine a shape for the plate so that if the plate is divided into any number of horizontal strips of equal height, the hydrostatic force on each strip is the same.

9. A uniform disk with radius 1 m is to be cut by a line so that the center of mass of the smaller piece lies halfway along a radius. How close to the center of the disk should the cut be made? (Express your answer correct to two decimal places.)

10. A triangle with area 30 cm^2 is cut from a corner of a square with side 10 cm, as shown in the figure. If the centroid of the remaining region is 4 cm from the right side of the square, how far is it from the bottom of the square?

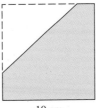

10 cm

11. In a famous 18th-century problem, known as *Buffon's needle problem*, a needle of length h is dropped onto a flat surface (for example, a table) on which parallel lines L units apart, $L \geqslant h$, have been drawn. The problem is to determine the probability that the needle will come to rest intersecting one of the lines. Assume that the lines run east-west, parallel to the x-axis in a rectangular coordinate system (as in the figure). Let y be the distance from the "southern" end of the needle to the nearest line to the north. (If the needle's southern end lies on a line, let $y = 0$. If the needle happens to lie east-west, let the "western" end be the "southern" end.) Let θ be the angle that the needle makes with a ray extending eastward from the "southern" end. Then $0 \leqslant y \leqslant L$ and $0 \leqslant \theta \leqslant \pi$. Note that the needle intersects one of the lines only when $y < h \sin \theta$. The total set of possibilities for the needle can be identified with the rectangular region $0 \leqslant y \leqslant L$, $0 \leqslant \theta \leqslant \pi$, and the proportion of times that the needle intersects a line is the ratio

$$\frac{\text{area under } y = h \sin \theta}{\text{area of rectangle}}$$

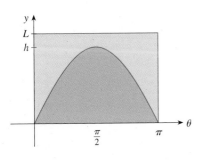

FIGURE FOR PROBLEM 11

This ratio is the probability that the needle intersects a line. Find the probability that the needle will intersect a line if $h = L$. What if $h = \frac{1}{2}L$?

12. If the needle in Problem 11 has length $h > L$, it's possible for the needle to intersect more than one line.
   (a) If $L = 4$, find the probability that a needle of length 7 will intersect at least one line. [*Hint:* Proceed as in Problem 11. Define y as before; then the total set of possibilities for the needle can be identified with the same rectangular region $0 \leqslant y \leqslant L$, $0 \leqslant \theta \leqslant \pi$. What portion of the rectangle corresponds to the needle intersecting a line?]
   (b) If $L = 4$, find the probability that a needle of length 7 will intersect *two* lines.
   (c) If $2L < h \leqslant 3L$, find a general formula for the probability that the needle intersects three lines.

13. Find the centroid of the region enclosed by the ellipse $x^2 + (x + y + 1)^2 = 1$.

Differential Equations

The relationship between populations of predators and prey (sharks and food fish, ladybugs and aphids, wolves and rabbits) is explored using pairs of differential equations in the last section of this chapter.

Perhaps the most important of all the applications of calculus is to differential equations. When physical scientists or social scientists use calculus, more often than not it is to analyze a differential equation that has arisen in the process of modeling some phenomenon that they are studying. Although it is often impossible to find an explicit formula for the solution of a differential equation, we will see that graphical and numerical approaches provide the needed information.

Modeling with Differential Equations

Now is a good time to read (or reread) the discussion of mathematical modeling on page 19.

In describing the process of modeling in Section 1.2, we talked about formulating a mathematical model of a real-world problem either through intuitive reasoning about the phenomenon or from a physical law based on evidence from experiments. The mathematical model often takes the form of a *differential equation,* that is, an equation that contains an unknown function and some of its derivatives. This is not surprising because in a real-world problem we often notice that changes occur and we want to predict future behavior on the basis of how current values change. Let's begin by examining several examples of how differential equations arise when we model physical phenomena.

Models of Population Growth

One model for the growth of a population is based on the assumption that the population grows at a rate proportional to the size of the population. That is a reasonable assumption for a population of bacteria or animals under ideal conditions (unlimited environment, adequate nutrition, absence of predators, immunity from disease).

Let's identify and name the variables in this model:

$$t = \text{time} \quad \text{(the independent variable)}$$

$$P = \text{the number of individuals in the population} \quad \text{(the dependent variable)}$$

The rate of growth of the population is the derivative dP/dt. So our assumption that the rate of growth of the population is proportional to the population size is written as the equation

$$\boxed{1} \qquad\qquad \frac{dP}{dt} = kP$$

where k is the proportionality constant. Equation 1 is our first model for population growth; it is a differential equation because it contains an unknown function P and its derivative dP/dt.

Having formulated a model, let's look at its consequences. If we rule out a population of 0, then $P(t) > 0$ for all t. So, if $k > 0$, then Equation 1 shows that $P'(t) > 0$ for all t. This means that the population is always increasing. In fact, as $P(t)$ increases, Equation 1 shows that dP/dt becomes larger. In other words, the growth rate increases as the population increases.

Let's try to think of a solution of Equation 1. This equation asks us to find a function whose derivative is a constant multiple of itself. We know from Chapter 6 that exponential functions have that property. In fact, if we let $P(t) = Ce^{kt}$, then

$$P'(t) = C(ke^{kt}) = k(Ce^{kt}) = kP(t)$$

Thus any exponential function of the form $P(t) = Ce^{kt}$ is a solution of Equation 1. In Section 9.4, we will see that there is no other solution.

Allowing C to vary through all the real numbers, we get the *family* of solutions $P(t) = Ce^{kt}$ whose graphs are shown in Figure 1. But populations have only positive values and so we are interested only in the solutions with $C > 0$. And we are probably con-

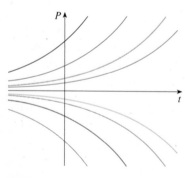

FIGURE 1
The family of solutions of $dP/dt = kP$

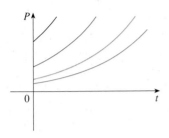

FIGURE 2

The family of solutions $P(t) = Ce^{kt}$
with $C > 0$ and $t \geqslant 0$

cerned only with values of t greater than the initial time $t = 0$. Figure 2 shows the physically meaningful solutions. Putting $t = 0$, we get $P(0) = Ce^{k(0)} = C$, so the constant C turns out to be the initial population, $P(0)$.

Equation 1 is appropriate for modeling population growth under ideal conditions, but we have to recognize that a more realistic model must reflect the fact that a given environment has limited resources. Many populations start by increasing in an exponential manner, but the population levels off when it approaches its *carrying capacity M* (or decreases toward M if it ever exceeds M). For a model to take into account both trends, we make two assumptions:

- $\dfrac{dP}{dt} \approx kP$ if P is small (Initially, the growth rate is proportional to P.)

- $\dfrac{dP}{dt} < 0$ if $P > M$ (P decreases if it ever exceeds M.)

A simple expression that incorporates both assumptions is given by the equation

$$\boxed{2} \qquad \frac{dP}{dt} = kP\left(1 - \frac{P}{M}\right)$$

Notice that if P is small compared with M, then P/M is close to 0 and so $dP/dt \approx kP$. If $P > M$, then $1 - P/M$ is negative and so $dP/dt < 0$.

Equation 2 is called the *logistic differential equation* and was proposed by the Dutch mathematical biologist Pierre-François Verhulst in the 1840s as a model for world population growth. We will develop techniques that enable us to find explicit solutions of the logistic equation in Section 9.4, but for now we can deduce qualitative characteristics of the solutions directly from Equation 2. We first observe that the constant functions $P(t) = 0$ and $P(t) = M$ are solutions because, in either case, one of the factors on the right side of Equation 2 is zero. (This certainly makes physical sense: If the population is ever either 0 or at the carrying capacity, it stays that way.) These two constant solutions are called *equilibrium solutions*.

If the initial population $P(0)$ lies between 0 and M, then the right side of Equation 2 is positive, so $dP/dt > 0$ and the population increases. But if the population exceeds the carrying capacity ($P > M$), then $1 - P/M$ is negative, so $dP/dt < 0$ and the population decreases. Notice that, in either case, if the population approaches the carrying capacity ($P \to M$), then $dP/dt \to 0$, which means the population levels off. So we expect that the solutions of the logistic differential equation have graphs that look something like the ones in Figure 3. Notice that the graphs move away from the equilibrium solution $P = 0$ and move toward the equilibrium solution $P = M$.

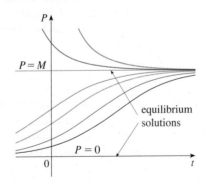

FIGURE 3

Solutions of the logistic equation

FIGURE 4

A Model for the Motion of a Spring

Let's now look at an example of a model from the physical sciences. We consider the motion of an object with mass m at the end of a vertical spring (as in Figure 4). In Section 5.4 we discussed Hooke's Law, which says that if the spring is stretched (or compressed) x units from its natural length, then it exerts a force that is proportional to x:

$$\text{restoring force} = -kx$$

where k is a positive constant (called the *spring constant*). If we ignore any external resisting forces (due to air resistance or friction) then, by Newton's Second Law (force equals mass times acceleration), we have

$$\boxed{3} \qquad\qquad m\frac{d^2x}{dt^2} = -kx$$

This is an example of what is called a *second-order differential equation* because it involves second derivatives. Let's see what we can guess about the form of the solution directly from the equation. We can rewrite Equation 3 in the form

$$\frac{d^2x}{dt^2} = -\frac{k}{m}x$$

which says that the second derivative of x is proportional to x but has the opposite sign. We know two functions with this property, the sine and cosine functions. In fact, it turns out that all solutions of Equation 3 can be written as combinations of certain sine and cosine functions (see Exercise 4). This is not surprising; we expect the spring to oscillate about its equilibrium position and so it is natural to think that trigonometric functions are involved.

General Differential Equations

In general, a **differential equation** is an equation that contains an unknown function and one or more of its derivatives. The **order** of a differential equation is the order of the highest derivative that occurs in the equation. Thus Equations 1 and 2 are first-order equations and Equation 3 is a second-order equation. In all three of those equations the independent variable is called t and represents time, but in general the independent variable doesn't have to represent time. For example, when we consider the differential equation

$$\boxed{4} \qquad\qquad y' = xy$$

it is understood that y is an unknown function of x.

A function f is called a **solution** of a differential equation if the equation is satisfied when $y = f(x)$ and its derivatives are substituted into the equation. Thus f is a solution of Equation 4 if

$$f'(x) = xf(x)$$

for all values of x in some interval.

When we are asked to *solve* a differential equation we are expected to find all possible solutions of the equation. We have already solved some particularly simple differential equations, namely, those of the form

$$y' = f(x)$$

For instance, we know that the general solution of the differential equation

$$y' = x^3$$

is given by

$$y = \frac{x^4}{4} + C$$

where C is an arbitrary constant.

But, in general, solving a differential equation is not an easy matter. There is no systematic technique that enables us to solve all differential equations. In Section 9.2, however, we will see how to draw rough graphs of solutions even when we have no explicit formula. We will also learn how to find numerical approximations to solutions.

V **EXAMPLE 1** Show that every member of the family of functions

$$y = \frac{1 + ce^t}{1 - ce^t}$$

is a solution of the differential equation $y' = \frac{1}{2}(y^2 - 1)$. ✗$y$

SOLUTION We use the Quotient Rule to differentiate the expression for y:

$$y' = \frac{(1 - ce^t)(ce^t) - (1 + ce^t)(-ce^t)}{(1 - ce^t)^2}$$

$$= \frac{ce^t - c^2e^{2t} + ce^t + c^2e^{2t}}{(1 - ce^t)^2} = \frac{2ce^t}{(1 - ce^t)^2}$$

The right side of the differential equation becomes

$$\frac{1}{2}(y^2 - 1) = \frac{1}{2}\left[\left(\frac{1 + ce^t}{1 - ce^t}\right)^2 - 1\right]$$

$$= \frac{1}{2}\left[\frac{(1 + ce^t)^2 - (1 - ce^t)^2}{(1 - ce^t)^2}\right]$$

$$= \frac{1}{2}\frac{4ce^t}{(1 - ce^t)^2} = \frac{2ce^t}{(1 - ce^t)^2}$$

Therefore, for every value of c, the given function is a solution of the differential equation.

Figure 5 shows graphs of seven members of the family in Example 1. The differential equation shows that if $y \approx \pm 1$, then $y' \approx 0$. That is borne out by the flatness of the graphs near $y = 1$ and $y = -1$.

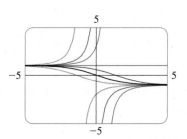

FIGURE 5

When applying differential equations, we are usually not as interested in finding a family of solutions (the *general solution*) as we are in finding a solution that satisfies some additional requirement. In many physical problems we need to find the particular solution that satisfies a condition of the form $y(t_0) = y_0$. This is called an **initial condition**, and the problem of finding a solution of the differential equation that satisfies the initial condition is called an **initial-value problem**.

Geometrically, when we impose an initial condition, we look at the family of solution curves and pick the one that passes through the point (t_0, y_0). Physically, this corresponds to measuring the state of a system at time t_0 and using the solution of the initial-value problem to predict the future behavior of the system.

V **EXAMPLE 2** Find a solution of the differential equation $y' = \frac{1}{2}(y^2 - 1)$ that satisfies the initial condition $y(0) = 2$.

SOLUTION Substituting the values $t = 0$ and $y = 2$ into the formula

$$y = \frac{1 + ce^t}{1 - ce^t}$$

from Example 1, we get

$$2 = \frac{1 + ce^0}{1 - ce^0} = \frac{1 + c}{1 - c}$$

Solving this equation for c, we get $2 - 2c = 1 + c$, which gives $c = \frac{1}{3}$. So the solution of the initial-value problem is

$$y = \frac{1 + \frac{1}{3}e^t}{1 - \frac{1}{3}e^t} = \frac{3 + e^t}{3 - e^t}$$

9.1 Exercises

Access end-of-section exercises and tools for visualizing calculus online at **www.webassign.net**   ~ENHANCED~ **Web**Assign

---

9.2 Direction Fields and Euler's Method

Unfortunately, it's impossible to solve most differential equations in the sense of obtaining an explicit formula for the solution. In this section we show that, despite the absence of an explicit solution, we can still learn a lot about the solution through a graphical approach (direction fields) or a numerical approach (Euler's method).

▩ Direction Fields

Suppose we are asked to sketch the graph of the solution of the initial-value problem

$$y' = x + y \qquad y(0) = 1$$

We don't know a formula for the solution, so how can we possibly sketch its graph? Let's think about what the differential equation means. The equation $y' = x + y$ tells us that the slope at any point (x, y) on the graph (called the *solution curve*) is equal to the sum of the x- and y-coordinates of the point (see Figure 1). In particular, because the curve passes through the point $(0, 1)$, its slope there must be $0 + 1 = 1$. So a small portion of the solution curve near the point $(0, 1)$ looks like a short line segment through $(0, 1)$ with slope 1. (See Figure 2.)

FIGURE 1
A solution of $y' = x + y$

FIGURE 2
Beginning of the solution curve through $(0, 1)$

As a guide to sketching the rest of the curve, let's draw short line segments at a number of points (x, y) with slope $x + y$. The result is called a *direction field* and is shown in Figure 3. For instance, the line segment at the point $(1, 2)$ has slope $1 + 2 = 3$. The direction field allows us to visualize the general shape of the solution curves by indicating the direction in which the curves proceed at each point.

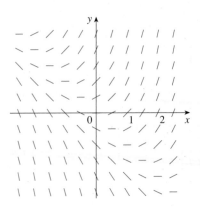

FIGURE 3
Direction field for $y' = x + y$

FIGURE 4
The solution curve through $(0, 1)$

Now we can sketch the solution curve through the point $(0, 1)$ by following the direction field as in Figure 4. Notice that we have drawn the curve so that it is parallel to nearby line segments.

In general, suppose we have a first-order differential equation of the form

$$y' = F(x, y)$$

where $F(x, y)$ is some expression in x and y. The differential equation says that the slope of a solution curve at a point (x, y) on the curve is $F(x, y)$. If we draw short line segments with slope $F(x, y)$ at several points (x, y), the result is called a **direction field** (or **slope field**). These line segments indicate the direction in which a solution curve is heading, so the direction field helps us visualize the general shape of these curves.

V EXAMPLE 1
(a) Sketch the direction field for the differential equation $y' = x^2 + y^2 - 1$.
(b) Use part (a) to sketch the solution curve that passes through the origin.

SOLUTION
(a) We start by computing the slope at several points in the following chart:

x	-2	-1	0	1	2	-2	-1	0	1	2	...
y	0	0	0	0	0	1	1	1	1	1	...
$y' = x^2 + y^2 - 1$	3	0	-1	0	3	4	1	0	1	4	...

Now we draw short line segments with these slopes at these points. The result is the direction field shown in Figure 5.

(b) We start at the origin and move to the right in the direction of the line segment (which has slope -1). We continue to draw the solution curve so that it moves parallel to the nearby line segments. The resulting solution curve is shown in Figure 6. Returning to the origin, we draw the solution curve to the left as well.

FIGURE 5

FIGURE 6

The more line segments we draw in a direction field, the clearer the picture becomes. Of course, it's tedious to compute slopes and draw line segments for a huge number of points by hand, but computers are well suited for this task. Figure 7 shows a more detailed, computer-drawn direction field for the differential equation in Example 1. It enables us to draw, with reasonable accuracy, the solution curves shown in Figure 8 with y-intercepts $-2, -1, 0, 1,$ and 2.

FIGURE 7

FIGURE 8

FIGURE 9

TEC Module 9.2A shows direction fields and solution curves for a variety of differential equations.

Now let's see how direction fields give insight into physical situations. The simple electric circuit shown in Figure 9 contains an electromotive force (usually a battery or generator) that produces a voltage of $E(t)$ volts (V) and a current of $I(t)$ amperes (A) at time t. The circuit also contains a resistor with a resistance of R ohms (Ω) and an inductor with an inductance of L henries (H).

Ohm's Law gives the drop in voltage due to the resistor as RI. The voltage drop due to the inductor is $L(dI/dt)$. One of Kirchhoff's laws says that the sum of the voltage drops is equal to the supplied voltage $E(t)$. Thus we have

$$\boxed{1} \qquad L\frac{dI}{dt} + RI = E(t)$$

which is a first-order differential equation that models the current I at time t.

V **EXAMPLE 2** Suppose that in the simple circuit of Figure 9 the resistance is 12 Ω, the inductance is 4 H, and a battery gives a constant voltage of 60 V.
(a) Draw a direction field for Equation 1 with these values.
(b) What can you say about the limiting value of the current?
(c) Identify any equilibrium solutions.
(d) If the switch is closed when $t = 0$ so the current starts with $I(0) = 0$, use the direction field to sketch the solution curve.

SOLUTION
(a) If we put $L = 4$, $R = 12$, and $E(t) = 60$ in Equation 1, we get

$$4\frac{dI}{dt} + 12I = 60 \qquad \text{or} \qquad \frac{dI}{dt} = 15 - 3I$$

The direction field for this differential equation is shown in Figure 10.

FIGURE 10

(b) It appears from the direction field that all solutions approach the value 5 A, that is,

$$\lim_{t \to \infty} I(t) = 5$$

(c) It appears that the constant function $I(t) = 5$ is an equilibrium solution. Indeed, we can verify this directly from the differential equation $dI/dt = 15 - 3I$. If $I(t) = 5$, then the left side is $dI/dt = 0$ and the right side is $15 - 3(5) = 0$.

(d) We use the direction field to sketch the solution curve that passes through $(0, 0)$, as shown in red in Figure 11.

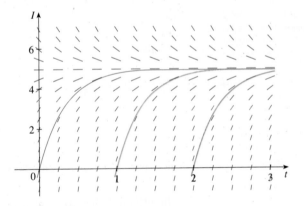

FIGURE 11

Notice from Figure 10 that the line segments along any horizontal line are parallel. That is because the independent variable t does not occur on the right side of the equation $I' = 15 - 3I$. In general, a differential equation of the form

$$y' = f(y)$$

in which the independent variable is missing from the right side, is called **autonomous**. For such an equation, the slopes corresponding to two different points with the same y-coordinate must be equal. This means that if we know one solution to an autonomous differential equation, then we can obtain infinitely many others just by shifting the graph of the known solution to the right or left. In Figure 11 we have shown the solutions that result from shifting the solution curve of Example 2 one and two time units (namely, seconds) to the right. They correspond to closing the switch when $t = 1$ or $t = 2$.

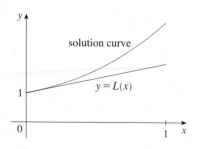

FIGURE 12

First Euler approximation

Euler's Method

The basic idea behind direction fields can be used to find numerical approximations to solutions of differential equations. We illustrate the method on the initial-value problem that we used to introduce direction fields:

$$y' = x + y \qquad y(0) = 1$$

The differential equation tells us that $y'(0) = 0 + 1 = 1$, so the solution curve has slope 1 at the point $(0, 1)$. As a first approximation to the solution we could use the linear approximation $L(x) = x + 1$. In other words, we could use the tangent line at $(0, 1)$ as a rough approximation to the solution curve (see Figure 12).

Euler's idea was to improve on this approximation by proceeding only a short distance along this tangent line and then making a midcourse correction by changing direction as indicated by the direction field. Figure 13 shows what happens if we start out along the tangent line but stop when $x = 0.5$. (This horizontal distance traveled is called the *step size*.) Since $L(0.5) = 1.5$, we have $y(0.5) \approx 1.5$ and we take $(0.5, 1.5)$ as the starting point for a new line segment. The differential equation tells us that $y'(0.5) = 0.5 + 1.5 = 2$, so we use the linear function

$$y = 1.5 + 2(x - 0.5) = 2x + 0.5$$

as an approximation to the solution for $x > 0.5$ (the green segment in Figure 13). If we decrease the step size from 0.5 to 0.25, we get the better Euler approximation shown in Figure 14.

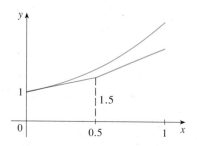

FIGURE 13

Euler approximation with step size 0.5

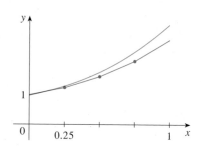

FIGURE 14

Euler approximation with step size 0.25

In general, Euler's method says to start at the point given by the initial value and proceed in the direction indicated by the direction field. Stop after a short time, look at the slope at the new location, and proceed in that direction. Keep stopping and changing direction according to the direction field. Euler's method does not produce the exact solution to an initial-value problem—it gives approximations. But by decreasing the step size (and therefore increasing the number of midcourse corrections), we obtain successively better approximations to the exact solution. (Compare Figures 12, 13, and 14.)

For the general first-order initial-value problem $y' = F(x, y)$, $y(x_0) = y_0$, our aim is to find approximate values for the solution at equally spaced numbers x_0, $x_1 = x_0 + h$, $x_2 = x_1 + h, \ldots$, where h is the step size. The differential equation tells us that the slope at (x_0, y_0) is $y' = F(x_0, y_0)$, so Figure 15 shows that the approximate value of the solution when $x = x_1$ is

$$y_1 = y_0 + hF(x_0, y_0)$$

Similarly,
$$y_2 = y_1 + hF(x_1, y_1)$$

In general,
$$y_n = y_{n-1} + hF(x_{n-1}, y_{n-1})$$

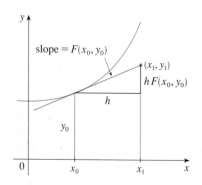

FIGURE 15

Euler's Method Approximate values for the solution of the initial-value problem $y' = F(x, y)$, $y(x_0) = y_0$, with step size h, at $x_n = x_{n-1} + h$, are

$$y_n = y_{n-1} + hF(x_{n-1}, y_{n-1}) \qquad n = 1, 2, 3, \ldots$$

EXAMPLE 3 Use Euler's method with step size 0.1 to construct a table of approximate values for the solution of the initial-value problem

$$y' = x + y \qquad y(0) = 1$$

SOLUTION We are given that $h = 0.1$, $x_0 = 0$, $y_0 = 1$, and $F(x, y) = x + y$. So we have

$$y_1 = y_0 + hF(x_0, y_0) = 1 + 0.1(0 + 1) = 1.1$$

$$y_2 = y_1 + hF(x_1, y_1) = 1.1 + 0.1(0.1 + 1.1) = 1.22$$

$$y_3 = y_2 + hF(x_2, y_2) = 1.22 + 0.1(0.2 + 1.22) = 1.362$$

This means that if $y(x)$ is the exact solution, then $y(0.3) \approx 1.362$.
Proceeding with similar calculations, we get the values in the table:

TEC Module 9.2B shows how Euler's method works numerically and visually for a variety of differential equations and step sizes.

n	x_n	y_n	n	x_n	y_n
1	0.1	1.100000	6	0.6	1.943122
2	0.2	1.220000	7	0.7	2.197434
3	0.3	1.362000	8	0.8	2.487178
4	0.4	1.528200	9	0.9	2.815895
5	0.5	1.721020	10	1.0	3.187485

For a more accurate table of values in Example 3 we could decrease the step size. But for a large number of small steps the amount of computation is considerable and so we need to program a calculator or computer to carry out these calculations. The following table shows the results of applying Euler's method with decreasing step size to the initial-value problem of Example 3.

Computer software packages that produce numerical approximations to solutions of differential equations use methods that are refinements of Euler's method. Although Euler's method is simple and not as accurate, it is the basic idea on which the more accurate methods are based.

Step size	Euler estimate of $y(0.5)$	Euler estimate of $y(1)$
0.500	1.500000	2.500000
0.250	1.625000	2.882813
0.100	1.721020	3.187485
0.050	1.757789	3.306595
0.020	1.781212	3.383176
0.010	1.789264	3.409628
0.005	1.793337	3.423034
0.001	1.796619	3.433848

Notice that the Euler estimates in the table seem to be approaching limits, namely, the true values of $y(0.5)$ and $y(1)$. Figure 16 (on page 468) shows graphs of the Euler approximations with step sizes 0.5, 0.25, 0.1, 0.05, 0.02, 0.01, and 0.005. They are approaching the exact solution curve as the step size h approaches 0.

Notice that the Euler estimates in the table seem to be approaching limits, namely, the true values of $y(0.5)$ and $y(1)$. Figure 16 shows graphs of the Euler approximations with step sizes 0.5, 0.25, 0.1, 0.05, 0.02, 0.01, and 0.005. They are approaching the exact solution curve as the step size h approaches 0.

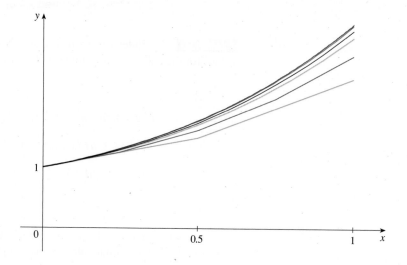

FIGURE 16
Euler approximations
approaching the exact solution

V **EXAMPLE 4** In Example 2 we discussed a simple electric circuit with resistance 12 Ω, inductance 4 H, and a battery with voltage 60 V. If the switch is closed when $t = 0$, we modeled the current I at time t by the initial-value problem

$$\frac{dI}{dt} = 15 - 3I \qquad I(0) = 0$$

Estimate the current in the circuit half a second after the switch is closed.

SOLUTION We use Euler's method with $F(t, I) = 15 - 3I$, $t_0 = 0$, $I_0 = 0$, and step size $h = 0.1$ second:

$$I_1 = 0 + 0.1(15 - 3 \cdot 0) = 1.5$$

$$I_2 = 1.5 + 0.1(15 - 3 \cdot 1.5) = 2.55$$

$$I_3 = 2.55 + 0.1(15 - 3 \cdot 2.55) = 3.285$$

$$I_4 = 3.285 + 0.1(15 - 3 \cdot 3.285) = 3.7995$$

$$I_5 = 3.7995 + 0.1(15 - 3 \cdot 3.7995) = 4.15965$$

So the current after 0.5 s is

$$I(0.5) \approx 4.16 \text{ A}$$

9.2 **Exercises**

Access end-of-section exercises and tools for visualizing calculus online at **www.webassign.net**   WebAssign

9.3 **Separable Equations**

We have looked at first-order differential equations from a geometric point of view (direction fields) and from a numerical point of view (Euler's method). What about the symbolic point of view? It would be nice to have an explicit formula for a solution of a differential equation. Unfortunately, that is not always possible. But in this section we examine a certain type of differential equation that *can* be solved explicitly.

A **separable equation** is a first-order differential equation in which the expression for dy/dx can be factored as a function of x times a function of y. In other words, it can be written in the form

$$\frac{dy}{dx} = g(x)f(y)$$

The name *separable* comes from the fact that the expression on the right side can be "separated" into a function of x and a function of y. Equivalently, if $f(y) \neq 0$, we could write

$$\boxed{1} \qquad \frac{dy}{dx} = \frac{g(x)}{h(y)}$$

where $h(y) = 1/f(y)$. To solve this equation we rewrite it in the differential form

$$h(y)\, dy = g(x)\, dx$$

so that all y's are on one side of the equation and all x's are on the other side. Then we integrate both sides of the equation:

The technique for solving separable differential equations was first used by James Bernoulli (in 1690) in solving a problem about pendulums and by Leibniz (in a letter to Huygens in 1691). John Bernoulli explained the general method in a paper published in 1694.

$$\boxed{2} \qquad \int h(y)\, dy = \int g(x)\, dx$$

Equation 2 defines y implicitly as a function of x. In some cases we may be able to solve for y in terms of x.

We use the Chain Rule to justify this procedure: If h and g satisfy $\boxed{2}$, then

$$\frac{d}{dx}\left(\int h(y)\, dy\right) = \frac{d}{dx}\left(\int g(x)\, dx\right)$$

so

$$\frac{d}{dy}\left(\int h(y)\, dy\right)\frac{dy}{dx} = g(x)$$

and

$$h(y)\frac{dy}{dx} = g(x)$$

Thus Equation 1 is satisfied.

EXAMPLE 1

(a) Solve the differential equation $\dfrac{dy}{dx} = \dfrac{x^2}{y^2}$.

(b) Find the solution of this equation that satisfies the initial condition $y(0) = 2$.

SOLUTION

(a) We write the equation in terms of differentials and integrate both sides:

$$y^2\,dy = x^2\,dx$$

$$\int y^2\,dy = \int x^2\,dx$$

$$\tfrac{1}{3}y^3 = \tfrac{1}{3}x^3 + C$$

Figure 1 shows graphs of several members of the family of solutions of the differential equation in Example 1. The solution of the initial-value problem in part (b) is shown in red.

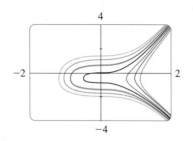

FIGURE 1

where C is an arbitrary constant. (We could have used a constant C_1 on the left side and another constant C_2 on the right side. But then we could combine these constants by writing $C = C_2 - C_1$.)

Solving for y, we get

$$y = \sqrt[3]{x^3 + 3C}$$

We could leave the solution like this or we could write it in the form

$$y = \sqrt[3]{x^3 + K}$$

where $K = 3C$. (Since C is an arbitrary constant, so is K.)

(b) If we put $x = 0$ in the general solution in part (a), we get $y(0) = \sqrt[3]{K}$. To satisfy the initial condition $y(0) = 2$, we must have $\sqrt[3]{K} = 2$ and so $K = 8$. Thus the solution of the initial-value problem is

$$y = \sqrt[3]{x^3 + 8}$$

Some computer algebra systems can plot curves defined by implicit equations. Figure 2 shows the graphs of several members of the family of solutions of the differential equation in Example 2. As we look at the curves from left to right, the values of C are 3, 2, 1, 0, −1, −2, and −3.

FIGURE 2

V **EXAMPLE 2** Solve the differential equation $\dfrac{dy}{dx} = \dfrac{6x^2}{2y + \cos y}$.

SOLUTION Writing the equation in differential form and integrating both sides, we have

$$(2y + \cos y)\,dy = 6x^2\,dx$$

$$\int (2y + \cos y)\,dy = \int 6x^2\,dx$$

$$\boxed{3} \qquad y^2 + \sin y = 2x^3 + C$$

where C is a constant. Equation 3 gives the general solution implicitly. In this case it's impossible to solve the equation to express y explicitly as a function of x.

EXAMPLE 3 Solve the equation $y' = x^2 y$.

SOLUTION First we rewrite the equation using Leibniz notation:

$$\frac{dy}{dx} = x^2 y$$

If a solution y is a function that satisfies $y(x) \neq 0$ for some x, it follows from a uniqueness theorem for solutions of differential equations that $y(x) \neq 0$ for all x.

If $y \neq 0$, we can rewrite it in differential notation and integrate:

$$\frac{dy}{y} = x^2 \, dx \qquad y \neq 0$$

$$\int \frac{dy}{y} = \int x^2 \, dx$$

$$\ln |y| = \frac{x^3}{3} + C$$

This equation defines y implicitly as a function of x. But in this case we can solve explicitly for y as follows:

$$|y| = e^{\ln |y|} = e^{(x^3/3)+C} = e^C e^{x^3/3}$$

so

$$y = \pm e^C e^{x^3/3}$$

We can easily verify that the function $y = 0$ is also a solution of the given differential equation. So we can write the general solution in the form

$$y = A e^{x^3/3}$$

where A is an arbitrary constant ($A = e^C$, or $A = -e^C$, or $A = 0$).

Figure 3 shows a direction field for the differential equation in Example 3. Compare it with Figure 4, in which we use the equation $y = A e^{x^3/3}$ to graph solutions for several values of A. If you use the direction field to sketch solution curves with y-intercepts 5, 2, 1, -1, and -2, they will resemble the curves in Figure 4.

FIGURE 3

FIGURE 4

FIGURE 5

▼ **EXAMPLE 4** In Section 9.2 we modeled the current $I(t)$ in the electric circuit shown in Figure 5 by the differential equation

$$L \frac{dI}{dt} + RI = E(t)$$

Find an expression for the current in a circuit where the resistance is 12 Ω, the inductance is 4 H, a battery gives a constant voltage of 60 V, and the switch is turned on when $t = 0$. What is the limiting value of the current?

SOLUTION With $L = 4$, $R = 12$, and $E(t) = 60$, the equation becomes

$$4 \frac{dI}{dt} + 12I = 60 \qquad \text{or} \qquad \frac{dI}{dt} = 15 - 3I$$

and the initial-value problem is

$$\frac{dI}{dt} = 15 - 3I \qquad I(0) = 0$$

We recognize this equation as being separable, and we solve it as follows:

$$\int \frac{dI}{15 - 3I} = \int dt \qquad (15 - 3I \neq 0)$$

$$-\tfrac{1}{3} \ln |15 - 3I| = t + C$$

$$|15 - 3I| = e^{-3(t+C)}$$

$$15 - 3I = \pm e^{-3C} e^{-3t} = Ae^{-3t}$$

$$I = 5 - \tfrac{1}{3} Ae^{-3t}$$

Figure 6 shows how the solution in Example 4 (the current) approaches its limiting value. Comparison with Figure 11 in Section 9.2 shows that we were able to draw a fairly accurate solution curve from the direction field.

FIGURE 6

Since $I(0) = 0$, we have $5 - \tfrac{1}{3}A = 0$, so $A = 15$ and the solution is

$$I(t) = 5 - 5e^{-3t}$$

The limiting current, in amperes, is

$$\lim_{t \to \infty} I(t) = \lim_{t \to \infty} (5 - 5e^{-3t}) = 5 - 5 \lim_{t \to \infty} e^{-3t} = 5 - 0 = 5$$

Orthogonal Trajectories

An **orthogonal trajectory** of a family of curves is a curve that intersects each curve of the family orthogonally, that is, at right angles (see Figure 7). For instance, each member of the family $y = mx$ of straight lines through the origin is an orthogonal trajectory of the family $x^2 + y^2 = r^2$ of concentric circles with center the origin (see Figure 8). We say that the two families are orthogonal trajectories of each other.

orthogonal
trajectory

FIGURE 7

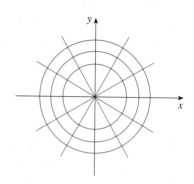

FIGURE 8

V EXAMPLE 5 Find the orthogonal trajectories of the family of curves $x = ky^2$, where k is an arbitrary constant.

SOLUTION The curves $x = ky^2$ form a family of parabolas whose axis of symmetry is the x-axis. The first step is to find a single differential equation that is satisfied by all members of the family. If we differentiate $x = ky^2$, we get

$$1 = 2ky \frac{dy}{dx} \quad \text{or} \quad \frac{dy}{dx} = \frac{1}{2ky}$$

This differential equation depends on k, but we need an equation that is valid for all values of k simultaneously. To eliminate k we note that, from the equation of the given general parabola $x = ky^2$, we have $k = x/y^2$ and so the differential equation can be written as

$$\frac{dy}{dx} = \frac{1}{2ky} = \frac{1}{2 \dfrac{x}{y^2} y}$$

or

$$\frac{dy}{dx} = \frac{y}{2x}$$

This means that the slope of the tangent line at any point (x, y) on one of the parabolas is $y' = y/(2x)$. On an orthogonal trajectory the slope of the tangent line must be the negative reciprocal of this slope. Therefore the orthogonal trajectories must satisfy the differential equation

$$\frac{dy}{dx} = -\frac{2x}{y}$$

This differential equation is separable, and we solve it as follows:

$$\int y \, dy = -\int 2x \, dx$$

$$\frac{y^2}{2} = -x^2 + C$$

$$\boxed{4} \qquad x^2 + \frac{y^2}{2} = C$$

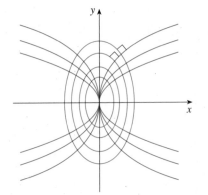

FIGURE 9

where C is an arbitrary positive constant. Thus the orthogonal trajectories are the family of ellipses given by Equation 4 and sketched in Figure 9. ▬▬

Orthogonal trajectories occur in various branches of physics. For example, in an electrostatic field the lines of force are orthogonal to the lines of constant potential. Also, the streamlines in aerodynamics are orthogonal trajectories of the velocity-equipotential curves.

▬ Mixing Problems

A typical mixing problem involves a tank of fixed capacity filled with a thoroughly mixed solution of some substance, such as salt. A solution of a given concentration enters the tank at a fixed rate and the mixture, thoroughly stirred, leaves at a fixed rate, which may differ from the entering rate. If $y(t)$ denotes the amount of substance in the tank at time t, then $y'(t)$ is the rate at which the substance is being added minus the rate at which it is being removed. The mathematical description of this situation often leads to a first-order separable differential equation. We can use the same type of reasoning to model a variety of phenomena: chemical reactions, discharge of pollutants into a lake, injection of a drug into the bloodstream.

EXAMPLE 6 A tank contains 20 kg of salt dissolved in 5000 L of water. Brine that contains 0.03 kg of salt per liter of water enters the tank at a rate of 25 L/min. The solution is kept thoroughly mixed and drains from the tank at the same rate. How much salt remains in the tank after half an hour?

SOLUTION Let $y(t)$ be the amount of salt (in kilograms) after t minutes. We are given that $y(0) = 20$ and we want to find $y(30)$. We do this by finding a differential equation satisfied by $y(t)$. Note that dy/dt is the rate of change of the amount of salt, so

$$\boxed{5} \qquad \frac{dy}{dt} = (\text{rate in}) - (\text{rate out})$$

where (rate in) is the rate at which salt enters the tank and (rate out) is the rate at which salt leaves the tank. We have

$$\text{rate in} = \left(0.03 \, \frac{\text{kg}}{\text{L}}\right)\left(25 \, \frac{\text{L}}{\text{min}}\right) = 0.75 \, \frac{\text{kg}}{\text{min}}$$

The tank always contains 5000 L of liquid, so the concentration at time t is $y(t)/5000$ (measured in kilograms per liter). Since the brine flows out at a rate of 25 L/min, we have

$$\text{rate out} = \left(\frac{y(t)}{5000} \, \frac{\text{kg}}{\text{L}}\right)\left(25 \, \frac{\text{L}}{\text{min}}\right) = \frac{y(t)}{200} \, \frac{\text{kg}}{\text{min}}$$

Thus, from Equation 5, we get

$$\frac{dy}{dt} = 0.75 - \frac{y(t)}{200} = \frac{150 - y(t)}{200}$$

Solving this separable differential equation, we obtain

$$\int \frac{dy}{150 - y} = \int \frac{dt}{200}$$

$$-\ln|150 - y| = \frac{t}{200} + C$$

Since $y(0) = 20$, we have $-\ln 130 = C$, so

$$-\ln|150 - y| = \frac{t}{200} - \ln 130$$

Therefore

$$|150 - y| = 130e^{-t/200}$$

Since $y(t)$ is continuous and $y(0) = 20$ and the right side is never 0, we deduce that $150 - y(t)$ is always positive. Thus $|150 - y| = 150 - y$ and so

$$y(t) = 150 - 130e^{-t/200}$$

The amount of salt after 30 min is

$$y(30) = 150 - 130e^{-30/200} \approx 38.1 \text{ kg}$$

Figure 10 shows the graph of the function $y(t)$ of Example 6. Notice that, as time goes by, the amount of salt approaches 150 kg.

FIGURE 10

9.3   Exercises

Access end-of-section exercises and tools for visualizing calculus online at **www.webassign.net**

APPLIED PROJECT   HOW FAST DOES A TANK DRAIN?

If water (or other liquid) drains from a tank, we expect that the flow will be greatest at first (when the water depth is greatest) and will gradually decrease as the water level decreases. But we need a more precise mathematical description of how the flow decreases in order to answer the kinds of questions that engineers ask: How long does it take for a tank to drain completely? How much water should a tank hold in order to guarantee a certain minimum water pressure for a sprinkler system?

Let $h(t)$ and $V(t)$ be the height and volume of water in a tank at time t. If water drains through a hole with area a at the bottom of the tank, then Torricelli's Law says that

$$\boxed{1} \qquad \frac{dV}{dt} = -a\sqrt{2gh}$$

where g is the acceleration due to gravity. So the rate at which water flows from the tank is proportional to the square root of the water height.

1. (a) Suppose the tank is cylindrical with height 6 ft and radius 2 ft and the hole is circular with radius 1 inch. If we take $g = 32$ ft/s^2, show that h satisfies the differential equation

$$\frac{dh}{dt} = -\frac{1}{72}\sqrt{h}$$

   (b) Solve this equation to find the height of the water at time t, assuming the tank is full at time $t = 0$.
   (c) How long will it take for the water to drain completely?

2. Because of the rotation and viscosity of the liquid, the theoretical model given by Equation 1 isn't quite accurate. Instead, the model

$$\boxed{2} \qquad \frac{dh}{dt} = k\sqrt{h}$$

is often used and the constant k (which depends on the physical properties of the liquid) is determined from data concerning the draining of the tank.
   (a) Suppose that a hole is drilled in the side of a cylindrical bottle and the height h of the water (above the hole) decreases from 10 cm to 3 cm in 68 seconds. Use Equation 2 to find an expression for $h(t)$. Evaluate $h(t)$ for $t = 10, 20, 30, 40, 50, 60$.
   (b) Drill a 4-mm hole near the bottom of the cylindrical part of a two-liter plastic soft-drink bottle. Attach a strip of masking tape marked in centimeters from 0 to 10, with 0 corresponding to the top of the hole. With one finger over the hole, fill the bottle with water to the 10-cm mark. Then take your finger off the hole and record the values of $h(t)$ for $t = 10, 20, 30, 40, 50, 60$ seconds. (You will probably find that it takes 68 seconds for the level to decrease to $h = 3$ cm.) Compare your data with the values of $h(t)$ from part (a). How well did the model predict the actual values?

3. In many parts of the world, the water for sprinkler systems in large hotels and hospitals is supplied by gravity from cylindrical tanks on or near the roofs of the buildings. Suppose such a tank has radius 10 ft and the diameter of the outlet is 2.5 inches. An engineer has to guarantee that the water pressure will be at least 2160 lb/ft^2 for a period of 10 minutes.

Problem 2(b) is best done as a classroom demonstration or as a group project with three students in each group: a timekeeper to call out seconds, a bottle keeper to estimate the height every 10 seconds, and a record keeper to record these values.

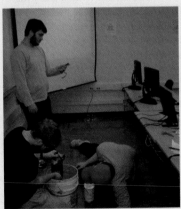

© Richard Le Borne, Dept. Mathematics, Tennessee Technological University

(When a fire happens, the electrical system might fail and it could take up to 10 minutes for the emergency generator and fire pump to be activated.) What height should the engineer specify for the tank in order to make such a guarantee? (Use the fact that the water pressure at a depth of d feet is $P = 62.5d$. See Section 8.3.)

4. Not all water tanks are shaped like cylinders. Suppose a tank has cross-sectional area $A(h)$ at height h. Then the volume of water up to height h is $V = \int_0^h A(u)\, du$ and so the Fundamental Theorem of Calculus gives $dV/dh = A(h)$. It follows that

$$\frac{dV}{dt} = \frac{dV}{dh}\frac{dh}{dt} = A(h)\frac{dh}{dt}$$

and so Torricelli's Law becomes

$$A(h)\frac{dh}{dt} = -a\sqrt{2gh}$$

(a) Suppose the tank has the shape of a sphere with radius 2 m and is initially half full of water. If the radius of the circular hole is 1 cm and we take $g = 10 \text{ m/s}^2$, show that h satisfies the differential equation

$$(4h - h^2)\frac{dh}{dt} = -0.0001\sqrt{20h}$$

(b) How long will it take for the water to drain completely?

APPLIED PROJECT

WHICH IS FASTER, GOING UP OR COMING DOWN?

Suppose you throw a ball into the air. Do you think it takes longer to reach its maximum height or to fall back to earth from its maximum height? We will solve the problem in this project but, before getting started, think about that situation and make a guess based on your physical intuition.

1. A ball with mass m is projected vertically upward from the earth's surface with a positive initial velocity v_0. We assume the forces acting on the ball are the force of gravity and a retarding force of air resistance with direction opposite to the direction of motion and with magnitude $p\,|v(t)|$, where p is a positive constant and $v(t)$ is the velocity of the ball at time t. In both the ascent and the descent, the total force acting on the ball is $-pv - mg$. [During ascent, $v(t)$ is positive and the resistance acts downward; during descent, $v(t)$ is negative and the resistance acts upward.] So, by Newton's Second Law, the equation of motion is

$$mv' = -pv - mg$$

Solve this differential equation to show that the velocity is

$$v(t) = \left(v_0 + \frac{mg}{p}\right)e^{-pt/m} - \frac{mg}{p}$$

2. Show that the height of the ball, until it hits the ground, is

$$y(t) = \left(v_0 + \frac{mg}{p}\right)\frac{m}{p}\left(1 - e^{-pt/m}\right) - \frac{mgt}{p}$$

In modeling force due to air resistance, various functions have been used, depending on the physical characteristics and speed of the ball. Here we use a linear model, $-pv$, but a quadratic model ($-pv^2$ on the way up and pv^2 on the way down) is another possibility for higher speeds (see Exercise 50 in Section 9.3). For a golf ball, experiments have shown that a good model is $-pv^{1.3}$ going up and $p\,|v|^{1.3}$ coming down. But no matter which force function $-f(v)$ is used [where $f(v) > 0$ for $v > 0$ and $f(v) < 0$ for $v < 0$], the answer to the question remains the same. See F. Brauer, "What Goes Up Must Come Down, Eventually," *Amer. Math. Monthly* 108 (2001), pp. 437–440.

⊞ Graphing calculator or computer required

3. Let t_1 be the time that the ball takes to reach its maximum height. Show that

$$t_1 = \frac{m}{p} \ln\left(\frac{mg + pv_0}{mg}\right)$$

Find this time for a ball with mass 1 kg and initial velocity 20 m/s. Assume the air resistance is $\frac{1}{10}$ of the speed.

4. Let t_2 be the time at which the ball falls back to earth. For the particular ball in Problem 3, estimate t_2 by using a graph of the height function $y(t)$. Which is faster, going up or coming down?

5. In general, it's not easy to find t_2 because it's impossible to solve the equation $y(t) = 0$ explicitly. We can, however, use an indirect method to determine whether ascent or descent is faster: we determine whether $y(2t_1)$ is positive or negative. Show that

$$y(2t_1) = \frac{m^2 g}{p^2}\left(x - \frac{1}{x} - 2 \ln x\right)$$

where $x = e^{pt_1/m}$. Then show that $x > 1$ and the function

$$f(x) = x - \frac{1}{x} - 2 \ln x$$

is increasing for $x > 1$. Use this result to decide whether $y(2t_1)$ is positive or negative. What can you conclude? Is ascent or descent faster?

9.4 **Models for Population Growth**

In this section we investigate differential equations that are used to model population growth: the law of natural growth, the logistic equation, and several others.

The Law of Natural Growth

One of the models for population growth that we considered in Section 9.1 was based on the assumption that the population grows at a rate proportional to the size of the population:

$$\frac{dP}{dt} = kP$$

Is that a reasonable assumption? Suppose we have a population (of bacteria, for instance) with size $P = 1000$ and at a certain time it is growing at a rate of $P' = 300$ bacteria per hour. Now let's take another 1000 bacteria of the same type and put them with the first population. Each half of the combined population was previously growing at a rate of 300 bacteria per hour. We would expect the total population of 2000 to increase at a rate of 600 bacteria per hour initially (provided there's enough room and nutrition). So if we double the size, we double the growth rate. It seems reasonable that the growth rate should be proportional to the size.

In general, if $P(t)$ is the value of a quantity y at time t and if the rate of change of P with respect to t is proportional to its size $P(t)$ at any time, then

1
$$\frac{dP}{dt} = kP$$

where k is a constant. Equation 1 is sometimes called the **law of natural growth**. If k is positive, then the population increases; if k is negative, it decreases.

Because Equation 1 is a separable differential equation, we can solve it by the methods of Section 9.3:

$$\int \frac{dP}{P} = \int k \, dt$$

$$\ln|P| = kt + C$$

$$|P| = e^{kt+C} = e^C e^{kt}$$

$$P = Ae^{kt}$$

where $A \; (= \pm e^C \text{ or } 0)$ is an arbitrary constant. To see the significance of the constant A, we observe that

$$P(0) = Ae^{k \cdot 0} = A$$

Therefore A is the initial value of the function.

Examples and exercises on the use of $\boxed{2}$ are given in Section 6.5.

$\boxed{2}$ The solution of the initial-value problem

$$\frac{dP}{dt} = kP \qquad P(0) = P_0$$

is $$P(t) = P_0 e^{kt}$$

Another way of writing Equation 1 is

$$\frac{1}{P} \frac{dP}{dt} = k$$

which says that the **relative growth rate** (the growth rate divided by the population size) is constant. Then $\boxed{2}$ says that a population with constant relative growth rate must grow exponentially.

We can account for emigration (or "harvesting") from a population by modifying Equation 1: If the rate of emigration is a constant m, then the rate of change of the population is modeled by the differential equation

$\boxed{3}$
$$\frac{dP}{dt} = kP - m$$

See Exercise 15 for the solution and consequences of Equation 3.

■ The Logistic Model

As we discussed in Section 9.1, a population often increases exponentially in its early stages but levels off eventually and approaches its carrying capacity because of limited resources. If $P(t)$ is the size of the population at time t, we assume that

$$\frac{dP}{dt} \approx kP \qquad \text{if } P \text{ is small}$$

This says that the growth rate is initially close to being proportional to size. In other words, the relative growth rate is almost constant when the population is small. But we also want to reflect the fact that the relative growth rate decreases as the population P increases and becomes negative if P ever exceeds its **carrying capacity** M, the maximum population that the environment is capable of sustaining in the long run. The simplest expression for the relative growth rate that incorporates these assumptions is

$$\frac{1}{P}\frac{dP}{dt} = k\left(1 - \frac{P}{M}\right)$$

Multiplying by P, we obtain the model for population growth known as the **logistic differential equation**:

$$\boxed{4} \qquad \boxed{\frac{dP}{dt} = kP\left(1 - \frac{P}{M}\right)}$$

Notice from Equation 4 that if P is small compared with M, then P/M is close to 0 and so $dP/dt \approx kP$. However, if $P \to M$ (the population approaches its carrying capacity), then $P/M \to 1$, so $dP/dt \to 0$. We can deduce information about whether solutions increase or decrease directly from Equation 4. If the population P lies between 0 and M, then the right side of the equation is positive, so $dP/dt > 0$ and the population increases. But if the population exceeds the carrying capacity ($P > M$), then $1 - P/M$ is negative, so $dP/dt < 0$ and the population decreases.

   Let's start our more detailed analysis of the logistic differential equation by looking at a direction field.

V EXAMPLE 1 Draw a direction field for the logistic equation with $k = 0.08$ and carrying capacity $M = 1000$. What can you deduce about the solutions?

SOLUTION In this case the logistic differential equation is

$$\frac{dP}{dt} = 0.08P\left(1 - \frac{P}{1000}\right)$$

A direction field for this equation is shown in Figure 1. We show only the first quadrant because negative populations aren't meaningful and we are interested only in what happens after $t = 0$.

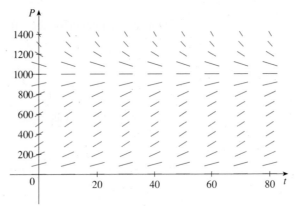

FIGURE 1
Direction field for the logistic equation in Example 1

The logistic equation is autonomous (dP/dt depends only on P, not on t), so the slopes are the same along any horizontal line. As expected, the slopes are positive for $0 < P < 1000$ and negative for $P > 1000$.

The slopes are small when P is close to 0 or 1000 (the carrying capacity). Notice that the solutions move away from the equilibrium solution $P = 0$ and move toward the equilibrium solution $P = 1000$.

In Figure 2 we use the direction field to sketch solution curves with initial populations $P(0) = 100$, $P(0) = 400$, and $P(0) = 1300$. Notice that solution curves that start below $P = 1000$ are increasing and those that start above $P = 1000$ are decreasing. The slopes are greatest when $P \approx 500$ and therefore the solution curves that start below $P = 1000$ have inflection points when $P \approx 500$. In fact we can prove that all solution curves that start below $P = 500$ have an inflection point when P is exactly 500. (See Exercise 11.)

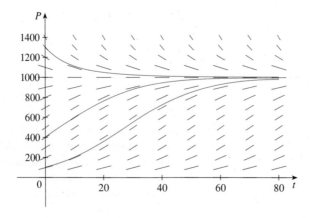

FIGURE 2
Solution curves for the logistic
equation in Example 1

The logistic equation $\boxed{4}$ is separable and so we can solve it explicitly using the method of Section 9.3. Since

$$\frac{dP}{dt} = kP\left(1 - \frac{P}{M}\right)$$

we have

$$\boxed{5} \qquad \int \frac{dP}{P(1 - P/M)} = \int k\, dt$$

To evaluate the integral on the left side, we write

$$\frac{1}{P(1 - P/M)} = \frac{M}{P(M - P)}$$

Using partial fractions (see Section 7.4), we get

$$\frac{M}{P(M - P)} = \frac{1}{P} + \frac{1}{M - P}$$

This enables us to rewrite Equation 5:

$$\int \left(\frac{1}{P} + \frac{1}{M - P} \right) dP = \int k \, dt$$

$$\ln |P| - \ln |M - P| = kt + C$$

$$\ln \left| \frac{M - P}{P} \right| = -kt - C$$

$$\left| \frac{M - P}{P} \right| = e^{-kt-C} = e^{-C}e^{-kt}$$

$$\boxed{6} \qquad \frac{M - P}{P} = Ae^{-kt}$$

where $A = \pm e^{-C}$. Solving Equation 6 for P, we get

$$\frac{M}{P} - 1 = Ae^{-kt} \qquad \Rightarrow \qquad \frac{P}{M} = \frac{1}{1 + Ae^{-kt}}$$

so

$$P = \frac{M}{1 + Ae^{-kt}}$$

We find the value of A by putting $t = 0$ in Equation 6. If $t = 0$, then $P = P_0$ (the initial population), so

$$\frac{M - P_0}{P_0} = Ae^0 = A$$

Thus the solution to the logistic equation is

$$\boxed{7} \qquad P(t) = \frac{M}{1 + Ae^{-kt}} \qquad \text{where } A = \frac{M - P_0}{P_0}$$

Using the expression for $P(t)$ in Equation 7, we see that

$$\lim_{t \to \infty} P(t) = M$$

which is to be expected.

EXAMPLE 2 Write the solution of the initial-value problem

$$\frac{dP}{dt} = 0.08P\left(1 - \frac{P}{1000} \right) \qquad P(0) = 100$$

and use it to find the population sizes $P(40)$ and $P(80)$. At what time does the population reach 900?

SOLUTION The differential equation is a logistic equation with $k = 0.08$, carrying capacity $M = 1000$, and initial population $P_0 = 100$. So Equation 7 gives the

population at time t as

$$P(t) = \frac{1000}{1 + Ae^{-0.08t}} \qquad \text{where } A = \frac{1000 - 100}{100} = 9$$

Thus

$$P(t) = \frac{1000}{1 + 9e^{-0.08t}}$$

So the population sizes when $t = 40$ and 80 are

$$P(40) = \frac{1000}{1 + 9e^{-3.2}} \approx 731.6 \qquad P(80) = \frac{1000}{1 + 9e^{-6.4}} \approx 985.3$$

The population reaches 900 when

$$\frac{1000}{1 + 9e^{-0.08t}} = 900$$

Solving this equation for t, we get

$$1 + 9e^{-0.08t} = \tfrac{10}{9}$$

$$e^{-0.08t} = \tfrac{1}{81}$$

$$-0.08t = \ln \tfrac{1}{81} = -\ln 81$$

$$t = \frac{\ln 81}{0.08} \approx 54.9$$

Compare the solution curve in Figure 3 with the lowest solution curve we drew from the direction field in Figure 2.

FIGURE 3

So the population reaches 900 when t is approximately 55. As a check on our work, we graph the population curve in Figure 3 and observe where it intersects the line $P = 900$. The cursor indicates that $t \approx 55$. ▬

Comparison of the Natural Growth and Logistic Models

In the 1930s the biologist G. F. Gause conducted an experiment with the protozoan *Paramecium* and used a logistic equation to model his data. The table gives his daily count of the population of protozoa. He estimated the initial relative growth rate to be 0.7944 and the carrying capacity to be 64.

t (days)	0	1	2	3	4	5	6	7	8	9	10	11	12	13	14	15	16
P (observed)	2	3	22	16	39	52	54	47	50	76	69	51	57	70	53	59	57

V EXAMPLE 3 Find the exponential and logistic models for Gause's data. Compare the predicted values with the observed values and comment on the fit.

SOLUTION Given the relative growth rate $k = 0.7944$ and the initial population $P_0 = 2$, the exponential model is

$$P(t) = P_0 e^{kt} = 2e^{0.7944t}$$

Gause used the same value of k for his logistic model. [This is reasonable because $P_0 = 2$ is small compared with the carrying capacity ($M = 64$). The equation

$$\frac{1}{P_0}\frac{dP}{dt}\bigg|_{t=0} = k\left(1 - \frac{2}{64}\right) \approx k$$

shows that the value of k for the logistic model is very close to the value for the exponential model.]

Then the solution of the logistic equation in Equation 7 gives

$$P(t) = \frac{M}{1 + Ae^{-kt}} = \frac{64}{1 + Ae^{-0.7944t}}$$

where

$$A = \frac{M - P_0}{P_0} = \frac{64 - 2}{2} = 31$$

So

$$P(t) = \frac{64}{1 + 31e^{-0.7944t}}$$

We use these equations to calculate the predicted values (rounded to the nearest integer) and compare them in the following table.

t (days)	0	1	2	3	4	5	6	7	8	9	10	11	12	13	14	15	16
P (observed)	2	3	22	16	39	52	54	47	50	76	69	51	57	70	53	59	57
P (logistic model)	2	4	9	17	28	40	51	57	61	62	63	64	64	64	64	64	64
P (exponential model)	2	4	10	22	48	106	$\cdots$										

We notice from the table and from the graph in Figure 4 that for the first three or four days the exponential model gives results comparable to those of the more sophisticated logistic model. For $t \geqslant 5$, however, the exponential model is hopelessly inaccurate, but the logistic model fits the observations reasonably well.

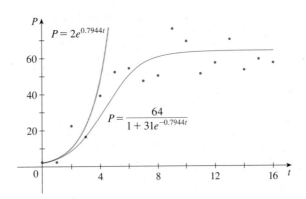

FIGURE 4

The exponential and logistic models for the *Paramecium* data

Many countries that formerly experienced exponential growth are now finding that their rates of population growth are declining and the logistic model provides a better model.

t	$B(t)$	t	$B(t)$
1980	9,847	1992	10,036
1982	9,856	1994	10,109
1984	9,855	1996	10,152
1986	9,862	1998	10,175
1988	9,884	2000	10,186
1990	9,962		

The table in the margin shows midyear values of $B(t)$, the population of Belgium, in thousands, at time t, from 1980 to 2000. Figure 5 shows these data points together with a shifted logistic function obtained from a calculator with the ability to fit a logistic function to these points by regression. We see that the logistic model provides a very good fit.

$$P = 9840 + \frac{350}{1 + 2.05e^{-0.48(t - 1990)}}$$

FIGURE 5

Logistic model for
the population of Belgium

Other Models for Population Growth

The Law of Natural Growth and the logistic differential equation are not the only equations that have been proposed to model population growth. In Exercise 20 we look at the Gompertz growth function and in Exercises 21 and 22 we investigate seasonal-growth models.

Two of the other models are modifications of the logistic model. The differential equation

$$\frac{dP}{dt} = kP\left(1 - \frac{P}{M}\right) - c$$

has been used to model populations that are subject to harvesting of one sort or another. (Think of a population of fish being caught at a constant rate.) This equation is explored in Exercises 17 and 18.

For some species there is a minimum population level m below which the species tends to become extinct. (Adults may not be able to find suitable mates.) Such populations have been modeled by the differential equation

$$\frac{dP}{dt} = kP\left(1 - \frac{P}{M}\right)\left(1 - \frac{m}{P}\right)$$

where the extra factor, $1 - m/P$, takes into account the consequences of a sparse population (see Exercise 19).

9.4 Exercises

9.5 Linear Equations

A first-order **linear** differential equation is one that can be put into the form

$$\boxed{1} \qquad \frac{dy}{dx} + P(x)y = Q(x)$$

where P and Q are continuous functions on a given interval. This type of equation occurs frequently in various sciences, as we will see.

An example of a linear equation is $xy' + y = 2x$ because, for $x \neq 0$, it can be written in the form

$$\boxed{2} \qquad y' + \frac{1}{x}y = 2$$

Notice that this differential equation is not separable because it's impossible to factor the expression for y' as a function of x times a function of y. But we can still solve the equation by noticing, by the Product Rule, that

$$xy' + y = (xy)'$$

and so we can rewrite the equation as

$$(xy)' = 2x$$

If we now integrate both sides of this equation, we get

$$xy = x^2 + C \qquad \text{or} \qquad y = x + \frac{C}{x}$$

If we had been given the differential equation in the form of Equation 2, we would have had to take the preliminary step of multiplying each side of the equation by x.

It turns out that every first-order linear differential equation can be solved in a similar fashion by multiplying both sides of Equation 1 by a suitable function $I(x)$ called an *integrating factor.* We try to find I so that the left side of Equation 1, when multiplied by $I(x)$, becomes the derivative of the product $I(x)y$:

$$\boxed{3} \qquad I(x)\big(y' + P(x)y\big) = \big(I(x)y\big)'$$

If we can find such a function I, then Equation 1 becomes

$$\big(I(x)y\big)' = I(x)\,Q(x)$$

Integrating both sides, we would have

$$I(x)y = \int I(x)\,Q(x)\,dx + C$$

so the solution would be

$$\boxed{4} \qquad y(x) = \frac{1}{I(x)}\left[\int I(x)\,Q(x)\,dx + C\right]$$

To find such an I, we expand Equation 3 and cancel terms:

$$I(x)y' + I(x)\,P(x)y = \big(I(x)y\big)' = I'(x)y + I(x)y'$$

$$I(x)\,P(x) = I'(x)$$

This is a separable differential equation for I, which we solve as follows:

$$\int \frac{dI}{I} = \int P(x)\,dx$$

$$\ln|I| = \int P(x)\,dx$$

$$I = Ae^{\int P(x)\,dx}$$

where $A = \pm e^C$. We are looking for a particular integrating factor, not the most general one, so we take $A = 1$ and use

5

$$I(x) = e^{\int P(x)\,dx}$$

Thus a formula for the general solution to Equation 1 is provided by Equation 4, where I is given by Equation 5. Instead of memorizing this formula, however, we just remember the form of the integrating factor.

> To solve the linear differential equation $y' + P(x)y = Q(x)$, multiply both sides by the **integrating factor** $I(x) = e^{\int P(x)\,dx}$ and integrate both sides.

V **EXAMPLE 1** Solve the differential equation $\dfrac{dy}{dx} + 3x^2y = 6x^2$.

SOLUTION The given equation is linear since it has the form of Equation 1 with $P(x) = 3x^2$ and $Q(x) = 6x^2$. An integrating factor is

$$I(x) = e^{\int 3x^2\,dx} = e^{x^3}$$

Multiplying both sides of the differential equation by e^{x^3}, we get

$$e^{x^3}\frac{dy}{dx} + 3x^2e^{x^3}y = 6x^2e^{x^3}$$

or

$$\frac{d}{dx}(e^{x^3}y) = 6x^2e^{x^3}$$

Integrating both sides, we have

$$e^{x^3}y = \int 6x^2e^{x^3}\,dx = 2e^{x^3} + C$$

$$y = 2 + Ce^{-x^3}$$

Figure 1 shows the graphs of several members of the family of solutions in Example 1. Notice that they all approach 2 as $x \to \infty$.

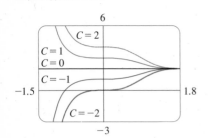

FIGURE 1

V **EXAMPLE 2** Find the solution of the initial-value problem

$$x^2y' + xy = 1 \qquad x > 0 \qquad y(1) = 2$$

SOLUTION We must first divide both sides by the coefficient of y' to put the differential equation into standard form:

$$\boxed{6} \qquad\qquad y' + \frac{1}{x}y = \frac{1}{x^2} \qquad x > 0$$

The integrating factor is

$$I(x) = e^{\int (1/x)\,dx} = e^{\ln x} = x$$

Multiplication of Equation 6 by x gives

$$xy' + y = \frac{1}{x} \qquad \text{or} \qquad (xy)' = \frac{1}{x}$$

Then

$$xy = \int \frac{1}{x}\,dx = \ln x + C$$

and so

$$y = \frac{\ln x + C}{x}$$

Since $y(1) = 2$, we have

$$2 = \frac{\ln 1 + C}{1} = C$$

Therefore the solution to the initial-value problem is

$$y = \frac{\ln x + 2}{x}$$

The solution of the initial-value problem in Example 2 is shown in Figure 2.

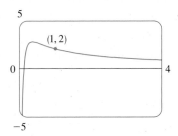

FIGURE 2

EXAMPLE 3 Solve $y' + 2xy = 1$.

SOLUTION The given equation is in the standard form for a linear equation. Multiplying by the integrating factor

$$e^{\int 2x\,dx} = e^{x^2}$$

we get

$$e^{x^2}y' + 2xe^{x^2}y = e^{x^2}$$

or

$$(e^{x^2}y)' = e^{x^2}$$

Therefore

$$e^{x^2}y = \int e^{x^2}\,dx + C$$

Recall from Section 7.5 that $\int e^{x^2}\,dx$ can't be expressed in terms of elementary functions. Nonetheless, it's a perfectly good function and we can leave the answer as

$$y = e^{-x^2}\int e^{x^2}\,dx + Ce^{-x^2}$$

Another way of writing the solution is

$$y = e^{-x^2}\int_0^x e^{t^2}\,dt + Ce^{-x^2}$$

(Any number can be chosen for the lower limit of integration.)

Even though the solutions of the differential equation in Example 3 are expressed in terms of an integral, they can still be graphed by a computer algebra system (Figure 3).

FIGURE 3

FIGURE 4

■ Application to Electric Circuits

In Section 9.2 we considered the simple electric circuit shown in Figure 4: An electromotive force (usually a battery or generator) produces a voltage of $E(t)$ volts (V) and a current of $I(t)$ amperes (A) at time t. The circuit also contains a resistor with a resistance of R ohms (Ω) and an inductor with an inductance of L henries (H).

Ohm's Law gives the drop in voltage due to the resistor as RI. The voltage drop due to the inductor is $L(dI/dt)$. One of Kirchhoff's laws says that the sum of the voltage drops is equal to the supplied voltage $E(t)$. Thus we have

$$\boxed{7} \qquad L\frac{dI}{dt} + RI = E(t)$$

which is a first-order linear differential equation. The solution gives the current I at time t.

V **EXAMPLE 4** Suppose that in the simple circuit of Figure 4 the resistance is 12 Ω and the inductance is 4 H. If a battery gives a constant voltage of 60 V and the switch is closed when $t = 0$ so the current starts with $I(0) = 0$, find (a) $I(t)$, (b) the current after 1 s, and (c) the limiting value of the current.

SOLUTION

The differential equation in Example 4 is both linear and separable, so an alternative method is to solve it as a separable equation (Example 4 in Section 9.3). If we replace the battery by a generator, however, we get an equation that is linear but not separable (Example 5).

(a) If we put $L = 4$, $R = 12$, and $E(t) = 60$ in Equation 7, we obtain the initial-value problem

$$4\frac{dI}{dt} + 12I = 60 \qquad I(0) = 0$$

or

$$\frac{dI}{dt} + 3I = 15 \qquad I(0) = 0$$

Multiplying by the integrating factor $e^{\int 3\,dt} = e^{3t}$, we get

$$e^{3t}\frac{dI}{dt} + 3e^{3t}I = 15e^{3t}$$

$$\frac{d}{dt}(e^{3t}I) = 15e^{3t}$$

$$e^{3t}I = \int 15e^{3t}\,dt = 5e^{3t} + C$$

$$I(t) = 5 + Ce^{-3t}$$

Since $I(0) = 0$, we have $5 + C = 0$, so $C = -5$ and

$$I(t) = 5(1 - e^{-3t})$$

(b) After 1 second the current is

$$I(1) = 5(1 - e^{-3}) \approx 4.75 \text{ A}$$

(c) The limiting value of the current is given by

$$\lim_{t\to\infty} I(t) = \lim_{t\to\infty} 5(1 - e^{-3t}) = 5 - 5\lim_{t\to\infty} e^{-3t} = 5 - 0 = 5 \qquad ■$$

Figure 5 shows how the current in Example 4 approaches its limiting value.

FIGURE 5

EXAMPLE 5 Suppose that the resistance and inductance remain as in Example 4 but, instead of the battery, we use a generator that produces a variable voltage of $E(t) = 60 \sin 30t$ volts. Find $I(t)$.

SOLUTION This time the differential equation becomes

$$4 \frac{dI}{dt} + 12I = 60 \sin 30t \qquad \text{or} \qquad \frac{dI}{dt} + 3I = 15 \sin 30t$$

The same integrating factor e^{3t} gives

$$\frac{d}{dt}(e^{3t}I) = e^{3t} \frac{dI}{dt} + 3e^{3t}I = 15e^{3t} \sin 30t$$

Figure 6 shows the graph of the current when the battery is replaced by a generator.

FIGURE 6

Using Formula 98 in the Table of Integrals, we have

$$e^{3t}I = \int 15e^{3t} \sin 30t \, dt = 15 \frac{e^{3t}}{909} (3 \sin 30t - 30 \cos 30t) + C$$

$$I = \tfrac{5}{101} (\sin 30t - 10 \cos 30t) + Ce^{-3t}$$

Since $I(0) = 0$, we get

$$-\tfrac{50}{101} + C = 0$$

so $\qquad I(t) = \tfrac{5}{101} (\sin 30t - 10 \cos 30t) + \tfrac{50}{101} e^{-3t}$

9.5 Exercises

Access end-of-section exercises and tools for visualizing calculus online at **www.webassign.net**   ENHANCED **WebAssign**

9.6 Predator-Prey Systems

We have looked at a variety of models for the growth of a single species that lives alone in an environment. In this section we consider more realistic models that take into account the interaction of two species in the same habitat. We will see that these models take the form of a pair of linked differential equations.

We first consider the situation in which one species, called the *prey,* has an ample food supply and the second species, called the *predators,* feeds on the prey. Examples of prey and predators include rabbits and wolves in an isolated forest, food fish and sharks, aphids and ladybugs, and bacteria and amoebas. Our model will have two dependent variables and both are functions of time. We let $R(t)$ be the number of prey (using R for rabbits) and $W(t)$ be the number of predators (with W for wolves) at time t.

In the absence of predators, the ample food supply would support exponential growth of the prey, that is,

$$\frac{dR}{dt} = kR \qquad \text{where } k \text{ is a positive constant}$$

In the absence of prey, we assume that the predator population would decline at a rate proportional to itself, that is,

$$\frac{dW}{dt} = -rW \qquad \text{where } r \text{ is a positive constant}$$

With both species present, however, we assume that the principal cause of death among the prey is being eaten by a predator, and the birth and survival rates of the predators depend on their available food supply, namely, the prey. We also assume that the two species encounter each other at a rate that is proportional to both populations and is therefore proportional to the product RW. (The more there are of either population, the more encounters there are likely to be.) A system of two differential equations that incorporates these assumptions is as follows:

W represents the predator.

R represents the prey.

$$\boxed{1} \qquad \frac{dR}{dt} = kR - aRW \qquad \frac{dW}{dt} = -rW + bRW$$

where k, r, a, and b are positive constants. Notice that the term $-aRW$ decreases the natural growth rate of the prey and the term bRW increases the natural growth rate of the predators.

The Lotka-Volterra equations were proposed as a model to explain the variations in the shark and food-fish populations in the Adriatic Sea by the Italian mathematician Vito Volterra (1860–1940).

The equations in $\boxed{1}$ are known as the **predator-prey equations**, or the **Lotka-Volterra equations**. A **solution** of this system of equations is a pair of functions $R(t)$ and $W(t)$ that describe the populations of prey and predator as functions of time. Because the system is coupled (R and W occur in both equations), we can't solve one equation and then the other; we have to solve them simultaneously. Unfortunately, it is usually impossible to find explicit formulas for R and W as functions of t. We can, however, use graphical methods to analyze the equations.

V **EXAMPLE 1** Suppose that populations of rabbits and wolves are described by the Lotka-Volterra equations $\boxed{1}$ with $k = 0.08$, $a = 0.001$, $r = 0.02$, and $b = 0.00002$. The time t is measured in months.
(a) Find the constant solutions (called the **equilibrium solutions**) and interpret the answer.
(b) Use the system of differential equations to find an expression for dW/dR.
(c) Draw a direction field for the resulting differential equation in the RW-plane. Then use that direction field to sketch some solution curves.
(d) Suppose that, at some point in time, there are 1000 rabbits and 40 wolves. Draw the corresponding solution curve and use it to describe the changes in both population levels.
(e) Use part (d) to make sketches of R and W as functions of t.

SOLUTION
(a) With the given values of k, a, r, and b, the Lotka-Volterra equations become

$$\frac{dR}{dt} = 0.08R - 0.001RW$$

$$\frac{dW}{dt} = -0.02W + 0.00002RW$$

Both R and W will be constant if both derivatives are 0, that is,

$$R' = R(0.08 - 0.001W) = 0$$
$$W' = W(-0.02 + 0.00002R) = 0$$

One solution is given by $R = 0$ and $W = 0$. (This makes sense: If there are no rabbits or wolves, the populations are certainly not going to increase.) The other constant solution is

$$W = \frac{0.08}{0.001} = 80 \qquad R = \frac{0.02}{0.00002} = 1000$$

So the equilibrium populations consist of 80 wolves and 1000 rabbits. This means that 1000 rabbits are just enough to support a constant wolf population of 80. There are neither too many wolves (which would result in fewer rabbits) nor too few wolves (which would result in more rabbits).

(b) We use the Chain Rule to eliminate t:

$$\frac{dW}{dt} = \frac{dW}{dR}\frac{dR}{dt}$$

so

$$\frac{dW}{dR} = \frac{\dfrac{dW}{dt}}{\dfrac{dR}{dt}} = \frac{-0.02W + 0.00002RW}{0.08R - 0.001RW}$$

(c) If we think of W as a function of R, we have the differential equation

$$\frac{dW}{dR} = \frac{-0.02W + 0.00002RW}{0.08R - 0.001RW}$$

We draw the direction field for this differential equation in Figure 1 and we use it to sketch several solution curves in Figure 2. If we move along a solution curve, we observe how the relationship between R and W changes as time passes. Notice that the curves appear to be closed in the sense that if we travel along a curve, we always return to the same point. Notice also that the point (1000, 80) is inside all the solution curves. That point is called an *equilibrium point* because it corresponds to the equilibrium solution $R = 1000$, $W = 80$.

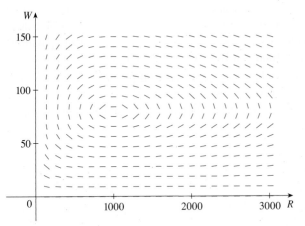

FIGURE 1 Direction field for the predator-prey system

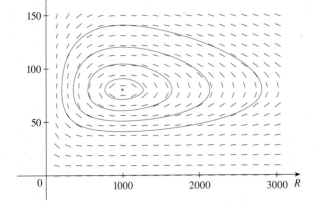

FIGURE 2 Phase portrait of the system

When we represent solutions of a system of differential equations as in Figure 2, we refer to the RW-plane as the **phase plane**, and we call the solution curves **phase trajectories**. So a phase trajectory is a path traced out by solutions (R, W) as time goes by. A **phase portrait** consists of equilibrium points and typical phase trajectories, as shown in Figure 2.

(d) Starting with 1000 rabbits and 40 wolves corresponds to drawing the solution curve through the point $P_0(1000, 40)$. Figure 3 shows this phase trajectory with the direction field removed. Starting at the point P_0 at time $t = 0$ and letting t increase, do we move clockwise or counterclockwise around the phase trajectory? If we put $R = 1000$ and $W = 40$ in the first differential equation, we get

$$\frac{dR}{dt} = 0.08(1000) - 0.001(1000)(40) = 80 - 40 = 40$$

Since $dR/dt > 0$, we conclude that R is increasing at P_0 and so we move counter-clockwise around the phase trajectory.

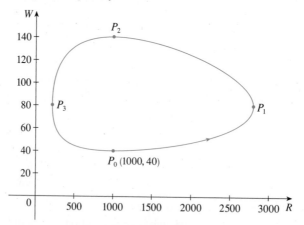

FIGURE 3
Phase trajectory through (1000, 40)

We see that at P_0 there aren't enough wolves to maintain a balance between the populations, so the rabbit population increases. That results in more wolves and eventually there are so many wolves that the rabbits have a hard time avoiding them. So the number of rabbits begins to decline (at P_1, where we estimate that R reaches its maximum population of about 2800). This means that at some later time the wolf population starts to fall (at P_2, where $R = 1000$ and $W \approx 140$). But this benefits the rabbits, so their population later starts to increase (at P_3, where $W = 80$ and $R \approx 210$). As a consequence, the wolf population eventually starts to increase as well. This happens when the populations return to their initial values of $R = 1000$ and $W = 40$, and the entire cycle begins again.

(e) From the description in part (d) of how the rabbit and wolf populations rise and fall, we can sketch the graphs of $R(t)$ and $W(t)$. Suppose the points P_1, P_2, and P_3 in Figure 3 are reached at times t_1, t_2, and t_3. Then we can sketch graphs of R and W as in Figure 4.

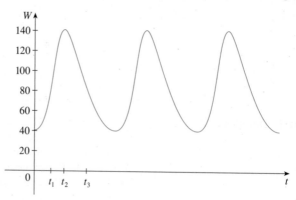

FIGURE 4   Graphs of the rabbit and wolf populations as functions of time

TEC In Module 9.6 you can change the coefficients in the Lotka-Volterra equations and observe the resulting changes in the phase trajectory and graphs of the rabbit and wolf populations.

To make the graphs easier to compare, we draw the graphs on the same axes but with different scales for R and W, as in Figure 5. Notice that the rabbits reach their maximum populations about a quarter of a cycle before the wolves.

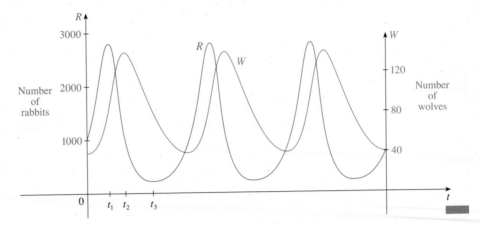

FIGURE 5

Comparison of the rabbit and wolf populations

An important part of the modeling process, as we discussed in Section 1.2, is to interpret our mathematical conclusions as real-world predictions and to test the predictions against real data. The Hudson's Bay Company, which started trading in animal furs in Canada in 1670, has kept records that date back to the 1840s. Figure 6 shows graphs of the number of pelts of the snowshoe hare and its predator, the Canada lynx, traded by the company over a 90-year period. You can see that the coupled oscillations in the hare and lynx populations predicted by the Lotka-Volterra model do actually occur and the period of these cycles is roughly 10 years.

FIGURE 6

Relative abundance of hare and lynx from Hudson's Bay Company records

Although the relatively simple Lotka-Volterra model has had some success in explaining and predicting coupled populations, more sophisticated models have also been proposed. One way to modify the Lotka-Volterra equations is to assume that, in the absence of predators, the prey grow according to a logistic model with carrying capacity M. Then the Lotka-Volterra equations [1] are replaced by the system of differential equations

$$\frac{dR}{dt} = kR\left(1 - \frac{R}{M}\right) - aRW \qquad \frac{dW}{dt} = -rW + bRW$$

This model is investigated in Exercises 11 and 12.

Models have also been proposed to describe and predict population levels of two or more species that compete for the same resources or cooperate for mutual benefit. Such models are explored in Exercises 2–4.

9.6 Exercises

Access end-of-section exercises and tools for visualizing calculus online at **www.webassign.net**

9 Review

Concept Check

1. (a) What is a differential equation?
   (b) What is the order of a differential equation?
   (c) What is an initial condition?

2. What can you say about the solutions of the equation $y' = x^2 + y^2$ just by looking at the differential equation?

3. What is a direction field for the differential equation $y' = F(x, y)$?

4. Explain how Euler's method works.

5. What is a separable differential equation? How do you solve it?

6. What is a first-order linear differential equation? How do you solve it?

7. (a) Write a differential equation that expresses the law of natural growth. What does it say in terms of relative growth rate?
   (b) Under what circumstances is this an appropriate model for population growth?
   (c) What are the solutions of this equation?

8. (a) Write the logistic equation.
   (b) Under what circumstances is this an appropriate model for population growth?

9. (a) Write Lotka-Volterra equations to model populations of food fish (F) and sharks (S).
   (b) What do these equations say about each population in the absence of the other?

True-False Quiz

Determine whether the statement is true or false. If it is true, explain why. If it is false, explain why or give an example that disproves the statement.

1. All solutions of the differential equation $y' = -1 - y^4$ are decreasing functions.

2. The function $f(x) = (\ln x)/x$ is a solution of the differential equation $x^2 y' + xy = 1$.

3. The equation $y' = x + y$ is separable.

4. The equation $y' = 3y - 2x + 6xy - 1$ is separable.

5. The equation $e^x y' = y$ is linear.

6. The equation $y' + xy = e^y$ is linear.

7. If y is the solution of the initial-value problem

$$\frac{dy}{dt} = 2y\left(1 - \frac{y}{5}\right) \qquad y(0) = 1$$

then $\lim_{t \to \infty} y = 5$.

Exercises

1. (a) A direction field for the differential equation
$y' = y(y - 2)(y - 4)$ is shown. Sketch the graphs of the
solutions that satisfy the given initial conditions.

   (i) $y(0) = -0.3$ (ii) $y(0) = 1$

   (iii) $y(0) = 3$ (iv) $y(0) = 4.3$

  (b) If the initial condition is $y(0) = c$, for what values of
c is $\lim_{t \to \infty} y(t)$ finite? What are the equilibrium solutions?

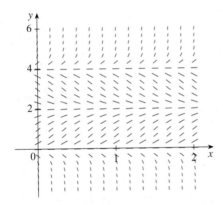

2. (a) Sketch a direction field for the differential equation
$y' = x/y$. Then use it to sketch the four solutions that
satisfy the initial conditions $y(0) = 1$, $y(0) = -1$,
$y(2) = 1$, and $y(-2) = 1$.

  (b) Check your work in part (a) by solving the differential
equation explicitly. What type of curve is each solution
curve?

3. (a) A direction field for the differential equation
$y' = x^2 - y^2$ is shown. Sketch the solution of the
initial-value problem

$$y' = x^2 - y^2 \qquad y(0) = 1$$

Use your graph to estimate the value of $y(0.3)$.

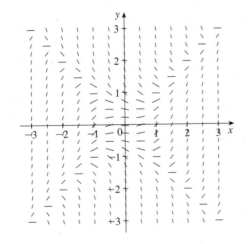

(b) Use Euler's method with step size 0.1 to estimate $y(0.3)$,
where $y(x)$ is the solution of the initial-value problem in
part (a). Compare with your estimate from part (a).

(c) On what lines are the centers of the horizontal line
segments of the direction field in part (a) located? What
happens when a solution curve crosses these lines?

4. (a) Use Euler's method with step size 0.2 to estimate $y(0.4)$,
where $y(x)$ is the solution of the initial-value problem

$$y' = 2xy^2 \qquad y(0) = 1$$

(b) Repeat part (a) with step size 0.1.

(c) Find the exact solution of the differential equation and
compare the value at 0.4 with the approximations in
parts (a) and (b).

5–8 Solve the differential equation.

5. $y' = xe^{-\sin x} - y \cos x$
    6. $\dfrac{dx}{dt} = 1 - t + x - tx$

7. $2ye^{y^2}y' = 2x + 3\sqrt{x}$
    8. $x^2y' - y = 2x^3e^{-1/x}$

9–11 Solve the initial-value problem.

9. $\dfrac{dr}{dt} + 2tr = r$, $r(0) = 5$

10. $(1 + \cos x)y' = (1 + e^{-y})\sin x$, $y(0) = 0$

11. $xy' - y = x \ln x$, $y(1) = 2$

12. Solve the initial-value problem $y' = 3x^2e^y$, $y(0) = 1$, and
graph the solution.

13–14 Find the orthogonal trajectories of the family of curves.

13. $y = ke^x$
    14. $y = e^{kx}$

15. (a) Write the solution of the initial-value problem

$$\frac{dP}{dt} = 0.1P\left(1 - \frac{P}{2000}\right) \qquad P(0) = 100$$

and use it to find the population when $t = 20$.

(b) When does the population reach 1200?

16. (a) The population of the world was 5.28 billion in 1990 and
6.07 billion in 2000. Find an exponential model for these
data and use the model to predict the world population in
the year 2020.

(b) According to the model in part (a), when will the world
population exceed 10 billion?

(c) Use the data in part (a) to find a logistic model for the population. Assume a carrying capacity of 100 billion. Then

use the logistic model to predict the population in 2020. Compare with your prediction from the exponential model.

(d) According to the logistic model, when will the world population exceed 10 billion? Compare with your prediction in part (b).

17. The von Bertalanffy growth model is used to predict the length $L(t)$ of a fish over a period of time. If L_∞ is the largest length for a species, then the hypothesis is that the rate of growth in length is proportional to $L_\infty - L$, the length yet to be achieved.

(a) Formulate and solve a differential equation to find an expression for $L(t)$.

(b) For the North Sea haddock it has been determined that $L_\infty = 53$ cm, $L(0) = 10$ cm, and the constant of proportionality is 0.2. What does the expression for $L(t)$ become with these data?

18. A tank contains 100 L of pure water. Brine that contains 0.1 kg of salt per liter enters the tank at a rate of 10 L/min. The solution is kept thoroughly mixed and drains from the tank at the same rate. How much salt is in the tank after 6 minutes?

19. One model for the spread of an epidemic is that the rate of spread is jointly proportional to the number of infected people and the number of uninfected people. In an isolated town of 5000 inhabitants, 160 people have a disease at the beginning of the week and 1200 have it at the end of the week. How long does it take for 80% of the population to become infected?

20. The Brentano-Stevens Law in psychology models the way that a subject reacts to a stimulus. It states that if R represents the reaction to an amount S of stimulus, then the relative rates of increase are proportional:

$$\frac{1}{R}\frac{dR}{dt} = \frac{k}{S}\frac{dS}{dt}$$

where k is a positive constant. Find R as a function of S.

21. The transport of a substance across a capillary wall in lung physiology has been modeled by the differential equation

$$\frac{dh}{dt} = -\frac{R}{V}\left(\frac{h}{k+h}\right)$$

where h is the hormone concentration in the bloodstream, t is time, R is the maximum transport rate, V is the volume of the capillary, and k is a positive constant that measures the affinity between the hormones and the enzymes that assist the process. Solve this differential equation to find a relationship between h and t.

22. Populations of birds and insects are modeled by the equations

$$\frac{dx}{dt} = 0.4x - 0.002xy$$

$$\frac{dy}{dt} = -0.2y + 0.000008xy$$

(a) Which of the variables, x or y, represents the bird population and which represents the insect population? Explain.

(b) Find the equilibrium solutions and explain their significance.

(c) Find an expression for dy/dx.

(d) The direction field for the differential equation in part (c) is shown. Use it to sketch the phase trajectory corresponding to initial populations of 100 birds and 40,000 insects. Then use the phase trajectory to describe how both populations change.

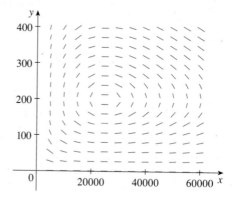

(e) Use part (d) to make rough sketches of the bird and insect populations as functions of time. How are these graphs related to each other?

23. Suppose the model of Exercise 22 is replaced by the equations

$$\frac{dx}{dt} = 0.4x(1 - 0.000005x) - 0.002xy$$

$$\frac{dy}{dt} = -0.2y + 0.000008xy$$

(a) According to these equations, what happens to the insect population in the absence of birds?

(b) Find the equilibrium solutions and explain their significance.

(c) The figure shows the phase trajectory that starts with 100 birds and 40,000 insects. Describe what eventually happens to the bird and insect populations.

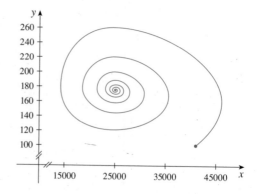

(d) Sketch graphs of the bird and insect populations as functions of time.

24. Barbara weighs 60 kg and is on a diet of 1600 calories per day, of which 850 are used automatically by basal metabolism. She spends about 15 cal/kg/day times her weight doing exercise. If 1 kg of fat contains 10,000 cal and we assume that the storage of calories in the form of fat is 100% efficient, formulate a differential equation and solve it to find her weight as a function of time. Does her weight ultimately approach an equilibrium weight?

25. When a flexible cable of uniform density is suspended between two fixed points and hangs of its own weight, the shape $y = f(x)$ of the cable must satisfy a differential equation of the form

$$\frac{d^2y}{dx^2} = k\sqrt{1 + \left(\frac{dy}{dx}\right)^2}$$

where k is a positive constant. Consider the cable shown in the figure.

(a) Let $z = dy/dx$ in the differential equation. Solve the resulting first-order differential equation (in z), and then integrate to find y.

(b) Determine the length of the cable.

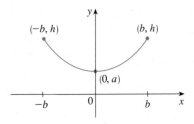

1. Find all functions f such that f' is continuous and

$$[f(x)]^2 = 100 + \int_0^x \{[f(t)]^2 + [f'(t)]^2\}\, dt \qquad \text{for all real } x$$

2. A student forgot the Product Rule for differentiation and made the mistake of thinking that $(fg)' = f'g'$. However, he was lucky and got the correct answer. The function f that he used was $f(x) = e^{x^2}$ and the domain of his problem was the interval $\left(\frac{1}{2}, \infty\right)$. What was the function g?

3. Let f be a function with the property that $f(0) = 1$, $f'(0) = 1$, and $f(a + b) = f(a)f(b)$ for all real numbers a and b. Show that $f'(x) = f(x)$ for all x and deduce that $f(x) = e^x$.

4. Find all functions f that satisfy the equation

$$\left(\int f(x)\, dx\right)\left(\int \frac{1}{f(x)}\, dx\right) = -1$$

5. Find the curve $y = f(x)$ such that $f(x) \ge 0$, $f(0) = 0$, $f(1) = 1$, and the area under the graph of f from 0 to x is proportional to the $(n + 1)$st power of $f(x)$.

6. A *subtangent* is a portion of the x-axis that lies directly beneath the segment of a tangent line from the point of contact to the x-axis. Find the curves that pass through the point $(c, 1)$ and whose subtangents all have length c.

7. A peach pie is taken out of the oven at 5:00 PM. At that time it is piping hot, 100°C. At 5:10 PM its temperature is 80°C; at 5:20 PM it is 65°C. What is the temperature of the room?

8. Snow began to fall during the morning of February 2 and continued steadily into the afternoon. At noon a snowplow began removing snow from a road at a constant rate. The plow traveled 6 km from noon to 1 PM but only 3 km from 1 PM to 2 PM. When did the snow begin to fall? [*Hints:* To get started, let t be the time measured in hours after noon; let $x(t)$ be the distance traveled by the plow at time t; then the speed of the plow is dx/dt. Let b be the number of hours before noon that it began to snow. Find an expression for the height of the snow at time t. Then use the given information that the rate of removal R (in m³/h) is constant.]

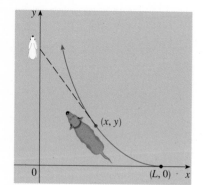

FIGURE FOR PROBLEM 9

9. A dog sees a rabbit running in a straight line across an open field and gives chase. In a rectangular coordinate system (as shown in the figure), assume:
  (i) The rabbit is at the origin and the dog is at the point $(L, 0)$ at the instant the dog first sees the rabbit.
  (ii) The rabbit runs up the y-axis and the dog always runs straight for the rabbit.
  (iii) The dog runs at the same speed as the rabbit.

  (a) Show that the dog's path is the graph of the function $y = f(x)$, where y satisfies the differential equation

$$x \frac{d^2 y}{dx^2} = \sqrt{1 + \left(\frac{dy}{dx}\right)^2}$$

  (b) Determine the solution of the equation in part (a) that satisfies the initial conditions $y = y' = 0$ when $x = L$. [*Hint:* Let $z = dy/dx$ in the differential equation and solve the resulting first-order equation to find z; then integrate z to find y.]
  (c) Does the dog ever catch the rabbit?

⊞ Graphing calculator or computer required

10. (a) Suppose that the dog in Problem 9 runs twice as fast as the rabbit. Find a differential equation for the path of the dog. Then solve it to find the point where the dog catches the rabbit.

(b) Suppose the dog runs half as fast as the rabbit. How close does the dog get to the rabbit? What are their positions when they are closest?

11. A planning engineer for a new alum plant must present some estimates to his company regarding the capacity of a silo designed to contain bauxite ore until it is processed into alum. The ore resembles pink talcum powder and is poured from a conveyor at the top of the silo. The silo is a cylinder 100 ft high with a radius of 200 ft. The conveyor carries ore at a rate of $60{,}000\pi$ ft^3/h and the ore maintains a conical shape whose radius is 1.5 times its height.

(a) If, at a certain time t, the pile is 60 ft high, how long will it take for the pile to reach the top of the silo?

(b) Management wants to know how much room will be left in the floor area of the silo when the pile is 60 ft high. How fast is the floor area of the pile growing at that height?

(c) Suppose a loader starts removing the ore at the rate of $20{,}000\pi$ ft^3/h when the height of the pile reaches 90 ft. Suppose, also, that the pile continues to maintain its shape. How long will it take for the pile to reach the top of the silo under these conditions?

12. Find the curve that passes through the point $(3, 2)$ and has the property that if the tangent line is drawn at any point P on the curve, then the part of the tangent line that lies in the first quadrant is bisected at P.

13. Recall that the normal line to a curve at a point P on the curve is the line that passes through P and is perpendicular to the tangent line at P. Find the curve that passes through the point $(3, 2)$ and has the property that if the normal line is drawn at any point on the curve, then the y-intercept of the normal line is always 6.

14. Find all curves with the property that if the normal line is drawn at any point P on the curve, then the part of the normal line between P and the x-axis is bisected by the y-axis.

15. Find all curves with the property that if a line is drawn from the origin to any point (x, y) on the curve, and then a tangent is drawn to the curve at that point and extended to meet the x-axis, the result is an isosceles triangle with equal sides meeting at (x, y).

16. (a) An outfielder fields a baseball 280 ft away from home plate and throws it directly to the catcher with an initial velocity of 100 ft/s. Assume that the velocity $v(t)$ of the ball after t seconds satisfies the differential equation $dv/dt = -\frac{1}{10}v$ because of air resistance. How long does it take for the ball to reach home plate? (Ignore any vertical motion of the ball.)

(b) The manager of the team wonders whether the ball will reach home plate sooner if it is relayed by an infielder. The shortstop can position himself directly between the outfielder and home plate, catch the ball thrown by the outfielder, turn, and throw the ball to the catcher with an initial velocity of 105 ft/s. The manager clocks the relay time of the shortstop (catching, turning, throwing) at half a second. How far from home plate should the shortstop position himself to minimize the total time for the ball to reach home plate? Should the manager encourage a direct throw or a relayed throw? What if the shortstop can throw at 115 ft/s?

(c) For what throwing velocity of the shortstop does a relayed throw take the same time as a direct throw?

10 Parametric Equations and Polar Coordinates

The Hale-Bopp comet, with its blue ion tail and white dust tail, appeared in the sky in March 1997. In Section 10.6 you will see how polar coordinates provide a convenient equation for the path of this comet.

© Dean Ketelsen

So far we have described plane curves by giving y as a function of x $[y = f(x)]$ or x as a function of y $[x = g(y)]$ or by giving a relation between x and y that defines y implicitly as a function of x $[f(x, y) = 0]$. In this chapter we discuss two new methods for describing curves.

Some curves, such as the cycloid, are best handled when both x and y are given in terms of a third variable t called a parameter $[x = f(t), y = g(t)]$. Other curves, such as the cardioid, have their most convenient description when we use a new coordinate system, called the polar coordinate system.

10.1 Curves Defined by Parametric Equations

FIGURE 1

Imagine that a particle moves along the curve C shown in Figure 1. It is impossible to describe C by an equation of the form $y = f(x)$ because C fails the Vertical Line Test. But the x- and y-coordinates of the particle are functions of time and so we can write $x = f(t)$ and $y = g(t)$. Such a pair of equations is often a convenient way of describing a curve and gives rise to the following definition.

Suppose that x and y are both given as functions of a third variable t (called a **param-eter**) by the equations

$$x = f(t) \qquad y = g(t)$$

(called **parametric equations**). Each value of t determines a point (x, y), which we can plot in a coordinate plane. As t varies, the point $(x, y) = (f(t), g(t))$ varies and traces out a curve C, which we call a **parametric curve**. The parameter t does not necessarily represent time and, in fact, we could use a letter other than t for the parameter. But in many applications of parametric curves, t does denote time and therefore we can interpret $(x, y) = (f(t), g(t))$ as the position of a particle at time t.

EXAMPLE 1 Sketch and identify the curve defined by the parametric equations

$$x = t^2 - 2t \qquad y = t + 1$$

SOLUTION Each value of t gives a point on the curve, as shown in the table. For instance, if $t = 0$, then $x = 0$, $y = 1$ and so the corresponding point is $(0, 1)$. In Figure 2 we plot the points (x, y) determined by several values of the parameter and we join them to produce a curve.

t	x	y
-2	8	-1
-1	3	0
0	0	1
1	-1	2
2	0	3
3	3	4
4	8	5

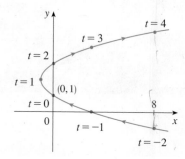

FIGURE 2

A particle whose position is given by the parametric equations moves along the curve in the direction of the arrows as t increases. Notice that the consecutive points marked on the curve appear at equal time intervals but not at equal distances. That is because the particle slows down and then speeds up as t increases.

It appears from Figure 2 that the curve traced out by the particle may be a parabola. This can be confirmed by eliminating the parameter t as follows. We obtain $t = y - 1$ from the second equation and substitute into the first equation. This gives

$$x = t^2 - 2t = (y - 1)^2 - 2(y - 1) = y^2 - 4y + 3$$

and so the curve represented by the given parametric equations is the parabola $x = y^2 - 4y + 3$. ▬

This equation in x and y describes *where* the particle has been, but it doesn't tell us *when* the particle was at a particular point. The parametric equations have an advantage—they tell us *when* the particle was at a point. They also indicate the *direction* of the motion.

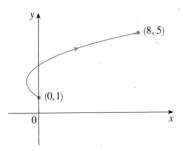

FIGURE 3

No restriction was placed on the parameter t in Example 1, so we assumed that t could be any real number. But sometimes we restrict t to lie in a finite interval. For instance, the parametric curve

$$x = t^2 - 2t \qquad y = t + 1 \qquad 0 \leqslant t \leqslant 4$$

shown in Figure 3 is the part of the parabola in Example 1 that starts at the point $(0, 1)$ and ends at the point $(8, 5)$. The arrowhead indicates the direction in which the curve is traced as t increases from 0 to 4.

In general, the curve with parametric equations

$$x = f(t) \qquad y = g(t) \qquad a \leqslant t \leqslant b$$

has **initial point** $(f(a), g(a))$ and **terminal point** $(f(b), g(b))$.

FIGURE 4

V EXAMPLE 2 What curve is represented by the following parametric equations?

$$x = \cos t \qquad y = \sin t \qquad 0 \leqslant t \leqslant 2\pi$$

SOLUTION If we plot points, it appears that the curve is a circle. We can confirm this impression by eliminating t. Observe that

$$x^2 + y^2 = \cos^2 t + \sin^2 t = 1$$

Thus the point (x, y) moves on the unit circle $x^2 + y^2 = 1$. Notice that in this example the parameter t can be interpreted as the angle (in radians) shown in Figure 4. As t increases from 0 to 2π, the point $(x, y) = (\cos t, \sin t)$ moves once around the circle in the counterclockwise direction starting from the point $(1, 0)$.

EXAMPLE 3 What curve is represented by the given parametric equations?

$$x = \sin 2t \qquad y = \cos 2t \qquad 0 \leqslant t \leqslant 2\pi$$

SOLUTION Again we have

$$x^2 + y^2 = \sin^2 2t + \cos^2 2t = 1$$

so the parametric equations again represent the unit circle $x^2 + y^2 = 1$. But as t increases from 0 to 2π, the point $(x, y) = (\sin 2t, \cos 2t)$ starts at $(0, 1)$ and moves *twice* around the circle in the clockwise direction as indicated in Figure 5.

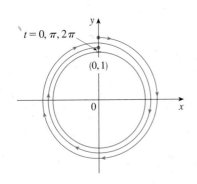

FIGURE 5

Examples 2 and 3 show that different sets of parametric equations can represent the same curve. Thus we distinguish between a *curve,* which is a set of points, and a *parametric curve,* in which the points are traced in a particular way.

EXAMPLE 4 Find parametric equations for the circle with center (h, k) and radius r.

SOLUTION If we take the equations of the unit circle in Example 2 and multiply the expressions for x and y by r, we get $x = r \cos t$, $y = r \sin t$. You can verify that these equations represent a circle with radius r and center the origin traced counterclockwise. We now shift h units in the x-direction and k units in the y-direction and obtain parametric equations of the circle (Figure 6) with center (h, k) and radius r:

$$x = h + r \cos t \qquad y = k + r \sin t \qquad 0 \leqslant t \leqslant 2\pi$$

FIGURE 6
$x = h + r \cos t, \ y = k + r \sin t$

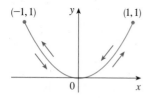

FIGURE 7

V **EXAMPLE 5** Sketch the curve with parametric equations $x = \sin t$, $y = \sin^2 t$.

SOLUTION Observe that $y = (\sin t)^2 = x^2$ and so the point (x, y) moves on the parabola $y = x^2$. But note also that, since $-1 \leqslant \sin t \leqslant 1$, we have $-1 \leqslant x \leqslant 1$, so the parametric equations represent only the part of the parabola for which $-1 \leqslant x \leqslant 1$. Since $\sin t$ is periodic, the point $(x, y) = (\sin t, \sin^2 t)$ moves back and forth infinitely often along the parabola from $(-1, 1)$ to $(1, 1)$. (See Figure 7.)

TEC Module 10.1A gives an animation of the relationship between motion along a parametric curve $x = f(t)$, $y = g(t)$ and motion along the graphs of f and g as functions of t. Clicking on TRIG gives you the family of parametric curves

$$x = a \cos bt \qquad y = c \sin dt$$

If you choose $a = b = c = d = 1$ and click on **animate**, you will see how the graphs of $x = \cos t$ and $y = \sin t$ relate to the circle in Example 2. If you choose $a = b = c = 1$, $d = 2$, you will see graphs as in Figure 8. By clicking on **animate** or moving the t-slider to the right, you can see from the color coding how motion along the graphs of $x = \cos t$ and $y = \sin 2t$ corresponds to motion along the parametric curve, which is called a **Lissajous figure**.

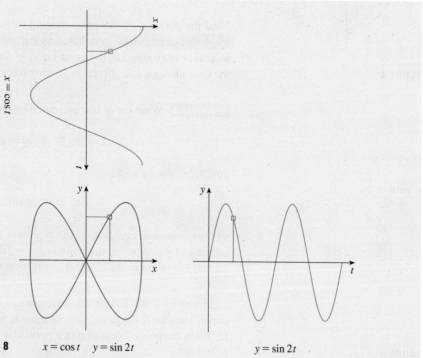

FIGURE 8 $x = \cos t$ $y = \sin 2t$ $y = \sin 2t$

Graphing Devices

Most graphing calculators and computer graphing programs can be used to graph curves defined by parametric equations. In fact, it's instructive to watch a parametric curve being drawn by a graphing calculator because the points are plotted in order as the corresponding parameter values increase.

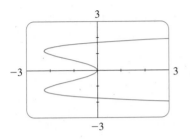

FIGURE 9

EXAMPLE 6 Use a graphing device to graph the curve $x = y^4 - 3y^2$.

SOLUTION If we let the parameter be $t = y$, then we have the equations

$$x = t^4 - 3t^2 \qquad y = t$$

Using these parametric equations to graph the curve, we obtain Figure 9. It would be possible to solve the given equation $(x = y^4 - 3y^2)$ for y as four functions of x and graph them individually, but the parametric equations provide a much easier method.

In general, if we need to graph an equation of the form $x = g(y)$, we can use the parametric equations

$$x = g(t) \qquad y = t$$

Notice also that curves with equations $y = f(x)$ (the ones we are most familiar with—graphs of functions) can also be regarded as curves with parametric equations

$$x = t \qquad y = f(t)$$

Graphing devices are particularly useful for sketching complicated curves. For instance, the curves shown in Figures 10, 11, and 12 would be virtually impossible to produce by hand.

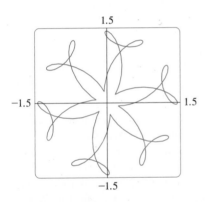

FIGURE 10
$x = \sin t + \frac{1}{2} \cos 5t + \frac{1}{4} \sin 13t$
$y = \cos t + \frac{1}{2} \sin 5t + \frac{1}{4} \cos 13t$

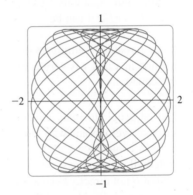

FIGURE 11
$x = \sin t - \sin 2.3t$
$y = \cos t$

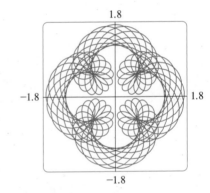

FIGURE 12
$x = \sin t + \frac{1}{2} \sin 5t + \frac{1}{4} \cos 2.3t$
$y = \cos t + \frac{1}{2} \cos 5t + \frac{1}{4} \sin 2.3t$

One of the most important uses of parametric curves is in computer-aided design (CAD). In the Laboratory Project after Section 10.2 we will investigate special parametric curves, called **Bézier curves**, that are used extensively in manufacturing, especially in the automotive industry. These curves are also employed in specifying the shapes of letters and other symbols in laser printers.

The Cycloid

TEC An animation in Module 10.1B shows how the cycloid is formed as the circle moves.

EXAMPLE 7 The curve traced out by a point P on the circumference of a circle as the circle rolls along a straight line is called a **cycloid** (see Figure 13). If the circle has radius r and rolls along the x-axis and if one position of P is the origin, find parametric equations for the cycloid.

FIGURE 13

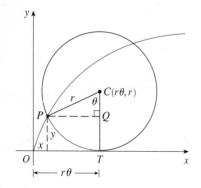

FIGURE 14

SOLUTION We choose as parameter the angle of rotation θ of the circle ($\theta = 0$ when P is at the origin). Suppose the circle has rotated through θ radians. Because the circle has been in contact with the line, we see from Figure 14 that the distance it has rolled from the origin is

$$|OT| = \text{arc } PT = r\theta$$

Therefore the center of the circle is $C(r\theta, r)$. Let the coordinates of P be (x, y). Then from Figure 14 we see that

$$x = |OT| - |PQ| = r\theta - r\sin\theta = r(\theta - \sin\theta)$$

$$y = |TC| - |QC| = r - r\cos\theta = r(1 - \cos\theta)$$

Therefore parametric equations of the cycloid are

$$\boxed{1} \qquad x = r(\theta - \sin\theta) \qquad y = r(1 - \cos\theta) \qquad \theta \in \mathbb{R}$$

One arch of the cycloid comes from one rotation of the circle and so is described by $0 \leq \theta \leq 2\pi$. Although Equations 1 were derived from Figure 14, which illustrates the case where $0 < \theta < \pi/2$, it can be seen that these equations are still valid for other values of θ (see Exercise 39).

Although it is possible to eliminate the parameter θ from Equations 1, the resulting Cartesian equation in x and y is very complicated and not as convenient to work with as the parametric equations. ▬

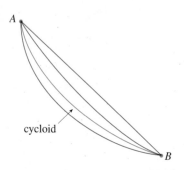

FIGURE 15

One of the first people to study the cycloid was Galileo, who proposed that bridges be built in the shape of cycloids and who tried to find the area under one arch of a cycloid. Later this curve arose in connection with the **brachistochrone problem**: Find the curve along which a particle will slide in the shortest time (under the influence of gravity) from a point A to a lower point B not directly beneath A. The Swiss mathematician John Bernoulli, who posed this problem in 1696, showed that among all possible curves that join A to B, as in Figure 15, the particle will take the least time sliding from A to B if the curve is part of an inverted arch of a cycloid.

The Dutch physicist Huygens had already shown that the cycloid is also the solution to the **tautochrone problem**; that is, no matter where a particle P is placed on an inverted cycloid, it takes the same time to slide to the bottom (see Figure 16). Huygens proposed that pendulum clocks (which he invented) should swing in cycloidal arcs because then the pendulum would take the same time to make a complete oscillation whether it swings through a wide or a small arc.

FIGURE 16

▬ Families of Parametric Curves

V   EXAMPLE 8   Investigate the family of curves with parametric equations

$$x = a + \cos t \qquad y = a\tan t + \sin t$$

What do these curves have in common? How does the shape change as a increases?

SOLUTION We use a graphing device to produce the graphs for the cases $a = -2, -1, -0.5, -0.2, 0, 0.5, 1,$ and 2 shown in Figure 17. Notice that all of these curves (except the case $a = 0$) have two branches, and both branches approach the vertical asymptote $x = a$ as x approaches a from the left or right.

FIGURE 17 Members of the family $x = a + \cos t$, $y = a \tan t + \sin t$, all graphed in the viewing rectangle $[-4, 4]$ by $[-4, 4]$

When $a < -1$, both branches are smooth; but when a reaches -1, the right branch acquires a sharp point, called a *cusp*. For a between -1 and 0 the cusp turns into a loop, which becomes larger as a approaches 0. When $a = 0$, both branches come together and form a circle (see Example 2). For a between 0 and 1, the left branch has a loop, which shrinks to become a cusp when $a = 1$. For $a > 1$, the branches become smooth again, and as a increases further, they become less curved. Notice that the curves with a positive are reflections about the y-axis of the corresponding curves with a negative.

These curves are called **conchoids of Nicomedes** after the ancient Greek scholar Nicomedes. He called them conchoids because the shape of their outer branches resembles that of a conch shell or mussel shell.

10.1   **Exercises**

Access end-of-section exercises and tools for visualizing calculus online at **www.webassign.net** ^{ENHANCED} WebAssign

---

LABORATORY PROJECT   📐 **RUNNING CIRCLES AROUND CIRCLES**

In this project we investigate families of curves, called *hypocycloids* and *epicycloids,* that are generated by the motion of a point on a circle that rolls inside or outside another circle.

1. A **hypocycloid** is a curve traced out by a fixed point P on a circle C of radius b as C rolls on the inside of a circle with center O and radius a. Show that if the initial position of P is $(a, 0)$ and the parameter θ is chosen as in the figure, then parametric equations of the hypocycloid are

$$x = (a - b)\cos\theta + b\cos\left(\frac{a-b}{b}\theta\right) \qquad y = (a - b)\sin\theta - b\sin\left(\frac{a-b}{b}\theta\right)$$

📐 Graphing calculator or computer required

2. Use a graphing device (or the interactive graphic in TEC Module 10.1B) to draw the graphs of hypocycloids with a a positive integer and $b = 1$. How does the value of a affect the graph? Show that if we take $a = 4$, then the parametric equations of the hypocycloid reduce to

$$x = 4\cos^3\theta \qquad y = 4\sin^3\theta$$

This curve is called a **hypocycloid of four cusps**, or an **astroid**.

3. Now try $b = 1$ and $a = n/d$, a fraction where n and d have no common factor. First let $n = 1$ and try to determine graphically the effect of the denominator d on the shape of the graph. Then let n vary while keeping d constant. What happens when $n = d + 1$?

4. What happens if $b = 1$ and a is irrational? Experiment with an irrational number like $\sqrt{2}$ or $e - 2$. Take larger and larger values for θ and speculate on what would happen if we were to graph the hypocycloid for all real values of θ.

5. If the circle C rolls on the *outside* of the fixed circle, the curve traced out by P is called an **epicycloid**. Find parametric equations for the epicycloid.

6. Investigate the possible shapes for epicycloids. Use methods similar to Problems 2–4.

10.2 Calculus with Parametric Curves

Having seen how to represent curves by parametric equations, we now apply the methods of calculus to these parametric curves. In particular, we solve problems involving tangents, area, arc length, and surface area.

Tangents

Suppose f and g are differentiable functions and we want to find the tangent line at a point on the curve where y is also a differentiable function of x. Then the Chain Rule gives

$$\frac{dy}{dt} = \frac{dy}{dx} \cdot \frac{dx}{dt}$$

If $dx/dt \neq 0$, we can solve for dy/dx:

If we think of the curve as being traced out by a moving particle, then dy/dt and dx/dt are the vertical and horizontal velocities of the particle and Formula 1 says that the slope of the tangent is the ratio of these velocities.

$$\boxed{1} \qquad \frac{dy}{dx} = \frac{\dfrac{dy}{dt}}{\dfrac{dx}{dt}} \qquad \text{if} \quad \frac{dx}{dt} \neq 0$$

Equation 1 (which you can remember by thinking of canceling the dt's) enables us to find the slope dy/dx of the tangent to a parametric curve without having to eliminate the parameter t. We see from $\boxed{1}$ that the curve has a horizontal tangent when $dy/dt = 0$ (provided that $dx/dt \neq 0$) and it has a vertical tangent when $dx/dt = 0$ (provided that $dy/dt \neq 0$). This information is useful for sketching parametric curves.

As we know from Chapter 4, it is also useful to consider d^2y/dx^2. This can be found by replacing y by dy/dx in Equation 1:

⊘ Note that $\dfrac{d^2y}{dx^2} \neq \dfrac{\dfrac{d^2y}{dt^2}}{\dfrac{d^2x}{dt^2}}$

$$\frac{d^2y}{dx^2} = \frac{d}{dx}\left(\frac{dy}{dx}\right) = \frac{\dfrac{d}{dt}\left(\dfrac{dy}{dx}\right)}{\dfrac{dx}{dt}}$$

EXAMPLE 1 A curve C is defined by the parametric equations $x = t^2$, $y = t^3 - 3t$.
(a) Show that C has two tangents at the point $(3, 0)$ and find their equations.
(b) Find the points on C where the tangent is horizontal or vertical.
(c) Determine where the curve is concave upward or downward.
(d) Sketch the curve.

SOLUTION
(a) Notice that $y = t^3 - 3t = t(t^2 - 3) = 0$ when $t = 0$ or $t = \pm\sqrt{3}$. Therefore the point $(3, 0)$ on C arises from two values of the parameter, $t = \sqrt{3}$ and $t = -\sqrt{3}$. This indicates that C crosses itself at $(3, 0)$. Since

$$\frac{dy}{dx} = \frac{dy/dt}{dx/dt} = \frac{3t^2 - 3}{2t} = \frac{3}{2}\left(t - \frac{1}{t}\right)$$

the slope of the tangent when $t = \pm\sqrt{3}$ is $dy/dx = \pm 6/(2\sqrt{3}) = \pm\sqrt{3}$, so the equations of the tangents at $(3, 0)$ are

$$y = \sqrt{3}\,(x - 3) \qquad \text{and} \qquad y = -\sqrt{3}\,(x - 3)$$

(b) C has a horizontal tangent when $dy/dx = 0$, that is, when $dy/dt = 0$ and $dx/dt \neq 0$. Since $dy/dt = 3t^2 - 3$, this happens when $t^2 = 1$, that is, $t = \pm 1$. The corresponding points on C are $(1, -2)$ and $(1, 2)$. C has a vertical tangent when $dx/dt = 2t = 0$, that is, $t = 0$. (Note that $dy/dt \neq 0$ there.) The corresponding point on C is $(0, 0)$.

(c) To determine concavity we calculate the second derivative:

$$\frac{d^2y}{dx^2} = \frac{\dfrac{d}{dt}\left(\dfrac{dy}{dx}\right)}{\dfrac{dx}{dt}} = \frac{\dfrac{3}{2}\left(1 + \dfrac{1}{t^2}\right)}{2t} = \frac{3(t^2 + 1)}{4t^3}$$

Thus the curve is concave upward when $t > 0$ and concave downward when $t < 0$.

(d) Using the information from parts (b) and (c), we sketch C in Figure 1.

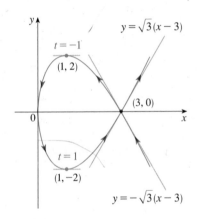

$y = \sqrt{3}(x - 3)$
$t = -1$
$(1, 2)$
$(3, 0)$
$t = 1$
$(1, -2)$
$y = -\sqrt{3}(x - 3)$

FIGURE 1

V EXAMPLE 2
(a) Find the tangent to the cycloid $x = r(\theta - \sin\theta)$, $y = r(1 - \cos\theta)$ at the point where $\theta = \pi/3$. (See Example 7 in Section 10.1.)
(b) At what points is the tangent horizontal? When is it vertical?

SOLUTION
(a) The slope of the tangent line is

$$\frac{dy}{dx} = \frac{dy/d\theta}{dx/d\theta} = \frac{r\sin\theta}{r(1 - \cos\theta)} = \frac{\sin\theta}{1 - \cos\theta}$$

When $\theta = \pi/3$, we have

$$x = r\left(\frac{\pi}{3} - \sin\frac{\pi}{3}\right) = r\left(\frac{\pi}{3} - \frac{\sqrt{3}}{2}\right) \qquad y = r\left(1 - \cos\frac{\pi}{3}\right) = \frac{r}{2}$$

and

$$\frac{dy}{dx} = \frac{\sin(\pi/3)}{1 - \cos(\pi/3)} = \frac{\sqrt{3}/2}{1 - \frac{1}{2}} = \sqrt{3}$$

Therefore the slope of the tangent is $\sqrt{3}$ and its equation is

$$y - \frac{r}{2} = \sqrt{3}\left(x - \frac{r\pi}{3} + \frac{r\sqrt{3}}{2}\right) \qquad \text{or} \qquad \sqrt{3}\,x - y = r\left(\frac{\pi}{\sqrt{3}} - 2\right)$$

The tangent is sketched in Figure 2.

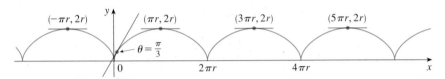

FIGURE 2

(b) The tangent is horizontal when $dy/dx = 0$, which occurs when $\sin\theta = 0$ and $1 - \cos\theta \neq 0$, that is, $\theta = (2n - 1)\pi$, n an integer. The corresponding point on the cycloid is $((2n - 1)\pi r, 2r)$.

When $\theta = 2n\pi$, both $dx/d\theta$ and $dy/d\theta$ are 0. It appears from the graph that there are vertical tangents at those points. We can verify this by using l'Hospital's Rule as follows:

$$\lim_{\theta \to 2n\pi^+} \frac{dy}{dx} = \lim_{\theta \to 2n\pi^+} \frac{\sin\theta}{1 - \cos\theta} = \lim_{\theta \to 2n\pi^+} \frac{\cos\theta}{\sin\theta} = \infty$$

A similar computation shows that $dy/dx \to -\infty$ as $\theta \to 2n\pi^-$, so indeed there are vertical tangents when $\theta = 2n\pi$, that is, when $x = 2n\pi r$.   ▬

Areas

We know that the area under a curve $y = F(x)$ from a to b is $A = \int_a^b F(x)\,dx$, where $F(x) \geq 0$. If the curve is traced out once by the parametric equations $x = f(t)$ and $y = g(t)$, $\alpha \leq t \leq \beta$, then we can calculate an area formula by using the Substitution Rule for Definite Integrals as follows:

$$A = \int_a^b y\,dx = \int_\alpha^\beta g(t)f'(t)\,dt \qquad \left[\text{or} \quad \int_\beta^\alpha g(t)f'(t)\,dt\right]$$

The limits of integration for t are found as usual with the Substitution Rule. When $x = a$, t is either α or β. When $x = b$, t is the remaining value.

V **EXAMPLE 3**   Find the area under one arch of the cycloid

$$x = r(\theta - \sin\theta) \qquad y = r(1 - \cos\theta)$$

(See Figure 3.)

FIGURE 3

SOLUTION   One arch of the cycloid is given by $0 \leq \theta \leq 2\pi$. Using the Substitution Rule with $y = r(1 - \cos\theta)$ and $dx = r(1 - \cos\theta)\,d\theta$, we have

$$A = \int_0^{2\pi r} y\,dx = \int_0^{2\pi} r(1 - \cos\theta)\,r(1 - \cos\theta)\,d\theta$$

$$= r^2 \int_0^{2\pi} (1 - \cos\theta)^2\,d\theta = r^2 \int_0^{2\pi} (1 - 2\cos\theta + \cos^2\theta)\,d\theta$$

$$= r^2 \int_0^{2\pi} \left[1 - 2\cos\theta + \tfrac{1}{2}(1 + \cos 2\theta)\right]\,d\theta$$

$$= r^2 \left[\tfrac{3}{2}\theta - 2\sin\theta + \tfrac{1}{4}\sin 2\theta\right]_0^{2\pi}$$

$$= r^2\left(\tfrac{3}{2} \cdot 2\pi\right) = 3\pi r^2$$

The result of Example 3 says that the area under one arch of the cycloid is three times the area of the rolling circle that generates the cycloid (see Example 7 in Section 10.1). Galileo guessed this result but it was first proved by the French mathematician Roberval and the Italian mathematician Torricelli.

■ Arc Length

We already know how to find the length L of a curve C given in the form $y = F(x)$, $a \leqslant x \leqslant b$. Formula 8.1.3 says that if F' is continuous, then

$$\boxed{2} \qquad L = \int_a^b \sqrt{1 + \left(\frac{dy}{dx}\right)^2} \, dx$$

Suppose that C can also be described by the parametric equations $x = f(t)$ and $y = g(t)$, $\alpha \leqslant t \leqslant \beta$, where $dx/dt = f'(t) > 0$. This means that C is traversed once, from left to right, as t increases from α to β and $f(\alpha) = a$, $f(\beta) = b$. Putting Formula 1 into Formula 2 and using the Substitution Rule, we obtain

$$L = \int_a^b \sqrt{1 + \left(\frac{dy}{dx}\right)^2} \, dx = \int_\alpha^\beta \sqrt{1 + \left(\frac{dy/dt}{dx/dt}\right)^2} \, \frac{dx}{dt} \, dt$$

Since $dx/dt > 0$, we have

$$\boxed{3} \qquad L = \int_\alpha^\beta \sqrt{\left(\frac{dx}{dt}\right)^2 + \left(\frac{dy}{dt}\right)^2} \, dt$$

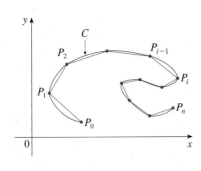

FIGURE 4

Even if C can't be expressed in the form $y = F(x)$, Formula 3 is still valid but we obtain it by polygonal approximations. We divide the parameter interval $[\alpha, \beta]$ into n subintervals of equal width Δt. If $t_0, t_1, t_2, \ldots, t_n$ are the endpoints of these subintervals, then $x_i = f(t_i)$ and $y_i = g(t_i)$ are the coordinates of points $P_i(x_i, y_i)$ that lie on C and the polygon with vertices $P_0, P_1, \ldots, P_n$ approximates C. (See Figure 4.)

As in Section 8.1, we define the length L of C to be the limit of the lengths of these approximating polygons as $n \to \infty$:

$$L = \lim_{n \to \infty} \sum_{i=1}^n |P_{i-1}P_i|$$

The Mean Value Theorem, when applied to f on the interval $[t_{i-1}, t_i]$, gives a number t_i^* in (t_{i-1}, t_i) such that

$$f(t_i) - f(t_{i-1}) = f'(t_i^*)(t_i - t_{i-1})$$

If we let $\Delta x_i = x_i - x_{i-1}$ and $\Delta y_i = y_i - y_{i-1}$, this equation becomes

$$\Delta x_i = f'(t_i^*) \, \Delta t$$

Similarly, when applied to g, the Mean Value Theorem gives a number t_i^{**} in (t_{i-1}, t_i) such that

$$\Delta y_i = g'(t_i^{**}) \, \Delta t$$

Therefore

$$|P_{i-1}P_i| = \sqrt{(\Delta x_i)^2 + (\Delta y_i)^2} = \sqrt{[f'(t_i^*) \, \Delta t]^2 + [g'(t_i^{**}) \, \Delta t]^2}$$
$$= \sqrt{[f'(t_i^*)]^2 + [g'(t_i^{**})]^2} \, \Delta t$$

and so

$$\boxed{4} \qquad L = \lim_{n \to \infty} \sum_{i=1}^n \sqrt{[f'(t_i^*)]^2 + [g'(t_i^{**})]^2} \, \Delta t$$

The sum in $\boxed{4}$ resembles a Riemann sum for the function $\sqrt{[f'(t)]^2 + [g'(t)]^2}$ but it is not exactly a Riemann sum because $t_i^* \neq t_i^{**}$ in general. Nevertheless, if f' and g' are continuous, it can be shown that the limit in $\boxed{4}$ is the same as if t_i^* and t_i^{**} were equal, namely,

$$L = \int_\alpha^\beta \sqrt{[f'(t)]^2 + [g'(t)]^2}\, dt$$

Thus, using Leibniz notation, we have the following result, which has the same form as Formula 3.

$\boxed{5}$ **Theorem** If a curve C is described by the parametric equations $x = f(t)$, $y = g(t)$, $\alpha \leqslant t \leqslant \beta$, where f' and g' are continuous on $[\alpha, \beta]$ and C is traversed exactly once as t increases from α to β, then the length of C is

$$L = \int_\alpha^\beta \sqrt{\left(\frac{dx}{dt}\right)^2 + \left(\frac{dy}{dt}\right)^2}\, dt$$

Notice that the formula in Theorem 5 is consistent with the general formulas $L = \int ds$ and $(ds)^2 = (dx)^2 + (dy)^2$ of Section 8.1.

EXAMPLE 4 If we use the representation of the unit circle given in Example 2 in Section 10.1,

$$x = \cos t \qquad y = \sin t \qquad 0 \leqslant t \leqslant 2\pi$$

then $dx/dt = -\sin t$ and $dy/dt = \cos t$, so Theorem 5 gives

$$L = \int_0^{2\pi} \sqrt{\left(\frac{dx}{dt}\right)^2 + \left(\frac{dy}{dt}\right)^2}\, dt = \int_0^{2\pi} \sqrt{\sin^2 t + \cos^2 t}\, dt = \int_0^{2\pi} dt = 2\pi$$

as expected. If, on the other hand, we use the representation given in Example 3 in Section 10.1,

$$x = \sin 2t \qquad y = \cos 2t \qquad 0 \leqslant t \leqslant 2\pi$$

then $dx/dt = 2 \cos 2t$, $dy/dt = -2 \sin 2t$, and the integral in Theorem 5 gives

$$\int_0^{2\pi} \sqrt{\left(\frac{dx}{dt}\right)^2 + \left(\frac{dy}{dt}\right)^2}\, dt = \int_0^{2\pi} \sqrt{4 \cos^2 2t + 4 \sin^2 2t}\, dt = \int_0^{2\pi} 2\, dt = 4\pi$$

⊘ Notice that the integral gives twice the arc length of the circle because as t increases from 0 to 2π, the point $(\sin 2t, \cos 2t)$ traverses the circle twice. In general, when finding the length of a curve C from a parametric representation, we have to be careful to ensure that C is traversed only once as t increases from α to β. ▬

V EXAMPLE 5 Find the length of one arch of the cycloid $x = r(\theta - \sin \theta)$, $y = r(1 - \cos \theta)$.

SOLUTION From Example 3 we see that one arch is described by the parameter interval $0 \leqslant \theta \leqslant 2\pi$. Since

$$\frac{dx}{d\theta} = r(1 - \cos \theta) \qquad \text{and} \qquad \frac{dy}{d\theta} = r \sin \theta$$

we have

$$L = \int_0^{2\pi} \sqrt{\left(\frac{dx}{d\theta}\right)^2 + \left(\frac{dy}{d\theta}\right)^2}\, d\theta$$

$$= \int_0^{2\pi} \sqrt{r^2(1 - \cos\theta)^2 + r^2 \sin^2\theta}\, d\theta$$

$$= \int_0^{2\pi} \sqrt{r^2(1 - 2\cos\theta + \cos^2\theta + \sin^2\theta)}\, d\theta$$

$$= r \int_0^{2\pi} \sqrt{2(1 - \cos\theta)}\, d\theta$$

The result of Example 5 says that the length of one arch of a cycloid is eight times the radius of the generating circle (see Figure 5). This was first proved in 1658 by Sir Christopher Wren, who later became the architect of St. Paul's Cathedral in London.

FIGURE 5

To evaluate this integral we use the identity $\sin^2 x = \frac{1}{2}(1 - \cos 2x)$ with $\theta = 2x$, which gives $1 - \cos\theta = 2\sin^2(\theta/2)$. Since $0 \le \theta \le 2\pi$, we have $0 \le \theta/2 \le \pi$ and so $\sin(\theta/2) \ge 0$. Therefore

$$\sqrt{2(1 - \cos\theta)} = \sqrt{4\sin^2(\theta/2)} = 2\,|\sin(\theta/2)| = 2\sin(\theta/2)$$

and so

$$L = 2r \int_0^{2\pi} \sin(\theta/2)\, d\theta = 2r\big[-2\cos(\theta/2)\big]_0^{2\pi}$$

$$= 2r[2 + 2] = 8r$$

Surface Area

In the same way as for arc length, we can adapt Formula 8.2.5 to obtain a formula for surface area. If the curve given by the parametric equations $x = f(t)$, $y = g(t)$, $\alpha \le t \le \beta$, is rotated about the x-axis, where f', g' are continuous and $g(t) \ge 0$, then the area of the resulting surface is given by

$$\boxed{6} \qquad S = \int_\alpha^\beta 2\pi y \sqrt{\left(\frac{dx}{dt}\right)^2 + \left(\frac{dy}{dt}\right)^2}\, dt$$

The general symbolic formulas $S = \int 2\pi y\, ds$ and $S = \int 2\pi x\, ds$ (Formulas 8.2.7 and 8.2.8) are still valid, but for parametric curves we use

$$ds = \sqrt{\left(\frac{dx}{dt}\right)^2 + \left(\frac{dy}{dt}\right)^2}\, dt$$

EXAMPLE 6 Show that the surface area of a sphere of radius r is $4\pi r^2$.

SOLUTION The sphere is obtained by rotating the semicircle

$$x = r\cos t \qquad y = r\sin t \qquad 0 \le t \le \pi$$

about the x-axis. Therefore, from Formula 6, we get

$$S = \int_0^\pi 2\pi r \sin t \sqrt{(-r\sin t)^2 + (r\cos t)^2}\, dt$$

$$= 2\pi \int_0^\pi r\sin t \sqrt{r^2(\sin^2 t + \cos^2 t)}\, dt = 2\pi \int_0^\pi r\sin t \cdot r\, dt$$

$$= 2\pi r^2 \int_0^\pi \sin t\, dt = 2\pi r^2(-\cos t)\big]_0^\pi = 4\pi r^2$$

10.2 Exercises

Access end-of-section exercises and tools for visualizing calculus online at **www.webassign.net**

LABORATORY PROJECT ⊞ BÉZIER CURVES

Bézier curves are used in computer-aided design and are named after the French mathematician Pierre Bézier (1910–1999), who worked in the automotive industry. A cubic Bézier curve is determined by four *control points*, $P_0(x_0, y_0)$, $P_1(x_1, y_1)$, $P_2(x_2, y_2)$, and $P_3(x_3, y_3)$, and is defined by the parametric equations

$$x = x_0(1 - t)^3 + 3x_1t(1 - t)^2 + 3x_2t^2(1 - t) + x_3t^3$$

$$y = y_0(1 - t)^3 + 3y_1t(1 - t)^2 + 3y_2t^2(1 - t) + y_3t^3$$

where $0 \le t \le 1$. Notice that when $t = 0$ we have $(x, y) = (x_0, y_0)$ and when $t = 1$ we have $(x, y) = (x_3, y_3)$, so the curve starts at P_0 and ends at P_3.

1. Graph the Bézier curve with control points $P_0(4, 1)$, $P_1(28, 48)$, $P_2(50, 42)$, and $P_3(40, 5)$. Then, on the same screen, graph the line segments P_0P_1, P_1P_2, and P_2P_3. (Exercise 31 in Section 10.1 shows how to do this.) Notice that the middle control points P_1 and P_2 don't lie on the curve; the curve starts at P_0, heads toward P_1 and P_2 without reaching them, and ends at P_3.

2. From the graph in Problem 1, it appears that the tangent at P_0 passes through P_1 and the tangent at P_3 passes through P_2. Prove it.

3. Try to produce a Bézier curve with a loop by changing the second control point in Problem 1.

4. Some laser printers use Bézier curves to represent letters and other symbols. Experiment with control points until you find a Bézier curve that gives a reasonable representation of the letter C.

5. More complicated shapes can be represented by piecing together two or more Bézier curves. Suppose the first Bézier curve has control points P_0, P_1, P_2, P_3 and the second one has control points P_3, P_4, P_5, P_6. If we want these two pieces to join together smoothly, then the tangents at P_3 should match and so the points P_2, P_3, and P_4 all have to lie on this common tangent line. Using this principle, find control points for a pair of Bézier curves that represent the letter S.

⊞ Graphing calculator or computer required

10.3 Polar Coordinates

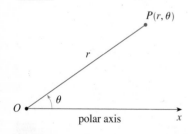

FIGURE 1

A coordinate system represents a point in the plane by an ordered pair of numbers called coordinates. Usually we use Cartesian coordinates, which are directed distances from two perpendicular axes. Here we describe a coordinate system introduced by Newton, called the **polar coordinate system**, which is more convenient for many purposes.

We choose a point in the plane that is called the **pole** (or origin) and is labeled O. Then we draw a ray (half-line) starting at O called the **polar axis**. This axis is usually drawn horizontally to the right and corresponds to the positive x-axis in Cartesian coordinates.

If P is any other point in the plane, let r be the distance from O to P and let θ be the angle (usually measured in radians) between the polar axis and the line OP as in Figure 1. Then

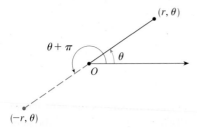

FIGURE 2

the point P is represented by the ordered pair (r, θ) and r, θ are called **polar coordinates** of P. We use the convention that an angle is positive if measured in the counterclockwise direction from the polar axis and negative in the clockwise direction. If $P = O$, then $r = 0$ and we agree that $(0, \theta)$ represents the pole for any value of θ.

We extend the meaning of polar coordinates (r, θ) to the case in which r is negative by agreeing that, as in Figure 2, the points $(-r, \theta)$ and (r, θ) lie on the same line through O and at the same distance $|r|$ from O, but on opposite sides of O. If $r > 0$, the point (r, θ) lies in the same quadrant as θ; if $r < 0$, it lies in the quadrant on the opposite side of the pole. Notice that $(-r, \theta)$ represents the same point as $(r, \theta + \pi)$.

EXAMPLE 1 Plot the points whose polar coordinates are given.
(a) $(1, 5\pi/4)$ (b) $(2, 3\pi)$ (c) $(2, -2\pi/3)$ (d) $(-3, 3\pi/4)$

SOLUTION The points are plotted in Figure 3. In part (d) the point $(-3, 3\pi/4)$ is located three units from the pole in the fourth quadrant because the angle $3\pi/4$ is in the second quadrant and $r = -3$ is negative.

FIGURE 3

In the Cartesian coordinate system every point has only one representation, but in the polar coordinate system each point has many representations. For instance, the point $(1, 5\pi/4)$ in Example 1(a) could be written as $(1, -3\pi/4)$ or $(1, 13\pi/4)$ or $(-1, \pi/4)$. (See Figure 4.)

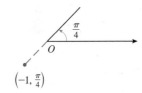

FIGURE 4

In fact, since a complete counterclockwise rotation is given by an angle 2π, the point represented by polar coordinates (r, θ) is also represented by

$$(r, \theta + 2n\pi) \qquad \text{and} \qquad (-r, \theta + (2n + 1)\pi)$$

where n is any integer.

The connection between polar and Cartesian coordinates can be seen from Figure 5, in which the pole corresponds to the origin and the polar axis coincides with the positive x-axis. If the point P has Cartesian coordinates (x, y) and polar coordinates (r, θ), then, from the figure, we have

$$\cos \theta = \frac{x}{r} \qquad \sin \theta = \frac{y}{r}$$

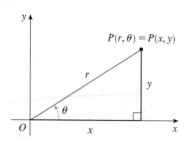

FIGURE 5

and so

$$\boxed{\mathbf{1}} \qquad \boxed{\quad x = r \cos \theta \qquad y = r \sin \theta \quad}$$

Although Equations 1 were deduced from Figure 5, which illustrates the case where $r > 0$ and $0 < \theta < \pi/2$, these equations are valid for all values of r and θ. (See the general definition of $\sin \theta$ and $\cos \theta$ in Appendix D.)

Equations 1 allow us to find the Cartesian coordinates of a point when the polar coordinates are known. To find r and θ when x and y are known, we use the equations

$$\boxed{\mathbf{2}} \qquad \boxed{\quad r^2 = x^2 + y^2 \qquad \tan \theta = \frac{y}{x} \quad}$$

which can be deduced from Equations 1 or simply read from Figure 5.

EXAMPLE 2 Convert the point $(2, \pi/3)$ from polar to Cartesian coordinates.

SOLUTION Since $r = 2$ and $\theta = \pi/3$, Equations 1 give

$$x = r \cos \theta = 2 \cos \frac{\pi}{3} = 2 \cdot \frac{1}{2} = 1$$

$$y = r \sin \theta = 2 \sin \frac{\pi}{3} = 2 \cdot \frac{\sqrt{3}}{2} = \sqrt{3}$$

Therefore the point is $\left(1, \sqrt{3}\right)$ in Cartesian coordinates.

EXAMPLE 3 Represent the point with Cartesian coordinates $(1, -1)$ in terms of polar coordinates.

SOLUTION If we choose r to be positive, then Equations 2 give

$$r = \sqrt{x^2 + y^2} = \sqrt{1^2 + (-1)^2} = \sqrt{2}$$

$$\tan \theta = \frac{y}{x} = -1$$

Since the point $(1, -1)$ lies in the fourth quadrant, we can choose $\theta = -\pi/4$ or $\theta = 7\pi/4$. Thus one possible answer is $\left(\sqrt{2}, -\pi/4\right)$; another is $\left(\sqrt{2}, 7\pi/4\right)$.

NOTE Equations 2 do not uniquely determine θ when x and y are given because, as θ increases through the interval $0 \le \theta < 2\pi$, each value of $\tan \theta$ occurs twice. Therefore, in converting from Cartesian to polar coordinates, it's not good enough just to find r and θ that satisfy Equations 2. As in Example 3, we must choose θ so that the point (r, θ) lies in the correct quadrant.

Polar Curves

The **graph of a polar equation** $r = f(\theta)$, or more generally $F(r, \theta) = 0$, consists of all points P that have at least one polar representation (r, θ) whose coordinates satisfy the equation.

V EXAMPLE 4 What curve is represented by the polar equation $r = 2$?

SOLUTION The curve consists of all points (r, θ) with $r = 2$. Since r represents the distance from the point to the pole, the curve $r = 2$ represents the circle with center O and radius 2. In general, the equation $r = a$ represents a circle with center O and radius $|a|$. (See Figure 6.)

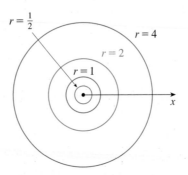

FIGURE 6

EXAMPLE 5 Sketch the polar curve $\theta = 1$.

SOLUTION This curve consists of all points (r, θ) such that the polar angle θ is 1 radian. It is the straight line that passes through O and makes an angle of 1 radian with the polar axis (see Figure 7). Notice that the points $(r, 1)$ on the line with $r > 0$ are in the first quadrant, whereas those with $r < 0$ are in the third quadrant.

FIGURE 7

EXAMPLE 6
(a) Sketch the curve with polar equation $r = 2 \cos \theta$.
(b) Find a Cartesian equation for this curve.

SOLUTION
(a) In Figure 8 we find the values of r for some convenient values of θ and plot the corresponding points (r, θ). Then we join these points to sketch the curve, which appears to be a circle. We have used only values of θ between 0 and π, since if we let θ increase beyond π, we obtain the same points again.

θ	$r = 2 \cos \theta$
0	2
$\pi/6$	$\sqrt{3}$
$\pi/4$	$\sqrt{2}$
$\pi/3$	1
$\pi/2$	0
$2\pi/3$	-1
$3\pi/4$	$-\sqrt{2}$
$5\pi/6$	$-\sqrt{3}$
π	-2

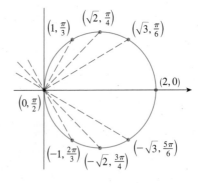

FIGURE 8
Table of values and
graph of $r = 2 \cos \theta$

(b) To convert the given equation to a Cartesian equation we use Equations 1 and 2. From $x = r \cos \theta$ we have $\cos \theta = x/r$, so the equation $r = 2 \cos \theta$ becomes $r = 2x/r$, which gives

$$2x = r^2 = x^2 + y^2 \qquad \text{or} \qquad x^2 + y^2 - 2x = 0$$

Completing the square, we obtain

$$(x - 1)^2 + y^2 = 1$$

which is an equation of a circle with center $(1, 0)$ and radius 1.

Figure 9 shows a geometrical illustration that the circle in Example 6 has the equation $r = 2 \cos \theta$. The angle OPQ is a right angle (Why?) and so $r/2 = \cos \theta$.

FIGURE 9

FIGURE 10
$r = 1 + \sin \theta$ in Cartesian coordinates,
$0 \leqslant \theta \leqslant 2\pi$

V **EXAMPLE 7** Sketch the curve $r = 1 + \sin \theta$.

SOLUTION Instead of plotting points as in Example 6, we first sketch the graph of $r = 1 + \sin \theta$ in *Cartesian* coordinates in Figure 10 by shifting the sine curve up one unit. This enables us to read at a glance the values of r that correspond to increasing values of θ. For instance, we see that as θ increases from 0 to $\pi/2$, r (the distance from O) increases from 1 to 2, so we sketch the corresponding part of the polar curve in Figure 11(a). As θ increases from $\pi/2$ to π, Figure 10 shows that r decreases from 2 to 1, so we sketch the next part of the curve as in Figure 11(b). As θ increases from π to $3\pi/2$, r decreases from 1 to 0 as shown in part (c). Finally, as θ increases from $3\pi/2$ to 2π, r increases from 0 to 1 as shown in part (d). If we let θ increase beyond 2π or decrease beyond 0, we would simply retrace our path. Putting together the parts of the curve from Figure 11(a)–(d), we sketch the complete curve in part (e). It is called a **cardioid** because it's shaped like a heart.

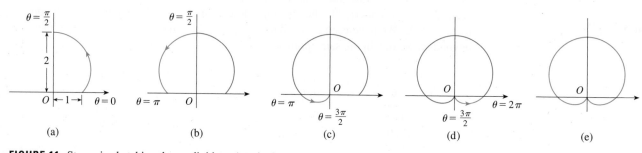

FIGURE 11 Stages in sketching the cardioid $r = 1 + \sin \theta$

EXAMPLE 8 Sketch the curve $r = \cos 2\theta$.

SOLUTION As in Example 7, we first sketch $r = \cos 2\theta$, $0 \leqslant \theta \leqslant 2\pi$, in Cartesian coordinates in Figure 12. As θ increases from 0 to $\pi/4$, Figure 12 shows that r decreases from 1 to 0 and so we draw the corresponding portion of the polar curve in Figure 13 (indicated by ①). As θ increases from $\pi/4$ to $\pi/2$, r goes from 0 to -1. This means that the distance from O increases from 0 to 1, but instead of being in the first quadrant this portion of the polar curve (indicated by ②) lies on the opposite side of the pole in the third quadrant. The remainder of the curve is drawn in a similar fashion, with the arrows

TEC Module 10.3 helps you see how polar curves are traced out by showing animations similar to Figures 10–13.

and numbers indicating the order in which the portions are traced out. The resulting curve has four loops and is called a **four-leaved rose**.

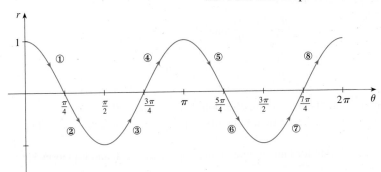

FIGURE 12
$r = \cos 2\theta$ in Cartesian coordinates

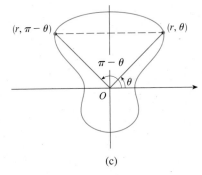

FIGURE 13
Four-leaved rose $r = \cos 2\theta$

Symmetry

When we sketch polar curves it is sometimes helpful to take advantage of symmetry. The following three rules are explained by Figure 14.

(a) If a polar equation is unchanged when θ is replaced by $-\theta$, the curve is symmetric about the polar axis.

(b) If the equation is unchanged when r is replaced by $-r$, or when θ is replaced by $\theta + \pi$, the curve is symmetric about the pole. (This means that the curve remains unchanged if we rotate it through 180° about the origin.)

(c) If the equation is unchanged when θ is replaced by $\pi - \theta$, the curve is symmetric about the vertical line $\theta = \pi/2$.

(a)

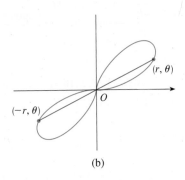

(b)

(c)

FIGURE 14

The curves sketched in Examples 6 and 8 are symmetric about the polar axis, since $\cos(-\theta) = \cos\theta$. The curves in Examples 7 and 8 are symmetric about $\theta = \pi/2$ because $\sin(\pi - \theta) = \sin\theta$ and $\cos 2(\pi - \theta) = \cos 2\theta$. The four-leaved rose is also symmetric about the pole. These symmetry properties could have been used in sketching the curves. For instance, in Example 6 we need only have plotted points for $0 \le \theta \le \pi/2$ and then reflected about the polar axis to obtain the complete circle.

Tangents to Polar Curves

To find a tangent line to a polar curve $r = f(\theta)$, we regard θ as a parameter and write its parametric equations as

$$x = r\cos\theta = f(\theta)\cos\theta \qquad y = r\sin\theta = f(\theta)\sin\theta$$

Then, using the method for finding slopes of parametric curves (Equation 10.2.1) and the Product Rule, we have

$$\boxed{3} \qquad \frac{dy}{dx} = \frac{\dfrac{dy}{d\theta}}{\dfrac{dx}{d\theta}} = \frac{\dfrac{dr}{d\theta}\sin\theta + r\cos\theta}{\dfrac{dr}{d\theta}\cos\theta - r\sin\theta}$$

We locate horizontal tangents by finding the points where $dy/d\theta = 0$ (provided that $dx/d\theta \neq 0$). Likewise, we locate vertical tangents at the points where $dx/d\theta = 0$ (provided that $dy/d\theta \neq 0$).

Notice that if we are looking for tangent lines at the pole, then $r = 0$ and Equation 3 simplifies to

$$\frac{dy}{dx} = \tan\theta \qquad \text{if} \quad \frac{dr}{d\theta} \neq 0$$

For instance, in Example 8 we found that $r = \cos 2\theta = 0$ when $\theta = \pi/4$ or $3\pi/4$. This means that the lines $\theta = \pi/4$ and $\theta = 3\pi/4$ (or $y = x$ and $y = -x$) are tangent lines to $r = \cos 2\theta$ at the origin.

EXAMPLE 9

(a) For the cardioid $r = 1 + \sin\theta$ of Example 7, find the slope of the tangent line when $\theta = \pi/3$.

(b) Find the points on the cardioid where the tangent line is horizontal or vertical.

SOLUTION Using Equation 3 with $r = 1 + \sin\theta$, we have

$$\frac{dy}{dx} = \frac{\dfrac{dr}{d\theta}\sin\theta + r\cos\theta}{\dfrac{dr}{d\theta}\cos\theta - r\sin\theta} = \frac{\cos\theta\sin\theta + (1 + \sin\theta)\cos\theta}{\cos\theta\cos\theta - (1 + \sin\theta)\sin\theta}$$

$$= \frac{\cos\theta\,(1 + 2\sin\theta)}{1 - 2\sin^2\theta - \sin\theta} = \frac{\cos\theta\,(1 + 2\sin\theta)}{(1 + \sin\theta)(1 - 2\sin\theta)}$$

(a) The slope of the tangent at the point where $\theta = \pi/3$ is

$$\left.\frac{dy}{dx}\right|_{\theta=\pi/3} = \frac{\cos(\pi/3)(1 + 2\sin(\pi/3))}{(1 + \sin(\pi/3))(1 - 2\sin(\pi/3))} = \frac{\tfrac{1}{2}\left(1 + \sqrt{3}\right)}{\left(1 + \sqrt{3}/2\right)\left(1 - \sqrt{3}\right)}$$

$$= \frac{1 + \sqrt{3}}{\left(2 + \sqrt{3}\right)\left(1 - \sqrt{3}\right)} = \frac{1 + \sqrt{3}}{-1 - \sqrt{3}} = -1$$

(b) Observe that

$$\frac{dy}{d\theta} = \cos\theta\,(1 + 2\sin\theta) = 0 \qquad \text{when } \theta = \frac{\pi}{2}, \frac{3\pi}{2}, \frac{7\pi}{6}, \frac{11\pi}{6}$$

$$\frac{dx}{d\theta} = (1 + \sin\theta)(1 - 2\sin\theta) = 0 \qquad \text{when } \theta = \frac{3\pi}{2}, \frac{\pi}{6}, \frac{5\pi}{6}$$

Therefore there are horizontal tangents at the points $(2, \pi/2)$, $\left(\frac{1}{2}, 7\pi/6\right)$, $\left(\frac{1}{2}, 11\pi/6\right)$ and vertical tangents at $\left(\frac{3}{2}, \pi/6\right)$ and $\left(\frac{3}{2}, 5\pi/6\right)$. When $\theta = 3\pi/2$, both $dy/d\theta$ and $dx/d\theta$ are 0, so we must be careful. Using l'Hospital's Rule, we have

$$\lim_{\theta \to (3\pi/2)^-} \frac{dy}{dx} = \left(\lim_{\theta \to (3\pi/2)^-} \frac{1 + 2\sin\theta}{1 - 2\sin\theta}\right)\left(\lim_{\theta \to (3\pi/2)^-} \frac{\cos\theta}{1 + \sin\theta}\right)$$

$$= -\frac{1}{3} \lim_{\theta \to (3\pi/2)^-} \frac{\cos\theta}{1 + \sin\theta} = -\frac{1}{3} \lim_{\theta \to (3\pi/2)^-} \frac{-\sin\theta}{\cos\theta} = \infty$$

By symmetry,

$$\lim_{\theta \to (3\pi/2)^+} \frac{dy}{dx} = -\infty$$

Thus there is a vertical tangent line at the pole (see Figure 15).

NOTE Instead of having to remember Equation 3, we could employ the method used to derive it. For instance, in Example 9 we could have written

$$x = r\cos\theta = (1 + \sin\theta)\cos\theta = \cos\theta + \tfrac{1}{2}\sin 2\theta$$

$$y = r\sin\theta = (1 + \sin\theta)\sin\theta = \sin\theta + \sin^2\theta$$

Then we have

$$\frac{dy}{dx} = \frac{dy/d\theta}{dx/d\theta} = \frac{\cos\theta + 2\sin\theta\cos\theta}{-\sin\theta + \cos 2\theta} = \frac{\cos\theta + \sin 2\theta}{-\sin\theta + \cos 2\theta}$$

which is equivalent to our previous expression.

FIGURE 15

Tangent lines for $r = 1 + \sin\theta$

Graphing Polar Curves with Graphing Devices

Although it's useful to be able to sketch simple polar curves by hand, we need to use a graphing calculator or computer when we are faced with a curve as complicated as the ones shown in Figures 16 and 17.

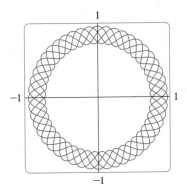

FIGURE 16

$r = \sin^2(2.4\theta) + \cos^4(2.4\theta)$

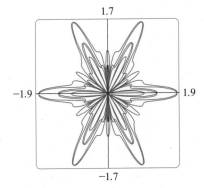

FIGURE 17

$r = \sin^2(1.2\theta) + \cos^3(6\theta)$

Some graphing devices have commands that enable us to graph polar curves directly. With other machines we need to convert to parametric equations first. In this case we take the polar equation $r = f(\theta)$ and write its parametric equations as

$$x = r\cos\theta = f(\theta)\cos\theta \qquad y = r\sin\theta = f(\theta)\sin\theta$$

Some machines require that the parameter be called t rather than θ.

EXAMPLE 10 Graph the curve $r = \sin(8\theta/5)$.

SOLUTION Let's assume that our graphing device doesn't have a built-in polar graphing command. In this case we need to work with the corresponding parametric equations, which are

$$x = r \cos\theta = \sin(8\theta/5) \cos\theta \qquad y = r \sin\theta = \sin(8\theta/5) \sin\theta$$

In any case we need to determine the domain for θ. So we ask ourselves: How many complete rotations are required until the curve starts to repeat itself? If the answer is n, then

$$\sin\frac{8(\theta + 2n\pi)}{5} = \sin\left(\frac{8\theta}{5} + \frac{16n\pi}{5}\right) = \sin\frac{8\theta}{5}$$

and so we require that $16n\pi/5$ be an even multiple of π. This will first occur when $n = 5$. Therefore we will graph the entire curve if we specify that $0 \leqslant \theta \leqslant 10\pi$. Switching from θ to t, we have the equations

$$x = \sin(8t/5) \cos t \qquad y = \sin(8t/5) \sin t \qquad 0 \leqslant t \leqslant 10\pi$$

and Figure 18 shows the resulting curve. Notice that this rose has 16 loops.   ▬

V EXAMPLE 11 Investigate the family of polar curves given by $r = 1 + c \sin\theta$. How does the shape change as c changes? (These curves are called **limaçons**, after a French word for snail, because of the shape of the curves for certain values of c.)

SOLUTION Figure 19 shows computer-drawn graphs for various values of c. For $c > 1$ there is a loop that decreases in size as c decreases. When $c = 1$ the loop disappears and the curve becomes the cardioid that we sketched in Example 7. For c between 1 and $\frac{1}{2}$ the cardioid's cusp is smoothed out and becomes a "dimple." When c decreases from $\frac{1}{2}$ to 0, the limaçon is shaped like an oval. This oval becomes more circular as $c \to 0$, and when $c = 0$ the curve is just the circle $r = 1$.

In Exercise 53 you are asked to prove analytically what we have discovered from the graphs in Figure 19.

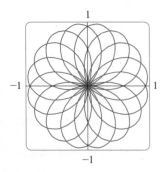

FIGURE 18
$r = \sin(8\theta/5)$

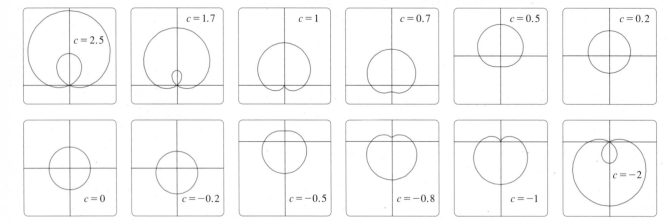

FIGURE 19
Members of the family of
limaçons $r = 1 + c \sin\theta$

The remaining parts of Figure 19 show that as c becomes negative, the shapes change in reverse order. In fact, these curves are reflections about the horizontal axis of the corresponding curves with positive c.   ▬

Limaçons arise in the study of planetary motion. In particular, the trajectory of Mars, as viewed from the planet Earth, has been modeled by a limaçon with a loop, as in the parts of Figure 19 with $|c| > 1$.

LABORATORY PROJECT FAMILIES OF POLAR CURVES

In this project you will discover the interesting and beautiful shapes that members of families of polar curves can take. You will also see how the shape of the curve changes when you vary the constants.

1. (a) Investigate the family of curves defined by the polar equations $r = \sin n\theta$, where n is a positive integer. How is the number of loops related to n?
    (b) What happens if the equation in part (a) is replaced by $r = |\sin n\theta|$?

2. A family of curves is given by the equations $r = 1 + c \sin n\theta$, where c is a real number and n is a positive integer. How does the graph change as n increases? How does it change as c changes? Illustrate by graphing enough members of the family to support your conclusions.

3. A family of curves has polar equations

$$r = \frac{1 - a \cos \theta}{1 + a \cos \theta}$$

Investigate how the graph changes as the number a changes. In particular, you should identify the transitional values of a for which the basic shape of the curve changes.

4. The astronomer Giovanni Cassini (1625–1712) studied the family of curves with polar equations

$$r^4 - 2c^2r^2 \cos 2\theta + c^4 - a^4 = 0$$

where a and c are positive real numbers. These curves are called the **ovals of Cassini** even though they are oval shaped only for certain values of a and c. (Cassini thought that these curves might represent planetary orbits better than Kepler's ellipses.) Investigate the variety of shapes that these curves may have. In particular, how are a and c related to each other when the curve splits into two parts?

 Graphing calculator or computer required

| **10.4** | **Areas and Lengths in Polar Coordinates** |

In this section we develop the formula for the area of a region whose boundary is given by a polar equation. We need to use the formula for the area of a sector of a circle:

> **1** $\qquad\qquad A = \frac{1}{2}r^2\theta$

where, as in Figure 1, r is the radius and θ is the radian measure of the central angle. Formula 1 follows from the fact that the area of a sector is proportional to its central angle: $A = (\theta/2\pi)\pi r^2 = \frac{1}{2}r^2\theta$. (See also Exercise 35 in Section 7.3.)

FIGURE 1

Let $\mathcal{R}$ be the region, illustrated in Figure 2 (on page 546), bounded by the polar curve $r = f(\theta)$ and by the rays $\theta = a$ and $\theta = b$, where f is a positive continuous function and

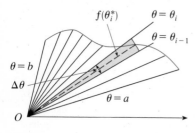

FIGURE 2

where $0 < b - a \leqslant 2\pi$. We divide the interval $[a, b]$ into subintervals with endpoints θ_0, $\theta_1, \theta_2, \ldots, \theta_n$ and equal width $\Delta\theta$. The rays $\theta = \theta_i$ then divide $\mathcal{R}$ into n smaller regions with central angle $\Delta\theta = \theta_i - \theta_{i-1}$. If we choose θ_i^* in the ith subinterval $[\theta_{i-1}, \theta_i]$, then the area ΔA_i of the ith region is approximated by the area of the sector of a circle with central angle $\Delta\theta$ and radius $f(\theta_i^*)$. (See Figure 3.)

Thus from Formula 1 we have

$$\Delta A_i \approx \tfrac{1}{2}[f(\theta_i^*)]^2 \Delta\theta$$

and so an approximation to the total area A of $\mathcal{R}$ is

$$\boxed{2} \qquad A \approx \sum_{i=1}^{n} \tfrac{1}{2}[f(\theta_i^*)]^2 \Delta\theta$$

It appears from Figure 3 that the approximation in $\boxed{2}$ improves as $n \to \infty$. But the sums in $\boxed{2}$ are Riemann sums for the function $g(\theta) = \tfrac{1}{2}[f(\theta)]^2$, so

$$\lim_{n\to\infty} \sum_{i=1}^{n} \tfrac{1}{2}[f(\theta_i^*)]^2 \Delta\theta = \int_a^b \tfrac{1}{2}[f(\theta)]^2 \, d\theta$$

It therefore appears plausible (and can in fact be proved) that the formula for the area A of the polar region $\mathcal{R}$ is

$$\boxed{3} \qquad A = \int_a^b \tfrac{1}{2}[f(\theta)]^2 \, d\theta$$

Formula 3 is often written as

$$\boxed{4} \qquad A = \int_a^b \tfrac{1}{2} r^2 \, d\theta$$

with the understanding that $r = f(\theta)$. Note the similarity between Formulas 1 and 4.

When we apply Formula 3 or 4 it is helpful to think of the area as being swept out by a rotating ray through O that starts with angle a and ends with angle b.

FIGURE 3

V **EXAMPLE 1** Find the area enclosed by one loop of the four-leaved rose $r = \cos 2\theta$.

SOLUTION The curve $r = \cos 2\theta$ was sketched in Example 8 in Section 10.3. Notice from Figure 4 that the region enclosed by the right loop is swept out by a ray that rotates from $\theta = -\pi/4$ to $\theta = \pi/4$. Therefore Formula 4 gives

$$A = \int_{-\pi/4}^{\pi/4} \tfrac{1}{2} r^2 \, d\theta = \tfrac{1}{2} \int_{-\pi/4}^{\pi/4} \cos^2 2\theta \, d\theta = \int_0^{\pi/4} \cos^2 2\theta \, d\theta$$

$$A = \int_0^{\pi/4} \tfrac{1}{2}(1 + \cos 4\theta) \, d\theta = \tfrac{1}{2}\left[\theta + \tfrac{1}{4}\sin 4\theta\right]_0^{\pi/4} = \frac{\pi}{8}$$

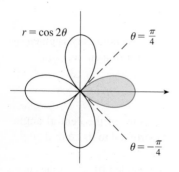

FIGURE 4

V **EXAMPLE 2** Find the area of the region that lies inside the circle $r = 3 \sin \theta$ and outside the cardioid $r = 1 + \sin \theta$.

SOLUTION The cardioid (see Example 7 in Section 10.3) and the circle are sketched in Figure 5 and the desired region is shaded. The values of a and b in Formula 4 are determined by finding the points of intersection of the two curves. They intersect when $3 \sin \theta = 1 + \sin \theta$, which gives $\sin \theta = \frac{1}{2}$, so $\theta = \pi/6, 5\pi/6$. The desired area can be found by subtracting the area inside the cardioid between $\theta = \pi/6$ and $\theta = 5\pi/6$ from the area inside the circle from $\pi/6$ to $5\pi/6$. Thus

$$A = \frac{1}{2} \int_{\pi/6}^{5\pi/6} (3 \sin \theta)^2 \, d\theta - \frac{1}{2} \int_{\pi/6}^{5\pi/6} (1 + \sin \theta)^2 \, d\theta$$

Since the region is symmetric about the vertical axis $\theta = \pi/2$, we can write

$$A = 2 \left[\frac{1}{2} \int_{\pi/6}^{\pi/2} 9 \sin^2 \theta \, d\theta - \frac{1}{2} \int_{\pi/6}^{\pi/2} (1 + 2 \sin \theta + \sin^2 \theta) \, d\theta \right]$$

$$= \int_{\pi/6}^{\pi/2} (8 \sin^2 \theta - 1 - 2 \sin \theta) \, d\theta$$

$$= \int_{\pi/6}^{\pi/2} (3 - 4 \cos 2\theta - 2 \sin \theta) \, d\theta \qquad [\text{because } \sin^2 \theta = \tfrac{1}{2}(1 - \cos 2\theta)]$$

$$= 3\theta - 2 \sin 2\theta + 2 \cos \theta \Big]_{\pi/6}^{\pi/2} = \pi \qquad \blacksquare$$

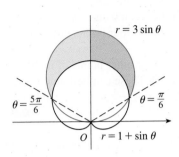

$r = 3 \sin \theta$

$\theta = \dfrac{5\pi}{6}$ $\theta = \dfrac{\pi}{6}$

O $r = 1 + \sin \theta$

FIGURE 5

Example 2 illustrates the procedure for finding the area of the region bounded by two polar curves. In general, let $\mathcal{R}$ be a region, as illustrated in Figure 6, that is bounded by curves with polar equations $r = f(\theta)$, $r = g(\theta)$, $\theta = a$, and $\theta = b$, where $f(\theta) \geq g(\theta) \geq 0$ and $0 < b - a \leq 2\pi$. The area A of $\mathcal{R}$ is found by subtracting the area inside $r = g(\theta)$ from the area inside $r = f(\theta)$, so using Formula 3 we have

$$A = \int_a^b \tfrac{1}{2} [f(\theta)]^2 \, d\theta - \int_a^b \tfrac{1}{2} [g(\theta)]^2 \, d\theta$$

$$= \tfrac{1}{2} \int_a^b ([f(\theta)]^2 - [g(\theta)]^2) \, d\theta$$

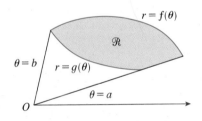

$r = f(\theta)$

$\mathcal{R}$

$\theta = b$ $r = g(\theta)$

$\theta = a$

O

FIGURE 6

⊘ **CAUTION** The fact that a single point has many representations in polar coordinates sometimes makes it difficult to find all the points of intersection of two polar curves. For instance, it is obvious from Figure 5 that the circle and the cardioid have three points of intersection; however, in Example 2 we solved the equations $r = 3 \sin \theta$ and $r = 1 + \sin \theta$ and found only two such points, $\left(\frac{3}{2}, \pi/6\right)$ and $\left(\frac{3}{2}, 5\pi/6\right)$. The origin is also a point of intersection, but we can't find it by solving the equations of the curves because the origin has no single representation in polar coordinates that satisfies both equations. Notice that, when represented as $(0, 0)$ or $(0, \pi)$, the origin satisfies $r = 3 \sin \theta$ and so it lies on the circle; when represented as $(0, 3\pi/2)$, it satisfies $r = 1 + \sin \theta$ and so it lies on the cardioid. Think of two points moving along the curves as the parameter value θ increases from 0 to 2π. On one curve the origin is reached at $\theta = 0$ and $\theta = \pi$; on the other curve it is reached at $\theta = 3\pi/2$. The points don't collide at the origin because they reach the origin at different times, but the curves intersect there nonetheless.

Thus, to find *all* points of intersection of two polar curves, it is recommended that you draw the graphs of both curves. It is especially convenient to use a graphing calculator or computer to help with this task.

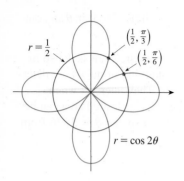

$r = \frac{1}{2}$

$\left(\frac{1}{2}, \frac{\pi}{3}\right)$

$\left(\frac{1}{2}, \frac{\pi}{6}\right)$

$r = \cos 2\theta$

FIGURE 7

EXAMPLE 3 Find all points of intersection of the curves $r = \cos 2\theta$ and $r = \frac{1}{2}$.

SOLUTION If we solve the equations $r = \cos 2\theta$ and $r = \frac{1}{2}$, we get $\cos 2\theta = \frac{1}{2}$ and, there-fore, $2\theta = \pi/3, 5\pi/3, 7\pi/3, 11\pi/3$. Thus the values of θ between 0 and 2π that satisfy both equations are $\theta = \pi/6, 5\pi/6, 7\pi/6, 11\pi/6$. We have found four points of inter-section: $\left(\frac{1}{2}, \pi/6\right)$, $\left(\frac{1}{2}, 5\pi/6\right)$, $\left(\frac{1}{2}, 7\pi/6\right)$, and $\left(\frac{1}{2}, 11\pi/6\right)$.

However, you can see from Figure 7 that the curves have four other points of inter-section—namely, $\left(\frac{1}{2}, \pi/3\right)$, $\left(\frac{1}{2}, 2\pi/3\right)$, $\left(\frac{1}{2}, 4\pi/3\right)$, and $\left(\frac{1}{2}, 5\pi/3\right)$. These can be found using symmetry or by noticing that another equation of the circle is $r = -\frac{1}{2}$ and then solving the equations $r = \cos 2\theta$ and $r = -\frac{1}{2}$. ∎

Arc Length

To find the length of a polar curve $r = f(\theta)$, $a \leqslant \theta \leqslant b$, we regard θ as a parameter and write the parametric equations of the curve as

$$x = r \cos \theta = f(\theta) \cos \theta \qquad y = r \sin \theta = f(\theta) \sin \theta$$

Using the Product Rule and differentiating with respect to θ, we obtain

$$\frac{dx}{d\theta} = \frac{dr}{d\theta} \cos \theta - r \sin \theta \qquad \frac{dy}{d\theta} = \frac{dr}{d\theta} \sin \theta + r \cos \theta$$

so, using $\cos^2\theta + \sin^2\theta = 1$, we have

$$\left(\frac{dx}{d\theta}\right)^2 + \left(\frac{dy}{d\theta}\right)^2 = \left(\frac{dr}{d\theta}\right)^2 \cos^2\theta - 2r \frac{dr}{d\theta} \cos \theta \sin \theta + r^2 \sin^2\theta$$

$$+ \left(\frac{dr}{d\theta}\right)^2 \sin^2\theta + 2r \frac{dr}{d\theta} \sin \theta \cos \theta + r^2 \cos^2\theta$$

$$= \left(\frac{dr}{d\theta}\right)^2 + r^2$$

Assuming that f' is continuous, we can use Theorem 10.2.5 to write the arc length as

$$L = \int_a^b \sqrt{\left(\frac{dx}{d\theta}\right)^2 + \left(\frac{dy}{d\theta}\right)^2} \, d\theta$$

Therefore the length of a curve with polar equation $r = f(\theta)$, $a \leqslant \theta \leqslant b$, is

5
$$L = \int_a^b \sqrt{r^2 + \left(\frac{dr}{d\theta}\right)^2} \, d\theta$$

▼ **EXAMPLE 4** Find the length of the cardioid $r = 1 + \sin \theta$.

SOLUTION The cardioid is shown in Figure 8. (We sketched it in Example 7 in Section 10.3.) Its full length is given by the parameter interval $0 \leqslant \theta \leqslant 2\pi$, so

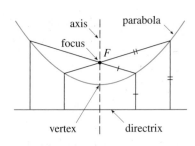

FIGURE 8

$r = 1 + \sin \theta$

Formula 5 gives

$$L = \int_0^{2\pi} \sqrt{r^2 + \left(\frac{dr}{d\theta}\right)^2}\ d\theta = \int_0^{2\pi} \sqrt{(1 + \sin \theta)^2 + \cos^2\theta}\ d\theta$$

$$= \int_0^{2\pi} \sqrt{2 + 2 \sin \theta}\ d\theta$$

We could evaluate this integral by multiplying and dividing the integrand by $\sqrt{2 - 2 \sin \theta}$, or we could use a computer algebra system. In any event, we find that the length of the cardioid is $L = 8$.

10.4 Exercises

Access end-of-section exercises and tools for visualizing calculus online at **www.webassign.net**

10.5 Conic Sections

In this section we give geometric definitions of parabolas, ellipses, and hyperbolas and derive their standard equations. They are called **conic sections**, or **conics**, because they result from intersecting a cone with a plane as shown in Figure 1.

ellipse   parabola   hyperbola

FIGURE 1

Conics

Parabolas

A **parabola** is the set of points in a plane that are equidistant from a fixed point F (called the **focus**) and a fixed line (called the **directrix**). This definition is illustrated by Figure 2. Notice that the point halfway between the focus and the directrix lies on the parabola; it is called the **vertex**. The line through the focus perpendicular to the directrix is called the **axis** of the parabola.

In the 16th century Galileo showed that the path of a projectile that is shot into the air at an angle to the ground is a parabola. Since then, parabolic shapes have been used in designing automobile headlights, reflecting telescopes, and suspension bridges. (See Problem 16 on page 153 for the reflection property of parabolas that makes them so useful.)

We obtain a particularly simple equation for a parabola if we place its vertex at the origin O and its directrix parallel to the x-axis as in Figure 3. If the focus is the point $(0, p)$, then the directrix has the equation $y = -p$. If $P(x, y)$ is any point on the parabola,

axis   parabola

focus

F

vertex   directrix

FIGURE 2

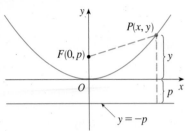

FIGURE 3

then the distance from P to the focus is

$$|PF| = \sqrt{x^2 + (y - p)^2}$$

and the distance from P to the directrix is $|y + p|$. (Figure 3 illustrates the case where $p > 0$.) The defining property of a parabola is that these distances are equal:

$$\sqrt{x^2 + (y - p)^2} = |y + p|$$

We get an equivalent equation by squaring and simplifying:

$$x^2 + (y - p)^2 = |y + p|^2 = (y + p)^2$$

$$x^2 + y^2 - 2py + p^2 = y^2 + 2py + p^2$$

$$x^2 = 4py$$

---

1   An equation of the parabola with focus $(0, p)$ and directrix $y = -p$ is

$$x^2 = 4py$$

If we write $a = 1/(4p)$, then the standard equation of a parabola **1** becomes $y = ax^2$. It opens upward if $p > 0$ and downward if $p < 0$ [see Figure 4, parts (a) and (b)]. The graph is symmetric with respect to the y-axis because **1** is unchanged when x is replaced by $-x$.

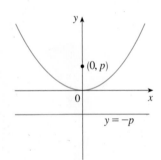

(a) $x^2 = 4py$, $p > 0$

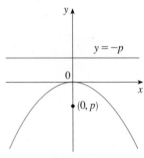

(b) $x^2 = 4py$, $p < 0$

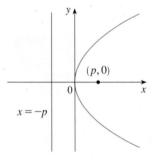

(c) $y^2 = 4px$, $p > 0$

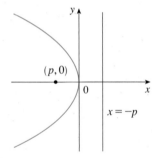

(d) $y^2 = 4px$, $p < 0$

FIGURE 4

If we interchange x and y in **1**, we obtain

2
$$y^2 = 4px$$

which is an equation of the parabola with focus $(p, 0)$ and directrix $x = -p$. (Interchanging x and y amounts to reflecting about the diagonal line $y = x$.) The parabola opens to the right if $p > 0$ and to the left if $p < 0$ [see Figure 4, parts (c) and (d)]. In both cases the graph is symmetric with respect to the x-axis, which is the axis of the parabola.

EXAMPLE 1   Find the focus and directrix of the parabola $y^2 + 10x = 0$ and sketch the graph.

SOLUTION   If we write the equation as $y^2 = -10x$ and compare it with Equation 2, we see that $4p = -10$, so $p = -\frac{5}{2}$. Thus the focus is $(p, 0) = \left(-\frac{5}{2}, 0\right)$ and the directrix is $x = \frac{5}{2}$. The sketch is shown in Figure 5.

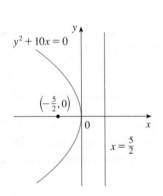

FIGURE 5

Ellipses

An **ellipse** is the set of points in a plane the sum of whose distances from two fixed points F_1 and F_2 is a constant (see Figure 6). These two fixed points are called the **foci** (plural of **focus**). One of Kepler's laws is that the orbits of the planets in the solar system are ellipses with the sun at one focus.

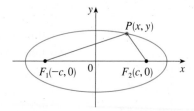

FIGURE 6   **FIGURE 7**

In order to obtain the simplest equation for an ellipse, we place the foci on the x-axis at the points $(-c, 0)$ and $(c, 0)$ as in Figure 7 so that the origin is halfway between the foci. Let the sum of the distances from a point on the ellipse to the foci be $2a > 0$. Then $P(x, y)$ is a point on the ellipse when

$$|PF_1| + |PF_2| = 2a$$

that is,

$$\sqrt{(x + c)^2 + y^2} + \sqrt{(x - c)^2 + y^2} = 2a$$

or

$$\sqrt{(x - c)^2 + y^2} = 2a - \sqrt{(x + c)^2 + y^2}$$

Squaring both sides, we have

$$x^2 - 2cx + c^2 + y^2 = 4a^2 - 4a\sqrt{(x + c)^2 + y^2} + x^2 + 2cx + c^2 + y^2$$

which simplifies to

$$a\sqrt{(x + c)^2 + y^2} = a^2 + cx$$

We square again:

$$a^2(x^2 + 2cx + c^2 + y^2) = a^4 + 2a^2cx + c^2x^2$$

which becomes

$$(a^2 - c^2)x^2 + a^2y^2 = a^2(a^2 - c^2)$$

From triangle F_1F_2P in Figure 7 we see that $2c < 2a$, so $c < a$ and therefore $a^2 - c^2 > 0$. For convenience, let $b^2 = a^2 - c^2$. Then the equation of the ellipse becomes $b^2x^2 + a^2y^2 = a^2b^2$ or, if both sides are divided by a^2b^2,

$$\boxed{3} \qquad \frac{x^2}{a^2} + \frac{y^2}{b^2} = 1$$

Since $b^2 = a^2 - c^2 < a^2$, it follows that $b < a$. The x-intercepts are found by setting $y = 0$. Then $x^2/a^2 = 1$, or $x^2 = a^2$, so $x = \pm a$. The corresponding points $(a, 0)$ and $(-a, 0)$ are called the **vertices** of the ellipse and the line segment joining the vertices is called the **major axis**. To find the y-intercepts we set $x = 0$ and obtain $y^2 = b^2$, so $y = \pm b$. The line segment joining $(0, b)$ and $(0, -b)$ is the **minor axis**. Equation 3 is unchanged if x is replaced by $-x$ or y is replaced by $-y$, so the ellipse is symmetric about both axes. Notice that if the foci coincide, then $c = 0$, so $a = b$ and the ellipse becomes a circle with radius $r = a = b$.

We summarize this discussion as follows (see also Figure 8).

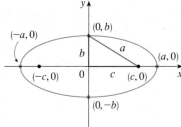

FIGURE 8

$$\frac{x^2}{a^2} + \frac{y^2}{b^2} = 1, \ a \geqslant b$$

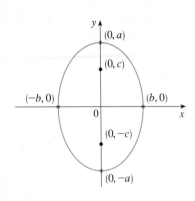

FIGURE 9
$\dfrac{x^2}{b^2} + \dfrac{y^2}{a^2} = 1,\ a \geqslant b$

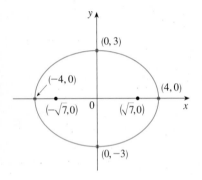

FIGURE 10
$9x^2 + 16y^2 = 144$

4 The ellipse

$$\frac{x^2}{a^2} + \frac{y^2}{b^2} = 1 \qquad a \geqslant b > 0$$

has foci $(\pm c, 0)$, where $c^2 = a^2 - b^2$, and vertices $(\pm a, 0)$.

If the foci of an ellipse are located on the y-axis at $(0, \pm c)$, then we can find its equation by interchanging x and y in $\boxed{4}$. (See Figure 9.)

5 The ellipse

$$\frac{x^2}{b^2} + \frac{y^2}{a^2} = 1 \qquad a \geqslant b > 0$$

has foci $(0, \pm c)$, where $c^2 = a^2 - b^2$, and vertices $(0, \pm a)$.

V EXAMPLE 2 Sketch the graph of $9x^2 + 16y^2 = 144$ and locate the foci.

SOLUTION Divide both sides of the equation by 144:

$$\frac{x^2}{16} + \frac{y^2}{9} = 1$$

The equation is now in the standard form for an ellipse, so we have $a^2 = 16$, $b^2 = 9$, $a = 4$, and $b = 3$. The x-intercepts are ± 4 and the y-intercepts are ± 3. Also, $c^2 = a^2 - b^2 = 7$, so $c = \sqrt{7}$ and the foci are $\left(\pm\sqrt{7}, 0\right)$. The graph is sketched in Figure 10. ■

V EXAMPLE 3 Find an equation of the ellipse with foci $(0, \pm 2)$ and vertices $(0, \pm 3)$.

SOLUTION Using the notation of $\boxed{5}$, we have $c = 2$ and $a = 3$. Then we obtain $b^2 = a^2 - c^2 = 9 - 4 = 5$, so an equation of the ellipse is

$$\frac{x^2}{5} + \frac{y^2}{9} = 1$$

Another way of writing the equation is $9x^2 + 5y^2 = 45$. ■

Like parabolas, ellipses have an interesting reflection property that has practical consequences. If a source of light or sound is placed at one focus of a surface with elliptical cross-sections, then all the light or sound is reflected off the surface to the other focus (see Exercise 65). This principle is used in *lithotripsy*, a treatment for kidney stones. A reflector with elliptical cross-section is placed in such a way that the kidney stone is at one focus. High-intensity sound waves generated at the other focus are reflected to the stone and destroy it without damaging surrounding tissue. The patient is spared the trauma of surgery and recovers within a few days.

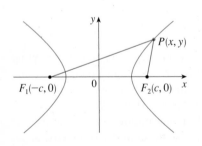

FIGURE 11

P is on the hyperbola when
$|PF_1| - |PF_2| = \pm 2a$.

Hyperbolas

A **hyperbola** is the set of all points in a plane the difference of whose distances from two fixed points F_1 and F_2 (the foci) is a constant. This definition is illustrated in Figure 11.

Hyperbolas occur frequently as graphs of equations in chemistry, physics, biology, and economics (Boyle's Law, Ohm's Law, supply and demand curves). A particularly significant application of hyperbolas is found in the navigation systems developed in World Wars I and II (see Exercise 51).

Notice that the definition of a hyperbola is similar to that of an ellipse; the only change is that the sum of distances has become a difference of distances. In fact, the derivation of the equation of a hyperbola is also similar to the one given earlier for an ellipse. It is left as Exercise 52 to show that when the foci are on the x-axis at $(\pm c, 0)$ and the difference of distances is $|PF_1| - |PF_2| = \pm 2a$, then the equation of the hyperbola is

$$\boxed{6} \qquad \frac{x^2}{a^2} - \frac{y^2}{b^2} = 1$$

where $c^2 = a^2 + b^2$. Notice that the x-intercepts are again $\pm a$ and the points $(a, 0)$ and $(-a, 0)$ are the **vertices** of the hyperbola. But if we put $x = 0$ in Equation 6 we get $y^2 = -b^2$, which is impossible, so there is no y-intercept. The hyperbola is symmetric with respect to both axes.

To analyze the hyperbola further, we look at Equation 6 and obtain

$$\frac{x^2}{a^2} = 1 + \frac{y^2}{b^2} \geq 1$$

This shows that $x^2 \geq a^2$, so $|x| = \sqrt{x^2} \geq a$. Therefore we have $x \geq a$ or $x \leq -a$. This means that the hyperbola consists of two parts, called its *branches*.

When we draw a hyperbola it is useful to first draw its **asymptotes**, which are the dashed lines $y = (b/a)x$ and $y = -(b/a)x$ shown in Figure 12. Both branches of the hyperbola approach the asymptotes; that is, they come arbitrarily close to the asymptotes. [See Exercise 73 in Section 4.5, where these lines are shown to be slant asymptotes.]

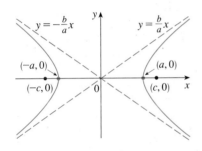

FIGURE 12
$\dfrac{x^2}{a^2} - \dfrac{y^2}{b^2} = 1$

$\boxed{7}$ The hyperbola

$$\frac{x^2}{a^2} - \frac{y^2}{b^2} = 1$$

has foci $(\pm c, 0)$, where $c^2 = a^2 + b^2$, vertices $(\pm a, 0)$, and asymptotes $y = \pm (b/a)x$.

If the foci of a hyperbola are on the y-axis, then by reversing the roles of x and y we obtain the following information, which is illustrated in Figure 13.

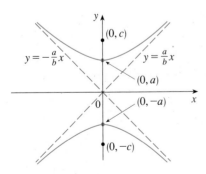

FIGURE 13
$\dfrac{y^2}{a^2} - \dfrac{x^2}{b^2} = 1$

$\boxed{8}$ The hyperbola

$$\frac{y^2}{a^2} - \frac{x^2}{b^2} = 1$$

has foci $(0, \pm c)$, where $c^2 = a^2 + b^2$, vertices $(0, \pm a)$, and asymptotes $y = \pm (a/b)x$.

EXAMPLE 4   Find the foci and asymptotes of the hyperbola $9x^2 - 16y^2 = 144$ and sketch its graph.

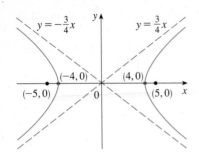

FIGURE 14
$9x^2 - 16y^2 = 144$

SOLUTION If we divide both sides of the equation by 144, it becomes

$$\frac{x^2}{16} - \frac{y^2}{9} = 1$$

which is of the form given in $\boxed{7}$ with $a = 4$ and $b = 3$. Since $c^2 = 16 + 9 = 25$, the foci are $(\pm 5, 0)$. The asymptotes are the lines $y = \frac{3}{4}x$ and $y = -\frac{3}{4}x$. The graph is shown in Figure 14.

EXAMPLE 5 Find the foci and equation of the hyperbola with vertices $(0, \pm 1)$ and asymptote $y = 2x$.

SOLUTION From $\boxed{8}$ and the given information, we see that $a = 1$ and $a/b = 2$. Thus $b = a/2 = \frac{1}{2}$ and $c^2 = a^2 + b^2 = \frac{5}{4}$. The foci are $\left(0, \pm\sqrt{5}/2\right)$ and the equation of the hyperbola is

$$y^2 - 4x^2 = 1$$

Shifted Conics

As discussed in Appendix C, we shift conics by taking the standard equations $\boxed{1}$, $\boxed{2}$, $\boxed{4}$, $\boxed{5}$, $\boxed{7}$, and $\boxed{8}$ and replacing x and y by $x - h$ and $y - k$.

EXAMPLE 6 Find an equation of the ellipse with foci $(2, -2)$, $(4, -2)$ and vertices $(1, -2)$, $(5, -2)$.

SOLUTION The major axis is the line segment that joins the vertices $(1, -2)$, $(5, -2)$ and has length 4, so $a = 2$. The distance between the foci is 2, so $c = 1$. Thus $b^2 = a^2 - c^2 = 3$. Since the center of the ellipse is $(3, -2)$, we replace x and y in $\boxed{4}$ by $x - 3$ and $y + 2$ to obtain

$$\frac{(x - 3)^2}{4} + \frac{(y + 2)^2}{3} = 1$$

as the equation of the ellipse.

V EXAMPLE 7 Sketch the conic $9x^2 - 4y^2 - 72x + 8y + 176 = 0$ and find its foci.

SOLUTION We complete the squares as follows:

$$4(y^2 - 2y) - 9(x^2 - 8x) = 176$$

$$4(y^2 - 2y + 1) - 9(x^2 - 8x + 16) = 176 + 4 - 144$$

$$4(y - 1)^2 - 9(x - 4)^2 = 36$$

$$\frac{(y - 1)^2}{9} - \frac{(x - 4)^2}{4} = 1$$

This is in the form $\boxed{8}$ except that x and y are replaced by $x - 4$ and $y - 1$. Thus $a^2 = 9$, $b^2 = 4$, and $c^2 = 13$. The hyperbola is shifted four units to the right and one unit upward. The foci are $\left(4, 1 + \sqrt{13}\right)$ and $\left(4, 1 - \sqrt{13}\right)$ and the vertices are $(4, 4)$ and $(4, -2)$. The asymptotes are $y - 1 = \pm\frac{3}{2}(x - 4)$. The hyperbola is sketched in Figure 15.

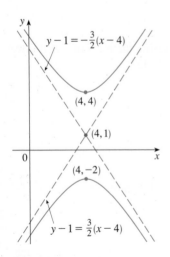

FIGURE 15
$9x^2 - 4y^2 - 72x + 8y + 176 = 0$

10.5 Exercises

Access end-of-section exercises and tools for visualizing calculus online at **www.webassign.net**

10.6 Conic Sections in Polar Coordinates

In the preceding section we defined the parabola in terms of a focus and directrix, but we defined the ellipse and hyperbola in terms of two foci. In this section we give a more unified treatment of all three types of conic sections in terms of a focus and directrix. Furthermore, if we place the focus at the origin, then a conic section has a simple polar equation, which provides a convenient description of the motion of planets, satellites, and comets.

---

1 Theorem Let F be a fixed point (called the **focus**) and l be a fixed line (called the **directrix**) in a plane. Let e be a fixed positive number (called the **eccentricity**). The set of all points P in the plane such that

$$\frac{|PF|}{|Pl|} = e$$

(that is, the ratio of the distance from F to the distance from l is the constant e) is a conic section. The conic is

(a) an ellipse if $e < 1$

(b) a parabola if $e = 1$

(c) a hyperbola if $e > 1$

PROOF Notice that if the eccentricity is $e = 1$, then $|PF| = |Pl|$ and so the given condition simply becomes the definition of a parabola as given in Section 10.5.

Let us place the focus F at the origin and the directrix parallel to the y-axis and d units to the right. Thus the directrix has equation $x = d$ and is perpendicular to the polar axis. If the point P has polar coordinates (r, θ), we see from Figure 1 that

$$|PF| = r \qquad |Pl| = d - r\cos\theta$$

Thus the condition $|PF|/|Pl| = e$, or $|PF| = e|Pl|$, becomes

2 $$r = e(d - r\cos\theta)$$

If we square both sides of this polar equation and convert to rectangular coordinates, we get

$$x^2 + y^2 = e^2(d - x)^2 = e^2(d^2 - 2dx + x^2)$$

or $$(1 - e^2)x^2 + 2de^2x + y^2 = e^2d^2$$

After completing the square, we have

3 $$\left(x + \frac{e^2d}{1 - e^2}\right)^2 + \frac{y^2}{1 - e^2} = \frac{e^2d^2}{(1 - e^2)^2}$$

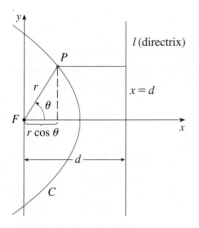

FIGURE 1

If $e < 1$, we recognize Equation 3 as the equation of an ellipse. In fact, it is of the form

$$\frac{(x - h)^2}{a^2} + \frac{y^2}{b^2} = 1$$

where

$$\boxed{4} \qquad h = -\frac{e^2 d}{1 - e^2} \qquad a^2 = \frac{e^2 d^2}{(1 - e^2)^2} \qquad b^2 = \frac{e^2 d^2}{1 - e^2}$$

In Section 10.5 we found that the foci of an ellipse are at a distance c from the center, where

$$\boxed{5} \qquad\qquad c^2 = a^2 - b^2 = \frac{e^4 d^2}{(1 - e^2)^2}$$

This shows that

$$c = \frac{e^2 d}{1 - e^2} = -h$$

and confirms that the focus as defined in Theorem 1 means the same as the focus defined in Section 10.5. It also follows from Equations 4 and 5 that the eccentricity is given by

$$e = \frac{c}{a}$$

If $e > 1$, then $1 - e^2 < 0$ and we see that Equation 3 represents a hyperbola. Just as we did before, we could rewrite Equation 3 in the form

$$\frac{(x - h)^2}{a^2} - \frac{y^2}{b^2} = 1$$

and see that

$$e = \frac{c}{a} \qquad \text{where} \quad c^2 = a^2 + b^2 \qquad\qquad \blacksquare$$

By solving Equation 2 for r, we see that the polar equation of the conic shown in Figure 1 can be written as

$$r = \frac{ed}{1 + e \cos\theta}$$

If the directrix is chosen to be to the left of the focus as $x = -d$, or if the directrix is chosen to be parallel to the polar axis as $y = \pm d$, then the polar equation of the conic is given by the following theorem, which is illustrated by Figure 2. (See Exercises 21–23.)

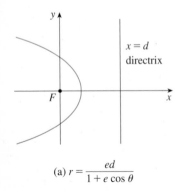

(a) $r = \dfrac{ed}{1 + e \cos\theta}$

(b) $r = \dfrac{ed}{1 - e \cos\theta}$

(c) $r = \dfrac{ed}{1 + e \sin\theta}$

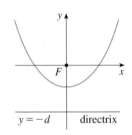

(d) $r = \dfrac{ed}{1 - e \sin\theta}$

FIGURE 2

Polar equations of conics

6 **Theorem** A polar equation of the form

$$r = \frac{ed}{1 \pm e \cos \theta} \qquad \text{or} \qquad r = \frac{ed}{1 \pm e \sin \theta}$$

represents a conic section with eccentricity e. The conic is an ellipse if $e < 1$, a parabola if $e = 1$, or a hyperbola if $e > 1$.

V EXAMPLE 1 Find a polar equation for a parabola that has its focus at the origin and whose directrix is the line $y = -6$.

SOLUTION Using Theorem 6 with $e = 1$ and $d = 6$, and using part (d) of Figure 2, we see that the equation of the parabola is

$$r = \frac{6}{1 - \sin \theta}$$

V EXAMPLE 2 A conic is given by the polar equation

$$r = \frac{10}{3 - 2 \cos \theta}$$

Find the eccentricity, identify the conic, locate the directrix, and sketch the conic.

SOLUTION Dividing numerator and denominator by 3, we write the equation as

$$r = \frac{\frac{10}{3}}{1 - \frac{2}{3} \cos \theta}$$

From Theorem 6 we see that this represents an ellipse with $e = \frac{2}{3}$. Since $ed = \frac{10}{3}$, we have

$$d = \frac{\frac{10}{3}}{e} = \frac{\frac{10}{3}}{\frac{2}{3}} = 5$$

so the directrix has Cartesian equation $x = -5$. When $\theta = 0$, $r = 10$; when $\theta = \pi$, $r = 2$. So the vertices have polar coordinates $(10, 0)$ and $(2, \pi)$. The ellipse is sketched in Figure 3.

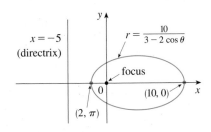

FIGURE 3

EXAMPLE 3 Sketch the conic $r = \dfrac{12}{2 + 4 \sin \theta}$.

SOLUTION Writing the equation in the form

$$r = \frac{6}{1 + 2 \sin \theta}$$

we see that the eccentricity is $e = 2$ and the equation therefore represents a hyperbola. Since $ed = 6$, $d = 3$ and the directrix has equation $y = 3$. The vertices occur when $\theta = \pi/2$ and $3\pi/2$, so they are $(2, \pi/2)$ and $(-6, 3\pi/2) = (6, \pi/2)$. It is also useful to plot the x-intercepts. These occur when $\theta = 0$, π; in both cases $r = 6$. For additional accuracy we could draw the asymptotes. Note that $r \to \pm\infty$ when $1 + 2 \sin \theta \to 0^+$ or

0^- and $1 + 2 \sin \theta = 0$ when $\sin \theta = -\frac{1}{2}$. Thus the asymptotes are parallel to the rays $\theta = 7\pi/6$ and $\theta = 11\pi/6$. The hyperbola is sketched in Figure 4.

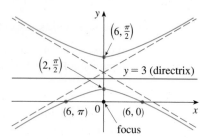

FIGURE 4

$$r = \frac{12}{2 + 4 \sin \theta}$$

When rotating conic sections, we find it much more convenient to use polar equations than Cartesian equations. We just use the fact (see Exercise 73 in Section 10.3) that the graph of $r = f(\theta - \alpha)$ is the graph of $r = f(\theta)$ rotated counterclockwise about the origin through an angle α.

 EXAMPLE 4 If the ellipse of Example 2 is rotated through an angle $\pi/4$ about the origin, find a polar equation and graph the resulting ellipse.

SOLUTION We get the equation of the rotated ellipse by replacing θ with $\theta - \pi/4$ in the equation given in Example 2. So the new equation is

$$r = \frac{10}{3 - 2\cos(\theta - \pi/4)}$$

We use this equation to graph the rotated ellipse in Figure 5. Notice that the ellipse has been rotated about its left focus.

FIGURE 5

In Figure 6 we use a computer to sketch a number of conics to demonstrate the effect of varying the eccentricity e. Notice that when e is close to 0 the ellipse is nearly circular, whereas it becomes more elongated as $e \to 1^-$. When $e = 1$, of course, the conic is a parabola.

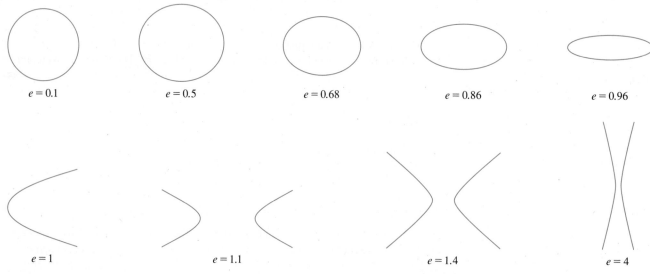

$e = 0.1$ $e = 0.5$ $e = 0.68$ $e = 0.86$ $e = 0.96$

$e = 1$ $e = 1.1$ $e = 1.4$ $e = 4$

FIGURE 6

Kepler's Laws

In 1609 the German mathematician and astronomer Johannes Kepler, on the basis of huge amounts of astronomical data, published the following three laws of planetary motion.

Kepler's Laws

1. A planet revolves around the sun in an elliptical orbit with the sun at one focus.
2. The line joining the sun to a planet sweeps out equal areas in equal times.
3. The square of the period of revolution of a planet is proportional to the cube of the length of the major axis of its orbit.

Although Kepler formulated his laws in terms of the motion of planets around the sun, they apply equally well to the motion of moons, comets, satellites, and other bodies that orbit subject to a single gravitational force. In Section 13.4 we will show how to deduce Kepler's Laws from Newton's Laws. Here we use Kepler's First Law, together with the polar equation of an ellipse, to calculate quantities of interest in astronomy.

For purposes of astronomical calculations, it's useful to express the equation of an ellipse in terms of its eccentricity e and its semimajor axis a. We can write the distance d from the focus to the directrix in terms of a if we use $\boxed{4}$:

$$a^2 = \frac{e^2 d^2}{(1 - e^2)^2} \quad \Rightarrow \quad d^2 = \frac{a^2(1 - e^2)^2}{e^2} \quad \Rightarrow \quad d = \frac{a(1 - e^2)}{e}$$

So $ed = a(1 - e^2)$. If the directrix is $x = d$, then the polar equation is

$$r = \frac{ed}{1 + e \cos \theta} = \frac{a(1 - e^2)}{1 + e \cos \theta}$$

$\boxed{7}$ The polar equation of an ellipse with focus at the origin, semimajor axis a, eccentricity e, and directrix $x = d$ can be written in the form

$$r = \frac{a(1 - e^2)}{1 + e \cos \theta}$$

The positions of a planet that are closest to and farthest from the sun are called its **perihelion** and **aphelion**, respectively, and correspond to the vertices of the ellipse. (See Figure 7.) The distances from the sun to the perihelion and aphelion are called the **perihelion distance** and **aphelion distance**, respectively. In Figure 1 the sun is at the focus F, so at perihelion we have $\theta = 0$ and, from Equation 7,

$$r = \frac{a(1 - e^2)}{1 + e \cos 0} = \frac{a(1 - e)(1 + e)}{1 + e} = a(1 - e)$$

Similarly, at aphelion $\theta = \pi$ and $r = a(1 + e)$.

FIGURE 7

$\boxed{8}$ The perihelion distance from a planet to the sun is $a(1 - e)$ and the aphelion distance is $a(1 + e)$.

EXAMPLE 5

(a) Find an approximate polar equation for the elliptical orbit of the earth around the sun (at one focus) given that the eccentricity is about 0.017 and the length of the major axis is about 2.99×10^8 km.

(b) Find the distance from the earth to the sun at perihelion and at aphelion.

SOLUTION

(a) The length of the major axis is $2a = 2.99 \times 10^8$, so $a = 1.495 \times 10^8$. We are given that $e = 0.017$ and so, from Equation 7, an equation of the earth's orbit around the sun is

$$r = \frac{a(1 - e^2)}{1 + e \cos \theta} = \frac{(1.495 \times 10^8)[1 - (0.017)^2]}{1 + 0.017 \cos \theta}$$

or, approximately,

$$r = \frac{1.49 \times 10^8}{1 + 0.017 \cos \theta}$$

(b) From $\boxed{8}$, the perihelion distance from the earth to the sun is

$$a(1 - e) \approx (1.495 \times 10^8)(1 - 0.017) \approx 1.47 \times 10^8 \text{ km}$$

and the aphelion distance is

$$a(1 + e) \approx (1.495 \times 10^8)(1 + 0.017) \approx 1.52 \times 10^8 \text{ km}$$

10.6 Exercises

Access end-of-section exercises and tools for visualizing calculus online at **www.webassign.net**

10 Review

Concept Check

1. (a) What is a parametric curve?
(b) How do you sketch a parametric curve?

2. (a) How do you find the slope of a tangent to a parametric curve?
(b) How do you find the area under a parametric curve?

3. Write an expression for each of the following:
(a) The length of a parametric curve
(b) The area of the surface obtained by rotating a parametric curve about the x-axis

4. (a) Use a diagram to explain the meaning of the polar coordinates (r, θ) of a point.

(b) Write equations that express the Cartesian coordinates (x, y) of a point in terms of the polar coordinates.
(c) What equations would you use to find the polar coordinates of a point if you knew the Cartesian coordinates?

5. (a) How do you find the slope of a tangent line to a polar curve?
(b) How do you find the area of a region bounded by a polar curve?
(c) How do you find the length of a polar curve?

6. (a) Give a geometric definition of a parabola.
(b) Write an equation of a parabola with focus $(0, p)$ and directrix $y = -p$. What if the focus is $(p, 0)$ and the directrix is $x = -p$?

7. (a) Give a definition of an ellipse in terms of foci.
   (b) Write an equation for the ellipse with foci $(\pm c, 0)$ and vertices $(\pm a, 0)$.

8. (a) Give a definition of a hyperbola in terms of foci.
   (b) Write an equation for the hyperbola with foci $(\pm c, 0)$ and vertices $(\pm a, 0)$.
   (c) Write equations for the asymptotes of the hyperbola in part (b).

9. (a) What is the eccentricity of a conic section?
   (b) What can you say about the eccentricity if the conic section is an ellipse? A hyperbola? A parabola?
   (c) Write a polar equation for a conic section with eccentricity e and directrix $x = d$. What if the directrix is $x = -d$? $y = d$? $y = -d$?

True-False Quiz

Determine whether the statement is true or false. If it is true, explain why. If it is false, explain why or give an example that disproves the statement.

1. If the parametric curve $x = f(t)$, $y = g(t)$ satisfies $g'(1) = 0$, then it has a horizontal tangent when $t = 1$.

2. If $x = f(t)$ and $y = g(t)$ are twice differentiable, then
$$\frac{d^2y}{dx^2} = \frac{d^2y/dt^2}{d^2x/dt^2}$$

3. The length of the curve $x = f(t)$, $y = g(t)$, $a \le t \le b$, is $\int_a^b \sqrt{[f'(t)]^2 + [g'(t)]^2}\, dt$.

4. If a point is represented by (x, y) in Cartesian coordinates (where $x \neq 0$) and (r, θ) in polar coordinates, then $\theta = \tan^{-1}(y/x)$.

5. The polar curves $r = 1 - \sin 2\theta$ and $r = \sin 2\theta - 1$ have the same graph.

6. The equations $r = 2$, $x^2 + y^2 = 4$, and $x = 2 \sin 3t$, $y = 2 \cos 3t$ $(0 \le t \le 2\pi)$ all have the same graph.

7. The parametric equations $x = t^2$, $y = t^4$ have the same graph as $x = t^3$, $y = t^6$.

8. The graph of $y^2 = 2y + 3x$ is a parabola.

9. A tangent line to a parabola intersects the parabola only once.

10. A hyperbola never intersects its directrix.

Exercises

1–4 Sketch the parametric curve and eliminate the parameter to find the Cartesian equation of the curve.

1. $x = t^2 + 4t$, $\quad y = 2 - t$, $\quad -4 \le t \le 1$

2. $x = 1 + e^{2t}$, $\quad y = e^t$

3. $x = \cos\theta$, $\quad y = \sec\theta$, $\quad 0 \le \theta < \pi/2$

4. $x = 2\cos\theta$, $\quad y = 1 + \sin\theta$

5. Write three different sets of parametric equations for the curve $y = \sqrt{x}$.

6. Use the graphs of $x = f(t)$ and $y = g(t)$ to sketch the parametric curve $x = f(t)$, $y = g(t)$. Indicate with arrows the direction in which the curve is traced as t increases.

7. (a) Plot the point with polar coordinates $(4, 2\pi/3)$. Then find its Cartesian coordinates.

   (b) The Cartesian coordinates of a point are $(-3, 3)$. Find two sets of polar coordinates for the point.

8. Sketch the region consisting of points whose polar coordinates satisfy $1 \le r < 2$ and $\pi/6 \le \theta \le 5\pi/6$.

9–16 Sketch the polar curve.

9. $r = 1 - \cos\theta$

10. $r = \sin 4\theta$

11. $r = \cos 3\theta$

12. $r = 3 + \cos 3\theta$

13. $r = 1 + \cos 2\theta$

14. $r = 2\cos(\theta/2)$

15. $r = \dfrac{3}{1 + 2\sin\theta}$

16. $r = \dfrac{3}{2 - 2\cos\theta}$

17–18 Find a polar equation for the curve represented by the given Cartesian equation.

17. $x + y = 2$

18. $x^2 + y^2 = 2$

19. The curve with polar equation $r = (\sin\theta)/\theta$ is called a **cochleoid**. Use a graph of r as a function of θ in Cartesian coordinates to sketch the cochleoid by hand. Then graph it with a machine to check your sketch.

Graphing calculator or computer required   CAS Computer algebra system required

20. Graph the ellipse $r = 2/(4 - 3 \cos \theta)$ and its directrix. Also graph the ellipse obtained by rotation about the origin through an angle $2\pi/3$.

21–24 Find the slope of the tangent line to the given curve at the point corresponding to the specified value of the parameter.

21. $x = \ln t, \ y = 1 + t^2; \quad t = 1$

22. $x = t^3 + 6t + 1, \quad y = 2t - t^2; \quad t = -1$

23. $r = e^{-\theta}; \quad \theta = \pi$

24. $r = 3 + \cos 3\theta; \quad \theta = \pi/2$

25–26 Find dy/dx and d^2y/dx^2.

25. $x = t + \sin t, \quad y = t - \cos t$

26. $x = 1 + t^2, \quad y = t - t^3$

27. Use a graph to estimate the coordinates of the lowest point on the curve $x = t^3 - 3t, \ y = t^2 + t + 1$. Then use calculus to find the exact coordinates.

28. Find the area enclosed by the loop of the curve in Exercise 27.

29. At what points does the curve

$$x = 2a \cos t - a \cos 2t \qquad y = 2a \sin t - a \sin 2t$$

have vertical or horizontal tangents? Use this information to help sketch the curve.

30. Find the area enclosed by the curve in Exercise 29.

31. Find the area enclosed by the curve $r^2 = 9 \cos 5\theta$.

32. Find the area enclosed by the inner loop of the curve $r = 1 - 3 \sin \theta$.

33. Find the points of intersection of the curves $r = 2$ and $r = 4 \cos \theta$.

34. Find the points of intersection of the curves $r = \cot \theta$ and $r = 2 \cos \theta$.

35. Find the area of the region that lies inside both of the circles $r = 2 \sin \theta$ and $r = \sin \theta + \cos \theta$.

36. Find the area of the region that lies inside the curve $r = 2 + \cos 2\theta$ but outside the curve $r = 2 + \sin \theta$.

37–40 Find the length of the curve.

37. $x = 3t^2, \quad y = 2t^3, \quad 0 \le t \le 2$

38. $x = 2 + 3t, \quad y = \cosh 3t, \quad 0 \le t \le 1$

39. $r = 1/\theta, \quad \pi \le \theta \le 2\pi$

40. $r = \sin^3(\theta/3), \quad 0 \le \theta \le \pi$

41–42 Find the area of the surface obtained by rotating the given curve about the x-axis.

41. $x = 4\sqrt{t}, \quad y = \dfrac{t^3}{3} + \dfrac{1}{2t^2}, \quad 1 \le t \le 4$

42. $x = 2 + 3t, \quad y = \cosh 3t, \quad 0 \le t \le 1$

43. The curves defined by the parametric equations

$$x = \frac{t^2 - c}{t^2 + 1} \qquad y = \frac{t(t^2 - c)}{t^2 + 1}$$

are called **strophoids** (from a Greek word meaning "to turn or twist"). Investigate how these curves vary as c varies.

44. A family of curves has polar equations $r^a = |\sin 2\theta|$ where a is a positive number. Investigate how the curves change as a changes.

45–48 Find the foci and vertices and sketch the graph.

45. $\dfrac{x^2}{9} + \dfrac{y^2}{8} = 1$

46. $4x^2 - y^2 = 16$

47. $6y^2 + x - 36y + 55 = 0$

48. $25x^2 + 4y^2 + 50x - 16y = 59$

49. Find an equation of the ellipse with foci $(\pm 4, 0)$ and vertices $(\pm 5, 0)$.

50. Find an equation of the parabola with focus $(2, 1)$ and directrix $x = -4$.

51. Find an equation of the hyperbola with foci $(0, \pm 4)$ and asymptotes $y = \pm 3x$.

52. Find an equation of the ellipse with foci $(3, \pm 2)$ and major axis with length 8.

53. Find an equation for the ellipse that shares a vertex and a focus with the parabola $x^2 + y = 100$ and that has its other focus at the origin.

54. Show that if m is any real number, then there are exactly two lines of slope m that are tangent to the ellipse $x^2/a^2 + y^2/b^2 = 1$ and their equations are $y = mx \pm \sqrt{a^2m^2 + b^2}$.

55. Find a polar equation for the ellipse with focus at the origin, eccentricity $\frac{1}{3}$, and directrix with equation $r = 4 \sec \theta$.

56. Show that the angles between the polar axis and the asymptotes of the hyperbola $r = ed/(1 - e \cos \theta)$, $e > 1$, are given by $\cos^{-1}(\pm 1/e)$.

57. A curve called the **folium of Descartes** is defined by the parametric equations

$$x = \frac{3t}{1 + t^3} \qquad y = \frac{3t^2}{1 + t^3}$$

(a) Show that if (a, b) lies on the curve, then so does (b, a); that is, the curve is symmetric with respect to the line $y = x$. Where does the curve intersect this line?

(b) Find the points on the curve where the tangent lines are horizontal or vertical.

(c) Show that the line $y = -x - 1$ is a slant asymptote.

(d) Sketch the curve.

(e) Show that a Cartesian equation of this curve is $x^3 + y^3 = 3xy$.

(f) Show that the polar equation can be written in the form

$$r = \frac{3 \sec \theta \tan \theta}{1 + \tan^3 \theta}$$

(g) Find the area enclosed by the loop of this curve.

(h) Show that the area of the loop is the same as the area that lies between the asymptote and the infinite branches of the curve. (Use a computer algebra system to evaluate the integral.)

1. A curve is defined by the parametric equations

$$x = \int_1^t \frac{\cos u}{u}\, du \qquad y = \int_1^t \frac{\sin u}{u}\, du$$

Find the length of the arc of the curve from the origin to the nearest point where there is a vertical tangent line.

2. (a) Find the highest and lowest points on the curve $x^4 + y^4 = x^2 + y^2$.
   (b) Sketch the curve. (Notice that it is symmetric with respect to both axes and both of the lines $y = \pm x$, so it suffices to consider $y \geqslant x \geqslant 0$ initially.)
   CAS (c) Use polar coordinates and a computer algebra system to find the area enclosed by the curve.

3. What is the smallest viewing rectangle that contains every member of the family of polar curves $r = 1 + c \sin \theta$, where $0 \leqslant c \leqslant 1$? Illustrate your answer by graphing several members of the family in this viewing rectangle.

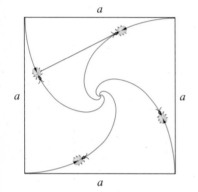

a

a

a

a

FIGURE FOR PROBLEM 4

4. Four bugs are placed at the four corners of a square with side length a. The bugs crawl counterclockwise at the same speed and each bug crawls directly toward the next bug at all times. They approach the center of the square along spiral paths.
   (a) Find the polar equation of a bug's path assuming the pole is at the center of the square. (Use the fact that the line joining one bug to the next is tangent to the bug's path.)
   (b) Find the distance traveled by a bug by the time it meets the other bugs at the center.

5. Show that any tangent line to a hyperbola touches the hyperbola halfway between the points of intersection of the tangent and the asymptotes.

6. A circle C of radius $2r$ has its center at the origin. A circle of radius r rolls without slipping in the counterclockwise direction around C. A point P is located on a fixed radius of the rolling circle at a distance b from its center, $0 < b < r$. [See parts (i) and (ii) of the figure.] Let L be the line from the center of C to the center of the rolling circle and let θ be the angle that L makes with the positive x-axis.
   (a) Using θ as a parameter, show that parametric equations of the path traced out by P are

$$x = b \cos 3\theta + 3r \cos \theta \qquad y = b \sin 3\theta + 3r \sin \theta$$

   Note: If $b = 0$, the path is a circle of radius $3r$; if $b = r$, the path is an *epicycloid*. The path traced out by P for $0 < b < r$ is called an *epitrochoid*.
   (b) Graph the curve for various values of b between 0 and r.
   (c) Show that an equilateral triangle can be inscribed in the epitrochoid and that its centroid is on the circle of radius b centered at the origin.

Note: This is the principle of the Wankel rotary engine. When the equilateral triangle rotates with its vertices on the epitrochoid, its centroid sweeps out a circle whose center is at the center of the curve.

   (d) In most rotary engines the sides of the equilateral triangles are replaced by arcs of circles centered at the opposite vertices as in part (iii) of the figure. (Then the diameter of the rotor is constant.) Show that the rotor will fit in the epitrochoid if $b \leqslant \frac{3}{2}(2 - \sqrt{3}\,)r$.

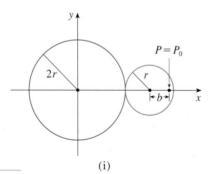

y

$2r$

r

$P = P_0$

$\leftarrow b \rightarrow$

x

(i)

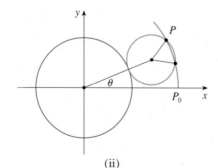

y

P

θ

P_0

x

(ii)

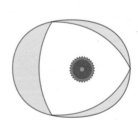

(iii)

11 Infinite Sequences and Series

In the last section of this chapter you are asked to use a series to derive a formula for the velocity of an ocean wave.

© Epic Stock / Shutterstock

Infinite sequences and series were introduced briefly in *A Preview of Calculus* in connection with Zeno's paradoxes and the decimal representation of numbers. Their importance in calculus stems from Newton's idea of representing functions as sums of infinite series. For instance, in finding areas he often integrated a function by first expressing it as a series and then integrating each term of the series. We will pursue his idea in Section 11.10 in order to integrate such functions as e^{-x^2}. (Recall that we have previously been unable to do this.) Many of the functions that arise in mathematical physics and chemistry, such as Bessel functions, are defined as sums of series, so it is important to be familiar with the basic concepts of convergence of infinite sequences and series.

Physicists also use series in another way, as we will see in Section 11.11. In studying fields as diverse as optics, special relativity, and electromagnetism, they analyze phenomena by replacing a function with the first few terms in the series that represents it.

11.1   Sequences

A **sequence** can be thought of as a list of numbers written in a definite order:

$$a_1, \ a_2, \ a_3, \ a_4, \ \ldots, \ a_n, \ldots$$

The number a_1 is called the *first term*, a_2 is the *second term*, and in general a_n is the *nth term*. We will deal exclusively with infinite sequences and so each term a_n will have a successor a_{n+1}.

Notice that for every positive integer n there is a corresponding number a_n and so a sequence can be defined as a function whose domain is the set of positive integers. But we usually write a_n instead of the function notation $f(n)$ for the value of the function at the number n.

NOTATION The sequence $\{a_1, a_2, a_3, \ldots\}$ is also denoted by

$$\{a_n\} \qquad \text{or} \qquad \{a_n\}_{n=1}^{\infty}$$

EXAMPLE 1 Some sequences can be defined by giving a formula for the nth term. In the following examples we give three descriptions of the sequence: one by using the preceding notation, another by using the defining formula, and a third by writing out the terms of the sequence. Notice that n doesn't have to start at 1.

(a) $\left\{ \dfrac{n}{n+1} \right\}_{n=1}^{\infty}$ $a_n = \dfrac{n}{n+1}$ $\left\{ \dfrac{1}{2}, \dfrac{2}{3}, \dfrac{3}{4}, \dfrac{4}{5}, \ldots, \dfrac{n}{n+1}, \ldots \right\}$

(b) $\left\{ \dfrac{(-1)^n(n+1)}{3^n} \right\}$ $a_n = \dfrac{(-1)^n(n+1)}{3^n}$ $\left\{ -\dfrac{2}{3}, \dfrac{3}{9}, -\dfrac{4}{27}, \dfrac{5}{81}, \ldots, \dfrac{(-1)^n(n+1)}{3^n}, \ldots \right\}$

(c) $\left\{ \sqrt{n-3} \right\}_{n=3}^{\infty}$ $a_n = \sqrt{n-3}, \ n \geqslant 3$ $\left\{ 0, 1, \sqrt{2}, \sqrt{3}, \ldots, \sqrt{n-3}, \ldots \right\}$

(d) $\left\{ \cos \dfrac{n\pi}{6} \right\}_{n=0}^{\infty}$ $a_n = \cos \dfrac{n\pi}{6}, \ n \geqslant 0$ $\left\{ 1, \dfrac{\sqrt{3}}{2}, \dfrac{1}{2}, 0, \ldots, \cos \dfrac{n\pi}{6}, \ldots \right\}$

V EXAMPLE 2 Find a formula for the general term a_n of the sequence

$$\left\{ \dfrac{3}{5}, -\dfrac{4}{25}, \dfrac{5}{125}, -\dfrac{6}{625}, \dfrac{7}{3125}, \ldots \right\}$$

assuming that the pattern of the first few terms continues.

SOLUTION We are given that

$$a_1 = \dfrac{3}{5} \qquad a_2 = -\dfrac{4}{25} \qquad a_3 = \dfrac{5}{125} \qquad a_4 = -\dfrac{6}{625} \qquad a_5 = \dfrac{7}{3125}$$

Notice that the numerators of these fractions start with 3 and increase by 1 whenever we go to the next term. The second term has numerator 4, the third term has numerator 5; in general, the nth term will have numerator $n + 2$. The denominators are the powers of 5, so a_n has denominator 5^n. The signs of the terms are alternately positive and negative, so we need to multiply by a power of -1. In Example 1(b) the factor $(-1)^n$ meant we started with a negative term. Here we want to start with a positive term and so we use $(-1)^{n-1}$ or $(-1)^{n+1}$. Therefore

$$a_n = (-1)^{n-1}\frac{n+2}{5^n}$$

EXAMPLE 3 Here are some sequences that don't have a simple defining equation.

(a) The sequence $\{p_n\}$, where p_n is the population of the world as of January 1 in the year n.

(b) If we let a_n be the digit in the nth decimal place of the number e, then $\{a_n\}$ is a well-defined sequence whose first few terms are

$$\{7, 1, 8, 2, 8, 1, 8, 2, 8, 4, 5, \ldots\}$$

(c) The **Fibonacci sequence** $\{f_n\}$ is defined recursively by the conditions

$$f_1 = 1 \qquad f_2 = 1 \qquad f_n = f_{n-1} + f_{n-2} \qquad n \geqslant 3$$

Each term is the sum of the two preceding terms. The first few terms are

$$\{1, 1, 2, 3, 5, 8, 13, 21, \ldots\}$$

This sequence arose when the 13th-century Italian mathematician known as Fibonacci solved a problem concerning the breeding of rabbits (see Exercise 83).

A sequence such as the one in Example 1(a), $a_n = n/(n+1)$, can be pictured either by plotting its terms on a number line, as in Figure 1, or by plotting its graph, as in Figure 2. Note that, since a sequence is a function whose domain is the set of positive integers, its graph consists of isolated points with coordinates

$$(1, a_1) \qquad (2, a_2) \qquad (3, a_3) \qquad \ldots \qquad (n, a_n) \qquad \ldots$$

From Figure 1 or Figure 2 it appears that the terms of the sequence $a_n = n/(n+1)$ are approaching 1 as n becomes large. In fact, the difference

$$1 - \frac{n}{n+1} = \frac{1}{n+1}$$

can be made as small as we like by taking n sufficiently large. We indicate this by writing

$$\lim_{n \to \infty} \frac{n}{n+1} = 1$$

In general, the notation

$$\lim_{n \to \infty} a_n = L$$

means that the terms of the sequence $\{a_n\}$ approach L as n becomes large. Notice that the following definition of the limit of a sequence is very similar to the definition of a limit of a function at infinity given in Section 3.4.

FIGURE 1

FIGURE 2

> **1** **Definition**   A sequence $\{a_n\}$ has the **limit** L and we write
>
> $$\lim_{n \to \infty} a_n = L \qquad \text{or} \qquad a_n \to L \text{ as } n \to \infty$$
>
> if we can make the terms a_n as close to L as we like by taking n sufficiently large. If $\lim_{n \to \infty} a_n$ exists, we say the sequence **converges** (or is **convergent**). Otherwise, we say the sequence **diverges** (or is **divergent**).

Figure 3 illustrates Definition 1 by showing the graphs of two sequences that have the limit L.

FIGURE 3
Graphs of two sequences with
$\lim\limits_{n \to \infty} a_n = L$

A more precise version of Definition 1 is as follows.

Compare this definition with Definition 3.4.5.

> **2** **Definition**   A sequence $\{a_n\}$ has the **limit** L and we write
>
> $$\lim_{n \to \infty} a_n = L \qquad \text{or} \qquad a_n \to L \text{ as } n \to \infty$$
>
> if for every $\varepsilon > 0$ there is a corresponding integer N such that
>
> $$\text{if} \qquad n > N \qquad \text{then} \qquad |a_n - L| < \varepsilon$$

Definition 2 is illustrated by Figure 4, in which the terms $a_1, a_2, a_3, \ldots$ are plotted on a number line. No matter how small an interval $(L - \varepsilon, L + \varepsilon)$ is chosen, there exists an N such that all terms of the sequence from a_{N+1} onward must lie in that interval.

FIGURE 4

Another illustration of Definition 2 is given in Figure 5. The points on the graph of $\{a_n\}$ must lie between the horizontal lines $y = L + \varepsilon$ and $y = L - \varepsilon$ if $n > N$. This picture must be valid no matter how small ε is chosen, but usually a smaller ε requires a larger N.

FIGURE 5

If you compare Definition 2 with Definition 3.4.5 you will see that the only difference between $\lim_{n \to \infty} a_n = L$ and $\lim_{x \to \infty} f(x) = L$ is that n is required to be an integer. Thus we have the following theorem, which is illustrated by Figure 6.

> **3** **Theorem** If $\lim_{x \to \infty} f(x) = L$ and $f(n) = a_n$ when n is an integer, then $\lim_{n \to \infty} a_n = L$.

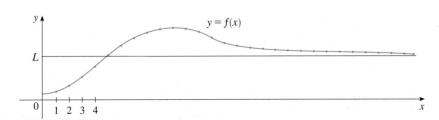

FIGURE 6

In particular, since we know that $\lim_{x \to \infty} (1/x^r) = 0$ when $r > 0$ (Theorem 3.4.4), we have

$$\boxed{4} \qquad \lim_{n \to \infty} \frac{1}{n^r} = 0 \qquad \text{if } r > 0$$

If a_n becomes large as n becomes large, we use the notation $\lim_{n \to \infty} a_n = \infty$. The following precise definition is similar to Definition 3.4.7.

> **5** **Definition** $\lim_{n \to \infty} a_n = \infty$ means that for every positive number M there is an integer N such that
>
> $$\text{if} \quad n > N \quad \text{then} \quad a_n > M$$

If $\lim_{n \to \infty} a_n = \infty$, then the sequence $\{a_n\}$ is divergent but in a special way. We say that $\{a_n\}$ diverges to ∞.

The Limit Laws given in Section 1.6 also hold for the limits of sequences and their proofs are similar.

Limit Laws for Sequences

> If $\{a_n\}$ and $\{b_n\}$ are convergent sequences and c is a constant, then
>
> $$\lim_{n \to \infty} (a_n + b_n) = \lim_{n \to \infty} a_n + \lim_{n \to \infty} b_n$$
>
> $$\lim_{n \to \infty} (a_n - b_n) = \lim_{n \to \infty} a_n - \lim_{n \to \infty} b_n$$
>
> $$\lim_{n \to \infty} c a_n = c \lim_{n \to \infty} a_n \qquad\qquad \lim_{n \to \infty} c = c$$
>
> $$\lim_{n \to \infty} (a_n b_n) = \lim_{n \to \infty} a_n \cdot \lim_{n \to \infty} b_n$$
>
> $$\lim_{n \to \infty} \frac{a_n}{b_n} = \frac{\lim_{n \to \infty} a_n}{\lim_{n \to \infty} b_n} \qquad \text{if } \lim_{n \to \infty} b_n \neq 0$$
>
> $$\lim_{n \to \infty} a_n^p = \left[\lim_{n \to \infty} a_n \right]^p \quad \text{if } p > 0 \text{ and } a_n > 0$$

The Squeeze Theorem can also be adapted for sequences as follows (see Figure 7).

Squeeze Theorem for Sequences

If $a_n \le b_n \le c_n$ for $n \ge n_0$ and $\lim\limits_{n \to \infty} a_n = \lim\limits_{n \to \infty} c_n = L$, then $\lim\limits_{n \to \infty} b_n = L$.

Another useful fact about limits of sequences is given by the following theorem, whose proof is left as Exercise 87.

6 Theorem   If $\lim\limits_{n \to \infty} |a_n| = 0$, then $\lim\limits_{n \to \infty} a_n = 0$.

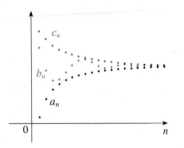

FIGURE 7

The sequence $\{b_n\}$ is squeezed between the sequences $\{a_n\}$ and $\{c_n\}$.

EXAMPLE 4 Find $\lim\limits_{n \to \infty} \dfrac{n}{n + 1}$.

SOLUTION The method is similar to the one we used in Section 3.4: Divide numerator and denominator by the highest power of n that occurs in the denominator and then use the Limit Laws.

$$\lim_{n \to \infty} \frac{n}{n + 1} = \lim_{n \to \infty} \frac{1}{1 + \dfrac{1}{n}} = \frac{\lim\limits_{n \to \infty} 1}{\lim\limits_{n \to \infty} 1 + \lim\limits_{n \to \infty} \dfrac{1}{n}}$$

This shows that the guess we made earlier from Figures 1 and 2 was correct.

$$= \frac{1}{1 + 0} = 1$$

Here we used Equation 4 with $r = 1$.   ▬

EXAMPLE 5 Is the sequence $a_n = \dfrac{n}{\sqrt{10 + n}}$ convergent or divergent?

SOLUTION As in Example 4, we divide numerator and denominator by n:

$$\lim_{n \to \infty} \frac{n}{\sqrt{10 + n}} = \lim_{n \to \infty} \frac{1}{\sqrt{\dfrac{10}{n^2} + \dfrac{1}{n}}} = \infty$$

because the numerator is constant and the denominator approaches 0. So $\{a_n\}$ is divergent.   ▬

EXAMPLE 6 Calculate $\lim\limits_{n \to \infty} \dfrac{\ln n}{n}$.

SOLUTION Notice that both numerator and denominator approach infinity as $n \to \infty$. We can't apply l'Hospital's Rule directly because it applies not to sequences but to functions of a real variable. However, we can apply l'Hospital's Rule to the related function $f(x) = (\ln x)/x$ and obtain

$$\lim_{x \to \infty} \frac{\ln x}{x} = \lim_{x \to \infty} \frac{1/x}{1} = 0$$

Therefore, by Theorem 3, we have

$$\lim_{n \to \infty} \frac{\ln n}{n} = 0$$ ▬

FIGURE 8

The graph of the sequence in Example 8 is shown in Figure 9 and supports our answer.

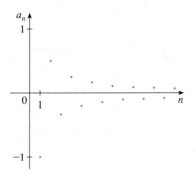

FIGURE 9

Creating Graphs of Sequences

Some computer algebra systems have special commands that enable us to create sequences and graph them directly. With most graphing calculators, however, sequences can be graphed by using parametric equations. For instance, the sequence in Example 10 can be graphed by entering the parametric equations

$$x = t \qquad y = t!/t^t$$

and graphing in dot mode, starting with $t = 1$ and setting the t-step equal to 1. The result is shown in Figure 10.

FIGURE 10

EXAMPLE 7 Determine whether the sequence $a_n = (-1)^n$ is convergent or divergent.

SOLUTION If we write out the terms of the sequence, we obtain

$$\{-1, 1, -1, 1, -1, 1, -1, \ldots\}$$

The graph of this sequence is shown in Figure 8. Since the terms oscillate between 1 and -1 infinitely often, a_n does not approach any number. Thus $\lim_{n \to \infty} (-1)^n$ does not exist; that is, the sequence $\{(-1)^n\}$ is divergent. ∎

EXAMPLE 8 Evaluate $\lim\limits_{n \to \infty} \dfrac{(-1)^n}{n}$ if it exists.

SOLUTION We first calculate the limit of the absolute value:

$$\lim_{n \to \infty} \left| \frac{(-1)^n}{n} \right| = \lim_{n \to \infty} \frac{1}{n} = 0$$

Therefore, by Theorem 6,

$$\lim_{n \to \infty} \frac{(-1)^n}{n} = 0$$ ∎

The following theorem says that if we apply a continuous function to the terms of a convergent sequence, the result is also convergent. The proof is left as Exercise 88.

> **7** **Theorem** If $\lim\limits_{n \to \infty} a_n = L$ and the function f is continuous at L, then
>
> $$\lim_{n \to \infty} f(a_n) = f(L)$$

EXAMPLE 9 Find $\lim\limits_{n \to \infty} \sin(\pi/n)$.

SOLUTION Because the sine function is continuous at 0, Theorem 7 enables us to write

$$\lim_{n \to \infty} \sin(\pi/n) = \sin\left(\lim_{n \to \infty} (\pi/n) \right) = \sin 0 = 0$$ ∎

V EXAMPLE 10 Discuss the convergence of the sequence $a_n = n!/n^n$, where $n! = 1 \cdot 2 \cdot 3 \cdot \cdots \cdot n$.

SOLUTION Both numerator and denominator approach infinity as $n \to \infty$ but here we have no corresponding function for use with l'Hospital's Rule ($x!$ is not defined when x is not an integer). Let's write out a few terms to get a feeling for what happens to a_n as n gets large:

$$a_1 = 1 \qquad a_2 = \frac{1 \cdot 2}{2 \cdot 2} \qquad a_3 = \frac{1 \cdot 2 \cdot 3}{3 \cdot 3 \cdot 3}$$

$$\boxed{8} \qquad a_n = \frac{1 \cdot 2 \cdot 3 \cdot \cdots \cdot n}{n \cdot n \cdot n \cdot \cdots \cdot n}$$

It appears from these expressions and the graph in Figure 10 that the terms are decreasing and perhaps approach 0. To confirm this, observe from Equation 8 that

$$a_n = \frac{1}{n} \left(\frac{2 \cdot 3 \cdot \cdots \cdot n}{n \cdot n \cdot \cdots \cdot n} \right)$$

Notice that the expression in parentheses is at most 1 because the numerator is less than (or equal to) the denominator. So

$$0 < a_n \leqslant \frac{1}{n}$$

We know that $1/n \to 0$ as $n \to \infty$. Therefore $a_n \to 0$ as $n \to \infty$ by the Squeeze Theorem.

V   EXAMPLE 11 For what values of r is the sequence $\{r^n\}$ convergent?

SOLUTION We know from Section 3.4 and the graphs of the exponential functions in Section 6.2 (or Section 6.4*) that $\lim_{x \to \infty} a^x = \infty$ for $a > 1$ and $\lim_{x \to \infty} a^x = 0$ for $0 < a < 1$. Therefore, putting $a = r$ and using Theorem 3, we have

$$\lim_{n \to \infty} r^n = \begin{cases} \infty & \text{if } r > 1 \\ 0 & \text{if } 0 < r < 1 \end{cases}$$

It is obvious that

$$\lim_{n \to \infty} 1^n = 1 \quad \text{and} \quad \lim_{n \to \infty} 0^n = 0$$

If $-1 < r < 0$, then $0 < |r| < 1$, so

$$\lim_{n \to \infty} |r^n| = \lim_{n \to \infty} |r|^n = 0$$

and therefore $\lim_{n \to \infty} r^n = 0$ by Theorem 6. If $r \leqslant -1$, then $\{r^n\}$ diverges as in Example 7. Figure 11 shows the graphs for various values of r. (The case $r = -1$ is shown in Figure 8.)

FIGURE 11
The sequence $a_n = r^n$

The results of Example 11 are summarized for future use as follows.

> **9** The sequence $\{r^n\}$ is convergent if $-1 < r \leqslant 1$ and divergent for all other values of r.
>
> $$\lim_{n \to \infty} r^n = \begin{cases} 0 & \text{if } -1 < r < 1 \\ 1 & \text{if } r = 1 \end{cases}$$

> **10** **Definition** A sequence $\{a_n\}$ is called **increasing** if $a_n < a_{n+1}$ for all $n \geqslant 1$, that is, $a_1 < a_2 < a_3 < \cdots$. It is called **decreasing** if $a_n > a_{n+1}$ for all $n \geqslant 1$. A sequence is **monotonic** if it is either increasing or decreasing.

EXAMPLE 12 The sequence $\left\{\dfrac{3}{n+5}\right\}$ is decreasing because

The right side is smaller because it has a larger denominator.

$$\frac{3}{n+5} > \frac{3}{(n+1)+5} = \frac{3}{n+6}$$

and so $a_n > a_{n+1}$ for all $n \geqslant 1$.

EXAMPLE 13 Show that the sequence $a_n = \dfrac{n}{n^2+1}$ is decreasing.

SOLUTION 1 We must show that $a_{n+1} < a_n$, that is,

$$\frac{n+1}{(n+1)^2+1} < \frac{n}{n^2+1}$$

This inequality is equivalent to the one we get by cross-multiplication:

$$\frac{n+1}{(n+1)^2+1} < \frac{n}{n^2+1} \iff (n+1)(n^2+1) < n[(n+1)^2+1]$$

$$\iff n^3 + n^2 + n + 1 < n^3 + 2n^2 + 2n$$

$$\iff 1 < n^2 + n$$

Since $n \geqslant 1$, we know that the inequality $n^2 + n > 1$ is true. Therefore $a_{n+1} < a_n$ and so $\{a_n\}$ is decreasing.

SOLUTION 2 Consider the function $f(x) = \dfrac{x}{x^2+1}$:

$$f'(x) = \frac{x^2+1-2x^2}{(x^2+1)^2} = \frac{1-x^2}{(x^2+1)^2} < 0 \qquad \text{whenever } x^2 > 1$$

Thus f is decreasing on $(1, \infty)$ and so $f(n) > f(n+1)$. Therefore $\{a_n\}$ is decreasing.

---

11 **Definition** A sequence $\{a_n\}$ is **bounded above** if there is a number M such that

$$a_n \leqslant M \qquad \text{for all } n \geqslant 1$$

It is **bounded below** if there is a number m such that

$$m \leqslant a_n \qquad \text{for all } n \geqslant 1$$

If it is bounded above and below, then $\{a_n\}$ is a **bounded sequence**.

For instance, the sequence $a_n = n$ is bounded below ($a_n > 0$) but not above. The sequence $a_n = n/(n+1)$ is bounded because $0 < a_n < 1$ for all n.

We know that not every bounded sequence is convergent [for instance, the sequence $a_n = (-1)^n$ satisfies $-1 \leqslant a_n \leqslant 1$ but is divergent from Example 7] and not every monotonic sequence is convergent ($a_n = n \to \infty$). But if a sequence is both bounded *and* monotonic, then it must be convergent. This fact is proved as Theorem 12, but intuitively you can understand why it is true by looking at Figure 12. If $\{a_n\}$ is increasing and $a_n \leq M$ for all n, then the terms are forced to crowd together and approach some number L.

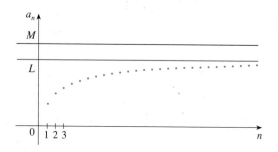

FIGURE 12

The proof of Theorem 12 is based on the **Completeness Axiom** for the set $\mathbb{R}$ of real numbers, which says that if S is a nonempty set of real numbers that has an upper bound M ($x \leq M$ for all x in S), then S has a **least upper bound** b. (This means that b is an upper bound for S, but if M is any other upper bound, then $b \leq M$.) The Completeness Axiom is an expression of the fact that there is no gap or hole in the real number line.

---

12   **Monotonic Sequence Theorem**   Every bounded, monotonic sequence is convergent.

PROOF   Suppose $\{a_n\}$ is an increasing sequence. Since $\{a_n\}$ is bounded, the set $S = \{a_n \mid n \geq 1\}$ has an upper bound. By the Completeness Axiom it has a least upper bound L. Given $\varepsilon > 0$, $L - \varepsilon$ is *not* an upper bound for S (since L is the *least* upper bound). Therefore

$$a_N > L - \varepsilon \qquad \text{for some integer } N$$

But the sequence is increasing so $a_n \geq a_N$ for every $n > N$. Thus if $n > N$, we have

$$a_n > L - \varepsilon$$

so

$$0 \leq L - a_n < \varepsilon$$

since $a_n \leq L$. Thus

$$\left| L - a_n \right| < \varepsilon \qquad \text{whenever } n > N$$

so $\lim_{n \to \infty} a_n = L$.

A similar proof (using the greatest lower bound) works if $\{a_n\}$ is decreasing.   ▬

The proof of Theorem 12 shows that a sequence that is increasing and bounded above is convergent. (Likewise, a decreasing sequence that is bounded below is convergent.) This fact is used many times in dealing with infinite series.

EXAMPLE 14 Investigate the sequence $\{a_n\}$ defined by the *recurrence relation*

$$a_1 = 2 \qquad a_{n+1} = \tfrac{1}{2}(a_n + 6) \qquad \text{for } n = 1, 2, 3, \ldots$$

SOLUTION We begin by computing the first several terms:

$$a_1 = 2 \qquad\qquad a_2 = \tfrac{1}{2}(2 + 6) = 4 \qquad a_3 = \tfrac{1}{2}(4 + 6) = 5$$

$$a_4 = \tfrac{1}{2}(5 + 6) = 5.5 \qquad a_5 = 5.75 \qquad\qquad a_6 = 5.875$$

$$a_7 = 5.9375 \qquad\qquad a_8 = 5.96875 \qquad\quad a_9 = 5.984375$$

Mathematical induction is often used in dealing with recursive sequences. See page 78 for a discussion of the Principle of Mathematical Induction.

These initial terms suggest that the sequence is increasing and the terms are approaching 6. To confirm that the sequence is increasing, we use mathematical induction to show that $a_{n+1} > a_n$ for all $n \geq 1$. This is true for $n = 1$ because $a_2 = 4 > a_1$. If we assume that it is true for $n = k$, then we have

$$a_{k+1} > a_k$$

so

$$a_{k+1} + 6 > a_k + 6$$

and

$$\tfrac{1}{2}(a_{k+1} + 6) > \tfrac{1}{2}(a_k + 6)$$

Thus

$$a_{k+2} > a_{k+1}$$

We have deduced that $a_{n+1} > a_n$ is true for $n = k + 1$. Therefore the inequality is true for all n by induction.

Next we verify that $\{a_n\}$ is bounded by showing that $a_n < 6$ for all n. (Since the sequence is increasing, we already know that it has a lower bound: $a_n \geq a_1 = 2$ for all n.) We know that $a_1 < 6$, so the assertion is true for $n = 1$. Suppose it is true for $n = k$. Then

$$a_k < 6$$

so

$$a_k + 6 < 12$$

and

$$\tfrac{1}{2}(a_k + 6) < \tfrac{1}{2}(12) = 6$$

Thus

$$a_{k+1} < 6$$

This shows, by mathematical induction, that $a_n < 6$ for all n.

Since the sequence $\{a_n\}$ is increasing and bounded, Theorem 12 guarantees that it has a limit. The theorem doesn't tell us what the value of the limit is. But now that we know $L = \lim_{n\to\infty} a_n$ exists, we can use the given recurrence relation to write

$$\lim_{n\to\infty} a_{n+1} = \lim_{n\to\infty} \tfrac{1}{2}(a_n + 6) = \tfrac{1}{2}\left(\lim_{n\to\infty} a_n + 6\right) = \tfrac{1}{2}(L + 6)$$

A proof of this fact is requested in Exercise 70.

Since $a_n \to L$, it follows that $a_{n+1} \to L$ too (as $n \to \infty$, $n + 1 \to \infty$ also). So we have

$$L = \tfrac{1}{2}(L + 6)$$

Solving this equation for L, we get $L = 6$, as we predicted.

11.1 Exercises

Access end-of-section exercises and tools for visualizing calculus online at **www.webassign.net**

LABORATORY PROJECT CAS LOGISTIC SEQUENCES

A sequence that arises in ecology as a model for population growth is defined by the **logistic difference equation**

$$p_{n+1} = kp_n(1 - p_n)$$

where p_n measures the size of the population of the nth generation of a single species. To keep the numbers manageable, p_n is a fraction of the maximal size of the population, so $0 \le p_n \le 1$. Notice that the form of this equation is similar to the logistic differential equation in Section 9.4. The discrete model—with sequences instead of continuous functions—is preferable for modeling insect populations, where mating and death occur in a periodic fashion.

An ecologist is interested in predicting the size of the population as time goes on, and asks these questions: Will it stabilize at a limiting value? Will it change in a cyclical fashion? Or will it exhibit random behavior?

Write a program to compute the first n terms of this sequence starting with an initial population p_0, where $0 < p_0 < 1$. Use this program to do the following.

1. Calculate 20 or 30 terms of the sequence for $p_0 = \frac{1}{2}$ and for two values of k such that $1 < k < 3$. Graph each sequence. Do the sequences appear to converge? Repeat for a different value of p_0 between 0 and 1. Does the limit depend on the choice of p_0? Does it depend on the choice of k?

2. Calculate terms of the sequence for a value of k between 3 and 3.4 and plot them. What do you notice about the behavior of the terms?

3. Experiment with values of k between 3.4 and 3.5. What happens to the terms?

4. For values of k between 3.6 and 4, compute and plot at least 100 terms and comment on the behavior of the sequence. What happens if you change p_0 by 0.001? This type of behavior is called *chaotic* and is exhibited by insect populations under certain conditions.

CAS Computer algebra system required

11.2 Series

What do we mean when we express a number as an infinite decimal? For instance, what does it mean to write

$$\pi = 3.14159\ 26535\ 89793\ 23846\ 26433\ 83279\ 50288 \ldots$$

The current record is that π has been computed to 2,576,980,370,000 (more than two trillion) decimal places by T. Daisuke and his team.

The convention behind our decimal notation is that any number can be written as an infinite sum. Here it means that

$$\pi = 3 + \frac{1}{10} + \frac{4}{10^2} + \frac{1}{10^3} + \frac{5}{10^4} + \frac{9}{10^5} + \frac{2}{10^6} + \frac{6}{10^7} + \frac{5}{10^8} + \cdots$$

where the three dots ($\cdots$) indicate that the sum continues forever, and the more terms we add, the closer we get to the actual value of π.

In general, if we try to add the terms of an infinite sequence $\{a_n\}_{n=1}^{\infty}$ we get an expression of the form

$$\boxed{1} \qquad a_1 + a_2 + a_3 + \cdots + a_n + \cdots$$

which is called an **infinite series** (or just a **series**) and is denoted, for short, by the symbol

$$\sum_{n=1}^{\infty} a_n \qquad \text{or} \qquad \sum a_n$$

Does it make sense to talk about the sum of infinitely many terms?

It would be impossible to find a finite sum for the series

$$1 + 2 + 3 + 4 + 5 + \cdots + n + \cdots$$

because if we start adding the terms we get the cumulative sums $1, 3, 6, 10, 15, 21, \ldots$ and, after the nth term, we get $n(n + 1)/2$, which becomes very large as n increases.

However, if we start to add the terms of the series

$$\frac{1}{2} + \frac{1}{4} + \frac{1}{8} + \frac{1}{16} + \frac{1}{32} + \frac{1}{64} + \cdots + \frac{1}{2^n} + \cdots$$

n	Sum of first n terms
1	0.50000000
2	0.75000000
3	0.87500000
4	0.93750000
5	0.96875000
6	0.98437500
7	0.99218750
10	0.99902344
15	0.99996948
20	0.99999905
25	0.99999997

we get $\frac{1}{2}, \frac{3}{4}, \frac{7}{8}, \frac{15}{16}, \frac{31}{32}, \frac{63}{64}, \ldots, 1 - 1/2^n, \ldots$. The table shows that as we add more and more terms, these *partial sums* become closer and closer to 1. (See also Figure 11 in *A Preview of Calculus,* page 6.) In fact, by adding sufficiently many terms of the series we can make the partial sums as close as we like to 1. So it seems reasonable to say that the sum of this infinite series is 1 and to write

$$\sum_{n=1}^{\infty} \frac{1}{2^n} = \frac{1}{2} + \frac{1}{4} + \frac{1}{8} + \frac{1}{16} + \cdots + \frac{1}{2^n} + \cdots = 1$$

We use a similar idea to determine whether or not a general series $\boxed{1}$ has a sum. We consider the **partial sums**

$$s_1 = a_1$$

$$s_2 = a_1 + a_2$$

$$s_3 = a_1 + a_2 + a_3$$

$$s_4 = a_1 + a_2 + a_3 + a_4$$

and, in general,

$$s_n = a_1 + a_2 + a_3 + \cdots + a_n = \sum_{i=1}^{n} a_i$$

These partial sums form a new sequence $\{s_n\}$, which may or may not have a limit. If $\lim_{n\to\infty} s_n = s$ exists (as a finite number), then, as in the preceding example, we call it the sum of the infinite series $\sum a_n$.

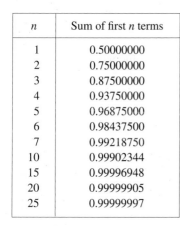

> **2 Definition** Given a series $\sum_{n=1}^{\infty} a_n = a_1 + a_2 + a_3 + \cdots$, let s_n denote its nth partial sum:
>
> $$s_n = \sum_{i=1}^{n} a_i = a_1 + a_2 + \cdots + a_n$$
>
> If the sequence $\{s_n\}$ is convergent and $\lim_{n \to \infty} s_n = s$ exists as a real number, then the series $\sum a_n$ is called **convergent** and we write
>
> $$a_1 + a_2 + \cdots + a_n + \cdots = s \qquad \text{or} \qquad \sum_{n=1}^{\infty} a_n = s$$
>
> The number s is called the **sum** of the series. If the sequence $\{s_n\}$ is divergent, then the series is called **divergent**.

Thus the sum of a series is the limit of the sequence of partial sums. So when we write $\sum_{n=1}^{\infty} a_n = s$, we mean that by adding sufficiently many terms of the series we can get as close as we like to the number s. Notice that

Compare with the improper integral

$$\int_{1}^{\infty} f(x)\, dx = \lim_{t \to \infty} \int_{1}^{t} f(x)\, dx$$

To find this integral we integrate from 1 to t and then let $t \to \infty$. For a series, we sum from 1 to n and then let $n \to \infty$.

$$\sum_{n=1}^{\infty} a_n = \lim_{n \to \infty} \sum_{i=1}^{n} a_i$$

EXAMPLE 1 Suppose we know that the sum of the first n terms of the series $\sum_{n=1}^{\infty} a_n$ is

$$s_n = a_1 + a_2 + \cdots + a_n = \frac{2n}{3n + 5}$$

Then the sum of the series is the limit of the sequence $\{s_n\}$:

$$\sum_{n=1}^{\infty} a_n = \lim_{n \to \infty} s_n = \lim_{n \to \infty} \frac{2n}{3n + 5} = \lim_{n \to \infty} \frac{2}{3 + \dfrac{5}{n}} = \frac{2}{3}$$　■

In Example 1 we were *given* an expression for the sum of the first n terms, but it's usually not easy to *find* such an expression. In Example 2, however, we look at a famous series for which we *can* find an explicit formula for s_n.

EXAMPLE 2 An important example of an infinite series is the **geometric series**

$$a + ar + ar^2 + ar^3 + \cdots + ar^{n-1} + \cdots = \sum_{n=1}^{\infty} ar^{n-1} \qquad a \neq 0$$

Each term is obtained from the preceding one by multiplying it by the **common ratio** r. (We have already considered the special case where $a = \frac{1}{2}$ and $r = \frac{1}{2}$ on page 577.)

If $r = 1$, then $s_n = a + a + \cdots + a = na \to \pm\infty$. Since $\lim_{n \to \infty} s_n$ doesn't exist, the geometric series diverges in this case.

If $r \neq 1$, we have

$$s_n = a + ar + ar^2 + \cdots + ar^{n-1}$$

and $\qquad\qquad rs_n = \qquad\quad ar + ar^2 + \cdots + ar^{n-1} + ar^n$

Figure 1 provides a geometric demonstration of the result in Example 2. If the triangles are constructed as shown and s is the sum of the series, then, by similar triangles,

$$\frac{s}{a} = \frac{a}{a - ar} \quad \text{so} \quad s = \frac{a}{1 - r}$$

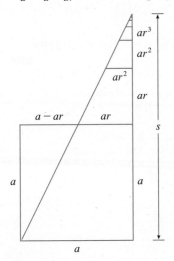

FIGURE 1

In words: The sum of a convergent geometric series is

$$\frac{\text{first term}}{1 - \text{common ratio}}$$

Subtracting these equations, we get

$$s_n - rs_n = a - ar^n$$

$$\boxed{3} \qquad\qquad s_n = \frac{a(1 - r^n)}{1 - r}$$

If $-1 < r < 1$, we know from (11.1.9) that $r^n \to 0$ as $n \to \infty$, so

$$\lim_{n \to \infty} s_n = \lim_{n \to \infty} \frac{a(1 - r^n)}{1 - r} = \frac{a}{1 - r} - \frac{a}{1 - r} \lim_{n \to \infty} r^n = \frac{a}{1 - r}$$

Thus when $|r| < 1$ the geometric series is convergent and its sum is $a/(1 - r)$.

If $r \leq -1$ or $r > 1$, the sequence $\{r^n\}$ is divergent by (11.1.9) and so, by Equation 3, $\lim_{n \to \infty} s_n$ does not exist. Therefore the geometric series diverges in those cases. ▬

We summarize the results of Example 2 as follows.

$\boxed{4}$ The geometric series

$$\sum_{n=1}^{\infty} ar^{n-1} = a + ar + ar^2 + \cdots$$

is convergent if $|r| < 1$ and its sum is

$$\sum_{n=1}^{\infty} ar^{n-1} = \frac{a}{1 - r} \qquad |r| < 1$$

If $|r| \geq 1$, the geometric series is divergent.

V EXAMPLE 3 Find the sum of the geometric series

$$5 - \frac{10}{3} + \frac{20}{9} - \frac{40}{27} + \cdots$$

SOLUTION The first term is $a = 5$ and the common ratio is $r = -\frac{2}{3}$. Since $|r| = \frac{2}{3} < 1$, the series is convergent by $\boxed{4}$ and its sum is

$$5 - \frac{10}{3} + \frac{20}{9} - \frac{40}{27} + \cdots = \frac{5}{1 - \left(-\frac{2}{3}\right)} = \frac{5}{\frac{5}{3}} = 3$$ ▬

What do we really mean when we say that the sum of the series in Example 3 is 3? Of course, we can't literally add an infinite number of terms, one by one. But, according to Definition 2, the total sum is the limit of the sequence of partial sums. So, by taking the sum of sufficiently many terms, we can get as close as we like to the number 3. The table shows the first ten partial sums s_n and the graph in Figure 2 shows how the sequence of partial sums approaches 3.

n	s_n
1	5.000000
2	1.666667
3	3.888889
4	2.407407
5	3.395062
6	2.736626
7	3.175583
8	2.882945
9	3.078037
10	2.947975

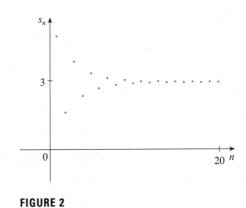

FIGURE 2

Another way to identify a and r is to write out the first few terms:
$$4 + \tfrac{16}{3} + \tfrac{64}{9} + \cdots$$

EXAMPLE 4 Is the series $\sum_{n=1}^{\infty} 2^{2n}3^{1-n}$ convergent or divergent?

SOLUTION Let's rewrite the nth term of the series in the form ar^{n-1}:

$$\sum_{n=1}^{\infty} 2^{2n}3^{1-n} = \sum_{n=1}^{\infty} (2^2)^n 3^{-(n-1)} = \sum_{n=1}^{\infty} \frac{4^n}{3^{n-1}} = \sum_{n=1}^{\infty} 4\left(\tfrac{4}{3}\right)^{n-1}$$

We recognize this series as a geometric series with $a = 4$ and $r = \tfrac{4}{3}$. Since $r > 1$, the series diverges by $\boxed{4}$.

V EXAMPLE 5 Write the number $2.3\overline{17} = 2.3171717\ldots$ as a ratio of integers.

SOLUTION

$$2.3171717\ldots = 2.3 + \frac{17}{10^3} + \frac{17}{10^5} + \frac{17}{10^7} + \cdots$$

After the first term we have a geometric series with $a = 17/10^3$ and $r = 1/10^2$. Therefore

$$2.3\overline{17} = 2.3 + \frac{\dfrac{17}{10^3}}{1 - \dfrac{1}{10^2}} = 2.3 + \frac{\dfrac{17}{1000}}{\dfrac{99}{100}}$$

$$= \frac{23}{10} + \frac{17}{990} = \frac{1147}{495}$$

EXAMPLE 6 Find the sum of the series $\sum_{n=0}^{\infty} x^n$, where $|x| < 1$.

SOLUTION Notice that this series starts with $n = 0$ and so the first term is $x^0 = 1$. (With series, we adopt the convention that $x^0 = 1$ even when $x = 0$.) Thus

$$\sum_{n=0}^{\infty} x^n = 1 + x + x^2 + x^3 + x^4 + \cdots$$

TEC Module 11.2 explores a series that depends on an angle θ in a triangle and enables you to see how rapidly the series converges when θ varies.

This is a geometric series with $a = 1$ and $r = x$. Since $|r| = |x| < 1$, it converges and $\boxed{4}$ gives

$$\boxed{5} \qquad \sum_{n=0}^{\infty} x^n = \frac{1}{1-x}$$

EXAMPLE 7 Show that the series $\sum_{n=1}^{\infty} \frac{1}{n(n+1)}$ is convergent, and find its sum.

SOLUTION This is not a geometric series, so we go back to the definition of a convergent series and compute the partial sums.

$$s_n = \sum_{i=1}^{n} \frac{1}{i(i+1)} = \frac{1}{1 \cdot 2} + \frac{1}{2 \cdot 3} + \frac{1}{3 \cdot 4} + \cdots + \frac{1}{n(n+1)}$$

We can simplify this expression if we use the partial fraction decomposition

$$\frac{1}{i(i+1)} = \frac{1}{i} - \frac{1}{i+1}$$

(see Section 7.4). Thus we have

$$s_n = \sum_{i=1}^{n} \frac{1}{i(i+1)} = \sum_{i=1}^{n} \left(\frac{1}{i} - \frac{1}{i+1} \right)$$

$$= \left(1 - \frac{1}{2} \right) + \left(\frac{1}{2} - \frac{1}{3} \right) + \left(\frac{1}{3} - \frac{1}{4} \right) + \cdots + \left(\frac{1}{n} - \frac{1}{n+1} \right)$$

$$= 1 - \frac{1}{n+1}$$

Notice that the terms cancel in pairs. This is an example of a **telescoping sum**: Because of all the cancellations, the sum collapses (like a pirate's collapsing telescope) into just two terms.

and so

$$\lim_{n \to \infty} s_n = \lim_{n \to \infty} \left(1 - \frac{1}{n+1} \right) = 1 - 0 = 1$$

Therefore the given series is convergent and

$$\sum_{n=1}^{\infty} \frac{1}{n(n+1)} = 1$$

Figure 3 illustrates Example 7 by showing the graphs of the sequence of terms $a_n = 1/[n(n+1)]$ and the sequence $\{s_n\}$ of partial sums. Notice that $a_n \to 0$ and $s_n \to 1$. See Exercises 76 and 77 for two geometric interpretations of Example 7.

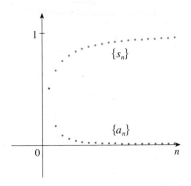

FIGURE 3

V **EXAMPLE 8** Show that the **harmonic series**

$$\sum_{n=1}^{\infty} \frac{1}{n} = 1 + \frac{1}{2} + \frac{1}{3} + \frac{1}{4} + \cdots$$

is divergent.

SOLUTION For this particular series it's convenient to consider the partial sums s_2, s_4, s_8, s_{16}, s_{32}, ... and show that they become large.

$$s_2 = 1 + \tfrac{1}{2}$$

$$s_4 = 1 + \tfrac{1}{2} + \left(\tfrac{1}{3} + \tfrac{1}{4} \right) > 1 + \tfrac{1}{2} + \left(\tfrac{1}{4} + \tfrac{1}{4} \right) = 1 + \tfrac{2}{2}$$

$$s_8 = 1 + \tfrac{1}{2} + \left(\tfrac{1}{3} + \tfrac{1}{4} \right) + \left(\tfrac{1}{5} + \tfrac{1}{6} + \tfrac{1}{7} + \tfrac{1}{8} \right)$$

$$> 1 + \tfrac{1}{2} + \left(\tfrac{1}{4} + \tfrac{1}{4} \right) + \left(\tfrac{1}{8} + \tfrac{1}{8} + \tfrac{1}{8} + \tfrac{1}{8} \right)$$

$$= 1 + \tfrac{1}{2} + \tfrac{1}{2} + \tfrac{1}{2} = 1 + \tfrac{3}{2}$$

$$s_{16} = 1 + \tfrac{1}{2} + \left(\tfrac{1}{3} + \tfrac{1}{4} \right) + \left(\tfrac{1}{5} + \cdots + \tfrac{1}{8} \right) + \left(\tfrac{1}{9} + \cdots + \tfrac{1}{16} \right)$$

$$> 1 + \tfrac{1}{2} + \left(\tfrac{1}{4} + \tfrac{1}{4} \right) + \left(\tfrac{1}{8} + \cdots + \tfrac{1}{8} \right) + \left(\tfrac{1}{16} + \cdots + \tfrac{1}{16} \right)$$

$$= 1 + \tfrac{1}{2} + \tfrac{1}{2} + \tfrac{1}{2} + \tfrac{1}{2} = 1 + \tfrac{4}{2}$$

Similarly, $s_{32} > 1 + \tfrac{5}{2}$, $s_{64} > 1 + \tfrac{6}{2}$, and in general

$$s_{2^n} > 1 + \frac{n}{2}$$

The method used in Example 8 for showing that the harmonic series diverges is due to the French scholar Nicole Oresme (1323–1382).

This shows that $s_{2^n} \to \infty$ as $n \to \infty$ and so $\{s_n\}$ is divergent. Therefore the harmonic series diverges.

---

6 **Theorem**   If the series $\displaystyle\sum_{n=1}^{\infty} a_n$ is convergent, then $\displaystyle\lim_{n \to \infty} a_n = 0$.

PROOF   Let $s_n = a_1 + a_2 + \cdots + a_n$. Then $a_n = s_n - s_{n-1}$. Since $\Sigma\, a_n$ is convergent, the sequence $\{s_n\}$ is convergent. Let $\lim_{n\to\infty} s_n = s$. Since $n - 1 \to \infty$ as $n \to \infty$, we also have $\lim_{n\to\infty} s_{n-1} = s$. Therefore

$$\lim_{n\to\infty} a_n = \lim_{n\to\infty} (s_n - s_{n-1}) = \lim_{n\to\infty} s_n - \lim_{n\to\infty} s_{n-1}$$
$$= s - s = 0$$

NOTE 1   With any *series* $\Sigma\, a_n$ we associate two *sequences:* the sequence $\{s_n\}$ of its partial sums and the sequence $\{a_n\}$ of its terms. If $\Sigma\, a_n$ is convergent, then the limit of the sequence $\{s_n\}$ is s (the sum of the series) and, as Theorem 6 asserts, the limit of the sequence $\{a_n\}$ is 0.

⊘   **NOTE 2**   The converse of Theorem 6 is not true in general. If $\lim_{n\to\infty} a_n = 0$, we cannot conclude that $\Sigma\, a_n$ is convergent. Observe that for the harmonic series $\Sigma\, 1/n$ we have $a_n = 1/n \to 0$ as $n \to \infty$, but we showed in Example 8 that $\Sigma\, 1/n$ is divergent.

---

7   **Test for Divergence**   If $\lim\limits_{n\to\infty} a_n$ does not exist or if $\lim\limits_{n\to\infty} a_n \neq 0$, then the series $\sum\limits_{n=1}^{\infty} a_n$ is divergent.

The Test for Divergence follows from Theorem 6 because, if the series is not divergent, then it is convergent, and so $\lim_{n\to\infty} a_n = 0$.

EXAMPLE 9   Show that the series $\sum\limits_{n=1}^{\infty} \dfrac{n^2}{5n^2 + 4}$ diverges.

SOLUTION

$$\lim_{n\to\infty} a_n = \lim_{n\to\infty} \frac{n^2}{5n^2 + 4} = \lim_{n\to\infty} \frac{1}{5 + 4/n^2} = \frac{1}{5} \neq 0$$

So the series diverges by the Test for Divergence.

NOTE 3   If we find that $\lim_{n\to\infty} a_n \neq 0$, we know that $\Sigma\, a_n$ is divergent. If we find that $\lim_{n\to\infty} a_n = 0$, we know *nothing* about the convergence or divergence of $\Sigma\, a_n$. Remember the warning in Note 2: If $\lim_{n\to\infty} a_n = 0$, the series $\Sigma\, a_n$ might converge or it might diverge.

---

8   **Theorem**   If $\Sigma\, a_n$ and $\Sigma\, b_n$ are convergent series, then so are the series $\Sigma\, ca_n$ (where c is a constant), $\Sigma\, (a_n + b_n)$, and $\Sigma\, (a_n - b_n)$, and

(i) $\sum\limits_{n=1}^{\infty} ca_n = c \sum\limits_{n=1}^{\infty} a_n$ 　　　　　　 (ii) $\sum\limits_{n=1}^{\infty} (a_n + b_n) = \sum\limits_{n=1}^{\infty} a_n + \sum\limits_{n=1}^{\infty} b_n$

(iii) $\sum\limits_{n=1}^{\infty} (a_n - b_n) = \sum\limits_{n=1}^{\infty} a_n - \sum\limits_{n=1}^{\infty} b_n$

These properties of convergent series follow from the corresponding Limit Laws for Sequences in Section 11.1. For instance, here is how part (ii) of Theorem 8 is proved: Let

$$s_n = \sum_{i=1}^{n} a_i \qquad s = \sum_{n=1}^{\infty} a_n \qquad t_n = \sum_{i=1}^{n} b_i \qquad t = \sum_{n=1}^{\infty} b_n$$

The nth partial sum for the series $\Sigma\,(a_n + b_n)$ is

$$u_n = \sum_{i=1}^{n} (a_i + b_i)$$

and, using Equation 4.2.10, we have

$$\lim_{n\to\infty} u_n = \lim_{n\to\infty} \sum_{i=1}^{n} (a_i + b_i) = \lim_{n\to\infty} \left(\sum_{i=1}^{n} a_i + \sum_{i=1}^{n} b_i \right)$$

$$= \lim_{n\to\infty} \sum_{i=1}^{n} a_i + \lim_{n\to\infty} \sum_{i=1}^{n} b_i$$

$$= \lim_{n\to\infty} s_n + \lim_{n\to\infty} t_n = s + t$$

Therefore $\Sigma\,(a_n + b_n)$ is convergent and its sum is

$$\sum_{n=1}^{\infty} (a_n + b_n) = s + t = \sum_{n=1}^{\infty} a_n + \sum_{n=1}^{\infty} b_n$$

EXAMPLE 10 Find the sum of the series $\displaystyle\sum_{n=1}^{\infty} \left(\frac{3}{n(n+1)} + \frac{1}{2^n} \right)$.

SOLUTION The series $\Sigma\, 1/2^n$ is a geometric series with $a = \frac{1}{2}$ and $r = \frac{1}{2}$, so

$$\sum_{n=1}^{\infty} \frac{1}{2^n} = \frac{\frac{1}{2}}{1 - \frac{1}{2}} = 1$$

In Example 7 we found that

$$\sum_{n=1}^{\infty} \frac{1}{n(n+1)} = 1$$

So, by Theorem 8, the given series is convergent and

$$\sum_{n=1}^{\infty} \left(\frac{3}{n(n+1)} + \frac{1}{2^n} \right) = 3 \sum_{n=1}^{\infty} \frac{1}{n(n+1)} + \sum_{n=1}^{\infty} \frac{1}{2^n}$$

$$= 3 \cdot 1 + 1 = 4$$

NOTE 4 A finite number of terms doesn't affect the convergence or divergence of a series. For instance, suppose that we were able to show that the series

$$\sum_{n=4}^{\infty} \frac{n}{n^3 + 1}$$

is convergent. Since

$$\sum_{n=1}^{\infty} \frac{n}{n^3 + 1} = \frac{1}{2} + \frac{2}{9} + \frac{3}{28} + \sum_{n=4}^{\infty} \frac{n}{n^3 + 1}$$

it follows that the entire series $\sum_{n=1}^{\infty} n/(n^3 + 1)$ is convergent. Similarly, if it is known that the series $\sum_{n=N+1}^{\infty} a_n$ converges, then the full series

$$\sum_{n=1}^{\infty} a_n = \sum_{n=1}^{N} a_n + \sum_{n=N+1}^{\infty} a_n$$

is also convergent.

11.2   **Exercises**

Access end-of-section exercises and tools for visualizing calculus online at **www.webassign.net**

11.3   **The Integral Test and Estimates of Sums**

In general, it is difficult to find the exact sum of a series. We were able to accomplish this for geometric series and the series $\Sigma\; 1/[n(n+1)]$ because in each of those cases we could find a simple formula for the nth partial sum s_n. But usually it isn't easy to discover such a formula. Therefore, in the next few sections, we develop several tests that enable us to determine whether a series is convergent or divergent without explicitly finding its sum. (In some cases, however, our methods will enable us to find good estimates of the sum.) Our first test involves improper integrals.

We begin by investigating the series whose terms are the reciprocals of the squares of the positive integers:

$$\sum_{n=1}^{\infty} \frac{1}{n^2} = \frac{1}{1^2} + \frac{1}{2^2} + \frac{1}{3^2} + \frac{1}{4^2} + \frac{1}{5^2} + \cdots$$

There's no simple formula for the sum s_n of the first n terms, but the computer-generated table of approximate values given in the margin suggests that the partial sums are approaching a number near 1.64 as $n \to \infty$ and so it looks as if the series is convergent.

We can confirm this impression with a geometric argument. Figure 1 shows the curve $y = 1/x^2$ and rectangles that lie below the curve. The base of each rectangle is an interval of length 1; the height is equal to the value of the function $y = 1/x^2$ at the right endpoint of the interval.

n	$s_n = \displaystyle\sum_{i=1}^{n} \frac{1}{i^2}$
5	1.4636
10	1.5498
50	1.6251
100	1.6350
500	1.6429
1000	1.6439
5000	1.6447

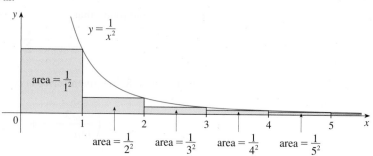

FIGURE 1

So the sum of the areas of the rectangles is

$$\frac{1}{1^2} + \frac{1}{2^2} + \frac{1}{3^2} + \frac{1}{4^2} + \frac{1}{5^2} + \cdots = \sum_{n=1}^{\infty} \frac{1}{n^2}$$

If we exclude the first rectangle, the total area of the remaining rectangles is smaller than the area under the curve $y = 1/x^2$ for $x \geqslant 1$, which is the value of the integral $\int_1^{\infty} (1/x^2)\, dx$. In Section 7.8 we discovered that this improper integral is convergent and has value 1. So the picture shows that all the partial sums are less than

$$\frac{1}{1^2} + \int_1^{\infty} \frac{1}{x^2}\, dx = 2$$

Thus the partial sums are bounded. We also know that the partial sums are increasing (because all the terms are positive). Therefore the partial sums converge (by the Monotonic Sequence Theorem) and so the series is convergent. The sum of the series (the limit of the partial sums) is also less than 2:

$$\sum_{n=1}^{\infty} \frac{1}{n^2} = \frac{1}{1^2} + \frac{1}{2^2} + \frac{1}{3^2} + \frac{1}{4^2} + \cdots < 2$$

[The exact sum of this series was found by the Swiss mathematician Leonhard Euler (1707–1783) to be $\pi^2/6$, but the proof of this fact is quite difficult. (See Problem 6 in the Problems Plus following Chapter 15.)]

Now let's look at the series

$$\sum_{n=1}^{\infty} \frac{1}{\sqrt{n}} = \frac{1}{\sqrt{1}} + \frac{1}{\sqrt{2}} + \frac{1}{\sqrt{3}} + \frac{1}{\sqrt{4}} + \frac{1}{\sqrt{5}} + \cdots$$

n	$s_n = \displaystyle\sum_{i=1}^{n} \frac{1}{\sqrt{i}}$
5	3.2317
10	5.0210
50	12.7524
100	18.5896
500	43.2834
1000	61.8010
5000	139.9681

The table of values of s_n suggests that the partial sums aren't approaching a finite number, so we suspect that the given series may be divergent. Again we use a picture for confirmation. Figure 2 shows the curve $y = 1/\sqrt{x}$, but this time we use rectangles whose tops lie *above* the curve.

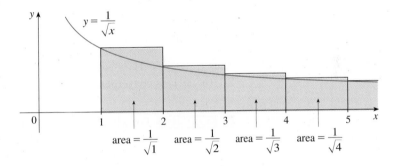

FIGURE 2

The base of each rectangle is an interval of length 1. The height is equal to the value of the function $y = 1/\sqrt{x}$ at the *left* endpoint of the interval. So the sum of the areas of all the rectangles is

$$\frac{1}{\sqrt{1}} + \frac{1}{\sqrt{2}} + \frac{1}{\sqrt{3}} + \frac{1}{\sqrt{4}} + \frac{1}{\sqrt{5}} + \cdots = \sum_{n=1}^{\infty} \frac{1}{\sqrt{n}}$$

This total area is greater than the area under the curve $y = 1/\sqrt{x}$ for $x \geqslant 1$, which is equal to the integral $\int_1^{\infty} (1/\sqrt{x})\, dx$. But we know from Section 7.8 that this improper integral is divergent. In other words, the area under the curve is infinite. So the sum of the series must be infinite; that is, the series is divergent.

The same sort of geometric reasoning that we used for these two series can be used to prove the following test. (The proof is given at the end of this section.)

The Integral Test   Suppose f is a continuous, positive, decreasing function on $[1, \infty)$ and let $a_n = f(n)$. Then the series $\sum_{n=1}^{\infty} a_n$ is convergent if and only if the improper integral $\int_1^{\infty} f(x)\, dx$ is convergent. In other words:

(i) If $\displaystyle\int_1^{\infty} f(x)\, dx$ is convergent, then $\displaystyle\sum_{n=1}^{\infty} a_n$ is convergent.

(ii) If $\displaystyle\int_1^{\infty} f(x)\, dx$ is divergent, then $\displaystyle\sum_{n=1}^{\infty} a_n$ is divergent.

NOTE When we use the Integral Test, it is not necessary to start the series or the integral at $n = 1$. For instance, in testing the series

$$\sum_{n=4}^{\infty} \frac{1}{(n-3)^2} \qquad \text{we use} \qquad \int_4^{\infty} \frac{1}{(x-3)^2}\, dx$$

Also, it is not necessary that f be *always* decreasing. What is important is that f be *ultimately* decreasing, that is, decreasing for x larger than some number N. Then $\sum_{n=N}^{\infty} a_n$ is convergent, so $\sum_{n=1}^{\infty} a_n$ is convergent by Note 4 of Section 11.2.

EXAMPLE 1 Test the series $\sum_{n=1}^{\infty} \frac{1}{n^2 + 1}$ for convergence or divergence.

SOLUTION The function $f(x) = 1/(x^2 + 1)$ is continuous, positive, and decreasing on $[1, \infty)$ so we use the Integral Test:

$$\int_1^{\infty} \frac{1}{x^2 + 1}\, dx = \lim_{t\to\infty} \int_1^t \frac{1}{x^2 + 1}\, dx = \lim_{t\to\infty} \tan^{-1}x \Big]_1^t$$

$$= \lim_{t\to\infty}\left(\tan^{-1}t - \frac{\pi}{4}\right) = \frac{\pi}{2} - \frac{\pi}{4} = \frac{\pi}{4}$$

Thus $\int_1^{\infty} 1/(x^2 + 1)\, dx$ is a convergent integral and so, by the Integral Test, the series $\sum 1/(n^2 + 1)$ is convergent.

V EXAMPLE 2 For what values of p is the series $\sum_{n=1}^{\infty} \frac{1}{n^p}$ convergent?

SOLUTION If $p < 0$, then $\lim_{n\to\infty} (1/n^p) = \infty$. If $p = 0$, then $\lim_{n\to\infty} (1/n^p) = 1$. In either case $\lim_{n\to\infty} (1/n^p) \neq 0$, so the given series diverges by the Test for Divergence (11.2.7).

If $p > 0$, then the function $f(x) = 1/x^p$ is clearly continuous, positive, and decreasing on $[1, \infty)$. We found in Chapter 7 [see (7.8.2)] that

$$\int_1^{\infty} \frac{1}{x^p}\, dx \quad \text{converges if } p > 1 \text{ and diverges if } p \leq 1$$

It follows from the Integral Test that the series $\sum 1/n^p$ converges if $p > 1$ and diverges if $0 < p \leq 1$. (For $p = 1$, this series is the harmonic series discussed in Example 8 in Section 11.2.)

The series in Example 2 is called the **p-series**. It is important in the rest of this chapter, so we summarize the results of Example 2 for future reference as follows.

> **1** The *p*-series $\sum_{n=1}^{\infty} \frac{1}{n^p}$ is convergent if $p > 1$ and divergent if $p \leq 1$.

EXAMPLE 3
(a) The series

$$\sum_{n=1}^{\infty} \frac{1}{n^3} = \frac{1}{1^3} + \frac{1}{2^3} + \frac{1}{3^3} + \frac{1}{4^3} + \cdots$$

is convergent because it is a *p*-series with $p = 3 > 1$.

In order to use the Integral Test we need to be able to evaluate $\int_1^{\infty} f(x)\, dx$ and therefore we have to be able to find an antiderivative of f. Frequently this is difficult or impossible, so we need other tests for convergence too.

(b) The series

$$\sum_{n=1}^{\infty} \frac{1}{n^{1/3}} = \sum_{n=1}^{\infty} \frac{1}{\sqrt[3]{n}} = 1 + \frac{1}{\sqrt[3]{2}} + \frac{1}{\sqrt[3]{3}} + \frac{1}{\sqrt[3]{4}} + \cdots$$

is divergent because it is a *p*-series with $p = \frac{1}{3} < 1$.

NOTE We should *not* infer from the Integral Test that the sum of the series is equal to the value of the integral. In fact,

$$\sum_{n=1}^{\infty} \frac{1}{n^2} = \frac{\pi^2}{6} \qquad \text{whereas} \qquad \int_1^{\infty} \frac{1}{x^2} \, dx = 1$$

Therefore, in general,

$$\sum_{n=1}^{\infty} a_n \neq \int_1^{\infty} f(x) \, dx$$

V **EXAMPLE 4** Determine whether the series $\displaystyle\sum_{n=1}^{\infty} \frac{\ln n}{n}$ converges or diverges.

SOLUTION The function $f(x) = (\ln x)/x$ is positive and continuous for $x > 1$ because the logarithm function is continuous. But it is not obvious whether or not f is decreasing, so we compute its derivative:

$$f'(x) = \frac{(1/x)x - \ln x}{x^2} = \frac{1 - \ln x}{x^2}$$

Thus $f'(x) < 0$ when $\ln x > 1$, that is, $x > e$. It follows that f is decreasing when $x > e$ and so we can apply the Integral Test:

$$\int_1^{\infty} \frac{\ln x}{x} \, dx = \lim_{t \to \infty} \int_1^t \frac{\ln x}{x} \, dx = \lim_{t \to \infty} \frac{(\ln x)^2}{2} \bigg]_1^t$$

$$= \lim_{t \to \infty} \frac{(\ln t)^2}{2} = \infty$$

Since this improper integral is divergent, the series $\sum (\ln n)/n$ is also divergent by the Integral Test.

Estimating the Sum of a Series

Suppose we have been able to use the Integral Test to show that a series $\sum a_n$ is convergent and we now want to find an approximation to the sum s of the series. Of course, any partial sum s_n is an approximation to s because $\lim_{n \to \infty} s_n = s$. But how good is such an approximation? To find out, we need to estimate the size of the **remainder**

$$R_n = s - s_n = a_{n+1} + a_{n+2} + a_{n+3} + \cdots$$

The remainder R_n is the error made when s_n, the sum of the first n terms, is used as an approximation to the total sum.

We use the same notation and ideas as in the Integral Test, assuming that f is decreasing on $[n, \infty)$. Comparing the areas of the rectangles with the area under $y = f(x)$ for $x > n$ in Figure 3, we see that

$$R_n = a_{n+1} + a_{n+2} + \cdots \leq \int_n^{\infty} f(x) \, dx$$

FIGURE 3

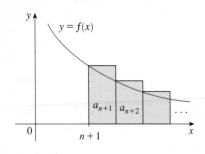

FIGURE 4

Similarly, we see from Figure 4 that

$$R_n = a_{n+1} + a_{n+2} + \cdots \geq \int_{n+1}^{\infty} f(x)\, dx$$

So we have proved the following error estimate.

---

2 **Remainder Estimate for the Integral Test** Suppose $f(k) = a_k$, where f is a continuous, positive, decreasing function for $x \geq n$ and $\Sigma\, a_n$ is convergent. If $R_n = s - s_n$, then

$$\int_{n+1}^{\infty} f(x)\, dx \leq R_n \leq \int_{n}^{\infty} f(x)\, dx$$

V EXAMPLE 5

(a) Approximate the sum of the series $\Sigma\, 1/n^3$ by using the sum of the first 10 terms. Estimate the error involved in this approximation.

(b) How many terms are required to ensure that the sum is accurate to within 0.0005?

SOLUTION In both parts (a) and (b) we need to know $\int_{n}^{\infty} f(x)\, dx$. With $f(x) = 1/x^3$, which satisfies the conditions of the Integral Test, we have

$$\int_{n}^{\infty} \frac{1}{x^3}\, dx = \lim_{t \to \infty} \left[-\frac{1}{2x^2} \right]_{n}^{t} = \lim_{t \to \infty} \left(-\frac{1}{2t^2} + \frac{1}{2n^2} \right) = \frac{1}{2n^2}$$

(a) Approximating the sum of the series by the 10th partial sum, we have

$$\sum_{n=1}^{\infty} \frac{1}{n^3} \approx s_{10} = \frac{1}{1^3} + \frac{1}{2^3} + \frac{1}{3^3} + \cdots + \frac{1}{10^3} \approx 1.1975$$

According to the remainder estimate in **2**, we have

$$R_{10} \leq \int_{10}^{\infty} \frac{1}{x^3}\, dx = \frac{1}{2(10)^2} = \frac{1}{200}$$

So the size of the error is at most 0.005.

(b) Accuracy to within 0.0005 means that we have to find a value of n such that $R_n \leq 0.0005$. Since

$$R_n \leq \int_{n}^{\infty} \frac{1}{x^3}\, dx = \frac{1}{2n^2}$$

we want

$$\frac{1}{2n^2} < 0.0005$$

Solving this inequality, we get

$$n^2 > \frac{1}{0.001} = 1000 \qquad \text{or} \qquad n > \sqrt{1000} \approx 31.6$$

We need 32 terms to ensure accuracy to within 0.0005.

If we add s_n to each side of the inequalities in **2**, we get

3
$$s_n + \int_{n+1}^{\infty} f(x)\, dx \leq s \leq s_n + \int_{n}^{\infty} f(x)\, dx$$

because $s_n + R_n = s$. The inequalities in $\boxed{3}$ give a lower bound and an upper bound for s. They provide a more accurate approximation to the sum of the series than the partial sum s_n does.

EXAMPLE 6 Use $\boxed{3}$ with $n = 10$ to estimate the sum of the series $\sum_{n=1}^{\infty} \dfrac{1}{n^3}$.

SOLUTION The inequalities in $\boxed{3}$ become

$$s_{10} + \int_{11}^{\infty} \frac{1}{x^3} \, dx \leq s \leq s_{10} + \int_{10}^{\infty} \frac{1}{x^3} \, dx$$

From Example 5 we know that

$$\int_{n}^{\infty} \frac{1}{x^3} \, dx = \frac{1}{2n^2}$$

so

$$s_{10} + \frac{1}{2(11)^2} \leq s \leq s_{10} + \frac{1}{2(10)^2}$$

Using $s_{10} \approx 1.197532$, we get

$$1.201664 \leq s \leq 1.202532$$

If we approximate s by the midpoint of this interval, then the error is at most half the length of the interval. So

$$\sum_{n=1}^{\infty} \frac{1}{n^3} \approx 1.2021 \qquad \text{with error} < 0.0005 \qquad \blacksquare$$

If we compare Example 6 with Example 5, we see that the improved estimate in $\boxed{3}$ can be much better than the estimate $s \approx s_n$. To make the error smaller than 0.0005 we had to use 32 terms in Example 5 but only 10 terms in Example 6.

Although Euler was able to calculate the exact sum of the p-series for $p = 2$, nobody has been able to find the exact sum for $p = 3$. In Example 6, however, we show how to *estimate* this sum.

■ **Proof of the Integral Test**

We have already seen the basic idea behind the proof of the Integral Test in Figures 1 and 2 for the series $\Sigma \, 1/n^2$ and $\Sigma \, 1/\sqrt{n}$. For the general series $\Sigma \, a_n$, look at Figures 5 and 6. The area of the first shaded rectangle in Figure 5 is the value of f at the right endpoint of $[1, 2]$, that is, $f(2) = a_2$. So, comparing the areas of the shaded rectangles with the area under $y = f(x)$ from 1 to n, we see that

$$\boxed{4} \qquad a_2 + a_3 + \cdots + a_n \leq \int_{1}^{n} f(x) \, dx$$

(Notice that this inequality depends on the fact that f is decreasing.) Likewise, Figure 6 shows that

$$\boxed{5} \qquad \int_{1}^{n} f(x) \, dx \leq a_1 + a_2 + \cdots + a_{n-1}$$

(i) If $\int_{1}^{\infty} f(x) \, dx$ is convergent, then $\boxed{4}$ gives

$$\sum_{i=2}^{n} a_i \leq \int_{1}^{n} f(x) \, dx \leq \int_{1}^{\infty} f(x) \, dx$$

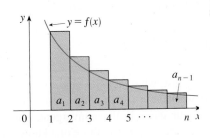

FIGURE 5

FIGURE 6

since $f(x) \geqslant 0$. Therefore

$$s_n = a_1 + \sum_{i=2}^{n} a_i \leqslant a_1 + \int_{1}^{\infty} f(x)\, dx = M, \text{ say}$$

Since $s_n \leqslant M$ for all n, the sequence $\{s_n\}$ is bounded above. Also

$$s_{n+1} = s_n + a_{n+1} \geqslant s_n$$

since $a_{n+1} = f(n + 1) \geqslant 0$. Thus $\{s_n\}$ is an increasing bounded sequence and so it is convergent by the Monotonic Sequence Theorem (11.1.12). This means that $\Sigma\, a_n$ is convergent.

   (ii) If $\int_{1}^{\infty} f(x)\, dx$ is divergent, then $\int_{1}^{n} f(x)\, dx \to \infty$ as $n \to \infty$ because $f(x) \geqslant 0$. But [5] gives

$$\int_{1}^{n} f(x)\, dx \leqslant \sum_{i=1}^{n-1} a_i = s_{n-1}$$

and so $s_{n-1} \to \infty$. This implies that $s_n \to \infty$ and so $\Sigma\, a_n$ diverges. ∎

11.3 Exercises

Access end-of-section exercises and tools for visualizing calculus online at **www.webassign.net**   ENHANCED   **WebAssign**

11.4 The Comparison Tests

In the comparison tests the idea is to compare a given series with a series that is known to be convergent or divergent. For instance, the series

[1]
$$\sum_{n=1}^{\infty} \frac{1}{2^n + 1}$$

reminds us of the series $\sum_{n=1}^{\infty} 1/2^n$, which is a geometric series with $a = \frac{1}{2}$ and $r = \frac{1}{2}$ and is therefore convergent. Because the series [1] is so similar to a convergent series, we have the feeling that it too must be convergent. Indeed, it is. The inequality

$$\frac{1}{2^n + 1} < \frac{1}{2^n}$$

shows that our given series [1] has smaller terms than those of the geometric series and therefore all its partial sums are also smaller than 1 (the sum of the geometric series). This means that its partial sums form a bounded increasing sequence, which is convergent. It also follows that the sum of the series is less than the sum of the geometric series:

$$\sum_{n=1}^{\infty} \frac{1}{2^n + 1} < 1$$

   Similar reasoning can be used to prove the following test, which applies only to series whose terms are positive. The first part says that if we have a series whose terms are *smaller* than those of a known *convergent* series, then our series is also convergent. The second part says that if we start with a series whose terms are *larger* than those of a known *divergent* series, then it too is divergent.

> **The Comparison Test** Suppose that $\Sigma\, a_n$ and $\Sigma\, b_n$ are series with positive terms.
>
> (i) If $\Sigma\, b_n$ is convergent and $a_n \leqslant b_n$ for all n, then $\Sigma\, a_n$ is also convergent.
>
> (ii) If $\Sigma\, b_n$ is divergent and $a_n \geqslant b_n$ for all n, then $\Sigma\, a_n$ is also divergent.

It is important to keep in mind the distinction between a sequence and a series. A sequence is a list of numbers, whereas a series is a sum. With every series $\Sigma\, a_n$ there are associated two sequences: the sequence $\{a_n\}$ of terms and the sequence $\{s_n\}$ of partial sums.

PROOF

(i) Let
$$s_n = \sum_{i=1}^{n} a_i \qquad t_n = \sum_{i=1}^{n} b_i \qquad t = \sum_{n=1}^{\infty} b_n$$

Since both series have positive terms, the sequences $\{s_n\}$ and $\{t_n\}$ are increasing $(s_{n+1} = s_n + a_{n+1} \geqslant s_n)$. Also $t_n \to t$, so $t_n \leqslant t$ for all n. Since $a_i \leqslant b_i$, we have $s_n \leqslant t_n$. Thus $s_n \leqslant t$ for all n. This means that $\{s_n\}$ is increasing and bounded above and therefore converges by the Monotonic Sequence Theorem. Thus $\Sigma\, a_n$ converges.

(ii) If $\Sigma\, b_n$ is divergent, then $t_n \to \infty$ (since $\{t_n\}$ is increasing). But $a_i \geqslant b_i$ so $s_n \geqslant t_n$. Thus $s_n \to \infty$. Therefore $\Sigma\, a_n$ diverges. ■

Standard Series for Use with the Comparison Test

In using the Comparison Test we must, of course, have some known series $\Sigma\, b_n$ for the purpose of comparison. Most of the time we use one of these series:

- A p-series $\left[\Sigma\, 1/n^p \text{ converges if } p > 1 \text{ and diverges if } p \leqslant 1; \text{ see } (11.3.1)\right]$

- A geometric series $\left[\Sigma\, ar^{n-1} \text{ converges if } |r| < 1 \text{ and diverges if } |r| \geqslant 1;\right.$ see (11.2.4)$\left.\right]$

V EXAMPLE 1 Determine whether the series $\displaystyle\sum_{n=1}^{\infty} \frac{5}{2n^2 + 4n + 3}$ converges or diverges.

SOLUTION For large n the dominant term in the denominator is $2n^2$, so we compare the given series with the series $\Sigma\, 5/(2n^2)$. Observe that

$$\frac{5}{2n^2 + 4n + 3} < \frac{5}{2n^2}$$

because the left side has a bigger denominator. (In the notation of the Comparison Test, a_n is the left side and b_n is the right side.) We know that

$$\sum_{n=1}^{\infty} \frac{5}{2n^2} = \frac{5}{2} \sum_{n=1}^{\infty} \frac{1}{n^2}$$

is convergent because it's a constant times a p-series with $p = 2 > 1$. Therefore

$$\sum_{n=1}^{\infty} \frac{5}{2n^2 + 4n + 3}$$

is convergent by part (i) of the Comparison Test. ■

NOTE 1 Although the condition $a_n \leqslant b_n$ or $a_n \geqslant b_n$ in the Comparison Test is given for all n, we need verify only that it holds for $n \geqslant N$, where N is some fixed integer, because the convergence of a series is not affected by a finite number of terms. This is illustrated in the next example.

V EXAMPLE 2 Test the series $\displaystyle\sum_{k=1}^{\infty} \frac{\ln k}{k}$ for convergence or divergence.

SOLUTION We used the Integral Test to test this series in Example 4 of Section 11.3, but we can also test it by comparing it with the harmonic series. Observe that $\ln k > 1$ for $k \geq 3$ and so

$$\frac{\ln k}{k} > \frac{1}{k} \qquad k \geq 3$$

We know that $\sum 1/k$ is divergent (p-series with $p = 1$). Thus the given series is divergent by the Comparison Test.

NOTE 2 The terms of the series being tested must be smaller than those of a convergent series or larger than those of a divergent series. If the terms are larger than the terms of a convergent series or smaller than those of a divergent series, then the Comparison Test doesn't apply. Consider, for instance, the series

$$\sum_{n=1}^{\infty} \frac{1}{2^n - 1}$$

The inequality

$$\frac{1}{2^n - 1} > \frac{1}{2^n}$$

is useless as far as the Comparison Test is concerned because $\sum b_n = \sum \left(\frac{1}{2}\right)^n$ is convergent and $a_n > b_n$. Nonetheless, we have the feeling that $\sum 1/(2^n - 1)$ ought to be convergent because it is very similar to the convergent geometric series $\sum \left(\frac{1}{2}\right)^n$. In such cases the following test can be used.

The Limit Comparison Test Suppose that $\sum a_n$ and $\sum b_n$ are series with positive terms. If

$$\lim_{n \to \infty} \frac{a_n}{b_n} = c$$

where c is a finite number and $c > 0$, then either both series converge or both diverge.

Exercises 40 and 41 deal with the cases $c = 0$ and $c = \infty$.

PROOF Let m and M be positive numbers such that $m < c < M$. Because a_n/b_n is close to c for large n, there is an integer N such that

$$m < \frac{a_n}{b_n} < M \qquad \text{when } n > N$$

and so

$$mb_n < a_n < Mb_n \qquad \text{when } n > N$$

If $\sum b_n$ converges, so does $\sum Mb_n$. Thus $\sum a_n$ converges by part (i) of the Comparison Test. If $\sum b_n$ diverges, so does $\sum mb_n$ and part (ii) of the Comparison Test shows that $\sum a_n$ diverges.

EXAMPLE 3 Test the series $\sum_{n=1}^{\infty} \dfrac{1}{2^n - 1}$ for convergence or divergence.

SOLUTION We use the Limit Comparison Test with

$$a_n = \frac{1}{2^n - 1} \qquad b_n = \frac{1}{2^n}$$

and obtain

$$\lim_{n \to \infty} \frac{a_n}{b_n} = \lim_{n \to \infty} \frac{1/(2^n - 1)}{1/2^n} = \lim_{n \to \infty} \frac{2^n}{2^n - 1} = \lim_{n \to \infty} \frac{1}{1 - 1/2^n} = 1 > 0$$

Since this limit exists and $\sum 1/2^n$ is a convergent geometric series, the given series converges by the Limit Comparison Test.

EXAMPLE 4 Determine whether the series $\sum_{n=1}^{\infty} \dfrac{2n^2 + 3n}{\sqrt{5 + n^5}}$ converges or diverges.

SOLUTION The dominant part of the numerator is $2n^2$ and the dominant part of the denominator is $\sqrt{n^5} = n^{5/2}$. This suggests taking

$$a_n = \frac{2n^2 + 3n}{\sqrt{5 + n^5}} \qquad b_n = \frac{2n^2}{n^{5/2}} = \frac{2}{n^{1/2}}$$

$$\lim_{n \to \infty} \frac{a_n}{b_n} = \lim_{n \to \infty} \frac{2n^2 + 3n}{\sqrt{5 + n^5}} \cdot \frac{n^{1/2}}{2} = \lim_{n \to \infty} \frac{2n^{5/2} + 3n^{3/2}}{2\sqrt{5 + n^5}}$$

$$= \lim_{n \to \infty} \frac{2 + \dfrac{3}{n}}{2\sqrt{\dfrac{5}{n^5} + 1}} = \frac{2 + 0}{2\sqrt{0 + 1}} = 1$$

Since $\sum b_n = 2 \sum 1/n^{1/2}$ is divergent (p-series with $p = \frac{1}{2} < 1$), the given series diverges by the Limit Comparison Test.

Notice that in testing many series we find a suitable comparison series $\sum b_n$ by keeping only the highest powers in the numerator and denominator.

Estimating Sums

If we have used the Comparison Test to show that a series $\sum a_n$ converges by comparison with a series $\sum b_n$, then we may be able to estimate the sum $\sum a_n$ by comparing remainders. As in Section 11.3, we consider the remainder

$$R_n = s - s_n = a_{n+1} + a_{n+2} + \cdots$$

For the comparison series $\sum b_n$ we consider the corresponding remainder

$$T_n = t - t_n = b_{n+1} + b_{n+2} + \cdots$$

Since $a_n \le b_n$ for all n, we have $R_n \le T_n$. If $\sum b_n$ is a p-series, we can estimate its remainder T_n as in Section 11.3. If $\sum b_n$ is a geometric series, then T_n is the sum of a geometric series and we can sum it exactly (see Exercises 35 and 36). In either case we know that R_n is smaller than T_n.

▼ **EXAMPLE 5** Use the sum of the first 100 terms to approximate the sum of the series $\Sigma\ 1/(n^3 + 1)$. Estimate the error involved in this approximation.

SOLUTION Since

$$\frac{1}{n^3 + 1} < \frac{1}{n^3}$$

the given series is convergent by the Comparison Test. The remainder T_n for the comparison series $\Sigma\ 1/n^3$ was estimated in Example 5 in Section 11.3 using the Remainder Estimate for the Integral Test. There we found that

$$T_n \le \int_n^\infty \frac{1}{x^3}\,dx = \frac{1}{2n^2}$$

Therefore the remainder R_n for the given series satisfies

$$R_n \le T_n \le \frac{1}{2n^2}$$

With $n = 100$ we have

$$R_{100} \le \frac{1}{2(100)^2} = 0.00005$$

Using a programmable calculator or a computer, we find that

$$\sum_{n=1}^{\infty} \frac{1}{n^3 + 1} \approx \sum_{n=1}^{100} \frac{1}{n^3 + 1} \approx 0.6864538$$

with error less than 0.00005. ∎

11.4   Exercises

Access end-of-section exercises and tools for visualizing calculus online at **www.webassign.net**   **WebAssign**

11.5   Alternating Series

The convergence tests that we have looked at so far apply only to series with positive terms. In this section and the next we learn how to deal with series whose terms are not necessarily positive. Of particular importance are *alternating series,* whose terms alternate in sign.

An **alternating series** is a series whose terms are alternately positive and negative. Here are two examples:

$$1 - \frac{1}{2} + \frac{1}{3} - \frac{1}{4} + \frac{1}{5} - \frac{1}{6} + \cdots = \sum_{n=1}^{\infty} (-1)^{n-1}\frac{1}{n}$$

$$-\frac{1}{2} + \frac{2}{3} - \frac{3}{4} + \frac{4}{5} - \frac{5}{6} + \frac{6}{7} - \cdots = \sum_{n=1}^{\infty} (-1)^{n}\frac{n}{n+1}$$

We see from these examples that the nth term of an alternating series is of the form

$$a_n = (-1)^{n-1}b_n \qquad \text{or} \qquad a_n = (-1)^{n}b_n$$

where b_n is a positive number. $\big($In fact, $b_n = |a_n|.\big)$

The following test says that if the terms of an alternating series decrease toward 0 in absolute value, then the series converges.

Alternating Series Test   If the alternating series

$$\sum_{n=1}^{\infty} (-1)^{n-1} b_n = b_1 - b_2 + b_3 - b_4 + b_5 - b_6 + \cdots \qquad b_n > 0$$

satisfies

(i) $b_{n+1} \leqslant b_n$ \qquad for all n

(ii) $\displaystyle\lim_{n \to \infty} b_n = 0$

then the series is convergent.

Before giving the proof let's look at Figure 1, which gives a picture of the idea behind the proof. We first plot $s_1 = b_1$ on a number line. To find s_2 we subtract b_2, so s_2 is to the left of s_1. Then to find s_3 we add b_3, so s_3 is to the right of s_2. But, since $b_3 < b_2$, s_3 is to the left of s_1. Continuing in this manner, we see that the partial sums oscillate back and forth. Since $b_n \to 0$, the successive steps are becoming smaller and smaller. The even partial sums $s_2, s_4, s_6, \ldots$ are increasing and the odd partial sums $s_1, s_3, s_5, \ldots$ are decreasing. Thus it seems plausible that both are converging to some number s, which is the sum of the series. Therefore we consider the even and odd partial sums separately in the following proof.

FIGURE 1

PROOF OF THE ALTERNATING SERIES TEST   We first consider the even partial sums:

$$s_2 = b_1 - b_2 \geqslant 0 \qquad\qquad \text{since } b_2 \leqslant b_1$$

$$s_4 = s_2 + (b_3 - b_4) \geqslant s_2 \qquad \text{since } b_4 \leqslant b_3$$

In general $s_{2n} = s_{2n-2} + (b_{2n-1} - b_{2n}) \geqslant s_{2n-2}$ \qquad since $b_{2n} \leqslant b_{2n-1}$

Thus $0 \leqslant s_2 \leqslant s_4 \leqslant s_6 \leqslant \cdots \leqslant s_{2n} \leqslant \cdots$

But we can also write

$$s_{2n} = b_1 - (b_2 - b_3) - (b_4 - b_5) - \cdots - (b_{2n-2} - b_{2n-1}) - b_{2n}$$

Every term in brackets is positive, so $s_{2n} \leqslant b_1$ for all n. Therefore the sequence $\{s_{2n}\}$ of even partial sums is increasing and bounded above. It is therefore convergent by the Monotonic Sequence Theorem. Let's call its limit s, that is,

$$\lim_{n \to \infty} s_{2n} = s$$

Now we compute the limit of the odd partial sums:

$$\lim_{n \to \infty} s_{2n+1} = \lim_{n \to \infty} (s_{2n} + b_{2n+1})$$

$$= \lim_{n \to \infty} s_{2n} + \lim_{n \to \infty} b_{2n+1}$$

$$= s + 0 \qquad\qquad \text{[by condition (ii)]}$$

$$= s$$

Since both the even and odd partial sums converge to s, we have $\lim_{n \to \infty} s_n = s$ [see Exercise 92(a) in Section 11.1] and so the series is convergent.

Figure 2 illustrates Example 1 by showing the graphs of the terms $a_n = (-1)^{n-1}/n$ and the partial sums s_n. Notice how the values of s_n zigzag across the limiting value, which appears to be about 0.7. In fact, it can be proved that the exact sum of the series is $\ln 2 \approx 0.693$ (see Exercise 36).

V **EXAMPLE 1** The alternating harmonic series

$$1 - \frac{1}{2} + \frac{1}{3} - \frac{1}{4} + \cdots = \sum_{n=1}^{\infty} \frac{(-1)^{n-1}}{n}$$

satisfies

(i) $b_{n+1} < b_n$ because $\dfrac{1}{n+1} < \dfrac{1}{n}$

(ii) $\displaystyle\lim_{n \to \infty} b_n = \lim_{n \to \infty} \frac{1}{n} = 0$

so the series is convergent by the Alternating Series Test.

FIGURE 2

V **EXAMPLE 2** The series $\displaystyle\sum_{n=1}^{\infty} \frac{(-1)^n 3n}{4n - 1}$ is alternating, but

$$\lim_{n \to \infty} b_n = \lim_{n \to \infty} \frac{3n}{4n - 1} = \lim_{n \to \infty} \frac{3}{4 - \dfrac{1}{n}} = \frac{3}{4}$$

so condition (ii) is not satisfied. Instead, we look at the limit of the nth term of the series:

$$\lim_{n \to \infty} a_n = \lim_{n \to \infty} \frac{(-1)^n 3n}{4n - 1}$$

This limit does not exist, so the series diverges by the Test for Divergence.

EXAMPLE 3 Test the series $\displaystyle\sum_{n=1}^{\infty} (-1)^{n+1} \frac{n^2}{n^3 + 1}$ for convergence or divergence.

SOLUTION The given series is alternating so we try to verify conditions (i) and (ii) of the Alternating Series Test.

Unlike the situation in Example 1, it is not obvious that the sequence given by $b_n = n^2/(n^3 + 1)$ is decreasing. However, if we consider the related function $f(x) = x^2/(x^3 + 1)$, we find that

$$f'(x) = \frac{x(2 - x^3)}{(x^3 + 1)^2}$$

Since we are considering only positive x, we see that $f'(x) < 0$ if $2 - x^3 < 0$, that is, $x > \sqrt[3]{2}$. Thus f is decreasing on the interval $(\sqrt[3]{2}, \infty)$. This means that $f(n + 1) < f(n)$

Instead of verifying condition (i) of the Alternating Series Test by computing a derivative, we could verify that $b_{n+1} < b_n$ directly by using the technique of Solution 1 of Example 13 in Section 11.1.

and therefore $b_{n+1} < b_n$ when $n \geq 2$. (The inequality $b_2 < b_1$ can be verified directly but all that really matters is that the sequence $\{b_n\}$ is eventually decreasing.)

Condition (ii) is readily verified:

$$\lim_{n \to \infty} b_n = \lim_{n \to \infty} \frac{n^2}{n^3 + 1} = \lim_{n \to \infty} \frac{\dfrac{1}{n}}{1 + \dfrac{1}{n^3}} = 0$$

Thus the given series is convergent by the Alternating Series Test.

Estimating Sums

A partial sum s_n of any convergent series can be used as an approximation to the total sum s, but this is not of much use unless we can estimate the accuracy of the approximation. The error involved in using $s \approx s_n$ is the remainder $R_n = s - s_n$. The next theorem says that for series that satisfy the conditions of the Alternating Series Test, the size of the error is smaller than b_{n+1}, which is the absolute value of the first neglected term.

You can see geometrically why the Alternating Series Estimation Theorem is true by looking at Figure 1 (on page 595). Notice that $s - s_4 < b_5$, $|s - s_5| < b_6$, and so on. Notice also that s lies between any two consecutive partial sums.

Alternating Series Estimation Theorem   If $s = \Sigma\, (-1)^{n-1} b_n$ is the sum of an alternating series that satisfies

(i) $b_{n+1} \leq b_n$ and   (ii) $\displaystyle\lim_{n \to \infty} b_n = 0$

then $|R_n| = |s - s_n| \leq b_{n+1}$

PROOF We know from the proof of the Alternating Series Test that s lies between any two consecutive partial sums s_n and s_{n+1}. (There we showed that s is larger than all even partial sums. A similar argument shows that s is smaller than all the odd sums.) It follows that

$$|s - s_n| \leq |s_{n+1} - s_n| = b_{n+1}$$

By definition, $0! = 1$.

▼ EXAMPLE 4 Find the sum of the series $\displaystyle\sum_{n=0}^{\infty} \frac{(-1)^n}{n!}$ correct to three decimal places.

SOLUTION We first observe that the series is convergent by the Alternating Series Test because

(i) $\dfrac{1}{(n+1)!} = \dfrac{1}{n!\,(n+1)} < \dfrac{1}{n!}$

(ii) $0 < \dfrac{1}{n!} < \dfrac{1}{n} \to 0$ so $\dfrac{1}{n!} \to 0$ as $n \to \infty$

To get a feel for how many terms we need to use in our approximation, let's write out the first few terms of the series:

$$s = \frac{1}{0!} - \frac{1}{1!} + \frac{1}{2!} - \frac{1}{3!} + \frac{1}{4!} - \frac{1}{5!} + \frac{1}{6!} - \frac{1}{7!} + \cdots$$

$$= 1 - 1 + \tfrac{1}{2} - \tfrac{1}{6} + \tfrac{1}{24} - \tfrac{1}{120} + \tfrac{1}{720} - \tfrac{1}{5040} + \cdots$$

Notice that
$$b_7 = \tfrac{1}{5040} < \tfrac{1}{5000} = 0.0002$$

and
$$s_6 = 1 - 1 + \tfrac{1}{2} - \tfrac{1}{6} + \tfrac{1}{24} - \tfrac{1}{120} + \tfrac{1}{720} \approx 0.368056$$

By the Alternating Series Estimation Theorem we know that

$$|s - s_6| \le b_7 < 0.0002$$

This error of less than 0.0002 does not affect the third decimal place, so we have $s \approx 0.368$ correct to three decimal places.

In Section 11.10 we will prove that $e^x = \sum_{n=0}^{\infty} x^n/n!$ for all x, so what we have obtained in Example 4 is actually an approximation to the number e^{-1}.

   NOTE The rule that the error (in using s_n to approximate s) is smaller than the first neglected term is, in general, valid only for alternating series that satisfy the conditions of the Alternating Series Estimation Theorem. The rule does not apply to other types of series.

11.5 Exercises

Access end-of-section exercises and tools for visualizing calculus online at **www.webassign.net**

11.6 Absolute Convergence and the Ratio and Root Tests

Given any series $\Sigma\, a_n$, we can consider the corresponding series

$$\sum_{n=1}^{\infty} |a_n| = |a_1| + |a_2| + |a_3| + \cdots$$

whose terms are the absolute values of the terms of the original series.

We have convergence tests for series with positive terms and for alternating series. But what if the signs of the terms switch back and forth irregularly? We will see in Example 3 that the idea of absolute convergence sometimes helps in such cases.

> **1** **Definition**   A series $\Sigma\, a_n$ is called **absolutely convergent** if the series of absolute values $\Sigma\, |a_n|$ is convergent.

Notice that if $\Sigma\, a_n$ is a series with positive terms, then $|a_n| = a_n$ and so absolute convergence is the same as convergence in this case.

EXAMPLE 1 The series

$$\sum_{n=1}^{\infty} \frac{(-1)^{n-1}}{n^2} = 1 - \frac{1}{2^2} + \frac{1}{3^2} - \frac{1}{4^2} + \cdots$$

is absolutely convergent because

$$\sum_{n=1}^{\infty} \left| \frac{(-1)^{n-1}}{n^2} \right| = \sum_{n=1}^{\infty} \frac{1}{n^2} = 1 + \frac{1}{2^2} + \frac{1}{3^2} + \frac{1}{4^2} + \cdots$$

is a convergent p-series ($p = 2$).

EXAMPLE 2 We know that the alternating harmonic series

$$\sum_{n=1}^{\infty} \frac{(-1)^{n-1}}{n} = 1 - \frac{1}{2} + \frac{1}{3} - \frac{1}{4} + \cdots$$

is convergent (see Example 1 in Section 11.5), but it is not absolutely convergent because the corresponding series of absolute values is

$$\sum_{n=1}^{\infty} \left| \frac{(-1)^{n-1}}{n} \right| = \sum_{n=1}^{\infty} \frac{1}{n} = 1 + \frac{1}{2} + \frac{1}{3} + \frac{1}{4} + \cdots$$

which is the harmonic series (*p*-series with $p = 1$) and is therefore divergent.   ▬

> **2**   **Definition**   A series $\Sigma\, a_n$ is called **conditionally convergent** if it is convergent but not absolutely convergent.

Example 2 shows that the alternating harmonic series is conditionally convergent. Thus it is possible for a series to be convergent but not absolutely convergent. However, the next theorem shows that absolute convergence implies convergence.

> **3**   **Theorem**   If a series $\Sigma\, a_n$ is absolutely convergent, then it is convergent.

PROOF   Observe that the inequality

$$0 \leq a_n + |a_n| \leq 2|a_n|$$

is true because $|a_n|$ is either a_n or $-a_n$. If $\Sigma\, a_n$ is absolutely convergent, then $\Sigma\, |a_n|$ is convergent, so $\Sigma\, 2|a_n|$ is convergent. Therefore, by the Comparison Test, $\Sigma\, (a_n + |a_n|)$ is convergent. Then

$$\sum a_n = \sum (a_n + |a_n|) - \sum |a_n|$$

is the difference of two convergent series and is therefore convergent.   ▬

V EXAMPLE 3 Determine whether the series

$$\sum_{n=1}^{\infty} \frac{\cos n}{n^2} = \frac{\cos 1}{1^2} + \frac{\cos 2}{2^2} + \frac{\cos 3}{3^2} + \cdots$$

is convergent or divergent.

SOLUTION   This series has both positive and negative terms, but it is not alternating. (The first term is positive, the next three are negative, and the following three are positive: The signs change irregularly.) We can apply the Comparison Test to the series of absolute values

$$\sum_{n=1}^{\infty} \left| \frac{\cos n}{n^2} \right| = \sum_{n=1}^{\infty} \frac{|\cos n|}{n^2}$$

Figure 1 shows the graphs of the terms a_n and partial sums s_n of the series in Example 3. Notice that the series is not alternating but has positive and negative terms.

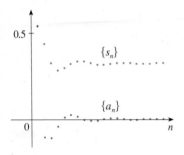

FIGURE 1

Since $|\cos n| \leq 1$ for all n, we have

$$\frac{|\cos n|}{n^2} \leq \frac{1}{n^2}$$

We know that $\Sigma \, 1/n^2$ is convergent (p-series with $p = 2$) and therefore $\Sigma \, |\cos n|/n^2$ is convergent by the Comparison Test. Thus the given series $\Sigma \, (\cos n)/n^2$ is absolutely convergent and therefore convergent by Theorem 3.   ▬

The following test is very useful in determining whether a given series is absolutely convergent.

The Ratio Test

(i) If $\displaystyle\lim_{n \to \infty} \left| \frac{a_{n+1}}{a_n} \right| = L < 1$, then the series $\displaystyle\sum_{n=1}^{\infty} a_n$ is absolutely convergent

    (and therefore convergent).

(ii) If $\displaystyle\lim_{n \to \infty} \left| \frac{a_{n+1}}{a_n} \right| = L > 1$ or $\displaystyle\lim_{n \to \infty} \left| \frac{a_{n+1}}{a_n} \right| = \infty$, then the series $\displaystyle\sum_{n=1}^{\infty} a_n$

    is divergent.

(iii) If $\displaystyle\lim_{n \to \infty} \left| \frac{a_{n+1}}{a_n} \right| = 1$, the Ratio Test is inconclusive; that is, no conclusion can

    be drawn about the convergence or divergence of $\Sigma \, a_n$.

PROOF

(i) The idea is to compare the given series with a convergent geometric series. Since $L < 1$, we can choose a number r such that $L < r < 1$. Since

$$\lim_{n \to \infty} \left| \frac{a_{n+1}}{a_n} \right| = L \qquad \text{and} \qquad L < r$$

the ratio $|a_{n+1}/a_n|$ will eventually be less than r; that is, there exists an integer N such that

$$\left| \frac{a_{n+1}}{a_n} \right| < r \qquad \text{whenever } n \geq N$$

or, equivalently,

$$\boxed{4} \qquad\qquad |a_{n+1}| < |a_n| r \qquad \text{whenever } n \geq N$$

Putting n successively equal to $N, N + 1, N + 2, \ldots$ in $\boxed{4}$, we obtain

$$|a_{N+1}| < |a_N| r$$

$$|a_{N+2}| < |a_{N+1}| r < |a_N| r^2$$

$$|a_{N+3}| < |a_{N+2}| r < |a_N| r^3$$

and, in general,

$$\boxed{5} \qquad\qquad |a_{N+k}| < |a_N|\, r^k \qquad \text{for all } k \geqslant 1$$

Now the series

$$\sum_{k=1}^{\infty} |a_N|\, r^k = |a_N|\, r + |a_N|\, r^2 + |a_N|\, r^3 + \cdots$$

is convergent because it is a geometric series with $0 < r < 1$. So the inequality $\boxed{5}$ together with the Comparison Test, shows that the series

$$\sum_{n=N+1}^{\infty} |a_n| = \sum_{k=1}^{\infty} |a_{N+k}| = |a_{N+1}| + |a_{N+2}| + |a_{N+3}| + \cdots$$

is also convergent. It follows that the series $\sum_{n=1}^{\infty} |a_n|$ is convergent. (Recall that a finite number of terms doesn't affect convergence.) Therefore $\sum a_n$ is absolutely convergent.

(ii) If $|a_{n+1}/a_n| \to L > 1$ or $|a_{n+1}/a_n| \to \infty$, then the ratio $|a_{n+1}/a_n|$ will eventually be greater than 1; that is, there exists an integer N such that

$$\left|\frac{a_{n+1}}{a_n}\right| > 1 \qquad \text{whenever } n \geqslant N$$

This means that $|a_{n+1}| > |a_n|$ whenever $n \geqslant N$ and so

$$\lim_{n \to \infty} a_n \neq 0$$

Therefore $\sum a_n$ diverges by the Test for Divergence. ▬

NOTE Part (iii) of the Ratio Test says that if $\lim_{n\to\infty} |a_{n+1}/a_n| = 1$, the test gives no information. For instance, for the convergent series $\sum 1/n^2$ we have

$$\left|\frac{a_{n+1}}{a_n}\right| = \frac{\dfrac{1}{(n+1)^2}}{\dfrac{1}{n^2}} = \frac{n^2}{(n+1)^2} = \frac{1}{\left(1 + \dfrac{1}{n}\right)^2} \to 1 \qquad \text{as } n \to \infty$$

whereas for the divergent series $\sum 1/n$ we have

$$\left|\frac{a_{n+1}}{a_n}\right| = \frac{\dfrac{1}{n+1}}{\dfrac{1}{n}} = \frac{n}{n+1} = \frac{1}{1 + \dfrac{1}{n}} \to 1 \qquad \text{as } n \to \infty$$

The Ratio Test is usually conclusive if the nth term of the series contains an exponential or a factorial, as we will see in Examples 4 and 5.

Therefore, if $\lim_{n\to\infty} |a_{n+1}/a_n| = 1$, the series $\sum a_n$ might converge or it might diverge. In this case the Ratio Test fails and we must use some other test.

EXAMPLE 4 Test the series $\displaystyle\sum_{n=1}^{\infty} (-1)^n \frac{n^3}{3^n}$ for absolute convergence.

Estimating Sums

In the last three sections we used various methods for estimating the sum of a series—the method depended on which test was used to prove convergence. What about series for which the Ratio Test works? There are two possibilities: If the series happens to be an alternating series, as in Example 4, then it is best to use the methods of Section 11.5. If the terms are all positive, then use the special methods explained in Exercise 38.

SOLUTION We use the Ratio Test with $a_n = (-1)^n n^3/3^n$:

$$\left| \frac{a_{n+1}}{a_n} \right| = \left| \frac{\dfrac{(-1)^{n+1}(n+1)^3}{3^{n+1}}}{\dfrac{(-1)^n n^3}{3^n}} \right| = \frac{(n+1)^3}{3^{n+1}} \cdot \frac{3^n}{n^3}$$

$$= \frac{1}{3} \left(\frac{n+1}{n} \right)^3 = \frac{1}{3} \left(1 + \frac{1}{n} \right)^3 \to \frac{1}{3} < 1$$

Thus, by the Ratio Test, the given series is absolutely convergent and therefore convergent.

V **EXAMPLE 5** Test the convergence of the series $\displaystyle\sum_{n=1}^{\infty} \frac{n^n}{n!}$.

SOLUTION Since the terms $a_n = n^n/n!$ are positive, we don't need the absolute value signs.

$$\frac{a_{n+1}}{a_n} = \frac{(n+1)^{n+1}}{(n+1)!} \cdot \frac{n!}{n^n}$$

$$= \frac{(n+1)(n+1)^n}{(n+1)n!} \cdot \frac{n!}{n^n}$$

$$= \left(\frac{n+1}{n} \right)^n = \left(1 + \frac{1}{n} \right)^n \to e \qquad \text{as } n \to \infty$$

(see Equation 6.4.9 or 6.4*.9). Since $e > 1$, the given series is divergent by the Ratio Test.

   NOTE Although the Ratio Test works in Example 5, an easier method is to use the Test for Divergence. Since

$$a_n = \frac{n^n}{n!} = \frac{n \cdot n \cdot n \cdot \cdots \cdot n}{1 \cdot 2 \cdot 3 \cdot \cdots \cdot n} \geq n$$

it follows that a_n does not approach 0 as $n \to \infty$. Therefore the given series is divergent by the Test for Divergence.

   The following test is convenient to apply when nth powers occur. Its proof is similar to the proof of the Ratio Test and is left as Exercise 41.

The Root Test

(i) If $\displaystyle\lim_{n\to\infty} \sqrt[n]{|a_n|} = L < 1$, then the series $\displaystyle\sum_{n=1}^{\infty} a_n$ is absolutely convergent
   (and therefore convergent).

(ii) If $\displaystyle\lim_{n\to\infty} \sqrt[n]{|a_n|} = L > 1$ or $\displaystyle\lim_{n\to\infty} \sqrt[n]{|a_n|} = \infty$, then the series $\displaystyle\sum_{n=1}^{\infty} a_n$ is divergent.

(iii) If $\displaystyle\lim_{n\to\infty} \sqrt[n]{|a_n|} = 1$, the Root Test is inconclusive.

   If $\lim_{n\to\infty} \sqrt[n]{|a_n|} = 1$, then part (iii) of the Root Test says that the test gives no information. The series $\sum a_n$ could converge or diverge. (If $L = 1$ in the Ratio Test, don't try the

Root Test because L will again be 1. And if $L = 1$ in the Root Test, don't try the Ratio Test because it will fail too.)

V EXAMPLE 6 Test the convergence of the series $\displaystyle\sum_{n=1}^{\infty} \left(\frac{2n+3}{3n+2}\right)^n$.

SOLUTION

$$a_n = \left(\frac{2n+3}{3n+2}\right)^n$$

$$\sqrt[n]{|a_n|} = \frac{2n+3}{3n+2} = \frac{2 + \dfrac{3}{n}}{3 + \dfrac{2}{n}} \to \frac{2}{3} < 1$$

Thus the given series converges by the Root Test.

Rearrangements

The question of whether a given convergent series is absolutely convergent or conditionally convergent has a bearing on the question of whether infinite sums behave like finite sums.

If we rearrange the order of the terms in a finite sum, then of course the value of the sum remains unchanged. But this is not always the case for an infinite series. By a **rearrangement** of an infinite series $\sum a_n$ we mean a series obtained by simply changing the order of the terms. For instance, a rearrangement of $\sum a_n$ could start as follows:

$$a_1 + a_2 + a_5 + a_3 + a_4 + a_{15} + a_6 + a_7 + a_{20} + \cdots$$

It turns out that

> if $\sum a_n$ is an absolutely convergent series with sum s,
> then any rearrangement of $\sum a_n$ has the same sum s.

However, any conditionally convergent series can be rearranged to give a different sum. To illustrate this fact let's consider the alternating harmonic series

$$\boxed{6} \qquad 1 - \tfrac{1}{2} + \tfrac{1}{3} - \tfrac{1}{4} + \tfrac{1}{5} - \tfrac{1}{6} + \tfrac{1}{7} - \tfrac{1}{8} + \cdots = \ln 2$$

(See Exercise 36 in Section 11.5.) If we multiply this series by $\tfrac{1}{2}$, we get

$$\tfrac{1}{2} - \tfrac{1}{4} + \tfrac{1}{6} - \tfrac{1}{8} + \cdots = \tfrac{1}{2}\ln 2$$

Inserting zeros between the terms of this series, we have

Adding these zeros does not affect the sum of the series; each term in the sequence of partial sums is repeated, but the limit is the same.

$$\boxed{7} \qquad 0 + \tfrac{1}{2} + 0 - \tfrac{1}{4} + 0 + \tfrac{1}{6} + 0 - \tfrac{1}{8} + \cdots = \tfrac{1}{2}\ln 2$$

Now we add the series in Equations 6 and 7 using Theorem 11.2.8:

$$\boxed{8} \qquad 1 + \tfrac{1}{3} - \tfrac{1}{2} + \tfrac{1}{5} + \tfrac{1}{7} - \tfrac{1}{4} + \cdots = \tfrac{3}{2}\ln 2$$

Notice that the series in $\boxed{8}$ contains the same terms as in $\boxed{6}$, but rearranged so that one negative term occurs after each pair of positive terms. The sums of these series, however, are different. In fact, Riemann proved that

> if $\sum a_n$ is a conditionally convergent series and r is any real number whatsoever, then there is a rearrangement of $\sum a_n$ that has a sum equal to r.

A proof of this fact is outlined in Exercise 44.

11.6 | Exercises

Access end-of-section exercises and tools for visualizing calculus online at **www.webassign.net**

11.7 | Strategy for Testing Series

We now have several ways of testing a series for convergence or divergence; the problem is to decide which test to use on which series. In this respect, testing series is similar to integrating functions. Again there are no hard and fast rules about which test to apply to a given series, but you may find the following advice of some use.

It is not wise to apply a list of the tests in a specific order until one finally works. That would be a waste of time and effort. Instead, as with integration, the main strategy is to classify the series according to its *form*.

1. If the series is of the form $\sum 1/n^p$, it is a *p*-series, which we know to be convergent if $p > 1$ and divergent if $p \leqslant 1$.

2. If the series has the form $\sum ar^{n-1}$ or $\sum ar^n$, it is a geometric series, which converges if $|r| < 1$ and diverges if $|r| \geqslant 1$. Some preliminary algebraic manipulation may be required to bring the series into this form.

3. If the series has a form that is similar to a *p*-series or a geometric series, then one of the comparison tests should be considered. In particular, if a_n is a rational function or an algebraic function of *n* (involving roots of polynomials), then the series should be compared with a *p*-series. Notice that most of the series in Exercises 11.4 have this form. (The value of *p* should be chosen as in Section 11.4 by keeping only the highest powers of *n* in the numerator and denominator.) The comparison tests apply only to series with positive terms, but if $\sum a_n$ has some negative terms, then we can apply the Comparison Test to $\sum |a_n|$ and test for absolute convergence.

4. If you can see at a glance that $\lim_{n\to\infty} a_n \neq 0$, then the Test for Divergence should be used.

5. If the series is of the form $\sum (-1)^{n-1} b_n$ or $\sum (-1)^n b_n$, then the Alternating Series Test is an obvious possibility.

6. Series that involve factorials or other products (including a constant raised to the *n*th power) are often conveniently tested using the Ratio Test. Bear in mind that $|a_{n+1}/a_n| \to 1$ as $n \to \infty$ for all *p*-series and therefore all rational or algebraic functions of *n*. Thus the Ratio Test should not be used for such series.

7. If a_n is of the form $(b_n)^n$, then the Root Test may be useful.

8. If $a_n = f(n)$, where $\int_1^\infty f(x)\, dx$ is easily evaluated, then the Integral Test is effective (assuming the hypotheses of this test are satisfied).

In the following examples we don't work out all the details but simply indicate which tests should be used.

V   EXAMPLE 1 $\displaystyle\sum_{n=1}^{\infty} \frac{n-1}{2n+1}$

Since $a_n \to \frac{1}{2} \neq 0$ as $n \to \infty$, we should use the Test for Divergence.

EXAMPLE 2 $\displaystyle\sum_{n=1}^{\infty} \frac{\sqrt{n^3 + 1}}{3n^3 + 4n^2 + 2}$

Since a_n is an algebraic function of n, we compare the given series with a p-series. The comparison series for the Limit Comparison Test is $\Sigma\, b_n$, where

$$b_n = \frac{\sqrt{n^3}}{3n^3} = \frac{n^{3/2}}{3n^3} = \frac{1}{3n^{3/2}}$$

V EXAMPLE 3 $\displaystyle\sum_{n=1}^{\infty} ne^{-n^2}$

Since the integral $\int_1^{\infty} xe^{-x^2}\,dx$ is easily evaluated, we use the Integral Test. The Ratio Test also works.

EXAMPLE 4 $\displaystyle\sum_{n=1}^{\infty} (-1)^n \frac{n^3}{n^4 + 1}$

Since the series is alternating, we use the Alternating Series Test.

V EXAMPLE 5 $\displaystyle\sum_{k=1}^{\infty} \frac{2^k}{k!}$

Since the series involves $k!$, we use the Ratio Test.

EXAMPLE 6 $\displaystyle\sum_{n=1}^{\infty} \frac{1}{2 + 3^n}$

Since the series is closely related to the geometric series $\Sigma\, 1/3^n$, we use the Comparison Test.

11.7 Exercises

Access end-of-section exercises and tools for visualizing calculus online at **www.webassign.net**

11.8 Power Series

A **power series** is a series of the form

$$\boxed{1} \qquad \sum_{n=0}^{\infty} c_n x^n = c_0 + c_1 x + c_2 x^2 + c_3 x^3 + \cdots$$

where x is a variable and the c_n's are constants called the **coefficients** of the series. For each fixed x, the series $\boxed{1}$ is a series of constants that we can test for convergence or divergence. A power series may converge for some values of x and diverge for other values of x. The sum of the series is a function

$$f(x) = c_0 + c_1 x + c_2 x^2 + \cdots + c_n x^n + \cdots$$

whose domain is the set of all x for which the series converges. Notice that f resembles a polynomial. The only difference is that f has infinitely many terms.

Trigonometric Series

A power series is a series in which each term is a power function. A **trigonometric series**

$$\sum_{n=0}^{\infty} (a_n \cos nx + b_n \sin nx)$$

is a series whose terms are trigonometric functions. This type of series is discussed on the website

www.stewartcalculus.com

Click on *Additional Topics* and then on *Fourier Series*.

For instance, if we take $c_n = 1$ for all n, the power series becomes the geometric series

$$\sum_{n=0}^{\infty} x^n = 1 + x + x^2 + \cdots + x^n + \cdots$$

which converges when $-1 < x < 1$ and diverges when $|x| \geq 1$. (See Equation 11.2.5.)

More generally, a series of the form

$$\boxed{2} \qquad \sum_{n=0}^{\infty} c_n(x - a)^n = c_0 + c_1(x - a) + c_2(x - a)^2 + \cdots$$

is called a **power series in $(x - a)$** or a **power series centered at a** or a **power series about a**. Notice that in writing out the term corresponding to $n = 0$ in Equations 1 and 2 we have adopted the convention that $(x - a)^0 = 1$ even when $x = a$. Notice also that when $x = a$ all of the terms are 0 for $n \geq 1$ and so the power series $\boxed{2}$ always converges when $x = a$.

V EXAMPLE 1 For what values of x is the series $\sum_{n=0}^{\infty} n!x^n$ convergent?

SOLUTION We use the Ratio Test. If we let a_n, as usual, denote the nth term of the series, then $a_n = n!x^n$. If $x \neq 0$, we have

Notice that
$$(n + 1)! = (n + 1)n(n - 1) \cdot \cdots \cdot 3 \cdot 2 \cdot 1$$
$$= (n + 1)n!$$

$$\lim_{n \to \infty} \left| \frac{a_{n+1}}{a_n} \right| = \lim_{n \to \infty} \left| \frac{(n + 1)!x^{n+1}}{n!x^n} \right|$$
$$= \lim_{n \to \infty} (n + 1)|x| = \infty$$

By the Ratio Test, the series diverges when $x \neq 0$. Thus the given series converges only when $x = 0$.

V EXAMPLE 2 For what values of x does the series $\sum_{n=1}^{\infty} \frac{(x - 3)^n}{n}$ converge?

SOLUTION Let $a_n = (x - 3)^n/n$. Then

$$\left| \frac{a_{n+1}}{a_n} \right| = \left| \frac{(x - 3)^{n+1}}{n + 1} \cdot \frac{n}{(x - 3)^n} \right|$$
$$= \frac{1}{1 + \dfrac{1}{n}} |x - 3| \to |x - 3| \qquad \text{as } n \to \infty$$

By the Ratio Test, the given series is absolutely convergent, and therefore convergent, when $|x - 3| < 1$ and divergent when $|x - 3| > 1$. Now

$$|x - 3| < 1 \quad \Longleftrightarrow \quad -1 < x - 3 < 1 \quad \Longleftrightarrow \quad 2 < x < 4$$

so the series converges when $2 < x < 4$ and diverges when $x < 2$ or $x > 4$.

The Ratio Test gives no information when $|x - 3| = 1$ so we must consider $x = 2$ and $x = 4$ separately. If we put $x = 4$ in the series, it becomes $\sum 1/n$, the harmonic series, which is divergent. If $x = 2$, the series is $\sum (-1)^n/n$, which converges by the Alternating Series Test. Thus the given power series converges for $2 \leq x < 4$.

Notice how closely the computer-generated model (which involves Bessel functions and cosine functions) matches the photograph of a vibrating rubber membrane.

We will see that the main use of a power series is that it provides a way to represent some of the most important functions that arise in mathematics, physics, and chemistry. In particular, the sum of the power series in the next example is called a **Bessel function**, after the German astronomer Friedrich Bessel (1784–1846), and the function given in Exercise 35 is another example of a Bessel function. In fact, these functions first arose when Bessel solved Kepler's equation for describing planetary motion. Since that time, these functions have been applied in many different physical situations, including the temperature distribution in a circular plate and the shape of a vibrating drumhead.

EXAMPLE 3 Find the domain of the Bessel function of order 0 defined by

$$J_0(x) = \sum_{n=0}^{\infty} \frac{(-1)^n x^{2n}}{2^{2n}(n!)^2}$$

SOLUTION Let $a_n = (-1)^n x^{2n}/[2^{2n}(n!)^2]$. Then

$$\left| \frac{a_{n+1}}{a_n} \right| = \left| \frac{(-1)^{n+1} x^{2(n+1)}}{2^{2(n+1)}[(n+1)!]^2} \cdot \frac{2^{2n}(n!)^2}{(-1)^n x^{2n}} \right|$$

$$= \frac{x^{2n+2}}{2^{2n+2}(n+1)^2(n!)^2} \cdot \frac{2^{2n}(n!)^2}{x^{2n}}$$

$$= \frac{x^2}{4(n+1)^2} \to 0 < 1 \qquad \text{for all } x$$

Thus, by the Ratio Test, the given series converges for all values of x. In other words, the domain of the Bessel function J_0 is $(-\infty, \infty) = \mathbb{R}$.

Recall that the sum of a series is equal to the limit of the sequence of partial sums. So when we define the Bessel function in Example 3 as the sum of a series we mean that, for every real number x,

$$J_0(x) = \lim_{n \to \infty} s_n(x) \qquad \text{where} \qquad s_n(x) = \sum_{i=0}^{n} \frac{(-1)^i x^{2i}}{2^{2i}(i!)^2}$$

The first few partial sums are

$$s_0(x) = 1 \qquad s_1(x) = 1 - \frac{x^2}{4} \qquad s_2(x) = 1 - \frac{x^2}{4} + \frac{x^4}{64}$$

$$s_3(x) = 1 - \frac{x^2}{4} + \frac{x^4}{64} - \frac{x^6}{2304} \qquad s_4(x) = 1 - \frac{x^2}{4} + \frac{x^4}{64} - \frac{x^6}{2304} + \frac{x^8}{147,456}$$

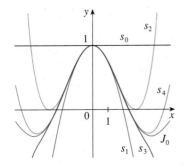

FIGURE 1
Partial sums of the Bessel function J_0

FIGURE 2

Figure 1 shows the graphs of these partial sums, which are polynomials. They are all appro ̶imations to the function J_0, but notice that the approximations become better when m ̶ terms are included. Figure 2 shows a more complete graph of the Bessel function.

For the power series that we have looked at so far, the set of values of x for which series is convergent has always turned out to be an interval [a finite interval for the geom ̶ series and the series in Example 2, the infinite interval $(-\infty, \infty)$ in Example 3, and a lapsed interval $[0, 0] = \{0\}$ in Example 1]. The following theorem, proved in Append says that this is true in general.

3   **Theorem**   For a given power series $\sum_{n=0}^{\infty} c_n(x-a)^n$ there are only three possibilities:

(i) The series converges only when $x = a$.

(ii) The series converges for all x.

(iii) There is a positive number R such that the series converges if $|x - a| < R$ and diverges if $|x - a| > R$.

The number R in case (iii) is called the **radius of convergence** of the power series. By convention, the radius of convergence is $R = 0$ in case (i) and $R = \infty$ in case (ii). The **interval of convergence** of a power series is the interval that consists of all values of x for which the series converges. In case (i) the interval consists of just a single point a. In case (ii) the interval is $(-\infty, \infty)$. In case (iii) note that the inequality $|x - a| < R$ can be rewritten as $a - R < x < a + R$. When x is an *endpoint* of the interval, that is, $x = a \pm R$, anything can happen—the series might converge at one or both endpoints or it might diverge at both endpoints. Thus in case (iii) there are four possibilities for the interval of convergence:

$$(a - R, a + R) \qquad (a - R, a + R] \qquad [a - R, a + R) \qquad [a - R, a + R]$$

The situation is illustrated in Figure 3.

FIGURE 3

We summarize here the radius and interval of convergence for each of the examples already considered in this section.

	Series	Radius of convergence	Interval of convergence
Geometric series	$\sum_{n=0}^{\infty} x^n$	$R = 1$	$(-1, 1)$
Example 1	$\sum_{n=0}^{\infty} n!\, x^n$	$R = 0$	$\{0\}$
Example 2	$\sum_{n=1}^{\infty} \dfrac{(x-3)^n}{n}$	$R = 1$	$[2, 4)$
Example 3	$\sum_{n=0}^{\infty} \dfrac{(-1)^n x^{2n}}{2^{2n}(n!)^2}$	$R = \infty$	$(-\infty, \infty)$

In general, the Ratio Test (or sometimes the Root Test) should be used to determine the radius of convergence R. The Ratio and Root Tests always fail when x is an endpoint of the interval of convergence, so the endpoints must be checked with some other test.

EXAMPLE 4 Find the radius of convergence and interval of convergence of the series

$$\sum_{n=0}^{\infty} \frac{(-3)^n x^n}{\sqrt{n+1}}$$

SOLUTION Let $a_n = (-3)^n x^n / \sqrt{n+1}$. Then

$$\left| \frac{a_{n+1}}{a_n} \right| = \left| \frac{(-3)^{n+1} x^{n+1}}{\sqrt{n+2}} \cdot \frac{\sqrt{n+1}}{(-3)^n x^n} \right| = \left| -3x \sqrt{\frac{n+1}{n+2}} \right|$$

$$= 3 \sqrt{\frac{1 + (1/n)}{1 + (2/n)}} |x| \rightarrow 3|x| \qquad \text{as } n \rightarrow \infty$$

By the Ratio Test, the given series converges if $3|x| < 1$ and diverges if $3|x| > 1$. Thus it converges if $|x| < \frac{1}{3}$ and diverges if $|x| > \frac{1}{3}$. This means that the radius of convergence is $R = \frac{1}{3}$.

We know the series converges in the interval $\left(-\frac{1}{3}, \frac{1}{3}\right)$, but we must now test for convergence at the endpoints of this interval. If $x = -\frac{1}{3}$, the series becomes

$$\sum_{n=0}^{\infty} \frac{(-3)^n \left(-\frac{1}{3}\right)^n}{\sqrt{n+1}} = \sum_{n=0}^{\infty} \frac{1}{\sqrt{n+1}} = \frac{1}{\sqrt{1}} + \frac{1}{\sqrt{2}} + \frac{1}{\sqrt{3}} + \frac{1}{\sqrt{4}} + \cdots$$

which diverges. (Use the Integral Test or simply observe that it is a p-series with $p = \frac{1}{2} < 1$.) If $x = \frac{1}{3}$, the series is

$$\sum_{n=0}^{\infty} \frac{(-3)^n \left(\frac{1}{3}\right)^n}{\sqrt{n+1}} = \sum_{n=0}^{\infty} \frac{(-1)^n}{\sqrt{n+1}}$$

which converges by the Alternating Series Test. Therefore the given power series converges when $-\frac{1}{3} < x \le \frac{1}{3}$, so the interval of convergence is $\left(-\frac{1}{3}, \frac{1}{3}\right]$.

V EXAMPLE 5 Find the radius of convergence and interval of convergence of the series

$$\sum_{n=0}^{\infty} \frac{n(x+2)^n}{3^{n+1}}$$

SOLUTION If $a_n = n(x+2)^n / 3^{n+1}$, then

$$\left| \frac{a_{n+1}}{a_n} \right| = \left| \frac{(n+1)(x+2)^{n+1}}{3^{n+2}} \cdot \frac{3^{n+1}}{n(x+2)^n} \right|$$

$$= \left(1 + \frac{1}{n}\right) \frac{|x+2|}{3} \rightarrow \frac{|x+2|}{3} \qquad \text{as } n \rightarrow \infty$$

Using the Ratio Test, we see that the series converges if $|x+2|/3 < 1$ and it diverges if $|x+2|/3 > 1$. So it converges if $|x+2| < 3$ and diverges if $|x+2| > 3$. Thus radius of convergence is $R = 3$.

The inequality $|x+2| < 3$ can be written as $-5 < x < 1$, so we test the series the endpoints -5 and 1. When $x = -5$, the series is

$$\sum_{n=0}^{\infty} \frac{n(-3)^n}{3^{n+1}} = \frac{1}{3} \sum_{n=0}^{\infty} (-1)^n n$$

which diverges by the Test for Divergence [$(-1)^n n$ doesn't converge to 0]. When $x = 1$, the series is

$$\sum_{n=0}^{\infty} \frac{n(3)^n}{3^{n+1}} = \tfrac{1}{3} \sum_{n=0}^{\infty} n$$

which also diverges by the Test for Divergence. Thus the series converges only when $-5 < x < 1$, so the interval of convergence is $(-5, 1)$.

11.8 Exercises

Access end-of-section exercises and tools for visualizing calculus online at **www.webassign.net**

11.9 Representations of Functions as Power Series

In this section we learn how to represent certain types of functions as sums of power series by manipulating geometric series or by differentiating or integrating such a series. You might wonder why we would ever want to express a known function as a sum of infinitely many terms. We will see later that this strategy is useful for integrating functions that don't have elementary antiderivatives, for solving differential equations, and for approximating functions by polynomials. (Scientists do this to simplify the expressions they deal with; computer scientists do this to represent functions on calculators and computers.)

We start with an equation that we have seen before:

$$\boxed{1} \qquad \frac{1}{1-x} = 1 + x + x^2 + x^3 + \cdots = \sum_{n=0}^{\infty} x^n \qquad |x| < 1$$

We first encountered this equation in Example 6 in Section 11.2, where we obtained it by observing that the series is a geometric series with $a = 1$ and $r = x$. But here our point of view is different. We now regard Equation 1 as expressing the function $f(x) = 1/(1-x)$ as a sum of a power series.

llustration of Equation 1 is shown cause the sum of a series is the uence of partial sums, we have

$$\frac{1}{-x} = \lim_{n \to \infty} s_n(x)$$

$+ x + x^2 + \cdots + x^n$

n. Notice that as n
mes a better approxi-
$1 < x < 1$.

es
s the

at

FIGURE 1

some partial sums

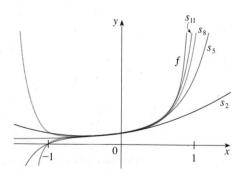

V EXAMPLE 1 Express $1/(1 + x^2)$ as the sum of a power series and find the interval of convergence.

SOLUTION Replacing x by $-x^2$ in Equation 1, we have

$$\frac{1}{1+x^2} = \frac{1}{1-(-x^2)} = \sum_{n=0}^{\infty} (-x^2)^n$$

$$= \sum_{n=0}^{\infty} (-1)^n x^{2n} = 1 - x^2 + x^4 - x^6 + x^8 - \cdots$$

Because this is a geometric series, it converges when $|-x^2| < 1$, that is, $x^2 < 1$, or $|x| < 1$. Therefore the interval of convergence is $(-1, 1)$. (Of course, we could have determined the radius of convergence by applying the Ratio Test, but that much work is unnecessary here.)

EXAMPLE 2 Find a power series representation for $1/(x+2)$.

SOLUTION In order to put this function in the form of the left side of Equation 1, we first factor a 2 from the denominator:

$$\frac{1}{2+x} = \frac{1}{2\left(1 + \dfrac{x}{2}\right)} = \frac{1}{2\left[1 - \left(-\dfrac{x}{2}\right)\right]}$$

$$= \frac{1}{2}\sum_{n=0}^{\infty}\left(-\frac{x}{2}\right)^n = \sum_{n=0}^{\infty}\frac{(-1)^n}{2^{n+1}} x^n$$

This series converges when $|-x/2| < 1$, that is, $|x| < 2$. So the interval of convergence is $(-2, 2)$.

EXAMPLE 3 Find a power series representation of $x^3/(x+2)$.

SOLUTION Since this function is just x^3 times the function in Example 2, all we have to do is to multiply that series by x^3:

It's legitimate to move x^3 across the sigma sign because it doesn't depend on n. [Use Theorem 11.2.8(i) with $c = x^3$.]

$$\frac{x^3}{x+2} = x^3 \cdot \frac{1}{x+2} = x^3\sum_{n=0}^{\infty}\frac{(-1)^n}{2^{n+1}} x^n = \sum_{n=0}^{\infty}\frac{(-1)^n}{2^{n+1}} x^{n+3}$$

$$= \tfrac{1}{2}x^3 - \tfrac{1}{4}x^4 + \tfrac{1}{8}x^5 - \tfrac{1}{16}x^6 + \cdots$$

Another way of writing this series is as follows:

$$\frac{x^3}{x+2} = \sum_{n=3}^{\infty}\frac{(-1)^{n-1}}{2^{n-2}} x^n$$

As in Example 2, the interval of convergence is $(-2, 2)$.

Differentiation and Integration of Power Series

The sum of a power series is a function $f(x) = \sum_{n=0}^{\infty} c_n(x-a)^n$ whose domain is the interval of convergence of the series. We would like to be able to differentiate and integrate such functions, and the following theorem (which we won't prove) says that we can do so by differentiating or integrating each individual term in the series, just as we would for a polynomial. This is called **term-by-term differentiation and integration**.

2 **Theorem** If the power series $\Sigma\, c_n(x - a)^n$ has radius of convergence $R > 0$, then the function f defined by

$$f(x) = c_0 + c_1(x - a) + c_2(x - a)^2 + \cdots = \sum_{n=0}^{\infty} c_n(x - a)^n$$

is differentiable (and therefore continuous) on the interval $(a - R, a + R)$ and

(i) $f'(x) = c_1 + 2c_2(x - a) + 3c_3(x - a)^2 + \cdots = \displaystyle\sum_{n=1}^{\infty} nc_n(x - a)^{n-1}$

In part (ii), $\int c_0\, dx = c_0 x + C_1$ is written as $c_0(x - a) + C$, where $C = C_1 + ac_0$, so all the terms of the series have the same form.

(ii) $\displaystyle\int f(x)\, dx = C + c_0(x - a) + c_1\frac{(x - a)^2}{2} + c_2\frac{(x - a)^3}{3} + \cdots$

$$= C + \sum_{n=0}^{\infty} c_n\frac{(x - a)^{n+1}}{n + 1}$$

The radii of convergence of the power series in Equations (i) and (ii) are both R.

NOTE 1 Equations (i) and (ii) in Theorem 2 can be rewritten in the form

(iii) $\displaystyle\frac{d}{dx}\left[\sum_{n=0}^{\infty} c_n(x - a)^n\right] = \sum_{n=0}^{\infty}\frac{d}{dx}\left[c_n(x - a)^n\right]$

(iv) $\displaystyle\int\left[\sum_{n=0}^{\infty} c_n(x - a)^n\right] dx = \sum_{n=0}^{\infty}\int c_n(x - a)^n\, dx$

We know that, for finite sums, the derivative of a sum is the sum of the derivatives and the integral of a sum is the sum of the integrals. Equations (iii) and (iv) assert that the same is true for infinite sums, provided we are dealing with *power series*. (For other types of series of functions the situation is not as simple; see Exercise 38.)

NOTE 2 Although Theorem 2 says that the radius of convergence remains the same when a power series is differentiated or integrated, this does not mean that the *interval* of convergence remains the same. It may happen that the original series converges at an endpoint, whereas the differentiated series diverges there. (See Exercise 39.)

NOTE 3 The idea of differentiating a power series term by term is the basis for a powerful method for solving differential equations. We will discuss this method in Chapter 17.

EXAMPLE 4 In Example 3 in Section 11.8 we saw that the Bessel function

$$J_0(x) = \sum_{n=0}^{\infty} \frac{(-1)^n x^{2n}}{2^{2n}(n!)^2}$$

is defined for all x. Thus, by Theorem 2, J_0 is differentiable for all x and its derivative is found by term-by-term differentiation as follows:

$$J_0'(x) = \sum_{n=0}^{\infty} \frac{d}{dx}\frac{(-1)^n x^{2n}}{2^{2n}(n!)^2} = \sum_{n=1}^{\infty} \frac{(-1)^n 2n x^{2n-1}}{2^{2n}(n!)^2}$$

V **EXAMPLE 5** Express $1/(1 - x)^2$ as a power series by differentiating Equation 1. What is the radius of convergence?

SOLUTION Differentiating each side of the equation

$$\frac{1}{1 - x} = 1 + x + x^2 + x^3 + \cdots = \sum_{n=0}^{\infty} x^n$$

we get

$$\frac{1}{(1 - x)^2} = 1 + 2x + 3x^2 + \cdots = \sum_{n=1}^{\infty} nx^{n-1}$$

If we wish, we can replace n by $n + 1$ and write the answer as

$$\frac{1}{(1 - x)^2} = \sum_{n=0}^{\infty} (n + 1)x^n$$

According to Theorem 2, the radius of convergence of the differentiated series is the same as the radius of convergence of the original series, namely, $R = 1$. ▬

EXAMPLE 6 Find a power series representation for $\ln(1 + x)$ and its radius of convergence.

SOLUTION We notice that the derivative of this function is $1/(1 + x)$. From Equation 1 we have

$$\frac{1}{1 + x} = \frac{1}{1 - (-x)} = 1 - x + x^2 - x^3 + \cdots \qquad |x| < 1$$

Integrating both sides of this equation, we get

$$\ln(1 + x) = \int \frac{1}{1 + x}\, dx = \int (1 - x + x^2 - x^3 + \cdots)\, dx$$

$$= x - \frac{x^2}{2} + \frac{x^3}{3} - \frac{x^4}{4} + \cdots + C$$

$$= \sum_{n=1}^{\infty} (-1)^{n-1} \frac{x^n}{n} + C \qquad |x| < 1$$

To determine the value of C we put $x = 0$ in this equation and obtain $\ln(1 + 0) = C$. Thus $C = 0$ and

$$\ln(1 + x) = x - \frac{x^2}{2} + \frac{x^3}{3} - \frac{x^4}{4} + \cdots = \sum_{n=1}^{\infty} (-1)^{n-1} \frac{x^n}{n} \qquad |x| < 1$$

The radius of convergence is the same as for the original series: $R = 1$. ▬

V **EXAMPLE 7** Find a power series representation for $f(x) = \tan^{-1}x$.

SOLUTION We observe that $f'(x) = 1/(1 + x^2)$ and find the required series by integrating the power series for $1/(1 + x^2)$ found in Example 1.

$$\tan^{-1}x = \int \frac{1}{1 + x^2}\, dx = \int (1 - x^2 + x^4 - x^6 + \cdots)\, dx$$

$$= C + x - \frac{x^3}{3} + \frac{x^5}{5} - \frac{x^7}{7} + \cdots$$

The power series for $\tan^{-1}x$ obtained in Example 7 is called *Gregory's series* after the Scottish mathematician James Gregory (1638–1675), who had anticipated some of Newton's discoveries. We have shown that Gregory's series is valid when $-1 < x < 1$, but it turns out (although it isn't easy to prove) that it is also valid when $x = \pm 1$. Notice that when $x = 1$ the series becomes

$$\frac{\pi}{4} = 1 - \frac{1}{3} + \frac{1}{5} - \frac{1}{7} + \cdots$$

This beautiful result is known as the Leibniz formula for π.

To find C we put $x = 0$ and obtain $C = \tan^{-1}0 = 0$. Therefore

$$\tan^{-1}x = x - \frac{x^3}{3} + \frac{x^5}{5} - \frac{x^7}{7} + \cdots = \sum_{n=0}^{\infty} (-1)^n \frac{x^{2n+1}}{2n+1}$$

Since the radius of convergence of the series for $1/(1 + x^2)$ is 1, the radius of convergence of this series for $\tan^{-1}x$ is also 1.

EXAMPLE 8

(a) Evaluate $\int [1/(1 + x^7)]\,dx$ as a power series.

(b) Use part (a) to approximate $\int_0^{0.5} [1/(1 + x^7)]\,dx$ correct to within 10^{-7}.

SOLUTION

(a) The first step is to express the integrand, $1/(1 + x^7)$, as the sum of a power series. As in Example 1, we start with Equation 1 and replace x by $-x^7$:

$$\frac{1}{1 + x^7} = \frac{1}{1 - (-x^7)} = \sum_{n=0}^{\infty} (-x^7)^n$$

$$= \sum_{n=0}^{\infty} (-1)^n x^{7n} = 1 - x^7 + x^{14} - \cdots$$

This example demonstrates one way in which power series representations are useful. Integrating $1/(1 + x^7)$ by hand is incredibly difficult. Different computer algebra systems return different forms of the answer, but they are all extremely complicated. (If you have a CAS, try it yourself.) The infinite series answer that we obtain in Example 8(a) is actually much easier to deal with than the finite answer provided by a CAS.

Now we integrate term by term:

$$\int \frac{1}{1 + x^7}\,dx = \int \sum_{n=0}^{\infty} (-1)^n x^{7n}\,dx = C + \sum_{n=0}^{\infty} (-1)^n \frac{x^{7n+1}}{7n+1}$$

$$= C + x - \frac{x^8}{8} + \frac{x^{15}}{15} - \frac{x^{22}}{22} + \cdots$$

This series converges for $|-x^7| < 1$, that is, for $|x| < 1$.

(b) In applying the Fundamental Theorem of Calculus, it doesn't matter which antiderivative we use, so let's use the antiderivative from part (a) with $C = 0$:

$$\int_0^{0.5} \frac{1}{1 + x^7}\,dx = \left[x - \frac{x^8}{8} + \frac{x^{15}}{15} - \frac{x^{22}}{22} + \cdots \right]_0^{1/2}$$

$$= \frac{1}{2} - \frac{1}{8 \cdot 2^8} + \frac{1}{15 \cdot 2^{15}} - \frac{1}{22 \cdot 2^{22}} + \cdots + \frac{(-1)^n}{(7n+1)2^{7n+1}} + \cdots$$

This infinite series is the exact value of the definite integral, but since it is an alternating series, we can approximate the sum using the Alternating Series Estimation Theorem. If we stop adding after the term with $n = 3$, the error is smaller than the term with $n = 4$:

$$\frac{1}{29 \cdot 2^{29}} \approx 6.4 \times 10^{-11}$$

So we have

$$\int_0^{0.5} \frac{1}{1 + x^7}\,dx \approx \frac{1}{2} - \frac{1}{8 \cdot 2^8} + \frac{1}{15 \cdot 2^{15}} - \frac{1}{22 \cdot 2^{22}} \approx 0.49951374$$

11.9   Exercises

Access end-of-section exercises and tools for visualizing calculus online at **www.webassign.net**

11.10   Taylor and Maclaurin Series

In the preceding section we were able to find power series representations for a certain restricted class of functions. Here we investigate more general problems: Which functions have power series representations? How can we find such representations?

We start by supposing that f is any function that can be represented by a power series

$$\boxed{1}\quad f(x) = c_0 + c_1(x - a) + c_2(x - a)^2 + c_3(x - a)^3 + c_4(x - a)^4 + \cdots \qquad |x - a| < R$$

Let's try to determine what the coefficients c_n must be in terms of f. To begin, notice that if we put $x = a$ in Equation 1, then all terms after the first one are 0 and we get

$$f(a) = c_0$$

By Theorem 11.9.2, we can differentiate the series in Equation 1 term by term:

$$\boxed{2}\quad f'(x) = c_1 + 2c_2(x - a) + 3c_3(x - a)^2 + 4c_4(x - a)^3 + \cdots \qquad |x - a| < R$$

and substitution of $x = a$ in Equation 2 gives

$$f'(a) = c_1$$

Now we differentiate both sides of Equation 2 and obtain

$$\boxed{3}\quad f''(x) = 2c_2 + 2 \cdot 3c_3(x - a) + 3 \cdot 4c_4(x - a)^2 + \cdots \qquad |x - a| < R$$

Again we put $x = a$ in Equation 3. The result is

$$f''(a) = 2c_2$$

Let's apply the procedure one more time. Differentiation of the series in Equation 3 gives

$$\boxed{4}\quad f'''(x) = 2 \cdot 3c_3 + 2 \cdot 3 \cdot 4c_4(x - a) + 3 \cdot 4 \cdot 5c_5(x - a)^2 + \cdots \qquad |x - a| < R$$

and substitution of $x = a$ in Equation 4 gives

$$f'''(a) = 2 \cdot 3c_3 = 3!c_3$$

By now you can see the pattern. If we continue to differentiate and substitute $x = a$, we obtain

$$f^{(n)}(a) = 2 \cdot 3 \cdot 4 \cdot \cdots \cdot nc_n = n!c_n$$

Solving this equation for the nth coefficient c_n, we get

$$c_n = \frac{f^{(n)}(a)}{n!}$$

This formula remains valid even for $n = 0$ if we adopt the conventions that $0! = 1$ and $f^{(0)} = f$. Thus we have proved the following theorem.

> **5 Theorem** If f has a power series representation (expansion) at a, that is, if
>
> $$f(x) = \sum_{n=0}^{\infty} c_n(x - a)^n \qquad |x - a| < R$$
>
> then its coefficients are given by the formula
>
> $$c_n = \frac{f^{(n)}(a)}{n!}$$

Substituting this formula for c_n back into the series, we see that *if f has a power series expansion at a, then it must be of the following form.*

> **6** $$f(x) = \sum_{n=0}^{\infty} \frac{f^{(n)}(a)}{n!} (x - a)^n$$
>
> $$= f(a) + \frac{f'(a)}{1!} (x - a) + \frac{f''(a)}{2!} (x - a)^2 + \frac{f'''(a)}{3!} (x - a)^3 + \cdots$$

Taylor and Maclaurin

The Taylor series is named after the English mathematician Brook Taylor (1685–1731) and the Maclaurin series is named in honor of the Scottish mathematician Colin Maclaurin (1698–1746) despite the fact that the Maclaurin series is really just a special case of the Taylor series. But the idea of representing particular functions as sums of power series goes back to Newton, and the general Taylor series was known to the Scottish mathematician James Gregory in 1668 and to the Swiss mathematician John Bernoulli in the 1690s. Taylor was apparently unaware of the work of Gregory and Bernoulli when he published his discoveries on series in 1715 in his book *Methodus incrementorum directa et inversa*. Maclaurin series are named after Colin Maclaurin because he popularized them in his calculus textbook *Treatise of Fluxions* published in 1742.

The series in Equation 6 is called the **Taylor series of the function f at a** (or **about a** or **centered at a**). For the special case $a = 0$ the Taylor series becomes

> **7** $$f(x) = \sum_{n=0}^{\infty} \frac{f^{(n)}(0)}{n!} x^n = f(0) + \frac{f'(0)}{1!} x + \frac{f''(0)}{2!} x^2 + \cdots$$

This case arises frequently enough that it is given the special name **Maclaurin series**.

NOTE We have shown that *if f can be represented as a power series about a, then f is equal to the sum of its Taylor series.* But there exist functions that are not equal to the sum of their Taylor series. An example of such a function is given in Exercise 74.

V EXAMPLE 1 Find the Maclaurin series of the function $f(x) = e^x$ and its radius of convergence.

SOLUTION If $f(x) = e^x$, then $f^{(n)}(x) = e^x$, so $f^{(n)}(0) = e^0 = 1$ for all n. Therefore the Taylor series for f at 0 (that is, the Maclaurin series) is

$$\sum_{n=0}^{\infty} \frac{f^{(n)}(0)}{n!} x^n = \sum_{n=0}^{\infty} \frac{x^n}{n!} = 1 + \frac{x}{1!} + \frac{x^2}{2!} + \frac{x^3}{3!} + \cdots$$

To find the radius of convergence we let $a_n = x^n/n!$. Then

$$\left| \frac{a_{n+1}}{a_n} \right| = \left| \frac{x^{n+1}}{(n+1)!} \cdot \frac{n!}{x^n} \right| = \frac{|x|}{n+1} \to 0 < 1$$

so, by the Ratio Test, the series converges for all x and the radius of convergence is $R = \infty$.

The conclusion we can draw from Theorem 5 and Example 1 is that *if* e^x *has a power series expansion at* 0, *then*

$$e^x = \sum_{n=0}^{\infty} \frac{x^n}{n!}$$

So how can we determine whether e^x *does* have a power series representation?

Let's investigate the more general question: Under what circumstances is a function equal to the sum of its Taylor series? In other words, if f has derivatives of all orders, when is it true that

$$f(x) = \sum_{n=0}^{\infty} \frac{f^{(n)}(a)}{n!} (x-a)^n$$

As with any convergent series, this means that $f(x)$ is the limit of the sequence of partial sums. In the case of the Taylor series, the partial sums are

$$T_n(x) = \sum_{i=0}^{n} \frac{f^{(i)}(a)}{i!} (x-a)^i$$

$$= f(a) + \frac{f'(a)}{1!}(x-a) + \frac{f''(a)}{2!}(x-a)^2 + \cdots + \frac{f^{(n)}(a)}{n!}(x-a)^n$$

Notice that T_n is a polynomial of degree n called the **nth-degree Taylor polynomial of f at a**. For instance, for the exponential function $f(x) = e^x$, the result of Example 1 shows that the Taylor polynomials at 0 (or Maclaurin polynomials) with $n = 1$, 2, and 3 are

$$T_1(x) = 1 + x \qquad T_2(x) = 1 + x + \frac{x^2}{2!} \qquad T_3(x) = 1 + x + \frac{x^2}{2!} + \frac{x^3}{3!}$$

The graphs of the exponential function and these three Taylor polynomials are drawn in Figure 1.

In general, $f(x)$ is the sum of its Taylor series if

$$f(x) = \lim_{n \to \infty} T_n(x)$$

If we let

$$R_n(x) = f(x) - T_n(x) \qquad \text{so that} \qquad f(x) = T_n(x) + R_n(x)$$

then $R_n(x)$ is called the **remainder** of the Taylor series. If we can somehow show that $\lim_{n \to \infty} R_n(x) = 0$, then it follows that

$$\lim_{n \to \infty} T_n(x) = \lim_{n \to \infty} [f(x) - R_n(x)] = f(x) - \lim_{n \to \infty} R_n(x) = f(x)$$

We have therefore proved the following theorem.

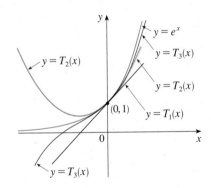

FIGURE 1

As n increases, $T_n(x)$ appears to approach e^x in Figure 1. This suggests that e^x is equal to the sum of its Taylor series.

> **8 Theorem** If $f(x) = T_n(x) + R_n(x)$, where T_n is the nth-degree Taylor polynomial of f at a and
> $$\lim_{n \to \infty} R_n(x) = 0$$
> for $|x - a| < R$, then f is equal to the sum of its Taylor series on the interval $|x - a| < R$.

In trying to show that $\lim_{n \to \infty} R_n(x) = 0$ for a specific function f, we usually use the following theorem.

> **9 Taylor's Inequality** If $|f^{(n+1)}(x)| \le M$ for $|x - a| \le d$, then the remainder $R_n(x)$ of the Taylor series satisfies the inequality
> $$|R_n(x)| \le \frac{M}{(n+1)!} |x - a|^{n+1} \qquad \text{for } |x - a| \le d$$

To see why this is true for $n = 1$, we assume that $|f''(x)| \le M$. In particular, we have $f''(x) \le M$, so for $a \le x \le a + d$ we have

$$\int_a^x f''(t)\, dt \le \int_a^x M\, dt$$

An antiderivative of f'' is f', so by Part 2 of the Fundamental Theorem of Calculus, we have

$$f'(x) - f'(a) \le M(x - a) \qquad \text{or} \qquad f'(x) \le f'(a) + M(x - a)$$

Thus

$$\int_a^x f'(t)\, dt \le \int_a^x [f'(a) + M(t - a)]\, dt$$

$$f(x) - f(a) \le f'(a)(x - a) + M \frac{(x - a)^2}{2}$$

$$f(x) - f(a) - f'(a)(x - a) \le \frac{M}{2}(x - a)^2$$

But $R_1(x) = f(x) - T_1(x) = f(x) - f(a) - f'(a)(x - a)$. So

$$R_1(x) \le \frac{M}{2}(x - a)^2$$

A similar argument, using $f''(x) \ge -M$, shows that

$$R_1(x) \ge -\frac{M}{2}(x - a)^2$$

So

$$|R_1(x)| \le \frac{M}{2}|x - a|^2$$

Although we have assumed that $x > a$, similar calculations show that this inequality is also true for $x < a$.

Formulas for the Taylor Remainder Term

As alternatives to Taylor's Inequality, we have the following formulas for the remainder term. If $f^{(n+1)}$ is continuous on an interval I and $x \in I$, then

$$R_n(x) = \frac{1}{n!} \int_a^x (x - t)^n f^{(n+1)}(t)\, dt$$

This is called the *integral form of the remainder term*. Another formula, called *Lagrange's form of the remainder term*, states that there is a number z between x and a such that

$$R_n(x) = \frac{f^{(n+1)}(z)}{(n+1)!} (x - a)^{n+1}$$

This version is an extension of the Mean Value Theorem (which is the case $n = 0$).

Proofs of these formulas, together with discussions of how to use them to solve the examples of Sections 11.10 and 11.11, are given on the website

www.stewartcalculus.com

Click on *Additional Topics* and then on *Formulas for the Remainder Term in Taylor series*.

This proves Taylor's Inequality for the case where $n = 1$. The result for any n is proved in a similar way by integrating $n + 1$ times. (See Exercise 73 for the case $n = 2$.)

NOTE In Section 11.11 we will explore the use of Taylor's Inequality in approximating functions. Our immediate use of it is in conjunction with Theorem 8.

In applying Theorems 8 and 9 it is often helpful to make use of the following fact.

$$\boxed{10} \qquad \lim_{n \to \infty} \frac{x^n}{n!} = 0 \qquad \text{for every real number } x$$

This is true because we know from Example 1 that the series $\Sigma \, x^n/n!$ converges for all x and so its nth term approaches 0.

V **EXAMPLE 2** Prove that e^x is equal to the sum of its Maclaurin series.

SOLUTION If $f(x) = e^x$, then $f^{(n+1)}(x) = e^x$ for all n. If d is any positive number and $|x| \le d$, then $|f^{(n+1)}(x)| = e^x \le e^d$. So Taylor's Inequality, with $a = 0$ and $M = e^d$, says that

$$|R_n(x)| \le \frac{e^d}{(n + 1)!} |x|^{n+1} \qquad \text{for } |x| \le d$$

Notice that the same constant $M = e^d$ works for every value of n. But, from Equation 10, we have

$$\lim_{n \to \infty} \frac{e^d}{(n + 1)!} |x|^{n+1} = e^d \lim_{n \to \infty} \frac{|x|^{n+1}}{(n + 1)!} = 0$$

It follows from the Squeeze Theorem that $\lim_{n \to \infty} |R_n(x)| = 0$ and therefore $\lim_{n \to \infty} R_n(x) = 0$ for all values of x. By Theorem 8, e^x is equal to the sum of its Maclaurin series, that is,

$$\boxed{11} \qquad e^x = \sum_{n=0}^{\infty} \frac{x^n}{n!} \qquad \text{for all } x$$

In particular, if we put $x = 1$ in Equation 11, we obtain the following expression for the number e as a sum of an infinite series:

In 1748 Leonhard Euler used Equation 12 to find the value of e correct to 23 digits. In 2007 Shigeru Kondo, again using the series in $\boxed{12}$, computed e to more than 100 billion decimal places. The special techniques employed to speed up the computation are explained on the website

numbers.computation.free.fr

$$\boxed{12} \qquad e = \sum_{n=0}^{\infty} \frac{1}{n!} = 1 + \frac{1}{1!} + \frac{1}{2!} + \frac{1}{3!} + \cdots$$

EXAMPLE 3 Find the Taylor series for $f(x) = e^x$ at $a = 2$.

SOLUTION We have $f^{(n)}(2) = e^2$ and so, putting $a = 2$ in the definition of a Taylor series $\boxed{6}$, we get

$$\sum_{n=0}^{\infty} \frac{f^{(n)}(2)}{n!} (x - 2)^n = \sum_{n=0}^{\infty} \frac{e^2}{n!} (x - 2)^n$$

Again it can be verified, as in Example 1, that the radius of convergence is $R = \infty$. As in Example 2 we can verify that $\lim_{n \to \infty} R_n(x) = 0$, so

$$\boxed{13} \qquad e^x = \sum_{n=0}^{\infty} \frac{e^2}{n!}(x - 2)^n \qquad \text{for all } x$$

We have two power series expansions for e^x, the Maclaurin series in Equation 11 and the Taylor series in Equation 13. The first is better if we are interested in values of x near 0 and the second is better if x is near 2.

EXAMPLE 4 Find the Maclaurin series for $\sin x$ and prove that it represents $\sin x$ for all x.

SOLUTION We arrange our computation in two columns as follows:

$$
\begin{array}{ll}
f(x) = \sin x & f(0) = 0 \\
f'(x) = \cos x & f'(0) = 1 \\
f''(x) = -\sin x & f''(0) = 0 \\
f'''(x) = -\cos x & f'''(0) = -1 \\
f^{(4)}(x) = \sin x & f^{(4)}(0) = 0
\end{array}
$$

Since the derivatives repeat in a cycle of four, we can write the Maclaurin series as follows:

$$f(0) + \frac{f'(0)}{1!}x + \frac{f''(0)}{2!}x^2 + \frac{f'''(0)}{3!}x^3 + \cdots$$

$$= x - \frac{x^3}{3!} + \frac{x^5}{5!} - \frac{x^7}{7!} + \cdots = \sum_{n=0}^{\infty}(-1)^n \frac{x^{2n+1}}{(2n+1)!}$$

Since $f^{(n+1)}(x)$ is $\pm\sin x$ or $\pm\cos x$, we know that $|f^{(n+1)}(x)| \le 1$ for all x. So we can take $M = 1$ in Taylor's Inequality:

$$\boxed{14} \qquad |R_n(x)| \le \frac{M}{(n+1)!}|x^{n+1}| = \frac{|x|^{n+1}}{(n+1)!}$$

By Equation 10 the right side of this inequality approaches 0 as $n \to \infty$, so $|R_n(x)| \to 0$ by the Squeeze Theorem. It follows that $R_n(x) \to 0$ as $n \to \infty$, so $\sin x$ is equal to the sum of its Maclaurin series by Theorem 8.

Figure 2 shows the graph of $\sin x$ together with its Taylor (or Maclaurin) polynomials

$$T_1(x) = x$$
$$T_3(x) = x - \frac{x^3}{3!}$$
$$T_5(x) = x - \frac{x^3}{3!} + \frac{x^5}{5!}$$

Notice that, as n increases, $T_n(x)$ becomes a better approximation to $\sin x$.

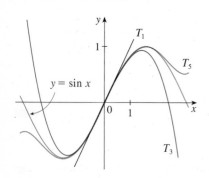

FIGURE 2

We state the result of Example 4 for future reference.

$$\boxed{15} \qquad \sin x = x - \frac{x^3}{3!} + \frac{x^5}{5!} - \frac{x^7}{7!} + \cdots$$

$$= \sum_{n=0}^{\infty}(-1)^n \frac{x^{2n+1}}{(2n+1)!} \qquad \text{for all } x$$

EXAMPLE 5 Find the Maclaurin series for $\cos x$.

SOLUTION We could proceed directly as in Example 4, but it's easier to differentiate the Maclaurin series for sin x given by Equation 15:

$$\cos x = \frac{d}{dx}(\sin x) = \frac{d}{dx}\left(x - \frac{x^3}{3!} + \frac{x^5}{5!} - \frac{x^7}{7!} + \cdots\right)$$

$$= 1 - \frac{3x^2}{3!} + \frac{5x^4}{5!} - \frac{7x^6}{7!} + \cdots = 1 - \frac{x^2}{2!} + \frac{x^4}{4!} - \frac{x^6}{6!} + \cdots$$

The Maclaurin series for e^x, sin x, and cos x that we found in Examples 2, 4, and 5 were discovered, using different methods, by Newton. These equations are remarkable because they say we know everything about each of these functions if we know all its derivatives at the single number 0.

Since the Maclaurin series for sin x converges for all x, Theorem 2 in Section 11.9 tells us that the differentiated series for cos x also converges for all x. Thus

16

$$\cos x = 1 - \frac{x^2}{2!} + \frac{x^4}{4!} - \frac{x^6}{6!} + \cdots$$

$$= \sum_{n=0}^{\infty}(-1)^n \frac{x^{2n}}{(2n)!} \qquad \text{for all } x$$

EXAMPLE 6 Find the Maclaurin series for the function $f(x) = x \cos x$.

SOLUTION Instead of computing derivatives and substituting in Equation 7, it's easier to multiply the series for cos x (Equation 16) by x:

$$x \cos x = x \sum_{n=0}^{\infty}(-1)^n \frac{x^{2n}}{(2n)!} = \sum_{n=0}^{\infty}(-1)^n \frac{x^{2n+1}}{(2n)!}$$

EXAMPLE 7 Represent $f(x) = \sin x$ as the sum of its Taylor series centered at $\pi/3$.

SOLUTION Arranging our work in columns, we have

$$f(x) = \sin x \qquad\qquad f\left(\frac{\pi}{3}\right) = \frac{\sqrt{3}}{2}$$

$$f'(x) = \cos x \qquad\qquad f'\left(\frac{\pi}{3}\right) = \frac{1}{2}$$

$$f''(x) = -\sin x \qquad\qquad f''\left(\frac{\pi}{3}\right) = -\frac{\sqrt{3}}{2}$$

$$f'''(x) = -\cos x \qquad\qquad f'''\left(\frac{\pi}{3}\right) = -\frac{1}{2}$$

We have obtained two different series representations for sin x, the Maclaurin series in Example 4 and the Taylor series in Example 7. It is best to use the Maclaurin series for values of x near 0 and the Taylor series for x near $\pi/3$. Notice that the third Taylor polynomial T_3 in Figure 3 is a good approximation to sin x near $\pi/3$ but not as good near 0. Compare it with the third Maclaurin polynomial T_3 in Figure 2, where the opposite is true.

and this pattern repeats indefinitely. Therefore the Taylor series at $\pi/3$ is

$$f\left(\frac{\pi}{3}\right) + \frac{f'\left(\frac{\pi}{3}\right)}{1!}\left(x - \frac{\pi}{3}\right) + \frac{f''\left(\frac{\pi}{3}\right)}{2!}\left(x - \frac{\pi}{3}\right)^2 + \frac{f'''\left(\frac{\pi}{3}\right)}{3!}\left(x - \frac{\pi}{3}\right)^3 + \cdots$$

$$= \frac{\sqrt{3}}{2} + \frac{1}{2 \cdot 1!}\left(x - \frac{\pi}{3}\right) - \frac{\sqrt{3}}{2 \cdot 2!}\left(x - \frac{\pi}{3}\right)^2 - \frac{1}{2 \cdot 3!}\left(x - \frac{\pi}{3}\right)^3 + \cdots$$

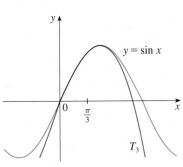

FIGURE 3

The proof that this series represents $\sin x$ for all x is very similar to that in Example 4. (Just replace x by $x - \pi/3$ in $\boxed{14}$.) We can write the series in sigma notation if we separate the terms that contain $\sqrt{3}$:

$$\sin x = \sum_{n=0}^{\infty} \frac{(-1)^n \sqrt{3}}{2(2n)!}\left(x - \frac{\pi}{3}\right)^{2n} + \sum_{n=0}^{\infty} \frac{(-1)^n}{2(2n+1)!}\left(x - \frac{\pi}{3}\right)^{2n+1} \qquad \blacksquare$$

The power series that we obtained by indirect methods in Examples 5 and 6 and in Section 11.9 are indeed the Taylor or Maclaurin series of the given functions because Theorem 5 asserts that, no matter how a power series representation $f(x) = \Sigma\, c_n(x - a)^n$ is obtained, it is always true that $c_n = f^{(n)}(a)/n!$. In other words, the coefficients are uniquely determined.

EXAMPLE 8 Find the Maclaurin series for $f(x) = (1 + x)^k$, where k is any real number.

SOLUTION Arranging our work in columns, we have

$$f(x) = (1 + x)^k \qquad\qquad\qquad f(0) = 1$$

$$f'(x) = k(1 + x)^{k-1} \qquad\qquad\qquad f'(0) = k$$

$$f''(x) = k(k - 1)(1 + x)^{k-2} \qquad\qquad f''(0) = k(k - 1)$$

$$f'''(x) = k(k - 1)(k - 2)(1 + x)^{k-3} \qquad f'''(0) = k(k - 1)(k - 2)$$

$$\vdots \qquad\qquad\qquad\qquad\qquad\qquad \vdots$$

$$f^{(n)}(x) = k(k - 1) \cdots (k - n + 1)(1 + x)^{k-n} \qquad f^{(n)}(0) = k(k - 1) \cdots (k - n + 1)$$

Therefore the Maclaurin series of $f(x) = (1 + x)^k$ is

$$\sum_{n=0}^{\infty} \frac{f^{(n)}(0)}{n!}x^n = \sum_{n=0}^{\infty} \frac{k(k - 1) \cdots (k - n + 1)}{n!}x^n$$

This series is called the **binomial series**. Notice that if k is a nonnegative integer, then the terms are eventually 0 and so the series is finite. For other values of k none of the terms is 0 and so we can try the Ratio Test. If the nth term is a_n, then

$$\left|\frac{a_{n+1}}{a_n}\right| = \left|\frac{k(k - 1) \cdots (k - n + 1)(k - n)x^{n+1}}{(n + 1)!} \cdot \frac{n!}{k(k - 1) \cdots (k - n + 1)x^n}\right|$$

$$= \frac{|k - n|}{n + 1}|x| = \frac{\left|1 - \dfrac{k}{n}\right|}{1 + \dfrac{1}{n}}|x| \to |x| \qquad \text{as } n \to \infty$$

Thus, by the Ratio Test, the binomial series converges if $|x| < 1$ and diverges if $|x| > 1$. $\qquad\qquad\qquad \blacksquare$

The traditional notation for the coefficients in the binomial series is

$$\binom{k}{n} = \frac{k(k - 1)(k - 2) \cdots (k - n + 1)}{n!}$$

and these numbers are called the **binomial coefficients**.

The following theorem states that $(1 + x)^k$ is equal to the sum of its Maclaurin series. It is possible to prove this by showing that the remainder term $R_n(x)$ approaches 0, but that turns out to be quite difficult. The proof outlined in Exercise 75 is much easier.

---

17 **The Binomial Series**   If k is any real number and $|x| < 1$, then

$$(1 + x)^k = \sum_{n=0}^{\infty} \binom{k}{n} x^n = 1 + kx + \frac{k(k-1)}{2!}x^2 + \frac{k(k-1)(k-2)}{3!}x^3 + \cdots$$

Although the binomial series always converges when $|x| < 1$, the question of whether or not it converges at the endpoints, ± 1, depends on the value of k. It turns out that the series converges at 1 if $-1 < k \leqslant 0$ and at both endpoints if $k \geqslant 0$. Notice that if k is a positive integer and $n > k$, then the expression for $\binom{k}{n}$ contains a factor $(k - k)$, so $\binom{k}{n} = 0$ for $n > k$. This means that the series terminates and reduces to the ordinary Binomial Theorem when k is a positive integer. (See Reference Page 1.)

V **EXAMPLE 9**   Find the Maclaurin series for the function $f(x) = \dfrac{1}{\sqrt{4 - x}}$ and its radius of convergence.

SOLUTION   We rewrite $f(x)$ in a form where we can use the binomial series:

$$\frac{1}{\sqrt{4-x}} = \frac{1}{\sqrt{4\left(1 - \dfrac{x}{4}\right)}} = \frac{1}{2\sqrt{1 - \dfrac{x}{4}}} = \frac{1}{2}\left(1 - \frac{x}{4}\right)^{-1/2}$$

Using the binomial series with $k = -\frac{1}{2}$ and with x replaced by $-x/4$, we have

$$\frac{1}{\sqrt{4-x}} = \frac{1}{2}\left(1 - \frac{x}{4}\right)^{-1/2} = \frac{1}{2}\sum_{n=0}^{\infty}\binom{-\frac{1}{2}}{n}\left(-\frac{x}{4}\right)^n$$

$$= \frac{1}{2}\left[1 + \left(-\frac{1}{2}\right)\left(-\frac{x}{4}\right) + \frac{\left(-\frac{1}{2}\right)\left(-\frac{3}{2}\right)}{2!}\left(-\frac{x}{4}\right)^2 + \frac{\left(-\frac{1}{2}\right)\left(-\frac{3}{2}\right)\left(-\frac{5}{2}\right)}{3!}\left(-\frac{x}{4}\right)^3\right.$$

$$\left. + \cdots + \frac{\left(-\frac{1}{2}\right)\left(-\frac{3}{2}\right)\left(-\frac{5}{2}\right)\cdots\left(-\frac{1}{2} - n + 1\right)}{n!}\left(-\frac{x}{4}\right)^n + \cdots\right]$$

$$= \frac{1}{2}\left[1 + \frac{1}{8}x + \frac{1 \cdot 3}{2!8^2}x^2 + \frac{1 \cdot 3 \cdot 5}{3!8^3}x^3 + \cdots + \frac{1 \cdot 3 \cdot 5 \cdot \cdots \cdot (2n - 1)}{n!8^n}x^n + \cdots\right]$$

We know from **17** that this series converges when $|-x/4| < 1$, that is, $|x| < 4$, so the radius of convergence is $R = 4$. ▬

We collect in the following table, for future reference, some important Maclaurin series that we have derived in this section and the preceding one.

TABLE 1

Important Maclaurin Series and
Their Radii of Convergence

$$\frac{1}{1-x} = \sum_{n=0}^{\infty} x^n = 1 + x + x^2 + x^3 + \cdots \qquad R = 1$$

$$e^x = \sum_{n=0}^{\infty} \frac{x^n}{n!} = 1 + \frac{x}{1!} + \frac{x^2}{2!} + \frac{x^3}{3!} + \cdots \qquad R = \infty$$

$$\sin x = \sum_{n=0}^{\infty} (-1)^n \frac{x^{2n+1}}{(2n+1)!} = x - \frac{x^3}{3!} + \frac{x^5}{5!} - \frac{x^7}{7!} + \cdots \qquad R = \infty$$

$$\cos x = \sum_{n=0}^{\infty} (-1)^n \frac{x^{2n}}{(2n)!} = 1 - \frac{x^2}{2!} + \frac{x^4}{4!} - \frac{x^6}{6!} + \cdots \qquad R = \infty$$

$$\tan^{-1}x = \sum_{n=0}^{\infty} (-1)^n \frac{x^{2n+1}}{2n+1} = x - \frac{x^3}{3} + \frac{x^5}{5} - \frac{x^7}{7} + \cdots \qquad R = 1$$

$$\ln(1+x) = \sum_{n=1}^{\infty} (-1)^{n-1} \frac{x^n}{n} = x - \frac{x^2}{2} + \frac{x^3}{3} - \frac{x^4}{4} + \cdots \qquad R = 1$$

$$(1+x)^k = \sum_{n=0}^{\infty} \binom{k}{n} x^n = 1 + kx + \frac{k(k-1)}{2!} x^2 + \frac{k(k-1)(k-2)}{3!} x^3 + \cdots \qquad R = 1$$

EXAMPLE 10   Find the sum of the series $\dfrac{1}{1 \cdot 2} - \dfrac{1}{2 \cdot 2^2} + \dfrac{1}{3 \cdot 2^3} - \dfrac{1}{4 \cdot 2^4} + \cdots$.

SOLUTION   With sigma notation we can write the given series as

$$\sum_{n=1}^{\infty} (-1)^{n-1} \frac{1}{n \cdot 2^n} = \sum_{n=1}^{\infty} (-1)^{n-1} \frac{\left(\frac{1}{2}\right)^n}{n}$$

Then from Table 1 we see that this series matches the entry for $\ln(1 + x)$ with $x = \frac{1}{2}$. So

$$\sum_{n=1}^{\infty} (-1)^{n-1} \frac{1}{n \cdot 2^n} = \ln\left(1 + \tfrac{1}{2}\right) = \ln \tfrac{3}{2} \qquad \blacksquare$$

TEC Module 11.10/11.11 enables you to see how successive Taylor polynomials approach the original function.

One reason that Taylor series are important is that they enable us to integrate functions that we couldn't previously handle. In fact, in the introduction to this chapter we mentioned that Newton often integrated functions by first expressing them as power series and then integrating the series term by term. The function $f(x) = e^{-x^2}$ can't be integrated by techniques discussed so far because its antiderivative is not an elementary function (see Section 7.5). In the following example we use Newton's idea to integrate this function.

V   EXAMPLE 11

(a) Evaluate $\int e^{-x^2} dx$ as an infinite series.

(b) Evaluate $\int_0^1 e^{-x^2} dx$ correct to within an error of 0.001.

SOLUTION

(a) First we find the Maclaurin series for $f(x) = e^{-x^2}$. Although it's possible to use the direct method, let's find it simply by replacing x with $-x^2$ in the series for e^x given in Table 1. Thus, for all values of x,

$$e^{-x^2} = \sum_{n=0}^{\infty} \frac{(-x^2)^n}{n!} = \sum_{n=0}^{\infty} (-1)^n \frac{x^{2n}}{n!} = 1 - \frac{x^2}{1!} + \frac{x^4}{2!} - \frac{x^6}{3!} + \cdots$$

Now we integrate term by term:

$$\int e^{-x^2}\, dx = \int \left(1 - \frac{x^2}{1!} + \frac{x^4}{2!} - \frac{x^6}{3!} + \cdots + (-1)^n \frac{x^{2n}}{n!} + \cdots \right) dx$$

$$= C + x - \frac{x^3}{3 \cdot 1!} + \frac{x^5}{5 \cdot 2!} - \frac{x^7}{7 \cdot 3!} + \cdots + (-1)^n \frac{x^{2n+1}}{(2n+1)n!} + \cdots$$

This series converges for all x because the original series for e^{-x^2} converges for all x.

(b) The Fundamental Theorem of Calculus gives

$$\int_0^1 e^{-x^2}\, dx = \left[x - \frac{x^3}{3 \cdot 1!} + \frac{x^5}{5 \cdot 2!} - \frac{x^7}{7 \cdot 3!} + \frac{x^9}{9 \cdot 4!} - \cdots \right]_0^1$$

$$= 1 - \tfrac{1}{3} + \tfrac{1}{10} - \tfrac{1}{42} + \tfrac{1}{216} - \cdots$$

$$\approx 1 - \tfrac{1}{3} + \tfrac{1}{10} - \tfrac{1}{42} + \tfrac{1}{216} \approx 0.7475$$

We can take $C = 0$ in the antiderivative in part (a).

The Alternating Series Estimation Theorem shows that the error involved in this approximation is less than

$$\frac{1}{11 \cdot 5!} = \frac{1}{1320} < 0.001 \qquad \blacksquare$$

Another use of Taylor series is illustrated in the next example. The limit could be found with l'Hospital's Rule, but instead we use a series.

EXAMPLE 12 Evaluate $\displaystyle\lim_{x \to 0} \frac{e^x - 1 - x}{x^2}$.

SOLUTION Using the Maclaurin series for e^x, we have

$$\lim_{x \to 0} \frac{e^x - 1 - x}{x^2} = \lim_{x \to 0} \frac{\left(1 + \dfrac{x}{1!} + \dfrac{x^2}{2!} + \dfrac{x^3}{3!} + \cdots \right) - 1 - x}{x^2}$$

$$= \lim_{x \to 0} \frac{\dfrac{x^2}{2!} + \dfrac{x^3}{3!} + \dfrac{x^4}{4!} + \cdots}{x^2}$$

Some computer algebra systems compute limits in this way.

$$= \lim_{x \to 0} \left(\frac{1}{2} + \frac{x}{3!} + \frac{x^2}{4!} + \frac{x^3}{5!} + \cdots \right) = \frac{1}{2}$$

because power series are continuous functions. $\qquad \blacksquare$

▪ Multiplication and Division of Power Series

If power series are added or subtracted, they behave like polynomials (Theorem 11.2.8 shows this). In fact, as the following example illustrates, they can also be multiplied and divided like polynomials. We find only the first few terms because the calculations for the later terms become tedious and the initial terms are the most important ones.

EXAMPLE 13   Find the first three nonzero terms in the Maclaurin series for (a) $e^x \sin x$ and (b) $\tan x$.

SOLUTION

(a) Using the Maclaurin series for e^x and $\sin x$ in Table 1, we have

$$e^x \sin x = \left(1 + \frac{x}{1!} + \frac{x^2}{2!} + \frac{x^3}{3!} + \cdots\right)\left(x - \frac{x^3}{3!} + \cdots\right)$$

We multiply these expressions, collecting like terms just as for polynomials:

$$
\begin{array}{r}
1 + x + \frac{1}{2}x^2 + \frac{1}{6}x^3 + \cdots \\
\times \qquad x \qquad\quad - \frac{1}{6}x^3 + \cdots \\
\hline
x + x^2 + \frac{1}{2}x^3 + \frac{1}{6}x^4 + \cdots \\
+ \qquad\qquad\quad - \frac{1}{6}x^3 - \frac{1}{6}x^4 - \cdots \\
\hline
x + x^2 + \frac{1}{3}x^3 + \cdots
\end{array}
$$

Thus
$$e^x \sin x = x + x^2 + \tfrac{1}{3}x^3 + \cdots$$

(b) Using the Maclaurin series in Table 1, we have

$$\tan x = \frac{\sin x}{\cos x} = \frac{x - \dfrac{x^3}{3!} + \dfrac{x^5}{5!} - \cdots}{1 - \dfrac{x^2}{2!} + \dfrac{x^4}{4!} - \cdots}$$

We use a procedure like long division:

$$
\begin{array}{r}
x + \frac{1}{3}x^3 + \frac{2}{15}x^5 + \cdots \\
1 - \frac{1}{2}x^2 + \frac{1}{24}x^4 - \cdots \overline{)x - \frac{1}{6}x^3 + \frac{1}{120}x^5 - \cdots} \\
\underline{x - \frac{1}{2}x^3 + \frac{1}{24}x^5 - \cdots} \\
\frac{1}{3}x^3 - \frac{1}{30}x^5 + \cdots \\
\underline{\frac{1}{3}x^3 - \frac{1}{6}x^5 + \cdots} \\
\frac{2}{15}x^5 + \cdots
\end{array}
$$

Thus
$$\tan x = x + \tfrac{1}{3}x^3 + \tfrac{2}{15}x^5 + \cdots$$

Although we have not attempted to justify the formal manipulations used in Example 13, they are legitimate. There is a theorem which states that if both $f(x) = \Sigma\, c_n x^n$ and $g(x) = \Sigma\, b_n x^n$ converge for $|x| < R$ and the series are multiplied as if they were polynomials, then the resulting series also converges for $|x| < R$ and represents $f(x)g(x)$. For division we require $b_0 \neq 0$; the resulting series converges for sufficiently small $|x|$.

11.10 Exercises

Access end-of-section exercises and tools for visualizing calculus online at **www.webassign.net**

LABORATORY PROJECT CAS AN ELUSIVE LIMIT

This project deals with the function

$$f(x) = \frac{\sin(\tan x) - \tan(\sin x)}{\arcsin(\arctan x) - \arctan(\arcsin x)}$$

1. Use your computer algebra system to evaluate $f(x)$ for $x = 1, 0.1, 0.01, 0.001$, and 0.0001. Does it appear that f has a limit as $x \to 0$?

2. Use the CAS to graph f near $x = 0$. Does it appear that f has a limit as $x \to 0$?

3. Try to evaluate $\lim_{x \to 0} f(x)$ with l'Hospital's Rule, using the CAS to find derivatives of the numerator and denominator. What do you discover? How many applications of l'Hospital's Rule are required?

4. Evaluate $\lim_{x \to 0} f(x)$ by using the CAS to find sufficiently many terms in the Taylor series of the numerator and denominator. (Use the command `taylor` in Maple or `Series` in Mathematica.)

5. Use the limit command on your CAS to find $\lim_{x \to 0} f(x)$ directly. (Most computer algebra systems use the method of Problem 4 to compute limits.)

6. In view of the answers to Problems 4 and 5, how do you explain the results of Problems 1 and 2?

CAS Computer algebra system required

WRITING PROJECT   HOW NEWTON DISCOVERED THE BINOMIAL SERIES

The Binomial Theorem, which gives the expansion of $(a + b)^k$, was known to Chinese mathematicians many centuries before the time of Newton for the case where the exponent k is a positive integer. In 1665, when he was 22, Newton was the first to discover the infinite series expansion of $(a + b)^k$ when k is a fractional exponent (positive or negative). He didn't publish his discovery, but he stated it and gave examples of how to use it in a letter (now called the *epistola prior*) dated June 13, 1676, that he sent to Henry Oldenburg, secretary of the Royal Society of London, to transmit to Leibniz. When Leibniz replied, he asked how Newton had discovered the binomial series. Newton wrote a second letter, the *epistola posterior* of October 24, 1676, in which he explained in great detail how he arrived at his discovery by a very indirect route. He was investigating the areas under the curves $y = (1 - x^2)^{n/2}$ from 0 to x for $n = 0, 1, 2, 3, 4, \ldots$. These are easy to calculate if n is even. By observing patterns and interpolating, Newton was able to guess the answers for odd values of n. Then he realized he could get the same answers by expressing $(1 - x^2)^{n/2}$ as an infinite series.

Write a report on Newton's discovery of the binomial series. Start by giving the statement of the binomial series in Newton's notation (see the *epistola prior* on page 285 of [4] or page 402 of [2]). Explain why Newton's version is equivalent to Theorem 17 on page 623. Then read

Newton's *epistola posterior* (page 287 in [4] or page 404 in [2]) and explain the patterns that Newton discovered in the areas under the curves $y = (1 - x^2)^{n/2}$. Show how he was able to guess the areas under the remaining curves and how he verified his answers. Finally, explain how these discoveries led to the binomial series. The books by Edwards [1] and Katz [3] contain commentaries on Newton's letters.

1. C. H. Edwards, *The Historical Development of the Calculus* (New York: Springer-Verlag, 1979), pp. 178–187.

2. John Fauvel and Jeremy Gray, eds., *The History of Mathematics: A Reader* (London: MacMillan Press, 1987).

3. Victor Katz, *A History of Mathematics: An Introduction* (New York: HarperCollins, 1993), pp. 463–466.

4. D. J. Struik, ed., *A Sourcebook in Mathematics, 1200–1800* (Princeton, NJ: Princeton University Press, 1969).

11.11   Applications of Taylor Polynomials

In this section we explore two types of applications of Taylor polynomials. First we look at how they are used to approximate functions—computer scientists like them because polynomials are the simplest of functions. Then we investigate how physicists and engineers use them in such fields as relativity, optics, blackbody radiation, electric dipoles, the velocity of water waves, and building highways across a desert.

Approximating Functions by Polynomials

Suppose that $f(x)$ is equal to the sum of its Taylor series at a:

$$f(x) = \sum_{n=0}^{\infty} \frac{f^{(n)}(a)}{n!} (x - a)^n$$

In Section 11.10 we introduced the notation $T_n(x)$ for the nth partial sum of this series and called it the nth-degree Taylor polynomial of f at a. Thus

$$T_n(x) = \sum_{i=0}^{n} \frac{f^{(i)}(a)}{i!} (x - a)^i$$

$$= f(a) + \frac{f'(a)}{1!} (x - a) + \frac{f''(a)}{2!} (x - a)^2 + \cdots + \frac{f^{(n)}(a)}{n!} (x - a)^n$$

Since f is the sum of its Taylor series, we know that $T_n(x) \to f(x)$ as $n \to \infty$ and so T_n can be used as an approximation to f: $f(x) \approx T_n(x)$.

Notice that the first-degree Taylor polynomial

$$T_1(x) = f(a) + f'(a)(x - a)$$

is the same as the linearization of f at a that we discussed in Section 2.9. Notice also that T_1 and its derivative have the same values at a that f and f' have. In general, it can be shown that the derivatives of T_n at a agree with those of f up to and including derivatives of order n.

To illustrate these ideas let's take another look at the graphs of $y = e^x$ and its first few Taylor polynomials, as shown in Figure 1. The graph of T_1 is the tangent line to $y = e^x$ at $(0, 1)$; this tangent line is the best linear approximation to e^x near $(0, 1)$. The graph

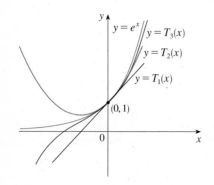

FIGURE 1

of T_2 is the parabola $y = 1 + x + x^2/2$, and the graph of T_3 is the cubic curve $y = 1 + x + x^2/2 + x^3/6$, which is a closer fit to the exponential curve $y = e^x$ than T_2. The next Taylor polynomial T_4 would be an even better approximation, and so on.

The values in the table give a numerical demonstration of the convergence of the Taylor polynomials $T_n(x)$ to the function $y = e^x$. We see that when $x = 0.2$ the convergence is very rapid, but when $x = 3$ it is somewhat slower. In fact, the farther x is from 0, the more slowly $T_n(x)$ converges to e^x.

	$x = 0.2$	$x = 3.0$
$T_2(x)$	1.220000	8.500000
$T_4(x)$	1.221400	16.375000
$T_6(x)$	1.221403	19.412500
$T_8(x)$	1.221403	20.009152
$T_{10}(x)$	1.221403	20.079665
e^x	1.221403	20.085537

When using a Taylor polynomial T_n to approximate a function f, we have to ask the questions: How good an approximation is it? How large should we take n to be in order to achieve a desired accuracy? To answer these questions we need to look at the absolute value of the remainder:

$$|R_n(x)| = |f(x) - T_n(x)|$$

There are three possible methods for estimating the size of the error:

1. If a graphing device is available, we can use it to graph $|R_n(x)|$ and thereby estimate the error.

2. If the series happens to be an alternating series, we can use the Alternating Series Estimation Theorem.

3. In all cases we can use Taylor's Inequality (Theorem 11.10.9), which says that if $|f^{(n+1)}(x)| \leq M$, then
$$|R_n(x)| \leq \frac{M}{(n+1)!}|x - a|^{n+1}$$

V EXAMPLE 1
(a) Approximate the function $f(x) = \sqrt[3]{x}$ by a Taylor polynomial of degree 2 at $a = 8$.
(b) How accurate is this approximation when $7 \leq x \leq 9$?

SOLUTION
(a)
$$f(x) = \sqrt[3]{x} = x^{1/3} \qquad f(8) = 2$$

$$f'(x) = \tfrac{1}{3}x^{-2/3} \qquad f'(8) = \tfrac{1}{12}$$

$$f''(x) = -\tfrac{2}{9}x^{-5/3} \qquad f''(8) = -\tfrac{1}{144}$$

$$f'''(x) = \tfrac{10}{27}x^{-8/3}$$

Thus the second-degree Taylor polynomial is

$$T_2(x) = f(8) + \frac{f'(8)}{1!}(x - 8) + \frac{f''(8)}{2!}(x - 8)^2$$

$$= 2 + \tfrac{1}{12}(x - 8) - \tfrac{1}{288}(x - 8)^2$$

The desired approximation is

$$\sqrt[3]{x} \approx T_2(x) = 2 + \tfrac{1}{12}(x - 8) - \tfrac{1}{288}(x - 8)^2$$

(b) The Taylor series is not alternating when $x < 8$, so we can't use the Alternating Series Estimation Theorem in this example. But we can use Taylor's Inequality with $n = 2$ and $a = 8$:

$$|R_2(x)| \leq \frac{M}{3!}|x - 8|^3$$

where $|f'''(x)| \leq M$. Because $x \geq 7$, we have $x^{8/3} \geq 7^{8/3}$ and so

$$f'''(x) = \frac{10}{27} \cdot \frac{1}{x^{8/3}} \leq \frac{10}{27} \cdot \frac{1}{7^{8/3}} < 0.0021$$

Therefore we can take $M = 0.0021$. Also $7 \leq x \leq 9$, so $-1 \leq x - 8 \leq 1$ and $|x - 8| \leq 1$. Then Taylor's Inequality gives

$$|R_2(x)| \leq \frac{0.0021}{3!} \cdot 1^3 = \frac{0.0021}{6} < 0.0004$$

Thus, if $7 \leq x \leq 9$, the approximation in part (a) is accurate to within 0.0004.　▬

Let's use a graphing device to check the calculation in Example 1. Figure 2 shows that the graphs of $y = \sqrt[3]{x}$ and $y = T_2(x)$ are very close to each other when x is near 8. Figure 3 shows the graph of $|R_2(x)|$ computed from the expression

$$|R_2(x)| = |\sqrt[3]{x} - T_2(x)|$$

We see from the graph that

$$|R_2(x)| < 0.0003$$

when $7 \leq x \leq 9$. Thus the error estimate from graphical methods is slightly better than the error estimate from Taylor's Inequality in this case.

FIGURE 2

FIGURE 3

V **EXAMPLE 2**
(a) What is the maximum error possible in using the approximation

$$\sin x \approx x - \frac{x^3}{3!} + \frac{x^5}{5!}$$

when $-0.3 \leq x \leq 0.3$? Use this approximation to find $\sin 12°$ correct to six decimal places.
(b) For what values of x is this approximation accurate to within 0.00005?

SOLUTION
(a) Notice that the Maclaurin series

$$\sin x = x - \frac{x^3}{3!} + \frac{x^5}{5!} - \frac{x^7}{7!} + \cdots$$

is alternating for all nonzero values of x, and the successive terms decrease in size because $|x| < 1$, so we can use the Alternating Series Estimation Theorem. The error in approximating $\sin x$ by the first three terms of its Maclaurin series is at most

$$\left| \frac{x^7}{7!} \right| = \frac{|x|^7}{5040}$$

If $-0.3 \leq x \leq 0.3$, then $|x| \leq 0.3$, so the error is smaller than

$$\frac{(0.3)^7}{5040} \approx 4.3 \times 10^{-8}$$

To find $\sin 12°$ we first convert to radian measure:

$$\sin 12° = \sin\left(\frac{12\pi}{180}\right) = \sin\left(\frac{\pi}{15}\right)$$

$$\approx \frac{\pi}{15} - \left(\frac{\pi}{15}\right)^3 \frac{1}{3!} + \left(\frac{\pi}{15}\right)^5 \frac{1}{5!} \approx 0.20791169$$

Thus, correct to six decimal places, $\sin 12° \approx 0.207912$.

(b) The error will be smaller than 0.00005 if

$$\frac{|x|^7}{5040} < 0.00005$$

Solving this inequality for x, we get

$$|x|^7 < 0.252 \qquad \text{or} \qquad |x| < (0.252)^{1/7} \approx 0.821$$

So the given approximation is accurate to within 0.00005 when $|x| < 0.82$. ▬

TEC Module 11.10/11.11 graphically shows the remainders in Taylor polynomial approximations.

What if we use Taylor's Inequality to solve Example 2? Since $f^{(7)}(x) = -\cos x$, we have $|f^{(7)}(x)| \leq 1$ and so

$$|R_6(x)| \leq \frac{1}{7!}|x|^7$$

So we get the same estimates as with the Alternating Series Estimation Theorem.

What about graphical methods? Figure 4 shows the graph of

$$|R_6(x)| = \left|\sin x - \left(x - \tfrac{1}{6}x^3 + \tfrac{1}{120}x^5\right)\right|$$

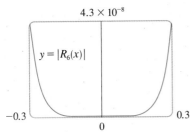

FIGURE 4

and we see from it that $|R_6(x)| < 4.3 \times 10^{-8}$ when $|x| \leq 0.3$. This is the same estimate that we obtained in Example 2. For part (b) we want $|R_6(x)| < 0.00005$, so we graph both $y = |R_6(x)|$ and $y = 0.00005$ in Figure 5. By placing the cursor on the right intersection point we find that the inequality is satisfied when $|x| < 0.82$. Again this is the same estimate that we obtained in the solution to Example 2.

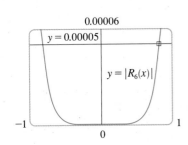

FIGURE 5

If we had been asked to approximate $\sin 72°$ instead of $\sin 12°$ in Example 2, it would have been wise to use the Taylor polynomials at $a = \pi/3$ (instead of $a = 0$) because they are better approximations to $\sin x$ for values of x close to $\pi/3$. Notice that 72° is close to 60° (or $\pi/3$ radians) and the derivatives of $\sin x$ are easy to compute at $\pi/3$.

Figure 6 shows the graphs of the Maclaurin polynomial approximations

$$T_1(x) = x \qquad\qquad T_3(x) = x - \frac{x^3}{3!}$$

$$T_5(x) = x - \frac{x^3}{3!} + \frac{x^5}{5!} \qquad T_7(x) = x - \frac{x^3}{3!} + \frac{x^5}{5!} - \frac{x^7}{7!}$$

to the sine curve. You can see that as n increases, $T_n(x)$ is a good approximation to $\sin x$ on a larger and larger interval.

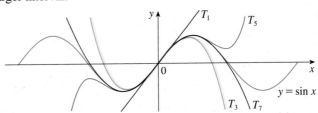

FIGURE 6

One use of the type of calculation done in Examples 1 and 2 occurs in calculators and computers. For instance, when you press the sin or e^x key on your calculator, or when a computer programmer uses a subroutine for a trigonometric or exponential or Bessel function, in many machines a polynomial approximation is calculated. The polynomial is often a Taylor polynomial that has been modified so that the error is spread more evenly throughout an interval.

Applications to Physics

Taylor polynomials are also used frequently in physics. In order to gain insight into an equation, a physicist often simplifies a function by considering only the first two or three terms in its Taylor series. In other words, the physicist uses a Taylor polynomial as an approximation to the function. Taylor's Inequality can then be used to gauge the accuracy of the approximation. The following example shows one way in which this idea is used in special relativity.

V EXAMPLE 3 In Einstein's theory of special relativity the mass of an object moving with velocity v is

$$m = \frac{m_0}{\sqrt{1 - v^2/c^2}}$$

where m_0 is the mass of the object when at rest and c is the speed of light. The kinetic energy of the object is the difference between its total energy and its energy at rest:

$$K = mc^2 - m_0c^2$$

(a) Show that when v is very small compared with c, this expression for K agrees with classical Newtonian physics: $K = \frac{1}{2}m_0v^2$.
(b) Use Taylor's Inequality to estimate the difference in these expressions for K when $|v| \leqslant 100$ m/s.

SOLUTION
(a) Using the expressions given for K and m, we get

$$K = mc^2 - m_0c^2 = \frac{m_0c^2}{\sqrt{1 - v^2/c^2}} - m_0c^2 = m_0c^2\left[\left(1 - \frac{v^2}{c^2}\right)^{-1/2} - 1\right]$$

With $x = -v^2/c^2$, the Maclaurin series for $(1 + x)^{-1/2}$ is most easily computed as a binomial series with $k = -\frac{1}{2}$. (Notice that $|x| < 1$ because $v < c$.) Therefore we have

$$(1 + x)^{-1/2} = 1 - \frac{1}{2}x + \frac{\left(-\frac{1}{2}\right)\left(-\frac{3}{2}\right)}{2!}x^2 + \frac{\left(-\frac{1}{2}\right)\left(-\frac{3}{2}\right)\left(-\frac{5}{2}\right)}{3!}x^3 + \cdots$$

$$= 1 - \frac{1}{2}x + \frac{3}{8}x^2 - \frac{5}{16}x^3 + \cdots$$

and

$$K = m_0c^2\left[\left(1 + \frac{1}{2}\frac{v^2}{c^2} + \frac{3}{8}\frac{v^4}{c^4} + \frac{5}{16}\frac{v^6}{c^6} + \cdots\right) - 1\right]$$

$$= m_0c^2\left(\frac{1}{2}\frac{v^2}{c^2} + \frac{3}{8}\frac{v^4}{c^4} + \frac{5}{16}\frac{v^6}{c^6} + \cdots\right)$$

If v is much smaller than c, then all terms after the first are very small when compared

The upper curve in Figure 7 is the graph of the expression for the kinetic energy K of an object with velocity v in special relativity. The lower curve shows the function used for K in classical Newtonian physics. When v is much smaller than the speed of light, the curves are practically identical.

FIGURE 7

with the first term. If we omit them, we get

$$K \approx m_0 c^2 \left(\frac{1}{2} \frac{v^2}{c^2} \right) = \tfrac{1}{2} m_0 v^2$$

(b) If $x = -v^2/c^2$, $f(x) = m_0 c^2 [(1 + x)^{-1/2} - 1]$, and M is a number such that $|f''(x)| \le M$, then we can use Taylor's Inequality to write

$$|R_1(x)| \le \frac{M}{2!} x^2$$

We have $f''(x) = \tfrac{3}{4} m_0 c^2 (1 + x)^{-5/2}$ and we are given that $|v| \le 100$ m/s, so

$$|f''(x)| = \frac{3 m_0 c^2}{4(1 - v^2/c^2)^{5/2}} \le \frac{3 m_0 c^2}{4(1 - 100^2/c^2)^{5/2}} \quad (= M)$$

Thus, with $c = 3 \times 10^8$ m/s,

$$|R_1(x)| \le \frac{1}{2} \cdot \frac{3 m_0 c^2}{4(1 - 100^2/c^2)^{5/2}} \cdot \frac{100^4}{c^4} < (4.17 \times 10^{-10}) m_0$$

So when $|v| \le 100$ m/s, the magnitude of the error in using the Newtonian expression for kinetic energy is at most $(4.2 \times 10^{-10}) m_0$. ▬

Another application to physics occurs in optics. Figure 8 is adapted from *Optics*, 4th ed., by Eugene Hecht (San Francisco, 2002), page 153. It depicts a wave from the point source S meeting a spherical interface of radius R centered at C. The ray SA is refracted toward P.

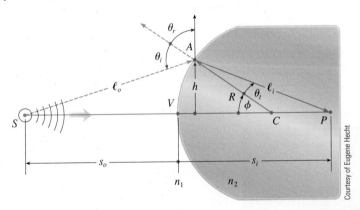

FIGURE 8

Refraction at a spherical interface

Using Fermat's principle that light travels so as to minimize the time taken, Hecht derives the equation

$$\boxed{1} \qquad \frac{n_1}{\ell_o} + \frac{n_2}{\ell_i} = \frac{1}{R} \left(\frac{n_2 s_i}{\ell_i} - \frac{n_1 s_o}{\ell_o} \right)$$

where n_1 and n_2 are indexes of refraction and ℓ_o, ℓ_i, s_o, and s_i are the distances indicated in Figure 8. By the Law of Cosines, applied to triangles ACS and ACP, we have

Here we use the identity

$$\cos(\pi - \phi) = -\cos \phi$$

$$\boxed{2} \qquad \ell_o = \sqrt{R^2 + (s_o + R)^2 - 2R(s_o + R) \cos \phi}$$

$$\ell_i = \sqrt{R^2 + (s_i - R)^2 + 2R(s_i - R) \cos \phi}$$

Because Equation 1 is cumbersome to work with, Gauss, in 1841, simplified it by using the linear approximation $\cos \phi \approx 1$ for small values of ϕ. (This amounts to using the Taylor polynomial of degree 1.) Then Equation 1 becomes the following simpler equation [as you are asked to show in Exercise 34(a)]:

$$\boxed{3} \qquad \frac{n_1}{s_o} + \frac{n_2}{s_i} = \frac{n_2 - n_1}{R}$$

The resulting optical theory is known as *Gaussian optics,* or *first-order optics,* and has become the basic theoretical tool used to design lenses.

A more accurate theory is obtained by approximating $\cos \phi$ by its Taylor polynomial of degree 3 (which is the same as the Taylor polynomial of degree 2). This takes into account rays for which ϕ is not so small, that is, rays that strike the surface at greater distances h above the axis. In Exercise 34(b) you are asked to use this approximation to derive the more accurate equation

$$\boxed{4} \qquad \frac{n_1}{s_o} + \frac{n_2}{s_i} = \frac{n_2 - n_1}{R} + h^2 \left[\frac{n_1}{2s_o} \left(\frac{1}{s_o} + \frac{1}{R} \right)^2 + \frac{n_2}{2s_i} \left(\frac{1}{R} - \frac{1}{s_i} \right)^2 \right]$$

The resulting optical theory is known as *third-order optics.*

Other applications of Taylor polynomials to physics and engineering are explored in Exercises 32, 33, 35, 36, 37, and 38, and in the following Applied Project.

11.11 Exercises

Access end-of-section exercises and tools for visualizing calculus online at **www.webassign.net**

APPLIED PROJECT   RADIATION FROM THE STARS

Any object emits radiation when heated. A *blackbody* is a system that absorbs all the radiation that falls on it. For instance, a matte black surface or a large cavity with a small hole in its wall (like a blastfurnace) is a blackbody and emits blackbody radiation. Even the radiation from the sun is close to being blackbody radiation.

Proposed in the late 19th century, the Rayleigh-Jeans Law expresses the energy density of blackbody radiation of wavelength λ as

$$f(\lambda) = \frac{8\pi kT}{\lambda^4}$$

where λ is measured in meters, T is the temperature in kelvins (K), and k is Boltzmann's constant. The Rayleigh-Jeans Law agrees with experimental measurements for long wavelengths but disagrees drastically for short wavelengths. [The law predicts that $f(\lambda) \to \infty$ as $\lambda \to 0^+$ but experiments have shown that $f(\lambda) \to 0$.] This fact is known as the *ultraviolet catastrophe.*

In 1900 Max Planck found a better model (known now as Planck's Law) for blackbody radiation:

$$f(\lambda) = \frac{8\pi hc \lambda^{-5}}{e^{hc/(\lambda kT)} - 1}$$

Graphing calculator or computer required

where λ is measured in meters, T is the temperature (in kelvins), and

$$h = \text{Planck's constant} = 6.6262 \times 10^{-34} \text{ J·s}$$

$$c = \text{speed of light} = 2.997925 \times 10^{8} \text{ m/s}$$

$$k = \text{Boltzmann's constant} = 1.3807 \times 10^{-23} \text{ J/K}$$

1. Use l'Hospital's Rule to show that

$$\lim_{\lambda \to 0^+} f(\lambda) = 0 \quad \text{and} \quad \lim_{\lambda \to \infty} f(\lambda) = 0$$

for Planck's Law. So this law models blackbody radiation better than the Rayleigh-Jeans Law for short wavelengths.

2. Use a Taylor polynomial to show that, for large wavelengths, Planck's Law gives approximately the same values as the Rayleigh-Jeans Law.

3. Graph f as given by both laws on the same screen and comment on the similarities and differences. Use $T = 5700$ K (the temperature of the sun). (You may want to change from meters to the more convenient unit of micrometers: 1 μm $= 10^{-6}$ m.)

4. Use your graph in Problem 3 to estimate the value of λ for which $f(\lambda)$ is a maximum under Planck's Law.

5. Investigate how the graph of f changes as T varies. (Use Planck's Law.) In particular, graph f for the stars Betelgeuse ($T = 3400$ K), Procyon ($T = 6400$ K), and Sirius ($T = 9200$ K), as well as the sun. How does the total radiation emitted (the area under the curve) vary with T? Use the graph to comment on why Sirius is known as a blue star and Betelgeuse as a red star.

---

11   Review

Concept Check

1. (a) What is a convergent sequence?
   (b) What is a convergent series?
   (c) What does $\lim_{n \to \infty} a_n = 3$ mean?
   (d) What does $\sum_{n=1}^{\infty} a_n = 3$ mean?

2. (a) What is a bounded sequence?
   (b) What is a monotonic sequence?
   (c) What can you say about a bounded monotonic sequence?

3. (a) What is a geometric series? Under what circumstances is it convergent? What is its sum?
   (b) What is a p-series? Under what circumstances is it convergent?

4. Suppose $\Sigma\, a_n = 3$ and s_n is the nth partial sum of the series. What is $\lim_{n \to \infty} a_n$? What is $\lim_{n \to \infty} s_n$?

5. State the following.
   (a) The Test for Divergence
   (b) The Integral Test
   (c) The Comparison Test

   (d) The Limit Comparison Test
   (e) The Alternating Series Test
   (f) The Ratio Test
   (g) The Root Test

6. (a) What is an absolutely convergent series?
   (b) What can you say about such a series?
   (c) What is a conditionally convergent series?

7. (a) If a series is convergent by the Integral Test, how do you estimate its sum?
   (b) If a series is convergent by the Comparison Test, how do you estimate its sum?
   (c) If a series is convergent by the Alternating Series Test, how do you estimate its sum?

8. (a) Write the general form of a power series.
   (b) What is the radius of convergence of a power series?
   (c) What is the interval of convergence of a power series?

9. Suppose $f(x)$ is the sum of a power series with radius of convergence R.
   (a) How do you differentiate f? What is the radius of convergence of the series for f'?
   (b) How do you integrate f? What is the radius of convergence of the series for $\int f(x)\,dx$?

10. (a) Write an expression for the nth-degree Taylor polynomial of f centered at a.
   (b) Write an expression for the Taylor series of f centered at a.
   (c) Write an expression for the Maclaurin series of f.

   (d) How do you show that $f(x)$ is equal to the sum of its Taylor series?
   (e) State Taylor's Inequality.

11. Write the Maclaurin series and the interval of convergence for each of the following functions.
   (a) $1/(1-x)$ (b) e^x
   (c) $\sin x$ (d) $\cos x$
   (e) $\tan^{-1}x$ (f) $\ln(1+x)$

12. Write the binomial series expansion of $(1+x)^k$. What is the radius of convergence of this series?

True-False Quiz

Determine whether the statement is true or false. If it is true, explain why. If it is false, explain why or give an example that disproves the statement.

1. If $\lim_{n\to\infty} a_n = 0$, then $\Sigma\, a_n$ is convergent.

2. The series $\sum_{n=1}^{\infty} n^{-\sin 1}$ is convergent.

3. If $\lim_{n\to\infty} a_n = L$, then $\lim_{n\to\infty} a_{2n+1} = L$.

4. If $\Sigma\, c_n 6^n$ is convergent, then $\Sigma\, c_n(-2)^n$ is convergent.

5. If $\Sigma\, c_n 6^n$ is convergent, then $\Sigma\, c_n(-6)^n$ is convergent.

6. If $\Sigma\, c_n x^n$ diverges when $x = 6$, then it diverges when $x = 10$.

7. The Ratio Test can be used to determine whether $\Sigma\, 1/n^3$ converges.

8. The Ratio Test can be used to determine whether $\Sigma\, 1/n!$ converges.

9. If $0 \le a_n \le b_n$ and $\Sigma\, b_n$ diverges, then $\Sigma\, a_n$ diverges.

10. $\displaystyle\sum_{n=0}^{\infty} \frac{(-1)^n}{n!} = \frac{1}{e}$

11. If $-1 < \alpha < 1$, then $\lim_{n\to\infty} \alpha^n = 0$.

12. If $\Sigma\, a_n$ is divergent, then $\Sigma\, |a_n|$ is divergent.

13. If $f(x) = 2x - x^2 + \frac{1}{3}x^3 - \cdots$ converges for all x, then $f'''(0) = 2$.

14. If $\{a_n\}$ and $\{b_n\}$ are divergent, then $\{a_n + b_n\}$ is divergent.

15. If $\{a_n\}$ and $\{b_n\}$ are divergent, then $\{a_n b_n\}$ is divergent.

16. If $\{a_n\}$ is decreasing and $a_n > 0$ for all n, then $\{a_n\}$ is convergent.

17. If $a_n > 0$ and $\Sigma\, a_n$ converges, then $\Sigma\, (-1)^n a_n$ converges.

18. If $a_n > 0$ and $\lim_{n\to\infty}(a_{n+1}/a_n) < 1$, then $\lim_{n\to\infty} a_n = 0$.

19. $0.99999\ldots = 1$

20. If $\lim_{n\to\infty} a_n = 2$, then $\lim_{n\to\infty}(a_{n+3} - a_n) = 0$.

21. If a finite number of terms are added to a convergent series, then the new series is still convergent.

22. If $\displaystyle\sum_{n=1}^{\infty} a_n = A$ and $\displaystyle\sum_{n=1}^{\infty} b_n = B$, then $\displaystyle\sum_{n=1}^{\infty} a_n b_n = AB$.

Exercises

1–8 Determine whether the sequence is convergent or divergent. If it is convergent, find its limit.

1. $a_n = \dfrac{2+n^3}{1+2n^3}$ **2.** $a_n = \dfrac{9^{n+1}}{10^n}$

3. $a_n = \dfrac{n^3}{1+n^2}$ **4.** $a_n = \cos(n\pi/2)$

5. $a_n = \dfrac{n \sin n}{n^2 + 1}$ **6.** $a_n = \dfrac{\ln n}{\sqrt{n}}$

7. $\{(1 + 3/n)^{4n}\}$ **8.** $\{(-10)^n/n!\}$

9. A sequence is defined recursively by the equations $a_1 = 1$, $a_{n+1} = \frac{1}{3}(a_n + 4)$. Show that $\{a_n\}$ is increasing and $a_n < 2$ for all n. Deduce that $\{a_n\}$ is convergent and find its limit.

10. Show that $\lim_{n\to\infty} n^4 e^{-n} = 0$ and use a graph to find the smallest value of N that corresponds to $\varepsilon = 0.1$ in the precise definition of a limit.

11–22 Determine whether the series is convergent or divergent.

11. $\displaystyle\sum_{n=1}^{\infty} \frac{n}{n^3 + 1}$ **12.** $\displaystyle\sum_{n=1}^{\infty} \frac{n^2+1}{n^3+1}$

13. $\displaystyle\sum_{n=1}^{\infty} \frac{n^3}{5^n}$ **14.** $\displaystyle\sum_{n=1}^{\infty} \frac{(-1)^n}{\sqrt{n+1}}$

Graphing calculator or computer required

15. $\displaystyle\sum_{n=2}^{\infty} \frac{1}{n\sqrt{\ln n}}$

16. $\displaystyle\sum_{n=1}^{\infty} \ln\left(\frac{n}{3n+1}\right)$

17. $\displaystyle\sum_{n=1}^{\infty} \frac{\cos 3n}{1+(1.2)^n}$

18. $\displaystyle\sum_{n=1}^{\infty} \frac{n^{2n}}{(1+2n^2)^n}$

19. $\displaystyle\sum_{n=1}^{\infty} \frac{1 \cdot 3 \cdot 5 \cdot \cdots \cdot (2n-1)}{5^n n!}$

20. $\displaystyle\sum_{n=1}^{\infty} \frac{(-5)^{2n}}{n^2 9^n}$

21. $\displaystyle\sum_{n=1}^{\infty} (-1)^{n-1} \frac{\sqrt{n}}{n+1}$

22. $\displaystyle\sum_{n=1}^{\infty} \frac{\sqrt{n+1}-\sqrt{n-1}}{n}$

23–26 Determine whether the series is conditionally convergent, absolutely convergent, or divergent.

23. $\displaystyle\sum_{n=1}^{\infty} (-1)^{n-1} n^{-1/3}$

24. $\displaystyle\sum_{n=1}^{\infty} (-1)^{n-1} n^{-3}$

25. $\displaystyle\sum_{n=1}^{\infty} \frac{(-1)^n (n+1)3^n}{2^{2n+1}}$

26. $\displaystyle\sum_{n=2}^{\infty} \frac{(-1)^n \sqrt{n}}{\ln n}$

27–31 Find the sum of the series.

27. $\displaystyle\sum_{n=1}^{\infty} \frac{(-3)^{n-1}}{2^{3n}}$

28. $\displaystyle\sum_{n=1}^{\infty} \frac{1}{n(n+3)}$

29. $\displaystyle\sum_{n=1}^{\infty} [\tan^{-1}(n+1) - \tan^{-1}n]$

30. $\displaystyle\sum_{n=0}^{\infty} \frac{(-1)^n \pi^n}{3^{2n}(2n)!}$

31. $1 - e + \dfrac{e^2}{2!} - \dfrac{e^3}{3!} + \dfrac{e^4}{4!} - \cdots$

32. Express the repeating decimal $4.17326326326\ldots$ as a fraction.

33. Show that $\cosh x \geq 1 + \frac{1}{2}x^2$ for all x.

34. For what values of x does the series $\sum_{n=1}^{\infty} (\ln x)^n$ converge?

35. Find the sum of the series $\displaystyle\sum_{n=1}^{\infty} \frac{(-1)^{n+1}}{n^5}$ correct to four decimal places.

36. (a) Find the partial sum s_5 of the series $\sum_{n=1}^{\infty} 1/n^6$ and estimate the error in using it as an approximation to the sum of the series.
(b) Find the sum of this series correct to five decimal places.

37. Use the sum of the first eight terms to approximate the sum of the series $\sum_{n=1}^{\infty} (2 + 5^n)^{-1}$. Estimate the error involved in this approximation.

38. (a) Show that the series $\displaystyle\sum_{n=1}^{\infty} \frac{n^n}{(2n)!}$ is convergent.

(b) Deduce that $\displaystyle\lim_{n\to\infty} \frac{n^n}{(2n)!} = 0$.

39. Prove that if the series $\sum_{n=1}^{\infty} a_n$ is absolutely convergent, then the series

$$\sum_{n=1}^{\infty} \left(\frac{n+1}{n}\right) a_n$$

is also absolutely convergent.

40–43 Find the radius of convergence and interval of convergence of the series.

40. $\displaystyle\sum_{n=1}^{\infty} (-1)^n \frac{x^n}{n^2 5^n}$

41. $\displaystyle\sum_{n=1}^{\infty} \frac{(x+2)^n}{n 4^n}$

42. $\displaystyle\sum_{n=1}^{\infty} \frac{2^n (x-2)^n}{(n+2)!}$

43. $\displaystyle\sum_{n=0}^{\infty} \frac{2^n (x-3)^n}{\sqrt{n+3}}$

44. Find the radius of convergence of the series

$$\sum_{n=1}^{\infty} \frac{(2n)!}{(n!)^2} x^n$$

45. Find the Taylor series of $f(x) = \sin x$ at $a = \pi/6$.

46. Find the Taylor series of $f(x) = \cos x$ at $a = \pi/3$.

47–54 Find the Maclaurin series for f and its radius of convergence. You may use either the direct method (definition of a Maclaurin series) or known series such as geometric series, binomial series, or the Maclaurin series for e^x, $\sin x$, $\tan^{-1}x$, and $\ln(1+x)$.

47. $f(x) = \dfrac{x^2}{1+x}$

48. $f(x) = \tan^{-1}(x^2)$

49. $f(x) = \ln(4-x)$

50. $f(x) = xe^{2x}$

51. $f(x) = \sin(x^4)$

52. $f(x) = 10^x$

53. $f(x) = 1/\sqrt[4]{16-x}$

54. $f(x) = (1-3x)^{-5}$

55. Evaluate $\displaystyle\int \frac{e^x}{x} \, dx$ as an infinite series.

56. Use series to approximate $\int_0^1 \sqrt{1+x^4} \, dx$ correct to two decimal places.

57–58
(a) Approximate f by a Taylor polynomial with degree n at the number a.
(b) Graph f and T_n on a common screen.
(c) Use Taylor's Inequality to estimate the accuracy of the approximation $f(x) \approx T_n(x)$ when x lies in the given interval.
(d) Check your result in part (c) by graphing $|R_n(x)|$.

57. $f(x) = \sqrt{x}, \quad a = 1, \quad n = 3, \quad 0.9 \leq x \leq 1.1$

58. $f(x) = \sec x, \quad a = 0, \quad n = 2, \quad 0 \leq x \leq \pi/6$

59. Use series to evaluate the following limit.

$$\lim_{x \to 0} \frac{\sin x - x}{x^3}$$

60. The force due to gravity on an object with mass m at a height h above the surface of the earth is

$$F = \frac{mgR^2}{(R + h)^2}$$

where R is the radius of the earth and g is the acceleration due to gravity.
(a) Express F as a series in powers of h/R.
(b) Observe that if we approximate F by the first term in the series, we get the expression $F \approx mg$ that is usually used

when h is much smaller than R. Use the Alternating Series Estimation Theorem to estimate the range of values of h for which the approximation $F \approx mg$ is accurate to within one percent. (Use $R = 6400$ km.)

61. Suppose that $f(x) = \sum_{n=0}^{\infty} c_n x^n$ for all x.
(a) If f is an odd function, show that

$$c_0 = c_2 = c_4 = \cdots = 0$$

(b) If f is an even function, show that

$$c_1 = c_3 = c_5 = \cdots = 0$$

62. If $f(x) = e^{x^2}$, show that $f^{(2n)}(0) = \frac{(2n)!}{n!}$.

Before you look at the solution of the example, cover it up and first try to solve the problem yourself.

EXAMPLE Find the sum of the series $\displaystyle\sum_{n=0}^{\infty} \frac{(x + 2)^n}{(n + 3)!}$.

SOLUTION The problem-solving principle that is relevant here is *recognizing something familiar.* Does the given series look anything like a series that we already know? Well, it does have some ingredients in common with the Maclaurin series for the exponential function:

$$e^x = \sum_{n=0}^{\infty} \frac{x^n}{n!} = 1 + x + \frac{x^2}{2!} + \frac{x^3}{3!} + \cdots$$

We can make this series look more like our given series by replacing x by $x + 2$:

$$e^{x+2} = \sum_{n=0}^{\infty} \frac{(x + 2)^n}{n!} = 1 + (x + 2) + \frac{(x + 2)^2}{2!} + \frac{(x + 2)^3}{3!} + \cdots$$

But here the exponent in the numerator matches the number in the denominator whose factorial is taken. To make that happen in the given series, let's multiply and divide by $(x + 2)^3$:

$$\sum_{n=0}^{\infty} \frac{(x + 2)^n}{(n + 3)!} = \frac{1}{(x + 2)^3} \sum_{n=0}^{\infty} \frac{(x + 2)^{n+3}}{(n + 3)!}$$

$$= (x + 2)^{-3} \left[\frac{(x + 2)^3}{3!} + \frac{(x + 2)^4}{4!} + \cdots \right]$$

We see that the series between brackets is just the series for e^{x+2} with the first three terms missing. So

$$\sum_{n=0}^{\infty} \frac{(x + 2)^n}{(n + 3)!} = (x + 2)^{-3} \left[e^{x+2} - 1 - (x + 2) - \frac{(x + 2)^2}{2!} \right]$$

—

Problems

1. If $f(x) = \sin(x^3)$, find $f^{(15)}(0)$.

2. A function f is defined by

$$f(x) = \lim_{n \to \infty} \frac{x^{2n} - 1}{x^{2n} + 1}$$

   Where is f continuous?

3. (a) Show that $\tan \frac{1}{2}x = \cot \frac{1}{2}x - 2 \cot x$.
   (b) Find the sum of the series

$$\sum_{n=1}^{\infty} \frac{1}{2^n} \tan \frac{x}{2^n}$$

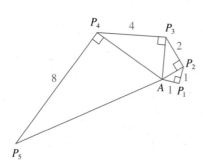

FIGURE FOR PROBLEM 4

4. Let $\{P_n\}$ be a sequence of points determined as in the figure. Thus $|AP_1| = 1$, $|P_n P_{n+1}| = 2^{n-1}$, and angle $AP_n P_{n+1}$ is a right angle. Find $\lim_{n \to \infty} \angle P_n A P_{n+1}$.

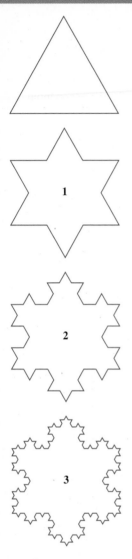

FIGURE FOR PROBLEM 5

5. To construct the **snowflake curve**, start with an equilateral triangle with sides of length 1. Step 1 in the construction is to divide each side into three equal parts, construct an equilateral triangle on the middle part, and then delete the middle part (see the figure). Step 2 is to repeat step 1 for each side of the resulting polygon. This process is repeated at each succeeding step. The snowflake curve is the curve that results from repeating this process indefinitely.

   (a) Let s_n, l_n, and p_n represent the number of sides, the length of a side, and the total length of the nth approximating curve (the curve obtained after step n of the construction), respectively. Find formulas for s_n, l_n, and p_n.

   (b) Show that $p_n \to \infty$ as $n \to \infty$.

   (c) Sum an infinite series to find the area enclosed by the snowflake curve.

   Note: Parts (b) and (c) show that the snowflake curve is infinitely long but encloses only a finite area.

6. Find the sum of the series

$$1 + \frac{1}{2} + \frac{1}{3} + \frac{1}{4} + \frac{1}{6} + \frac{1}{8} + \frac{1}{9} + \frac{1}{12} + \cdots$$

   where the terms are the reciprocals of the positive integers whose only prime factors are 2s and 3s.

7. (a) Show that for $xy \neq -1$,

$$\arctan x - \arctan y = \arctan \frac{x - y}{1 + xy}$$

   if the left side lies between $-\pi/2$ and $\pi/2$.

   (b) Show that $\arctan \frac{120}{119} - \arctan \frac{1}{239} = \pi/4$.

   (c) Deduce the following formula of John Machin (1680–1751):

$$4 \arctan \tfrac{1}{5} - \arctan \tfrac{1}{239} = \frac{\pi}{4}$$

   (d) Use the Maclaurin series for arctan to show that

$$0.1973955597 < \arctan \tfrac{1}{5} < 0.1973955616$$

   (e) Show that

$$0.004184075 < \arctan \tfrac{1}{239} < 0.004184077$$

   (f) Deduce that, correct to seven decimal places, $\pi \approx 3.1415927$.

   Machin used this method in 1706 to find π correct to 100 decimal places. Recently, with the aid of computers, the value of π has been computed to increasingly greater accuracy. In 2009 T. Daisuke and his team computed the value of π to more than two trillion decimal places!

8. (a) Prove a formula similar to the one in Problem 7(a) but involving arccot instead of arctan.

   (b) Find the sum of the series $\sum_{n=0}^{\infty} \operatorname{arccot}(n^2 + n + 1)$.

9. Find the interval of convergence of $\sum_{n=1}^{\infty} n^3 x^n$ and find its sum.

10. If $a_0 + a_1 + a_2 + \cdots + a_k = 0$, show that

$$\lim_{n \to \infty} \left(a_0 \sqrt{n} + a_1 \sqrt{n+1} + a_2 \sqrt{n+2} + \cdots + a_k \sqrt{n+k} \right) = 0$$

   If you don't see how to prove this, try the problem-solving strategy of *using analogy* (see page 77). Try the special cases $k = 1$ and $k = 2$ first. If you can see how to prove the assertion for these cases, then you will probably see how to prove it in general.

11. Find the sum of the series $\sum_{n=2}^{\infty} \ln\left(1 - \frac{1}{n^2} \right)$.

FIGURE FOR PROBLEM 12

12. Suppose you have a large supply of books, all the same size, and you stack them at the edge of a table, with each book extending farther beyond the edge of the table than the one beneath it. Show that it is possible to do this so that the top book extends entirely beyond the table. In fact, show that the top book can extend any distance at all beyond the edge of the table if the stack is high enough. Use the following method of stacking: The top book extends half its length beyond the second book. The second book extends a quarter of its length beyond the third. The third extends one-sixth of its length beyond the fourth, and so on. (Try it yourself with a deck of cards.) Consider centers of mass.

13. If the curve $y = e^{-x/10} \sin x$, $x \geq 0$, is rotated about the x-axis, the resulting solid looks like an infinite decreasing string of beads.
    (a) Find the exact volume of the nth bead. (Use either a table of integrals or a computer algebra system.)
    (b) Find the total volume of the beads.

14. If $p > 1$, evaluate the expression

$$\frac{1 + \dfrac{1}{2^p} + \dfrac{1}{3^p} + \dfrac{1}{4^p} + \cdots}{1 - \dfrac{1}{2^p} + \dfrac{1}{3^p} - \dfrac{1}{4^p} + \cdots}$$

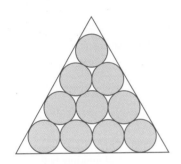

FIGURE FOR PROBLEM 15

15. Suppose that circles of equal diameter are packed tightly in n rows inside an equilateral triangle. (The figure illustrates the case $n = 4$.) If A is the area of the triangle and A_n is the total area occupied by the n rows of circles, show that

$$\lim_{n \to \infty} \frac{A_n}{A} = \frac{\pi}{2\sqrt{3}}$$

16. A sequence $\{a_n\}$ is defined recursively by the equations

$$a_0 = a_1 = 1 \qquad n(n-1)a_n = (n-1)(n-2)a_{n-1} - (n-3)a_{n-2}$$

    Find the sum of the series $\sum_{n=0}^{\infty} a_n$.

17. Taking the value of x^x at 0 to be 1 and integrating a series term by term, show that

$$\int_0^1 x^x \, dx = \sum_{n=1}^{\infty} \frac{(-1)^{n-1}}{n^n}$$

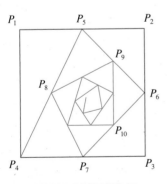

FIGURE FOR PROBLEM 18

18. Starting with the vertices $P_1(0, 1)$, $P_2(1, 1)$, $P_3(1, 0)$, $P_4(0, 0)$ of a square, we construct further points as shown in the figure: P_5 is the midpoint of P_1P_2, P_6 is the midpoint of P_2P_3, P_7 is the midpoint of P_3P_4, and so on. The polygonal spiral path $P_1P_2P_3P_4P_5P_6P_7 \ldots$ approaches a point P inside the square.
    (a) If the coordinates of P_n are (x_n, y_n), show that $\frac{1}{2}x_n + x_{n+1} + x_{n+2} + x_{n+3} = 2$ and find a similar equation for the y-coordinates.
    (b) Find the coordinates of P.

19. Find the sum of the series $\displaystyle\sum_{n=1}^{\infty} \frac{(-1)^n}{(2n+1)3^n}$.

20. Carry out the following steps to show that

$$\frac{1}{1 \cdot 2} + \frac{1}{3 \cdot 4} + \frac{1}{5 \cdot 6} + \frac{1}{7 \cdot 8} + \cdots = \ln 2$$

    (a) Use the formula for the sum of a finite geometric series (11.2.3) to get an expression for

$$1 - x + x^2 - x^3 + \cdots + x^{2n-2} - x^{2n-1}$$

641

(b) Integrate the result of part (a) from 0 to 1 to get an expression for

$$1 - \frac{1}{2} + \frac{1}{3} - \frac{1}{4} + \cdots + \frac{1}{2n-1} - \frac{1}{2n}$$

as an integral.

(c) Deduce from part (b) that

$$\left| \frac{1}{1 \cdot 2} + \frac{1}{3 \cdot 4} + \frac{1}{5 \cdot 6} + \cdots + \frac{1}{(2n-1)(2n)} - \int_0^1 \frac{dx}{1+x} \right| < \int_0^1 x^{2n}\, dx$$

(d) Use part (c) to show that the sum of the given series is ln 2.

21. Find all the solutions of the equation

$$1 + \frac{x}{2!} + \frac{x^2}{4!} + \frac{x^3}{6!} + \frac{x^4}{8!} + \cdots = 0$$

Hint: Consider the cases $x \geqslant 0$ and $x < 0$ separately.

22. Right-angled triangles are constructed as in the figure. Each triangle has height 1 and its base is the hypotenuse of the preceding triangle. Show that this sequence of triangles makes indefinitely many turns around P by showing that $\Sigma\, \theta_n$ is a divergent series.

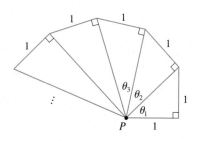

FIGURE FOR PROBLEM 22

23. Consider the series whose terms are the reciprocals of the positive integers that can be written in base 10 notation without using the digit 0. Show that this series is convergent and the sum is less than 90.

24. (a) Show that the Maclaurin series of the function

$$f(x) = \frac{x}{1-x-x^2} \qquad \text{is} \qquad \sum_{n=1}^{\infty} f_n x^n$$

where f_n is the nth Fibonacci number, that is, $f_1 = 1$, $f_2 = 1$, and $f_n = f_{n-1} + f_{n-2}$ for $n \geqslant 3$. [*Hint:* Write $x/(1-x-x^2) = c_0 + c_1 x + c_2 x^2 + \cdots$ and multiply both sides of this equation by $1 - x - x^2$.]

(b) By writing $f(x)$ as a sum of partial fractions and thereby obtaining the Maclaurin series in a different way, find an explicit formula for the nth Fibonacci number.

25. Let

$$u = 1 + \frac{x^3}{3!} + \frac{x^6}{6!} + \frac{x^9}{9!} + \cdots$$

$$v = x + \frac{x^4}{4!} + \frac{x^7}{7!} + \frac{x^{10}}{10!} + \cdots$$

$$w = \frac{x^2}{2!} + \frac{x^5}{5!} + \frac{x^8}{8!} + \cdots$$

Show that $u^3 + v^3 + w^3 - 3uvw = 1$.

26. Prove that if $n > 1$, the nth partial sum of the harmonic series is not an integer.

Hint: Let 2^k be the largest power of 2 that is less than or equal to n and let M be the product of all odd integers that are less than or equal to n. Suppose that $s_n = m$, an integer. Then $M2^k s_n = M2^k m$. The right side of this equation is even. Prove that the left side is odd by showing that each of its terms is an even integer, except for the last one.

Appendixes

A Numbers, Inequalities, and Absolute Values

B Coordinate Geometry and Lines

C Graphs of Second-Degree Equations

D Trigonometry

E Sigma Notation

F Proofs of Theorems

G Graphing Calculators and Computers

H Complex Numbers

I Answers to Odd-Numbered Exercises

A   Numbers, Inequalities, and Absolute Values

Calculus is based on the real number system. We start with the **integers**:

$$\ldots, \quad -3, \quad -2, \quad -1, \quad 0, \quad 1, \quad 2, \quad 3, \quad 4, \quad \ldots$$

Then we construct the **rational numbers**, which are ratios of integers. Thus any rational number r can be expressed as

$$r = \frac{m}{n} \qquad \text{where } m \text{ and } n \text{ are integers and } n \neq 0$$

Examples are

$$\tfrac{1}{2} \qquad -\tfrac{3}{7} \qquad 46 = \tfrac{46}{1} \qquad 0.17 = \tfrac{17}{100}$$

(Recall that division by 0 is always ruled out, so expressions like $\frac{3}{0}$ and $\frac{0}{0}$ are undefined.) Some real numbers, such as $\sqrt{2}$, can't be expressed as a ratio of integers and are therefore called **irrational numbers**. It can be shown, with varying degrees of difficulty, that the following are also irrational numbers:

$$\sqrt{3} \qquad \sqrt{5} \qquad \sqrt[3]{2} \qquad \pi \qquad \sin 1° \qquad \log_{10} 2$$

The set of all real numbers is usually denoted by the symbol $\mathbb{R}$. When we use the word *number* without qualification, we mean "real number."

Every number has a decimal representation. If the number is rational, then the corresponding decimal is repeating. For example,

$$\tfrac{1}{2} = 0.5000\ldots = 0.5\overline{0} \qquad\qquad \tfrac{2}{3} = 0.66666\ldots = 0.\overline{6}$$

$$\tfrac{157}{495} = 0.317171717\ldots = 0.3\overline{17} \qquad \tfrac{9}{7} = 1.285714285714\ldots = 1.\overline{285714}$$

(The bar indicates that the sequence of digits repeats forever.) On the other hand, if the number is irrational, the decimal is nonrepeating:

$$\sqrt{2} = 1.414213562373095\ldots \qquad \pi = 3.141592653589793\ldots$$

If we stop the decimal expansion of any number at a certain place, we get an approximation to the number. For instance, we can write

$$\pi \approx 3.14159265$$

where the symbol $\approx$ is read "is approximately equal to." The more decimal places we retain, the better the approximation we get.

The real numbers can be represented by points on a line as in Figure 1. The positive direction (to the right) is indicated by an arrow. We choose an arbitrary reference point O, called the **origin**, which corresponds to the real number 0. Given any convenient unit of measurement, each positive number x is represented by the point on the line a distance of x units to the right of the origin, and each negative number $-x$ is represented by the point x units to the left of the origin. Thus every real number is represented by a point on the line, and every point P on the line corresponds to exactly one real number. The number associated with the point P is called the **coordinate** of P and the line is then called a **coordinate**

line, or a **real number line**, or simply a **real line**. Often we identify the point with its coordinate and think of a number as being a point on the real line.

FIGURE 1

The real numbers are ordered. We say *a is less than b* and write $a < b$ if $b - a$ is a positive number. Geometrically this means that a lies to the left of b on the number line. (Equivalently, we say *b is greater than a* and write $b > a$.) The symbol $a \le b$ (or $b \ge a$) means that either $a < b$ or $a = b$ and is read "a is less than or equal to b." For instance, the following are true inequalities:

$$7 < 7.4 < 7.5 \qquad -3 > -\pi \qquad \sqrt{2} < 2 \qquad \sqrt{2} \le 2 \qquad 2 \le 2$$

In what follows we need to use *set notation*. A **set** is a collection of objects, and these objects are called the **elements** of the set. If S is a set, the notation $a \in S$ means that a is an element of S, and $a \notin S$ means that a is not an element of S. For example, if Z represents the set of integers, then $-3 \in Z$ but $\pi \notin Z$. If S and T are sets, then their **union** $S \cup T$ is the set consisting of all elements that are in S or T (or in both S and T). The **intersection** of S and T is the set $S \cap T$ consisting of all elements that are in both S and T. In other words, $S \cap T$ is the common part of S and T. The empty set, denoted by $\varnothing$, is the set that contains no element.

Some sets can be described by listing their elements between braces. For instance, the set A consisting of all positive integers less than 7 can be written as

$$A = \{1, 2, 3, 4, 5, 6\}$$

We could also write A in *set-builder notation* as

$$A = \{x \mid x \text{ is an integer and } 0 < x < 7\}$$

which is read "A is the set of x such that x is an integer and $0 < x < 7$."

▰ Intervals

Certain sets of real numbers, called **intervals,** occur frequently in calculus and correspond geometrically to line segments. For example, if $a < b$, the **open interval** from a to b consists of all numbers between a and b and is denoted by the symbol (a, b). Using set-builder notation, we can write

$$(a, b) = \{x \mid a < x < b\}$$

Notice that the endpoints of the interval—namely, a and b—are excluded. This is indicated by the round brackets $(\ )$ and by the open dots in Figure 2. The **closed interval** from a to b is the set

$$[a, b] = \{x \mid a \le x \le b\}$$

Here the endpoints of the interval are included. This is indicated by the square brackets $[\ ]$ and by the solid dots in Figure 3. It is also possible to include only one endpoint in an interval, as shown in Table 1.

FIGURE 2
Open interval (a, b)

FIGURE 3
Closed interval $[a, b]$

Table 1 lists the nine possible types of intervals. When these intervals are discussed, it is always assumed that $a < b$.

Notation	Set description	Picture
(a, b)	$\{x \mid a < x < b\}$	
$[a, b]$	$\{x \mid a \leq x \leq b\}$	
$[a, b)$	$\{x \mid a \leq x < b\}$	
$(a, b]$	$\{x \mid a < x \leq b\}$	
(a, ∞)	$\{x \mid x > a\}$	
$[a, \infty)$	$\{x \mid x \geq a\}$	
$(-\infty, b)$	$\{x \mid x < b\}$	
$(-\infty, b]$	$\{x \mid x \leq b\}$	
$(-\infty, \infty)$	$\mathbb{R}$ (set of all real numbers)	

We also need to consider infinite intervals such as

$$(a, \infty) = \{x \mid x > a\}$$

This does not mean that ∞ ("infinity") is a number. The notation (a, ∞) stands for the set of all numbers that are greater than a, so the symbol ∞ simply indicates that the interval extends indefinitely far in the positive direction.

Inequalities

When working with inequalities, note the following rules.

2 | **Rules for Inequalities**

1. If $a < b$, then $a + c < b + c$.

2. If $a < b$ and $c < d$, then $a + c < b + d$.

3. If $a < b$ and $c > 0$, then $ac < bc$.

4. If $a < b$ and $c < 0$, then $ac > bc$.

5. If $0 < a < b$, then $1/a > 1/b$.

Rule 1 says that we can add any number to both sides of an inequality, and Rule 2 says that two inequalities can be added. However, we have to be careful with multiplication. Rule 3 says that we can multiply both sides of an inequality by a *positive* number, but Rule 4 says that if we multiply both sides of an inequality by a negative number, then we reverse the direction of the inequality. For example, if we take the inequality $3 < 5$ and multiply by 2, we get $6 < 10$, but if we multiply by -2, we get $-6 > -10$. Finally, Rule 5 says that if we take reciprocals, then we reverse the direction of an inequality (provided the numbers are positive).

EXAMPLE 1 Solve the inequality $1 + x < 7x + 5$.

SOLUTION The given inequality is satisfied by some values of x but not by others. To *solve* an inequality means to determine the set of numbers x for which the inequality is true. This is called the *solution set*.

First we subtract 1 from each side of the inequality (using Rule 1 with $c = -1$):

$$x < 7x + 4$$

Then we subtract $7x$ from both sides (Rule 1 with $c = -7x$):

$$-6x < 4$$

Now we divide both sides by -6 $\left(\text{Rule 4 with } c = -\frac{1}{6}\right)$:

$$x > -\frac{4}{6} = -\frac{2}{3}$$

These steps can all be reversed, so the solution set consists of all numbers greater than $-\frac{2}{3}$. In other words, the solution of the inequality is the interval $\left(-\frac{2}{3}, \infty\right)$. ■

EXAMPLE 2 Solve the inequalities $4 \leqslant 3x - 2 < 13$.

SOLUTION Here the solution set consists of all values of x that satisfy both inequalities. Using the rules given in $\boxed{2}$, we see that the following inequalities are equivalent:

$$4 \leqslant 3x - 2 < 13$$

$$6 \leqslant 3x < 15 \qquad \text{(add 2)}$$

$$2 \leqslant x < 5 \qquad \text{(divide by 3)}$$

Therefore the solution set is $[2, 5)$. ■

EXAMPLE 3 Solve the inequality $x^2 - 5x + 6 \leqslant 0$.

SOLUTION First we factor the left side:

$$(x - 2)(x - 3) \leqslant 0$$

We know that the corresponding equation $(x - 2)(x - 3) = 0$ has the solutions 2 and 3. The numbers 2 and 3 divide the real line into three intervals:

$$(-\infty, 2) \qquad (2, 3) \qquad (3, \infty)$$

On each of these intervals we determine the signs of the factors. For instance,

$$x \in (-\infty, 2) \quad \Rightarrow \quad x < 2 \quad \Rightarrow \quad x - 2 < 0$$

Then we record these signs in the following chart:

Interval	$x - 2$	$x - 3$	$(x - 2)(x - 3)$
$x < 2$	$-$	$-$	$+$
$2 < x < 3$	$+$	$-$	$-$
$x > 3$	$+$	$+$	$+$

Another method for obtaining the information in the chart is to use *test values*. For instance, if we use the test value $x = 1$ for the interval $(-\infty, 2)$, then substitution in $x^2 - 5x + 6$ gives

$$1^2 - 5(1) + 6 = 2$$

A visual method for solving Example 3 is to use a graphing device to graph the parabola $y = x^2 - 5x + 6$ (as in Figure 4) and observe that the curve lies on or below the x-axis when $2 \leqslant x \leqslant 3$.

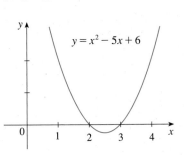

$y = x^2 - 5x + 6$

FIGURE 4

FIGURE 5

The polynomial $x^2 - 5x + 6$ doesn't change sign inside any of the three intervals, so we conclude that it is positive on $(-\infty, 2)$.

Then we read from the chart that $(x - 2)(x - 3)$ is negative when $2 < x < 3$. Thus the solution of the inequality $(x - 2)(x - 3) \leq 0$ is

$$\{x \mid 2 \leq x \leq 3\} = [2, 3]$$

Notice that we have included the endpoints 2 and 3 because we are looking for values of x such that the product is either negative or zero. The solution is illustrated in Figure 5.

EXAMPLE 4 Solve $x^3 + 3x^2 > 4x$.

SOLUTION First we take all nonzero terms to one side of the inequality sign and factor the resulting expression:

$$x^3 + 3x^2 - 4x > 0 \qquad \text{or} \qquad x(x - 1)(x + 4) > 0$$

As in Example 3 we solve the corresponding equation $x(x - 1)(x + 4) = 0$ and use the solutions $x = -4$, $x = 0$, and $x = 1$ to divide the real line into four intervals $(-\infty, -4)$, $(-4, 0)$, $(0, 1)$, and $(1, \infty)$. On each interval the product keeps a constant sign as shown in the following chart:

Interval	x	$x - 1$	$x + 4$	$x(x - 1)(x + 4)$
$x < -4$	$-$	$-$	$-$	$-$
$-4 < x < 0$	$-$	$-$	$+$	$+$
$0 < x < 1$	$+$	$-$	$+$	$-$
$x > 1$	$+$	$+$	$+$	$+$

Then we read from the chart that the solution set is

$$\{x \mid -4 < x < 0 \text{ or } x > 1\} = (-4, 0) \cup (1, \infty)$$

FIGURE 6

The solution is illustrated in Figure 6.

Absolute Value

The **absolute value** of a number a, denoted by $|a|$, is the distance from a to 0 on the real number line. Distances are always positive or 0, so we have

$$|a| \geq 0 \qquad \text{for every number } a$$

For example,

$$|3| = 3 \qquad |-3| = 3 \qquad |0| = 0 \qquad |\sqrt{2} - 1| = \sqrt{2} - 1 \qquad |3 - \pi| = \pi - 3$$

In general, we have

Remember that if a is negative, then $-a$ is positive.

3

$$|a| = a \qquad \text{if } a \geq 0$$

$$|a| = -a \qquad \text{if } a < 0$$

EXAMPLE 5 Express $|3x - 2|$ without using the absolute-value symbol.

SOLUTION

$$|3x - 2| = \begin{cases} 3x - 2 & \text{if } 3x - 2 \geqslant 0 \\ -(3x - 2) & \text{if } 3x - 2 < 0 \end{cases}$$

$$= \begin{cases} 3x - 2 & \text{if } x \geqslant \frac{2}{3} \\ 2 - 3x & \text{if } x < \frac{2}{3} \end{cases}$$

Recall that the symbol $\sqrt{}$ means "the positive square root of." Thus $\sqrt{r} = s$ means $s^2 = r$ and $s \geqslant 0$. Therefore the equation $\sqrt{a^2} = a$ is not always true. It is true only when $a \geqslant 0$. If $a < 0$, then $-a > 0$, so we have $\sqrt{a^2} = -a$. In view of $\boxed{3}$, we then have the equation

$$\boxed{4} \qquad \boxed{\sqrt{a^2} = |a|}$$

which is true for all values of a.

Hints for the proofs of the following properties are given in the exercises.

$\boxed{5}$ **Properties of Absolute Values** Suppose a and b are any real numbers and n is an integer. Then

1. $|ab| = |a||b|$ **2.** $\left|\dfrac{a}{b}\right| = \dfrac{|a|}{|b|}$ $(b \neq 0)$ **3.** $|a^n| = |a|^n$

For solving equations or inequalities involving absolute values, it's often very helpful to use the following statements.

$\boxed{6}$ Suppose $a > 0$. Then

4. $|x| = a$ if and only if $x = \pm a$

5. $|x| < a$ if and only if $-a < x < a$

6. $|x| > a$ if and only if $x > a$ or $x < -a$

For instance, the inequality $|x| < a$ says that the distance from x to the origin is less than a, and you can see from Figure 7 that this is true if and only if x lies between $-a$ and a.

If a and b are any real numbers, then the distance between a and b is the absolute value of the difference, namely, $|a - b|$, which is also equal to $|b - a|$. (See Figure 8.)

FIGURE 7

EXAMPLE 6 Solve $|2x - 5| = 3$.

SOLUTION By Property 4 of $\boxed{6}$, $|2x - 5| = 3$ is equivalent to

$$2x - 5 = 3 \qquad \text{or} \qquad 2x - 5 = -3$$

So $2x = 8$ or $2x = 2$. Thus $x = 4$ or $x = 1$.

FIGURE 8

Length of a line segment $= |a - b|$

EXAMPLE 7 Solve $|x - 5| < 2$.

SOLUTION 1 By Property 5 of $\boxed{6}$, $|x - 5| < 2$ is equivalent to

$$-2 < x - 5 < 2$$

Therefore, adding 5 to each side, we have

$$3 < x < 7$$

and the solution set is the open interval $(3, 7)$.

SOLUTION 2 Geometrically the solution set consists of all numbers x whose distance from 5 is less than 2. From Figure 9 we see that this is the interval $(3, 7)$.

FIGURE 9

EXAMPLE 8 Solve $|3x + 2| \geq 4$.

SOLUTION By Properties 4 and 6 of $\boxed{6}$, $|3x + 2| \geq 4$ is equivalent to

$$3x + 2 \geq 4 \quad \text{or} \quad 3x + 2 \leq -4$$

In the first case $3x \geq 2$, which gives $x \geq \frac{2}{3}$. In the second case $3x \leq -6$, which gives $x \leq -2$. So the solution set is

$$\left\{ x \mid x \leq -2 \ \text{or} \ x \geq \tfrac{2}{3} \right\} = (-\infty, -2] \cup \left[\tfrac{2}{3}, \infty\right)$$

Another important property of absolute value, called the Triangle Inequality, is used frequently not only in calculus but throughout mathematics in general.

> $\boxed{7}$ **The Triangle Inequality** If a and b are any real numbers, then
> $$|a + b| \leq |a| + |b|$$

Observe that if the numbers a and b are both positive or both negative, then the two sides in the Triangle Inequality are actually equal. But if a and b have opposite signs, the left side involves a subtraction and the right side does not. This makes the Triangle Inequality seem reasonable, but we can prove it as follows.

Notice that

$$-|a| \leq a \leq |a|$$

is always true because a equals either $|a|$ or $-|a|$. The corresponding statement for b is

$$-|b| \leq b \leq |b|$$

Adding these inequalities, we get

$$-(|a| + |b|) \leq a + b \leq |a| + |b|$$

If we now apply Properties 4 and 5 $\big($with x replaced by $a + b$ and a by $|a| + |b|\big)$, we obtain

$$|a + b| \leq |a| + |b|$$

which is what we wanted to show.

EXAMPLE 9 If $|x - 4| < 0.1$ and $|y - 7| < 0.2$, use the Triangle Inequality to estimate $|(x + y) - 11|$.

SOLUTION In order to use the given information, we use the Triangle Inequality with $a = x - 4$ and $b = y - 7$:

$$|(x + y) - 11| = |(x - 4) + (y - 7)|$$
$$\leqslant |x - 4| + |y - 7|$$
$$< 0.1 + 0.2 = 0.3$$

Thus

$$|(x + y) - 11| < 0.3$$

A Exercises

Access end-of-section exercises and tools for visualizing calculus online at **www.webassign.net**   ENHANCED Web**Assign**

B Coordinate Geometry and Lines

Just as the points on a line can be identified with real numbers by assigning them coordinates, as described in Appendix A, so the points in a plane can be identified with ordered pairs of real numbers. We start by drawing two perpendicular coordinate lines that intersect at the origin O on each line. Usually one line is horizontal with positive direction to the right and is called the *x*-axis; the other line is vertical with positive direction upward and is called the *y*-axis.

Any point P in the plane can be located by a unique ordered pair of numbers as follows. Draw lines through P perpendicular to the *x*- and *y*-axes. These lines intersect the axes in points with coordinates a and b as shown in Figure 1. Then the point P is assigned the ordered pair (a, b). The first number a is called the **x-coordinate** of P; the second number b is called the **y-coordinate** of P. We say that P is the point with coordinates (a, b), and we denote the point by the symbol $P(a, b)$. Several points are labeled with their coordinates in Figure 2.

FIGURE 1

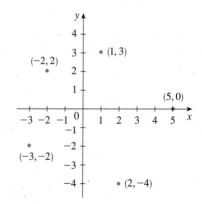

FIGURE 2

By reversing the preceding process we can start with an ordered pair (a, b) and arrive at the corresponding point P. Often we identify the point P with the ordered pair (a, b) and refer to "the point (a, b)." [Although the notation used for an open interval (a, b) is the same as the notation used for a point (a, b), you will be able to tell from the context which meaning is intended.]

This coordinate system is called the **rectangular coordinate system** or the **Cartesian coordinate system** in honor of the French mathematician René Descartes (1596–1650), even though another Frenchman, Pierre Fermat (1601–1665), invented the principles of analytic geometry at about the same time as Descartes. The plane supplied with this coordinate system is called the **coordinate plane** or the **Cartesian plane** and is denoted by $\mathbb{R}^2$.

The x- and y-axes are called the **coordinate axes** and divide the Cartesian plane into four quadrants, which are labeled I, II, III, and IV in Figure 1. Notice that the first quadrant consists of those points whose x- and y-coordinates are both positive.

EXAMPLE 1 Describe and sketch the regions given by the following sets.

(a) $\{(x, y) \mid x \geqslant 0\}$ (b) $\{(x, y) \mid y = 1\}$ (c) $\{(x, y) \mid |y| < 1\}$

SOLUTION

(a) The points whose x-coordinates are 0 or positive lie on the y-axis or to the right of it as indicated by the shaded region in Figure 3(a).

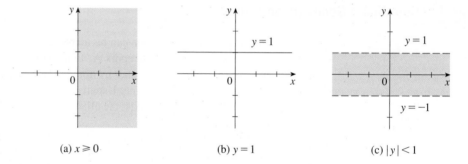

FIGURE 3   (a) $x \geqslant 0$ (b) $y = 1$ (c) $|y| < 1$

(b) The set of all points with y-coordinate 1 is a horizontal line one unit above the x-axis [see Figure 3(b)].

(c) Recall from Appendix A that

$$|y| < 1 \qquad \text{if and only if} \qquad -1 < y < 1$$

The given region consists of those points in the plane whose y-coordinates lie between -1 and 1. Thus the region consists of all points that lie between (but not on) the horizontal lines $y = 1$ and $y = -1$. [These lines are shown as dashed lines in Figure 3(c) to indicate that the points on these lines don't lie in the set.]

Recall from Appendix A that the distance between points a and b on a number line is $|a - b| = |b - a|$. Thus the distance between points $P_1(x_1, y_1)$ and $P_3(x_2, y_1)$ on a horizontal line must be $|x_2 - x_1|$ and the distance between $P_2(x_2, y_2)$ and $P_3(x_2, y_1)$ on a vertical line must be $|y_2 - y_1|$. (See Figure 4.)

To find the distance $|P_1P_2|$ between any two points $P_1(x_1, y_1)$ and $P_2(x_2, y_2)$, we note that triangle $P_1P_2P_3$ in Figure 4 is a right triangle, and so by the Pythagorean Theorem we have

$$|P_1P_2| = \sqrt{|P_1P_3|^2 + |P_2P_3|^2} = \sqrt{|x_2 - x_1|^2 + |y_2 - y_1|^2}$$

$$= \sqrt{(x_2 - x_1)^2 + (y_2 - y_1)^2}$$

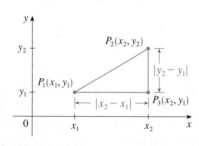

FIGURE 4

> **1** **Distance Formula** The distance between the points $P_1(x_1, y_1)$ and $P_2(x_2, y_2)$ is
>
> $$|P_1P_2| = \sqrt{(x_2 - x_1)^2 + (y_2 - y_1)^2}$$

EXAMPLE 2 The distance between $(1, -2)$ and $(5, 3)$ is

$$\sqrt{(5 - 1)^2 + [3 - (-2)]^2} = \sqrt{4^2 + 5^2} = \sqrt{41}$$

Lines

We want to find an equation of a given line L; such an equation is satisfied by the coordinates of the points on L and by no other point. To find the equation of L we use its *slope,* which is a measure of the steepness of the line.

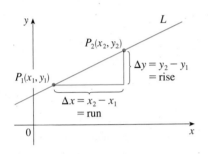

FIGURE 5

> **2** **Definition** The **slope** of a nonvertical line that passes through the points $P_1(x_1, y_1)$ and $P_2(x_2, y_2)$ is
>
> $$m = \frac{\Delta y}{\Delta x} = \frac{y_2 - y_1}{x_2 - x_1}$$
>
> The slope of a vertical line is not defined.

Thus the slope of a line is the ratio of the change in y, Δy, to the change in x, Δx. (See Figure 5.) The slope is therefore the rate of change of y with respect to x. The fact that the line is straight means that the rate of change is constant.

Figure 6 shows several lines labeled with their slopes. Notice that lines with positive slope slant upward to the right, whereas lines with negative slope slant downward to the right. Notice also that the steepest lines are the ones for which the absolute value of the slope is largest, and a horizontal line has slope 0.

Now let's find an equation of the line that passes through a given point $P_1(x_1, y_1)$ and has slope m. A point $P(x, y)$ with $x \neq x_1$ lies on this line if and only if the slope of the line through P_1 and P is equal to m; that is,

$$\frac{y - y_1}{x - x_1} = m$$

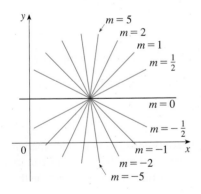

FIGURE 6

This equation can be rewritten in the form

$$y - y_1 = m(x - x_1)$$

and we observe that this equation is also satisfied when $x = x_1$ and $y = y_1$. Therefore it is an equation of the given line.

> **3** **Point-Slope Form of the Equation of a Line** An equation of the line passing through the point $P_1(x_1, y_1)$ and having slope m is
>
> $$y - y_1 = m(x - x_1)$$

EXAMPLE 3   Find an equation of the line through $(1, -7)$ with slope $-\frac{1}{2}$.

SOLUTION   Using $\boxed{3}$ with $m = -\frac{1}{2}$, $x_1 = 1$, and $y_1 = -7$, we obtain an equation of the line as

$$y + 7 = -\tfrac{1}{2}(x - 1)$$

which we can rewrite as

$$2y + 14 = -x + 1 \quad \text{or} \quad x + 2y + 13 = 0$$

EXAMPLE 4   Find an equation of the line through the points $(-1, 2)$ and $(3, -4)$.

SOLUTION   By Definition 2 the slope of the line is

$$m = \frac{-4 - 2}{3 - (-1)} = -\frac{3}{2}$$

Using the point-slope form with $x_1 = -1$ and $y_1 = 2$, we obtain

$$y - 2 = -\tfrac{3}{2}(x + 1)$$

which simplifies to $\qquad\qquad 3x + 2y = 1$

Suppose a nonvertical line has slope m and y-intercept b. (See Figure 7.) This means it intersects the y-axis at the point $(0, b)$, so the point-slope form of the equation of the line, with $x_1 = 0$ and $y_1 = b$, becomes

$$y - b = m(x - 0)$$

This simplifies as follows.

FIGURE 7

$\boxed{4}$ **Slope-Intercept Form of the Equation of a Line**   An equation of the line with slope m and y-intercept b is

$$y = mx + b$$

In particular, if a line is horizontal, its slope is $m = 0$, so its equation is $y = b$, where b is the y-intercept (see Figure 8). A vertical line does not have a slope, but we can write its equation as $x = a$, where a is the x-intercept, because the x-coordinate of every point on the line is a.

Observe that the equation of every line can be written in the form

FIGURE 8

$\boxed{5}$ $\qquad\qquad\qquad$ $Ax + By + C = 0$

because a vertical line has the equation $x = a$ or $x - a = 0$ ($A = 1$, $B = 0$, $C = -a$) and a nonvertical line has the equation $y = mx + b$ or $-mx + y - b = 0$ ($A = -m$, $B = 1$, $C = -b$). Conversely, if we start with a general first-degree equation, that is, an equation of the form $\boxed{5}$, where A, B, and C are constants and A and B are not both 0, then we can show that it is the equation of a line. If $B = 0$, the equation becomes $Ax + C = 0$ or $x = -C/A$, which represents a vertical line with x-intercept $-C/A$. If $B \neq 0$, the equation can be rewritten by solving for y:

$$y = -\frac{A}{B}x - \frac{C}{B}$$

and we recognize this as being the slope-intercept form of the equation of a line ($m = -A/B$, $b = -C/B$). Therefore an equation of the form $\boxed{5}$ is called a **linear equation** or the **general equation of a line**. For brevity, we often refer to "the line $Ax + By + C = 0$" instead of "the line whose equation is $Ax + By + C = 0$."

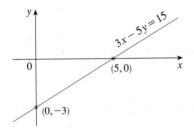

FIGURE 9

EXAMPLE 5 Sketch the graph of the equation $3x - 5y = 15$.

SOLUTION Since the equation is linear, its graph is a line. To draw the graph, we can simply find two points on the line. It's easiest to find the intercepts. Substituting $y = 0$ (the equation of the x-axis) in the given equation, we get $3x = 15$, so $x = 5$ is the x-intercept. Substituting $x = 0$ in the equation, we see that the y-intercept is -3. This allows us to sketch the graph as in Figure 9.

EXAMPLE 6 Graph the inequality $x + 2y > 5$.

SOLUTION We are asked to sketch the graph of the set $\{(x, y) \mid x + 2y > 5\}$ and we begin by solving the inequality for y:

$$x + 2y > 5$$
$$2y > -x + 5$$
$$y > -\tfrac{1}{2}x + \tfrac{5}{2}$$

FIGURE 10

Compare this inequality with the equation $y = -\frac{1}{2}x + \frac{5}{2}$, which represents a line with slope $-\frac{1}{2}$ and y-intercept $\frac{5}{2}$. We see that the given graph consists of points whose y-coordinates are *larger* than those on the line $y = -\frac{1}{2}x + \frac{5}{2}$. Thus the graph is the region that lies *above* the line, as illustrated in Figure 10.

Parallel and Perpendicular Lines

Slopes can be used to show that lines are parallel or perpendicular. The following facts are proved, for instance, in *Precalculus: Mathematics for Calculus, Sixth Edition* by Stewart, Redlin, and Watson (Belmont, CA, 2012).

$\boxed{6}$ **Parallel and Perpendicular Lines**

1. Two nonvertical lines are parallel if and only if they have the same slope.

2. Two lines with slopes m_1 and m_2 are perpendicular if and only if $m_1 m_2 = -1$; that is, their slopes are negative reciprocals:

$$m_2 = -\frac{1}{m_1}$$

EXAMPLE 7 Find an equation of the line through the point $(5, 2)$ that is parallel to the line $4x + 6y + 5 = 0$.

SOLUTION The given line can be written in the form

$$y = -\tfrac{2}{3}x - \tfrac{5}{6}$$

which is in slope-intercept form with $m = -\tfrac{2}{3}$. Parallel lines have the same slope, so the required line has slope $-\tfrac{2}{3}$ and its equation in point-slope form is

$$y - 2 = -\tfrac{2}{3}(x - 5)$$

We can write this equation as $2x + 3y = 16$.

EXAMPLE 8 Show that the lines $2x + 3y = 1$ and $6x - 4y - 1 = 0$ are perpendicular.

SOLUTION The equations can be written as

$$y = -\tfrac{2}{3}x + \tfrac{1}{3} \qquad \text{and} \qquad y = \tfrac{3}{2}x - \tfrac{1}{4}$$

from which we see that the slopes are

$$m_1 = -\tfrac{2}{3} \qquad \text{and} \qquad m_2 = \tfrac{3}{2}$$

Since $m_1 m_2 = -1$, the lines are perpendicular.

B Exercises

Access end-of-section exercises and tools for visualizing calculus online at **www.webassign.net**

C Graphs of Second-Degree Equations

In Appendix B we saw that a first-degree, or linear, equation $Ax + By + C = 0$ represents a line. In this section we discuss second-degree equations such as

$$x^2 + y^2 = 1 \qquad y = x^2 + 1 \qquad \frac{x^2}{9} + \frac{y^2}{4} = 1 \qquad x^2 - y^2 = 1$$

which represent a circle, a parabola, an ellipse, and a hyperbola, respectively.

The graph of such an equation in x and y is the set of all points (x, y) that satisfy the equation; it gives a visual representation of the equation. Conversely, given a curve in the xy-plane, we may have to find an equation that represents it, that is, an equation satisfied by the coordinates of the points on the curve and by no other point. This is the other half of the basic principle of analytic geometry as formulated by Descartes and Fermat. The idea is that if a geometric curve can be represented by an algebraic equation, then the rules of algebra can be used to analyze the geometric problem.

Circles

As an example of this type of problem, let's find an equation of the circle with radius r and center (h, k). By definition, the circle is the set of all points $P(x, y)$ whose distance from the center $C(h, k)$ is r. (See Figure 1.) Thus P is on the circle if and only if $|PC| = r$. From the distance formula, we have

$$\sqrt{(x - h)^2 + (y - k)^2} = r$$

or equivalently, squaring both sides, we get

$$(x - h)^2 + (y - k)^2 = r^2$$

This is the desired equation.

FIGURE 1

> **1 Equation of a Circle** An equation of the circle with center (h, k) and radius r is
>
> $$(x - h)^2 + (y - k)^2 = r^2$$
>
> In particular, if the center is the origin $(0, 0)$, the equation is
>
> $$x^2 + y^2 = r^2$$

EXAMPLE 1 Find an equation of the circle with radius 3 and center $(2, -5)$.

SOLUTION From Equation 1 with $r = 3$, $h = 2$, and $k = -5$, we obtain

$$(x - 2)^2 + (y + 5)^2 = 9$$

EXAMPLE 2 Sketch the graph of the equation $x^2 + y^2 + 2x - 6y + 7 = 0$ by first showing that it represents a circle and then finding its center and radius.

SOLUTION We first group the x-terms and y-terms as follows:

$$(x^2 + 2x) + (y^2 - 6y) = -7$$

Then we complete the square within each grouping, adding the appropriate constants (the squares of half the coefficients of x and y) to both sides of the equation:

$$(x^2 + 2x + 1) + (y^2 - 6y + 9) = -7 + 1 + 9$$

or
$$(x + 1)^2 + (y - 3)^2 = 3$$

Comparing this equation with the standard equation of a circle 1, we see that $h = -1$, $k = 3$, and $r = \sqrt{3}$, so the given equation represents a circle with center $(-1, 3)$ and radius $\sqrt{3}$. It is sketched in Figure 2.

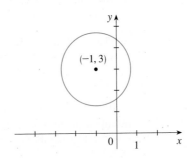

FIGURE 2
$x^2 + y^2 + 2x - 6y + 7 = 0$

■ Parabolas

The geometric properties of parabolas are reviewed in Section 10.5. Here we regard a parabola as a graph of an equation of the form $y = ax^2 + bx + c$.

EXAMPLE 3 Draw the graph of the parabola $y = x^2$.

SOLUTION We set up a table of values, plot points, and join them by a smooth curve to obtain the graph in Figure 3.

x	$y = x^2$
0	0
$\pm\frac{1}{2}$	$\frac{1}{4}$
± 1	1
± 2	4
± 3	9

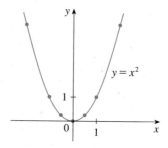

FIGURE 3

Figure 4 shows the graphs of several parabolas with equations of the form $y = ax^2$ for various values of the number a. In each case the *vertex*, the point where the parabola changes direction, is the origin. We see that the parabola $y = ax^2$ opens upward if $a > 0$ and downward if $a < 0$ (as in Figure 5).

FIGURE 4

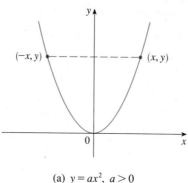

(a) $y = ax^2$, $a > 0$

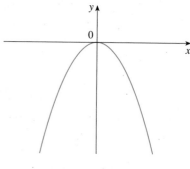

(b) $y = ax^2$, $a < 0$

FIGURE 5

Notice that if (x, y) satisfies $y = ax^2$, then so does $(-x, y)$. This corresponds to the geometric fact that if the right half of the graph is reflected about the y-axis, then the left half of the graph is obtained. We say that the graph is **symmetric with respect to the y-axis**.

> The graph of an equation is symmetric with respect to the y-axis if the equation is unchanged when x is replaced by $-x$.

If we interchange x and y in the equation $y = ax^2$, the result is $x = ay^2$, which also represents a parabola. (Interchanging x and y amounts to reflecting about the diagonal line $y = x$.) The parabola $x = ay^2$ opens to the right if $a > 0$ and to the left if $a < 0$. (See

Figure 6.) This time the parabola is symmetric with respect to the x-axis because if (x, y) satisfies $x = ay^2$, then so does $(x, -y)$.

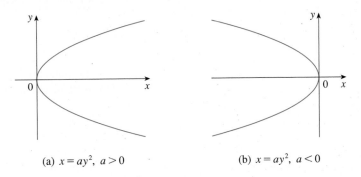

FIGURE 6

(a) $x = ay^2$, $a > 0$

(b) $x = ay^2$, $a < 0$

> The graph of an equation is symmetric with respect to the x-axis if the equation is unchanged when y is replaced by $-y$.

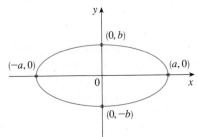

FIGURE 7

EXAMPLE 4 Sketch the region bounded by the parabola $x = y^2$ and the line $y = x - 2$.

SOLUTION First we find the points of intersection by solving the two equations. Substituting $x = y + 2$ into the equation $x = y^2$, we get $y + 2 = y^2$, which gives

$$0 = y^2 - y - 2 = (y - 2)(y + 1)$$

so $y = 2$ or -1. Thus the points of intersection are $(4, 2)$ and $(1, -1)$, and we draw the line $y = x - 2$ passing through these points. We then sketch the parabola $x = y^2$ by referring to Figure 6(a) and having the parabola pass through $(4, 2)$ and $(1, -1)$. The region bounded by $x = y^2$ and $y = x - 2$ means the finite region whose boundaries are these curves. It is sketched in Figure 7.

Ellipses

The curve with equation

2
$$\frac{x^2}{a^2} + \frac{y^2}{b^2} = 1$$

where a and b are positive numbers, is called an **ellipse** in standard position. (Geometric properties of ellipses are discussed in Section 10.5.) Observe that Equation 2 is unchanged if x is replaced by $-x$ or y is replaced by $-y$, so the ellipse is symmetric with respect to both axes. As a further aid to sketching the ellipse, we find its intercepts.

> The **x-intercepts** of a graph are the x-coordinates of the points where the graph intersects the x-axis. They are found by setting $y = 0$ in the equation of the graph.
>
> The **y-intercepts** are the y-coordinates of the points where the graph intersects the y-axis. They are found by setting $x = 0$ in its equation.

FIGURE 8
$$\frac{x^2}{a^2} + \frac{y^2}{b^2} = 1$$

If we set $y = 0$ in Equation 2, we get $x^2 = a^2$ and so the x-intercepts are $\pm a$. Setting $x = 0$, we get $y^2 = b^2$, so the y-intercepts are $\pm b$. Using this information, together with symmetry, we sketch the ellipse in Figure 8. If $a = b$, the ellipse is a circle with radius a.

EXAMPLE 5 Sketch the graph of $9x^2 + 16y^2 = 144$.

SOLUTION We divide both sides of the equation by 144:

$$\frac{x^2}{16} + \frac{y^2}{9} = 1$$

The equation is now in the standard form for an ellipse $\boxed{2}$, so we have $a^2 = 16$, $b^2 = 9$, $a = 4$, and $b = 3$. The x-intercepts are ± 4; the y-intercepts are ± 3. The graph is sketched in Figure 9.

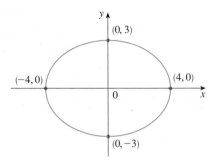

FIGURE 9

$9x^2 + 16y^2 = 144$

Hyperbolas

The curve with equation

$$\boxed{3} \qquad \frac{x^2}{a^2} - \frac{y^2}{b^2} = 1$$

is called a **hyperbola** in standard position. Again, Equation 3 is unchanged when x is replaced by $-x$ or y is replaced by $-y$, so the hyperbola is symmetric with respect to both axes. To find the x-intercepts we set $y = 0$ and obtain $x^2 = a^2$ and $x = \pm a$. However, if we put $x = 0$ in Equation 3, we get $y^2 = -b^2$, which is impossible, so there is no y-intercept. In fact, from Equation 3 we obtain

$$\frac{x^2}{a^2} = 1 + \frac{y^2}{b^2} \geq 1$$

which shows that $x^2 \geq a^2$ and so $|x| = \sqrt{x^2} \geq a$. Therefore we have $x \geq a$ or $x \leq -a$. This means that the hyperbola consists of two parts, called its *branches*. It is sketched in Figure 10.

In drawing a hyperbola it is useful to draw first its *asymptotes,* which are the lines $y = (b/a)x$ and $y = -(b/a)x$ shown in Figure 10. Both branches of the hyperbola approach the asymptotes; that is, they come arbitrarily close to the asymptotes. This involves the idea of a limit, which is discussed in Chapter 1. (See also Exercise 57 in Section 3.5.)

By interchanging the roles of x and y we get an equation of the form

$$\frac{y^2}{a^2} - \frac{x^2}{b^2} = 1$$

which also represents a hyperbola and is sketched in Figure 11.

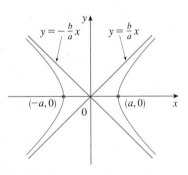

FIGURE 10

The hyperbola $\dfrac{x^2}{a^2} - \dfrac{y^2}{b^2} = 1$

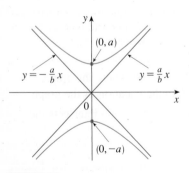

FIGURE 11

The hyperbola $\dfrac{y^2}{a^2} - \dfrac{x^2}{b^2} = 1$

EXAMPLE 6   Sketch the curve $9x^2 - 4y^2 = 36$.

SOLUTION   Dividing both sides by 36, we obtain

$$\frac{x^2}{4} - \frac{y^2}{9} = 1$$

which is the standard form of the equation of a hyperbola (Equation 3). Since $a^2 = 4$, the x-intercepts are ± 2. Since $b^2 = 9$, we have $b = 3$ and the asymptotes are $y = \pm\left(\frac{3}{2}\right)x$. The hyperbola is sketched in Figure 12.

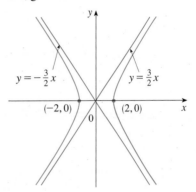

FIGURE 12
The hyperbola $9x^2 - 4y^2 = 36$

If $b = a$, a hyperbola has the equation $x^2 - y^2 = a^2$ (or $y^2 - x^2 = a^2$) and is called an *equilateral hyperbola* [see Figure 13(a)]. Its asymptotes are $y = \pm x$, which are perpendicular. If an equilateral hyperbola is rotated by $45°$, the asymptotes become the x- and y-axes, and it can be shown that the new equation of the hyperbola is $xy = k$, where k is a constant [see Figure 13(b)].

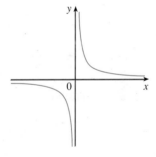

FIGURE 13
Equilateral hyperbolas

(a) $x^2 - y^2 = a^2$

(b) $xy = k$ $(k > 0)$

Shifted Conics

Recall that an equation of the circle with center the origin and radius r is $x^2 + y^2 = r^2$, but if the center is the point (h, k), then the equation of the circle becomes

$$(x - h)^2 + (y - k)^2 = r^2$$

Similarly, if we take the ellipse with equation

4

$$\frac{x^2}{a^2} + \frac{y^2}{b^2} = 1$$

and translate it (shift it) so that its center is the point (h, k), then its equation becomes

$$\boxed{5} \qquad \frac{(x - h)^2}{a^2} + \frac{(y - k)^2}{b^2} = 1$$

(See Figure 14.)

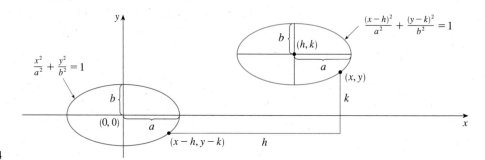

FIGURE 14

Notice that in shifting the ellipse, we replaced x by $x - h$ and y by $y - k$ in Equation 4 to obtain Equation 5. We use the same procedure to shift the parabola $y = ax^2$ so that its vertex (the origin) becomes the point (h, k) as in Figure 15. Replacing x by $x - h$ and y by $y - k$, we see that the new equation is

$$y - k = a(x - h)^2 \qquad \text{or} \qquad y = a(x - h)^2 + k$$

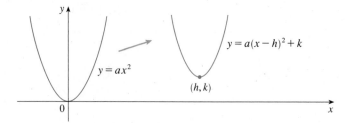

FIGURE 15

EXAMPLE 7 Sketch the graph of the equation $y = 2x^2 - 4x + 1$.

SOLUTION First we complete the square:

$$y = 2(x^2 - 2x) + 1 = 2(x - 1)^2 - 1$$

In this form we see that the equation represents the parabola obtained by shifting $y = 2x^2$ so that its vertex is at the point $(1, -1)$. The graph is sketched in Figure 16.

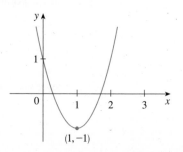

FIGURE 16
$y = 2x^2 - 4x + 1$

EXAMPLE 8   Sketch the curve $x = 1 - y^2$.

SOLUTION   This time we start with the parabola $x = -y^2$ (as in Figure 6 with $a = -1$) and shift one unit to the right to get the graph of $x = 1 - y^2$. (See Figure 17.)

FIGURE 17

(a) $x = -y^2$

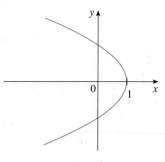

(b) $x = 1 - y^2$

C | Exercises

Access end-of-section exercises and tools for visualizing calculus online at **www.webassign.net**   ENHANCED   **WebAssign**

D | Trigonometry

Angles

Angles can be measured in degrees or in radians (abbreviated as rad). The angle given by a complete revolution contains 360°, which is the same as 2π rad. Therefore

$$\boxed{1} \qquad \pi \text{ rad} = 180°$$

and

$$\boxed{2} \qquad 1 \text{ rad} = \left(\frac{180}{\pi}\right)^{\circ} \approx 57.3° \qquad 1° = \frac{\pi}{180} \text{ rad} \approx 0.017 \text{ rad}$$

EXAMPLE 1
(a) Find the radian measure of 60°.   (b) Express $5\pi/4$ rad in degrees.

SOLUTION
(a) From Equation 1 or 2 we see that to convert from degrees to radians we multiply by $\pi/180$. Therefore

$$60° = 60\left(\frac{\pi}{180}\right) = \frac{\pi}{3} \text{ rad}$$

(b) To convert from radians to degrees we multiply by $180/\pi$. Thus

$$\frac{5\pi}{4} \text{ rad} = \frac{5\pi}{4}\left(\frac{180}{\pi}\right) = 225°$$

In calculus we use radians to measure angles except when otherwise indicated. The following table gives the correspondence between degree and radian measures of some common angles.

Degrees	0°	30°	45°	60°	90°	120°	135°	150°	180°	270°	360°
Radians	0	$\dfrac{\pi}{6}$	$\dfrac{\pi}{4}$	$\dfrac{\pi}{3}$	$\dfrac{\pi}{2}$	$\dfrac{2\pi}{3}$	$\dfrac{3\pi}{4}$	$\dfrac{5\pi}{6}$	π	$\dfrac{3\pi}{2}$	2π

Figure 1 shows a sector of a circle with central angle θ and radius r subtending an arc with length a. Since the length of the arc is proportional to the size of the angle, and since the entire circle has circumference $2\pi r$ and central angle 2π, we have

$$\frac{\theta}{2\pi} = \frac{a}{2\pi r}$$

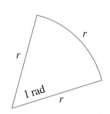

FIGURE 1

Solving this equation for θ and for a, we obtain

$$\boxed{3} \qquad \boxed{\theta = \frac{a}{r}} \qquad\qquad \boxed{a = r\theta}$$

Remember that Equations 3 are valid only when θ is measured in radians.

In particular, putting $a = r$ in Equation 3, we see that an angle of 1 rad is the angle subtended at the center of a circle by an arc equal in length to the radius of the circle (see Figure 2).

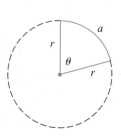

FIGURE 2

EXAMPLE 2
(a) If the radius of a circle is 5 cm, what angle is subtended by an arc of 6 cm?
(b) If a circle has radius 3 cm, what is the length of an arc subtended by a central angle of $3\pi/8$ rad?

SOLUTION
(a) Using Equation 3 with $a = 6$ and $r = 5$, we see that the angle is

$$\theta = \tfrac{6}{5} = 1.2 \text{ rad}$$

(b) With $r = 3$ cm and $\theta = 3\pi/8$ rad, the arc length is

$$a = r\theta = 3\left(\frac{3\pi}{8}\right) = \frac{9\pi}{8} \text{ cm}$$

The **standard position** of an angle occurs when we place its vertex at the origin of a coordinate system and its initial side on the positive x-axis as in Figure 3. A **positive** angle is obtained by rotating the initial side counterclockwise until it coincides with the terminal side. Likewise, **negative** angles are obtained by clockwise rotation as in Figure 4.

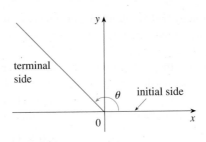

FIGURE 3 $\theta \geqslant 0$

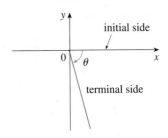

FIGURE 4 $\theta < 0$

Figure 5 shows several examples of angles in standard position. Notice that different angles can have the same terminal side. For instance, the angles $3\pi/4$, $-5\pi/4$, and $11\pi/4$ have the same initial and terminal sides because

$$\frac{3\pi}{4} - 2\pi = -\frac{5\pi}{4} \qquad \frac{3\pi}{4} + 2\pi = \frac{11\pi}{4}$$

and 2π rad represents a complete revolution.

FIGURE 5

Angles in standard position

◼ The Trigonometric Functions

For an acute angle θ the six trigonometric functions are defined as ratios of lengths of sides of a right triangle as follows (see Figure 6).

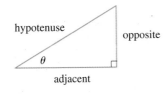

FIGURE 6

$$\boxed{4} \qquad \sin\theta = \frac{\text{opp}}{\text{hyp}} \qquad \csc\theta = \frac{\text{hyp}}{\text{opp}}$$

$$\cos\theta = \frac{\text{adj}}{\text{hyp}} \qquad \sec\theta = \frac{\text{hyp}}{\text{adj}}$$

$$\tan\theta = \frac{\text{opp}}{\text{adj}} \qquad \cot\theta = \frac{\text{adj}}{\text{opp}}$$

This definition doesn't apply to obtuse or negative angles, so for a general angle θ in standard position we let $P(x, y)$ be any point on the terminal side of θ and we let r be the distance $|OP|$ as in Figure 7. Then we define

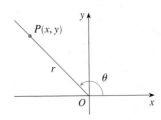

FIGURE 7

$$\boxed{5} \qquad \sin\theta = \frac{y}{r} \qquad \csc\theta = \frac{r}{y}$$

$$\cos\theta = \frac{x}{r} \qquad \sec\theta = \frac{r}{x}$$

$$\tan\theta = \frac{y}{x} \qquad \cot\theta = \frac{x}{y}$$

If we put $r = 1$ in Definition 5 and draw a unit circle with center the origin and label θ as in Figure 8, then the coordinates of P are $(\cos \theta, \sin \theta)$.

FIGURE 8

FIGURE 9

FIGURE 10

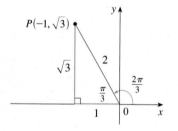

FIGURE 11

Since division by 0 is not defined, $\tan \theta$ and $\sec \theta$ are undefined when $x = 0$ and $\csc \theta$ and $\cot \theta$ are undefined when $y = 0$. Notice that the definitions in $\boxed{4}$ and $\boxed{5}$ are consistent when θ is an acute angle.

If θ is a number, the convention is that $\sin \theta$ means the sine of the angle whose *radian* measure is θ. For example, the expression $\sin 3$ implies that we are dealing with an angle of 3 rad. When finding a calculator approximation to this number, we must remember to set our calculator in radian mode, and then we obtain

$$\sin 3 \approx 0.14112$$

If we want to know the sine of the angle 3° we would write $\sin 3°$ and, with our calculator in degree mode, we find that

$$\sin 3° \approx 0.05234$$

The exact trigonometric ratios for certain angles can be read from the triangles in Figure 9. For instance,

$$\sin \frac{\pi}{4} = \frac{1}{\sqrt{2}} \qquad \sin \frac{\pi}{6} = \frac{1}{2} \qquad \sin \frac{\pi}{3} = \frac{\sqrt{3}}{2}$$

$$\cos \frac{\pi}{4} = \frac{1}{\sqrt{2}} \qquad \cos \frac{\pi}{6} = \frac{\sqrt{3}}{2} \qquad \cos \frac{\pi}{3} = \frac{1}{2}$$

$$\tan \frac{\pi}{4} = 1 \qquad \tan \frac{\pi}{6} = \frac{1}{\sqrt{3}} \qquad \tan \frac{\pi}{3} = \sqrt{3}$$

The signs of the trigonometric functions for angles in each of the four quadrants can be remembered by means of the rule "All Students Take Calculus" shown in Figure 10.

EXAMPLE 3 Find the exact trigonometric ratios for $\theta = 2\pi/3$.

SOLUTION From Figure 11 we see that a point on the terminal line for $\theta = 2\pi/3$ is $P(-1, \sqrt{3})$. Therefore, taking

$$x = -1 \qquad y = \sqrt{3} \qquad r = 2$$

in the definitions of the trigonometric ratios, we have

$$\sin \frac{2\pi}{3} = \frac{\sqrt{3}}{2} \qquad \cos \frac{2\pi}{3} = -\frac{1}{2} \qquad \tan \frac{2\pi}{3} = -\sqrt{3}$$

$$\csc \frac{2\pi}{3} = \frac{2}{\sqrt{3}} \qquad \sec \frac{2\pi}{3} = -2 \qquad \cot \frac{2\pi}{3} = -\frac{1}{\sqrt{3}} \qquad \blacksquare$$

The following table gives some values of $\sin \theta$ and $\cos \theta$ found by the method of Example 3.

θ	0	$\dfrac{\pi}{6}$	$\dfrac{\pi}{4}$	$\dfrac{\pi}{3}$	$\dfrac{\pi}{2}$	$\dfrac{2\pi}{3}$	$\dfrac{3\pi}{4}$	$\dfrac{5\pi}{6}$	π	$\dfrac{3\pi}{2}$	2π
$\sin \theta$	0	$\dfrac{1}{2}$	$\dfrac{1}{\sqrt{2}}$	$\dfrac{\sqrt{3}}{2}$	1	$\dfrac{\sqrt{3}}{2}$	$\dfrac{1}{\sqrt{2}}$	$\dfrac{1}{2}$	0	-1	0
$\cos \theta$	1	$\dfrac{\sqrt{3}}{2}$	$\dfrac{1}{\sqrt{2}}$	$\dfrac{1}{2}$	0	$-\dfrac{1}{2}$	$-\dfrac{1}{\sqrt{2}}$	$-\dfrac{\sqrt{3}}{2}$	-1	0	1

FIGURE 12

FIGURE 13

EXAMPLE 4 If $\cos \theta = \frac{2}{5}$ and $0 < \theta < \pi/2$, find the other five trigonometric functions of θ.

SOLUTION Since $\cos \theta = \frac{2}{5}$, we can label the hypotenuse as having length 5 and the adjacent side as having length 2 in Figure 12. If the opposite side has length x, then the Pythagorean Theorem gives $x^2 + 4 = 25$ and so $x^2 = 21$, $x = \sqrt{21}$. We can now use the diagram to write the other five trigonometric functions:

$$\sin \theta = \frac{\sqrt{21}}{5} \qquad \tan \theta = \frac{\sqrt{21}}{2}$$

$$\csc \theta = \frac{5}{\sqrt{21}} \qquad \sec \theta = \frac{5}{2} \qquad \cot \theta = \frac{2}{\sqrt{21}}$$

EXAMPLE 5 Use a calculator to approximate the value of x in Figure 13.

SOLUTION From the diagram we see that

$$\tan 40° = \frac{16}{x}$$

Therefore

$$x = \frac{16}{\tan 40°} \approx 19.07$$

■ Trigonometric Identities

A trigonometric identity is a relationship among the trigonometric functions. The most elementary are the following, which are immediate consequences of the definitions of the trigonometric functions.

$$\boxed{6} \qquad \csc \theta = \frac{1}{\sin \theta} \qquad \sec \theta = \frac{1}{\cos \theta} \qquad \cot \theta = \frac{1}{\tan \theta}$$

$$\tan \theta = \frac{\sin \theta}{\cos \theta} \qquad \cot \theta = \frac{\cos \theta}{\sin \theta}$$

For the next identity we refer back to Figure 7. The distance formula (or, equivalently, the Pythagorean Theorem) tells us that $x^2 + y^2 = r^2$. Therefore

$$\sin^2\theta + \cos^2\theta = \frac{y^2}{r^2} + \frac{x^2}{r^2} = \frac{x^2 + y^2}{r^2} = \frac{r^2}{r^2} = 1$$

We have therefore proved one of the most useful of all trigonometric identities:

$$\boxed{7} \qquad \qquad \sin^2\theta + \cos^2\theta = 1$$

If we now divide both sides of Equation 7 by $\cos^2\theta$ and use Equations 6, we get

$$\boxed{8} \qquad \qquad \tan^2\theta + 1 = \sec^2\theta$$

Similarly, if we divide both sides of Equation 7 by $\sin^2\theta$, we get

$$\boxed{1 + \cot^2\theta = \csc^2\theta}$$

9

The identities

10a

10b

$$\boxed{\begin{aligned}\sin(-\theta) &= -\sin\theta \\ \cos(-\theta) &= \cos\theta\end{aligned}}$$

Odd functions and even functions are discussed in Section 1.1.

show that sine is an odd function and cosine is an even function. They are easily proved by drawing a diagram showing θ and $-\theta$ in standard position (see Exercise 39).

Since the angles θ and $\theta + 2\pi$ have the same terminal side, we have

11

$$\boxed{\sin(\theta + 2\pi) = \sin\theta \qquad \cos(\theta + 2\pi) = \cos\theta}$$

These identities show that the sine and cosine functions are periodic with period 2π.

The remaining trigonometric identities are all consequences of two basic identities called the **addition formulas**:

12a

12b

$$\boxed{\begin{aligned}\sin(x + y) &= \sin x \cos y + \cos x \sin y \\ \cos(x + y) &= \cos x \cos y - \sin x \sin y\end{aligned}}$$

The proofs of these addition formulas are outlined in Exercises 85, 86, and 87.

By substituting $-y$ for y in Equations 12a and 12b and using Equations 10a and 10b, we obtain the following **subtraction formulas**:

13a

13b

$$\boxed{\begin{aligned}\sin(x - y) &= \sin x \cos y - \cos x \sin y \\ \cos(x - y) &= \cos x \cos y + \sin x \sin y\end{aligned}}$$

Then, by dividing the formulas in Equations 12 or Equations 13, we obtain the corresponding formulas for $\tan(x \pm y)$:

14a

14b

$$\boxed{\begin{aligned}\tan(x + y) &= \frac{\tan x + \tan y}{1 - \tan x \tan y} \\ \tan(x - y) &= \frac{\tan x - \tan y}{1 + \tan x \tan y}\end{aligned}}$$

If we put $y = x$ in the addition formulas $\boxed{12}$, we get the **double-angle formulas**:

15a
$$\sin 2x = 2 \sin x \cos x$$

15b
$$\cos 2x = \cos^2 x - \sin^2 x$$

Then, by using the identity $\sin^2 x + \cos^2 x = 1$, we obtain the following alternate forms of the double-angle formulas for $\cos 2x$:

16a
$$\cos 2x = 2 \cos^2 x - 1$$

16b
$$\cos 2x = 1 - 2 \sin^2 x$$

If we now solve these equations for $\cos^2 x$ and $\sin^2 x$, we get the following **half-angle formulas**, which are useful in integral calculus:

17a
$$\cos^2 x = \frac{1 + \cos 2x}{2}$$

17b
$$\sin^2 x = \frac{1 - \cos 2x}{2}$$

Finally, we state the **product formulas**, which can be deduced from Equations 12 and 13:

18a
$$\sin x \cos y = \tfrac{1}{2}[\sin(x + y) + \sin(x - y)]$$

18b
$$\cos x \cos y = \tfrac{1}{2}[\cos(x + y) + \cos(x - y)]$$

18c
$$\sin x \sin y = \tfrac{1}{2}[\cos(x - y) - \cos(x + y)]$$

There are many other trigonometric identities, but those we have stated are the ones used most often in calculus. If you forget any of the identities 13–18, remember that they can all be deduced from Equations 12a and 12b.

EXAMPLE 6 Find all values of x in the interval $[0, 2\pi]$ such that $\sin x = \sin 2x$.

SOLUTION Using the double-angle formula (15a), we rewrite the given equation as

$$\sin x = 2 \sin x \cos x \qquad \text{or} \qquad \sin x (1 - 2 \cos x) = 0$$

Therefore there are two possibilities:

$$\sin x = 0 \qquad \text{or} \qquad 1 - 2 \cos x = 0$$

$$x = 0, \pi, 2\pi \qquad\qquad \cos x = \tfrac{1}{2}$$

$$x = \frac{\pi}{3}, \frac{5\pi}{3}$$

The given equation has five solutions: 0, $\pi/3$, π, $5\pi/3$, and 2π.

Graphs of the Trigonometric Functions

The graph of the function $f(x) = \sin x$, shown in Figure 14(a), is obtained by plotting points for $0 \le x \le 2\pi$ and then using the periodic nature of the function (from Equation 11) to complete the graph. Notice that the zeros of the sine function occur at the integer multiples of π, that is,

$$\sin x = 0 \qquad \text{whenever } x = n\pi, \quad n \text{ an integer}$$

(a) $f(x) = \sin x$

(b) $g(x) = \cos x$

FIGURE 14

Because of the identity

$$\cos x = \sin\left(x + \frac{\pi}{2}\right)$$

(which can be verified using Equation 12a), the graph of cosine is obtained by shifting the graph of sine by an amount $\pi/2$ to the left [see Figure 14(b)]. Note that for both the sine and cosine functions the domain is $(-\infty, \infty)$ and the range is the closed interval $[-1, 1]$. Thus, for all values of x, we have

$$-1 \le \sin x \le 1 \qquad -1 \le \cos x \le 1$$

The graphs of the remaining four trigonometric functions are shown in Figure 15 and their domains are indicated there. Notice that tangent and cotangent have range $(-\infty, \infty)$, whereas cosecant and secant have range $(-\infty, -1] \cup [1, \infty)$. All four functions are periodic: tangent and cotangent have period π, whereas cosecant and secant have period 2π.

(a) $y = \tan x$

(b) $y = \cot x$

(c) $y = \csc x$

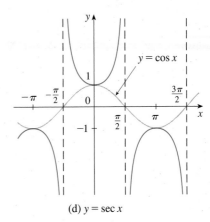

(d) $y = \sec x$

FIGURE 15

D **Exercises**

Access end-of-section exercises and tools for visualizing calculus online at **www.webassign.net**

E **Sigma Notation**

A convenient way of writing sums uses the Greek letter Σ (capital sigma, corresponding to our letter S) and is called **sigma notation**.

This tells us to end with $i = n$.

This tells us to add. $\displaystyle\sum_{i=m}^{n} a_i$

This tells us to start with $i = m$.

> **1** **Definition** If $a_m, a_{m+1}, \ldots, a_n$ are real numbers and m and n are integers such that $m \leq n$, then
>
> $$\sum_{i=m}^{n} a_i = a_m + a_{m+1} + a_{m+2} + \cdots + a_{n-1} + a_n$$

With function notation, Definition 1 can be written as

$$\sum_{i=m}^{n} f(i) = f(m) + f(m + 1) + f(m + 2) + \cdots + f(n - 1) + f(n)$$

Thus the symbol $\sum_{i=m}^{n}$ indicates a summation in which the letter i (called the **index of summation**) takes on consecutive integer values beginning with m and ending with n, that is, $m, m + 1, \ldots, n$. Other letters can also be used as the index of summation.

EXAMPLE 1

(a) $\displaystyle\sum_{i=1}^{4} i^2 = 1^2 + 2^2 + 3^2 + 4^2 = 30$

(b) $\displaystyle\sum_{i=3}^{n} i = 3 + 4 + 5 + \cdots + (n - 1) + n$

(c) $\displaystyle\sum_{j=0}^{5} 2^j = 2^0 + 2^1 + 2^2 + 2^3 + 2^4 + 2^5 = 63$

(d) $\displaystyle\sum_{k=1}^{n} \frac{1}{k} = 1 + \frac{1}{2} + \frac{1}{3} + \cdots + \frac{1}{n}$

(e) $\displaystyle\sum_{i=1}^{3} \frac{i - 1}{i^2 + 3} = \frac{1 - 1}{1^2 + 3} + \frac{2 - 1}{2^2 + 3} + \frac{3 - 1}{3^2 + 3} = 0 + \frac{1}{7} + \frac{1}{6} = \frac{13}{42}$

(f) $\displaystyle\sum_{i=1}^{4} 2 = 2 + 2 + 2 + 2 = 8$

EXAMPLE 2   Write the sum $2^3 + 3^3 + \cdots + n^3$ in sigma notation.

SOLUTION   There is no unique way of writing a sum in sigma notation. We could write

$$2^3 + 3^3 + \cdots + n^3 = \sum_{i=2}^{n} i^3$$

or

$$2^3 + 3^3 + \cdots + n^3 = \sum_{j=1}^{n-1} (j + 1)^3$$

or

$$2^3 + 3^3 + \cdots + n^3 = \sum_{k=0}^{n-2} (k + 2)^3$$

The following theorem gives three simple rules for working with sigma notation.

---

2   **Theorem**   If c is any constant (that is, it does not depend on i), then

(a) $\displaystyle\sum_{i=m}^{n} ca_i = c \sum_{i=m}^{n} a_i$ (b) $\displaystyle\sum_{i=m}^{n} (a_i + b_i) = \sum_{i=m}^{n} a_i + \sum_{i=m}^{n} b_i$

(c) $\displaystyle\sum_{i=m}^{n} (a_i - b_i) = \sum_{i=m}^{n} a_i - \sum_{i=m}^{n} b_i$

PROOF   To see why these rules are true, all we have to do is write both sides in expanded form. Rule (a) is just the distributive property of real numbers:

$$ca_m + ca_{m+1} + \cdots + ca_n = c(a_m + a_{m+1} + \cdots + a_n)$$

Rule (b) follows from the associative and commutative properties:

$$(a_m + b_m) + (a_{m+1} + b_{m+1}) + \cdots + (a_n + b_n)$$
$$= (a_m + a_{m+1} + \cdots + a_n) + (b_m + b_{m+1} + \cdots + b_n)$$

Rule (c) is proved similarly.

EXAMPLE 3 Find $\displaystyle\sum_{i=1}^{n} 1$.

SOLUTION
$$\sum_{i=1}^{n} 1 = \underbrace{1 + 1 + \cdots + 1}_{n \text{ terms}} = n$$

EXAMPLE 4 Prove the formula for the sum of the first n positive integers:

$$\sum_{i=1}^{n} i = 1 + 2 + 3 + \cdots + n = \frac{n(n+1)}{2}$$

SOLUTION This formula can be proved by mathematical induction (see page 78) or by the following method used by the German mathematician Karl Friedrich Gauss (1777–1855) when he was ten years old.

Write the sum S twice, once in the usual order and once in reverse order:

$$S = 1 + \quad 2 \quad + \quad 3 \quad + \cdots + (n-1) + n$$
$$S = n + (n-1) + (n-2) + \cdots + \quad 2 \quad + 1$$

Adding all columns vertically, we get

$$2S = (n+1) + (n+1) + (n+1) + \cdots + (n+1) + (n+1)$$

On the right side there are n terms, each of which is $n + 1$, so

$$2S = n(n+1) \qquad \text{or} \qquad S = \frac{n(n+1)}{2}$$

EXAMPLE 5 Prove the formula for the sum of the squares of the first n positive integers:

$$\sum_{i=1}^{n} i^2 = 1^2 + 2^2 + 3^2 + \cdots + n^2 = \frac{n(n+1)(2n+1)}{6}$$

SOLUTION 1 Let S be the desired sum. We start with the *telescoping sum* (or collapsing sum):

Most terms cancel in pairs.

$$\sum_{i=1}^{n} [(1+i)^3 - i^3] = (2^3 - 1^3) + (3^3 - 2^3) + (4^3 - 3^3) + \cdots + [(n+1)^3 - n^3]$$
$$= (n+1)^3 - 1^3 = n^3 + 3n^2 + 3n$$

On the other hand, using Theorem 2 and Examples 3 and 4, we have

$$\sum_{i=1}^{n} [(1+i)^3 - i^3] = \sum_{i=1}^{n} [3i^2 + 3i + 1] = 3\sum_{i=1}^{n} i^2 + 3\sum_{i=1}^{n} i + \sum_{i=1}^{n} 1$$
$$= 3S + 3\frac{n(n+1)}{2} + n = 3S + \tfrac{3}{2}n^2 + \tfrac{5}{2}n$$

Thus we have

$$n^3 + 3n^2 + 3n = 3S + \tfrac{3}{2}n^2 + \tfrac{5}{2}n$$

Solving this equation for S, we obtain

$$3S = n^3 + \tfrac{3}{2}n^2 + \tfrac{1}{2}n$$

or
$$S = \frac{2n^3 + 3n^2 + n}{6} = \frac{n(n+1)(2n+1)}{6}$$

Principle of Mathematical Induction

Let S_n be a statement involving the positive integer n. Suppose that

1. S_1 is true.
2. If S_k is true, then S_{k+1} is true.

Then S_n is true for all positive integers n.

See pages 78 and 80 for a more thorough discussion of mathematical induction.

SOLUTION 2 Let S_n be the given formula.

1. S_1 is true because
$$1^2 = \frac{1(1+1)(2 \cdot 1 + 1)}{6}$$

2. Assume that S_k is true; that is,

$$1^2 + 2^2 + 3^2 + \cdots + k^2 = \frac{k(k+1)(2k+1)}{6}$$

Then

$$1^2 + 2^2 + 3^2 + \cdots + (k+1)^2 = (1^2 + 2^2 + 3^2 + \cdots + k^2) + (k+1)^2$$

$$= \frac{k(k+1)(2k+1)}{6} + (k+1)^2$$

$$= (k+1)\frac{k(2k+1) + 6(k+1)}{6}$$

$$= (k+1)\frac{2k^2 + 7k + 6}{6}$$

$$= \frac{(k+1)(k+2)(2k+3)}{6}$$

$$= \frac{(k+1)[(k+1)+1][2(k+1)+1]}{6}$$

So S_{k+1} is true.

By the Principle of Mathematical Induction, S_n is true for all n. ▬

We list the results of Examples 3, 4, and 5 together with a similar result for cubes (see Exercises 37–40) as Theorem 3. These formulas are needed for finding areas and evaluating integrals in Chapter 4.

---

3 **Theorem** Let c be a constant and n a positive integer. Then

(a) $\displaystyle\sum_{i=1}^{n} 1 = n$

(b) $\displaystyle\sum_{i=1}^{n} c = nc$

(c) $\displaystyle\sum_{i=1}^{n} i = \frac{n(n+1)}{2}$

(d) $\displaystyle\sum_{i=1}^{n} i^2 = \frac{n(n+1)(2n+1)}{6}$

(e) $\displaystyle\sum_{i=1}^{n} i^3 = \left[\frac{n(n+1)}{2}\right]^2$

EXAMPLE 6 Evaluate $\sum_{i=1}^{n} i(4i^2 - 3)$.

SOLUTION Using Theorems 2 and 3, we have

$$\sum_{i=1}^{n} i(4i^2 - 3) = \sum_{i=1}^{n} (4i^3 - 3i) = 4\sum_{i=1}^{n} i^3 - 3\sum_{i=1}^{n} i$$

$$= 4\left[\frac{n(n+1)}{2}\right]^2 - 3\frac{n(n+1)}{2}$$

$$= \frac{n(n+1)[2n(n+1) - 3]}{2}$$

$$= \frac{n(n+1)(2n^2 + 2n - 3)}{2}$$

EXAMPLE 7 Find $\displaystyle\lim_{n\to\infty} \sum_{i=1}^{n} \frac{3}{n}\left[\left(\frac{i}{n}\right)^2 + 1\right]$.

The type of calculation in Example 7 arises in Chapter 4 when we compute areas.

SOLUTION

$$\lim_{n\to\infty} \sum_{i=1}^{n} \frac{3}{n}\left[\left(\frac{i}{n}\right)^2 + 1\right] = \lim_{n\to\infty} \sum_{i=1}^{n} \left[\frac{3}{n^3}i^2 + \frac{3}{n}\right]$$

$$= \lim_{n\to\infty} \left[\frac{3}{n^3}\sum_{i=1}^{n} i^2 + \frac{3}{n}\sum_{i=1}^{n} 1\right]$$

$$= \lim_{n\to\infty} \left[\frac{3}{n^3}\frac{n(n+1)(2n+1)}{6} + \frac{3}{n}\cdot n\right]$$

$$= \lim_{n\to\infty} \left[\frac{1}{2}\cdot\frac{n}{n}\cdot\left(\frac{n+1}{n}\right)\left(\frac{2n+1}{n}\right) + 3\right]$$

$$= \lim_{n\to\infty} \left[\frac{1}{2}\cdot 1\left(1 + \frac{1}{n}\right)\left(2 + \frac{1}{n}\right) + 3\right]$$

$$= \tfrac{1}{2}\cdot 1\cdot 1\cdot 2 + 3 = 4$$

E **Exercises**

Access end-of-section exercises and tools for visualizing calculus online at **www.webassign.net** WebAssign

F Proofs of Theorems

In this appendix we present proofs of several theorems that are stated in the main body of the text. The sections in which they occur are indicated in the margin.

Section 1.6

Limit Laws   Suppose that c is a constant and the limits

$$\lim_{x \to a} f(x) = L \quad \text{and} \quad \lim_{x \to a} g(x) = M$$

exist. Then

1. $\lim_{x \to a} [f(x) + g(x)] = L + M$ 2. $\lim_{x \to a} [f(x) - g(x)] = L - M$

3. $\lim_{x \to a} [cf(x)] = cL$ 4. $\lim_{x \to a} [f(x)g(x)] = LM$

5. $\lim_{x \to a} \dfrac{f(x)}{g(x)} = \dfrac{L}{M}$ if $M \neq 0$

PROOF OF LAW 4   Let $\varepsilon > 0$ be given. We want to find $\delta > 0$ such that

$$\text{if} \quad 0 < |x - a| < \delta \quad \text{then} \quad |f(x)g(x) - LM| < \varepsilon$$

In order to get terms that contain $|f(x) - L|$ and $|g(x) - M|$, we add and subtract $Lg(x)$ as follows:

$$\begin{aligned}
|f(x)g(x) - LM| &= |f(x)g(x) - Lg(x) + Lg(x) - LM| \\
&= |[f(x) - L]g(x) + L[g(x) - M]| \\
&\leq |[f(x) - L]g(x)| + |L[g(x) - M]| \quad \text{(Triangle Inequality)} \\
&= |f(x) - L||g(x)| + |L||g(x) - M|
\end{aligned}$$

We want to make each of these terms less than $\varepsilon/2$.

Since $\lim_{x \to a} g(x) = M$, there is a number $\delta_1 > 0$ such that

$$\text{if} \quad 0 < |x - a| < \delta_1 \quad \text{then} \quad |g(x) - M| < \frac{\varepsilon}{2(1 + |L|)}$$

Also, there is a number $\delta_2 > 0$ such that if $0 < |x - a| < \delta_2$, then

$$|g(x) - M| < 1$$

and therefore

$$|g(x)| = |g(x) - M + M| \leq |g(x) - M| + |M| < 1 + |M|$$

Since $\lim_{x \to a} f(x) = L$, there is a number $\delta_3 > 0$ such that

$$\text{if} \quad 0 < |x - a| < \delta_3 \quad \text{then} \quad |f(x) - L| < \frac{\varepsilon}{2(1 + |M|)}$$

Let $\delta = \min\{\delta_1, \delta_2, \delta_3\}$. If $0 < |x - a| < \delta$, then we have $0 < |x - a| < \delta_1$, $0 < |x - a| < \delta_2$, and $0 < |x - a| < \delta_3$, so we can combine the inequalities to obtain

$$|f(x)g(x) - LM| \le |f(x) - L||g(x)| + |L||g(x) - M|$$

$$< \frac{\varepsilon}{2(1 + |M|)}(1 + |M|) + |L|\frac{\varepsilon}{2(1 + |L|)}$$

$$< \frac{\varepsilon}{2} + \frac{\varepsilon}{2} = \varepsilon$$

This shows that $\lim_{x \to a} [f(x)g(x)] = LM$. ▪

PROOF OF LAW 3 If we take $g(x) = c$ in Law 4, we get

$$\lim_{x \to a} [cf(x)] = \lim_{x \to a} [g(x)f(x)] = \lim_{x \to a} g(x) \cdot \lim_{x \to a} f(x)$$

$$= \lim_{x \to a} c \cdot \lim_{x \to a} f(x)$$

$$= c \lim_{x \to a} f(x) \qquad \text{(by Law 7)} \qquad ▪$$

PROOF OF LAW 2 Using Law 1 and Law 3 with $c = -1$, we have

$$\lim_{x \to a} [f(x) - g(x)] = \lim_{x \to a} [f(x) + (-1)g(x)] = \lim_{x \to a} f(x) + \lim_{x \to a} (-1)g(x)$$

$$= \lim_{x \to a} f(x) + (-1)\lim_{x \to a} g(x) = \lim_{x \to a} f(x) - \lim_{x \to a} g(x) \qquad ▪$$

PROOF OF LAW 5 First let us show that

$$\lim_{x \to a} \frac{1}{g(x)} = \frac{1}{M}$$

To do this we must show that, given $\varepsilon > 0$, there exists $\delta > 0$ such that

$$\text{if} \qquad 0 < |x - a| < \delta \qquad \text{then} \qquad \left| \frac{1}{g(x)} - \frac{1}{M} \right| < \varepsilon$$

Observe that

$$\left| \frac{1}{g(x)} - \frac{1}{M} \right| = \frac{|M - g(x)|}{|Mg(x)|}$$

We know that we can make the numerator small. But we also need to know that the denominator is not small when x is near a. Since $\lim_{x \to a} g(x) = M$, there is a number $\delta_1 > 0$ such that, whenever $0 < |x - a| < \delta_1$, we have

$$|g(x) - M| < \frac{|M|}{2}$$

and therefore

$$|M| = |M - g(x) + g(x)| \le |M - g(x)| + |g(x)|$$

$$< \frac{|M|}{2} + |g(x)|$$

This shows that

$$\text{if} \quad 0 < |x - a| < \delta_1 \quad \text{then} \quad |g(x)| > \frac{|M|}{2}$$

and so, for these values of x,

$$\frac{1}{|Mg(x)|} = \frac{1}{|M||g(x)|} < \frac{1}{|M|} \cdot \frac{2}{|M|} = \frac{2}{M^2}$$

Also, there exists $\delta_2 > 0$ such that

$$\text{if} \quad 0 < |x - a| < \delta_2 \quad \text{then} \quad |g(x) - M| < \frac{M^2}{2}\varepsilon$$

Let $\delta = \min\{\delta_1, \delta_2\}$. Then, for $0 < |x - a| < \delta$, we have

$$\left| \frac{1}{g(x)} - \frac{1}{M} \right| = \frac{|M - g(x)|}{|Mg(x)|} < \frac{2}{M^2} \frac{M^2}{2} \varepsilon = \varepsilon$$

It follows that $\lim_{x \to a} 1/g(x) = 1/M$. Finally, using Law 4, we obtain

$$\lim_{x \to a} \frac{f(x)}{g(x)} = \lim_{x \to a} f(x) \left(\frac{1}{g(x)} \right) = \lim_{x \to a} f(x) \lim_{x \to a} \frac{1}{g(x)} = L \cdot \frac{1}{M} = \frac{L}{M} \qquad \blacksquare$$

---

2   **Theorem**   If $f(x) \le g(x)$ for all x in an open interval that contains a (except possibly at a) and

$$\lim_{x \to a} f(x) = L \quad \text{and} \quad \lim_{x \to a} g(x) = M$$

then $L \le M$.

PROOF   We use the method of proof by contradiction. Suppose, if possible, that $L > M$. Law 2 of limits says that

$$\lim_{x \to a} [g(x) - f(x)] = M - L$$

Therefore, for any $\varepsilon > 0$, there exists $\delta > 0$ such that

$$\text{if} \quad 0 < |x - a| < \delta \quad \text{then} \quad |[g(x) - f(x)] - (M - L)| < \varepsilon$$

In particular, taking $\varepsilon = L - M$ (noting that $L - M > 0$ by hypothesis), we have a number $\delta > 0$ such that

$$\text{if} \quad 0 < |x - a| < \delta \quad \text{then} \quad |[g(x) - f(x)] - (M - L)| < L - M$$

Since $a \le |a|$ for any number a, we have

$$\text{if} \quad 0 < |x - a| < \delta \quad \text{then} \quad [g(x) - f(x)] - (M - L) < L - M$$

which simplifies to

$$\text{if} \quad 0 < |x - a| < \delta \quad \text{then} \quad g(x) < f(x)$$

But this contradicts $f(x) \le g(x)$. Thus the inequality $L > M$ must be false. Therefore $L \le M$. $\blacksquare$

---

3 **The Squeeze Theorem** If $f(x) \leqslant g(x) \leqslant h(x)$ for all x in an open interval that contains a (except possibly at a) and

$$\lim_{x \to a} f(x) = \lim_{x \to a} h(x) = L$$

then
$$\lim_{x \to a} g(x) = L$$

PROOF Let $\varepsilon > 0$ be given. Since $\lim_{x \to a} f(x) = L$, there is a number $\delta_1 > 0$ such that

$$\text{if} \qquad 0 < |x - a| < \delta_1 \qquad \text{then} \qquad |f(x) - L| < \varepsilon$$

that is,

$$\text{if} \qquad 0 < |x - a| < \delta_1 \qquad \text{then} \qquad L - \varepsilon < f(x) < L + \varepsilon$$

Since $\lim_{x \to a} h(x) = L$, there is a number $\delta_2 > 0$ such that

$$\text{if} \qquad 0 < |x - a| < \delta_2 \qquad \text{then} \qquad |h(x) - L| < \varepsilon$$

that is,

$$\text{if} \qquad 0 < |x - a| < \delta_2 \qquad \text{then} \qquad L - \varepsilon < h(x) < L + \varepsilon$$

Let $\delta = \min\{\delta_1, \delta_2\}$. If $0 < |x - a| < \delta$, then $0 < |x - a| < \delta_1$ and $0 < |x - a| < \delta_2$, so

$$L - \varepsilon < f(x) \leqslant g(x) \leqslant h(x) < L + \varepsilon$$

In particular,
$$L - \varepsilon < g(x) < L + \varepsilon$$

and so $|g(x) - L| < \varepsilon$. Therefore $\lim_{x \to a} g(x) = L$. ▪

Section 1.8

---

8 **Theorem** If f is continuous at b and $\lim_{x \to a} g(x) = b$, then

$$\lim_{x \to a} f\big(g(x)\big) = f(b)$$

PROOF Let $\varepsilon > 0$ be given. We want to find a number $\delta > 0$ such that

$$\text{if} \qquad 0 < |x - a| < \delta \qquad \text{then} \qquad \big|f(g(x)) - f(b)\big| < \varepsilon$$

Since f is continuous at b, we have

$$\lim_{y \to b} f(y) = f(b)$$

and so there exists $\delta_1 > 0$ such that

$$\text{if} \qquad 0 < |y - b| < \delta_1 \qquad \text{then} \qquad |f(y) - f(b)| < \varepsilon$$

Since $\lim_{x \to a} g(x) = b$, there exists $\delta > 0$ such that

$$\text{if} \qquad 0 < |x - a| < \delta \qquad \text{then} \qquad |g(x) - b| < \delta_1$$

Combining these two statements, we see that whenever $0 < |x - a| < \delta$ we have $|g(x) - b| < \delta_1$, which implies that $|f(g(x)) - f(b)| < \varepsilon$. Therefore we have proved that $\lim_{x \to a} f(g(x)) = f(b)$. ▬

Section 2.4

The proof of the following result was promised when we proved that $\lim_{\theta \to 0} \dfrac{\sin \theta}{\theta} = 1$.

> **Theorem**   If $0 < \theta < \pi/2$, then $\theta \leqslant \tan \theta$.

PROOF   Figure 1 shows a sector of a circle with center O, central angle θ, and radius 1. Then

$$|AD| = |OA| \tan \theta = \tan \theta$$

We approximate the arc AB by an inscribed polygon consisting of n equal line segments and we look at a typical segment PQ. We extend the lines OP and OQ to meet AD in the points R and S. Then we draw $RT \parallel PQ$ as in Figure 2. Observe that

$$\angle RTO = \angle PQO < 90°$$

and so $\angle RTS > 90°$. Therefore we have

$$|PQ| < |RT| < |RS|$$

If we add n such inequalities, we get

$$L_n < |AD| = \tan \theta$$

where L_n is the length of the inscribed polygon. Thus, by Theorem 1.6.2, we have

$$\lim_{n \to \infty} L_n \leqslant \tan \theta$$

But the arc length is defined in Equation 8.1.1 as the limit of the lengths of inscribed polygons, so

$$\theta = \lim_{n \to \infty} L_n \leqslant \tan \theta$$ ▬

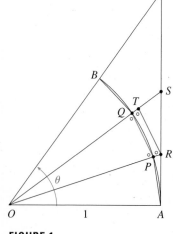

FIGURE 1

Section 3.3

> **Concavity Test**
> (a) If $f''(x) > 0$ for all x in I, then the graph of f is concave upward on I.
> (b) If $f''(x) < 0$ for all x in I, then the graph of f is concave downward on I.

PROOF OF (a)   Let a be any number in I. We must show that the curve $y = f(x)$ lies above the tangent line at the point $(a, f(a))$. The equation of this tangent is

$$y = f(a) + f'(a)(x - a)$$

So we must show that

$$f(x) > f(a) + f'(a)(x - a)$$

FIGURE 2

whenever $x \in I$ $(x \neq a)$. (See Figure 2.)

First let us take the case where $x > a$. Applying the Mean Value Theorem to f on the interval $[a, x]$, we get a number c, with $a < c < x$, such that

$$\boxed{1} \qquad\qquad f(x) - f(a) = f'(c)(x - a)$$

Since $f'' > 0$ on I, we know from the Increasing/Decreasing Test that f' is increasing on I. Thus, since $a < c$, we have

$$f'(a) < f'(c)$$

and so, multiplying this inequality by the positive number $x - a$, we get

$$\boxed{2} \qquad\qquad f'(a)(x - a) < f'(c)(x - a)$$

Now we add $f(a)$ to both sides of this inequality:

$$f(a) + f'(a)(x - a) < f(a) + f'(c)(x - a)$$

But from Equation 1 we have $f(x) = f(a) + f'(c)(x - a)$. So this inequality becomes

$$\boxed{3} \qquad\qquad f(x) > f(a) + f'(a)(x - a)$$

which is what we wanted to prove.

For the case where $x < a$ we have $f'(c) < f'(a)$, but multiplication by the negative number $x - a$ reverses the inequality, so we get $\boxed{2}$ and $\boxed{3}$ as before.

Section 6.1

> **Theorem** If f is a one-to-one continuous function defined on an interval (a, b), then its inverse function f^{-1} is also continuous.

PROOF First we show that if f is both one-to-one and continuous on (a, b), then it must be either increasing or decreasing on (a, b). If it were neither increasing nor decreasing, then there would exist numbers x_1, x_2, and x_3 in (a, b) with $x_1 < x_2 < x_3$ such that $f(x_2)$ does not lie between $f(x_1)$ and $f(x_3)$. There are two possibilities: either (1) $f(x_3)$ lies between $f(x_1)$ and $f(x_2)$ or (2) $f(x_1)$ lies between $f(x_2)$ and $f(x_3)$. (Draw a picture.) In case (1) we apply the Intermediate Value Theorem to the continuous function f to get a number c between x_1 and x_2 such that $f(c) = f(x_3)$. In case (2) the Intermediate Value Theorem gives a number c between x_2 and x_3 such that $f(c) = f(x_1)$. In either case we have contradicted the fact that f is one-to-one.

Let us assume, for the sake of definiteness, that f is increasing on (a, b). We take any number y_0 in the domain of f^{-1} and we let $f^{-1}(y_0) = x_0$; that is, x_0 is the number in (a, b) such that $f(x_0) = y_0$. To show that f^{-1} is continuous at y_0 we take any $\varepsilon > 0$ such that the interval $(x_0 - \varepsilon, x_0 + \varepsilon)$ is contained in the interval (a, b). Since f is increasing, it maps the numbers in the interval $(x_0 - \varepsilon, x_0 + \varepsilon)$ onto the numbers in the interval $(f(x_0 - \varepsilon), f(x_0 + \varepsilon))$ and f^{-1} reverses the correspondence. If we let δ denote the smaller of the numbers $\delta_1 = y_0 - f(x_0 - \varepsilon)$ and $\delta_2 = f(x_0 + \varepsilon) - y_0$, then the interval $(y_0 - \delta, y_0 + \delta)$ is contained in the interval $(f(x_0 - \varepsilon), f(x_0 + \varepsilon))$ and so is mapped into the interval $(x_0 - \varepsilon, x_0 + \varepsilon)$ by f^{-1}. (See the arrow diagram in Figure 3 on page A40.) We have therefore found a number $\delta > 0$ such that

$$\text{if} \qquad |y - y_0| < \delta \qquad \text{then} \qquad |f^{-1}(y) - f^{-1}(y_0)| < \varepsilon$$

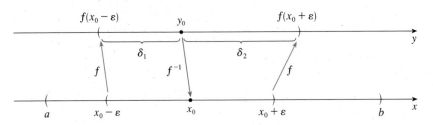

FIGURE 3

This shows that $\lim_{y \to y_0} f^{-1}(y) = f^{-1}(y_0)$ and so f^{-1} is continuous at any number y_0 in its domain. ∎

Section 11.8

In order to prove Theorem 11.8.3, we first need the following results.

Theorem

1. If a power series $\Sigma\, c_n x^n$ converges when $x = b$ (where $b \neq 0$), then it converges whenever $|x| < |b|$.

2. If a power series $\Sigma\, c_n x^n$ diverges when $x = d$ (where $d \neq 0$), then it diverges whenever $|x| > |d|$.

PROOF OF 1 Suppose that $\Sigma\, c_n b^n$ converges. Then, by Theorem 11.2.6, we have $\lim_{n \to \infty} c_n b^n = 0$. According to Definition 11.1.2 with $\varepsilon = 1$, there is a positive integer N such that $|c_n b^n| < 1$ whenever $n \geqslant N$. Thus, for $n \geqslant N$, we have

$$|c_n x^n| = \left| \frac{c_n b^n x^n}{b^n} \right| = |c_n b^n| \left| \frac{x}{b} \right|^n < \left| \frac{x}{b} \right|^n$$

If $|x| < |b|$, then $|x/b| < 1$, so $\Sigma\, |x/b|^n$ is a convergent geometric series. Therefore, by the Comparison Test, the series $\Sigma_{n=N}^{\infty} |c_n x^n|$ is convergent. Thus the series $\Sigma\, c_n x^n$ is absolutely convergent and therefore convergent. ∎

PROOF OF 2 Suppose that $\Sigma\, c_n d^n$ diverges. If x is any number such that $|x| > |d|$, then $\Sigma\, c_n x^n$ cannot converge because, by part 1, the convergence of $\Sigma\, c_n x^n$ would imply the convergence of $\Sigma\, c_n d^n$. Therefore $\Sigma\, c_n x^n$ diverges whenever $|x| > |d|$. ∎

Theorem For a power series $\Sigma\, c_n x^n$ there are only three possibilities:

1. The series converges only when $x = 0$.

2. The series converges for all x.

3. There is a positive number R such that the series converges if $|x| < R$ and diverges if $|x| > R$.

PROOF Suppose that neither case 1 nor case 2 is true. Then there are nonzero numbers b and d such that $\Sigma\, c_n x^n$ converges for $x = b$ and diverges for $x = d$. Therefore the set $S = \{x \mid \Sigma\, c_n x^n \text{ converges}\}$ is not empty. By the preceding theorem, the series diverges if $|x| > |d|$, so $|x| \leqslant |d|$ for all $x \in S$. This says that $|d|$ is an upper bound for the set S. Thus, by the Completeness Axiom (see Section 11.1), S has a least upper bound R. If $|x| > R$, then $x \notin S$, so $\Sigma\, c_n x^n$ diverges. If $|x| < R$, then $|x|$ is not an upper bound for S and so there exists $b \in S$ such that $b > |x|$. Since $b \in S$, $\Sigma\, c_n b^n$ converges, so by the preceding theorem $\Sigma\, c_n x^n$ converges. ∎

---

3 **Theorem** For a power series $\sum c_n(x - a)^n$ there are only three possibilities:

1. The series converges only when $x = a$.

2. The series converges for all x.

3. There is a positive number R such that the series converges if $|x - a| < R$ and diverges if $|x - a| > R$.

PROOF If we make the change of variable $u = x - a$, then the power series becomes $\sum c_n u^n$ and we can apply the preceding theorem to this series. In case 3 we have convergence for $|u| < R$ and divergence for $|u| > R$. Thus we have convergence for $|x - a| < R$ and divergence for $|x - a| > R$. ∎

G | Graphing Calculators and Computers

In this appendix we assume that you have access to a graphing calculator or a computer with graphing software. We see that the use of such a device enables us to graph more complicated functions and to solve more complex problems than would otherwise be possible. We also point out some of the pitfalls that can occur with these machines.

Graphing calculators and computers can give very accurate graphs of functions. But we will see in Chapter 3 that only through the use of calculus can we be sure that we have uncovered all the interesting aspects of a graph.

A graphing calculator or computer displays a rectangular portion of the graph of a function in a **display window** or **viewing screen**, which we refer to as a **viewing rectangle**. The default screen often gives an incomplete or misleading picture, so it is important to choose the viewing rectangle with care. If we choose the x-values to range from a minimum value of $Xmin = a$ to a maximum value of $Xmax = b$ and the y-values to range from a minimum of $Ymin = c$ to a maximum of $Ymax = d$, then the visible portion of the graph lies in the rectangle

$$[a, b] \times [c, d] = \{(x, y) \mid a \le x \le b, c \le y \le d\}$$

shown in Figure 1. We refer to this rectangle as the $[a, b]$ *by* $[c, d]$ *viewing rectangle.*

FIGURE 1
The viewing rectangle $[a, b]$ by $[c, d]$

The machine draws the graph of a function f much as you would. It plots points of the form $(x, f(x))$ for a certain number of equally spaced values of x between a and b. If an x-value is not in the domain of f, or if $f(x)$ lies outside the viewing rectangle, it moves on to the next x-value. The machine connects each point to the preceding plotted point to form a representation of the graph of f.

EXAMPLE 1 Draw the graph of the function $f(x) = x^2 + 3$ in each of the following viewing rectangles.

(a) $[-2, 2]$ by $[-2, 2]$ (b) $[-4, 4]$ by $[-4, 4]$

(c) $[-10, 10]$ by $[-5, 30]$ (d) $[-50, 50]$ by $[-100, 1000]$

SOLUTION For part (a) we select the range by setting $Xmin = -2$, $Xmax = 2$, $Ymin = -2$, and $Ymax = 2$. The resulting graph is shown in Figure 2(a). The display window is blank! A moment's thought provides the explanation: Notice that $x^2 \geq 0$ for all x, so $x^2 + 3 \geq 3$ for all x. Thus the range of the function $f(x) = x^2 + 3$ is $[3, \infty)$. This means that the graph of f lies entirely outside the viewing rectangle $[-2, 2]$ by $[-2, 2]$.

The graphs for the viewing rectangles in parts (b), (c), and (d) are also shown in Figure 2. Observe that we get a more complete picture in parts (c) and (d), but in part (d) it is not clear that the y-intercept is 3.

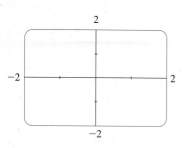

(a) $[-2, 2]$ by $[-2, 2]$

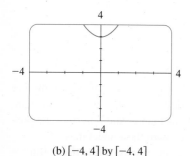

(b) $[-4, 4]$ by $[-4, 4]$

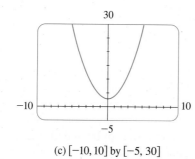

(c) $[-10, 10]$ by $[-5, 30]$

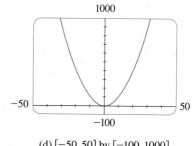

(d) $[-50, 50]$ by $[-100, 1000]$

FIGURE 2 Graphs of $f(x) = x^2 + 3$

We see from Example 1 that the choice of a viewing rectangle can make a big difference in the appearance of a graph. Often it's necessary to change to a larger viewing rectangle to obtain a more complete picture, a more global view, of the graph. In the next example we see that knowledge of the domain and range of a function sometimes provides us with enough information to select a good viewing rectangle.

EXAMPLE 2 Determine an appropriate viewing rectangle for the function $f(x) = \sqrt{8 - 2x^2}$ and use it to graph f.

SOLUTION The expression for $f(x)$ is defined when

$$8 - 2x^2 \geq 0 \iff 2x^2 \leq 8 \iff x^2 \leq 4$$
$$\iff |x| \leq 2 \iff -2 \leq x \leq 2$$

Therefore the domain of f is the interval $[-2, 2]$. Also,

$$0 \leq \sqrt{8 - 2x^2} \leq \sqrt{8} = 2\sqrt{2} \approx 2.83$$

so the range of f is the interval $[0, 2\sqrt{2}]$.

We choose the viewing rectangle so that the x-interval is somewhat larger than the domain and the y-interval is larger than the range. Taking the viewing rectangle to be $[-3, 3]$ by $[-1, 4]$, we get the graph shown in Figure 3.

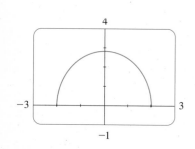

FIGURE 3
$f(x) = \sqrt{8 - 2x^2}$

EXAMPLE 3 Graph the function $y = x^3 - 150x$.

SOLUTION Here the domain is $\mathbb{R}$, the set of all real numbers. That doesn't help us choose a viewing rectangle. Let's experiment. If we start with the viewing rectangle $[-5, 5]$ by

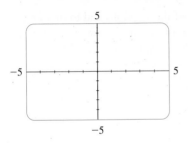

5

−5   5

−5

FIGURE 4

$[-5, 5]$, we get the graph in Figure 4. It appears blank, but actually the graph is so nearly vertical that it blends in with the y-axis.

If we change the viewing rectangle to $[-20, 20]$ by $[-20, 20]$, we get the picture shown in Figure 5(a). The graph appears to consist of vertical lines, but we know that can't be correct. If we look carefully while the graph is being drawn, we see that the graph leaves the screen and reappears during the graphing process. This indicates that we need to see more in the vertical direction, so we change the viewing rectangle to $[-20, 20]$ by $[-500, 500]$. The resulting graph is shown in Figure 5(b). It still doesn't quite reveal all the main features of the function, so we try $[-20, 20]$ by $[-1000, 1000]$ in Figure 5(c). Now we are more confident that we have arrived at an appropriate viewing rectangle. In Chapter 4 we will be able to see that the graph shown in Figure 5(c) does indeed reveal all the main features of the function.

  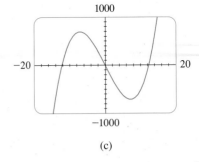

(a)   (b)   (c)

FIGURE 5 Graphs of $y = x^3 - 150x$

V EXAMPLE 4 Graph the function $f(x) = \sin 50x$ in an appropriate viewing rectangle.

SOLUTION Figure 6(a) shows the graph of f produced by a graphing calculator using the viewing rectangle $[-12, 12]$ by $[-1.5, 1.5]$. At first glance the graph appears to be reasonable. But if we change the viewing rectangle to the ones shown in the following parts of Figure 6, the graphs look very different. Something strange is happening.

(a)   (b)

(c)   (d)

The appearance of the graphs in Figure 6 depends on the machine used. The graphs you get with your own graphing device might not look like these figures, but they will also be quite inaccurate.

FIGURE 6
Graphs of $f(x) = \sin 50x$
in four viewing rectangles

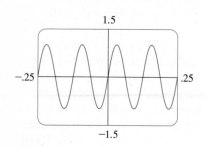

FIGURE 7

$f(x) = \sin 50x$

In order to explain the big differences in appearance of these graphs and to find an appropriate viewing rectangle, we need to find the period of the function $y = \sin 50x$. We know that the function $y = \sin x$ has period 2π and the graph of $y = \sin 50x$ is shrunk horizontally by a factor of 50, so the period of $y = \sin 50x$ is

$$\frac{2\pi}{50} = \frac{\pi}{25} \approx 0.126$$

This suggests that we should deal only with small values of x in order to show just a few oscillations of the graph. If we choose the viewing rectangle $[-0.25, 0.25]$ by $[-1.5, 1.5]$, we get the graph shown in Figure 7.

Now we see what went wrong in Figure 6. The oscillations of $y = \sin 50x$ are so rapid that when the calculator plots points and joins them, it misses most of the maximum and minimum points and therefore gives a very misleading impression of the graph. ▬

We have seen that the use of an inappropriate viewing rectangle can give a misleading impression of the graph of a function. In Examples 1 and 3 we solved the problem by changing to a larger viewing rectangle. In Example 4 we had to make the viewing rectangle smaller. In the next example we look at a function for which there is no single viewing rectangle that reveals the true shape of the graph.

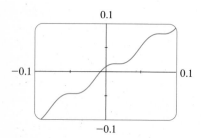

FIGURE 8

V **EXAMPLE 5** Graph the function $f(x) = \sin x + \frac{1}{100}\cos 100x$.

SOLUTION Figure 8 shows the graph of f produced by a graphing calculator with viewing rectangle $[-6.5, 6.5]$ by $[-1.5, 1.5]$. It looks much like the graph of $y = \sin x$, but perhaps with some bumps attached. If we zoom in to the viewing rectangle $[-0.1, 0.1]$ by $[-0.1, 0.1]$, we can see much more clearly the shape of these bumps in Figure 9. The reason for this behavior is that the second term, $\frac{1}{100}\cos 100x$, is very small in comparison with the first term, $\sin x$. Thus we really need two graphs to see the true nature of this function. ▬

EXAMPLE 6 Draw the graph of the function $y = \dfrac{1}{1-x}$.

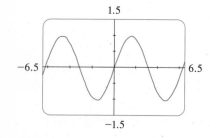

FIGURE 9

SOLUTION Figure 10(a) shows the graph produced by a graphing calculator with viewing rectangle $[-9, 9]$ by $[-9, 9]$. In connecting successive points on the graph, the calculator produced a steep line segment from the top to the bottom of the screen. That line segment is not truly part of the graph. Notice that the domain of the function $y = 1/(1 - x)$ is $\{x \mid x \neq 1\}$. We can eliminate the extraneous near-vertical line by experimenting with a change of scale. When we change to the smaller viewing rectangle $[-4.7, 4.7]$ by $[-4.7, 4.7]$ on this particular calculator, we obtain the much better graph in Figure 10(b).

Another way to avoid the extraneous line is to change the graphing mode on the calculator so that the dots are not connected.

(a)

(b)

FIGURE 10

EXAMPLE 7 Graph the function $y = \sqrt[3]{x}$.

SOLUTION Some graphing devices display the graph shown in Figure 11, whereas others produce a graph like that in Figure 12. We know from Section 1.2 (Figure 13) that the graph in Figure 12 is correct, so what happened in Figure 11? The explanation is that some machines compute the cube root of x using a logarithm, which is not defined if x is negative, so only the right half of the graph is produced.

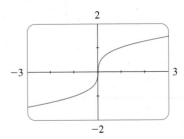

FIGURE 11   **FIGURE 12**

You can get the correct graph with Maple if you first type

    with(RealDomain);

You should experiment with your own machine to see which of these two graphs is produced. If you get the graph in Figure 11, you can obtain the correct picture by graphing the function

$$f(x) = \frac{x}{|x|} \cdot |x|^{1/3}$$

Notice that this function is equal to $\sqrt[3]{x}$ (except when $x = 0$).

To understand how the expression for a function relates to its graph, it's helpful to graph a **family of functions**, that is, a collection of functions whose equations are related. In the next example we graph members of a family of cubic polynomials.

TEC In Visual G you can see an animation of Figure 13.

V **EXAMPLE 8** Graph the function $y = x^3 + cx$ for various values of the number c. How does the graph change when c is changed?

SOLUTION Figure 13 shows the graphs of $y = x^3 + cx$ for $c = 2, 1, 0, -1$, and -2. We see that, for positive values of c, the graph increases from left to right with no maximum or minimum points (peaks or valleys). When $c = 0$, the curve is flat at the origin. When c is negative, the curve has a maximum point and a minimum point. As c decreases, the maximum point becomes higher and the minimum point lower.

(a) $y = x^3 + 2x$ (b) $y = x^3 + x$ (c) $y = x^3$ (d) $y = x^3 - x$ (e) $y = x^3 - 2x$

FIGURE 13
Several members of the family of functions $y = x^3 + cx$, all graphed in the viewing rectangle $[-2, 2]$ by $[-2.5, 2.5]$

EXAMPLE 9 Find the solution of the equation $\cos x = x$ correct to two decimal places.

SOLUTION The solutions of the equation $\cos x = x$ are the x-coordinates of the points of intersection of the curves $y = \cos x$ and $y = x$. From Figure 14(a) we see that there is only one solution and it lies between 0 and 1. Zooming in to the viewing rectangle $[0, 1]$ by $[0, 1]$, we see from Figure 14(b) that the root lies between 0.7 and 0.8. So we zoom in further to the viewing rectangle $[0.7, 0.8]$ by $[0.7, 0.8]$ in Figure 14(c). By moving the cursor to the intersection point of the two curves, or by inspection and the fact that the x-scale is 0.01, we see that the solution of the equation is about 0.74. (Many calculators have a built-in intersection feature.)

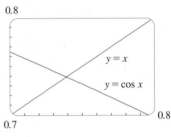

FIGURE 14
Locating the roots
of $\cos x = x$

(a) $[-5, 5]$ by $[-1.5, 1.5]$
x-scale = 1

(b) $[0, 1]$ by $[0, 1]$
x-scale = 0.1

(c) $[0.7, 0.8]$ by $[0.7, 0.8]$
x-scale = 0.01

G   Exercises

Access end-of-section exercises and tools for visualizing calculus online at **www.webassign.net**   **WebAssign**

H   Complex Numbers

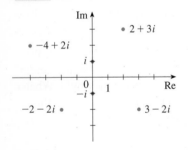

FIGURE 1
Complex numbers as points in
the Argand plane

A **complex number** can be represented by an expression of the form $a + bi$, where a and b are real numbers and i is a symbol with the property that $i^2 = -1$. The complex number $a + bi$ can also be represented by the ordered pair (a, b) and plotted as a point in a plane (called the Argand plane) as in Figure 1. Thus the complex number $i = 0 + 1 \cdot i$ is identified with the point $(0, 1)$.

The **real part** of the complex number $a + bi$ is the real number a and the **imaginary part** is the real number b. Thus the real part of $4 - 3i$ is 4 and the imaginary part is -3. Two complex numbers $a + bi$ and $c + di$ are **equal** if $a = c$ and $b = d$, that is, their real parts are equal and their imaginary parts are equal. In the Argand plane the horizontal axis is called the real axis and the vertical axis is called the imaginary axis.

The sum and difference of two complex numbers are defined by adding or subtracting their real parts and their imaginary parts:

$$(a + bi) + (c + di) = (a + c) + (b + d)i$$

$$(a + bi) - (c + di) = (a - c) + (b - d)i$$

For instance,

$$(1 - i) + (4 + 7i) = (1 + 4) + (-1 + 7)i = 5 + 6i$$

The product of complex numbers is defined so that the usual commutative and distributive laws hold:

$$(a + bi)(c + di) = a(c + di) + (bi)(c + di)$$

$$= ac + adi + bci + bdi^2$$

Since $i^2 = -1$, this becomes

$$(a + bi)(c + di) = (ac - bd) + (ad + bc)i$$

EXAMPLE 1

$$(-1 + 3i)(2 - 5i) = (-1)(2 - 5i) + 3i(2 - 5i)$$
$$= -2 + 5i + 6i - 15(-1) = 13 + 11i$$

Division of complex numbers is much like rationalizing the denominator of a rational expression. For the complex number $z = a + bi$, we define its **complex conjugate** to be $\bar{z} = a - bi$. To find the quotient of two complex numbers we multiply numerator and denominator by the complex conjugate of the denominator.

EXAMPLE 2 Express the number $\dfrac{-1 + 3i}{2 + 5i}$ in the form $a + bi$.

SOLUTION We multiply numerator and denominator by the complex conjugate of $2 + 5i$, namely $2 - 5i$, and we take advantage of the result of Example 1:

$$\frac{-1 + 3i}{2 + 5i} = \frac{-1 + 3i}{2 + 5i} \cdot \frac{2 - 5i}{2 - 5i} = \frac{13 + 11i}{2^2 + 5^2} = \frac{13}{29} + \frac{11}{29}i$$

The geometric interpretation of the complex conjugate is shown in Figure 2: $\bar{z}$ is the reflection of z in the real axis. We list some of the properties of the complex conjugate in the following box. The proofs follow from the definition and are requested in Exercise 18.

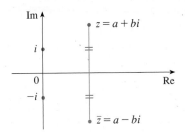

FIGURE 2

Properties of Conjugates

$$\overline{z + w} = \bar{z} + \bar{w} \qquad \overline{zw} = \bar{z}\,\bar{w} \qquad \overline{z^n} = \bar{z}^n$$

The **modulus**, or **absolute value**, $|z|$ of a complex number $z = a + bi$ is its distance from the origin. From Figure 3 we see that if $z = a + bi$, then

$$\boxed{|z| = \sqrt{a^2 + b^2}}$$

FIGURE 3

Notice that

$$z\bar{z} = (a + bi)(a - bi) = a^2 + abi - abi - b^2i^2 = a^2 + b^2$$

and so

$$\boxed{z\bar{z} = |z|^2}$$

This explains why the division procedure in Example 2 works in general:

$$\frac{z}{w} = \frac{z\bar{w}}{w\bar{w}} = \frac{z\bar{w}}{|w|^2}$$

Since $i^2 = -1$, we can think of i as a square root of -1. But notice that we also have $(-i)^2 = i^2 = -1$ and so $-i$ is also a square root of -1. We say that i is the **principal square root** of -1 and write $\sqrt{-1} = i$. In general, if c is any positive number, we write

$$\sqrt{-c} = \sqrt{c}\, i$$

With this convention, the usual derivation and formula for the roots of the quadratic equation $ax^2 + bx + c = 0$ are valid even when $b^2 - 4ac < 0$:

$$x = \frac{-b \pm \sqrt{b^2 - 4ac}}{2a}$$

EXAMPLE 3 Find the roots of the equation $x^2 + x + 1 = 0$.

SOLUTION Using the quadratic formula, we have

$$x = \frac{-1 \pm \sqrt{1^2 - 4 \cdot 1}}{2} = \frac{-1 \pm \sqrt{-3}}{2} = \frac{-1 \pm \sqrt{3}\ i}{2}$$

We observe that the solutions of the equation in Example 3 are complex conjugates of each other. In general, the solutions of any quadratic equation $ax^2 + bx + c = 0$ with real coefficients a, b, and c are always complex conjugates. (If z is real, $\bar{z} = z$, so z is its own conjugate.)

We have seen that if we allow complex numbers as solutions, then every quadratic equation has a solution. More generally, it is true that every polynomial equation

$$a_n x^n + a_{n-1} x^{n-1} + \cdots + a_1 x + a_0 = 0$$

of degree at least one has a solution among the complex numbers. This fact is known as the Fundamental Theorem of Algebra and was proved by Gauss.

Polar Form

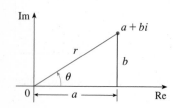

FIGURE 4

We know that any complex number $z = a + bi$ can be considered as a point (a, b) and that any such point can be represented by polar coordinates (r, θ) with $r \geq 0$. In fact,

$$a = r \cos \theta \qquad b = r \sin \theta$$

as in Figure 4. Therefore we have

$$z = a + bi = (r \cos \theta) + (r \sin \theta)i$$

Thus we can write any complex number z in the form

$$\boxed{z = r(\cos \theta + i \sin \theta)}$$

where $\qquad r = |z| = \sqrt{a^2 + b^2} \qquad$ and $\qquad \tan \theta = \dfrac{b}{a}$

The angle θ is called the **argument** of z and we write $\theta = \arg(z)$. Note that $\arg(z)$ is not unique; any two arguments of z differ by an integer multiple of 2π.

EXAMPLE 4 Write the following numbers in polar form.
(a) $z = 1 + i$
(b) $w = \sqrt{3} - i$

SOLUTION
(a) We have $r = |z| = \sqrt{1^2 + 1^2} = \sqrt{2}$ and $\tan \theta = 1$, so we can take $\theta = \pi/4$.

Therefore the polar form is

$$z = \sqrt{2}\left(\cos\frac{\pi}{4} + i\sin\frac{\pi}{4}\right)$$

(b) Here we have $r = |w| = \sqrt{3 + 1} = 2$ and $\tan\theta = -1/\sqrt{3}$. Since w lies in the fourth quadrant, we take $\theta = -\pi/6$ and

$$w = 2\left[\cos\left(-\frac{\pi}{6}\right) + i\sin\left(-\frac{\pi}{6}\right)\right]$$

The numbers z and w are shown in Figure 5.

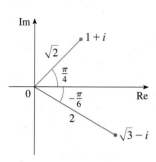

FIGURE 5

The polar form of complex numbers gives insight into multiplication and division. Let

$$z_1 = r_1(\cos\theta_1 + i\sin\theta_1) \qquad z_2 = r_2(\cos\theta_2 + i\sin\theta_2)$$

be two complex numbers written in polar form. Then

$$z_1 z_2 = r_1 r_2(\cos\theta_1 + i\sin\theta_1)(\cos\theta_2 + i\sin\theta_2)$$

$$= r_1 r_2[(\cos\theta_1\cos\theta_2 - \sin\theta_1\sin\theta_2) + i(\sin\theta_1\cos\theta_2 + \cos\theta_1\sin\theta_2)]$$

Therefore, using the addition formulas for cosine and sine, we have

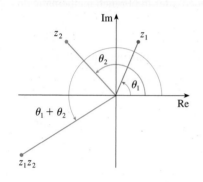

FIGURE 6

$$\boxed{1} \qquad \boxed{z_1 z_2 = r_1 r_2[\cos(\theta_1 + \theta_2) + i\sin(\theta_1 + \theta_2)]}$$

This formula says that *to multiply two complex numbers we multiply the moduli and add the arguments*. (See Figure 6.)

A similar argument using the subtraction formulas for sine and cosine shows that *to divide two complex numbers we divide the moduli and subtract the arguments*.

$$\boxed{\frac{z_1}{z_2} = \frac{r_1}{r_2}[\cos(\theta_1 - \theta_2) + i\sin(\theta_1 - \theta_2)] \qquad z_2 \neq 0}$$

In particular, taking $z_1 = 1$ and $z_2 = z$ (and therefore $\theta_1 = 0$ and $\theta_2 = \theta$), we have the following, which is illustrated in Figure 7.

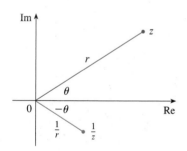

FIGURE 7

$$\boxed{\text{If}\quad z = r(\cos\theta + i\sin\theta), \quad\text{then}\quad \frac{1}{z} = \frac{1}{r}(\cos\theta - i\sin\theta).}$$

EXAMPLE 5 Find the product of the complex numbers $1 + i$ and $\sqrt{3} - i$ in polar form.

SOLUTION From Example 4 we have

$$1 + i = \sqrt{2}\left(\cos\frac{\pi}{4} + i\sin\frac{\pi}{4}\right)$$

and

$$\sqrt{3} - i = 2\left[\cos\left(-\frac{\pi}{6}\right) + i\sin\left(-\frac{\pi}{6}\right)\right]$$

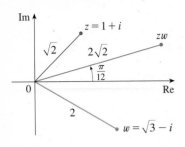

FIGURE 8

So, by Equation 1,

$$(1 + i)(\sqrt{3} - i) = 2\sqrt{2}\left[\cos\left(\frac{\pi}{4} - \frac{\pi}{6}\right) + i\sin\left(\frac{\pi}{4} - \frac{\pi}{6}\right)\right]$$

$$= 2\sqrt{2}\left(\cos\frac{\pi}{12} + i\sin\frac{\pi}{12}\right)$$

This is illustrated in Figure 8.

Repeated use of Formula 1 shows how to compute powers of a complex number. If

$$z = r(\cos\theta + i\sin\theta)$$

then

$$z^2 = r^2(\cos 2\theta + i\sin 2\theta)$$

and

$$z^3 = zz^2 = r^3(\cos 3\theta + i\sin 3\theta)$$

In general, we obtain the following result, which is named after the French mathematician Abraham De Moivre (1667–1754).

---

2 De Moivre's Theorem If $z = r(\cos\theta + i\sin\theta)$ and n is a positive integer, then

$$z^n = [r(\cos\theta + i\sin\theta)]^n = r^n(\cos n\theta + i\sin n\theta)$$

This says that *to take the nth power of a complex number we take the nth power of the modulus and multiply the argument by n.*

EXAMPLE 6 Find $\left(\frac{1}{2} + \frac{1}{2}i\right)^{10}$.

SOLUTION Since $\frac{1}{2} + \frac{1}{2}i = \frac{1}{2}(1 + i)$, it follows from Example 4(a) that $\frac{1}{2} + \frac{1}{2}i$ has the polar form

$$\frac{1}{2} + \frac{1}{2}i = \frac{\sqrt{2}}{2}\left(\cos\frac{\pi}{4} + i\sin\frac{\pi}{4}\right)$$

So by De Moivre's Theorem,

$$\left(\frac{1}{2} + \frac{1}{2}i\right)^{10} = \left(\frac{\sqrt{2}}{2}\right)^{10}\left(\cos\frac{10\pi}{4} + i\sin\frac{10\pi}{4}\right)$$

$$= \frac{2^5}{2^{10}}\left(\cos\frac{5\pi}{2} + i\sin\frac{5\pi}{2}\right) = \frac{1}{32}i$$

De Moivre's Theorem can also be used to find the nth roots of complex numbers. An nth root of the complex number z is a complex number w such that

$$w^n = z$$

Writing these two numbers in trigonometric form as

$$w = s(\cos\phi + i\sin\phi) \qquad \text{and} \qquad z = r(\cos\theta + i\sin\theta)$$

and using De Moivre's Theorem, we get

$$s^n(\cos n\phi + i\sin n\phi) = r(\cos\theta + i\sin\theta)$$

The equality of these two complex numbers shows that

$$s^n = r \quad\text{or}\quad s = r^{1/n}$$

and $$\cos n\phi = \cos\theta \quad\text{and}\quad \sin n\phi = \sin\theta$$

From the fact that sine and cosine have period 2π it follows that

$$n\phi = \theta + 2k\pi \quad\text{or}\quad \phi = \frac{\theta + 2k\pi}{n}$$

Thus $$w = r^{1/n}\left[\cos\left(\frac{\theta + 2k\pi}{n}\right) + i\sin\left(\frac{\theta + 2k\pi}{n}\right)\right]$$

Since this expression gives a different value of w for $k = 0, 1, 2, \ldots, n-1$, we have the following.

---

3 **Roots of a Complex Number** Let $z = r(\cos\theta + i\sin\theta)$ and let n be a positive integer. Then z has the n distinct nth roots

$$w_k = r^{1/n}\left[\cos\left(\frac{\theta + 2k\pi}{n}\right) + i\sin\left(\frac{\theta + 2k\pi}{n}\right)\right]$$

where $k = 0, 1, 2, \ldots, n-1$.

Notice that each of the nth roots of z has modulus $|w_k| = r^{1/n}$. Thus all the nth roots of z lie on the circle of radius $r^{1/n}$ in the complex plane. Also, since the argument of each successive nth root exceeds the argument of the previous root by $2\pi/n$, we see that the nth roots of z are equally spaced on this circle.

EXAMPLE 7 Find the six sixth roots of $z = -8$ and graph these roots in the complex plane.

SOLUTION In trigonometric form, $z = 8(\cos\pi + i\sin\pi)$. Applying Equation 3 with $n = 6$, we get

$$w_k = 8^{1/6}\left(\cos\frac{\pi + 2k\pi}{6} + i\sin\frac{\pi + 2k\pi}{6}\right)$$

We get the six sixth roots of -8 by taking $k = 0, 1, 2, 3, 4, 5$ in this formula:

$$w_0 = 8^{1/6}\left(\cos\frac{\pi}{6} + i\sin\frac{\pi}{6}\right) = \sqrt{2}\left(\frac{\sqrt{3}}{2} + \frac{1}{2}i\right)$$

$$w_1 = 8^{1/6}\left(\cos\frac{\pi}{2} + i\sin\frac{\pi}{2}\right) = \sqrt{2}\,i$$

$$w_2 = 8^{1/6}\left(\cos\frac{5\pi}{6} + i\sin\frac{5\pi}{6}\right) = \sqrt{2}\left(-\frac{\sqrt{3}}{2} + \frac{1}{2}i\right)$$

$$w_3 = 8^{1/6}\left(\cos\frac{7\pi}{6} + i\sin\frac{7\pi}{6}\right) = \sqrt{2}\left(-\frac{\sqrt{3}}{2} - \frac{1}{2}i\right)$$

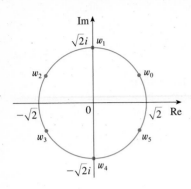

FIGURE 9
The six sixth roots of $z = -8$

$$w_4 = 8^{1/6}\left(\cos\frac{3\pi}{2} + i\sin\frac{3\pi}{2}\right) = -\sqrt{2}\,i$$

$$w_5 = 8^{1/6}\left(\cos\frac{11\pi}{6} + i\sin\frac{11\pi}{6}\right) = \sqrt{2}\left(\frac{\sqrt{3}}{2} - \frac{1}{2}i\right)$$

All these points lie on the circle of radius $\sqrt{2}$ as shown in Figure 9.

Complex Exponentials

We also need to give a meaning to the expression e^z when $z = x + iy$ is a complex number. The theory of infinite series as developed in Chapter 11 can be extended to the case where the terms are complex numbers. Using the Taylor series for e^x (11.10.11) as our guide, we define

4
$$e^z = \sum_{n=0}^{\infty} \frac{z^n}{n!} = 1 + z + \frac{z^2}{2!} + \frac{z^3}{3!} + \cdots$$

and it turns out that this complex exponential function has the same properties as the real exponential function. In particular, it is true that

5
$$e^{z_1 + z_2} = e^{z_1}e^{z_2}$$

If we put $z = iy$, where y is a real number, in Equation 4, and use the facts that

$$i^2 = -1, \quad i^3 = i^2 i = -i, \quad i^4 = 1, \quad i^5 = i, \quad \ldots$$

we get
$$e^{iy} = 1 + iy + \frac{(iy)^2}{2!} + \frac{(iy)^3}{3!} + \frac{(iy)^4}{4!} + \frac{(iy)^5}{5!} + \cdots$$

$$= 1 + iy - \frac{y^2}{2!} - i\frac{y^3}{3!} + \frac{y^4}{4!} + i\frac{y^5}{5!} + \cdots$$

$$= \left(1 - \frac{y^2}{2!} + \frac{y^4}{4!} - \frac{y^6}{6!} + \cdots\right) + i\left(y - \frac{y^3}{3!} + \frac{y^5}{5!} - \cdots\right)$$

$$= \cos y + i\sin y$$

Here we have used the Taylor series for $\cos y$ and $\sin y$ (Equations 11.10.16 and 11.10.15). The result is a famous formula called **Euler's formula**:

6
$$\boxed{\,e^{iy} = \cos y + i\sin y\,}$$

Combining Euler's formula with Equation 5, we get

7
$$e^{x+iy} = e^x e^{iy} = e^x(\cos y + i\sin y)$$

EXAMPLE 8 Evaluate: (a) $e^{i\pi}$ (b) $e^{-1+i\pi/2}$

SOLUTION
(a) From Euler's equation **6** we have

$$e^{i\pi} = \cos\pi + i\sin\pi = -1 + i(0) = -1$$

We could write the result of Example 8(a) as

$$e^{i\pi} + 1 = 0$$

This equation relates the five most famous numbers in all of mathematics: 0, 1, e, i, and π.

(b) Using Equation 7 we get

$$e^{-1+i\pi/2} = e^{-1}\left(\cos\frac{\pi}{2} + i\sin\frac{\pi}{2}\right) = \frac{1}{e}[0 + i(1)] = \frac{i}{e}$$

Finally, we note that Euler's equation provides us with an easier method of proving De Moivre's Theorem:

$$[r(\cos\theta + i\sin\theta)]^n = (re^{i\theta})^n = r^n e^{in\theta} = r^n(\cos n\theta + i\sin n\theta)$$

H Exercises

Access end-of-section exercises and tools for visualizing calculus online at **www.webassign.net**

Answers to Odd-Numbered Exercises

CHAPTER 1

True-False Quiz

1. False   **3.** False   **5.** True   **7.** False   **9.** True
11. False   **13.** True   **15.** True   **17.** False   **19.** True
21. True   **23.** True   **25.** False

Exercises

1. (a) 2.7   (b) 2.3, 5.6   (c) $[-6, 6]$ (d) $[-4, 4]$
(e) $[-4, 4]$ (f) Odd; its graph is symmetric about the origin.
3. $2a + h - 2$ **5.** $\left(-\infty, \frac{1}{3}\right) \cup \left(\frac{1}{3}, \infty\right)$, $(-\infty, 0) \cup (0, \infty)$
7. $\mathbb{R}, [0, 2]$
9. (a) Shift the graph 8 units upward.
(b) Shift the graph 8 units to the left.
(c) Stretch the graph vertically by a factor of 2, then shift it
1 unit upward.
(d) Shift the graph 2 units to the right and 2 units downward.
(e) Reflect the graph about the x-axis.
(f) Reflect the graph about the x-axis, then shift 3 units upward.
11.

$y = -\sin 2x$

13.

$y = \frac{1}{2}x^3 + 1$

15.

$x = -2$, $y = \dfrac{1}{x + 2}$

17. (a) Neither   (b) Odd   (c) Even   (d) Neither
19. (a) $(f \circ g)(x) = \sqrt{\sin x}$, $\{x \mid x \in [2n\pi, \pi + 2n\pi], n \text{ an integer}\}$
(b) $(g \circ f)(x) = \sin \sqrt{x}$, $[0, \infty)$
(c) $(f \circ f)(x) = \sqrt[4]{x}$, $[0, \infty)$
(d) $(g \circ g)(x) = \sin(\sin x)$, $\mathbb{R}$
21. $y = 0.2493x - 423.4818$; about 77.6 years
23. (a) (i) 3   (ii) 0   (iii) Does not exist   (iv) 2
(v) ∞ (vi) $-\infty$ (b) $x = 0, x = 2$ (c) $-3, 0, 2, 4$
25. 1   **27.** $\frac{3}{2}$ **29.** 3   **31.** ∞ **33.** $\frac{4}{7}$
35. $-\frac{1}{8}$ **37.** 0   **39.** 1

45. (a) (i) 3   (ii) 0   (iii) Does not exist   (iv) 0   (v) 0   (vi) 0
(b) At 0 and 3   (c)

47. $[0, \infty)$

1.

3. $f_n(x) = x^{2^{n+1}}$ **5.** $\frac{2}{3}$ **7.** -4
9. (a) Does not exist   (b) 1
11. $a = \frac{1}{2} \pm \frac{1}{2}\sqrt{5}$ **13.** $\frac{3}{4}$ **15.** (b) Yes   (c) Yes; no

CHAPTER 2

True-False Quiz

1. False   **3.** False   **5.** True   **7.** False
9. True   **11.** False

Exercises

1. (a) (i) 3 m/s   (ii) 2.75 m/s   (iii) 2.625 m/s
(iv) 2.525 m/s   (b) 2.5 m/s
3.

5. $a = f, b = f'', c = f'$

7. (a) The rate at which the cost changes with respect to the
interest rate; dollars/(percent per year)
(b) As the interest rate increases past 10%, the cost is increasing
at a rate of $1200/(percent per year).
(c) Always positive
9. The rate at which the total value of US currency in circulation is
changing in billions of dollars per year; $22.2 billion/year
11. $f'(x) = 3x^2 + 5$ **13.** $4x^7(x + 1)^3(3x + 2)$
15. $\frac{3}{2}\sqrt{x} - \dfrac{1}{2\sqrt{x}} - \dfrac{1}{\sqrt{x^3}}$ **17.** $x(\pi x \cos \pi x + 2 \sin \pi x)$

19. $\dfrac{8t^3}{(t^4 + 1)^2}$ **21.** $-\dfrac{\sec^2\sqrt{1 - x}}{2\sqrt{1 - x}}$

23. $\dfrac{1 - y^4 - 2xy}{4xy^3 + x^2 - 3}$ **25.** $\dfrac{2\sec 2\theta\,(\tan 2\theta - 1)}{(1 + \tan 2\theta)^2}$

27. $-(x - 1)^{-2}$ **29.** $\dfrac{2x - y\cos(xy)}{x\cos(xy) + 1}$ **31.** $-6x\csc^2(3x^2 + 5)$

33. $\dfrac{\cos\sqrt{x} - \sqrt{x}\sin\sqrt{x}}{2\sqrt{x}}$ **35.** $2\cos\theta\tan(\sin\theta)\sec^2(\sin\theta)$

37. $\tfrac{1}{5}(x\tan x)^{-4/5}(\tan x + x\sec^2 x)$

39. $\cos(\tan\sqrt{1 + x^3})(\sec^2\sqrt{1 + x^3})\dfrac{3x^2}{2\sqrt{1 + x^3}}$

41. $-\tfrac{4}{27}$ **43.** $-5x^4/y^{11}$ **45.** 1

47. $y = 2\sqrt{3}x + 1 - \pi\sqrt{3}/3$ **49.** $y = 2x + 1,\ y = -\tfrac{1}{2}x + 1$

51. (a) $\dfrac{10 - 3x}{2\sqrt{5 - x}}$ (b) $y = \tfrac{7}{4}x + \tfrac{1}{4},\ y = -x + 8$

(c)

53. $\left(\pi/4, \sqrt{2}\right), \left(5\pi/4, -\sqrt{2}\right)$ **55.** $y = -\tfrac{2}{3}x^2 + \tfrac{14}{3}x$

59. (a) 2   (b) 44

61. $2xg(x) + x^2 g'(x)$ **63.** $2g(x)g'(x)$ **65.** $g'(g(x))g'(x)$

67. $f'(x) = g'(\sin x)\cdot\cos x$

69. $\dfrac{f'(x)[g(x)]^2 + g'(x)\,[f(x)]^2}{[f(x) + g(x)]^2}$

71. $f'(g(\sin 4x))g'(\sin 4x)(\cos 4x)(4)$

73. (a) $v(t) = 3t^2 - 12;\ a(t) = 6t$ (b) $t > 2;\ 0 \le t < 2$
(c) 23   (d)   (e) $t > 2;\ 0 < t < 2$

75. 4 kg/m   **77.** $\tfrac{4}{3}$ cm²/min

79. 13 ft/s   **81.** 400 ft/h

83. (a) $L(x) = 1 + x;\ \sqrt[3]{1 + 3x} \approx 1 + x;\ \sqrt[3]{1.03} \approx 1.01$
(b) $-0.235 < x < 0.401$

85. $12 + \tfrac{3}{2}\pi \approx 16.7$ cm²   **87.** $\tfrac{1}{32}$ **89.** $\tfrac{1}{4}$ **91.** $\tfrac{1}{8}x^2$

PROBLEMS PLUS ▪ PAGE 151

1. $\left(\pm\sqrt{3}/2, \tfrac{1}{4}\right)$ **5.** $3\sqrt{2}$ **9.** $\left(0, \tfrac{5}{4}\right)$

11. (a) $4\pi\sqrt{3}/\sqrt{11}$ rad/s   (b) $40\left(\cos\theta + \sqrt{8 + \cos^2\theta}\right)$ cm
(c) $-480\pi\sin\theta\left(1 + \cos\theta/\sqrt{8 + \cos^2\theta}\right)$ cm/s

13. $x_T \in (3, \infty),\ y_T \in (2, \infty),\ x_N \in \left(0, \tfrac{5}{3}\right),\ y_N \in \left(-\tfrac{5}{2}, 0\right)$

15. (b) (i) $53°$ (or $127°$)   (ii) $63°$ (or $117°$)

17. R approaches the midpoint of the radius AO.

19. $-\sin a$ **21.** $(1, -2), (-1, 0)$

23. $\sqrt{29}/58$ **25.** $2 + \tfrac{375}{128}\pi \approx 11.204$ cm³/min

---

CHAPTER 3

CHAPTER 3 REVIEW ▪ PAGE 213

True-False Quiz

1. False   **3.** False   **5.** True   **7.** False   **9.** True
11. True   **13.** False   **15.** True   **17.** True   **19.** True

Exercises

1. Abs max $f(4) = 5$, abs and loc min $f(3) = 1$
3. Abs max $f(2) = \tfrac{2}{5}$, abs and loc min $f\left(-\tfrac{1}{3}\right) = -\tfrac{9}{2}$
5. Abs and loc max $f(\pi/6) = \pi/6 + \sqrt{3}$,
abs min $f(-\pi) = -\pi - 2$, loc min $f(5\pi/6) = 5\pi/6 - \sqrt{3}$
7. $\tfrac{1}{2}$ **9.** $-\tfrac{2}{3}$ **11.** $\tfrac{3}{4}$

13.

15.

17. A. $\mathbb{R}$ B. y-int 2
C. None   D. None
E. Dec on $(-\infty, \infty)$ F. None
G. CU on $(-\infty, 0)$;
CD on $(0, \infty)$; IP $(0, 2)$
H. See graph at right.

19. A. $\mathbb{R}$ B. y-int 0; x-int 0, 1
C. None   D. None
E. Inc on $\left(\tfrac{1}{4}, \infty\right)$, dec on $\left(-\infty, \tfrac{1}{4}\right)$
F. Loc min $f\left(\tfrac{1}{4}\right) = -\tfrac{27}{256}$
G. CU on $\left(-\infty, \tfrac{1}{2}\right), (1, \infty)$;
CD on $\left(\tfrac{1}{2}, 1\right)$; IP $\left(\tfrac{1}{2}, -\tfrac{1}{16}\right), (1, 0)$
H. See graph at right.

21. A. $\{x \mid x \neq 0, 3\}$
B. None   C. None
D. HA $y = 0$; VA $x = 0, x = 3$
E. Inc on $(1, 3)$; dec on $(-\infty, 0)$,
$(0, 1), (3, \infty)$
F. Loc min $f(1) = \tfrac{1}{4}$
G. CU on $(0, 3), (3, \infty)$; CD on $(-\infty, 0)$
H. See graph at right.

23. A. $\{x \mid x \neq -8\}$
B. y-int 0, x-int 0   C. None
D. VA $x = -8$; SA $y = x - 8$
E. Inc on $(-\infty, -16)$, $(0, \infty)$;
dec on $(-16, -8)$, $(-8, 0)$
F. Loc max $f(-16) = -32$;
loc min $f(0) = 0$
G. CU on $(-8, \infty)$; CD on $(-\infty, -8)$
H. See graph at right.

25. A. $[-2, \infty)$
B. y-int 0; x-int -2, 0
C. None   D. None
E. Inc on $\left(-\frac{4}{3}, \infty\right)$, dec on $\left(-2, -\frac{4}{3}\right)$
F. Loc min $f\left(-\frac{4}{3}\right) = -\frac{4}{9}\sqrt{6}$
G. CU on $(-2, \infty)$
H. See graph at right.

27. A. $\mathbb{R}$ B. y-int -2
C. About y-axis, period 2π D. None
E. Inc on $(2n\pi, (2n + 1)\pi)$, n an integer; dec on $((2n - 1)\pi, 2n\pi)$
F. Loc max $f((2n + 1)\pi) = 2$; loc min $f(2n\pi) = -2$
G. CU on $(2n\pi - (\pi/3), 2n\pi + (\pi/3))$;
CD on $(2n\pi + (\pi/3), 2n\pi + (5\pi/3))$; IP $\left(2n\pi \pm (\pi/3), -\frac{1}{4}\right)$
H.

29. Inc on $\left(-\sqrt{3}, 0\right)$, $\left(0, \sqrt{3}\right)$;
dec on $\left(-\infty, -\sqrt{3}\right)$, $\left(\sqrt{3}, \infty\right)$;
loc max $f\left(\sqrt{3}\right) = \frac{2}{9}\sqrt{3}$,
loc min $f\left(-\sqrt{3}\right) = -\frac{2}{9}\sqrt{3}$;
CU on $\left(-\sqrt{6}, 0\right)$, $\left(\sqrt{6}, \infty\right)$;
CD on $\left(-\infty, -\sqrt{6}\right)$, $\left(0, \sqrt{6}\right)$;
IP $\left(\sqrt{6}, \frac{5}{36}\sqrt{6}\right)$, $\left(-\sqrt{6}, -\frac{5}{36}\sqrt{6}\right)$

31. Inc on $(-0.23, 0)$, $(1.62, \infty)$; dec on $(-\infty, -0.23)$, $(0, 1.62)$;
loc max $f(0) = 2$; loc min $f(-0.23) \approx 1.96$, $f(1.62) \approx -19.2$;
CU on $(-\infty, -0.12)$, $(1.24, \infty)$;
CD on $(-0.12, 1.24)$; IP $(-0.12, 1.98)$, $(1.24, -12.1)$

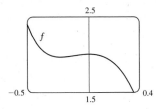

37. (a) 0   (b) CU on $\mathbb{R}$ **41.** $3\sqrt{3}\,r^2$
43. $4/\sqrt{3}$ cm from D **45.** $L = C$ **47.** \$11.50
49. 1.297383   **51.** 1.16718557

53. $f(x) = \frac{2}{5}x^{5/2} + \frac{3}{5}x^{5/3} + C$
55. $f(t) = t^2 + 3\cos t + 2$
57. $f(x) = \frac{1}{2}x^2 - x^3 + 4x^4 + 2x + 1$
59. $s(t) = t^2 + \cos t + 2$
61.

63. No
65. (b) About 8.5 in. by 2 in.   (c) $20/\sqrt{3}$ in., $20\sqrt{2/3}$ in.

PROBLEMS PLUS ▪ PAGE 218

5. $(-2, 4)$, $(2, -4)$ **7.** $\frac{4}{3}$ **9.** $(m/2, m^2/4)$
11. $-3.5 < a < -2.5$
13. (a) $x/(x^2 + 1)$ (b) $\frac{1}{2}$
15. (a) $-\tan\theta \left[\dfrac{1}{c}\dfrac{dc}{dt} + \dfrac{1}{b}\dfrac{db}{dt}\right]$

(b) $\dfrac{b\dfrac{db}{dt} + c\dfrac{dc}{dt} - \left(b\dfrac{dc}{dt} + c\dfrac{db}{dt}\right)\sec\theta}{\sqrt{b^2 + c^2 - 2bc\cos\theta}}$

17. (a) $T_1 = D/c_1$, $T_2 = (2h\sec\theta)/c_1 + (D - 2h\tan\theta)/c_2$,
$T_3 = \sqrt{4h^2 + D^2}/c_1$
(c) $c_1 \approx 3.85$ km/s, $c_2 \approx 7.66$ km/s, $h \approx 0.42$ km
21. $3/\left(\sqrt[3]{2} - 1\right) \approx 11\frac{1}{2}$ h

CHAPTER 4

CHAPTER 4 REVIEW ▪ PAGE 264

True-False Quiz

1. True   **3.** True   **5.** False   **7.** True   **9.** True
11. False   **13.** True   **15.** False   **17.** False

Exercises

1. (a) 8   (b) 5.7

3. $\frac{1}{2} + \pi/4$ **5.** 3   **7.** f is c, f' is b, $\int_0^x f(t)\,dt$ is a
9. 37   **11.** $\frac{9}{10}$ **13.** -76 **15.** $\frac{21}{4}$ **17.** Does not exist
19. $\frac{1}{3}\sin 1$ **21.** 0   **23.** $\dfrac{1}{2\pi}\sin^2\pi t + C$ **25.** $\frac{1}{2}\sqrt{2} - \frac{1}{2}$

27. $\frac{23}{3}$ **29.** $2\sqrt{1 + \sin x} + C$ **31.** $\frac{64}{5}$

33. $F'(x) = x^2/(1 + x^3)$ **35.** $g'(x) = 4x^3 \cos(x^8)$

37. $y' = \dfrac{2 \cos x - \cos \sqrt{x}}{2x}$ **39.** $4 \leqslant \int_1^3 \sqrt{x^2 + 3}\, dx \leqslant 4\sqrt{3}$

43. 0.280981

45. Number of barrels of oil consumed from Jan. 1, 2000, through Jan. 1, 2008

47. 72,400   **49.** 3   **51.** $(1 + x^2)(x \cos x + \sin x)/x^2$

PROBLEMS PLUS ■ PAGE 269

1. $\pi/2$ **3.** $f(x) = \frac{1}{2}x$ **5.** -1 **7.** $[-1, 2]$

9. (a) $\frac{1}{2}(n - 1)n$ (b) $\frac{1}{2}[\![b]\!](2b - [\![b]\!] - 1) - \frac{1}{2}[\![a]\!](2a - [\![a]\!] - 1)$

15. $2(\sqrt{2} - 1)$

CHAPTER 5

CHAPTER 5 REVIEW ■ PAGE 297

Exercises

1. $\frac{8}{3}$ **3.** $\frac{7}{12}$ **5.** $\frac{4}{3} + 4/\pi$ **7.** $64\pi/15$ **9.** $1656\pi/5$

11. $\frac{4}{3}\pi(2ah + h^2)^{3/2}$ **13.** $\int_{-\pi/3}^{\pi/3} 2\pi(\pi/2 - x)(\cos^2 x - \frac{1}{4})\, dx$

15. (a) $2\pi/15$ (b) $\pi/6$ (c) $8\pi/15$

17. (a) 0.38   (b) 0.87

19. Solid obtained by rotating the region $0 \leqslant y \leqslant \cos x$, $0 \leqslant x \leqslant \pi/2$ about the y-axis

21. Solid obtained by rotating the region $0 \leqslant x \leqslant \pi$, $0 \leqslant y \leqslant 2 - \sin x$ about the x-axis

23. 36   **25.** $\frac{125}{3}\sqrt{3}$ m³   **27.** 3.2 J

29. (a) $8000\pi/3 \approx 8378$ ft-lb   (b) 2.1 ft

31. $f(x)$

PROBLEMS PLUS ■ PAGE 299

1. (a) $f(t) = 3t^2$ (b) $f(x) = \sqrt{2x/\pi}$ **3.** $\frac{32}{27}$

5. (b) 0.2261   (c) 0.6736 m

(d) (i) $1/(105\pi) \approx 0.003$ in/s   (ii) $370\pi/3$ s ≈ 6.5 min

9. $y = \frac{32}{9}x^2$

11. (a) $V = \int_0^h \pi[f(y)]^2\, dy$

(c) $f(y) = \sqrt{kA/(\pi C)}\, y^{1/4}$. Advantage: the markings on the container are equally spaced.

13. $b = 2a$

CHAPTER 6

CHAPTER 6 REVIEW ■ PAGE 378

True-False Quiz

1. True   **3.** False   **5.** True   **7.** True   **9.** False

11. False   **13.** False   **15.** True   **17.** True

Exercises

1. No   **3.** (a) 7   (b) $\frac{1}{8}$

5.

7.

9.

11. (a) 9   (b) 2

13. $e^{1/3}$

15. $\ln \ln 17$ **17.** $\sqrt{1 + e}$

19. $\tan 1$ **21.** $f'(t) = t + 2t \ln t$

23. $h'(\theta) = 2 \sec^2(2\theta)e^{\tan 2\theta}$ **25.** $y' = 5 \sec 5x$

27. $y' = \dfrac{4x}{1 + 16x^2} + \tan^{-1}(4x)$ **29.** $y' = 2 \tan x$

31. $y' = -\dfrac{e^{1/x}(1 + 2x)}{x^4}$ **33.** $y' = 3^{x \ln x}(\ln 3)(1 + \ln x)$

35. $H'(v) = \dfrac{v}{1 + v^2} + \tan^{-1}v$

37. $y' = 2x^2 \cosh(x^2) + \sinh(x^2)$ **39.** $y' = \cot x - \sin x \cos x$

41. $y' = -(1/x)[1 + 1/(\ln x)^2]$

43. $y' = 3 \tanh 3x$ **45.** $y' = (\cosh x)/\sqrt{\sinh^2 x - 1}$

47. $y' = \dfrac{-3 \sin(e^{\sqrt{\tan 3x}})e^{\sqrt{\tan 3x}} \sec^2(3x)}{2\sqrt{\tan 3x}}$ **49.** $e^{g(x)}g'(x)$

51. $g'(x)/g(x)$ **53.** $2^x(\ln 2)^n$ **57.** $y = -x + 2$ **59.** $(-3, 0)$

61. (a) $y = \frac{1}{4}x + \frac{1}{4}(\ln 4 + 1)$ (b) $y = ex$

63. 0   **65.** 0   **67.** $-\infty$ **69.** -1

71. 1   **73.** 8   **75.** 0   **77.** $\frac{1}{2}$

79. A. $[-\pi, \pi]$ B. y-int 0; x-int $-\pi, 0, \pi$

C. None   D. None

E. Inc on $(-\pi/4, 3\pi/4)$; dec on $(-\pi, -\pi/4)$, $(3\pi/4, \pi)$

F. Loc max $f(3\pi/4) = \frac{1}{2}\sqrt{2}e^{3\pi/4}$, loc min $f(-\pi/4) = -\frac{1}{2}\sqrt{2}e^{3\pi/4}$

G. CU on $(-\pi/2, \pi/2)$; CD on $(-\pi, -\pi/2)$, $(\pi/2, \pi)$;

IP $(-\pi/2, -e^{-\pi/2})$, $(\pi/2, e^{\pi/2})$

H.

81. A. $(0, \infty)$ B. x-int 1
C. None   D. None
E. Inc on $(1/e, \infty)$; dec on $(0, 1/e)$
F. Loc min $f(1/e) = -1/e$
G. CU on $(0, \infty)$
H. See graph at right.

83. A. $\mathbb{R}$
B. y-int -2; x-int 2
C. None   D. HA $y = 0$
E. Inc on $(-\infty, 3)$; dec on $(3, \infty)$
F. Loc max $f(3) = e^{-3}$
G. CU on $(4, \infty)$; CD on $(-\infty, 4)$;
IP $(4, 2e^{-4})$
H. See graph at right.

85.

For $c > 0$, $\lim_{x \to \infty} f(x) = 0$ and $\lim_{x \to -\infty} f(x) = -\infty$.
For $c < 0$, $\lim_{x \to \infty} f(x) = \infty$ and $\lim_{x \to -\infty} f(x) = 0$.
As $|c|$ increases, the max and min points and the IPs get closer
to the origin.
87. $v(t) = -Ae^{-ct}[c \cos(\omega t + \delta) + \omega \sin(\omega t + \delta)]$,
$a(t) = Ae^{-ct}[(c^2 - \omega^2) \cos(\omega t + \delta) + 2c\omega \sin(\omega t + \delta)]$
89. (a) $200(3.24)^t$ (b) $\approx 22,040$
(c) $\approx 25,910$ bacteria/h   (d) $(\ln 50)/(\ln 3.24) \approx 3.33$ h
91. 4.32 days   **93.** $\frac{1}{4}(1 - e^{-2})$ **95.** $\arctan e - \pi/4$
97. $2e^{\sqrt{x}} + C$ **99.** $\frac{1}{2}\ln|x^2 + 2x| + C$
101. $-\frac{1}{2}[\ln(\cos x)]^2 + C$ **103.** $2^{\tan \theta}/\ln 2 + C$
105. $-(1/x) - 2\ln|x| + x + C$ **109.** $e^{\sqrt{x}}/(2x)$
111. $\frac{1}{3}\ln 4$ **113.** $\pi^2/4$ **115.** $\frac{2}{3}$ **117.** $2/e$
121. $e^{2x}(1 + 2x)/(1 - e^{-x})$

PROBLEMS PLUS ▪ PAGE 383

3. Abs max $f(-5) = e^{45}$, no abs min   **9.** $1/\sqrt{2}$ **11.** $a = \frac{1}{2}$
15. $2\sqrt{e}$ **17.** $a \le e^{1/e}$

CHAPTER 7

CHAPTER 7 REVIEW ▪ PAGE 437

True-False Quiz

1. False   **3.** False   **5.** False   **7.** False
9. (a) True   (b) False   **11.** False   **13.** False

Exercises

1. $\frac{7}{2} + \ln 2$ **3.** $e - 1$
5. $\ln|2t + 1| - \ln|t + 1| + C$
7. $\frac{2}{15}$ **9.** $-\cos(\ln t) + C$ **11.** $\sqrt{3} - \frac{1}{3}\pi$
13. $3e^{\sqrt[3]{x}}(x^{2/3} - 2x^{1/3} + 2) + C$
15. $-\frac{1}{2}\ln|x| + \frac{3}{2}\ln|x + 2| + C$
17. $x \sec x - \ln|\sec x + \tan x| + C$
19. $\frac{1}{18}\ln(9x^2 + 6x + 5) + \frac{1}{9}\tan^{-1}[\frac{1}{2}(3x + 1)] + C$
21. $\ln|x - 2 + \sqrt{x^2 - 4x}| + C$
23. $\ln\left|\dfrac{\sqrt{x^2 + 1} - 1}{x}\right| + C$
25. $\frac{3}{2}\ln(x^2 + 1) - 3\tan^{-1}x + \sqrt{2}\tan^{-1}(x/\sqrt{2}) + C$
27. $\frac{2}{5}$ **29.** 0   **31.** $6 - \frac{3}{2}\pi$
33. $\dfrac{x}{\sqrt{4 - x^2}} - \sin^{-1}\left(\dfrac{x}{2}\right) + C$
35. $4\sqrt{1 + \sqrt{x}} + C$
37. $\frac{1}{2}\sin 2x - \frac{1}{8}\cos 4x + C$
39. $\frac{1}{8}e - \frac{1}{4}$ **41.** $\frac{1}{36}$ **43.** D
45. $4\ln 4 - 8$ **47.** $-\frac{4}{3}$ **49.** $\pi/4$
51. $(x + 1)\ln(x^2 + 2x + 2) + 2\arctan(x + 1) - 2x + C$
53. 0
55. $\frac{1}{4}(2x - 1)\sqrt{4x^2 - 4x - 3} -$
$$\ln|2x - 1 + \sqrt{4x^2 - 4x - 3}| + C$$
57. $\frac{1}{2}\sin x\sqrt{4 + \sin^2 x} + 2\ln(\sin x + \sqrt{4 + \sin^2 x}) + C$
61. No
63. (a) 1.925444   (b) 1.920915   (c) 1.922470
65. (a) $0.01348, n \ge 368$ (b) $0.00674, n \ge 260$
67. 8.6 mi
69. (a) 3.8   (b) 1.7867, 0.000646   (c) $n \ge 30$
71. (a) D   (b) C
73. 2   **75.** $\frac{3}{16}\pi^2$

PROBLEMS PLUS ▪ PAGE 441

1. About 1.85 inches from the center   **3.** 0
7. $f(\pi) = -\pi/2$ **11.** $(b^b a^{-a})^{1/(b-a)}e^{-1}$ **13.** $\frac{1}{8}\pi - \frac{1}{12}$
15. $2 - \sin^{-1}(2/\sqrt{5})$

CHAPTER 8

CHAPTER 8 REVIEW ▪ PAGE 474

Exercises

1. $\frac{15}{2}$ **3.** (a) $\frac{21}{16}$ (b) $\frac{41}{10}\pi$
5. 3.8202   **7.** $\frac{124}{5}$ **9.** ≈ 458 lb   **11.** $(\frac{8}{5}, 1)$

13. $\left(2, \frac{2}{3}\right)$ **15.** $2\pi^2$ **17.** $7166.67

19. (a) $f(x) \geq 0$ for all x and $\int_{-\infty}^{\infty} f(x)\,dx = 1$
(b) ≈ 0.3455 (c) 5, yes

21. (a) $1 - e^{-3/8} \approx 0.31$ (b) $e^{-5/4} \approx 0.29$
(c) $8 \ln 2 \approx 5.55$ min

PROBLEMS PLUS ■ **PAGE 476**

1. $\frac{2}{3}\pi - \frac{1}{2}\sqrt{3}$

3. (a) $2\pi r(r \pm d)$ (b) $\approx 3.36 \times 10^6$ mi^2
(d) $\approx 7.84 \times 10^7$ mi^2

5. (a) $P(z) = P_0 + g \int_0^z \rho(x)\,dx$
(b) $(P_0 - \rho_0 gH)(\pi r^2) + \rho_0 g H e^{L/H} \int_{-r}^{r} e^{x/H} \cdot 2\sqrt{r^2 - x^2}\,dx$

7. Height $\sqrt{2}\,b$, volume $\left(\frac{28}{27}\sqrt{6} - 2\right)\pi b^3$ **9.** 0.14 m

11. $2/\pi, 1/\pi$ **13.** $(0, -1)$

CHAPTER 9

CHAPTER 9 REVIEW ■ **PAGE 516**

True-False Quiz

1. True   **3.** False   **5.** True   **7.** True

Exercises

1. (a)

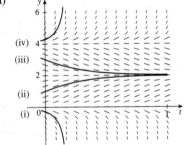

(b) $0 \leq c \leq 4$; $y = 0$, $y = 2$, $y = 4$

3. (a) $y(0.3) \approx 0.8$

(b) 0.75676
(c) $y = x$ and $y = -x$; there is a loc max or loc min

5. $y = \left(\frac{1}{2}x^2 + C\right)e^{-\sin x}$

7. $y = \pm\sqrt{\ln(x^2 + 2x^{3/2} + C)}$

9. $r(t) = 5e^{t - t^2}$ **11.** $y = \frac{1}{2}x(\ln x)^2 + 2x$ **13.** $x = C - \frac{1}{2}y^2$

15. (a) $P(t) = \dfrac{2000}{1 + 19e^{-0.1t}}$; ≈ 560 (b) $t = -10 \ln \frac{2}{57} \approx 33.5$

17. (a) $L(t) = L_\infty - [L_\infty - L(0)]e^{-kt}$ (b) $L(t) = 53 - 43e^{-0.2t}$

19. 15 days   **21.** $k \ln h + h = (-R/V)t + C$

23. (a) Stabilizes at 200,000
(b) (i) $x = 0$, $y = 0$: Zero populations
(ii) $x = 200,000$, $y = 0$: In the absence of birds, the insect
population is always 200,000.
(iii) $x = 25,000$, $y = 175$: Both populations are stable.
(c) The populations stabilize at 25,000 insects and 175 birds.
(d)

25. (a) $y = (1/k) \cosh kx + a - 1/k$ or
$y = (1/k) \cosh kx - (1/k) \cosh kb + h$ (b) $(2/k) \sinh kb$

PROBLEMS PLUS ■ **PAGE 520**

1. $f(x) = \pm 10e^x$ **5.** $y = x^{1/n}$ **7.** $20°C$

9. (b) $f(x) = \dfrac{x^2 - L^2}{4L} - \frac{1}{2}L \ln\left(\dfrac{x}{L}\right)$ (c) No

11. (a) 9.8 h   (b) $31,900\pi$ ft^2; 2000π ft^2/h
(c) 5.1 h

13. $x^2 + (y - 6)^2 = 25$

15. $y = K/x$, $K \neq 0$

CHAPTER 10

CHAPTER 10 REVIEW ■ **PAGE 561**

True-False Quiz

1. False   **3.** False   **5.** True   **7.** False   **9.** True

Exercises

1. $x = y^2 - 8y + 12$ **3.** $y = 1/x$

5. $x = t$, $y = \sqrt{t}$; $x = t^4$, $y = t^2$;
$x = \tan^2 t$, $y = \tan t$, $0 \leq t < \pi/2$

7. (a)

$\left(4, \frac{2\pi}{3}\right)$

(b) $\left(3\sqrt{2}, 3\pi/4\right)$, $\left(-3\sqrt{2}, 7\pi/4\right)$

$\left(-2, 2\sqrt{3}\right)$

9.

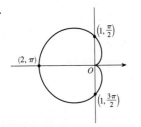

$\left(1, \frac{\pi}{2}\right)$, $(2, \pi)$, $\left(1, \frac{3\pi}{2}\right)$

11.

$\theta = \frac{\pi}{6}$, $(1, 0)$

13.

$(2, \pi)$, $(2, 0)$

15.

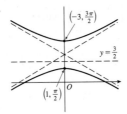

$\left(-3, \frac{3\pi}{2}\right)$, $y = \frac{3}{2}$, $\left(1, \frac{\pi}{2}\right)$

17. $r = \dfrac{2}{\cos\theta + \sin\theta}$

19.

$r = \dfrac{\sin\theta}{\theta}$

21. 2   **23.** −1

25. $\dfrac{1 + \sin t}{1 + \cos t}$, $\dfrac{1 + \cos t + \sin t}{(1 + \cos t)^3}$ **27.** $\left(\frac{11}{8}, \frac{3}{4}\right)$

29. Vertical tangent at $\left(\frac{3}{2}a, \pm\frac{1}{2}\sqrt{3}\,a\right)$, $(-3a, 0)$; horizontal tangent at $(a, 0)$, $\left(-\frac{1}{2}a, \pm\frac{3}{2}\sqrt{3}\,a\right)$

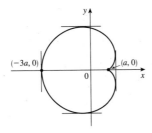

$(-3a, 0)$, $(a, 0)$

31. 18   **33.** $(2, \pm\pi/3)$ **35.** $\frac{1}{2}(\pi - 1)$
37. $2\left(5\sqrt{5} - 1\right)$

39. $\dfrac{2\sqrt{\pi^2 + 1} - \sqrt{4\pi^2 + 1}}{2\pi} + \ln\left(\dfrac{2\pi + \sqrt{4\pi^2 + 1}}{\pi + \sqrt{\pi^2 + 1}}\right)$

41. $471{,}295\pi/1024$

43. All curves have the vertical asymptote $x = 1$. For $c < -1$, the curve bulges to the right. At $c = -1$, the curve is the line $x = 1$. For $-1 < c < 0$, it bulges to the left. At $c = 0$ there is a cusp at $(0, 0)$. For $c > 0$, there is a loop.

45. $(\pm1, 0)$, $(\pm3, 0)$ **47.** $\left(-\frac{25}{24}, 3\right)$, $(-1, 3)$

49. $\dfrac{x^2}{25} + \dfrac{y^2}{9} = 1$ **51.** $\dfrac{y^2}{72/5} - \dfrac{x^2}{8/5} = 1$

53. $\dfrac{x^2}{25} + \dfrac{(8y - 399)^2}{160{,}801} = 1$ **55.** $r = \dfrac{4}{3 + \cos\theta}$

57. (a) At $(0, 0)$ and $\left(\frac{3}{2}, \frac{3}{2}\right)$
(b) Horizontal tangents at $(0, 0)$ and $\left(\sqrt[3]{2}, \sqrt[3]{4}\right)$; vertical tangents at $(0, 0)$ and $\left(\sqrt[3]{4}, \sqrt[3]{2}\right)$
(d)   **(g)** $\frac{3}{2}$

$y = -x - 1$

PROBLEMS PLUS ▪ **PAGE 564**

1. $\ln(\pi/2)$ **3.** $\left[-\frac{3}{4}\sqrt{3}, \frac{3}{4}\sqrt{3}\right] \times [-1, 2]$

CHAPTER 11

CHAPTER 11 REVIEW ▪ **PAGE 636**

True-False Quiz
1. False   **3.** True   **5.** False   **7.** False   **9.** False
11. True   **13.** True   **15.** False   **17.** True
19. True   **21.** True

Exercises
1. $\frac{1}{2}$ **3.** D   **5.** 0   **7.** e^{12} **9.** 2   **11.** C   **13.** C
15. D   **17.** C   **19.** C   **21.** C   **23.** CC   **25.** AC
27. $\frac{1}{11}$ **29.** $\pi/4$ **31.** e^{-e} **35.** 0.9721

37. 0.18976224, error $< 6.4 \times 10^{-7}$

41. $4, [-6, 2)$ **43.** $0.5, [2.5, 3.5)$

45. $\dfrac{1}{2} \displaystyle\sum_{n=0}^{\infty} (-1)^n \left[\dfrac{1}{(2n)!} \left(x - \dfrac{\pi}{6}\right)^{2n} + \dfrac{\sqrt{3}}{(2n+1)!} \left(x - \dfrac{\pi}{6}\right)^{2n+1} \right]$

47. $\displaystyle\sum_{n=0}^{\infty} (-1)^n x^{n+2}, R = 1$

49. $\ln 4 - \displaystyle\sum_{n=1}^{\infty} \dfrac{x^n}{n 4^n}, R = 4$

51. $\displaystyle\sum_{n=0}^{\infty} (-1)^n \dfrac{x^{8n+4}}{(2n+1)!}, R = \infty$

53. $\dfrac{1}{2} + \displaystyle\sum_{n=1}^{\infty} \dfrac{1 \cdot 5 \cdot 9 \cdot \,\cdots\, \cdot (4n-3)}{n! \, 2^{6n+1}} x^n, R = 16$

55. $C + \ln |x| + \displaystyle\sum_{n=1}^{\infty} \dfrac{x^n}{n \cdot n!}$

57. (a) $1 + \dfrac{1}{2}(x-1) - \dfrac{1}{8}(x-1)^2 + \dfrac{1}{16}(x-1)^3$

(b)

(c) 0.000006

59. $-\dfrac{1}{6}$

PROBLEMS PLUS ▪ PAGE 639

1. $15!/5! = 10{,}897{,}286{,}400$

3. (b) 0 if $x = 0$, $(1/x) - \cot x$ if $x \neq k\pi$, k an integer

5. (a) $s_n = 3 \cdot 4^n$, $l_n = 1/3^n$, $p_n = 4^n/3^{n-1}$ (c) $\dfrac{2}{5}\sqrt{3}$

9. $(-1, 1)$, $\dfrac{x^3 + 4x^2 + x}{(1 - x)^4}$

11. $\ln \dfrac{1}{2}$ **13.** (a) $\dfrac{250}{101} \pi (e^{-(n-1)\pi/5} - e^{-n\pi/5})$ (b) $\dfrac{250}{101} \pi$

19. $\dfrac{\pi}{2\sqrt{3}} - 1$

21. $-\left(\dfrac{\pi}{2} - \pi k\right)^2$ where k is a positive integer

Index

RP denotes Reference Page numbers.
Numbers followed by asterisks denote pages in Sections 6.2*, 6.3*, and 6.4*.

Abel, Niels, 126
absolute maximum and minimum
    values, 156
absolute value, 16, A6, A47
absolute value function, 16
absolutely convergent series, 598
acceleration as a rate of change, 98, 128,
    254
Achilles and the tortoise, 5
adaptive numerical integration, 428
addition formulas for sine and cosine, A26
algebraic function, 27
alternating harmonic series, 596, 599
alternating series, 594
Alternating Series Estimation
    Theorem, 597
Alternating Series Test, 595
analytic geometry, A9
angle(s), A21
    between curves, 152
    of deviation, 162
    negative or positive, A22
    standard position, A22
antiderivative, 208
antidifferentiation formulas, 210
aphelion, 559
approach path of an aircraft, 122
approximate integration, 420
approximating cylinder, 280
approximating surface, 450
approximation
    by differentials, 143
    to e, 315, 316, 334*
    linear, 141
    by the Midpoint Rule, 238, 421
    by Newton's method, 205
    by an nth-degree Taylor polynomial, 145
    quadratic, 145
    by Riemann sums, 232
    by Simpson's Rule, 424, 426
    tangent line, 141

    by Taylor polynomials, 145, 628
    by Taylor's Inequality, 618, 629
    by the Trapezoidal Rule, 421
Archimedes, 257, 458
Archimedes' Principle, 300
arc length, 444
    of a parametric curve, 533
    of a polar curve, 548
arc length contest, 449
arc length formula, 445
arc length function, 447
arcsine function, 357
area, 2, 222
    of a circle, 398
    under a curve, 222, 22
    between curves, 272, 275
    of an ellipse, 398
    by exhaustion, 2, 51
    enclosed by a parametric curve, 532
    in polar coordinates, 545
    of a sector of a circle, 545
    surface, 535
    of a surface of a revolution, 450, 454
area function, 242
area problem, 2, 222
argument of a complex number, A48
arrow diagram, 11
astroid, 530
asymptote(s), 187
    in graphing, 187
    horizontal, 177, 187
    of a hyperbola, 553, A18
    slant, 187, 190
    vertical, 47, 187
autonomous differential equation, 487
average rate of change, 88, 128
average value of a function, 294, 470
average velocity, 4, 38, 86
axes, coordinate, A10
axes of an ellipse, A17
axis of a parabola, 549

bacterial growth, 499, 504
Barrow, Isaac, 3, 51, 91, 243, 257
base of a cylinder, 278
base of a logarithm, 319, 348*
    change of, 322, 349*
baseball and calculus, 296. 521
Bernoulli, James, 491
Bernoulli, John, 370, 377, 491, 528, 616
Bessel, Friedrich, 607
Bessel function, 607
Bézier, Pierre, 536
Bézier curves, 527, 536
binomial coefficients, 622
binomial series, 622, 623
    discovery by Newton, 627
binomial theorem, 101, RP1
blackbody radiation, 634
blood flow, 134, 466
boundary curve, 918
bounded sequence, 573
brachistochrone problem, 528
Brahe, Tycho, 710
branches of a hyperbola, 553, A18
Buffon's needle problem, 477

cable (hanging), 365
calculator, graphing, 191, 526, 543, A41.
    See also computer algebra system
calculus, 8
    differential, 3
    integral, 2
    invention of, 8, 257
    of inverse functions, 308
cancellation equations
    for inverse functions, 306
    for inverse trigonometric functions, 357,
        359
    for logarithms, 319, 348*
    for natural exponential functon, 339*
cans, minimizing manufacturing cost of, 203
cardiac output, 467

cardioid, 540
carrying capacity, 172, 481, 501
Cartesian coordinate system, A10
Cartesian plane, A10
Cassini, Giovanni, 545
CAS. *See* computer algebra system
catenary, 365
Cauchy, Augustin-Louis, 60, 375
Cauchy's Mean Value Theorem, 376
Cavalieri, 426
center of gravity. *See* center of mass
center of mass, 458
    of a plate, 461
centroid of a plane region, 460, 461, 462
Chain Rule, 116, 119
change of base, formula for, 322, 349*
change of variables in integration, 258
charge, electric, 131
chemical reaction, 131
circle, area of, 398
circle, equation of, A14
circular cylinder, 278
closed interval, A3
Closed Interval Method, 160
cochleoid, 561
coefficient(s)
    binomial, 622
    of inequality, 277
    of a polynomial, 23
    of a power series, 605
coffee cups, volumes of, 464
combinations of functions, 33
common ratio, 578
comparison properties of the integral, 241
comparison test for improper integrals, 435
Comparison Test for series, 591
Comparison Theorem for integrals, 435
Completeness Axiom, 574
complex conjugate, A47
complex exponentials, A52
complex number(s), A46
    addition and subtraction of, A46
    argument of, A48
    division of, A47, A49
    equality of, A46
    imaginary part of, A46
    modulus of, A47
    multiplication of, A46, A49
    polar form, A48
    powers of, A50
    principal square root of, A47
    real part of, A46
    roots of, A51
composition of functions, 33
    continuity of, 70
    derivative of, 118

compound interest, 355
compressibility, 132
computer algebra system, 43, 526
    for integration, 408, 416, 614
    pitfalls of using, 43
computer algebra system, graphing
        with, A41
    a curve, 191
    parametric equations, 526
    polar curve, 543
    sequence, 571
concavity, 171, 188
Concavity Test, 172, A38
concentration, 132
conchoid of Nicomedes, 529
conditionally convergent series, 599
conic section, 549, 555
    directrix, 549, 555
    eccentricity, 555
    focus (foci), 549, 551, 555
    polar equation, 556
    shifted, 554, A19
    vertex (vertices), 549
conjugates, properties of, A47
constant force, 290
constant function, 99
Constant Multiple Law of limits, 49
Constant Multiple Rule, 101
continuity
    of a function, 64, 66
    on an interval, 66
    from the left or right, 66
continuous compounding of interest, 355
continuous random variable, 468
convergence
    absolute, 598
    conditional, 599
    of an improper integral, 430, 433
    interval of, 608
    radius of, 608
    of a sequence, 568, 572
    of a series, 578, 585
convergent improper integral, 430, 433
convergent sequence, 568, 572
convergent series, 578, 585
    properties of, 582
coordinate axes, A10
coordinate plane(s), A10
coordinate system, A2
    Cartesian, A10
    polar, 537
    retangular, A10
cosine function, A23
    derivative of, 113
    graph of, 27, A28
    power series for, 621, 624

cost function, 135, 202
critical number, 159
cross-section, 278
cubic function, 24
current, 131
curve(s)
    Bézier, 527, 536
    demand, 465
    epicycloid, 530
    folium of Descartes, 123
    initial point, 525
    ovals of Cassini, 545
    parametric, 524
    polar, 538
    smooth, 444
    strophoid, 562
    terminal point, 525
curve fitting, 21
curve-sketching procedure, 186
cusp, 529
cycloid, 527
cylinder, 278
cylindrical shell, 287

Daisuke, T., 576
decay, law of natural, 351
decay, radioactive, 353
decomposition of a function, 34
decreasing function, 19
decreasing sequence, 572
definite integral, 231
    properties of, 239
    Substitution Rule for, 261
definite integration
    by parts, 386, 389
    by substitution, 261
degree of a polynomial, 23
delta (Δ) notation, 88
demand curve, 465
demand function, 202, 465
De Moivre, Abraham, A50
De Moivre's Theorem, A50
density
    linear, 130, 131, 254
    liquid, 456
    mass vs. weight, 456
dependent variable, 10
derivative(s), 87, 91
    of a composite function, 116
    of a constant function, 99
    domain of, 91
    of exponential functions, 314, 315, 341*
    as a function, 87, 91
    higher, 97
    of hyperbolic functions, 366
    of an integral, 246

of an inverse function, 324
of inverse hyperbolic functions, 368
of inverse trigonometric functions, 361
of logarithmic functions, 324, 328, 332*, 335*
notation, 94
of a power function, 100, 106, 107
of a power series, 612
of a product, 103
of a quotient, 105
as a rate of change, 88
second, 97
as the slope of a tangent, 87
third, 98
of trigonometric functions, 110, 113
Descartes, René, A10
descent of aircraft, determining start of, 122
Difference Law of limits, 49
difference quotient, 12
Difference Rule, 102
differentiable function, 94
differential, 143
differential calculus, 3
differential equation, 210, 351, 480, 482
autonomous, 487
family of solutions, 480, 483
first-order, 480, 507
general solution of, 483
linear, 507
logistic, 481, 501
order of, 482
second-order, 482
separable, 491
solution of, 482
differentiation, 94
formulas for, 109, RP5
implicit, 122, 123
logarithmic, 329, 338*
of a function defined as a power series, 612
term-by-term, 611
differentiation operator, 94
Direct Substitution Property, 51
direction field, 484, 485
directrix, 549, 555
discontinuity, 64, 66
discontinuous function, 64, 66
discontinuous integrand, 433
disk method for approximating volume, 280
dispersion, 163
displacement, 86, 255
distance
between points in a plane, A10
between real numbers, A7
distance formula, A11

distance problem, 229
divergence
of an improper integral, 430, 433
of a series, 578, 585
of a sequence, 568, 572
Divergence, Test for, 582
divergent improper integral, 430, 433
divergent sequence, 568, 572
divergent series, 578, 585
division of power series, 625
domain of a function, 10, 187
convention for, 14
double-angle formulas, A27
dye dilution method, 467

e (the number), 315, 334*
as a limit, 331, 350*
as a sum of an infinite series, 619
eccentricity, 555
electric circuit, 486, 493, 510
electric current to a flash bulb, 37
elementary function, integrability of, 413
element of a set, A3
ellipse, 551, 555, A17
area, 397
directrix, 555
eccentricity, 555
equation of, 552
foci, 551, 555
major axis, 551
minor axis, 551
polar equation, 556, 559
reflection property, 552
vertices, 551
empirical model, 21
energy, kinetic, 296
epicycloid, 530
equation(s)
of a circle, A15
differential (see differential equation)
of an ellipse, 552, 557, A17
of a graph, A14, A15
of a hyperbola, 553, 557, A18
of a line, A11, A12, A13
linear, A13
logistic difference, 576
logistic differential, 481
Lotka-Volterra, 512
nth-degree, 126
of a parabola, 550, 557, A16
parametric, 524
point-slope, 17, A11
polar, 537, 556
predator-prey, 512
second-degree, A14
slope-intercept, A12

equilateral hyperbola, A19
equilibrium point, 513
equilibrium solution, 481, 512
error
in approximate integration, 422, 423
percentage, 145
relative, 145
in Taylor approximation, 629
error bounds, 423, 428
error estimate
for alternating series, 597
for the Midpoint Rule, 423
for Simpson's Rule, 428
for the Trapezoidal Rule, 423
estimate of the sum of a series, 587, 593, 597, 602
Euclid, 51
Eudoxus, 2, 51, 257
Euler, Leonhard, 488, 585, 619
Euler's formula, A52
Euler's Method, 488, 489
even function, 17, 187
exponential decay, 351
exponential function(s), 28, 310, 315, RP4
with base a, 310, 328, 344*
defined as a limit, 311
derivative, of 314, 318, 341*, 344*
graphs of, 311, 315, 317, 345*
integration of, 237, 318, 347*, 624
limits of, 312, 317, 340*
power series for, 616, 620, 624
properties of, 312, 344*
exponential graph, 315, 317
exponential growth, 316, 504
exponential integral, 347*
exponents, laws of, 312, 341*, 344*
extrapolation, 23
extreme value, 156
Extreme Value Theorem, 157

family
of epicycloids and hypocycloids, 529
of exponential functions, 311
of functions, 195, 196, A45
of implicit curves, 127
of parametric curves, 528
of polar curves, 545
of solutions, 480, 483
fat circles, 127
Fermat, Pierre, 3, 91, 158, 257, A10
Fermat's Theorem, 158
Fibonacci, 567
Fibonacci sequence, 567
First Derivative Test, 169
for Absolute Extreme Values, 200

first-order linear differential equation, 480, 507

first-order optics, 634

fixed point of a function, 82

flash bulb, current to, 37

flux, 466

FM synthesis, 195

foci, 551

focus, 549, 555
    of a conic section, 555
    of an ellipse, 551, 555
    of a hyperbola, 553
    of a parabola, 549

folium of Descartes, 123

foot-pount (unit of work), 291

force
    constant, 290
    exerted by fluid, 456

Fourier, Joseph, 137

four-leaved rose, 541

fractions (partial), 403

Fresnel, Augustin, 246

Fresnel function, 246

function(s), 10
    absolute value, 16
    algebraic, 27
    arc length, 447
    arcsine, 357
    area, 242
    arrow diagram of, 11
    average value of, 294, 470
    Bessel, 607
    combinations of, 33
    composite, 33, 116
    constant, 99
    continuity of, 64, 66
    cost, 135, 202
    cubic, 24
    decreasing, 19
    demand, 202, 465
    derivative of, 87
    differentiability of, 94
    discontinuous, 64
    domain of, 10
    elementary, 413
    even, 17, 187
    exponential, 28, 310, 315, RP4
    extreme values of, 156
    family of, 195, 197, A45
    fixed point of, 82
    Fresnel, 246
    Gompertz, 506
    graph of, 11
    greatest integer, 54
    Heaviside, 44
    hyperbolic, 364
    implicit, 122

increasing, 19
    inverse, 304, 305
    inverse cosine, 358
    inverse hyperbolic, 366, 367
    inverse sine, 357
    inverse tangent, 359
    inverse trigonometric, 356
    limit of, 39, 56
    linear, 20
    logarithmic, 29, 319, 348*
    machine diagram of, 11
    marginal cost, 89, 135, 202, 254
    marginal profit, 203
    marginal revenue, 203
    maximum and minimum values of, 156
    natural exponential, 315, 339*
    natural logarithmic, 321, 332*
    nondifferentiable, 96
    odd, 18, 187
    one-to-one, 304
    periodic, 187
    piecewise defined, 16
    polynomial, 23
    position, 86
    power, 25, 99, 106
    probability density, 468
    profit, 203
    quadratic, 23
    range of, 10
    rational, 26, 402
    reciprocal, 26
    reflected, 30
    representation as a power series, 610
    representations of, 10, 12
    revenue, 203
    root, 26
    shifted, 29
    smooth, 444
    step, 17
    stretched, 30
    tabular, 13
    transformation of, 29
    translation of, 29
    trigonometric, 27, A23
    value of, 10

Fundamental Theorem of Calculus, 243, 245, 248, 250

Galileo, 528, 5332, 549

Galois, Evariste, 126

Gause, G. F., 504

Gauss, Karl Friedrich, A31

Gaussian optics, 634

geometric series, 578
    convergence or divergence of, 579
    sum of, 579

Gini, Corrado, 277

Gini coefficient, 277

Gini index, 277

global maximum and minimum, 156

Gompertz function, 506

graph(s)
    of an equation, A14, A15
    of exponential functions, 311, 315, 317, 345*, RP4
    of a function, 11
    of logarithmic functions, 320, 322
    polar, 538, 543
    of power functions, 25, RP3
    of a sequence, 571
    of trigonometric functions, 27, A28, RP2

graphing calculator, 191, 526, 543, A41

graphing device. *See* computer algebra system

gravitational acceleration, 291

greatest integer function, 54

Gregory, James, 127, 394, 426, 614, 616

Gregory's series, 614

growth, law of natural, 351, 500

growth rate, 133, 254
    relative, 351, 500

half-angle formulas, A27

half-life, 351

hare-lynx system, 515

harmonic series, 581
    alternating, 596

Heaviside, Oliver, 44

Heaviside function, 44

Hecht, Eugene, 143, 633

higher derivatives, 97

Hooke's Law, 292

horizontal asymptote, 177, 187

horizontal line, equation of, A12

Horizontal Line Test, 304

horizontal translation of a graph, 29

Hubble Space Telescope, 161

Huygens, Christiaan, 528

hydrostatic pressure and force, 456

hyperbola, 552, 555, A18
    asymptotes, 553, A18
    branches, 553, A18
    directrix, 555
    eccentricity, 555
    equation, 553, 557, A18
    equilateral, A19
    foci, 553, 555
    polar equation, 557
    vertices, 553

hyperbolic function(s), 364
    derivatives of, 366
    inverse, 366, 367

hyperbolic identities, 365
hyperbolic substitution, 400
hypocycloid, 529

i (imaginary number), A46
I/D Test, 168
implicit differentiation, 157
implicit function, 122
improper integral, 429
    convergence or divergence of, 430, 433
impulse of a force, 296
increasing function, 19
increasing sequence, 572
Increasing/Decreasing Test, 168
increment, 88
indefinite integral(s), 251
    table of, 252
independent variable, 10
indeterminate difference, 374
indeterminate forms of limits, 369
indeterminate power, 374
indeterminate product, 373
index of summation, A30
inequalities, rules for, A4
infinite discontinuity, 66
infinite interval, 429
infinite limit, 45, 63, 182
infinite sequence. *See* sequence
infinite series. *See* series
inflection point, 172, 188
initial condition, 483
initial point of a parametric curve, 525
initial-value problem, 483
instantaneous rate of change, 88, 128
instantaneous rate of growth, 133
instantaneous rate of reaction, 132
instantaneous velocity, 38, 86, 128
integer, A2
integral(s)
    approximations to, 238
    change of variables in, 258
    comparison properties of, 241
    definite, 231
    derivative of, 245
    evaluating, 234
    improper, 429
    indefinite, 251
    patterns in, 419
    properties of, 239
    of symmetric functions, 262
    table of, 410, 414, RP6–10
    units for, 256
integral calculus, 2
Integral Test, 585
integrand, 232
    discontinuous, 433

integration, 232
    approximate, 420
    by computer algebra system, 416
    of exponential functions, 237, 318, 343*, 347*
    formulas, 410, 414, RP6–10
    indefinite, 251
    limits of, 232
    of logarithmic functions, 329
    numerical, 420
    by partial fractions, 402
    by parts, 386
    of a function defined as a power series, 612
    of rational functions, 402
    by a rationalizing substitution, 409
    strategy for, 410
    substitution in, 258
    tables, use of, 414
    term-by-term, 611
    of trignometric functions, 390
intercepts, 187, A17
interest compunded continuously, 355
Intermediate Value Theorem, 71
interpolation, 23
intersection
    of polar graphs, area of, 547
    of sets, A3
interval, A3
interval, derivative of, 368
interval of convergence, 608
interval of increase or decrease, 187
inverse cosine function, 358
inverse function(s), 304, 305
inverse hyperbolic functions, 366, 367
inverse sine function, 357
inverse substitution, 396
inverse tangent function, 359
    limits involving, 360
inverse trigonometric functions, 356, 361
    derivatives of, 361
irrational number, A2
isothermal compressibility, 132

jerk, 99
joule, 291
jump discontinuity, 66

Kepler, Johannes, 559
Kepler's Laws, 559
kinetic energy, 296
Kirchhoff's Laws, 486
Kondo, Shigeru, 619

Lagrange, Joseph-Louis, 164, 166
lamina, 460
lattice point, 153

law of laminar flow, 134, 466
law of natural growth or decay, 351
laws of exponents, 312, 341*
laws of logarithms, 321, 332*
least squares method, 22
least upper bound, 574
left-hand limit, 44, 60
Leibniz, Gottfried Wilhelm, 3, 94, 243, 257, 491, 627
Leibniz notation, 94
length
    of a curve, 444
    of a line segment, A7, A11
    of a parametric curve, 533
    of a polar curve, 548
l'Hospital, Marquis de, 370, 377
l'Hospital's Rule, 370
    origins of, 377
limaçon, 544
limit(s), 2, 39, 40
    calculating, 48
    e (the number) as, 331, 350*
    of exponential functions, 312, 317, 340*
    of a function, 39, 56
    infinite, 45, 63, 182
    at infinity, 175, 176, 177, 182
    of integration, 232
    left-hand, 44, 60
    of logarithmic functions, 320, 323
    one-sided, 44, 53
    precise definitions, 56, 183
    properties of, 48
    right-hand, 44, 60
    of a sequence, 5, 222, 568
    involving sine and cosine functions, 111, 112
Limit Comparison Test, 592
Limit Laws, 48, A34
    for sequences, 569
linear approximation, 141
linear density, 130, 131, 254
linear differential equation, 507
linear equation, A13
linear function, 20
linearization, 141
linear model, 20
linear regression, 22
line(s), A11
    equation of, A11, A12, A13
    horizontal, A12
    normal, 108
    parallel, A13
    perpendicular, A13
    secant, 35, 46
    slope of, A11
    tangent, 35, 84

liquid force, 426
Lissajous figure, 526
lithotripsy, 552
local maximum and minimum values, 156, 188
logarithm(s), 29, 319
  laws of, 321, 333*
  natural, 321, 332*
  notation for, 321, 349*
logarithmic differentiation, 329, 338*
logarithmic function(s), 29, 319, 332*, 349*
  with base a, 319
  derivative of, 324, 328, 332*, 335*, 336*
  graph of, 320, 322
  integration of, 329, 337*
  limits of, 320, 323
  properties of, 320, 333*
logistic difference equation, 576
logistic differential equation, 481, 501
logistic model, 481, 500
logistic sequence, 576
Lorenz curve, 277
Lotka-Volterra equations, 512
lower sum, 227

machine diagram of a function, 11
Maclaurin, Colin, 616
Maclaurin series, 615, 616
  table of, 624
major axis of ellipse, 551
marginal cost function, 89, 135, 202, 254
marginal profit function, 203
marginal revenue function, 203
mass, center of. *See* center of mass
mathematical induction, 78, 80, 575
  principle of, 78, 80, A32
mathematical model. *See* model(s), mathematical
maximum and minimum values, 156, 188
mean of a probability density function, 471
Mean Value Theorem, 163, 165
  for integrals, 295
mean waiting time, 470
median of a probability density function, 472
method of cylindrical shells, 287
method of exhaustion, 2, 51
method of least squares, 22
Midpoint Rule, 238, 421
  error in using, 422, 423
minor axis of ellipse, 551
mixing problems, 495
model(s), mathematical, 13, 19
  comparison of natural growth vs. logistic, 504
  of electric current, 486, 493

empirical, 21
exponential, 28, 313, 351
Gompertz function, 506
linear, 20
logarithmic, 29
polynomial, 24
for population growth, 313, 346*, 480, 499, 500, 504
power function, 25
predator-prey, 511
rational function, 26
trigonometric, 28
for vibration of membrane, 607
von Bertalanffy, 518
modeling
  with differential equations, 480
  force due to air resistance, 498
  motion of a spring, 482
  population growth, 313, 346*, 480, 499, 500, 504
modulus, A47
moment(s)
  about an axis, 458
  of a lamina, 460
  of a mass, 458
  of a system of particles, 459
momentum of an object, 296
monotonic sequence, 572
Monotonic Sequence Theorem, 574
movie theater seating, 363
multiplication of power series, 625

natural exponential function, 315, 339*
  derivative of, 315, 341*
  graph of, 315, 317
  power series for, 616, 620
  properties of, 317, 340*
natural growth law, 351, 500
natural logarithm function, 321, 332*
  derivative of, 332*, 335*, 336*
  graph of, 322, 334*
  limits of, 323, 334*
  properties of, 323, 334*
negative angle, A22
net area, 233
Net Change Theorem, 254
newton (unit of force), 290
Newton, Sir Isaac, 3, 8, 51, 60, 91, 94, 243, 257, 627
Newton's Law of Cooling, 354
Newton's method, 204, 205
Newton's Second Law of Motion, 290, 296, 442
Nicomedes, 529
nondifferentiable function, 96
normal distribution, 472
normal line, 108

nth-degree equation, finding roots of, 126
nth-degree Taylor polynomial, 145, 617
number
  complex, A46
  integer, A2
  irrational, A2
  rational, A2
  real, A2
numerical integration, 420

O (origin), A2, A9
odd function, 18, 187
one-sided limits, 44, 60
one-to-one function, 304
open interval, A3
optics
  first-order, 634
  Gaussian, 634
  third-order, 634
optimization problems, 156, 197
order of a differential equation, 482
ordered pair, A9
Oresme, Nicole, 581
origin, A2, A10
orthogonal trajectory, 494
ovals of Cassini, 545

Pappus, Theorem of, 463
Pappus of Alexandria, 463
parabola, 549, 555, A16
  axis, 549
  directrix, 549, 555
  equation, 550, 557
  focus, 549, 522
  polar equation, 557
  reflection property, 153
  vertex, 549
paradoxes of Zeno, 5
parallel lines, A13
parallelepiped, 278
parameter, 534
parametric curve(s), 534
  arc length of, 533
  area under, 532
  family of, 528
  slope of tangent line to, 530
parametric equations, 534
paraxial rays, 143
partial fractions, 403
partial integration, 386, 389
partial sum of a series, 577
parts, integration by, 386, 389
pascal (unit of pressure), 456
patterns in integrals, 419
pendulum, approximating the period of, 143
percentage error, 145

perihelion, 559
period, 187
periodic function, 187
perpendicular lines, A13
phase plane, 513
phase portrait, 513
phase trajectory, 513
piecewise defined function, 16
Planck's Law, 634
planetary motion, laws of, 559
point of inflection, 172, 188
point-slope equation of a line, A11
Poiseuille, Jean-Louis-Marie, 134
Poiseuille's Laws, 467
polar axis, 536
polar coordinate system, 536
    area in, 545
    conic sections in, 555
    conversion equations for Cartesian
        coordinates, 537
polar curve(s), 538
    arc length of, 548
    family of, 545
    graph of, 538
    symmetry in, 541
    tangent line to, 541
polar equation, graph of, 538
polar equation of a conic, 556
polar form of a complex number, A48
polar graph, 538
polar region, area of, 545
pole, 536
polynomial, 23
polynomial function, 23
    continuity of, 68, 69
population growth, 351, 499
    of bacteria, 134, 499, 504
    models, 480
    world, 313, 346*, 351
position function, 86
positive angle, A22
pound (unit of force), 291
power, 90
power consumption, approximation
    of, 256
power function(s), 25, 99, 106
    derviative of, 100
Power Law of limits, 50
Power Rule, 100, 107, 119, 330
power series, 605
    centered at a, 606
    coefficients of, 605
    for cosine and sine, 620, 621
    differentiation of, 612
    division of, 625
    for exponential function, 616, 620
    integration of, 612

interval of convergence, 608
    multiplication of, 625
    radius of convergence, 608
    representations of functions as, 610
predator-prey model, 511, 512
pressure exerted by a fluid, 456
prime notation, 87, 102, 105
principal square root of a complex
        number, A47
principle of mathematical induction, 78,
        80, A32
probability, 468
probability density function, 468
problem-solving principles, 77
    uses of, 217, 268, 382, 440
product formulas, A27
Product Law of limits, 49
Product Rule, 103
profit function, 203
p-series, 586

quadrant, A10
quadratic approximation, 145
quadratic function, 23
Quotient Law of limits, 49
Quotient Rule, 105

radian measure, 110, A21
radiation from stars, 634
radioactive decay, 353
radius of convergence, 608
rainbow, formation and location of, 162
rainbow angle, 162
range of a function, 10
rate of change
    average, 88, 128
    derivative as, 88
    instantaneous, 38, 88, 128
rate of growth, 133, 254
rate of reaction, 90, 132, 254
rates, related, 137
rational function, 26, 402
    continuity of, 68, 69
    integration of, 402
rational number, A2
rationalizing substitution for
        integration, 409
Ratio Test, 600
Rayleigh-Jeans Law, 634
real line, A3
real number, A2
rearrangement of a series, 603
reciprocal function, 26
rectangular coordinate system, A10
rectilinear motion, 211
reduction formula, 389
reflecting a function, 30

reflection property of an ellipse, 552
reflection property of a parabola, 153
region between two graphs, 272
region under a graph, 222, 227
regression, linear, 22
related rates, 137
relative error, 145
relative growth rate, 351, 500
relative maximum or minimum, 156
remainder estimates
    for the Alternating Series, 597
    for the Integral Test, 588
remainder of the Taylor series, 617
removable discontinuity, 66
representation(s) of a function, 12
    as a power series, 610
revenue function, 203
revolution, solid of, 283
revolution, surface of, 450
Riemann, Georg Bernhard, 232
Riemann sum(s), 232
right circular cylinder, 278
right-hand limit, 44, 60
Roberval, Gilles de, 249, 532
Rolle, Michel, 163
roller coaster, design of, 109
Rolle's Theorem, 163
root function, 26
Root Law of limits, 50
Root Test, 602
roots of a complex number, A51
roots of an nth-degree equation, 126
rubber membrane, vibration of, 607
rumors, rate of spread, 136

sample point, 227, 231
scatter plot, 13
secant function, A23
    derivative of, 113
    integration of, 394
    graph of, A29
secant line, 3, 35, 39
secant vector, 872
second derivative, 97
Second Derivative Test, 173
second-order differential equation,
        482
sector of a circle, area of, 545
separable differential equation, 491
sequence, 5, 566
    bounded, 573
    convergent, 568
    decreasing, 572
    divergent, 568
    Fibonacci, 567
    graph of, 571
    increasing, 572

sequence (*continued*)
   limit of, 5, 224, 568
   logistic, 576
   monotonic, 572
   of partial sums, 577
   term of, 566
series, 6, 576
   absolutely convergent, 598
   alternating, 594
   alternating harmonic, 596, 599
   binomial, 622, 623
   coefficients of, 605
   conditionally convergent, 599
   convergent, 578
   divergent, 578
   geometric, 578
   Gregory's, 614
   harmonic, 581
   infinite, 577
   Maclaurin, 615, 616
   p-, 586
   partial sum of, 577
   power, 605
   rearrangement of, 603
   strategy for testing, 604
   sum of, 6, 578
   Taylor, 615, 616
   term of, 577
   trigonometric, 606
set notation, A3
shell method for approximating
   volume, 287
shift of a function, 29
shifted conic, 554, A19
sigma notation, 228, A29
Simpson, Thomas, 426
Simpson's Rule, 424, 426
   error bounds for, 428
sine function, A23
   derivative of, 112
   graph of, 27, A28
   power series for, 620, 624
slant asymptote, 187, 190
slope, A11
   of a curve, 84
slope field, 485
slope-intercept equation of a line, A12
smooth curve, 444
smooth function, 444
snowflake curve, 640
solid, 278
   volume of, 278, 279, 283, 288
solid of revolution, 283
   rotated on a slant, 454
   volume of, 283, 286, 288, 454

solution curve, 484
solution of a differential equation, 482
solutions of predator-prey
   equations, 512
speed of a particle, 89
spherical zone, 476
spring constant, 292, 482
Squeeze Theorem, 55, A37
   for sequences, 570
standard deviation, 472
standard position of an angle, A22
step function, 17
strategy
   for integration, 411
   for optimization problems, 197
   for problem solving, 77
   for related rates, 139
   for testing series, 604
   for trigonometric integrals,
      392, 394
stretching of a function, 30
strophoid, 562
substitution, inverse, 396
Substitution Rule, 258
   for definite integrals, 261
subtraction formulas for sine
   and cosine, A26
sum
   of a geometric series, 579
   of a series, 578
   lower or upper, 227
   of partial fractions, 403
   Riemann, 232
   telescoping, 581
Sum Law of limits, 49
Sum Rule, 102
summation notation, A30
surface area, 451
   of a parametric surface, 535
surface of revolution, 450
   surface area of, 451
symmetric functions, integrals of, 262
symmetry, 17, 187, 262
   in polar graphs, 541
symmetry principle, 460

table of differentiation formulas,
   109, RP5
tables of integrals, 410, RP6–10
   use of, 414
tabular function, 13
tangent function, A23
   derivative of, 113
   graph of, 280

tangent line(s), 84
   to a curve, 3, 35, 84
   early methods of finding, 91
   to a parametric curve, 530
   to a polar curve, 541
   vertical, 97
tangent line approximation, 141
tangent problem, 2, 35, 84
tank, rate of drainage of, 497
tautochrone problem, 528
Taylor, Brook, 616
Taylor polynomial, 146, 617
   applications of, 628
Taylor series, 615, 616
   remainder of, 617
Taylor's Inequality, 618
techniques of integration, summary, 411
telescoping sum, 582
term of a sequence, 566
term of a series, 577
term-by-term differentiation and
   integration, 611
terminal point of a parametric curve, 525
Test for Divergence, 581
tests for convergence and divergence
   of series
      Alternating Series Test, 595
      Comparison Test, 591
      Integral Test, 585
      Limit Comparison Test, 592
      Ratio Test, 600
      Root Test, 602
      summary of tests, 604
Theorem of Pappus, 463
third derivative, 98
third-order optics, 634
Torricelli, Evangelista, 532
total fertility rate, 148
transformation of a function, 29
translation of a function, 29
Trapezoidal Rule, 421
   error in, 422, 423
Triangle Inequality, 62, A8
trigonometric functions, 27, A23
   continuity of, 69
   derivatives of, 110, 113
   graphs of, 27, 28,0
   integrals of, 252, 390
   inverse, 356
   limits involving, 111, 112
trigonometric identities, A25
trigonometric integrals, 390
   strategy for evaluating, 392, 394
trigonometric series, 606
trigonometric substitution(s), 396, 397

ultraviolet catastrophe, 634
union of sets, A3
upper sum, 227

value of a function, 10
variable(s)
    change of, 258
    continuous random, 468
    dependent, 10
    independent, 10
variables, change of, 258
velocity, 3, 37, 86, 128, 254
    average, 4, 38, 86
    instantaneous, 38, 86, 128
velocity gradient, 134
velocity problem, 37, 85
Verhulst, Pierre-François, 481
vertex of a parabola, 549
vertical asymptote, 47, 187

vertical line, A12
Vertical Line Test, 15
vertical tangent line, 97
vertical translation of a graph, 29
vertices of an ellipse, 551
vertices of a hyperbola, 553
vibration of a rubber membrane, 607
viewing rectangle, A41
visual representations of a function,
    11, 12
volume, 279
    by cross-sections, 272, 273, 466
    by cylindrical shells, 287
    by disks, 280, 283
    of a solid, 279
    of a solid of revolution, 283, 454
    of a solid on a slant, 454
    by washers, 282, 283
Volterra, Vito, 512

Wallis, John, 3
washer method, 282, 283
Weierstrass, Karl, 60
weight (force), 290
work (force), 290, 291
Wren, Sir Christopher, 535

x-axis, A9
x-coordinate, A9
x-intercept, A12, A17

y-axis, A9
y-coordinate, A9
y-intercept, A12, A17

Zeno, 5
Zeno's paradoxes, 5

SPECIAL FUNCTIONS

Power Functions $f(x) = x^a$

(i) $f(x) = x^n$, n a positive integer

n even

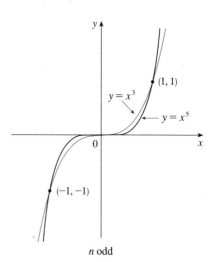

n odd

(ii) $f(x) = x^{1/n} = \sqrt[n]{x}$, n a positive integer

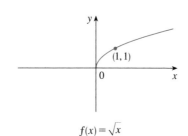

$f(x) = \sqrt{x}$

$f(x) = \sqrt[3]{x}$

(iii) $f(x) = x^{-1} = \dfrac{1}{x}$

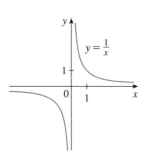

$y = \dfrac{1}{x}$

Inverse Trigonometric Functions

$\arcsin x = \sin^{-1}x = y \iff \sin y = x \quad$ and $\quad -\dfrac{\pi}{2} \le y \le \dfrac{\pi}{2}$

$\arccos x = \cos^{-1}x = y \iff \cos y = x \quad$ and $\quad 0 \le y \le \pi$

$\arctan x = \tan^{-1}x = y \iff \tan y = x \quad$ and $\quad -\dfrac{\pi}{2} < y < \dfrac{\pi}{2}$

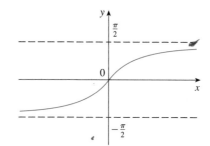

$y = \tan^{-1}x = \arctan x$

$$\lim_{x \to -\infty} \tan^{-1}x = -\dfrac{\pi}{2}$$

$$\lim_{x \to \infty} \tan^{-1}x = \dfrac{\pi}{2}$$

SPECIAL FUNCTIONS

Exponential and Logarithmic Functions

$\log_a x = y \iff a^y = x$

$\ln x = \log_e x, \quad \text{where} \quad \ln e = 1$

$\ln x = y \iff e^y = x$

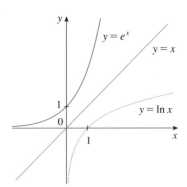

Cancellation Equations

$\log_a(a^x) = x \qquad a^{\log_a x} = x$

$\ln(e^x) = x \qquad e^{\ln x} = x$

Laws of Logarithms

1. $\log_a(xy) = \log_a x + \log_a y$

2. $\log_a\left(\dfrac{x}{y}\right) = \log_a x - \log_a y$

3. $\log_a(x^r) = r \log_a x$

$\lim_{x \to -\infty} e^x = 0 \qquad \lim_{x \to \infty} e^x = \infty$

$\lim_{x \to 0^+} \ln x = -\infty \qquad \lim_{x \to \infty} \ln x = \infty$

Exponential functions

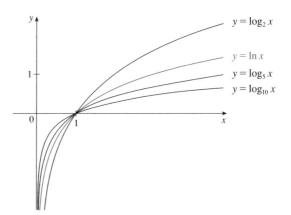

Logarithmic functions

Hyperbolic Functions

$\sinh x = \dfrac{e^x - e^{-x}}{2} \qquad\qquad \operatorname{csch} x = \dfrac{1}{\sinh x}$

$\cosh x = \dfrac{e^x + e^{-x}}{2} \qquad\qquad \operatorname{sech} x = \dfrac{1}{\cosh x}$

$\tanh x = \dfrac{\sinh x}{\cosh x} \qquad\qquad \coth x = \dfrac{\cosh x}{\sinh x}$

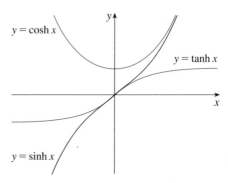

Inverse Hyperbolic Functions

$y = \sinh^{-1} x \iff \sinh y = x \qquad\qquad \sinh^{-1} x = \ln\left(x + \sqrt{x^2 + 1}\right)$

$y = \cosh^{-1} x \iff \cosh y = x \quad \text{and} \quad y \geq 0 \qquad \cosh^{-1} x = \ln\left(x + \sqrt{x^2 - 1}\right)$

$y = \tanh^{-1} x \iff \tanh y = x \qquad\qquad \tanh^{-1} x = \tfrac{1}{2} \ln\left(\dfrac{1 + x}{1 - x}\right)$

DIFFERENTIATION RULES

General Formulas

1. $\dfrac{d}{dx}(c) = 0$

2. $\dfrac{d}{dx}[cf(x)] = cf'(x)$

3. $\dfrac{d}{dx}[f(x) + g(x)] = f'(x) + g'(x)$

4. $\dfrac{d}{dx}[f(x) - g(x)] = f'(x) - g'(x)$

5. $\dfrac{d}{dx}[f(x)g(x)] = f(x)g'(x) + g(x)f'(x)$ (Product Rule)

6. $\dfrac{d}{dx}\left[\dfrac{f(x)}{g(x)}\right] = \dfrac{g(x)f'(x) - f(x)g'(x)}{[g(x)]^2}$ (Quotient Rule)

7. $\dfrac{d}{dx}f(g(x)) = f'(g(x))g'(x)$ (Chain Rule)

8. $\dfrac{d}{dx}(x^n) = nx^{n-1}$ (Power Rule)

Exponential and Logarithmic Functions

9. $\dfrac{d}{dx}(e^x) = e^x$

10. $\dfrac{d}{dx}(a^x) = a^x \ln a$

11. $\dfrac{d}{dx}\ln|x| = \dfrac{1}{x}$

12. $\dfrac{d}{dx}(\log_a x) = \dfrac{1}{x \ln a}$

Trigonometric Functions

13. $\dfrac{d}{dx}(\sin x) = \cos x$

14. $\dfrac{d}{dx}(\cos x) = -\sin x$

15. $\dfrac{d}{dx}(\tan x) = \sec^2 x$

16. $\dfrac{d}{dx}(\csc x) = -\csc x \cot x$

17. $\dfrac{d}{dx}(\sec x) = \sec x \tan x$

18. $\dfrac{d}{dx}(\cot x) = -\csc^2 x$

Inverse Trigonometric Functions

19. $\dfrac{d}{dx}(\sin^{-1}x) = \dfrac{1}{\sqrt{1 - x^2}}$

20. $\dfrac{d}{dx}(\cos^{-1}x) = -\dfrac{1}{\sqrt{1 - x^2}}$

21. $\dfrac{d}{dx}(\tan^{-1}x) = \dfrac{1}{1 + x^2}$

22. $\dfrac{d}{dx}(\csc^{-1}x) = -\dfrac{1}{x\sqrt{x^2 - 1}}$

23. $\dfrac{d}{dx}(\sec^{-1}x) = \dfrac{1}{x\sqrt{x^2 - 1}}$

24. $\dfrac{d}{dx}(\cot^{-1}x) = -\dfrac{1}{1 + x^2}$

Hyperbolic Functions

25. $\dfrac{d}{dx}(\sinh x) = \cosh x$

26. $\dfrac{d}{dx}(\cosh x) = \sinh x$

27. $\dfrac{d}{dx}(\tanh x) = \text{sech}^2 x$

28. $\dfrac{d}{dx}(\text{csch}\, x) = -\text{csch}\, x \coth x$

29. $\dfrac{d}{dx}(\text{sech}\, x) = -\text{sech}\, x \tanh x$

30. $\dfrac{d}{dx}(\coth x) = -\text{csch}^2 x$

Inverse Hyperbolic Functions

31. $\dfrac{d}{dx}(\sinh^{-1}x) = \dfrac{1}{\sqrt{1 + x^2}}$

32. $\dfrac{d}{dx}(\cosh^{-1}x) = \dfrac{1}{\sqrt{x^2 - 1}}$

33. $\dfrac{d}{dx}(\tanh^{-1}x) = \dfrac{1}{1 - x^2}$

34. $\dfrac{d}{dx}(\text{csch}^{-1}x) = -\dfrac{1}{|x|\sqrt{x^2 + 1}}$

35. $\dfrac{d}{dx}(\text{sech}^{-1}x) = -\dfrac{1}{x\sqrt{1 - x^2}}$

36. $\dfrac{d}{dx}(\coth^{-1}x) = \dfrac{1}{1 - x^2}$

TABLE OF INTEGRALS

Basic Forms

1. $\displaystyle\int u\,dv = uv - \int v\,du$

2. $\displaystyle\int u^n\,du = \frac{u^{n+1}}{n+1} + C, \quad n \neq -1$

3. $\displaystyle\int \frac{du}{u} = \ln|u| + C$

4. $\displaystyle\int e^u\,du = e^u + C$

5. $\displaystyle\int a^u\,du = \frac{a^u}{\ln a} + C$

6. $\displaystyle\int \sin u\,du = -\cos u + C$

7. $\displaystyle\int \cos u\,du = \sin u + C$

8. $\displaystyle\int \sec^2 u\,du = \tan u + C$

9. $\displaystyle\int \csc^2 u\,du = -\cot u + C$

10. $\displaystyle\int \sec u\,\tan u\,du = \sec u + C$

11. $\displaystyle\int \csc u\,\cot u\,du = -\csc u + C$

12. $\displaystyle\int \tan u\,du = \ln|\sec u| + C$

13. $\displaystyle\int \cot u\,du = \ln|\sin u| + C$

14. $\displaystyle\int \sec u\,du = \ln|\sec u + \tan u| + C$

15. $\displaystyle\int \csc u\,du = \ln|\csc u - \cot u| + C$

16. $\displaystyle\int \frac{du}{\sqrt{a^2 - u^2}} = \sin^{-1}\frac{u}{a} + C, \quad a > 0$

17. $\displaystyle\int \frac{du}{a^2 + u^2} = \frac{1}{a}\tan^{-1}\frac{u}{a} + C$

18. $\displaystyle\int \frac{du}{u\sqrt{u^2 - a^2}} = \frac{1}{a}\sec^{-1}\frac{u}{a} + C$

19. $\displaystyle\int \frac{du}{a^2 - u^2} = \frac{1}{2a}\ln\left|\frac{u+a}{u-a}\right| + C$

20. $\displaystyle\int \frac{du}{u^2 - a^2} = \frac{1}{2a}\ln\left|\frac{u-a}{u+a}\right| + C$

Forms Involving $\sqrt{a^2 + u^2}$, $a > 0$

21. $\displaystyle\int \sqrt{a^2 + u^2}\,du = \frac{u}{2}\sqrt{a^2 + u^2} + \frac{a^2}{2}\ln\left(u + \sqrt{a^2 + u^2}\right) + C$

22. $\displaystyle\int u^2\sqrt{a^2 + u^2}\,du = \frac{u}{8}(a^2 + 2u^2)\sqrt{a^2 + u^2} - \frac{a^4}{8}\ln\left(u + \sqrt{a^2 + u^2}\right) + C$

23. $\displaystyle\int \frac{\sqrt{a^2 + u^2}}{u}\,du = \sqrt{a^2 + u^2} - a\ln\left|\frac{a + \sqrt{a^2 + u^2}}{u}\right| + C$

24. $\displaystyle\int \frac{\sqrt{a^2 + u^2}}{u^2}\,du = -\frac{\sqrt{a^2 + u^2}}{u} + \ln\left(u + \sqrt{a^2 + u^2}\right) + C$

25. $\displaystyle\int \frac{du}{\sqrt{a^2 + u^2}} = \ln\left(u + \sqrt{a^2 + u^2}\right) + C$

26. $\displaystyle\int \frac{u^2\,du}{\sqrt{a^2 + u^2}} = \frac{u}{2}\sqrt{a^2 + u^2} - \frac{a^2}{2}\ln\left(u + \sqrt{a^2 + u^2}\right) + C$

27. $\displaystyle\int \frac{du}{u\sqrt{a^2 + u^2}} = -\frac{1}{a}\ln\left|\frac{\sqrt{a^2 + u^2} + a}{u}\right| + C$

28. $\displaystyle\int \frac{du}{u^2\sqrt{a^2 + u^2}} = -\frac{\sqrt{a^2 + u^2}}{a^2 u} + C$

29. $\displaystyle\int \frac{du}{(a^2 + u^2)^{3/2}} = \frac{u}{a^2\sqrt{a^2 + u^2}} + C$

TABLE OF INTEGRALS

Forms Involving $\sqrt{a^2 - u^2}$, $a > 0$

30. $\displaystyle\int \sqrt{a^2 - u^2}\, du = \frac{u}{2}\sqrt{a^2 - u^2} + \frac{a^2}{2}\sin^{-1}\frac{u}{a} + C$

31. $\displaystyle\int u^2\sqrt{a^2 - u^2}\, du = \frac{u}{8}(2u^2 - a^2)\sqrt{a^2 - u^2} + \frac{a^4}{8}\sin^{-1}\frac{u}{a} + C$

32. $\displaystyle\int \frac{\sqrt{a^2 - u^2}}{u}\, du = \sqrt{a^2 - u^2} - a\ln\left|\frac{a + \sqrt{a^2 - u^2}}{u}\right| + C$

33. $\displaystyle\int \frac{\sqrt{a^2 - u^2}}{u^2}\, du = -\frac{1}{u}\sqrt{a^2 - u^2} - \sin^{-1}\frac{u}{a} + C$

34. $\displaystyle\int \frac{u^2\, du}{\sqrt{a^2 - u^2}} = -\frac{u}{2}\sqrt{a^2 - u^2} + \frac{a^2}{2}\sin^{-1}\frac{u}{a} + C$

35. $\displaystyle\int \frac{du}{u\sqrt{a^2 - u^2}} = -\frac{1}{a}\ln\left|\frac{a + \sqrt{a^2 - u^2}}{u}\right| + C$

36. $\displaystyle\int \frac{du}{u^2\sqrt{a^2 - u^2}} = -\frac{1}{a^2 u}\sqrt{a^2 - u^2} + C$

37. $\displaystyle\int (a^2 - u^2)^{3/2}\, du = -\frac{u}{8}(2u^2 - 5a^2)\sqrt{a^2 - u^2} + \frac{3a^4}{8}\sin^{-1}\frac{u}{a} + C$

38. $\displaystyle\int \frac{du}{(a^2 - u^2)^{3/2}} = \frac{u}{a^2\sqrt{a^2 - u^2}} + C$

Forms Involving $\sqrt{u^2 - a^2}$, $a > 0$

39. $\displaystyle\int \sqrt{u^2 - a^2}\, du = \frac{u}{2}\sqrt{u^2 - a^2} - \frac{a^2}{2}\ln\left|u + \sqrt{u^2 - a^2}\right| + C$

40. $\displaystyle\int u^2\sqrt{u^2 - a^2}\, du = \frac{u}{8}(2u^2 - a^2)\sqrt{u^2 - a^2} - \frac{a^4}{8}\ln\left|u + \sqrt{u^2 - a^2}\right| + C$

41. $\displaystyle\int \frac{\sqrt{u^2 - a^2}}{u}\, du = \sqrt{u^2 - a^2} - a\cos^{-1}\frac{a}{|u|} + C$

42. $\displaystyle\int \frac{\sqrt{u^2 - a^2}}{u^2}\, du = -\frac{\sqrt{u^2 - a^2}}{u} + \ln\left|u + \sqrt{u^2 - a^2}\right| + C$

43. $\displaystyle\int \frac{du}{\sqrt{u^2 - a^2}} = \ln\left|u + \sqrt{u^2 - a^2}\right| + C$

44. $\displaystyle\int \frac{u^2\, du}{\sqrt{u^2 - a^2}} = \frac{u}{2}\sqrt{u^2 - a^2} + \frac{a^2}{2}\ln\left|u + \sqrt{u^2 - a^2}\right| + C$

45. $\displaystyle\int \frac{du}{u^2\sqrt{u^2 - a^2}} = \frac{\sqrt{u^2 - a^2}}{a^2 u} + C$

46. $\displaystyle\int \frac{du}{(u^2 - a^2)^{3/2}} = -\frac{u}{a^2\sqrt{u^2 - a^2}} + C$

TABLE OF INTEGRALS

Forms Involving $a + bu$

47. $\displaystyle \int \frac{u \, du}{a + bu} = \frac{1}{b^2} \left(a + bu - a \ln |a + bu| \right) + C$

48. $\displaystyle \int \frac{u^2 \, du}{a + bu} = \frac{1}{2b^3} \left[(a + bu)^2 - 4a(a + bu) + 2a^2 \ln |a + bu| \right] + C$

49. $\displaystyle \int \frac{du}{u(a + bu)} = \frac{1}{a} \ln \left| \frac{u}{a + bu} \right| + C$

50. $\displaystyle \int \frac{du}{u^2(a + bu)} = -\frac{1}{au} + \frac{b}{a^2} \ln \left| \frac{a + bu}{u} \right| + C$

51. $\displaystyle \int \frac{u \, du}{(a + bu)^2} = \frac{a}{b^2(a + bu)} + \frac{1}{b^2} \ln |a + bu| + C$

52. $\displaystyle \int \frac{du}{u(a + bu)^2} = \frac{1}{a(a + bu)} - \frac{1}{a^2} \ln \left| \frac{a + bu}{u} \right| + C$

53. $\displaystyle \int \frac{u^2 \, du}{(a + bu)^2} = \frac{1}{b^3} \left(a + bu - \frac{a^2}{a + bu} - 2a \ln |a + bu| \right) + C$

54. $\displaystyle \int u \sqrt{a + bu} \, du = \frac{2}{15b^2} (3bu - 2a)(a + bu)^{3/2} + C$

55. $\displaystyle \int \frac{u \, du}{\sqrt{a + bu}} = \frac{2}{3b^2} (bu - 2a) \sqrt{a + bu} + C$

56. $\displaystyle \int \frac{u^2 \, du}{\sqrt{a + bu}} = \frac{2}{15b^3} (8a^2 + 3b^2u^2 - 4abu) \sqrt{a + bu} + C$

57. $\displaystyle \int \frac{du}{u \sqrt{a + bu}} = \frac{1}{\sqrt{a}} \ln \left| \frac{\sqrt{a + bu} - \sqrt{a}}{\sqrt{a + bu} + \sqrt{a}} \right| + C, \quad \text{if } a > 0$

$\displaystyle \qquad\qquad = \frac{2}{\sqrt{-a}} \tan^{-1} \sqrt{\frac{a + bu}{-a}} + C, \qquad \text{if } a < 0$

58. $\displaystyle \int \frac{\sqrt{a + bu}}{u} \, du = 2\sqrt{a + bu} + a \int \frac{du}{u \sqrt{a + bu}}$

59. $\displaystyle \int \frac{\sqrt{a + bu}}{u^2} \, du = -\frac{\sqrt{a + bu}}{u} + \frac{b}{2} \int \frac{du}{u \sqrt{a + bu}}$

60. $\displaystyle \int u^n \sqrt{a + bu} \, du = \frac{2}{b(2n + 3)} \left[u^n(a + bu)^{3/2} - na \int u^{n-1} \sqrt{a + bu} \, du \right]$

61. $\displaystyle \int \frac{u^n \, du}{\sqrt{a + bu}} = \frac{2u^n \sqrt{a + bu}}{b(2n + 1)} - \frac{2na}{b(2n + 1)} \int \frac{u^{n-1} \, du}{\sqrt{a + bu}}$

62. $\displaystyle \int \frac{du}{u^n \sqrt{a + bu}} = -\frac{\sqrt{a + bu}}{a(n - 1)u^{n-1}} - \frac{b(2n - 3)}{2a(n - 1)} \int \frac{du}{u^{n-1} \sqrt{a + bu}}$

TABLE OF INTEGRALS

Trigonometric Forms

63. $\int \sin^2 u \, du = \frac{1}{2}u - \frac{1}{4}\sin 2u + C$

64. $\int \cos^2 u \, du = \frac{1}{2}u + \frac{1}{4}\sin 2u + C$

65. $\int \tan^2 u \, du = \tan u - u + C$

66. $\int \cot^2 u \, du = -\cot u - u + C$

67. $\int \sin^3 u \, du = -\frac{1}{3}(2 + \sin^2 u)\cos u + C$

68. $\int \cos^3 u \, du = \frac{1}{3}(2 + \cos^2 u)\sin u + C$

69. $\int \tan^3 u \, du = \frac{1}{2}\tan^2 u + \ln|\cos u| + C$

70. $\int \cot^3 u \, du = -\frac{1}{2}\cot^2 u - \ln|\sin u| + C$

71. $\int \sec^3 u \, du = \frac{1}{2}\sec u \tan u + \frac{1}{2}\ln|\sec u + \tan u| + C$

72. $\int \csc^3 u \, du = -\frac{1}{2}\csc u \cot u + \frac{1}{2}\ln|\csc u - \cot u| + C$

73. $\int \sin^n u \, du = -\frac{1}{n}\sin^{n-1} u \cos u + \frac{n-1}{n}\int \sin^{n-2} u \, du$

74. $\int \cos^n u \, du = \frac{1}{n}\cos^{n-1} u \sin u + \frac{n-1}{n}\int \cos^{n-2} u \, du$

75. $\int \tan^n u \, du = \frac{1}{n-1}\tan^{n-1} u - \int \tan^{n-2} u \, du$

76. $\int \cot^n u \, du = \frac{-1}{n-1}\cot^{n-1} u - \int \cot^{n-2} u \, du$

77. $\int \sec^n u \, du = \frac{1}{n-1}\tan u \sec^{n-2} u + \frac{n-2}{n-1}\int \sec^{n-2} u \, du$

78. $\int \csc^n u \, du = \frac{-1}{n-1}\cot u \csc^{n-2} u + \frac{n-2}{n-1}\int \csc^{n-2} u \, du$

79. $\int \sin au \sin bu \, du = \frac{\sin(a-b)u}{2(a-b)} - \frac{\sin(a+b)u}{2(a+b)} + C$

80. $\int \cos au \cos bu \, du = \frac{\sin(a-b)u}{2(a-b)} + \frac{\sin(a+b)u}{2(a+b)} + C$

81. $\int \sin au \cos bu \, du = -\frac{\cos(a-b)u}{2(a-b)} - \frac{\cos(a+b)u}{2(a+b)} + C$

82. $\int u \sin u \, du = \sin u - u \cos u + C$

83. $\int u \cos u \, du = \cos u + u \sin u + C$

84. $\int u^n \sin u \, du = -u^n \cos u + n \int u^{n-1} \cos u \, du$

85. $\int u^n \cos u \, du = u^n \sin u - n \int u^{n-1} \sin u \, du$

86. $\int \sin^n u \cos^m u \, du = -\frac{\sin^{n-1} u \cos^{m+1} u}{n+m} + \frac{n-1}{n+m}\int \sin^{n-2} u \cos^m u \, du$

$\qquad = \frac{\sin^{n+1} u \cos^{m-1} u}{n+m} + \frac{m-1}{n+m}\int \sin^n u \cos^{m-2} u \, du$

Inverse Trigonometric Forms

87. $\int \sin^{-1} u \, du = u \sin^{-1} u + \sqrt{1 - u^2} + C$

88. $\int \cos^{-1} u \, du = u \cos^{-1} u - \sqrt{1 - u^2} + C$

89. $\int \tan^{-1} u \, du = u \tan^{-1} u - \frac{1}{2}\ln(1 + u^2) + C$

90. $\int u \sin^{-1} u \, du = \frac{2u^2 - 1}{4}\sin^{-1} u + \frac{u\sqrt{1 - u^2}}{4} + C$

91. $\int u \cos^{-1} u \, du = \frac{2u^2 - 1}{4}\cos^{-1} u - \frac{u\sqrt{1 - u^2}}{4} + C$

92. $\int u \tan^{-1} u \, du = \frac{u^2 + 1}{2}\tan^{-1} u - \frac{u}{2} + C$

93. $\int u^n \sin^{-1} u \, du = \frac{1}{n+1}\left[u^{n+1}\sin^{-1} u - \int \frac{u^{n+1}\, du}{\sqrt{1 - u^2}} \right], \quad n \neq -1$

94. $\int u^n \cos^{-1} u \, du = \frac{1}{n+1}\left[u^{n+1}\cos^{-1} u + \int \frac{u^{n+1}\, du}{\sqrt{1 - u^2}} \right], \quad n \neq -1$

95. $\int u^n \tan^{-1} u \, du = \frac{1}{n+1}\left[u^{n+1}\tan^{-1} u - \int \frac{u^{n+1}\, du}{1 + u^2} \right], \quad n \neq -1$

TABLE OF INTEGRALS

Exponential and Logarithmic Forms

96. $\displaystyle\int ue^{au}\,du = \frac{1}{a^2}(au - 1)e^{au} + C$

97. $\displaystyle\int u^n e^{au}\,du = \frac{1}{a}u^n e^{au} - \frac{n}{a}\int u^{n-1}e^{au}\,du$

98. $\displaystyle\int e^{au}\sin bu\,du = \frac{e^{au}}{a^2 + b^2}(a\sin bu - b\cos bu) + C$

99. $\displaystyle\int e^{au}\cos bu\,du = \frac{e^{au}}{a^2 + b^2}(a\cos bu + b\sin bu) + C$

100. $\displaystyle\int \ln u\,du = u\ln u - u + C$

101. $\displaystyle\int u^n \ln u\,du = \frac{u^{n+1}}{(n+1)^2}[(n+1)\ln u - 1] + C$

102. $\displaystyle\int \frac{1}{u\ln u}\,du = \ln|\ln u| + C$

Hyperbolic Forms

103. $\displaystyle\int \sinh u\,du = \cosh u + C$

104. $\displaystyle\int \cosh u\,du = \sinh u + C$

105. $\displaystyle\int \tanh u\,du = \ln\cosh u + C$

106. $\displaystyle\int \coth u\,du = \ln|\sinh u| + C$

107. $\displaystyle\int \operatorname{sech} u\,du = \tan^{-1}|\sinh u| + C$

108. $\displaystyle\int \operatorname{csch} u\,du = \ln\left|\tanh\tfrac{1}{2}u\right| + C$

109. $\displaystyle\int \operatorname{sech}^2 u\,du = \tanh u + C$

110. $\displaystyle\int \operatorname{csch}^2 u\,du = -\coth u + C$

111. $\displaystyle\int \operatorname{sech} u\tanh u\,du = -\operatorname{sech} u + C$

112. $\displaystyle\int \operatorname{csch} u\coth u\,du = -\operatorname{csch} u + C$

Forms Involving $\sqrt{2au - u^2}$, $a > 0$

113. $\displaystyle\int \sqrt{2au - u^2}\,du = \frac{u - a}{2}\sqrt{2au - u^2} + \frac{a^2}{2}\cos^{-1}\left(\frac{a - u}{a}\right) + C$

114. $\displaystyle\int u\sqrt{2au - u^2}\,du = \frac{2u^2 - au - 3a^2}{6}\sqrt{2au - u^2} + \frac{a^3}{2}\cos^{-1}\left(\frac{a - u}{a}\right) + C$

115. $\displaystyle\int \frac{\sqrt{2au - u^2}}{u}\,du = \sqrt{2au - u^2} + a\cos^{-1}\left(\frac{a - u}{a}\right) + C$

116. $\displaystyle\int \frac{\sqrt{2au - u^2}}{u^2}\,du = -\frac{2\sqrt{2au - u^2}}{u} - \cos^{-1}\left(\frac{a - u}{a}\right) + C$

117. $\displaystyle\int \frac{du}{\sqrt{2au - u^2}} = \cos^{-1}\left(\frac{a - u}{a}\right) + C$

118. $\displaystyle\int \frac{u\,du}{\sqrt{2au - u^2}} = -\sqrt{2au - u^2} + a\cos^{-1}\left(\frac{a - u}{a}\right) + C$

119. $\displaystyle\int \frac{u^2\,du}{\sqrt{2au - u^2}} = -\frac{(u + 3a)}{2}\sqrt{2au - u^2} + \frac{3a^2}{2}\cos^{-1}\left(\frac{a - u}{a}\right) + C$

120. $\displaystyle\int \frac{du}{u\sqrt{2au - u^2}} = -\frac{\sqrt{2au - u^2}}{au} + C$